IET ENERGY ENGINEERING SERIES 217

Advances in Power System Modelling, Control and Stability Analysis

Other volumes in this series:

Volume 1	**Power Circuit Breaker Theory and Design** C.H. Flurscheim (Editor)
Volume 4	**Industrial Microwave Heating** A.C. Metaxas and R.J. Meredith
Volume 7	**Insulators for High Voltages** J.S.T. Looms
Volume 8	**Variable Frequency AC Motor Drive Systems** D. Finney
Volume 10	**SF_6 Switchgear** H.M. Ryan and G.R. Jones
Volume 11	**Conduction and Induction Heating** E.J. Davies
Volume 13	**Statistical Techniques for High Voltage Engineering** W. Hauschild and W. Mosch
Volume 14	**Uninterruptible Power Supplies** J. Platts and J.D. St Aubyn (Editors)
Volume 15	**Digital Protection for Power Systems** A.T. Johns and S.K. Salman
Volume 16	**Electricity Economics and Planning** T.W. Berrie
Volume 18	**Vacuum Switchgear** A. Greenwood
Volume 19	**Electrical Safety: A guide to causes and prevention of hazards** J.M. Adams
Volume 21	**Electricity Distribution Network Design, 2nd Edition** E. Lakervi and E.J. Holmes
Volume 22	**Artificial Intelligence Techniques in Power Systems** K. Warwick, A.O. Ekwue and R. Aggarwal (Editors)
Volume 24	**Power System Commissioning and Maintenance Practice** K. Harker
Volume 25	**Engineers' Handbook of Industrial Microwave Heating** R.J. Meredith
Volume 26	**Small Electric Motors** H. Moczala, J. Draeger, H. Krauss, H. Schock and S. Tillner
Volume 27	**AC-DC Power System Analysis** J. Arrillaga and B.C. Smith
Volume 29	**High Voltage Direct Current Transmission, 2nd Edition** J. Arrillaga
Volume 30	**Flexible AC Transmission Systems (FACTS)** Y.-H. Song (Editor)
Volume 31	**Embedded generation** N. Jenkins, R. Allan, P. Crossley, D. Kirschen and G. Strbac
Volume 32	**High Voltage Engineering and Testing, 2nd Edition** H.M. Ryan (Editor)
Volume 33	**Overvoltage Protection of Low-Voltage Systems, Revised Edition** P. Hasse
Volume 36	**Voltage Quality in Electrical Power Systems** J. Schlabbach, D. Blume and T. Stephanblome
Volume 37	**Electrical Steels for Rotating Machines** P. Beckley
Volume 38	**The Electric Car: Development and future of battery, hybrid and fuel-cell cars** M. Westbrook
Volume 39	**Power Systems Electromagnetic Transients Simulation** J. Arrillaga and N. Watson
Volume 40	**Advances in High Voltage Engineering** M. Haddad and D. Warne
Volume 41	**Electrical Operation of Electrostatic Precipitators** K. Parker
Volume 43	**Thermal Power Plant Simulation and Control** D. Flynn
Volume 44	**Economic Evaluation of Projects in the Electricity Supply Industry** H. Khatib
Volume 45	**Propulsion Systems for Hybrid Vehicles** J. Miller
Volume 46	**Distribution Switchgear** S. Stewart
Volume 47	**Protection of Electricity Distribution Networks, 2nd Edition** J. Gers and E. Holmes
Volume 48	**Wood Pole Overhead Lines** B. Wareing
Volume 49	**Electric Fuses, 3rd Edition** A. Wright and G. Newbery
Volume 50	**Wind Power Integration: Connection and system operational aspects** B. Fox, L. Bryans, D. Flynn, N. Jenkins, D. Milborrow, M. O'Malley, R. Watson and O. Anaya-Lara
Volume 51	**Short Circuit Currents** J. Schlabbach
Volume 52	**Nuclear Power** J. Wood
Volume 53	**Condition Assessment of High Voltage Insulation in Power System Equipment** R.E. James and Q. Su
Volume 55	**Local Energy: Distributed generation of heat and power** J. Wood
Volume 56	**Condition Monitoring of Rotating Electrical Machines** P. Tavner, L. Ran, J. Penman and H. Sedding
Volume 57	**The Control Techniques Drives and Controls Handbook, 2nd Edition** B. Drury
Volume 58	**Lightning Protection** V. Cooray (Editor)
Volume 59	**Ultracapacitor Applications** J.M. Miller
Volume 62	**Lightning Electromagnetics** V. Cooray
Volume 63	**Energy Storage for Power Systems, 2nd Edition** A. Ter-Gazarian
Volume 65	**Protection of Electricity Distribution Networks, 3rd Edition** J. Gers
Volume 66	**High Voltage Engineering Testing, 3rd Edition** H. Ryan (Editor)
Volume 67	**Multicore Simulation of Power System Transients** F.M. Uriate
Volume 68	**Distribution System Analysis and Automation** J. Gers
Volume 69	**The Lightening Flash, 2nd Edition** V. Cooray (Editor)
Volume 70	**Economic Evaluation of Projects in the Electricity Supply Industry,** *3rd Edition* H. Khatib
Volume 72	**Control Circuits in Power Electronics: Practical issues in design and implementation** M. Castilla (Editor)
Volume 73	**Wide Area Monitoring, Protection and Control Systems: The enabler for** *Smarter Grids* A. Vaccaro and A. Zobaa (Editors)

Volume 74 **Power Electronic Converters and Systems: Frontiers and applications** A.M. Trzynadlowski (Editor)
Volume 75 **Power Distribution Automation** B. Das (Editor)
Volume 76 **Power System Stability: Modelling, analysis and control** A.A. Sallam and O.P. Malik
Volume 78 **Numerical Analysis of Power System Transients and Dynamics** A. Ametani (Editor)
Volume 79 **Vehicle-to-Grid: Linking electric vehicles to the smart grid** J. Lu and J. Hossain (Editors)
Volume 81 **Cyber-Physical-Social Systems and Constructs in Electric Power Engineering** S. Suryanarayanan, R. Roche and T.M. Hansen (Editors)
Volume 82 **Periodic Control of Power Electronic Converters** F. Blaabjerg, K. Zhou, D. Wang and Y. Yang
Volume 86 **Advances in Power System Modelling, Control and Stability Analysis** F. Milano (Editor)
Volume 87 **Cogeneration: Technologies, optimisation and implentation** C.A. Frangopoulos (Editor)
Volume 88 **Smarter Energy: from Smart Metering to the Smart Grid** H. Sun, N. Hatziargyriou, H.V. Poor, L. Carpanini and M.A. Sánchez Fornié (Editors)
Volume 89 **Hydrogen Production, Separation and Purification for Energy** A. Basile, F. Dalena, J. Tong and T.N. Veziroğlu (Editors)
Volume 90 **Clean Energy Microgrids** S. Obara and J. Morel (Editors)
Volume 91 **Fuzzy Logic Control in Energy Systems with Design Applications in Matlab®/Simulink®** I.H. Altaş
Volume 92 **Power Quality in Future Electrical Power Systems** A.F. Zobaa and S.H.E.A. Aleem (Editors)
Volume 93 **Cogeneration and District Energy Systems: Modelling, Analysis and & Optimization** M.A. Rosen and S. Koohi-Fayegh
Volume 94 **Introduction to the Smart Grid: Concepts, technologies and evolution** S.K. Salman
Volume 95 **Communication, Control and Security Challenges for the Smart Grid** S.M. Muyeen and S. Rahman (Editors)
Volume 96 **Industrial Power Systems with Distributed and Embedded Generation** R. Belu
Volume 97 **Synchronized Phasor Measurements for Smart Grids** M.J.B. Reddy and D.K. Mohanta (Editors)
Volume 98 **Large Scale Grid Integration of Renewable Energy Sources** A. Moreno-Munoz (Editor)
Volume 100 **Modeling and Dynamic Behaviour of Hydropower Plants** N. Kishor and J. Fraile-Ardanuy (Editors)
Volume 101 **Methane and Hydrogen for Energy Storage** R. Carriveau and D.S.-K. Ting
Volume 104 **Power Transformer Condition Monitoring and Diagnosis** A. Abu-Siada (Editor)
Volume 106 **Surface Passivation of Industrial Crystalline Silicon Solar Cells** J. John (Editor)
Volume 107 **Bifacial Photovoltaics: Technology, applications and economics** J. Libal and R. Kopecek (Editors)
Volume 108 **Fault Diagnosis of Induction Motors** J. Faiz, V. Ghorbanian and G. Joksimović
Volume 109 **Cooling of Rotating Electrical Machines:** Fundamentals, modelling, testing and design D. Staton, E. Chong, S. Pickering and A. Boglietti
Volume 110 **High Voltage Power Network Construction** K. Harker
Volume 111 **Energy Storage at Different Voltage Levels: Technology, integration, and market aspects** A.F. Zobaa, P.F. Ribeiro, S.H.A. Aleem and S.N. Afifi (Editors)
Volume 112 **Wireless Power Transfer: Theory, Technology and Application** N. Shinohara
Volume 114 **Lightning-Induced Effects in Electrical and Telecommunication Systems** Y. Baba and V.A. Rakov
Volume 115 **DC Distribution Systems and Microgrids** T. Dragičević, F. Blaabjerg and P. Wheeler
Volume 116 **Modelling and Simulation of HVDC Transmission** M. Han (Editor)
Volume 117 **Structural Control and Fault Detection of Wind Turbine Systems** H.R. Karimi
Volume 118 **Modelling and Simulation of Complex Power Systems** A. Monti and A. Benigni
Volume 119 **Thermal & Power Plant Control and Instrumentation: The control of boilers and HRSGs, 2nd Edition** D. Lindsley, J. Grist and D. Parker
Volume 120 **Fault Diagnosis for Robust Inverter Power Drives** A. Ginart (Editor)
Volume 121 Monitoring and Control using Synchrophasors in Power Systems with Renewables I. Kamwa and C. Lu (Editors)
Volume 123 **Power Systems Electromagnetic Transients Simulation, 2nd Edition** N. Watson and J. Arrillaga
Volume 124 **Power Market Transformation** B. Murray
Volume 125 **Wind Energy Modeling and Simulation, Volume 1: Atmosphere and plant** P. Veers (Editor)
Volume 126 **Diagnosis and Fault Tolerance of Electrical Machines, Power Electronics and Drives** A.J.M. Cardoso
Volume 128 **Characterization of Wide Bandgap Power Semiconductor Devices** F. Wang, Z. Zhang and E.A. Jones
Volume 129 **Renewable Energy from the Oceans: From wave, tidal and gradient systems to offshore wind and solar** D. Coiro and T. Sant (Editors)
Volume 130 **Wind and Solar Based Energy Systems for Communities** R. Carriveau and D.S.-K. Ting (Editors)
Volume 131 **Metaheuristic Optimization in Power Engineering** J. Radosavljević

Volume 132	**Power Line Communication Systems for Smart Grids** I.R.S. Casella and A. Anpalagan
Volume 134	**Hydrogen Passivation and Laser Doping for Silicon Solar Cells** B. Hallam and C. Chan (Editors)
Volume 139	**Variability, Scalability and Stability of Microgrids** S.M. Muyeen, S.M. Islam and F. Blaabjerg (Editors)
Volume 142	**Wind Turbine System Design: Volume 1: Nacelles, drive trains and verification** J. Wenske (Editor)
Volume 143	**Medium Voltage DC System Architectures** B. Grainger and R.D. Doncker (Editors)
Volume 145	**Condition Monitoring of Rotating Electrical Machines** P. Tavner, L. Ran, C. Crabtree
Volume 146	**Energy Storage for Power Systems, 3rd Edition** A.G. Ter-Gazarian
Volume 147	**Distribution Systems Analysis and Automation, 2nd Edition** J. Gers
Volume 151	**SiC Power Module Design: Performance, robustness and reliability** A. Castellazzi and A. Irace (Editors)
Volume 152	**Power Electronic Devices: Applications, failure mechanisms and reliability** F. Iannuzzo (Editor)
Volume 153	**Signal Processing for Fault Detection and Diagnosis in Electric Machines and Systems** M. Benbouzid (Editor)
Volume 155	**Energy Generation and Efficiency Technologies for Green Residential Buildings** D. Ting and R. Carriveau (Editors)
Volume 156	**Lithium-ion Batteries Enabled by Silicon Anodes** C. Ban and K. Xu (Editors)
Volume 157	**Electrical Steels, 2 Volumes** A. Moses, K. Jenkins, P. Anderson and H. Stanbury
Volume 158	**Advanced Dielectric Materials for Electrostatic Capacitors** Q. Li (Editor)
Volume 159	**Transforming the Grid Towards Fully Renewable Energy** O. Probst, S. Castellanos and R. Palacios (Editors)
Volume 160	**Microgrids for Rural Areas: Research and case studies** R.K. Chauhan, K. Chauhan and S.N. Singh (Editors)
Volume 161	**Artificial Intelligence for Smarter Power Systems: Fuzzy Logic and Neural Networks** M.G. Simoes
Volume 165	**Digital Protection for Power Systems, 2nd Edition** S.K. Salman
Volume 166	**Advanced Characterization of Thin Film Solar Cells** N. Haegel and M. Al-Jassim (Editors)
Volume 167	**Power Grids with Renewable Energy:** Storage, integration and digitalization A.A. Sallam and O.P. Malik
Volume 169	**Small Wind and Hydrokinetic Turbines** P. Clausen, J. Whale and D. Wood (Editors)
Volume 170	**Reliability of Power Electronics Converters for Solar Photovoltaic Applications** F. Blaabjerg, A.l Haque, H. Wang, Z. Abdin Jaffery and Y. Yang (Editors)
Volume 171	**Utility-scale Wind Turbines and Wind Farms** A. Vasel-Be-Hagh and D.S.-K. Ting
Volume 172	**Lighting interaction with Power Systems, 2 volumes** A. Piantini (Editor)
Volume 174	**Silicon Solar Cell Metallization and Module Technology** T. Dullweber (Editor)
Volume 180	**Protection of Electricity Distribution Networks, 4th Edition** J. Gers and E. Holmes
Volume 182	**Surge Protection for Low Voltage Systems** A. Rousseau (Editor)
Volume 184	**Compressed Air Energy Storage: Types, systems and applications** D. Ting and J. Stagner
Volume 186	**Synchronous Reluctance Machines: Analysis, optimization and applications** N. Bianchi, C. Babetto and G. Bacco
Volume 191	**Electric Fuses: Fundamentals and new applications, 4th Edition** N. Nurse, A. Wright and P.G. Newbery
Volume 193	**Overhead Electric Power Lines: Theory and practice** S. Chattopadhyay and A. Das
Volume 194	**Offshore Wind Power Reliability, availability and maintenance, 2nd edition** P. Tavner
Volume 196	**Cyber Security for Microgrids** S. Sahoo, F. Blaajberg and T. Dragicevic
Volume 198	**Battery Management Systems and Inductive Balancing** A. Van den Bossche and A. Farzan Moghaddam
Volume 199	**Model Predictive Control for Microgrids: From power electronic converters to energy management** J. Hu, J. M. Guerrero and S. Islam
Volume 204	**Electromagnetic Transients in Large HV Cable Networks: Modeling and calculations** A. Ametani, H. Xue, T. Ohno and H. Khalilnezhad
Volume 208	**Nanogrids and Picogrids and their Integration with Electric Vehicles** S. Chattopadhyay
Volume 210	**Superconducting Magnetic Energy Storage in Power Grids** M.H. Ali
Volume 211	**Blockchain Technology for Smart Grids: Implementation, management and security** H.L. Gururaj, K.V. Ravi, F. Flammini, H. Lin, B. Goutham, K.B.R. Sunil and C. Sivapragash
Volume 212	**Battery State Estimation: Methods and Models** S. Wang
Volume 213	**Wide Area Monitoring of Interconnected Power Systems, 2nd Edition** A.R. Messina
Volume 215	**Industrial Demand Response: Methods, best practices, case studies, and applications** H. H. Alhelou, A. Moreno-Muñoz and P. Siano (Editors)
Volume 217	**Advances in Power System Modelling, Control and Stability Analysis, 2nd Edition** F. Milano (Editor)
Volume 225	**Fusion-Fission Hybrid Nuclear Reactors: For enhanced nuclear fuel utilization and radioactive waste reduction** W. M. Stacey
Volume 238	**AI for Status Monitoring of Utility Scale Batteries** S. Wang, K. Liu, Y. Wang, D.-I. Stroe, C. Fernandez and J.M. Guerrero
Volume 905	**Power System Protection, 4 volumes** The Electricity Training Association

Advances in Power System Modelling, Control and Stability Analysis

2nd Edition

Edited by
Federico Milano

The Institution of Engineering and Technology

Published by The Institution of Engineering and Technology, London, United Kingdom

The Institution of Engineering and Technology is registered as a Charity in England & Wales (no. 211014) and Scotland (no. SC038698).

First published 2022

The Institution of Engineering and Technology
Futures Place
Kings Way, Stevenage
Hertfordshire, SG1 2UA, United Kingdom

www.theiet.org

British Library Cataloguing in Publication Data
A catalogue record for this product is available from the British Library

ISBN 978-1-83953-575-8 (hardback)
ISBN 978-1-83953-576-5 (PDF)

Typeset in India by MPS Limited
Printed in the UK by CPI Group (UK) Ltd, Croydon

Cover image: Chris McLoughlin via Getty Images, overlaid image is the editor's own.

Contents

About the editors

Federico Milano is a professor at the University College Dublin, Ireland. He has authored 8 books and about 250 scientific papers. He was elevated to IEEE fellow in 2016, IET fellow in 2017, and IEEE PES distinguished lecturer in 2020. He has been an associate editor of several international journals published by IEEE, IET, and other publishers, including the *IEEE Transactions of Power Systems.* In 2022, he became Co-Editor in Chief of the *IET Generation, Transmission & Distribution.*

Muhammad Adeen received B.E. electrical engineering degree from National University of Sciences and Technology, Pakistan in 2013 and M.Sc. degree in electrical engineering from Politecnico di Milano, Italy in 2017. Since May 2018, he is a Ph.D. candidate with the Department of Electrical Engineering in University College Dublin, Ireland. His current research interests include stochastic processes, power system modelling and dynamic analysis.

Vittorio Arcidiacono received the Ph.D. degree (with honours) in Electronic Engineering from the Politecnico of Torino in 1966. After a first work experience in designing special electronic equipment, in 1967 he joined ENEL (former National Electricity Board in Italy) at the Automatica Research Centre (CRA). There he has emerged as one of the leading experts in the fields of excitation and turbine control of generators, secondary and tertiary voltage control of EHV transmission system, power system dynamics and stability, mathematical modelling, high-voltage direct current transmission systems (HVDC), advanced power electronics. He is author of over two hundred scientific publications in the fields mentioned above. He has been the creator and the designer of the secondary and tertiary voltage control of the Italian EHV transmission system. At ENEL, he has been appointed manager in 1978, Head of Systems Department in 1984, Vice Director in 1993 and Director in 1998 of the Automatica Research Centre. Since March 2001, date of his retirement, he cooperates with MAI Control Systems Ltd., Milan, overseeing all the R&D activities.

Alberto Berizzi received his M.Sc. degree (1990) and his Ph.D. degree (1994) both in Electrical Engineering from the Politecnico di Milano. He is now a professor at the Energy Department of the Politecnico di Milano. His areas of research include power system analysis and optimization and power system dynamics.

Federico Bizzarri received the master's degree (magna cum laude) in electronic engineering and the Ph.D. degree in electrical engineering from the University of Genova, Italy, in 1998 and 2001, respectively. Since October 2018, he has been an associate professor at the Electronic and Information Department of the Politecnico di Milano, Italy, where he had been a research assistant since June 2010. From 2002 to 2008, he was a post-doctoral research assistant in the Biophysical and Electronic Engineering Department of the University of Genova. In 2009, he was a post-doctoral research assistant in Advanced Research Center on Electronic Systems for Information and Communication Technologies "E. De Castro" (ARCES) at the University of Bologna, Italy. His main research interests are in the area of circuit theory and simulation, power system theory and simulation, nonlinear dynamical systems, and electrical engineering. He is the author or co-author of about 120 scientific papers, more than half of which have been published in international journals. He is a research fellow of the Advanced Research Center on Electronic Systems for Information and Communication Technologies "E. De Castro" (ARCES) at the University of Bologna, and an IEEE Senior Member since 2014. He served as an Associate Editor of the IEEE Transactions on Circuits and Systems-I and he received the 2012–2013 Best Associate Editor Award of that journal. He is currently serving as an associate editor for the IEEE Open Journal of Circuits and Systems and the NOLTA Journal.

Alberto Borghetti graduated (with honours) in electrical engineering from the University of Bologna, Italy, in 1992. Since then, he has been working with the power system group of the same University, where he was appointed as a researcher in 1994 and associate professor in 2004. His research and teaching activities are in the areas of power system analysis, with reference to voltage collapse, power system restoration after black-out, electromagnetic transients, optimal generation scheduling, and distribution system operation. He is a fellow of the Institute of Electrical and Electronics Engineers (class 2015) for contributions to modeling of power distribution systems under transient conditions. He has served as Technical Program Committee chairperson of the 30th International Conference on Lightning Protection (Cagliari, Italy, 2010). Currently, he serves as an editor of IEEE Transactions on Smart Grid and is on the editorial board of Electric Power System Research.

Daniele Bosich is an assistant professor of Microgrids for the Sustainable Energy at the University of Trieste (Italy). He received the M.Sc. degree (Hon) in electrical engineering at the University of Trieste (Italy) in 2010, and the Ph.D. degree in energy engineering at the University of Padua (Italy) in 2014. He is the author of more than 60 papers in the field of marine shipboard power systems, microgrid modeling, voltage control and nonlinear systems analysis. Dr Bosich is an IEEE Senior Member (PES, IAS, IES and VTS societies).

Cristian Bovo received his M.Sc. degree (1998) and his Ph.D. degree (2002) in Electrical Engineering from the Politecnico di Milano, where he is now an associate professor at the Department of Energy. His areas of research include power system analysis and control, voltage stability, optimization, and electricity markets.

Angelo Brambilla received the Dr Ing. degree in electronics engineering cum Laude from the University of Pavia, Pavia, Italy, in 1986. Since 2020, he is a full professor at the Dipartimento di Elettronica, Informazione e Bioingegneria,Politecnico di Milano,

Milano, Italy. His main research areas are circuit analysis, simulation, and modeling of both micro-electronics circuits and power systems. He was an associate editor of IEEE Transaction on Circuits and Systems II: Express Brief (TCASII) from 2008 to 2011.

Sergio Bruno received his Ph.D. in Electrotechical Engineering from the Politecnico di Bari in 2004. He was a consultant for energy and environment for the Strategic Plan of the Bari Metropolitan Area and is currently involved in research activities with the Politecnico di Bari. He is an adjunct professor of Power Systems at the Politecnico di Bari. Most recent studies deal with power system dynamics & control, monitoring and control of smart distribution grids, and energy hubs.

Claudio Cañizares is a University Professor, Hydro One Chair, and the WISE Executive Director at the University of Waterloo, where he has been since 1993. His highly-cited research focuses on modeling, simulation, computation, stability, control, and optimization of power and energy systems. He is the IEEE Trans. Smart Gird EIC; Director of the IEEE and PES Boards; a Fellow of the IEEE, the Royal Society of Canada, and the Canadian Academy of Engineering; and has received the 2017 IEEE PES Outstanding Educator Award, the 2016 IEEE Canada Electric Power Medal, and multiple awards and recognitions from PES technical committees.

Massimiliano Chiandone was graduated from University of Udine (Italy) in Computer Science (M.Sc.), from University of Trieste (Italy) in Electrical Engineering (B.Sc.) and obtained his PhD in Industrial Engineering from University of Padova (Italy). He has been working for several years as software developer and system administrator at MSC Software Corporation and at Elettra Synchrotron Light Source research facility. He is a researcher at University of Trieste, his main interests are in real-time control systems and applications to electric power system.

Konstantina Christakou was graduated from the National Technical University of Athens in 2010. She obtained her Ph.D. in 2015 from École Polytechnique Fédérale de Lausanne (EPFL), Switzerland where she worked under the joint supervision of Professors Jean-Yves Le Boudec (LCA2) and Mario Paolone (DESL). Since February 2016, she is a postdoctoral researcher in the group of Professor Mario Paolone (DESL). Her research interests include control and real-time operation of electrical distribution grids with reference to voltage control and lines congestion management, demand side management and storage applications for active distribution networks, optimal power flow algorithms and optimization under uncertainty.

Emanuele Crisostomi received a B.Sc. degree in computer science engineering, M.Sc. degree in automatic control, and Ph.D. degree in automatics, robotics, and bioengineering from the University of Pisa, Italy, in 2002, 2005, and 2009, respectively. He is currently an associate professor of Electrotechnics with the Department of Energy, Systems, Territory, and Constructions Engineering, University of Pisa. His research interests include control and optimization of large-scale systems, including smart grids, and green mobility networks. He is a co-author of the recently published book on "Electric and Plug-in Vehicle Networks: Optimization and Control", CRC Press, Series: Automation and Control Engineering, 2017, and an Editor of the Springer Nature book on "Analytics for the Sharing Economy: Mathematics, Engineering and Business Perspectives", 2020.

Salvatore D'Arco received the M.Sc. and Ph.D. degrees in electrical engineering from the University of Naples "Federico II," Naples, Italy, in 2002 and 2005, respectively. From 2006 to 2007, he was a postdoctoral researcher at the University of South Carolina, Columbia, SC, USA. In 2008, he joined ASML, Veldhoven, the Netherlands, as a Power Electronics Designer consultant, where he worked until 2010. From 2010 to 2012, he was a postdoctoral researcher in the Department of Electric Power Engineering at the Norwegian University of Science and Technology (NTNU), Trondheim, Norway. In 2012, he joined SINTEF Energy Research where he currently works as a senior research scientist. He is the author of more than 130 scientific papers and is the holder of one patent. His main research activities are related to control and analysis of power-electronic conversion systems for power system applications, including real-time simulation and rapid prototyping of converter control systems.

Ioannis Dassios received his Ph.D. in Applied Mathematics from the Department of Mathematics, University of Athens, Greece, in 2013. He worked as a postdoctoral research and teaching fellow in optimization at the School of Mathematics, University of Edinburgh, UK. He also worked as a research associate at the Modelling and Simulation Centre, University of Manchester, UK, and as a research fellow at MACSI, University of Limerick, Ireland. He is currently a UCD research fellow at UCD, Ireland.

Giovanni De Carne received the B.Sc. and M.Sc. degrees in electrical engineering from the Polytechnic University of Bari, Italy, in 2011 and 2013, respectively, and the Ph.D. degree from the Chair of Power Electronics, Kiel University, Germany, in 2018. He is currently the head of the ìReal Time System Integrationî Group and head of the "Power Hardware In the Loop Lab" at the Institute for Technical Physics at the Karlsruhe Institute of Technology. In 2020, Dr-Ing. De Carne has been awarded with the Helmholtz "Young Investigator Group" for the project "Hybrid Networks: a multi-modal design for the future energy system." He has authored/coauthored more than 60 peer-reviewed scientific papers. His research interests include power electronics transformers, real-time modeling, and power hardware in the loop. He is the Chairman of the IEEE PES Task Force on "Solid State Transformer integration in distribution grids." He is an associate editor of the IEEE Industrial Electronics Magazine and IEEE Open Journal of Power Electronics.

Davide del Giudice received the M.S. degree (cum laude) and the Ph.D. degree in electrical engineering from Politecnico di Milano, Italy, in 2017 and 2022 respectively. His main research activities are related to simulation techniques specifically tailored to electric power systems characterized by high penetration of renewable energy sources, such as HVDC systems.

Ekaterina Dudkina received her M.Sc. degree in energy management from Peter the Great St. Petersburg Polytechnic University, Saint-Petersburg, Russia, in 2014. After that, she continued her education and obtained a Master in Electrical Engineering for Smart Grids in Grenoble INP, Grenoble, France, in 2016. Currently, she is pursuing a Ph.D. degree from the University of Pisa, Pisa, Italy. Her current research interests are on peer-to-peer energy markets and graph theory.

Aboutaleb Haddadi received the Ph.D. degree in electrical and computer engineering from McGill University, Montréal, QC, Canada, in 2015. From 2015 to 2020, he was a postdoctoral fellow and research associate with Polytechnique Montréal, Montréal, QC, Canada. Since 2020, he has been with the Transmission Operations and Planning R&D group of the Electric Power Research Institute (EPRI), USA, as a Senior Engineer Scientist (2020–2022) and a technical leader (since 2022). He is the EPRI Manager of Advanced Grid Innovation Lab for Energy (AGILe) at New York Power Authority (NYPA), White Plains, NY, USA. Dr Haddadi is an IEEE Senior Member and Chair of CIGRE working group C4.60 on Generic EMT Modeling of Inverter-Based Resources for Long-Term Planning Studies. He is further active in several working groups of North American Electric Reliability Corporation (NERC), IEEE Power & Energy Society (PES), and IEEE Power System Relaying and Control Committee (PSRC). He has consulting experience in power system modeling and simulation. Dr Haddadi has (co-)authored more than 60 technical publications which have been recognized as best IEEE conference papers and top-list CIGRE technical brochure. His research interests include renewable energy integration, transmission system protection and the impact of renewables, and power system modeling and simulation.

Tobias Heins received the M.Sc. degree in electrical engineering from RWTH Aachen University in 2020. He is currently a research associate with the Institute for Automation of Complex Power Systems, EON Energy Research Center, RWTH Aachen University. His research interests include analysis and control of converter-interfaced power generation and power systems.

Keijo Jacobs received the B.Sc. degree in electrical engineering, information technology, and computer engineering and the M.Sc. degree in electrical power engineering from RWTH Aachen University, Aachen, Germany, in 2011 and 2015, respectively. He is currently working toward the Ph.D. degree in the field of silicon carbide semiconductor devices and HVdc converters with the Division of Electrical Power and Energy Systems, KTH Royal Institute of Technology, Stockholm, Sweden. Since 2015, he has been a part of the Division of Electrical Power and Energy Systems, KTH Royal Institute of Technology, Stockholm.

Guðrún Margrét Jónsdóttir received a B.Sc. in Electrical and Computer Eng. for the University of Iceland in 2013, a M.Sc. in Electric Power Eng. from KTH, Stockholm, Sweden in 2015 and a Ph.D. from University College Dublin in Ireland. She is currently working for power systems engineering consultant at Norconsult Iceland.

Ulas Karaagac received the B.Sc. and M.Sc. degrees in electrical and electronics engineering from the Middle East Technical University, Ankara, Turkey, in 1999 and 2002, respectively, and the Ph.D. degree in electrical engineering from Polytechnique Montréal, Montréal, Canada, in 2011. He was a Research and Development Engineer with the Information Technology and Electronics Research Institute (BILTEN), Scientific and Technical Research Council of Turkey (TUBI-TAK), Ankara, Turkey, from 1999 to 2007. He was also a postdoctoral fellow with Polytechnique Montréal, from 2011 to 2013, and a research associate, from 2013 to 2016. In 2017, he joined the Department of Electrical Engineering, The Hong Kong Polytechnic University, China, where he is currently an assistant professor. His research interests include integration of large-scale renewables into power grids, modeling and simulation of large-scale power systems, and power system dynamics and control.

Houshang Karimi received the Ph.D. degree in electrical engineering from the University of Toronto, Toronto, ON, Canada, in 2007. He was an assistant professor with the Department of Electrical Engineering, Sharif University of Technology, Tehran, Iran, from 2009 to 2012. In 2013, he joined the Department of Electrical Engineering, Polytechnique Montreal, Montreal, QC, Canada, where he is currently a full professor. His research interests include control systems, microgrid control, and smart grids.

Massimo La Scala received his Ph.D. in Electrical Engineering from the University of Bari, Italy, in 1989. In 1987, he joined ENEL. He is currently a professor of power systems at the Politecnico di Bari, Italy. His research interests are in the areas of power system analysis and control. In 2007, he was appointed as a fellow member of the IEEE PES for his contributions to computationally efficient power system dynamic performance simulation and control.

Jean-Yves Le Boudec is a professor at EPFL and fellow of the IEEE. He graduated from Ecole Normale Supérieure de Saint-Cloud, Paris, where he obtained the Agrégation in Mathematics in 1980 and received his doctorate in 1984 from the University of Rennes, France. From 1984 to 1987, he was with INSA/IRISA, Rennes. In 1987, he joined Bell Northern Research, Ottawa, Canada, as a member of scientific staff in the Network and Product Traffic Design Department. In 1988, he joined the IBM Zurich Research Laboratory where he was a manager of the Customer Premises Network Department. In 1994, he became an associate professor at EPFL. His interests are in the performance and architecture of communication systems and smart grids. He co-authored a book on network calculus, which forms a foundation to many traffic control concepts in the internet, an introductory textbook on Information Sciences and is also the author of the book "Performance Evaluation."

Daniele Linaro received his M.Sc. in electronic engineering from the University of Genoa (Italy) in 2007 and a Ph.D. in electrical engineering from the same university in 2011. In the same year, he was awarded a fellowship from the Flemish Research Foundation – FWO to conduct postdoctoral work in the Laboratory of Theoretical Neurobiology and Neuroengineering at the University of Antwerp, under the supervision of Prof. Michele Giugliano, where he used dynamic-clamp to elucidate the computational capabilities of different cell types in the rodent cortex. In 2015, he moved to the Laboratory of Cortical Development at VIB (Leuven, Belgium), under the supervision of Prof. Pierre Vanderhaeghen, where he studied the protracted electrophysiological and morphological development of human cortical neurons grafted in the rodent cortex. From 2014 until 2018 he held a position as a visiting scientist at Janelia Research Campus in collaboration with the laboratory of Dr Nelson Spruston, where he investigated the network properties of different cell types in the rodent hippocampus. Since 2018, he is an assistant professor in the Department of Electronics, Information Technology and Bioengineering at the Polytechnic of Milan. His main research interests are currently in the area of circuit theory and non-linear dynamical systems, with applications to electronic oscillators and power systems and in the field of computational neuroscience, in particular biophysically realistic single-cell models of neuronal cells.

Marco Liserre received the M.Sc. and Ph.D. degrees in electrical engineering from the Polytechnic University of Bari, Bari, Italy, in 1998 and 2002, respectively. He has been an associate professor with the Polytechnic University of Bari. Since 2012, he has been a professor of reliable power electronics with Aalborg University, Aalborg, Denmark. Since 2013, he has been a full professor and holds the Chair of Power Electronics at Kiel University, Kiel, Germany. In 2022, he joined, part-time, Fraunhofer ISIT, Itzehoe, Germany, as the Deputy Director and the Director of a new Center for Electronic Energy Systems funded for five million Euros. He has published more than 600 technical papers (one-third of them in international peer-reviewed journals), a book, and two granted patents, with more under evaluation, some of them involving companies. These works have received more than 45,000 citations. He was listed in ISI Thomson Report "The World's Most Influential Scientific Minds" in 2014. Prof. Liserre is also a member of the IEEE Industry Applications Society (IAS), the IEEE Power Electronics Society (PELS), the IEEE Power and Energy Society (PES), and the IEEE Industrial Electronics Society (IES); he has been serving all these societies in different capacities. He received the IES 2009 Early Career Award, the IES 2011 Anthony J. Hornfeck Service Award, the 2014 Dr Bimal Bose Energy Systems Award, the 2017 IEEE PELS Sustainable Energy Systems Technical Achievement Award, the 2018 IEEE IES Mittelmann Achievement Award, and six IEEE Journal Awards. In PELS, he is an AdCom Member (second mandate), a Co-Editor of the *IEEE Open Journal of Power Electronics*, an associate editor of *IEEE Transactions on Power Electronics*, and *IEEE Journal of Emerging and Selected Topics in Power Electronics* (JESTPE), a guest editor of several special issues of JESTPE, the Technical Committee Chairperson of the new Committee on Electronic Power Grid Systems, a member of the IEEE Digital Committee, the IESLiaison responsible, the eGrid 2021 Workshop Co-Chairperson, and the PEDG 2022 Co-Chairperson.

Muyang Liu received the M.E. and Ph.D. in electrical energy engineering from University College Dublin, Ireland, in 2016 and 2019. From 2019 to 2020, she was a senior researcher with University College Dublin. She is currently working in Xinjiang University, Urumqi, China, where she is currently an associate professor of

Power System Modelling and Simulation. Her research interests include the monitor, modelling and control of low-inertia power system.

Jean Mahseredjian received the Ph.D. degree in electrical engineering from Polytechnique Montréal, Montréal, Canada, in 1991. From 1987 to 2004, he was with IREQ (Hydro-Quebec), Montréal, Canada, where he was involved in research and development activities related to the simulation and analysis of electromagnetic transients. In December 2004, he joined the Faculty of Electrical Engineering, Polytechnique Montréal. Jean Mahseredjian is IEEE fellow. Jean Mahseredjian is the creator and lead-developer of the EMTP® software.

Federico Milano received from the University of Genoa, Italy, the Electrical Engineer degree and the Ph.D. in Electrical Engineering in March 1999 and June 2003, respectively. From September 2001 to December 2002, he worked at the Electrical & Computer Engineering Department of the University of Waterloo, Canada, as a visiting scholar. He was with the Department of Electrical Engineering of University of Castilla-La Mancha, Spain, from September 2003 to May 2013. In June 2013, he joined the UCD School of Electrical and Electronic Engineering, where he is currently Professor of Power Systems Control and Protections. In January 2016, he was elevated to IEEE fellow for his contributions to power system modelling and simulation. In December 2017, he was elevated to IET fellow. In February 2020, he joined the IEEE PES Distinguished Lecturer Program. He is or has been an editor of several international journals published by IEEE, IET, Elsevier, and Springer, including the *IEEE Transactions on Power Systems and IET Generation, Transmission & Distribution*.

Antonello Monti received the M.Sc. (summa cum laude) and Ph.D. degrees in electrical engineering from the Politecnico di Milano, Milano, Italy, in 1989 and 1994, respectively. He started his career with Ansaldo Industria and then moved in 1995 to the Politecnico di Milano as an assistant professor. In 2000, he joined the Department of Electrical Engineering, University of South Carolina, Columbia, SC, USA, as an associate professor and later became a full professor. Since 2008, he has been the Director of the Institute for Automation of Complex Power System, E.ON Energy Research Center, RWTH Aachen University, Aachen, Germany. Since 2019 he also has a joined appointment at Fraunhofer FIT within the Center for Digital Energy Aachen. He has authored or co-authored more than 400 peer-reviewed papers published in international journals and in the proceedings of international conferences. Dr Monti is an associate editor for the IEEE Electrification Magazine. He is a member of the Editorial Board of the Sustainable Energy, Grids and Networks, and a member of the Founding Board of Springer Energy Informatics. In 2017, he was the recipient of the IEEE Innovation in Societal Infrastructure Award.

Mohammed Ahsan Adib Murad received B.Sc. degree in electrical engineering from IUT, Bangladesh, in 2009, and a double M.Sc. degree in smart electrical networks and systems from KU Leuven, Belgium and KTH, Sweden in 2015. He received a Ph.D.

degree in electrical engineering from University College Dublin, Ireland and is currently working with DIgSILENT GmbH, Germany, as an application developer. His research interests include power system modeling, simulation, and dynamic analysis.

Fabio Napolitano is an assistant professor at Department of Electrical, Electronic and Information Engineering of the University of Bologna, Italy. From the same university, he received the degree (with hons.) in electrical engineering in 2003 and the Ph.D. degree in 2009. Since his M.Sc. graduation, he collaborated with the Power Systems group of the University of Bologna on the analysis of power systems transients, in particular those due to indirect lightning strikes, and lightning protection systems.

Carlo Alberto Nucci is a full professor and head of the Power Systems Laboratory of the Department of Electrical, Electronic and Information Engineering "Guglielmo Marconi" of the University of Bologna, Italy. He is author or co-author of over 300 scientific papers published on peer-reviewed journals or on proceedings of international conferences, of five book chapters edited by IEE (two), Kluwer, Rumanian Academy of Science and WIT press and of a couple of IEEE Standards and some CIGRE technical brochures. He is a fellow of the IEEE and of the IET, CIGRE honorary member and has received some best paper/technical international awards, including the CIGRE Technical Committee Award and the ICLP Golde Award. From January 2006 to September 2012, he has served as Chairman of Cigré Study Committee C4 "System Technical Performance." Since January 2010, he is serving as an editor in chief of the Electric Power Systems Research journal, Elsevier. Prof. Nucci is doctor *honoris causa* of the University Politehnica of Bucharest and corresponding member of the Bologna Science Academy. He has served as the President of the Italian Group of University Professors of Electrical Power Systems (GUSEE) from 2012 to 2015. He is an advisor of the Global Resource Management Program of Doshisha University, Kyoto, supported by the Japanese Ministry of Education and Science. He is the coordinator of the Working Group "Smart City" of the University of Bologna, has been serving as member of the EU Smart City Stakeholder Platform since 2013, and since 2014 is representing PES in the IEEE Smart City Initiatives Program.

Luca Papini received the bachelor's (Hons.) and master's (Hons.) degrees in Electrical Engineering from the University of Pisa, Italy, in 2009 and 2011, respectively, and the Ph.D. degree from the University of Nottingham, UK, in 2018. He has been a research assistant with the University of Nottingham, since 2013. He was JSPS Fellow in 2018 and is currently a senior researcher with the University of Pisa. His research interests include high-speed, high power density electric machines, machine control, and levitating systems.

Mario Paolone received the M.Sc. (Hons.) and Ph.D. degrees in electrical engineering from the University of Bologna, Italy, in 1998 and 2002, respectively. In 2005, he was an assistant professor of Power Systems with the University of Bologna, where he was with the Power Systems Laboratory until 2011. Since 2011, he has been with the Swiss Federal Institute of Technology, Lausanne, Switzerland, where he is a full professor and the Chair of the Distributed Electrical Systems Laboratory. His research interests focus on power systems with reference to real-time monitoring and operational aspects, power system protections, dynamics, and transients. He has authored or co-authored over 300 papers published in mainstream journals and international conferences in the area of energy and power systems that received numerous awards, including the IEEE EMC Technical Achievement Award, two IEEE Transactions on EMC Best Paper Awards, the IEEE Power System Dynamic Performance Committee's Prize Paper Award, and the Basil Papadias Best Paper Award at the 2013 IEEE PowerTech. He was the founder editor-in-chief of the Sustainable Energy, Grids and Networks (Elsevier). He was elevated IEEE Fellow in 2022.

Antonio Pepiciello received his B.S., M.S., and Ph.D. in energy engineering from University of Sannio, Benevento, Italy, where he is currently a postdoctoral scholar. His research interests include integration of renewable energy sources in power systems, power system dynamics, decision making under uncertainty, and time synchronization of sensor networks.

Farhad Rachidi received the M.Sc. degree in electrical engineering and the Ph.D. degree from the Swiss Federal Institute of Technology, Lausanne, in 1986 and 1991, respectively. He worked at the Power Systems Laboratory of the same institute until 1996. In 1997, he joined the Lightning Research Laboratory of the University of Toronto in Canada and from April 1998 until September 1999, he was with Montena EMC in Switzerland. He is currently a Titular Professor and the head of the EMC Laboratory at the Swiss Federal Institute of Technology (EPFL), Lausanne, Switzerland. Dr Rachidi served as the Vice-Chair of the European COST Action on the Physics of Lightning Flash and its Effects (2005–2009), the Chairman of the 2008 European Electromagnetics International Symposium (EUROEM), the President of the International Conference on Lightning Protection (2008–2014), and the Editor-in-Chief of the IEEE Transactions on Electromagnetic Compatibility (2013–2016). He is currently the President of the Swiss National Committee of the International Union of Radio Science (URSI) and a member of the Advisory Board of the IEEE Transactions on Electromagnetic Compatibility. Farhad Rachidi is the author or co-author of 140 scientific papers published in peer-reviewed journals and over 350 papers presented at international conferences. In 2005, he was the recipient of the IEEE Technical Achievement Award and the Cigré Technical Committee Award. In 2006, he was awarded the Blondel Medal from the French Association of Electrical Engineering, Electronics, Information Technology and Communication (SEE). In 2014, Farhad Rachidi was conferred the title of Honorary Professor of the Xi'an Jiaotong University in China.

Paolo Romano received his B.Sc. and M.Sc. degrees (with honours) in electronics engineering from the University of Genova, Italy, in 2008 and 2011, respectively, and is Ph.D. from the Swiss Federal Institute of Technology of Lausanne in 2016. He is currently postdoctoral fellow at the Distributed Electrical Systems Laboratory (DESL) of the Swiss Federal Institute of Technology of Lausanne. His research interests refer to the synchrophasor area and particularly to the development of advanced Phasor Measurement Units for the real-time monitoring of active distribution networks.

Marcos Rubinstein received the M.Sc. and Ph.D. degrees in electrical engineering from the University of Florida, Gainesville, FL, USA, in 1986 and 1991, respectively. In 1992, he joined the Swiss Federal Institute of Technology, Lausanne, Switzerland, where he was involved in the fields of electromagnetic compatibility and lightning. In 1995, he was with Swisscom, where he worked in numerical electromagnetics and EMC. In 2001, he moved to the University of Applied Sciences of Western Switzerland HES-SO, Yverdon-les-Bains, where he is currently a full professor, Head of the advanced Communication Technologies Group and a member of the IICT Institute Team. He is the author or co-author of more than 200 scientific publications in reviewed journals and international conferences. He is also the co-author of six book chapters. He is the Chairman of the International Project on Electromagnetic Radiation from Lightning to Tall structures and the editor-in-chief of the *Open Atmospheric Science Journal* and serves as an associate editor of the IEEE Transactions on Electromagnetic Compatibility. Prof. Rubinstein received the best Master's Thesis award from the University of Florida. He received the IEEE achievement award, and he is a co-recipient of the NASA's Recognition for Innovative Technological Work award. He is a fellow of the IEEE and of the SUMMA Foundation, a member of the Swiss Academy of Sciences and of the International Union of Radio Science.

Roberto S. Salgado received his B.Sc. degree in electrical engineering from Federal University of Pará in 1976, and the M.Sc. and Ph.D. degrees in electrical engineering from Federal University of Santa Catarina, in 1981, and from University of Manchester Institute of Science and Technology (UMIST), UK, in 1989, respectively. Since 1978, he has been with Federal University of Santa Catarina. His main research interests are in the field of power system analysis, particularly in optimal power flow, state estimation, voltage control, and stability and security analysis.

Styliani Sarri received the diploma degree from the Electrical and Computer Engineering Department, Aristotle University of Thessaloniki, Thessaloniki, Greece, in 2011, and the Ph.D. degree from the École Polytechnique Fédérale de Lausanne, Lausanne, Switzerland, in 2016. Her current research interests include real-time electrical network monitoring and control with particular focus on state estimation of active distribution networks, real-time operation of electrical networks, and development of real-time models.

Younes Seyedi received the Ph.D. degree in electrical engineering from Polytechnique Montreal, Montreal, QC, Canada, in 2017. He is currently a postdoctoral research fellow in the Department of Electrical Engineering at Polytechnique Montreal. His current research is focused on smart grids, data analytics and networked systems.

Robert Shorten is a professor of Cyberphysical Systems Design at Imperial College London. He was a co-founder of the Hamilton Institute at Maynooth University, and led the Optimisation and Control team at IBM Research Smart Cities Research

Lab in Dublin Ireland. He has been a visiting professor at TU Berlin and a research visitor at Yale University and Technion. He is the Irish member of the European Union Control Association assembly, a member of the IEEE Control Systems Society Technical Group on Smart Cities, and a member of the IFAC Technical Committees for Automotive Control, and for Discrete Event and Hybrid Systems. He is a co-author of the recently published books AIMD Dynamics and Distributed Resource Allocation (SIAM 2016) and Electric and Plug-in Vehicle Networks: Optimisation and Control (CRC Press, Taylor and Francis Group, 2017).

Rifat Sipahi received the B.Sc. degree in mechanical engineering from Istanbul Technical University, Istanbul, Turkey, in 2000, and the M.Sc. and Ph.D. degrees in mechanical engineering from the University of Connecticut, Storrs, CT, USA, in 2003 and 2005, respectively. He was a post-doctoral fellow with the HeuDiaSyC (CNRS) Laboratory, Université de Technologie de Compiègne, Compiègne, France, from 2005 to 2006. In 2006, he joined the Department of Mechanical and Industrial Engineering, Northeastern University, Boston, MA, USA, where he is currently a professor. His current research interests include stability, stabilization of dynamical systems at the interplay between multiple time delays and network graphs, human–machine systems, and human–robotic interactions. He is a fellow of ASME, Senior Member of IEEE, and an associate editor of Automatica.

Giorgio Sulligoi earned the Ph.D. (University of Padua, 2005) and the M.Sc. (University of Trieste, 2001, Hon), both in Electrical Engineering. He is the founder and director of the Digital Energy Transformation & Electrification Facility (D-ETEF) at the Department of Engineering and Architecture of the University of Trieste. He is a full professor of electric power generation and control and appointed Full Professor of Shipboard Electrical Power Systems. He is the author of more than 150 scientific papers in the fields of shipboard power systems, all-electric ships, generators modeling and voltage control.

Jon Are Suul received the M.Sc. degree in energy and environmental engineering and the Ph.D. degree in electric power engineering from the Norwegian University of Science and Technology (NTNU), Trondheim, Norway, in 2006 and 2012, respectively. From 2006 to 2007, he was with SINTEF Energy Research, Trondheim, where he was working with simulation of power electronic converters and marine propulsion systems until starting his Ph.D. studies. Since 2012, he has been a Research Scientist with SINTEF Energy Research, first in a part-time position while working as a part-time postdoctoral researcher with the Department of Electric Power Engineering of NTNU until 2016. Since August 2017, he has been an adjunct associate professor with the Department of Engineering Cybernetics, NTNU. His research interests are mainly related to modeling, analysis, and control of power electronic converters in power systems, renewable energy applications, and electrification of transport.

Dan-Cristian Tomozei received undergraduate degrees at the École Polytechnique, Paris, France, and received the Ph.D. degree from the University "Pierre et Marie Curie" (UPMC), Paris, France, in 2011. During the Ph.D. degree, he was with the

Technicolor Paris Research Lab. He developed distributed algorithms for congestion control and content recommendation in peer-to-peer networks. Since March 2011, he has been a postdoctoral researcher at the École Polytechnique Fédérale de Lausanne (EPFL), Switzerland. He is working in the group of Prof. J.-Y. Le Boudec (LCA2) on communication and control mechanisms for the smart grid.

Georgios Tzounas is a postdoctoral researcher at ETH Zurich, Switzerland. He received from the National Technical University of Athens, Greece, the Diploma (ME) in Electrical and Computer Engineering in 2017, and the PhD in Electrical Engineering from University College Dublin, Ireland, in 2021. In 2020, he was for 4 months a visiting student researcher at Northeastern University, Boston, MA, USA. After his Ph.D. and before joining ETH Zurich, he worked as a senior power systems researcher with University College Dublin. His research interests include modelling, stability analysis, and automatic control of power systems.

Alfredo Vaccaro received the M.Sc. (Hons.) degree in electronic engineering from the University of Salerno, Salerno, Italy, and the Ph.D. degree in electrical and computer engineering from the University of Waterloo, Waterloo, ON, Canada. From 1999 to 2002, he was an assistant researcher with the Department of Electrical and Electronic Engineering, University of Salerno. From March 2002 to October 2015, he was an assistant professor of electric power systems with the Department of Engineering, University of Sannio, Benevento, Italy, where he is currently an associate professor of electrical power system. His research interests include soft computing and interval-based method for uncertain power system analysis, and decentralized architectures for smart grids computing.

Felix Wald received his bachelor's degree from Berlin University of Applied Sciences in 2019 and his master's degree from the Karlsruher Institute of Technology in 2021, both in the field of in electrical engineering. He is currently working towards his Ph.D. degree as part of the "Real Time System Integration" group and "Power Hardware-in-the-Loop" Lab at the Institute for Technical Physics at the Karlsruhe Institute of Technology. His research interests power hardware-in-the-loop as well as the technical and economic investigation of power electronic transformers.

Lorenzo Zanni received the B.Sc. and M.Sc. degrees, both with honors, in electrical engineering from the University of Bologna, Bologna, Italy, in 2010 and 2012, respectively. He is currently pursuing the Ph.D. degree with the Distributed Electrical System Laboratory, École Polytechnique Fédérale de Lausanne, Lausanne, Switzerland. His current research interests include real-time monitoring and control of active distribution networks with particular focus on state estimation using phasor measurement units, and synchrophasor-based fault location.

Rafael Zárate-Miñano received the electrical engineering degree and the Ph.D. degree from the University of Castilla-La Mancha, Ciudad Real, Spain, in April 2005 and July 2010, respectively. He is currently an associate professor at the University of Castilla-La Mancha, Almadén, Spain. His research interests include power system modelling, operations and stability, as well as optimization and numerical methods.

Weilin Zhong received the B.E. degree from Hunan University, China, in 2016, and M.Sc. in Advanced Control and System Engineering from the University of Manchester, UK, in 2017. In February 2022, he obtained the Ph.D. degree in Electrical Engineering from the University College Dublin, Ireland. His current research interests include inertia estimation and frequency control of virtual power plants, stability analysis and control of distributed energy resources, and co-simulation for power systems and communication networks.

Preface to the 1st edition

Electric energy is a fundamental component of the productive processes of any economic sector. The mission of modern and future power systems is to supply electric energy satisfying conflicting requirements: reliability/security of supply, economy, and finally, environmental protection. With this regard, security, quality, and stability of the electric energy supply are key aspects to maintain the productivity of the industrial sector. On the one hand, the progressive increasing concern about climate change and the effects of energy production in greenhouse gas emissions has led to the wide integration of renewable energy sources with obvious advantages in the environmental behaviour of power systems. On the other hand, the integration of communication systems has led to the redefinition of standards and practices of transmission and distribution systems and to the new concept of "smart grid". In this context, the integration of new technologies passes through the definition and validation of advanced techniques for the modelling, planning, monitoring, and control of power systems. These technical innovations point out the need either to reformulate some conventional modelling and control aspects with a modern perspective, or to address new aspects and phenomena related to issues that have not been considered in the past.

The aim of this book is to provide a collection of studies that, while focusing on specific topics, are able to capture the variety of new methodologies and technologies that are changing the way modern electric power systems are modelled, simulated and operated. The approach of the book mixes theoretical aspects with practical considerations, as well as benchmarks test systems and real-world applications. With this aim, the book is divided into three parts, namely modelling, control and stability analysis. Part I presents research works on power system modelling and includes applications of telegrapher equations, power flow analysis with inclusion of uncertainty, discrete Fourier transformation and stochastic differential equations. Part II focuses on power system operation and control and presents insights on optimal power flow, real-time control and state estimation techniques. In this part, special attention is devoted to distribution systems. Finally, Part III describes advances in the stability analysis of power systems and covers voltage stability, transient stability, time delays, and limit cycles.

The book provides a glance on the state-of-the-art of the research that has been carried out in the last decades by the authors of each chapter. The common background of lead authors is the unifying thread of the whole book. In particular, the lead authors of each chapter have obtained their degree in Electrical Engineering in Italy, which provides solid theoretical basis on modelling and stability analysis of non-linear systems and, in most of the cases, have extended their knowledge and improved their skills with visiting periods in prestigious European and North American universities. These experiences have led to several fruitful international collaborations as well as career opportunities, as shown by the biographies included in the list of contributors. We believe that this unique cultural mix provides an added value to the book, which, as a whole, offers the reader an unconventional viewpoint on current research on electric power systems.

Federico Milano
Dublin, February 2016

Preface to the 2nd edition

About a year ago, in the middle of the pandemic, I was contacted by Christoph von Friedeburg, the IET Senior Commissioning Editor that helped me prepare the first edition of this book in 2016. He suggested to start thinking of a new edition since already 5 years had passed. This kind invitation was a chance for me to go back to this book with a new perspective: how many of the advances that were timely and state-of-the-art in 2016 remained so? What was missing?

Answering these questions led me to realise that in the first edition there were two kinds of chapters: "classics," which provided a milestone on a certain topic; and "work in progress," which focussed on a topic still object of intense research. Moreover, I noticed that the first edition did not, surprisingly enough, cover appropriately the modelling, control, and stability analysis of converter-interfaced devices. The objectives of this new edition, thus, are twofold: updating chapters that were "in progress" in 2016 and filling the gap on power electronic converters.

Four chapters of the first edition, namely Chapters 2 and 4 in Part I (Modelling) and Chapters 11 and 12 in Part III (Stability Analysis), have been updated with recent findings. These updates reflect the rapid evolving of the research that focus on the dynamic performance of power systems.

The reader will then find eight new chapters, namely Chapters 5 and 6 in Part I (Modelling), Chapters 11–14 in Part II (Control), and Chapters 19 and 20 in Part III (Stability Analysis). The authors of all chapters are widely well-known and respected authorities in their fields of expertise. Moreover, in the same vein of the first edition, wherever it was possible, new chapters are co-authored by a senior author who has obtained their degree in Italy, which gives to the matter a characteristic theoretical cut. These new chapters cover several relevant and timely aspects of the dynamic behaviour of converters and the resources that are connected to them. These include emerging topics, such as electric vehicle chargers, smart transformers, and microgrids. Particular attention has been devoted to the modeling as well as numerical simulation aspects of power electronic converters, topics that are currently object of intense discussions in the power system community.

We trust that the reader will find in this new edition the same spirit of the first one, namely an unconventional, rigorous, and timely viewpoint on current research on electric power systems.

Federico Milano
Dublin, March 2022

Part I

Modelling

Chapter 1
Telegrapher's equations for field-to-transmission line interaction

A. Borghetti[1], F. Napolitano[1], C.A. Nucci[1], F. Rachidi[2] and M. Rubinstein[3]

In this chapter, we discuss the transmission line theory and its application to the problem of external electromagnetic field coupling to transmission lines, with particular reference to lightning-induced overvoltages on overhead power lines. After a short discussion on the underlying assumptions of the transmission line theory, we provide the derivation of field-to-transmission line coupling equations for the case of a single-wire line above a perfectly conducting ground. We also describe three seemingly different but completely equivalent approaches that have been proposed in the literature to describe the coupling of electromagnetic fields to transmission lines. The derived equations are extended to deal with the presence of losses and multiple conductors. The time-domain representation of the field-to-transmission line coupling equations, which allows for a straightforward treatment of non-linear phenomena as well as the variation in the line topology, is also described. Solution methods in the time domain are presented. The description of the main modelling features of an advanced computer code for the calculation of lightning originated voltages, i.e., the LIOV-EMTP-RV code, is given. The application of the illustrated theory and relevant computer codes to the case of a typical medium-voltage multi-conductor distribution feeder, which includes transformers and surge protection devices, is presented. The lightning performance assessment of traditional and compact overhead lines is dealt with as well.

1.1 Transmission line approximation

The problem of an external electromagnetic field coupling to an overhead line can be solved using different approaches. One such approach makes use of antenna

[1]Department of Electrical, Electronic, and Information Engineering, "Guglielmo Marconi", University of Bologna, Italy
[2]School of Engineering, École Polytechnique Fédérale de Lausanne (EPFL), Switzerland
[3]IICT Institute, University of Applied Sciences of Western Switzerland HES-SO, Switzerland

theory, a general methodology based on Maxwell's equations* [1]. When electrically long lines are involved, however, the antenna theory approach implies prohibitively long computational times and high computer resources. On the other hand, the less resource hungry quasi-static approximation [1], in which propagation is neglected and coupling is described by means of lumped elements, can be adopted only when the overall dimensions of the circuit are smaller than the minimum significant wavelength of the electromagnetic field. For many practical cases, however, this condition is not satisfied. As an example, let us consider the case of power lines illuminated by a lightning electromagnetic pulse (LEMP) originated by an indirect lightning, namely, a lightning hitting the ground near the line. Power networks extend, in general, over distances of several kilometres, which are much larger than the minimum wavelength λ associated with LEMP, as significant portions of the frequency spectrum of LEMP extend to frequencies up to a few megahertz and beyond, corresponding to wavelengths of about 100 m or less [2]. Therefore, the quasi-static approach cannot be applied for such a case.

A third approach is known as transmission line (TL) theory, or TL approximation. The main assumptions for this approach, which we shall use in this chapter, are the following:

(i) Propagation occurs along the line axis.
(ii) The sum of the line currents at any cross section of the line is zero. In other words, the ground – i.e., the reference conductor – is the return path for the currents in the n overhead conductors.
(iii) The response of the line to the coupled electromagnetic fields is quasi-transverse electromagnetic (quasi-TEM). In other words, the electromagnetic field produced by the electric charges and currents along the line is essentially confined to the transverse plane and perpendicular to the line axis.

Assumption (i) can be considered to be a good approximation if the cross-sectional dimensions of the line are electrically small since, in that case, propagation can indeed be assumed to occur essentially along the line axis only.

Assumption (ii) is satisfied if the ground plane exhibits infinite conductivity since, in that case, the currents and voltages can be obtained making use of the method of images, which guarantees currents of equal amplitude and opposite direction in the ground. It is worth noting that, by assuming that the sum of all the currents is equal to zero, we are considering only "transmission line mode" currents and neglecting the so-called antenna-mode currents [1,3]. If we wish to compute the load responses of the line, this assumption is adequate, because the antenna mode current response is small near the ends of the line. Along the line, however, and even for electrically small line cross sections, the presence of antenna-mode currents implies that the sum of the currents at a cross section is not necessarily equal to zero [1,4]. However, the quasi-symmetry due to the presence of the ground plane, even of finite conductivity, results in a very small contribution of antenna mode currents and, consequently, the predominant mode on the line will be transmission line [1].

*Methods based on this approach generally assume that the wire's cross section is smaller than the minimum significant wavelength (thin-wire approximation).

Assumption (iii) is satisfied only up to a threshold frequency, above which higher-order modes begin to appear [1]. For some cases, such as infinite parallel plates or coaxial lines, it is possible to derive an exact expression for a cut-off frequency below which only the TEM mode exists [4]. For other line structures (i.e., multiple conductors above a ground plane), the TEM mode response is generally satisfied as long as the line cross section is electrically small [4].

Under these conditions, the line can be represented by a distributed-parameter structure along its axis.

For uniform transmission lines with electrically small cross-sectional dimensions (not exceeding about one-tenth of the minimum significant wavelength of the exciting electromagnetic field), a number of theoretical and experimental studies have shown a fairly good agreement between results obtained using the TL approximation and results obtained either by means of antenna theory approaches or experiments (see, e.g., Reference 2). A more detailed discussion of the validity of the basic assumptions of the TL theory is beyond the scope of this chapter.

It is finally worth noting that a number of recent studies have attempted to enhance the classical TL theory to take into account high-frequency radiation effects (e.g., References 5–8). These models, which are beyond the scope of this chapter, are referred to as enhanced TL theory, or full-wave TL theory.

1.2 Single-wire line above a perfectly conducting ground

We will consider first the case of a single-wire, lossless line above a perfectly conducting ground. This simple case will allow us to introduce various coupling models and to discuss a number of concepts essential to the understanding of the electromagnetic field coupling phenomenon. Later in this chapter (in Sections 1.4 and 1.5), we will cover the cases of lossy and multi-conductor lines. The transmission line is defined by its geometrical parameters (wire radius a and height above ground h) and its terminations Z_A and Z_B, as illustrated in Figure 1.1, where the line is illuminated by an external electromagnetic field. The problem of interest is the calculation of the induced voltages and currents along the line and at the terminations.

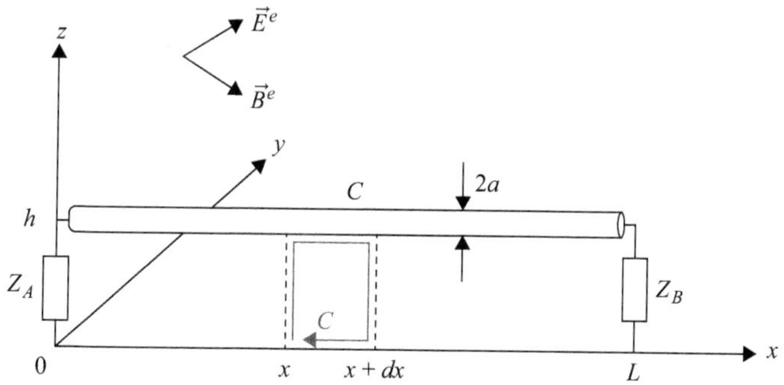

Figure 1.1 Geometry of the problem

It is worth noting that the external exciting electric and magnetic fields $\vec{E}^e$ and $\vec{B}^e$ are defined as the sum of the incident fields, $\vec{E}^i$ and $\vec{B}^i$, and the ground-reflected fields, $\vec{E}^r$ and $\vec{B}^r$, determined in absence of the line conductor. The total fields $\vec{E}$ and $\vec{B}$ at a given point in space are given by the sum of the excitation fields and the scattered fields from the line, the latter being denoted as $\vec{E}^s$ and $\vec{B}^s$. The scattered fields are those generated by the currents and charges flowing in the line conductor and by the corresponding currents and charges induced in the ground.

Three seemingly different but completely equivalent approaches were proposed to describe the coupling of electromagnetic fields to transmission lines. In what follows, we will present each one of them in turn in chronological order of publication, starting in the next subsection with a detailed derivation of the field-to-transmission line coupling equations† published by Taylor *et al.* [9] in 1965. For the other two models, the model by Agrawal *et al.* [10] and the Rachidi's model [11], we will present the corresponding telegrapher's equations and equivalent circuit representations in Sections 1.2.2 and 1.2.3.

1.2.1 Taylor, Satterwhite and Harrison model

1.2.1.1 Derivation of the first field-to-transmission line coupling (generalized telegrapher's) equation

Consider the single-conductor transmission line of height h in Figure 1.1. Applying Stokes' theorem to Maxwell's equation $\nabla \times \vec{E} = -j\omega\vec{B}$ for the area enclosed by the closed contour C yields

$$\oint_C \vec{E} \cdot d\vec{l} = -j\omega \int_S \vec{B} \cdot d\vec{S} \tag{1.1}$$

where surface S has C as boundary.

Since the contour has a differential width Δx, (1.1) can be written as‡

$$\int_0^h [E_z(x,z) - E_z(x+\Delta x, z)]dz + \int_x^{x+\Delta x} [E_x(x,h) - E_x(x,0)]dx$$
$$= -j\omega \int_0^h \int_x^{x+\Delta x} B_y(x,z)dxdz \tag{1.2}$$

Dividing by Δx and taking the limit as Δx approaches zero yields

$$-\frac{\partial}{\partial x}\int_0^h E_z(x,z)dz + E_x(x,h) - E_x(x,0) = -j\omega \int_0^h B_y(x,z)dz \tag{1.3}$$

†The field-to-transmission line coupling equations are sometimes referred to as the generalized telegrapher's equations.
‡The coordinate y will be implicitly assumed to be 0 and, for the sake of clarity, we will omit the y-dependence unless the explicit inclusion is important for the discussion.

Since the wire and the ground are assumed to be perfect conductors, the total tangential electric fields, $E_x(x,h)$ and $E_x(x,0)$, are zero. Defining also the total transverse voltage $V(x)$ in the quasi-static sense (since $h \ll \lambda$) as

$$V(x) = -\int_0^h E_z(x,z)dz \tag{1.4}$$

Equation (1.3) becomes

$$\frac{dV(x)}{dx} = -j\omega \int_0^h B_y(x,z)dz = -j\omega \int_0^h B_y^e(x,z)dz - j\omega \int_0^h B_y^s(x,z)dz \tag{1.5}$$

where we have decomposed the B-field into the excitation and scattered components.

The last integral in (1.5) represents the magnetic flux between the conductor and the ground produced by the current $I(x)$ flowing in the conductor.

Now, Ampère–Maxwell's equation in integral form is given by

$$\oint_{C'} \vec{B} \cdot d\vec{l} = I + j\omega \int_{S'} \vec{D} \cdot d\vec{S} \tag{1.6}$$

If we use a path C' and a surface S' in the transverse plane, defined by a constant x in such a manner that the conductor goes through it, we can write

$$\oint_{C'} \vec{B}_T^s(x,y,z) \cdot d\vec{l} = I(x) + j\omega \int_{S'} D_x^s(x,y,z)dS \tag{1.7}$$

where the subscript T is used to indicate that the field is in the transverse direction, and where we have explicitly included the dependence of the fields on the three Cartesian coordinates.

If the response of the wire is TEM, the scattered electric flux density in the x-direction, D_x^s, is zero and (1.7) can be written as

$$\oint_{C'} \vec{B}_T^s(x,y,z) \cdot d\vec{l} = I(x) \tag{1.8}$$

Clearly, $I(x)$ is the only source of $\vec{B}_T^s(x,y,z)$. Further, it is apparent from (1.8) that $\vec{B}_T^s(x,y,z)$ is directly proportional to $I(x)$. Indeed, if $I(x)$ is multiplied by a constant multiplicative factor which, in general, can be complex, $\vec{B}_T^s(x,y,z)$ too will be multiplied by that factor. Further, the proportionality factor for a uniform cross section line must be independent of x.

Let us now concentrate on the y component of $\vec{B}_T^s(x,y,z)$ for points in the plane $y=0$. Using the facts we just established that $I(x)$ and $\vec{B}_T^s(x,y,z)$ are proportional and that the proportionality factor is independent of x, we can now write

$$B_y^s(x, y=0, z) = k(y=0, z)I(x) \tag{1.9}$$

where $k(y,z)$ is the proportionality constant.

With this result, we now go back to the last integral in (1.5)

$$\int_0^h B_y^s(x,z)dz$$

Note that, although the value of y is not explicitly given, $y=0$. The integral represents the per-unit-length magnetic flux under the line. Substituting (1.9) into it, we obtain

$$\int_0^h B_y^s(x,z)dz = \int_0^h k(y=0,z)I(x)dz \tag{1.10}$$

We can rewrite (1.10) as follows:

$$\int_0^h B_y^s(x,z)dz = I(x)\int_0^h k(y=0,z)dz \tag{1.11}$$

Equation (1.11) implies that the per-unit-length scattered magnetic flux under the line at any point along it is proportional to the current at that point. The proportionality constant, given by $\int_0^h k(y=0,z)dz$, is the per-unit-length inductance of the line.

This result in the well-known linear relationship between the magnetic flux and the line current, the proportionality constant being the line per-unit-length inductance:

$$\int_0^h B_y^s(x,z)dz = L'I(x) \tag{1.12}$$

Assuming that the transverse dimension of the line is much greater than the height of the line, $(a \ll h)$, the magnetic flux density can be calculated using Ampere's law and the integral can be evaluated analytically [1].

Inserting (1.12) into (1.5), we obtain the first generalized telegrapher's equation

$$\frac{dV(x)}{dx} + j\omega L'I(x) = -j\omega\int_0^h B_y^e(x,z)dz \tag{1.13}$$

Note that, unlike the classical telegrapher's equations in which no external excitation is considered, the presence of an external field results in a forcing function expressed in terms of the exciting magnetic flux. This forcing function can be viewed as a distributed voltage source along the line.

Attention must paid to the fact that the voltage $V(x)$ in (1.13) depends on the integration path in (1.4) since it is obtained by integration of an electric field whose curl is not necessarily zero.

1.2.1.2 Derivation of the second field-to-transmission line coupling equation

To derive the second telegrapher's equation, we will assume that the medium surrounding the line is air ($\varepsilon = \varepsilon_o$) and we will start from the second Maxwell's equation $\nabla \times \vec{H} = \vec{J} + j\omega\varepsilon_o\vec{E}$. Rearranging the terms and writing it in Cartesian coordinates for the z-component, we get:

$$j\omega E_z(x,z) = \frac{1}{\varepsilon_o\mu_o}\left[\frac{\partial B_y(x,z)}{\partial x} - \frac{\partial B_x(x,z)}{\partial y}\right] - \frac{J_z}{\varepsilon_o} \tag{1.14}$$

The current density can be related to the E-field using Ohm's law, $\vec{J} = \sigma_{\text{air}}\vec{E}$, where σ_{air} is the air conductivity. Since the air conductivity is generally low, we will assume here that $\sigma_{\text{air}} = 0$ and will therefore neglect this term.[§]

Integrating (1.14) along the z-axis from 0 to h, and making use of (1.4), we obtain

$$\begin{aligned} -j\omega V(x) &= \frac{1}{\varepsilon_o\mu_o}\int_0^h \left[\frac{\partial B_y^e(x,z)}{\partial x} - \frac{\partial B_x^e(x,z)}{\partial y}\right] dz \\ &+ \frac{1}{\varepsilon_o\mu_o}\int_0^h \left[\frac{\partial B_y^s(x,z)}{\partial x} - \frac{\partial B_x^s(x,z)}{\partial y}\right] dz \end{aligned} \tag{1.15}$$

in which we have decomposed the magnetic flux density field into the excitation and scattered components.

Applying (1.14) to the components of the excitation electromagnetic field and integrating along z from 0 to h yields

$$\frac{1}{\varepsilon_o\mu_o}\int_0^h \left[\frac{\partial B_y^e}{\partial x} - \frac{\partial B_x^e}{\partial y}\right] dz = j\omega\int_0^h E_z^e dz. \tag{1.16}$$

Using (1.12), (1.16) and given that $B_x^s = 0$ by virtue of the assumed TEM nature of the line response, (1.15) becomes

$$\frac{dI(x)}{dx} + j\omega C'V(x) = -j\omega C'\int_0^h E_z^e(x,z)dz \tag{1.17}$$

where C' is the per-unit-length line capacitance related to the per-unit-length inductance through $\varepsilon_o\mu_o = L'C'$. Equation (1.17) is the second field-to-transmission line coupling equation.

For a line of finite length, such as the one represented in Figure 1.1, the boundary conditions for the load currents and voltages must be enforced. They are simply given by

$$V(0) = -Z_A I(0) \tag{1.18}$$

$$V(L) = Z_B I(L) \tag{1.19}$$

[§]This term will eventually result in an equivalent transverse conductance in the coupling equation (see Section 1.4).

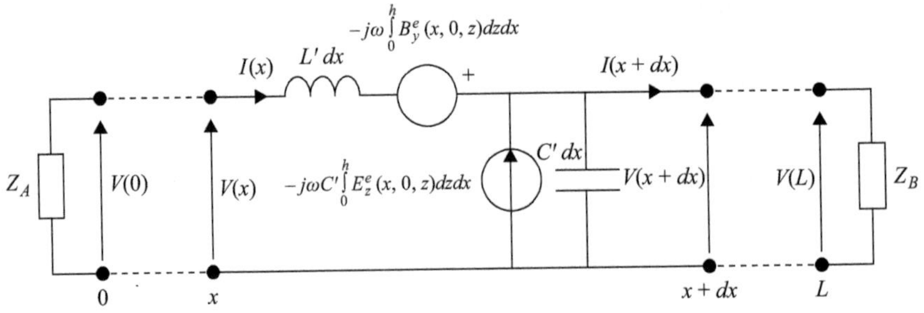

Figure 1.2 Equivalent circuit of a lossless single-wire overhead line excited by an electromagnetic field. Taylor et al. model

1.2.1.3 Equivalent circuit

Equations (1.13) and (1.17) are referred to as the Taylor *et al.* model. They can be represented using an equivalent circuit, as shown in Figure 1.2. The forcing functions (source terms) in (1.13) and (1.17) are included as a set of distributed series voltage and parallel current sources along the line.

1.2.2 Agrawal, Price and Gurbaxani model

An equivalent formulation of the field-to-transmission line coupling equations was proposed in 1980 by Agrawal, Price and Gurbaxani [10]. This model is commonly referred to as the Agrawal model. We will call it the Agrawal *et al.* model hereafter.

The basis for the derivation of the Agrawal *et al.* model can be described as follows: The excitation fields produce a line response that is TEM. This response is expressed in terms of a scattered voltage $V^s(x)$, which is defined in terms of the line integral of the scattered electric field from the ground to the line, and a scattered current $I^s(x)$ which flows in the line. The total voltage $V(x)$ and the total current $I(x)$ (the quantities that are actually measurable) are computed as the sum of the excitation and the scattered voltages and currents.

The coupling equations in the model of Agrawal *et al.* are used to obtain the scattered voltage and the scattered current only. Specific components of the incident fields either appear as a source term in the coupling equations or are used to compute the total voltage $V(x)$, which corresponds to that used in the model of Taylor *et al.* In the model of Agrawal *et al.*, the total current $I(x)$ is identical to the scattered current and it is obtained directly from the coupling equations since it is assumed that the line is not energized and, therefore, only induced currents flow in it. As we will see in the next section, when we present *Rachidi*'s model [11], it is possible to define a distinct excitation current $I_e(x)$.

In the model of Agrawal *et al.*, the rationale behind the writing of the telegrapher's equations in terms of the scattered quantities only is that, whereas the incident fields are arbitrary (they are of course constrained to satisfy Maxwell's equations and the ground boundary conditions), the scattered response is TEM, which allows for them to be calculated using TL theory.

The total voltage can be obtained from the scattered voltage through

$$V(x) = V^s(x) + V^e(x) = V^s(x) - \int_0^h E_z^e(x, z)dz \tag{1.20}$$

The field-to-transmission line coupling equations as derived by *Agrawal et al.* [10] are given by

$$\frac{dV^s(x)}{dx} + j\omega L' I(x) = E_x^e(x, h) \tag{1.21}$$

$$\frac{dI(x)}{dx} + j\omega C' V^s(x) = 0 \tag{1.22}$$

Note that, in this model, only one source term is present (in the first equation) and it is simply expressed in terms of the exciting electric field tangential to the line conductor $E_x^e(x, h)$.

The boundary conditions, in terms of the scattered voltage and the total current as used in (1.21) and (1.22), are given by

$$V^s(0) = -Z_A I(0) + \int_0^h E_z^e(0, z)dz \tag{1.23}$$

$$V^s(L) = Z_B I(L) + \int_0^h E_z^e(L, z)dz \tag{1.24}$$

The equivalent circuit representation of this model (equations (1.21)–(1.24)) is shown in Figure 1.3. For this model, the forcing function (the exciting electric field tangential to the line conductor) is represented by distributed voltage sources along the line. In accordance with boundary conditions (1.23) and (1.24), two lumped voltage sources (equal to the line integral of the exciting vertical electric field) are inserted at the line terminations.

It is also interesting to note that this model involves only electric field components of the exciting field and the exciting magnetic field does not appear explicitly as a source term in the coupling equations.

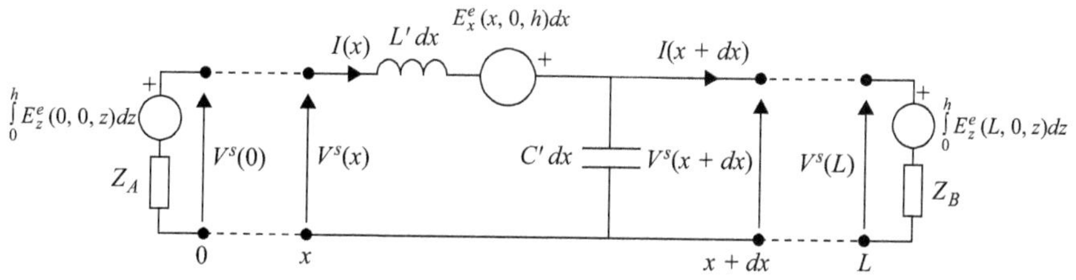

Figure 1.3 Equivalent circuit of a lossless single-wire overhead line excited by an electromagnetic field. Agrawal et al. model

Note that in Reference 12, a derivation of Agrawal *et al.* model is presented starting from the so-called full-wave field-to-transmission line coupling equations.

1.2.3 Rachidi model

Another form of coupling equations, equivalent to the Agrawal *et al.* and to the Taylor *et al.* models, has been derived by Rachidi [11]. In this model, only the exciting magnetic field components appear explicitly as forcing functions in the equations:

$$\frac{dV(x)}{dx} + j\omega L' I^s(x) = 0 \tag{1.25}$$

$$\frac{dI^s(x)}{dx} + j\omega C' V(x) = \frac{1}{L'}\int_0^h \frac{\partial B_x^e(x,z)}{\partial y} dz \tag{1.26}$$

in which $I^s(x)$ is the so-called scattered current related to the total current by

$$I(x) = I^s(x) + I^e(x) \tag{1.27}$$

where the excitation current $I^e(x)$ is defined as

$$I^e(x) = -\frac{1}{L'}\int_0^h B_y^e(x,z)dz \tag{1.28}$$

The boundary conditions corresponding to this formulation are

$$I^s(0) = -\frac{V(0)}{Z_A} + \frac{1}{L'}\int_0^h B_y^e(0,z)dz \tag{1.29}$$

$$I^s(L) = \frac{V(L)}{Z_B} + \frac{1}{L'}\int_0^h B_y^e(L,z)dz \tag{1.30}$$

The equivalent circuit corresponding to the above equivalent pair of coupling equations is shown in Figure 1.4. Note that the equivalent circuit associated with the *Rachidi model* could be seen as the dual circuit – in the sense of electrical network theory – of the one corresponding to the Agrawal *et al.* model (Figure 1.3).

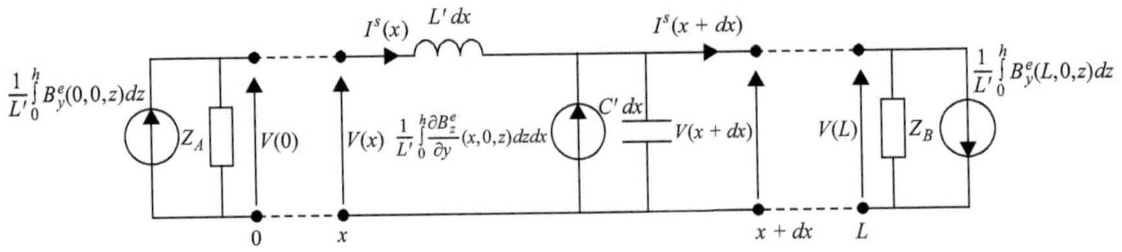

Figure 1.4 Equivalent circuit of a lossless single-wire overhead line excited by an electromagnetic field. Rachidi model

1.3 Contribution of the different electromagnetic field components

Nucci and Rachidi [13] showed, on the basis of a specific numerical example that, as predicted theoretically, the total induced voltage waveforms obtained using the three coupling models presented in Sections 1.2.1–1.2.3 are identical. However, the contribution of a given component of the exciting electromagnetic field to the total induced voltage and current varies depending on the adopted coupling model. Indeed, the three coupling models are different but fully equivalent approaches that predict identical results in terms of total voltages and total currents, in spite of the fact that they take into account the electromagnetic coupling in different ways. In other words, the three models are different expressions of the same equations, cast in terms of different combinations of the various electromagnetic field components, which are related through Maxwell's equations.[‖]

1.4 Inclusion of losses

In the calculation of lightning-induced voltages, losses are, in principle, to be taken into account both in the wire and in the ground. Losses due to the finite ground conductivity are the most important ones, and they affect both the electromagnetic field and the surge propagation along the line [15].

Let us make reference to the same geometry of Figure 1.1, and let us now take into account losses both in the wire and in the ground plane. The wire conductivity and relative permittivity will be denoted σ_w and ε_{rw}, respectively, and the ground, assumed to be homogeneous, is characterized by its conductivity σ_g and its relative permittivity ε_{rg}. The Agrawal *et al.* coupling equations extended to the present case of

[‖]Nucci and Rachidi [13] have highlighted that the three popular forms of coupling equations discussed so far are not the only possible ones. Other coupling formulations can be derived by algebraically re-arranging some of the above equations and by taking into account the link given by Maxwell's equations among the various electromagnetic field components. For instance, it is possible, making use of the following relation, which can be derived from Maxwell's equations [14]

$$j\omega \int_0^h B_y^e(x,z)dz = E_z^e(h,z) - \frac{\partial}{\partial z}\int_0^h E_x^e(x,z)dx \tag{1.31}$$

to express the coupling equations in terms of the total electric field only, namely

$$\frac{dV(z)}{dz} + j\omega L' I(z) = E_z^e(h,z) - \frac{\partial}{\partial z}\int_0^h E_x^e(x,z)dx \tag{1.32}$$

$$\frac{dI(z)}{dz} + j\omega C' V(z) = -j\omega C' \int_0^h E_x^e(x,z)dx \tag{1.33}$$

a wire above an imperfectly conducting ground can be written as (for a step-by-step derivation see Reference 1)

$$\frac{dV^s(x)}{dx} + Z'I(x) = E_x^e(x, h) \tag{1.34}$$

$$\frac{dI(x)}{dx} + Y'V^s(x) = 0 \tag{1.35}$$

where Z' and Y' are the longitudinal and transverse per-unit-length impedance and admittance, respectively, given by [1,15]¶

$$Z' = j\omega L' + Z_w' + Z_g' \tag{1.36}$$

$$Y' = \frac{(G' + j\omega C')Y_g'}{G' + j\omega C' + Y_g'} \tag{1.37}$$

in which

– L', C' and G' are the per-unit-length longitudinal inductance, transverse capacitance and transverse conductance, respectively, calculated for a lossless wire above a perfectly conducting ground:

$$L' = \frac{\mu_o}{2\pi}\cosh^{-1}\left(\frac{h}{a}\right) \cong \frac{\mu_o}{2\pi}\ln\left(\frac{2h}{a}\right) \quad \text{for } h \gg a \tag{1.38}$$

$$C' = \frac{2\pi\varepsilon_o}{\cosh^{-1}(h/a)} \cong \frac{2\pi\varepsilon_o}{\ln(2h/a)} \quad \text{for } h \gg a \tag{1.39}$$

$$G' = \frac{\sigma_{\text{air}}}{\varepsilon_o}C' \tag{1.40}$$

– Z_w' is the per-unit-length internal impedance of the wire; assuming a round wire and an axial symmetry for the current, the following expression can be derived for the wire internal impedance (e.g., Reference 16):

$$Z_w' = \frac{\gamma_w \mathrm{I}_0(\gamma_w a)}{2\pi a\sigma_w \mathrm{I}_1(\gamma_w a)} \tag{1.41}$$

where $\gamma_w = \sqrt{j\omega\mu_o(\sigma_w + j\omega\varepsilon_o\varepsilon_{rw})}$ is the propagation constant in the wire and I_0 and I_1 are the modified Bessel functions of zero and first order, respectively;

– Z_g' is the per-unit-length ground impedance, which is defined as [2,17]

$$Z_g' = \frac{j\omega \int_{-\infty}^{h} B_y^s(x,z)dx}{I} - j\omega L' \tag{1.42}$$

where B_y^s is the y-component of the scattered magnetic induction field.

¶In Reference 1, the per-unit-length transverse conductance was disregarded.

Several expressions for the ground impedance have been proposed in the literature, see for instance Reference 2 for a survey. Sunde [18] derived a general expression for the ground impedance, which is given by

$$Z'_g = \frac{j\omega\mu_o}{\pi}\int_0^\infty \frac{e^{-2hx}}{\sqrt{x^2+\gamma_g^2}+x}dx \tag{1.43}$$

where $\gamma_g = \sqrt{j\omega\mu_o(\sigma_g + j\omega\varepsilon_o\varepsilon_{rg})}$ is the propagation constant in the ground.

As noted in Reference 19, *Sunde*'s expression (1.43) is directly connected to the general expressions obtained from scattering theory. Indeed, it is shown in Reference 1 that the general expression for the ground impedance derived using scattering theory reduces to the *Sunde* approximation when considering the transmission line approximation. Also, the results obtained using (1.43) are shown to be accurate within the limits of the transmission line approximation [1].

The general expression (1.43) is not suitable for a numerical evaluation since it involves an integral over an infinitely long interval. Several approximations for the expression of the ground impedance of a single-wire line have been proposed in the literature (see Reference 2 for a survey). One of the simplest and most accurate was proposed by *Sunde* himself and is given by the following logarithmic function:

$$Z'_g \cong \frac{j\omega\mu_o}{2\pi}\ln\left(\frac{1+\gamma_g h}{\gamma_g h}\right) \tag{1.44}$$

It has been shown [2] that the above logarithmic expression represents an excellent approximation to the general expression (1.43) over the frequency range of interest.

Finally, Y'_g is the so-called ground admittance, given by [1]

$$Y'_g \cong \frac{\gamma_g^2}{Z'_g} \tag{1.45}$$

1.5 Case of multi-conductor line

Making reference to the geometry of Figure 1.5, the field-to-transmission line coupling equations for the case of a multi-wire system along the x-axis above an imperfectly conducting ground and in the presence of an external electromagnetic excitation, based on the model of Agrawal *et al.*, are given by [1,2,19]

$$\frac{d}{dx}[V_i^s(x)] + j\omega[L'_{ij}][I_i(x)] + [Z'_{gij}][I_i(x)] = [E_x^e(x, h_i)] \tag{1.46}$$

$$\frac{d}{dx}[I_i(x)] + [G'_{ij}][V_i^s(x)] + j\omega[C'_{ij}][V_i^s(x)] = [0] \tag{1.47}$$

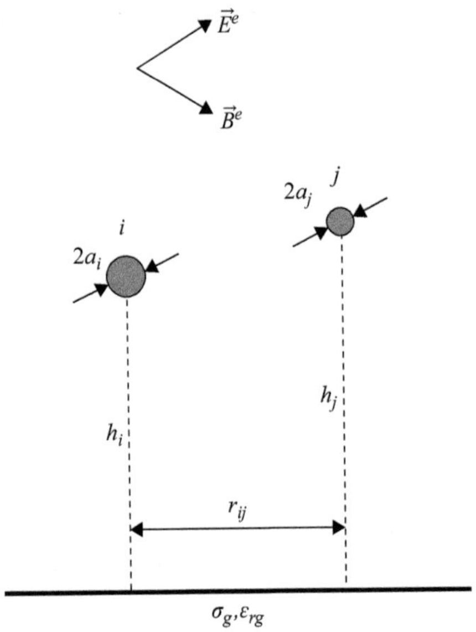

Figure 1.5 Cross-sectional geometry of a multi-conductor line in presence of an external electromagnetic field

in which

- $[V_i^s(x)]$ and $[I_i(x)]$ are frequency-domain vectors of the scattered voltage and the current along the line;
- $[E_x^e(x, h_i)]$ is the vector of the exciting electric field tangential to the line conductors;
- [0] is the zero vector (all elements are equal to zero);
- $[L'_{ij}]$ is the per-unit-length line inductance matrix. Assuming that the distances between conductors are much larger than their radii, the general expression for the mutual inductance between two conductors i and j is given by [1]

$$L'_{ij} = \frac{\mu_o}{2\pi} \ln \frac{r_{ij}^2 + (h_i + h_j)^2}{r_{ij}^2 + (h_i - h_j)^2} \tag{1.48}$$

The self-inductance for conductor i is given by

$$L'_{ii} = \frac{\mu_o}{2\pi} \ln\left(\frac{2h_i}{r_{ii}}\right) \tag{1.49}$$

- $[C'_{ij}]$ is the per-unit-length line capacitance matrix, which can be evaluated directly from the inductance matrix using the following expression [1]:

$$[C'_{ij}] = \varepsilon_o \mu_o [L'_{ij}]^{-1} \tag{1.50}$$

- $[G'_{ij}]$ is the per-unit-length transverse conductance matrix. The transverse conductance matrix elements can be evaluated starting either from the capacitance matrix or the inductance matrix using the following relations:

$$[G'_{ij}] = \frac{\sigma_{\text{air}}}{\varepsilon_o}[C'_{ij}] = \sigma_{\text{air}} \mu_o [L'_{ij}]^{-1} \tag{1.51}$$

In most practical cases, the transverse conductance matrix elements G'_{ij} are much smaller than $j\omega C'_{ij}$ [4] and can therefore be neglected in the computation of the induced overvoltages.

– Finally, $[Z'_{g_{ij}}]$ is the ground impedance matrix. The general expression for the mutual ground impedance between two conductors i and j derived by Sunde is given by [18]

$$Z'_{g_{ij}} = \frac{j\omega\mu_o}{\pi}\int_0^{\infty}\frac{e^{-(h_i+h_j)x}}{\sqrt{x^2+\gamma_g^2}+x}\cos(r_{ij}x)dx \tag{1.52}$$

In a similar way to that proposed by Sunde for the case of a single-wire line, an accurate logarithmic approximation is proposed by Rachidi *et al.* [19] which is given by

$$Z'_{g_{ij}} \cong \frac{j\omega\mu_o}{\pi}\ln\frac{\left(1+\gamma_g\frac{h_i+h_j}{2}\right)^2+\left(\gamma_g\frac{r_{ij}}{2}\right)^2}{\left(\gamma_g\frac{h_i+h_j}{2}\right)^2+\left(\gamma_g\frac{r_{ij}}{2}\right)^2} \tag{1.53}$$

Note that in (1.46) and (1.47), the terms corresponding to the wire impedance and the so-called ground admittance have been neglected. This approximation is valid for typical overhead power lines [15].

The boundary conditions for the two line terminations are given by

$$[V_i^s(0)] = -[Z_A][I_i(0)] + \left[\int_0^h E_z^e(0,z)dz\right] \tag{1.54}$$

$$[V_i^s(L)] = [Z_B][I_i(L)] + \left[\int_0^h E_z^e(L,z)dz\right] \tag{1.55}$$

in which $[Z_A]$ and $[Z_B]$ are the impedance matrices at the two line terminations.

1.6 Time-domain representation of the coupling equations

A time-domain representation of the field-to-transmission line coupling equations is sometimes preferable because it allows for the straightforward treatment of non-linear phenomena as well as the variation in the line topology [2]. On the other hand, frequency-dependent parameters, such as the ground impedance, need to be represented using convolution integrals.

The field-to-transmission line coupling equations (1.46) and (1.47) can be converted into the time domain to obtain the following expressions:

$$\frac{\partial}{\partial x}[v_i^s(x,t)] + [L'_{ij}]\frac{\partial}{\partial t}[i_i(x,t)] + [\xi'_{g_{ij}}]\otimes\frac{\partial}{\partial t}[i_i(x,t)] = [E_x^e(x,h_i,t)] \tag{1.56}$$

$$\frac{\partial}{\partial x}[i_i(x,t)] + [G'_{ij}][v_i^s(x,t)] + [C'_{ij}]\frac{\partial}{\partial t}[v_i^s(x,t)] = [0] \tag{1.57}$$

in which ⊗ denotes convolution product and the matrix $[\xi'_{gij}]$ is called the transient ground resistance matrix; its elements are defined as

$$[\xi'_{gij}] = F^{-1}\left\{\frac{[Z'_{gij}]}{j\omega}\right\} \tag{1.58}$$

The inverse Fourier transforms of the boundary conditions written, for simplicity, for resistive terminal loads read

$$[v_i^s(0,t)] = -[R_A][i_i(0,t)] + \left[\int_0^{h_i} E_z^e(0,z,t)dz\right] \tag{1.59}$$

$$[v_i^s(L,t)] = [R_B][i_i(0,t)] + \left[\int_0^{h_i} E_z^e(L,z,t)dz\right] \tag{1.60}$$

where $[R_A]$ and $[R_B]$ are the matrices of the resistive loads at the two line terminals.

Some clarification on the subject of the response of multi-conductor lines to lightning-originated electromagnetic fields, with and without taking into account the ground losses in the surge propagation, are provided in Reference 20.

The general expression for the ground impedance matrix terms in the frequency domain (1.52) does not have an analytical inverse Fourier transform. Thus, the elements of the transient ground resistance matrix in the time domain are to be, in general, determined using a numerical inverse Fourier transform algorithm. However, analytical expressions have been derived which are shown to be reasonable approximations to the numerical values obtained using an inverse FFT [17].

Recently, in Reference 21, Tossani *et al.* proposed a new analytical approach for calculating the transient ground resistance in the time domain that is based on the very accurate *Sunde*'s and *Rachidi*'s logarithmic expressions (1.44) and (1.53) for the ground impedance matrix of an overhead multi-conductor line.

1.7 Solutions with particular reference to time-domain numerical solutions

Different approaches can be employed to find solutions to the presented coupling equations. Reference 22 presents some solution methods in the frequency domain and in the time domain. Time-domain solutions are particularly convenient when the theory we are illustrating is applied to the problem of estimating the response of overhead lines to the excitation of external fields, such as LEMP (which will be the focus of the next section). As a matter of fact, the presence of surge protection devices along the line and the operation of line circuit breakers that can change the line topology/geometry are two aspects of the problem that make preferable the use of time-domain approaches. Several methods can be used to solve the coupling equations in the time domain [1,4]. Among the proposed methods, the finite-difference time-domain (FDTD) technique is particularly convenient (e.g., Reference 23). Such a technique was indeed used by Agrawal *et al.* in Reference 10 when presenting their field-to-transmission line coupling equations. In Reference 10, partial time and space derivatives were approximated using a first-order FDTD scheme. Here, we chose to use the second-order FDTD scheme based on the Lax–Wendroff algorithm [24], which is more suitable for the problem of interest [25].

To simplify the notation, we shall make reference to the case of a lossless single-conductor overhead line. For such a case, the time-domain coupling equations (1.56) and (1.57) read

$$\frac{\partial v^s(x,t)}{\partial x} + L'\frac{\partial i(x,t)}{\partial t} = E_x^e(x,h,t) \tag{1.61}$$

$$\frac{\partial i(x,t)}{\partial x} + C'\frac{\partial v^s(x,t)}{\partial t} = 0 \tag{1.62}$$

The expansion of both line currents and scattered voltages of the Agrawal *et al.* model using Taylor series applied to the variable t up to the second-order term yields

$$v^s(x,t) = v^s(x,t_0) + \frac{\partial v^s(x,t)}{\partial t}\Delta t + \frac{\partial^2 v^s(x,t)}{\partial t^2}\frac{\Delta t^2}{2} + \mathrm{O}(\Delta t^3) \tag{1.63}$$

$$i(x,t) = i(x,t_0) + \frac{\partial i(x,t)}{\partial t}\Delta t + \frac{\partial^2 i(x,t)}{\partial t^2}\frac{\Delta t^2}{2} + \mathrm{O}(\Delta t^3) \tag{1.64}$$

where $\mathrm{O}(\Delta t^3)$ is the remainder term.

By differentiating (1.61) and (1.62) with respect to x and t and after substitution, we obtain

$$\frac{\partial^2 i(x,t)}{\partial x^2} - L'C'\frac{\partial^2 i(x,t)}{\partial t^2} = -C'\frac{\partial E_x^e(x,h,t)}{\partial t} \tag{1.65}$$

$$\frac{\partial^2 v^s(x,t)}{\partial x^2} - L'C'\frac{\partial^2 v^s(x,t)}{\partial t^2} = \frac{\partial E_x^e(x,h,t)}{\partial t} \tag{1.66}$$

The substitution of the time derivatives of v^s and i from (1.61), (1.62), (1.65) and (1.66) into (1.63) and (1.64) yields

$$\begin{aligned} v^s(x,t) &= v^s(x,t_0) - \frac{\Delta t}{C'}\frac{\partial i(x,t)}{\partial x} - \frac{\Delta t^2}{2L'C'}\left(\frac{\partial E_x^e(x,h,t)}{\partial x} - \frac{\partial^2 v^2(x,t)}{\partial x^2}\right) \\ &\quad + \mathrm{O}(\Delta t^3) \end{aligned} \tag{1.67}$$

$$\begin{aligned} i(x,t) &= i(x,t_0) - \frac{\Delta t}{L'}\left(\frac{\partial V^s(x,t)}{\partial x} - E_x^e(x,h,t)\right) \\ &\quad + \frac{\Delta t^2}{2L'C'}\left(\frac{\partial^2 i(x,t)}{\partial x^2} + C'\frac{\partial E_x^e(x,h,t)}{\partial t}\right) + \mathrm{O}(\Delta t^3) \end{aligned} \tag{1.68}$$

In order to represent (1.67) and (1.68) within the FDTD integration scheme, one can proceed with the time and space discretization as follows:

$$v^s(x,t) = v^s(k\Delta x, n\Delta t) = v_k^n \tag{1.69}$$

$$i(x,t) = i(k\Delta x, n\Delta t) = i_k^n \tag{1.70}$$

$$E_x^e(x,h,t) = E_x^e(k\Delta x, h, n\Delta t) = Eh_k^n \tag{1.71}$$

where

- Δx is the spatial integration step;
- Δt is time integration step;
- $k = 0, 1, 2, \ldots, k_{\max}$ is the spatial discretization index ($k_{\max} = (L/\Delta x) + 1$, where L is the line length);
- $n = 0, 1, 2, \ldots, n_{\max}$ is the time discretization index.

In the integration scheme, the scattered voltage and current at time step n are assumed as known for all spatial nodes. Therefore, (1.67) and (1.68) allow the scattered voltages and currents to be calculated at time step $n+1$.

The discrete spatial derivatives of the scattered voltages, line currents and horizontal electric fields can be written as follows:

$$\left.\frac{\partial v^s(x,t)}{\partial x}\right|_{t=n\Delta t} \cong \frac{v_{k+1}^n - v_{k-1}^n}{2\Delta x} \tag{1.72}$$

$$\left.\frac{\partial i(x,t)}{\partial x}\right|_{t=n\Delta t} \cong \frac{i_{k+1}^n - i_{k-1}^n}{2\Delta x} \tag{1.73}$$

$$\left.\frac{\partial E_x^e(x,h,t)}{\partial x}\right|_{t=n\Delta t} \cong \frac{Eh_{k+1}^n - Eh_{k-1}^n}{2\Delta x} \tag{1.74}$$

On the other hand, the discrete time derivatives of the horizontal electric field read

$$\left.\frac{\partial E_x^e(x,h,t)}{\partial t}\right|_{t=n\Delta t} \cong \frac{Eh_k^{n+1} - Eh_k^{n-1}}{2\Delta t} \tag{1.75}$$

The discrete second-order spatial derivatives of the scattered voltages and currents can be written as

$$\left.\frac{\partial^2 v^s(x,t)}{\partial x^2}\right|_{t=n\Delta t} \cong \frac{v_{k+1}^n + v_{k-1}^n - 2v_k^n}{\Delta x^2} \tag{1.76}$$

$$\left.\frac{\partial^2 i(x,t)}{\partial x^2}\right|_{t=n\Delta t} \cong \frac{i_{k+1}^n + i_{k-1}^n - 2i_k^n}{\Delta x^2} \tag{1.77}$$

Inserting (1.69)–(1.77) into (1.67) and (1.68), the equations of the second-order FDTD integration scheme for the internal nodes of the line, namely $k = 1, 2, \ldots, k_{\max-1}$, finally become

$$v_k^{n+1} = v_k^n - \frac{\Delta t}{C'}\frac{i_{k+1}^n - i_{k-1}^n}{2\Delta x} - \frac{\Delta t^2}{2L'C'}\frac{E_{k+1}^n - E_{k-1}^n}{2\Delta x} + \frac{\Delta t^2}{2L'C'}\frac{v_{k+1}^n - 2v_k^n + v_{k-1}^n}{\Delta x^2} \tag{1.78}$$

$$i_k^{n+1} = i_k^n - \frac{\Delta t}{L'}\frac{v_{k+1}^n - v_{k-1}^n}{2\Delta x} + \frac{\Delta t}{L'}\frac{E_k^{n+1} + 2E_k^n - E_k^{n-1}}{2} + \frac{\Delta t^2}{2L'C'}\frac{i_{k+1}^n - 2i_k^n + i_{k-1}^n}{\Delta x^2} \tag{1.79}$$

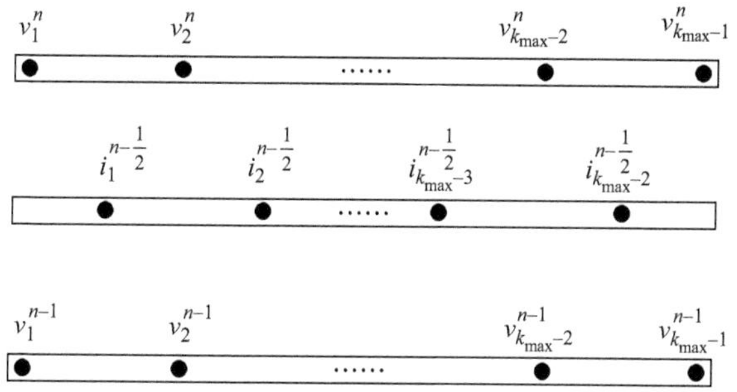

Figure 1.6 Interlaced FDTD discretization scheme for the internal points of the line

By expanding currents and scattered voltages in a Taylor series up to the first-order term, the first-order recursive solutions are obtained. Note that when the first-order FDTD solution is adopted, voltage and currents points are usually interlaced in order to improve stability and accuracy. As a consequence, proximate current and voltage discretization points are separated by $\Delta x/2$ in space and $\Delta t/2$ in time, as represented in Figure 1.6.

The first-order FDTD recursive solution is given by

$$v_k^n = v_k^{n-1} - \frac{\Delta t}{C'} \frac{i_{k+1}^{n-\frac{1}{2}} - i_{k-1}^{n-\frac{1}{2}}}{\Delta x} \tag{1.80}$$

$$i_k^{n+\frac{1}{2}} = i_k^{n-\frac{1}{2}} - \frac{\Delta t}{L'} \frac{v_{k+1}^n - v_{k-1}^n}{\Delta x} + \frac{\Delta t}{L'} E_k^n \tag{1.81}$$

In order to account for the boundary conditions at the terminal ends of the line, a proper implementation of the relationships Γ_0 and $\Gamma_{k_{max}}$ between currents and total voltages at the terminal ends is needed.

The boundary conditions of the LEMP-coupled line can be written as:

$$v^s(0,t) = -\Gamma_0 i(0,t) + \int_0^h E_z^e(0,z,t)dz \tag{1.82}$$

$$v^s(k_{max}\Delta x, t) = \Gamma_{k_{max}} i(k_{max}\Delta x, t) + \int_0^h E_z^e(k_{max}\Delta x, z, t)dz \tag{1.83}$$

The equations relevant to the line terminations provide the boundary conditions of the model. They are described in Section 1.8.2, which will be devoted to the description of the interface between the popular ElectroMagnetic Transient Program (EMTP) [26,27] and the LIOV computer code for the calculation of lightning-induced voltages on overhead lines [2], which is described below, in Section 1.8.1.

1.8 Application of theory to the case of lightning-induced voltages on distribution overhead lines

1.8.1 The LIOV code

LIOV (lightning-induced overvoltages) [2,28] is a computer code which allows for the calculation of lightning-induced overvoltages on multi-conductor overhead lines as a function of lightning current waveshape (amplitude, front steepness, time to half-width), return-stroke velocity, line geometry (height, length, number and position of conductors, presence of shield wire and relevant grounding), stroke location with respect to the line, ground resistivity, ground relative permittivity and termination impedances. Corona effects on the induced voltages [29,30] and the induction effects of downward leader electric fields [31] can be also dealt with.

LIOV has been experimentally validated by means of tests carried out on reduced scale set-ups with NEMP (nuclear electromagnetic pulse) simulators [32], LEMP simulators [33] and on full-scale set-ups illuminated by artificially initiated lightning [34].

The code is based on the field-to-transmission line coupling formulation of Agrawal *et al.* presented in Section 1.2.2. The electromagnetic field radiated by the lightning channel, assumed as a straight vertical antenna, is calculated using the field equations in the form given by *Master* and *Uman* [35] with the extension to the case of lossy ground introduced by Cooray [36] and Rubinstein [37]. Concerning the model of the lightning current along the channel, the LIOV code adopts an engineering return stroke model, namely an expression that describes the spatial and temporal distributions of the return stroke current along the lightning channel, as a function of the current waveshape at the base of the channel and one or two additional parameters. Different return stroke models are available in the literature (see, e.g., References 38, 39) and they are implemented in the LIOV code. The channel base current waveshape can be represented by means of *Heidler* functions [40] or by a piecewise straight line having a linear front and a flat top. The Heidler function is given by

$$I(t) = \frac{I_0}{\eta} \frac{(t/\tau_1)^n}{(t/\tau_1)^n + 1} \exp(-t/\tau_2) \tag{1.84}$$

where

$$\eta = \exp[-(\tau_1/\tau_2)(n\tau_2/\tau_1)^{1/n}] \tag{1.85}$$

while the parameters I_0, τ_1, τ_2 are selected to approximate the desired current waveform. To obtain realistic induced overvoltage waveshapes, the sum of two Heidler functions can be adopted. A simplified representation of the channel base current, i.e., a linear front and a flat top, can be adopted when only the amplitude of the induced voltage is of interest.

In indirect lightning performance studies, usually performed by means of statistical procedures (see, e.g., References 41, 42), the computational burden is a crucial feature, in which case the TL return stroke model by Uman and McLain [43] can be adopted, in order to take advantage of the calculation speed provided by the analytical formulation proposed by Napolitano in Reference 44. Such analytical formulation can be adopted also with channel base currents represented by Heidler functions, provided the current waveshape is suitably approximated by means of piecewise straight lines. The default return stroke engineering model in LIOV is the modified transmission line with exponential decay (MTLE) (Nucci *et al.* [28,45]; Rachidi and Nucci [46]), in which the lightning return stroke current is assumed to decay exponentially as it travels up the lightning channel. The amplitude decay constant of the MTLE model, λ, is fixed at 2 km, as inferred by Nucci and Rachidi in Reference 47 by means of simultaneously measured electromagnetic fields at two different distances from the lightning channel. The vertical electric field radiated by the lightning channel is calculated assuming the ground as a perfectly conducting plane, since such a field component is not affected significantly by the soil resistivity in the frequency and distance range of interest [37,48]. For the calculation of the horizontal electric field component, the earlier-mentioned approximate formula proposed by Cooray and Rubinstein is adopted. It is worth noting that Cooray [49] has proposed an improved version of such a formula, taking into account remarks by Wait [50]. We assume in what follows that, for the adopted values for ground conductivity, our results will not be significantly affected by the adoption of one expression instead of the other.

The Agrawal et al. transmission line equations are adopted in LIOV, where, as seen in Section 1.2.2, the forcing functions are expressed in terms of the vertical and horizontal electric field components discussed above. The Agrawal *et al.* model is solved by means of a one-dimensional finite-difference time-domain (FDTD) technique described earlier, according to which the line length is discretized into a finite number of nodes, at which scattered voltages, currents and the horizontal component of the external electromagnetic field are evaluated at each time step. In general, space and time steps in FDTD schemes need to be suitably chosen in order to obtain accurate results with reasonable computational times; moreover, they have to satisfy the Courant stability condition [10].

LIOV was developed in the framework of an international collaboration involving the University of Bologna (Department of Electrical Engineering, now Department of Electrical, Electronic and Information Engineering), the Swiss Federal Institute of Technology (Power Systems Laboratory, now EMC Laboratory), and the University of Rome "La Sapienza" (Department of Electrical Engineering). A free version of the LIOV code can be downloaded from the following web address: http://www.liov.ing.unibo.it/.

1.8.2 The LIOV-EMTP-RV code

In order to take into account the presence of complex types of terminations, i.e., surge arresters, groundings of shielding wires, transformers or other power components, as well as of complex network topologies, the LIOV code has been interfaced with electromagnetic transient programs, namely, EMTP [51], MATLAB®-Simulink [32] and EMTP-RV [52], as described in detail in this section. At each time step of the time-domain transient calculation, LIOV performs: (1) the LEMP calculation, (2) the field-to-line coupling solution and (3) the update of the variables to be exchanged with EMTP for the solution of the boundary conditions at the line terminations.

The distribution network is assumed as consisting of a number of illuminated lines (henceforth called LIOV lines, as their model is implemented by the LIOV code) connected to each other through the nodal analysis equations solved by EMTP. The LIOV code has the task of calculating the response of the various LEMP-coupled lines connecting two nodes while the EMTP has the task of solving the line model boundary conditions through the nodal analysis equations.

The link between the LIOV line and the EMTP is realized by numerical processing of the travelling wave solution known as the Bergeron method or the method of characteristics [53].

According to the interface described in References 32, 51, the terminal ends are connected to the LIOV line by means of short lossless line of length $\Delta l = c\Delta t$, not illuminated by any electromagnetic field. The scheme of the interface is represented in Figure 1.7, with reference to a single-phase line for simplicity.

The application of the Bergeron's method gives

$$v_{t0}^{n+1} - Zi_{t0}^{n+1} = G_0 \tag{1.86}$$

$$v_{tk\max}^{n+1} + Zi_{tk\max}^{n+1} = G_{k\max} \tag{1.87}$$

$$v_0^{n+1} + Zi_0^{n+1} = G_0' \tag{1.88}$$

$$v_{k\max}^{n+1} - Zi_{k\max}^{n+1} = G_{k\max}' \tag{1.89}$$

where Z is the surge impedance of the line.

The Bergeron equivalent generators can be estimated as follows:

$$G_0 = v_0^n - Ev_0^n h - Zi_0^n \tag{1.90}$$

$$G_{k\max} = v_{k\max}^n - Ev_{k\max}^n h + Zi_{k\max}^n \tag{1.91}$$

$$G_0' = v_{t0}^{n+1} - Zi_{t0}^{n+1} Ev_0^{n+1} h \tag{1.92}$$

$$G_{k\max}' = v_{tk\max}^{n+1} - Zi_{tk\max}^{n+1} + Ev_{k\max}^{n+1} h \tag{1.93}$$

where

$$Ev_k^n = \int_0^h E_z^e(k\Delta x, z, n\Delta t)dz \tag{1.94}$$

is the incident voltage, that is added or subtracted according to the line type (LIOV or Bergeron) the generator belongs to. Indeed, in EMTP, the voltages are the sum of the scattered and incident voltages, while in LIOV lines, only scattered voltages are accounted for.

The terminal voltages and currents of the LIOV lines can be obtained by (1.88) and (1.89) together with the second coupling equation, which gives

$$i_0^{n+1} = \frac{\frac{\Delta t}{C'\Delta x} i_1^{n+1} - v_0^n + G_0'}{\frac{\Delta t}{C'\Delta x} + Z} \tag{1.95}$$

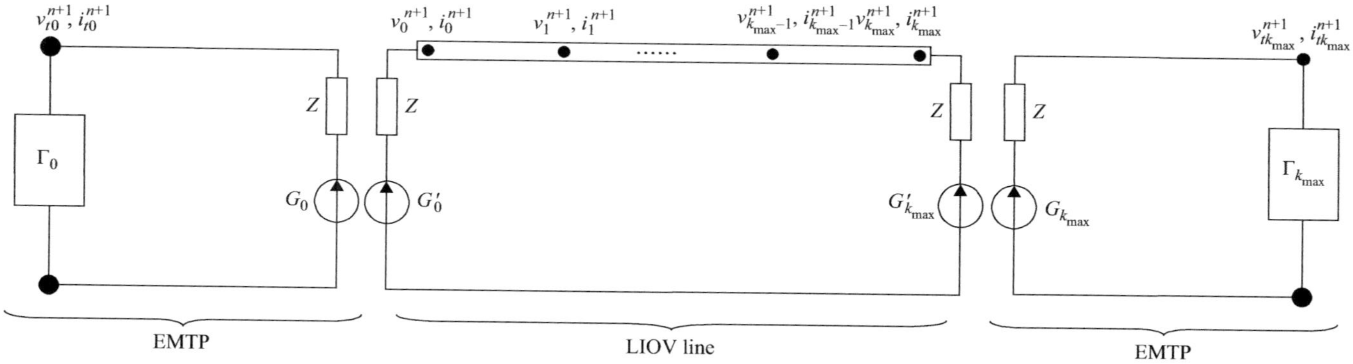

Figure 1.7 Interface between LIOV and EMTP by means of not illuminated Bergeron lines

$$v_0^{n+1} = G_0' - Zi_0^{n+1} \tag{1.96}$$

$$i_{k\max}^{n+1} = \frac{\frac{\Delta t}{C'\Delta x} i_{k\max-1}^{n+1} + v_{k\max}^{n} - G_{k\max}'}{\frac{\Delta t}{C'\Delta x} + Z} \tag{1.97}$$

$$v_{k\max}^{n+1} = G_{k\max}' + Zi_{k\max}^{n+1} \tag{1.98}$$

Note that the presence of external fictitious Bergeron lines, and the corresponding propagation time on them, introduces a delay at each simulation time step, that is, to some extent, reduced by neglecting propagation in (1.92) and (1.93). On the other hand, this approach allows to choose independently the spatial and the temporal integration steps Δx and Δt, provided the Courant's criterion is satisfied.

This delay can be avoided by adopting the interface scheme described in Reference 52 and adopted in the interface between LIOV and EMTP-RV, in which the spatial and temporal steps Δx and Δt are taken at the limit of Courant's criterion, i.e., $\Delta x = c\Delta t$. This gives the first and last spatial discretization of the LIOV line the role to provide the interface LIOV with EMTP, provided a suitable numerical treatment accounts for the presence of the exciting LEMP on them. In what follows, we shall call them illuminated Bergeron lines.

In the presence of an exciting LEMP field and using the Agrawal *et al.* coupling model, the equations that describe the travelling waves on the illuminated Bergeron lines read

$$v^s(0,t) - Zi(0,t) = v^s(c\Delta t, t-\Delta t) - Zi(c\Delta t, t-\Delta t) - \int_0^{c\Delta t} E_x^e(x,h,t)dx \tag{1.99}$$

$$v^s(L,t) + Zi(L,t) = v^s(L - c\Delta t, t-\Delta t) + Zi(L - c\Delta t, t-\Delta t) + \int_{L-c\Delta t}^{L} E_x^e(x,h,t)dx \tag{1.100}$$

where L is the length of the line.

By applying the FDTD method, (1.99) and (1.100) become:

$$v_0^{n+1} - Zi_0^{n+1} = v_1^n - Zi_1^n - \frac{\Delta x}{2}\left(Eh_0^{n+1} + Eh_1^n\right) \tag{1.101}$$

$$v_{k\max}^{n+1} + Zi_{k\max}^{n+1} = v_{k\max-1}^n + Zi_{k\max-1}^n + \frac{\Delta x}{2}\left(Eh_{k\max}^{n+1} + Eh_{k\max-1}^n\right) \tag{1.102}$$

These equations provide the link between the LIOV line and EMTP-RV represented in Figure 1.8 and described by the following equations:

$$v_{t0}^{n+1} - Zi_0^{n+1} = G_1 \tag{1.103}$$

$$v_{tk\max}^{n+1} + Zi_{k\max}^{n+1} = G_{k\max-1} \tag{1.104}$$

$$v_1^{n+1} + Zi_1^{n+1} = G_1' \tag{1.105}$$

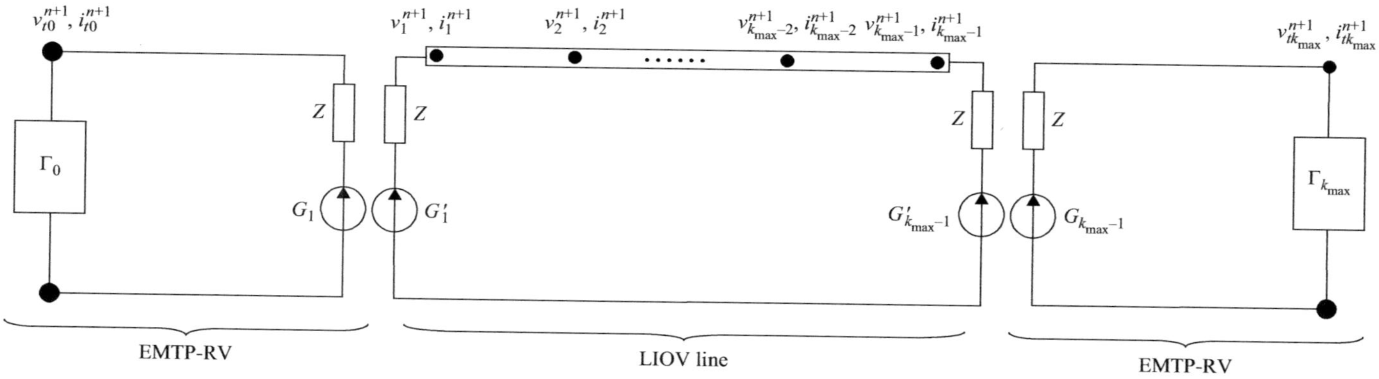

Figure 1.8 Interface between LIOV and EMTP by means of illuminated Bergeron lines

$$v^{n+1}_{k\max-1} - Zi^{n+1}_{k\max-1} = G'_{k\max} \tag{1.106}$$

$$G_1 = v^n_1 - Ev^{n+1}_0 h - Zi^n_1 - \frac{\Delta x}{2}\left(Eh^{n+1}_0 + Eh^n_1\right) \tag{1.107}$$

$$G_{k\max-1} = v^n_{k\max-1} - Ev^{n+1}_{k\max} + Zi^n_{k_{\max}-1} + \frac{\Delta x}{2}\left(Eh^{n+1}_{k_{\max}} + Eh^n_{k_{\max}-1}\right) \tag{1.108}$$

Equations (1.107) and (1.108) are calculated at each time step by means of a dynamic link library (DLL) called within the EMTP-RV simulation environment.

Similar expressions may also be obtained for G'_1 and $G'_{k\max}$, indeed their calculation is not necessary as v^{n+1}_1, i^{n+1}_1, $v^{n+1}_{k\max-1}$ and $i^{n+1}_{k\max-1}$ can be updated by using (1.78) and (1.79).

G_1 and G_{kmax-1} are voltage sources added to the nodal analysis solved by the EMTP-RV. Therefore, it is necessary to define them in the solution provided by the augmented nodal analysis formulation [27]. In particular, each voltage source is defined by adding one row to the augmented nodal admittance matrix and, in the added row, the unknown quantity is the current of the added voltage source, while the known voltage is inserted in the known coefficients column, namely the column of the history currents sources. Together with the additional row, an auxiliary column is also added, where a coefficient equal to one is suitably inserted in order to satisfy *Kirchhoff's* laws in the loop containing the voltage source.

As described in Reference 54, the equations that provide the interface between LIOV and EMTP can be modified to account for the presence of the voltage at the utility frequency. As the time horizon usually adopted for induced voltage calculations is of the order of some tens of microseconds, the 50 or 60 Hz voltage is assumed equal to a value V, for simplicity constant along the line (positive sequence in three-phase lines). The value V is added to the left side of both (1.90) and (1.91) (or (1.107) and (1.108)).

In induced voltage calculations, the overhead line of finite length is often assumed to be terminated in its surge impedance at one or both ends. This simplifies the analysis of the results as the effects of the reflections of the travelling surge waves at the line terminations are avoided.

In the case of unenergized lines, a matched line is represented by defining Γ_0 and Γ_{kmax} equal to the surge impedance Z of the line (or a set of branches of coupled impedances in case of multi-conductor lines). Then, total voltages v_{t0} and $v_{tk\max}$ are equal to one-half of the value of the sources G_0 and $G_{k\max}$ or G_1 and $G_{k\max-1}$.

In order to preserve the possibility to represent a matched line at one or both ends also when a stationary voltage V is taken into account, the line termination is kept open in the EMTP circuit. Sources G_0 and/or $G_{k\max}$ are set equal to half of the value given by (1.90) and (1.91) plus V.

Sources G'_0 and/or $G'_{k\max}$ are taken as null to simulate the absence of a voltage wave reflected inward towards the line at the matched termination.

Analogously, in the linking method with illuminated Bergeron lines G_1 and $G_{k\max-1}$ are set equal to half of the value given by (1.107) and (1.108) plus V, while the currents at the matched terminations are calculated as follows:

$$i_0^{n+1} = -\frac{G_1 - V}{Z} \tag{1.109}$$

$$i_{k_{\max}}^{n+1} = \frac{G_{k\max-1} - V}{Z} \tag{1.110}$$

1.8.3 LEMP response of electrical distribution systems

The theory illustrated so far finds several applications in the area of electric power systems. One of these is the classical problem that electric engineers have solve when they need to choose the insulation level of a distribution system and the number and location of protection devices in order to minimize the number of flashovers – and relevant power supply interruptions – due to lightning strikes. For the case of distribution systems, the attention is often concentrated only on indirect events, as the height of surrounding building and/or elevated objects, such as trees, is in general higher than that of the overhead conductors and, therefore, direct events are expected to occur on these structures and not on the line. What needs to be inferred is the so-called lighting performance of a distribution line (or system) consisting of a plot in an $x - -y$ system having, in the abscissa, the voltage (or the critical flashover voltage – CFO of the insulating system) and, in the vertical scale, the number of lightning-induced events exceeding the given voltage (or the number of flashover for that CFO), as illustrated in References 41, 42, 55. Such a plot is inferred by means of a large number of lightning-induced voltage computations as a function of the line design/configuration and of the annual ground flash density. These procedures start from the statistical distribution of lightning current parameters, assume a lightning incidence model to distinguish between direct and indirect strokes [55], make use of a model for the calculation of the lightning-induced voltages and, by means of an iterative/statistical process in which the strike location and lightning current parameters are randomly varied, infer the number of lightning-induced voltages capable of resulting in line flashover. The above-mentioned computations are rather CPU time consuming, and, for this reason, procedures have been proposed in the literature to optimize them, especially when distribution systems with complex topologies have to be assessed (e.g., Reference 56).

An example is given in Figure 1.9, adapted from Reference 41, in which the lightning performance of a single conductor, 10 m high, overhead line was evaluated.

By examining Figure 1.9, one of the conclusions that one can immediately draw is that the lightning performance of the assumed line is worsened by the ground resistivity, a somewhat unexpected result since, in general, higher line losses result in increased amplitude attenuation of the induced surges. To be able to understand the reasons for that, one has first to analyse the waveshapes of the induced voltages for some specific instances. This can be done in a straightforward manner using LIOV for overhead lines or, for the case of more topologically complex systems, LIOV-EMTP. LIOV was indeed used in Reference 57 to explain, for the first time,

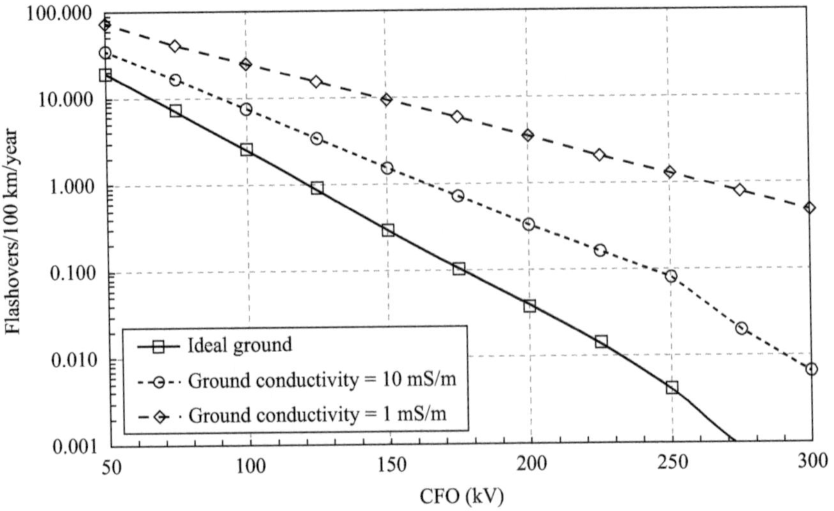

Figure 1.9 Number of induced-voltage flashovers versus distribution-line insulation level. © 2011 IEEE. Reprinted, with permission, from Reference 41

how the ground resistivity can result in an enhancement of the amplitude of the lightning-induced overvoltages, a result that has been further discussed in subsequent publications (e.g., References 2, 22).

In view of the above, we feel it useful in this section to show some examples of application of LIOV-EMTP through several case studies.

The first case study, which we will number (i), involves the calculation of overvoltages induced on a distribution line, showing the effect of the finite conductivity of the ground.

The second case study, numbered (ii), illustrates a calculation of the overvoltages induced on a distribution line, transferred from the medium voltage (MV) to the low voltage (LV) side of a distribution transformer.

The third case study to be illustrated, which we will number (iii), shows the effects of the utility frequency voltage on the calculation of the lightning-induced overvoltages.

The final test case, (iv), is an application of the LIOV code for the lightning performance assessment of a distribution line, carried out by means of the statistical simulation procedure [42] based on the Monte Carlo method.

(i) Overvoltages induced on a distribution line and transferred from the MV to LV side of a distribution transformer

Lightning-induced overvoltages are calculated at multiple observation points along a distribution line as shown in Figure 1.10. The line has the following characteristic: conductor diameter of 1 cm, height of 10.2 m from the ground surface, distance

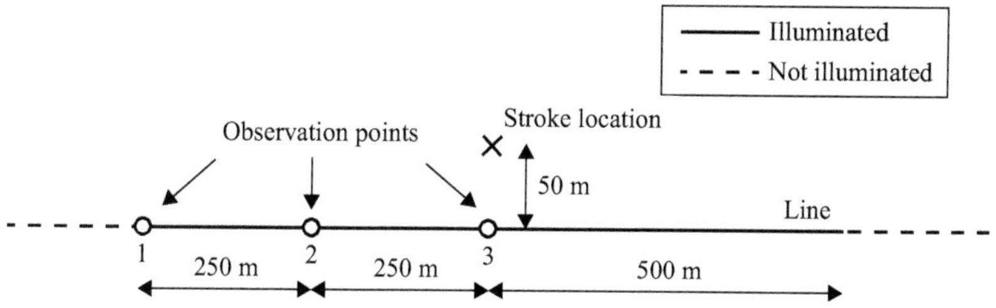

Figure 1.10 Geometry of the first case study

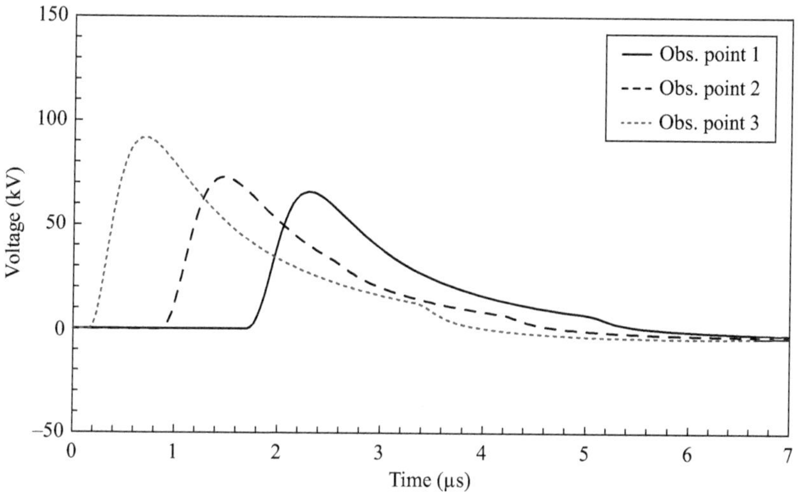

Figure 1.11 Induced overvoltages for the first case study with lossless ground

between the outside conductors of 2.24 m, distance between the central conductor and the nearest outside one equal to 1.07 m. The adopted engineering return stroke model is the TL [35]. The return stroke speed is 0.15 km/μs. The lightning current at the channel base has a peak amplitude equal to 12 kA and a maximum time derivative of 40 kA/μs. The current waveshape is represented by the sum of two Heidler functions with the same parameters adopted in Reference 58, i.e.: $I_{01} = 10.7$ kA, $\tau_{11} = 0.25$ μs, $\tau_{21} = 2.5$ μs, $n_1 = 2$, for the first function, $I_{02} = 6.5$ kA, $\tau_{12} = 2.1$ μs, $\tau_{22} = 230$ μs, $n_2 = 2$, for the second function.

Two ground conditions are considered: one with an infinite conductivity and a second with a finite conductivity. In the latter case, the conductivity is set equal to 1 mS/m and the relative permittivity to 10. The overvoltages for the lossless and the finite conductivity ground are reported in Figures 1.11 and 1.12, respectively. The effect of the finite conductivity of the ground on the transient propagation is neglected.

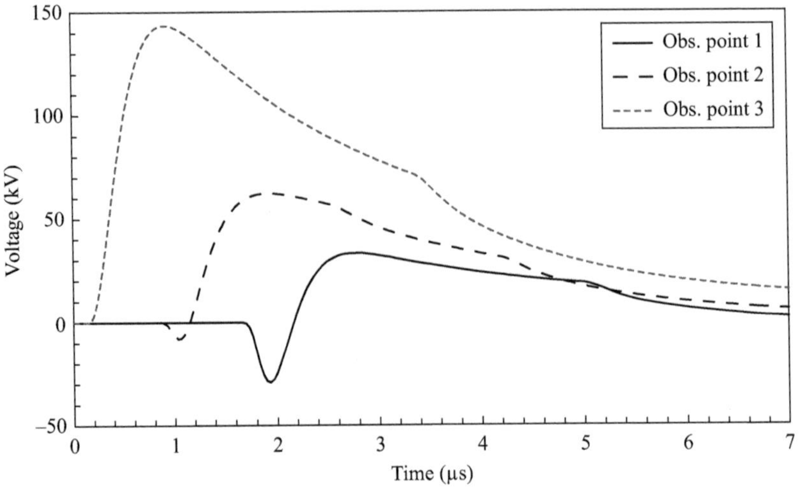

Figure 1.12 Induced overvoltages for the first case study with lossy ground

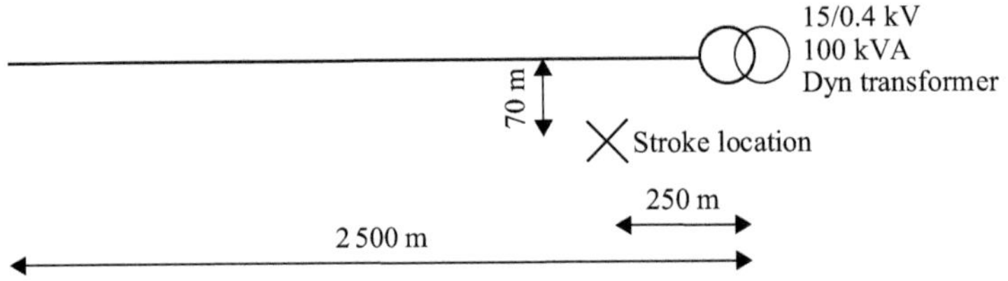

Figure 1.13 Geometry of the distribution system with a no load transformer. © 2009 IEEE. Reprinted, with permission, from Reference 59

(ii) Overvoltages induced on a distribution line and transferred from the medium voltage to the low voltage side of a distribution transformer

The geometry of the system is sketched in Figure 1.13 and it is composed of a three-phase 2.5 km-long MV overhead line terminated with the no load MV–LV transformer at one end. The lightning strike location is 70 m from the line and it is 250 m far from the MV side of the transformer.

The line is overhead and each wire has a diameter of 1 cm. The two outside conductors are suspended by the poles 10 m above ground, while the central one is located 10.8 m above ground. The horizontal distance between the conductors is 0.675 m.

The values of the transformer model have been identified from laboratory measurements on a 100 kVA 15/0.4 kV three-phase Dyn transformer. As described in Reference 59, three different high-frequency transformer models were identified and compared. The first model consists of a π capacitance network; the second has the π capacitance network in parallel with a model at the utility frequency (the BCTRAN model provided by EMTP and identified on the basis of the manufacturer's test data);

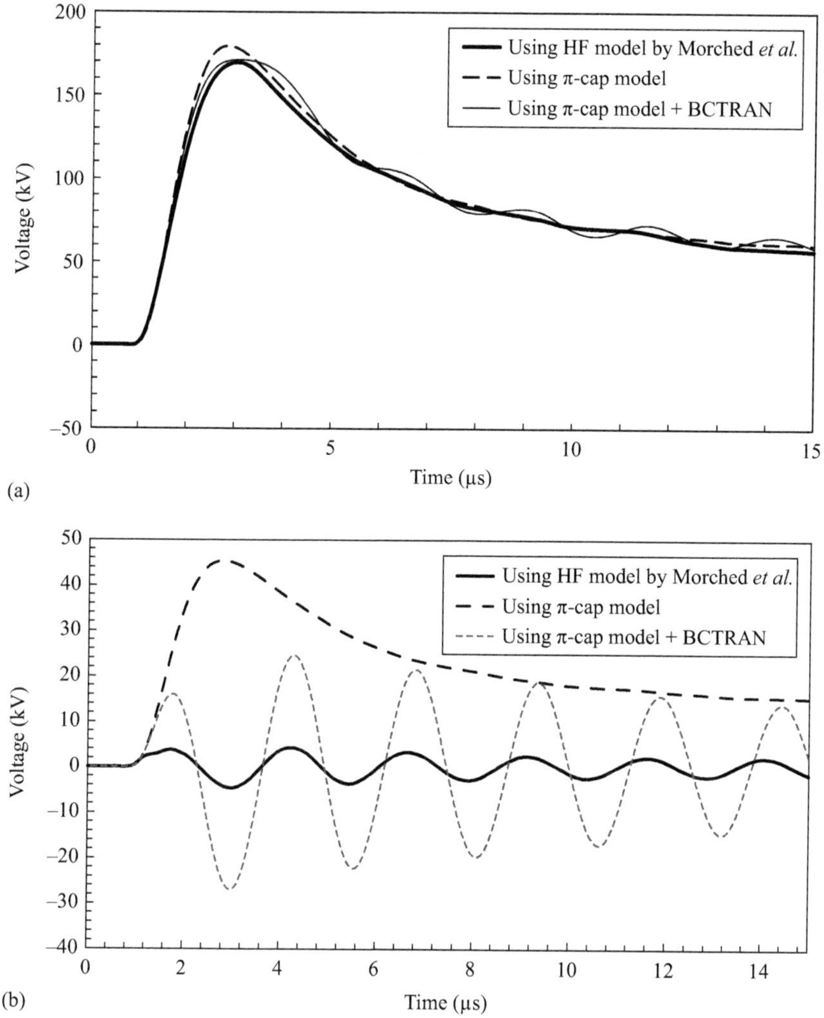

Figure 1.14 Comparison between the voltage transients calculated for the distribution system configuration of Figure 1.13 by adopting the π-capacitance transformer model (with and without the BCTRAN model), and the high-frequency transformer model by Morched et al. (a) Induced voltage at an MV terminal of the transformer. (b) At an LV terminal. © 2009 IEEE. Reprinted, with permission, from Reference 59

the third, more complex and accurate, is the frequency-dependent multiterminal-equivalent model presented by Morched *et al.* in Reference 60 and identified by using the vector-fitting technique [61].

Figure 1.14 shows the induced voltages at the MV and LV terminals of the transformer, obtained by using the considered high-frequency models. The results are obtained for the configuration of Figure 1.13 and for a channel base current waveform with amplitude equal to 31 kA and maximum time derivative of 40 kA/μs. The adopted engineering return stroke model is the MTLE [45]. The finite ground conductivity is taken into account and assumed equal to 0.01 S/m. The effect of the finite ground conductivity is taken into account only in the horizontal component of the lightning-generated electromagnetic field and not in the surge propagation along the lines.

The results of Figure 1.14 show that the considered models predict similar induced voltages at the MV terminals. On the other hand, for the simulation of the overvoltages transferred through the distribution transformer, the detailed high-frequency model by Morched *et al.* predicts different results from the other ones, which are expected to be more accurate, given the comparison between model-predicted results and measurements reported in Reference 58.

(iii) Overvoltages induced on a distribution line – effect of the utility frequency voltage

The test case hereafter illustrated shows the effects of the utility frequency voltage on the calculation of the lightning-induced voltages; it has been presented in Reference 54. We consider a 13.8 kV 2 km-long overhead line with the following characteristics: wire diameter of 1 cm, height of 9.3 m from the ground surface, distance between the outside conductors of 2.2 m, distance between the central conductor and the nearest outside conductor equal to 0.7 m.

The ground conductivity value is assumed equal to 1 mS/m. We assume two stroke locations, both 40-m far from the nearest point of the line: strike location A is equidistant from the line terminations and facing the observation point, stroke location B is near the right termination. The channel base current is the same of case study (i). The calculation is repeated with surge arresters located at the pole in the middle of the line and without surge arresters. The following figures show, for each case, the voltage waveforms observed at the line centre. Figure 1.15 shows the induced voltages calculated with the line assumed to be unenergized.

In Figures 1.16 and 1.17, the three phases are characterized by stationary voltages equal to $V_a = 9.8$ kV, $V_b = -9.8$ kV, and $V_c = 0$, respectively. Figure 1.17 compares the results by assuming the open or matched terminations.

(iv) Lightning performance assessment

The calculation procedure adopted for the estimation of the lightning performance to obtain the results of this subsection is based on the application of the Monte Carlo method, and it is described in Reference 42. Such a procedure can be summarized as follows. A large number n_{tot} of lightning events is randomly generated; each event is characterized by four parameters: lightning current amplitude I_p, time to peak t_f and stroke location with coordinates x and y.

The lightning current parameters are assumed to follow the Cigré log-normal probability distributions [62, 63] for negative first strokes, with a correlation coefficient between t_f and I_p equal to 0.47. The effects of the presence of positive flashes and of subsequent strokes in negative flashes on the lightning performance of the feeder are assumed to be negligible.

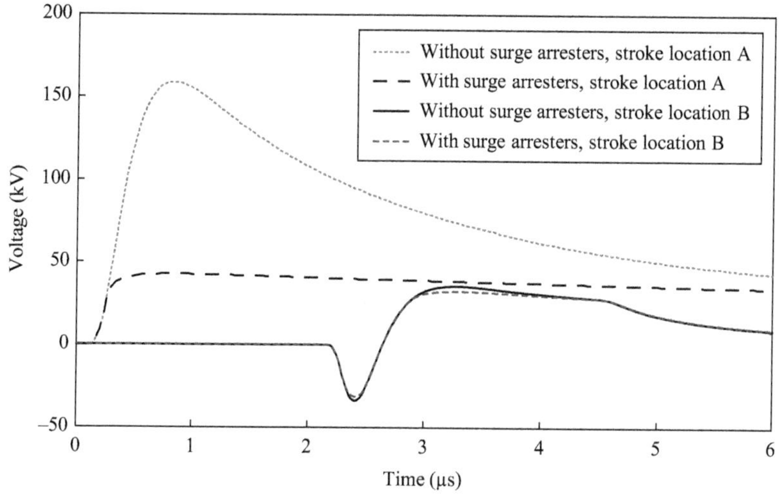

Figure 1.15 Lightning-induced voltages at the line centre of an unenergized line with and without three surge arresters at the line centre. © 2014 Elsevier. Reprinted, with permission, from Reference 54

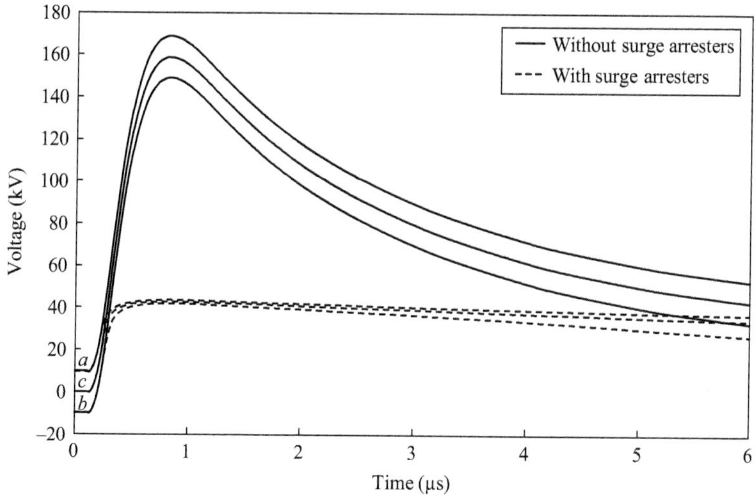

Figure 1.16 Lightning-induced voltages at the line centre of an energized line with and without surge arresters for strike location A. © 2014 Elsevier. Reprinted, with permission, from Reference 54

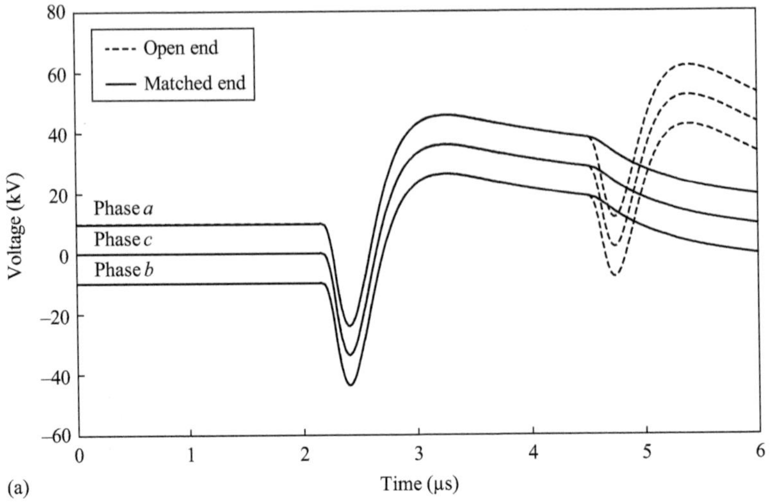

(a)

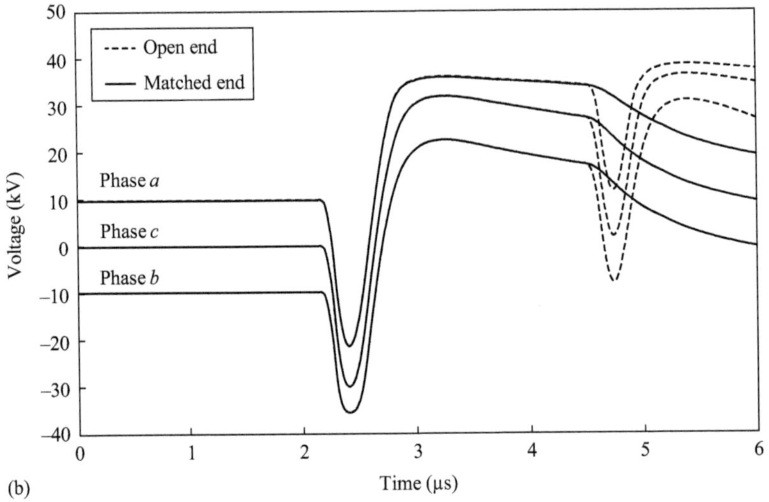

(b)

Figure 1.17 Lightning-induced voltages at the line centre of on an energized line for strike location B: (a) without surge arresters and (b) with surge arresters. © 2014 Elsevier. Reprinted, with permission, from Reference 54

The strike locations are assumed to be uniformly distributed in a striking area, having a size large enough to contain the entire network and all the lightning events that could cause voltages larger than the minimum voltage value of interest for the analysis.

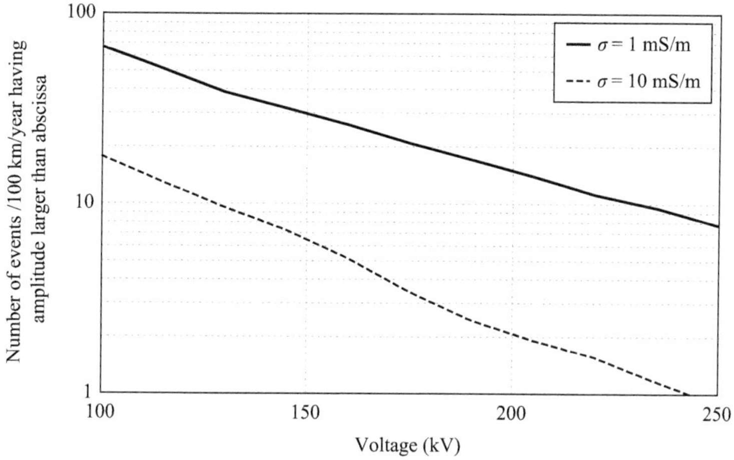

Figure 1.18 Indirect lightning performance of a conventional line for different values of the soil conductivity σ. © 2014 IEEJ. Reprinted, with permission, from Reference 64

As mentioned earlier, the lightning performance is expressed by means of a curve providing the expected annual numbers of lightning events F_p that cause voltages with amplitude larger than the insulation voltage value reported in the abscissa:

$$F_p = \frac{n}{n_{tot}} A_s N_g \tag{1.111}$$

where n is the number of events causing overvoltages higher than the considered insulation level, A_s is the striking area and N_g is the annual ground flash density (assumed here to be equal to 1 flash/km^2/year).

If referred to a single straight line, the lightning performance is usually expressed in terms of number of events per year per unit length of line, e.g., 100 km of line as usually done for transmission lines.

We consider 30 000 lightning events, with stroke locations uniformly distributed over a striking area A_s that extends up to 2 km from a 2-km long line in order to include all the lightning events that could produce induced voltages larger than the minimum withstand voltage considered in the analysis. The line geometrical configuration is the same of test case (iii).

Figure 1.18 shows the graph of the annual number of lightning events that cause a voltage larger than the value in the abscissa in at least one pole of the line; the results make reference to the same 2 km long line of test case (iii), without surge arresters. The calculation was repeated for two different ground conductivities, namely $\sigma = 1$ mS/m and $\sigma = 10$ mS/m. As already known, the higher the soil resistivity the higher the expected maximum amplitude of induced voltages [15].

The calculations are now repeated for a typical modern compact line design, characterized by insulated covered conductors (without shield) and a close upper

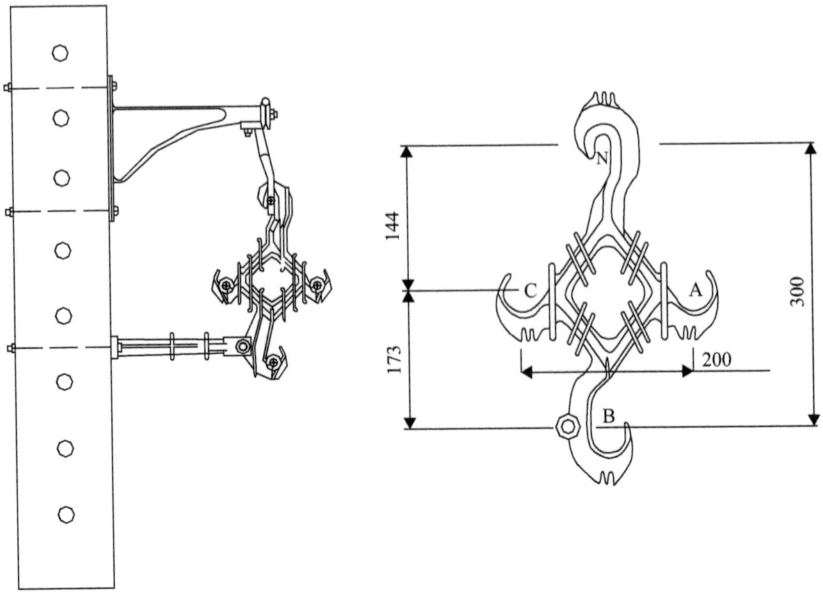

Figure 1.19 *Compact configuration of overhead distribution lines with insulation class equal to 15 kV (dimensions in millimetres; adapted from Reference 66)*

unenergized wire that has the main function of sustaining periodical spacers of the phase conductors (e.g., References 65, 66). The adoption of a compact design for overhead lines is increasing due to the advantages of covered conductors, especially in the regions where contacts with tree branches represent a non-negligible cause of outages and high maintenance costs. As the unenergized wire is periodically grounded, the issue of assessing its effectiveness in reducing the overvoltages due to lightning has been investigated [64]. The specific configuration of the poles and the spacers dealt with in this section is shown in Figure 1.19, with the upper wire located at 9.3 m above ground.

The insulation between the upper wire and covered conductors is around 215 kV and it reduces to 95 kV in case of damaged insulation. Therefore, the presence of the upper wire is not expected to provide a reduction of the effects of direct strikes. We focus instead on its contribution to the reduction of the induced effects by indirect strikes.

Figure 1.20 compares the results obtained for two values of the distance d between two succeeding groundings, namely, 50 m and 200 m. The first value roughly corresponds to grounding at each pole while the second is often assumed as a compromise between effectiveness and costs [67].

The results show that the lightning performance of the compact line is substantially improved by the periodical grounding of the upper wire. Both the distance

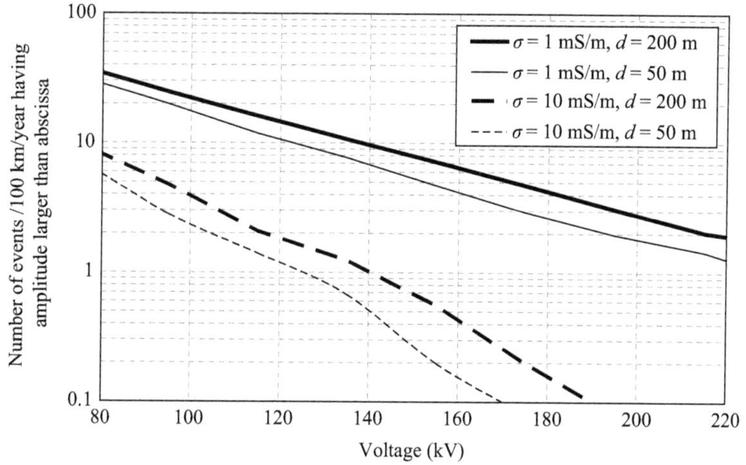

Figure 1.20 Indirect lightning performance of the compact line with the resistance of upper wire groundings equal to 50 Ω, for different values of soil conductivity σ and of spacing d between adjacent groundings. © 2014 IEEJ. Reprinted, with permission, from Reference 64

between groundings and the grounding resistance have a significant impact also for poor ground conductivity.

1.9 Summary and concluding remarks

This chapter discussed the TL theory and its application to the problem of external electromagnetic field coupling to transmission lines. After a short discussion on the underlying assumptions of the TL theory, the field-to-transmission line coupling equations were derived for the case of a single-wire line above a perfectly conducting ground. Three different but completely equivalent approaches that have been proposed to describe the coupling of electromagnetic fields to transmission lines were also presented and discussed. The derived equations were extended to deal with the presence of losses (soil and conductors) and with the case of multiple conductors. The time-domain representation of the field-to-transmission line coupling equations, which allows for a straightforward treatment of both non-linear phenomena and the change in the line topology was also described. Solution methods in the time domain were presented. Computer tools that implement the above equations and solutions were illustrated as well, and applied to several test cases. This has allowed us to show the importance of the illustrated theory and the relevant computer codes when advanced insulation coordination practice is required. As a matter of fact, this requires the accurate assessment of lightning-induced voltages on overhead lines and, in particular, the estimation of the lightning performance of distribution systems. Such systems are characterized by increasing complexity, which needs to, and can be taken into account, thanks to adequate computer tools, such as those described in this chapter.

Acknowledgements

The material presented in this chapter reports some of the results of a joint Italo-Swiss research cooperation carried out during the last three decades involving Silvia Guerreri, Michel Ianoz, Carlo Mazzetti, Mario Paolone and, more recently, Fabio Tossani. Their contribution and support are gratefully acknowledged. The authors wish to express their gratitude to Manuel Martinez, Alexander Piantini, Vladimir A. Rakov, Fred M. Tesche and Martin A. Uman for their precious cooperation throughout these years.

Bibliography

[1] Tesche FM, Ianoz M, Karlsson T. *EMC Analysis Methods and Computational Models*. New York: Wiley Interscience; 1997.

[2] Nucci CA, Rachidi F. "Interaction of electromagnetic fields generated by lightning with overhead electrical networks". In: Cooray V, editor. *The Lightning Flash*, 2nd ed. London: The IET; 2014. p. 559–610.

[3] Vukicevic A, Rachidi F, Rubinstein M, Tkachenko SV. "On the evaluation of antenna-mode currents along transmission lines". *IEEE Trans Electromagn Compat*. 2006 Nov;48(4):693–700.

[4] Paul CR. *Analysis of Multiconductor Transmission Lines*. New York, NY: John Wiley & Sons; 1994.

[5] Maffucci A, Miano G, Villone F. "An enhanced transmission line model for conducting wires". *IEEE Trans Electromagn Compat*. 2004 Nov;46(4): 512–528.

[6] Rachidi F, Tkachenko SV, editors. *Electromagnetic Field Interaction with Transmission Lines: From Classical Theory to HF Radiation Effects*. Southampton: WIT Press; 2008.

[7] Nitsch J, Tkachenko SV. "High-frequency multiconductor transmission-line theory". *Found Phys*. 2010 Mar;40(9):1231–1252.

[8] Lugrin G, Tkachenko SV, Rachidi F, Rubinstein M, Cherkaoui R. "High-frequency electromagnetic coupling to multiconductor transmission lines of finite length". *IEEE Trans Electromagn Compat*. 2015 Dec;57(6):1714–1723.

[9] Taylor CD, Satterwhite RS, Harrison CW. "The response of a terminated two-wire transmission line excited by a nonuniform electromagnetic field". *IEEE Trans Antennas Propag*. 1965 Nov;13(6):987–989.

[10] Agrawal AK, Price HJ, Gurbaxani SH. "Transient response of multiconductor transmission lines excited by a nonuniform electromagnetic field". *IEEE Trans Electromagn Compat*. 1980 May;22(2):119–129.

[11] Rachidi F. "Formulation of the field-to-transmission line coupling equations in terms of magnetic excitation field". *IEEE Trans Electromagn Compat*. 1993 Aug;35(3):404–407.

[12] Rachidi F. "A review of field-to-transmission line coupling models with special emphasis to lightning-induced voltages on overhead lines". *IEEE Trans Electromagn Compat.* 2012 Aug;54(4):898–911.
[13] Nucci CA, Rachidi F. "On the contribution of the electromagnetic field components in field-to-transmission lines interaction". *IEEE Trans Electromagn Compat.* 1995 Nov;37(4):505–508.
[14] Diendorfer G. "Induced voltage on an overhead line due to nearby lightning". *IEEE Trans Electromagn Compat.* 1990 Nov;32(4):292–299.
[15] Rachidi F, Nucci CA, Ianoz M, Mazzetti C. "Influence of a lossy ground on lightning induced voltages on overhead lines". *IEEE Trans Electromagn Compat.* 1996 Aug;38(3):250–264.
[16] Ramo S, Whinnery JR, Van Duzer T. *Fields and Waves in Communication Electronics*, 3rd ed. New York, NY: John Wiley & Sons; 1994.
[17] Tesche FM. "Comparison of the transmission line and scattering models for computing the HEMP response of overhead cables". *IEEE Trans Electromagn Compat.* 1992 May;34(2):93–99.
[18] Sunde ED. *Earth Conduction Effects in Transmission Systems*. New York, NY: D. Van Nostrand Company; 1949.
[19] Rachidi F, Nucci CA, Ianoz M. "Transient analysis of multiconductor lines above a lossy ground". *IEEE Trans Power Deliv.* 1999 Jan;14(1):294–302.
[20] Napolitano F, Tossani F, Nucci CA, Rachidi F. "On the transmission-line approach for the evaluation of LEMP coupling to multiconductor lines". *IEEE Trans Power Deliv*. 2015 Apr;30(2):861–869.
[21] Tossani F, Napolitano F, Rachidi F, Nucci CA. "An improved approach for the calculation of the transient ground resistance matrix of multiconductor lines". *IEEE Trans Power Deliv*. 2016 May;31(3):1142–1149.
[22] Nucci CA, Rachidi F, Rubinstein M. "Interaction of lightning-generated electromagnetic fields with overhead and underground cables". In: Cooray V, editor. *Lightning Electromagnetics*. London: The IET; 2012. p. 687–718.
[23] Taflove A. *Computational Electrodynamics: The Finite-Difference Time-Domain Method*. Norwood, MA: Artech House; 1995.
[24] Lax P, Wendroff B. "System of conservations laws". *Commun Pure Appl Math.* 1960;13:217–237.
[25] Paolone M, Nucci CA, Rachidi F. "A new finite difference time domain scheme for the evaluation of lightning induced overvoltages on multiconductor overhead lines". In: *Proceedings of Fifth international Conference on Power System Transients*, Rio de Janeiro, Brazil; 2001.
[26] Dommel DH. *Electromagnetic Transient Program Reference Manual (EMTP Theory Book)*. Portland: Bonneville Power Administration; 1986.
[27] Mahseredjian J, Dennetière S, Dubé L, Khodabakhchian B, Gérin-Lajoie L. "On a new approach for the simulation of transients in power systems". *Electr Power Syst Res*. 2007 Sep;77(11):1514–1520.
[28] Nucci CA, Rachidi F, Ianoz M, Mazzetti C. "Lightning-induced voltages on overhead lines". *IEEE Trans Electromagn Compat.* 1993 Feb;35(1):75–86.

[29] Nucci CA, Guerrieri S, Correia De Barros MT, Rachidi F. "Influence of corona on the voltages induced by nearby lightning on overhead distribution lines". *IEEE Trans Power Deliv*. 2000 Oct;15(4):1265–1273.

[30] Dragan G, Florea G, Nucci CA, Paolone M. "On the influence of corona on lightning-induced overvoltages". In: *Proceedings of 30th International Conference on Lightning Protection*, Cagliari, Italy; 2010.

[31] Rachidi F, Rubinstein M, Guerrieri S, Nucci CA. "Voltages induced on overhead lines by dart leaders and subsequent return strokes in natural and rocket-triggered lightning". *IEEE Trans Electromagn Compat*. 1997 May; 39(2):160–166.

[32] Borghetti A, Gutierrez JA, Nucci CA, Paolone M, Petrache E, Rachidi F. "Lightning-induced voltages on complex distribution systems: models, advanced software tools and experimental validation". *J Electrostat*. 2004;60(2–4):163–174.

[33] Piantini A, Janiszewski JM, Borghetti A, Nucci CA, Paolone M. "A scale model for the study of the LEMP response of complex power distribution networks". *IEEE Trans Power Deliv*. 2007 Jan;22(1):710–720.

[34] Paolone M, Rachidi F, Borghetti A, *et al*. "Lightning electromagnetic field coupling to overhead lines: theory, numerical simulations, and experimental validation". *IEEE Trans Electromagn Compat*. 2009 Aug;51(3):532–547.

[35] Master MJ, Uman MA. "Transient electric and magnetic fields associated with establishing a finite electrostatic dipole". *Am J Phys*. 1983 Feb;51(2): 118–126.

[36] Cooray V. "Horizontal fields generated by return strokes". *Radio Sci*. 1992 Jul;27(4):529–537.

[37] Rubinstein M. "An approximate formula for the calculation of the horizontal electric field from lightning at close, intermediate, and long range". *IEEE Trans Electromagn Compat*. 1996 Aug;38(3):531–535.

[38] Nucci CA, Diendorfer G, Uman MA, Rachidi F, Ianoz M, Mazzetti C. "Lightning return stroke current models with specified channel-base current: a review and comparison". *J Geophys Res*. 1990 Nov;95(D12):20395–20408.

[39] Rakov VA, Uman MA. "Review and evaluation of lightning return stroke models including some aspects of their application". *IEEE Trans Electromagn Compat*. 1998 Nov;40(4):403–426.

[40] Heidler F. "Analytische blitzstromfunktion zur LEMP-berechnung". In: *Proceedings of 18th International Conference on Lightning Protection*, Munich, Germany; 1985. p. 63–66.

[41] IEEE guide for improving the lightning performance of electric power overhead distribution lines. IEEE Std 1410-2010. 2011.

[42] Borghetti A, Nucci CA, Paolone M. "An improved procedure for the assessment of overhead line indirect lightning performance and its comparison with the IEEE Std. 1410 method". *IEEE Trans Power Deliv*. 2007 Jan;22(1):684–692.

[43] Uman MA, McLain DK. "Magnetic field of lightning return stroke". *J Geophys Res*. 1969 Dec;74(28):6899–6910.

[44] Napolitano F. "An analytical formulation of the electromagnetic field generated by lightning return strokes". *IEEE Trans Electromagn Compat.* 2011 Feb;53(1):108–113.

[45] Nucci CA, Mazzetti C, Rachidi F, Ianoz M. "On lightning return stroke models for LEMP calculations". In: *Proceedings of 19th International Conference on Lightning Protection*, Graz, Austria; 1988. p. 464–470.

[46] Rachidi F, Nucci CA. "On the Master, Uman, Lin, Standler and the Modified Transmission Line Lightning return stroke current models". *J Geophys Res.* 1990;95(D12):20389.

[47] Nucci CA, Rachidi F. "Experimental validation of a modification to the transmission line model for LEMP calculations". In: *Proceedings of Eighth International Symposium on Electromagnetic Compatibility*, Zurich, Switzerland; 1989. p. 389–394.

[48] Cooray V, Scuka V. "Lightning-induced overvoltages in power lines: validity of various approximations made in overvoltage calculations". *IEEE Trans Electromagn Compat.* 1998 Nov;40(4):355–363.

[49] Cooray V. "Some considerations on the 'Cooray–Rubinstein' formulation used in deriving the horizontal electric field of lightning return strokes over finitely conducting ground". *IEEE Trans Electromagn Compat.* 2002 Nov;44(4): 560–566.

[50] Wait JR. "Concerning the horizontal electric field of lightning". *IEEE Trans Electromagn Compat.* 1997 May;39(2):186.

[51] Nucci CA, Bardazzi V, Iorio R, Mansoldo A, Porrino A. "A code for the calculation of lightning-induced overvoltages and its interface with the electromagnetic transient program". In: *Proceedings of 22nd International Conference on Lightning Protection*, Budapest, Hungary; 1994.

[52] Napolitano F, Borghetti A, Nucci CA, Paolone M, Rachidi F, Mahseredjian J. "An advanced interface between the LIOV code and the EMTP-RV". In: *Proceedings of 29th International Conference on Lightning Protection*, Uppsala, Sweden; 2008.

[53] Bergeron L. *Du Coupde Bélier en Hydraulique au Coupde Foudre en Électricité, Paris, France, Dunod, 1949*. Paris: Dunod; 1949.

[54] Napolitano F, Borghetti A, Nucci CA, Martinez MLB, Lopes GP, Dos Santos GJG. "Protection against lightning overvoltages in resonant grounded power distribution networks". *Electr Power Syst Res.* 2014;113: 121–128.

[55] Nucci CA, Rachidi F. "Lightning protection of medium voltage lines". In: Cooray V, editor. *Lightning Protection.* London: The IET; 2010.

[56] Borghetti A, Nucci CA, Paolone M. "Indirect-lightning performance of overhead distribution networks with complex topology". *IEEE Trans Power Deliv.* 2009 Oct;24(4):2206–2213.

[57] Guerrieri S, Nucci CA, Rachidi F. "Influence of the ground resistivity on the polarity and intensity of lightning induced voltages". In: *Proceedings of 10th International Symposium on High Voltage Engineering*, Montreal, Canada; 1997.

[58] Ianoz M, Mazzetti C, Nucci CA, Rachidi F. "Induced overvoltages on overhead transmission lines by indirect lightning return strokes: a sensitivity analysis". In: *Proceedings of 20th International Conference on Lightning Protection*, Interlaken, Switzerland; 1990.

[59] Borghetti A, Morched AS, Napolitano F, Nucci CA, Paolone M. "Lightning-induced overvoltages transferred through distribution power transformers". *IEEE Trans Power Deliv*. 2009 Jan;24(1):360–372.

[60] Morched AS, Martì L, Ottavengers J. "A high frequency transformer model for the EMTP". *IEEE Trans Power Deliv*. 1993 Jul;8(3):1615–1626.

[61] Gustavsen B, Semlyen G. "Rational approximation of frequency domain responses by vector fitting". *IEEE Trans Power Deliv*. 1999 Jul;14(3): 1052–1061.

[62] Anderson RB, Eriksson AJ. "Lightning parameters for engineering applications". *Electra*. 1980;69:65–102.

[63] Cigré Working Group 01 of SC 33. Guide to procedures for estimating the lightning performance of transmission lines. 1991 Oct;63.

[64] Napolitano F, Borghetti A, Messori D, *et al*. "Assessment of the lightning performance of compact overhead distribution lines". *IEEJ Trans Power Energy*. 2013;133(12):987–993.

[65] Oosthuizen M, Lategan R. *Hendrix Covered Conductor Manual*. Western Power. 2010. Available from http://www.westernpower.com.au/documents/WE_n5523521_v7_HENDRIX_COVERED_CONDUCTOR.pdf

[66] AES SUL. "Especificações técnicas de materiais: ferragens para rede compacta". 2005; NTD 004.005.

[67] Piantini A. "Lightning protection of overhead power distribution lines". In: *Proceedings of 29th International Conference on Lightning Protection ICLP*, Uppsala, Sweden; 2008.

Chapter 2

Reliable solutions of uncertain optimal power flow problems by affine arithmetic

Alfredo Vaccaro[1], Claudio A. Cañizares[2] and Antonio Pepiciello[1]

An affine arithmetic (AA)-based computing paradigm aimed at achieving more efficient computational processes and better enclosures of uncertain power flow (PF) and optimal PF (OPF) solution sets is presented in this chapter. The main idea is to formulate a generic mathematical programming problem under uncertainty by means of deterministic problems, based on equivalent AA minimization, equality, inequality operators. Compared to existing solution paradigms, the described formulation presents a different approach to handle uncertainty, yielding adequate and meaningful PF and OPF solution enclosures. Detailed numerical results are presented and discussed using a variety of test systems, demonstrating the effectiveness of the explained AA-based methodology and comparing it to previously proposed techniques for uncertain PF and OPF analyses. Finally, the concept of second-order AA is introduced for mitigating the over-conservatism of the conventional AA-based operators in solving uncertain programming problems.

2.1 Introduction

Optimal power system operation requires intensive numerical analysis to study and improve system security and reliability. In this context, power system operators need to understand and reduce the impact of system uncertainties. To address this issue, power flow (PF) and optimal PF (OPF) analyses are some of the most important tools, as these represent the mathematical foundation of many power engineering applications such as state estimation, network optimization, voltage control, generation dispatch, and market studies.

For the most common formalization of the PF and OPF problems, all input data are specified using deterministic variables resulting either from a snapshot of the system or defined by the analyst based on several assumptions about the system

[1]Department of Engineering, University of Sannio, Italy
[2]Department of Electrical and Computer Engineering, University of Waterloo, Canada

under study (e.g., expected/desired generation/load profiles). This approach can be employed to compute solutions for a single system state that is deemed representative of the limited set of system conditions corresponding to the data assumptions. Thus, when the input conditions are uncertain, numerous scenarios need to be analyzed. These uncertainties are due to several internal and external sources in power systems. The most relevant uncertainties are related to the complex dynamics of the active and reactive power supply and demand, which may vary due to, for example [1]:

- the variable nature of generation patterns due to competition;
- the increasing number of smaller, geographically dispersed generation that could sensibly affect power transactions;
- the difficulties arising in predicting and modeling market participants' behavior, governed mainly by unpredictable economic dynamics, which introduce considerable uncertainty in short-term power system operation;
- the high penetration of non-dispatchable generation units powered by renewable energy sources that induce considerable uncertainty in power system operation.

Since uncertainties can affect the PF and OPF solutions to a considerable extent, reliable solution paradigms, incorporating the effect of data uncertainties, are required. Such algorithms can be employed by analysts to estimate both the data tolerance (i.e., uncertainties characterization) and the solution tolerance (i.e., uncertainty propagation assessment), thus providing insight into the level of confidence of the solutions. Furthermore, these methodologies could effectively support sensitivity analysis of large variables variations to estimate the rate of change in the solution with respect to changes in input data. Hence, to address this problem, this chapter presents a solution methodology based on the use of AA, which is an enhanced model for self-validated numerical analysis in which the quantities of interest are represented as affine combinations of certain primitive variables representing the sources of uncertainty in the data or approximations made during computations.

An enhanced method aimed at computing reliable enclosures of the PF equations by using second-order affine forms is also discussed. In the presented method, the endogenous uncertainty introduced by the approximation of nonlinear functions is not modeled by new noise symbols, but instead it is represented through the partial deviations of the primitive noise symbols. The affine forms resulting from this computing process are called second-order affine forms, which can be processed by proper mathematical operators in order to reliably solve uncertain programming problems.

2.2 Overview of existing approaches

Conventional methodologies available in the literature propose the use of sampling, analytic, and approximate methods for OPF analysis [2, 3], accounting for the variability and stochastic nature of the input data used. A critical review of the most relevant papers proposing these solution methodologies is presented in the following subsections.

2.2.1 *Sampling methods*

Uncertainty propagation studies based on sampling-based methods, such as Monte Carlo, require several model runs that sample various combinations of input values. In particular, the most popular Monte Carlo-based algorithm adopted to solve PF and OPF problems is simple random sampling, in which a large number of samples are randomly generated from the probability distribution functions of the input uncertain variables. Although this technique can provide highly accurate results, it has the drawback of requiring high computation resources needed for the large number of repeated solutions [4]. This hinders the application of this solution algorithm, especially for large-scale power system analysis, where the number of simulations may be rather large and thus the needed computational resources could be prohibitively expensive [5, 6].

The need to reduce the computational costs of Monte Carlo simulations has stimulated the research for improved sampling techniques aimed at reducing the number of model runs, at the cost of accuracy, which requires accepting some level of risk. For example, in [6], the uncertain OPF problem is formulated as a chance-constrained programming model, and the stochastic features of its solutions are obtained by combining Monte Carlo-based simulations with deterministic optimization models. Also, the Bootstrap method, which is a statistical method that empirically assesses the parametric confidence intervals of a sampling distribution based on resampling theory, has been applied to uncertain PF analyses to significantly reduce the sampling space and improving computational performance [7].

Although the application of the aforementioned techniques lowers the computational burden of sampling-based approaches, they reduce the accuracy of the estimation of uncertainty regions of PF and OPF solutions. Therefore, the dichotomy between accuracy and computational efficiency is still an open problem that requires further investigation.

2.2.2 *Analytical methods*

Analytical methods are computationally more effective, but they require some mathematical assumptions in order to simplify the problem and obtain an effective characterization of the output random variables [8]. These assumptions are typically based on model multi-linearization, convolution techniques, and fast Fourier transforms. For example, the cumulant method has been applied to solve the probabilistic OPF problem in [8]; the performance of this method is enhanced by combining it with the Gram–Charlier expansion in [9], and by integrating the Von Mises functions in [10], to handle discrete distributions.

In [6], a novel OPF formulation based on a chance-constrained programming model is proposed to explore the stochastic features of the OPF solution by means of a Monte Carlo-based probabilistic model, whose parameters are identified by solving a deterministic optimization problem. However, the application of these techniques to solve OPF problems is not straightforward and requires a back-mapping approach and a linear approximation of the non-linear PF equations [11]; this is mainly due to the non-linearities, the multiple uncertain variables, and the multiple output constraints characterizing OPF problems.

Analytical techniques present various shortcomings, as discussed in, for example, [12], such as the need to assume statistical independence of the input data, and the problems associated with accurately identifying probability distributions for some input data. This is a problem for PF and OPF analyses, since it is not always feasible to translate imprecise knowledge into probability distributions, as in the case of power generated by wind or photovoltaic generators, due to the inherently qualitative knowledge of the phenomena and the lack of sufficient data to estimate the required probability density distributions. To address this issue, the assumptions of normality and statistical independence of the input variables are often made, but experimental results show that these assumptions are not always supported by empirical evidence. These drawbacks may limit the usefulness of analytical methods in practical applications, especially for the study of large-scale power networks.

2.2.3 Approximate methods

In order to overcome some of the aforementioned limitations of sampling and analytical methods, the use of approximate methods, such as the first-order second-moment method and point estimate methods, has been proposed in the literature [13]. Rather than computing the exact OPF solution, these methods aim at approximating the statistical proprieties of the output random variables by means of a probability distribution fitting algorithm. In particular, the application of the first-order second-moment method allows to compute the first two moments of the OPF solution by propagating the moments of the input variables by the Taylor-series expansion of the model equations.

Point estimate methods represent a more effective strategy, especially if the input parameters uncertainties can be directly estimated or measured. The application of these solution algorithms allows to estimate the statistical moments of the OPF solution by properly amalgamating the solutions of $2p$ deterministic problems, where p is the number of uncertain parameters. This feature could be further enhanced by deploying more sophisticated point estimation schemes, based, for example, on Hong's point estimate method [14]. The application of this enhanced solution strategy allows solving the OPF problem in the presence of multiple uncertainty sources characterized by both normal and binomial distributions, which could be particularly useful in modeling generator outages. These papers demonstrate that point estimate methods allow to effectively approximate the OPF solution while keeping low the computational burden, which is confirmed in [15], where a comparison between the two-point estimate method proposed in [16] and a cumulant method proposed in [8] for solving the OPF problem in the presence of multiple data uncertainty is presented. The results obtained in this paper show that both approaches give similar results in most cases and are accurate provided that the OPF has a feasible solution. It also observed that the cumulant method exhibits better performances for higher uncertainty in the input variables; however, since it is based on a linearization around an operation point, its performances rapidly decrease when this approximation is no longer valid. Both of these methods are shown to be computationally significantly faster than a standard sampling-based approach, since they solve a reduced number of deterministic problems.

The application of the aforementioned solution methods presents several shortcomings. In particular, two-point estimate methods are not suitable to solve large-scale problems, since they typically do not provide acceptable results in the presence of a large number of input random variables. Moreover, the identification of the most effective scheme that should be adopted to select the number of estimated points is still an open problem that requires further investigations [17]; this is a critical issue, since a limited number of estimated points do not allow for an accurate and reliable exploration of the solution space, especially for input uncertainties characterized by relatively large standard deviations, such as in the case of lognormal or exponential distributions. On the other hand, an increased number of estimated points reduce the computational benefits deriving by the application of point estimated methods, which could degenerate into a standard Monte Carlo solution approach.

2.2.4 Non-probabilistic methods

Recent research has enriched the spectrum of available techniques to deal with uncertainty in OPF by proposing non-probabilistic formalisms, which are commonly adopted when uncertainty does not originate from unpredictable numerical measurements but stems from imprecise human knowledge about the system; as a consequence, only imprecise estimates of values and relations between variables are available. For example, wind can be locally measured, but it is difficult to estimate the spatial distribution of wind speed in a geographical area using probabilities; also, weather forecasts provide qualitatively information about environmental variables that can hardly be represented in a probabilistic form. Hence, the availability of modeling and simulation tools able to deal with non-probabilistic knowledge can be useful to analysts for PF and OPF studies.

The application of fuzzy set theory to represent imprecise information in PF and OPF studies, rather than using uncertainty associated with a frequency of occurrence, has been proposed in several papers [18–20]. In this paradigm, the input data and the inequality constraints are modeled by fuzzy numbers, which are special instances of fuzzy sets [20], and the problem solution is computed by deploying efficient linear programming solution algorithms based on Dantzig–Wolfe decomposition and dual simplex [19].

Other studies reported in the literature have proposed the employment of self-validated computing for uncertainty representation in PF and OPF analyses. The main advantage of self-validated computation is that the algorithm itself keeps track of the accuracy of the computed quantities, as part of the process of computing them, without requiring information about the type of uncertainty in the variables. The simplest and most popular of these models is interval mathematics (IM), which allows for numerical computation where each quantity is represented by an interval of real numbers without a probability structure; such intervals are added, subtracted, and/or multiplied in such a way that each computed interval is guaranteed to contain the unknown value of the quantity it represents. The application of "standard" IM, referred here as interval arithmetic (IA), to PF analysis has been investigated by various authors (e.g., [12, 21]). However, the adoption of this solution technique presents many drawbacks derived mainly by the so-called "dependency problem" and "wrapping effect"; as a consequence, the solution provided by an IA method for PF solution is not always as informative as expected. Thus, in [22], it is shown that the use of IA for the solution of the PF equations may easily yield aberrant solutions, due to the fact that the IA formalism is unable to represent the correlations that the PF equations establishes between the power systems state variables; therefore, at each algorithm step, spurious values are added to the solutions, which could converge to large domains that include the correct solution.

2.2.5 AA-based methods

To overcome the aforementioned limitations in IA, in [22], the employment of more effective self-validated paradigms based on AA is explored. In this approach, each state variable is approximated by a first degree polynomial composed by a central value, i.e., the nameplate value, and a number of partial deviations that represent the correlation among various variables. The adoption of AA for uncertainty representation allows expressing the PF and OPF equations in a more convenient formalism, so that a reliable estimation of the solution hull can be computed taking into account the parameter uncertainty inter-dependencies, as well as the diversity of uncertainty sources. These benefits have been confirmed in [23] and in [24], where AA-based methods have been proposed to solve uncertain OPF problems, which allows to determine operating margins for thermal generators in systems with uncertain parameters, by representing all the state and control variables with affine forms accounting for forecast, model error, and other sources of uncertainty, without the need to assume a probability density function. These methodologies have been recently recognized as a promising alternative for stochastic information management in bulk generation and transmission systems for smart grids [25].

Based on the work reported in [22], several papers have explored the application of AA-based computing in power system analysis. In particular, in [26], the state estimation problem in the presence of a mixed phasor and conventional power measurements has been addressed, considering the effect of network parameters uncertainty by an iterative weight least square algorithm based on IA and AA processing. In [27], an AA-based model of the uncertain PF problem is proposed, using complementarity conditions to properly represent generator bus voltage controls, including reactive power limits and voltage recovery; the model is then used to obtain operational intervals for the PF variables considering active and reactive power demand uncertainties. In [28], a non-iterative solution scheme based on AA is proposed to estimate the bounds of the uncertain PF solutions by solving an uncertain PF problem, which is formalized by an interval PF problem and solved by quadratic programming optimization models. More recent applications include the analysis of multi-energy systems [29, 30], optimal scheduling of microgrids [31], optimal dispatch of energy storage [32, 33], and wind power integration [34].

The benefits deriving from the application of AA-based computing to power system planning and operation in the presence of data uncertainty have been assessed in [35], which confirms that AA represents a fast and reliable computing paradigm that allows planners and operators to cope with high levels of renewable energy penetration, electric vehicle load integration, and other uncertain sources. Moreover, as confirmed in [23, 24, 36, 37], AA allows the analyst to narrow the gap between the upper and lower bounds of the OPF solutions, avoiding the overestimation of bounds resulting from correlation of variables in IA.

In this chapter, an AA-based computing paradigm proposed in [38] to efficiently determine better enclosures of uncertain PF and OPF solution sets is presented. The main idea is to formulate a generic mathematical programming problem under uncertainty by means of equivalent deterministic problems, based on a set of AA-based minimization, equality, inequality operators, and second-order affine forms to express nonlinearities resulting from the multiplication of two affine forms. Compared to previously proposed AA solution paradigms, this formulation presents greater flexibility, as it allows to find feasible solutions and inclusion of multiple equality and inequality constraints, while reducing the approximation errors to obtain better PF and OPF solution enclosures.

2.3 Mathematical background

2.3.1 PF analysis

PF analysis deals mainly with the calculation of the steady-state voltage phasor angle and magnitude for each network bus, for a given set of variables such as load demand and real-power generation, under certain assumptions such as balanced system operation. Based on this information, the network operating conditions, in particular, real and reactive PFs on each branch, power losses, and generator reactive power outputs, can be determined. Thus, the input (output) variables of the PF problem are typically:

- the real and reactive power (voltage magnitude and angle) at each load bus, i.e., PQ buses;
- the real power generated and the voltage magnitude (reactive power generated and voltage angle) at each generation bus, i.e., PV buses;
- the voltage magnitude and angle (the real and reactive power generated) at the reference or slack bus.

The equations typically used to solve the PF problem are the real-power balance equations at the generation and load buses, and the reactive power balance at the load buses. These equations can be written in a polar form as:

$$\begin{aligned} P_i^F &= V_i \textstyle\sum_{b=1}^{N} V_b Y_{ib} \cos(\delta_i - \delta_b - \theta_{ib}) \quad \forall i \in \mathcal{N}_P \\ Q_j^F &= V_j \textstyle\sum_{b=1}^{N} V_b Y_{jb} \sin\left(\delta_j - \delta_k - \theta_{jb}\right) \quad \forall j \in \mathcal{N}_Q \end{aligned} \tag{2.1}$$

and in a rectangular form as:

$$\begin{aligned} P_i^F &= \sum_{b=1}^{N} G_{ib}(V_{Ri}V_{Rb} + V_{Ii}V_{Ib}) + B_{ib}(V_{Ii}V_{Rb} - V_{Ri}V_{Ib}) \quad \forall i \in \mathcal{N}_P \\ Q_j^F &= \sum_{b=1}^{N} G_{jb}(V_{Ij}V_{Rb} - V_{Rj}V_{Ib}) - B_{jb}(V_{Ij}V_{Ib} + V_{Rj}V_{Rb}) \quad \forall j \in \mathcal{N}_Q \end{aligned} \tag{2.2}$$

where:

- $\mathcal{N}$ is the set of all the N buses of the system;
- $\mathcal{N}_P$ is the set of the buses in which the active power is specified;
- $\mathcal{N}_Q$ is the set of the buses in which the reactive power is specified;
- P_i^F and Q_j^F are the real and reactive power injections specified at the ith and jth bus;
- V_{Rb} is the real part of the voltage of the bth bus in a rectangular form;
- V_{Ib} is the imaginary part of the voltage of the bth bus in a rectangular form;
- $V_g\angle\delta_g$ is the unknown gth bus voltage in a polar form;
- and $Y_{ij}\angle\theta_{ij}$ is the ijth element of the bus admittance matrix, with real and imaginary terms G_{ij} and B_{ij}, respectively.

Due to the nonlinear nature of these equations, the solution is not unique, and numerical algorithms, mainly based on Newton–Raphson or fast-decoupled methods, are employed to obtain a solution that is within an acceptable tolerance. These algorithms aim at approximating the non-linear PF equations by linearized Jacobian-matrix equations, which are solved by means of numerical iteration algorithms and sparse factorization techniques.

The PF solution should take into account the limits on certain variables, in particular max/min values of the reactive power at generation buses, to properly model the generator voltage controls. To address this particular issue, the typical solution strategy is to carry out a bus-type "switching," which consists on converting a PV-bus into a PQ-bus with the reactive power set at the limiting value, if the corresponding limits are violated. If at any consequent iteration, the voltage magnitude at that bus is below or above its original set point, depending on whether the generator is, respectively, underexcited or overexcited, the bus is then reverted back to a PV-bus. An alternative and more effective strategy to represent generator bus voltage controls, including reactive power limits and voltage recovery processes, has been proposed in [39], where an OPF-based model of the PF problem using complementarity conditions is proposed; this is the approach used here to solve the AA-based PF problem.

2.3.2 Optimal PF analysis

OPF analysis aims at computing the power system operation state according to, for example, cost, planning, or reliability criteria without violating system and equipment operating limits. The solution of this problem yields for identifying the optimal asset of the control/decision variables **u** that minimizes a scalar objective function f (e.g., thermal generation costs), subject to a number of nonlinear equality g_l (e.g., power flow equations) and inequality h_r (e.g., generator reactive power limits) constraints, where f, g_l, and h_r are continuous and differentiable functions. Hence, this problem can be formalized in general by the following constrained, non-linear multi-objective programming problem:

$$\min_{(\mathbf{x},\mathbf{u})} \quad f(\mathbf{x},\mathbf{u}) \tag{2.3}$$

$$\text{s.t.} \quad g_l(\mathbf{x},\mathbf{u}) = 0 \qquad \forall l \in (1,\ldots,n) \tag{2.4}$$

$$h_r(\mathbf{x},\mathbf{u}) < 0 \qquad \forall r \in (1,\ldots,m) \tag{2.5}$$

where $\mathbf{x}$ is the vector of dependent variables, n is the number of equality constraints, and m is the number of inequality constraints. These equations can be expressed in a more compact vectorial form as follows:

$$\min_{(\mathbf{x},\mathbf{u})} \quad f(\mathbf{x},\mathbf{u}) \tag{2.6}$$

$$\text{s.t.} \quad \mathbf{g}(\mathbf{x},\mathbf{u}) = 0 \tag{2.7}$$

$$\mathbf{h}(\mathbf{x},\mathbf{u}) < 0 \tag{2.8}$$

where $\mathbf{g}(.)$ and $\mathbf{h}(.)$ are the n-dimensional and m-dimensional vectors representing the equality and inequality constraints, respectively.

The control/decision variables in (2.6) depend on the specific application domain. These can include both real-valued variables, such as the active power generated by the available generators (i.e., optimal power dispatch), the set points of the primary voltage controllers (i.e., secondary voltage regulation), the optimal location of control/generator resources (i.e., planning studies), the maximum loading factor (i.e., voltage stability analysis), and integer variables, such as the set of the available generators (i.e., unit commitment). As a consequence, the OPF can be in general classified as a non-convex mixed integer non-linear programming (MINLP) problem.

The dependent variables include the voltage magnitude and phase angle at PQ buses, the voltage phase angle and the reactive power generated at the PV buses, and the active and reactive power generated at the slack bus. The inequality constraints include the maximum allowable PFs for the power lines, the minimum and maximum allowable limits for most control/decision variables, i.e., $u_{\text{min,i}} \leq u_i \leq u_{\text{max,i}}$, $\forall i \in (1,\ldots,n_u)$, such as generator voltages, and for some dependent variables, i.e., $x_{\text{min,i}} \leq x_i \leq x_{\text{max,i}}$, $\forall i \in (1,\ldots,n_x)$, such as load bus voltage limits. In addition, the control/decision and the dependent variables should satisfy the PF equations (2.1), which are the main equality constraints for (2.6).

The objective functions $f(.)$ could integrate both technical and economic criteria, including the minimization of the production costs, the minimization of the transmission line losses, the minimization of the voltage deviations, etc. Because these design objectives are typically competing, and due to its non-convexity and non-linear characteristics, the OPF problem has no unique solution, and thus a suitable trade-off between the objectives needs to be identified. To deal with a multi-objective OPF problem, one of the most common solution approaches used is the weighted global criterion method, in which all objective functions, which are assumed to be non-negative, are combined to form a single utility function expressed as:

$$f(\mathbf{x}, \mathbf{u}) = \sum_{\gamma=1}^{q} \omega_\gamma f_\gamma(\mathbf{x}, \mathbf{u}) \tag{2.9}$$

where the normalized weights ω_γ, so that $\sum_{\gamma=1}^{q} \omega_\gamma = 1, \omega_\gamma > 0$, are typically set by the analyst depending on the relative importance of the objective functions.

Many classes of programming algorithms, such as nonlinear programming, quadratic programming, and linear programming, have been proposed to solve the OPF problem [40]. Some methods formalize the problem's Karush–Kuhn–Tucker (KKT) optimality conditions, which are a set of nonlinear equations that can be solved by using an iterative Newton-based algorithm. These methods can handle both equality and inequality constraints, with the latter being added as quadratic penalty terms to the objective function and multiplied by proper penalty multipliers. Another useful paradigm to handle inequality constraints is based on the Interior Point method, i.e., barrier method. This approach converts the inequality constraints into equalities by the introduction of non-negative slack variables; a self-concordant barrier function (e.g., logarithmic) of these slack variables is then added to the objective function and multiplied by a barrier parameter, which is gradually reduced to zero during the solution process.

2.3.3 *Self-validated computing*

2.3.3.1 IA

The most intuitive approach to the numerical solution of uncertain PF and OPF problems consists of extending the numerical algorithms for the solution of the corresponding deterministic problems using IA operators. IA is a range-based formalism for numerical computation, where each real quantity θ is assumed to be "unknown but bounded" in an interval of real numbers $\Theta = [\theta_{inf}, \theta_{sup}]$, also known as the tolerance of θ. The key element of IA based computing is based on the following theorem [41].

Theorem 2.1. (Fundamental invariant of range analysis for IA) $\forall\ \Gamma : \Re^P \to \Re^q$, *globally Lipschitz with bounded slope, there exists an interval extension* $\Gamma^I : \Re^P \to \Re^q$, *such that:* $\forall\ (\theta_1, .., \theta_P) \in (\Theta_1, .., \Theta_P) \Rightarrow \Gamma\ (\theta_1, .., \theta_P) \in \Gamma^I\ (\Theta_1, .., \Theta_P)$.

The implementation of the interval extension Γ^I is generally straightforward for elementary operations, such as sums, products, square roots, since it requires only the identification of the maximum and minimum values of $\Gamma(\theta_1, .., \theta_P)$, when the corresponding arguments vary independently over specified intervals. Examples of simple arithmetic operations between two intervals $\Theta_1 = [\theta_{1,inf}, \theta_{2,sup}]$ and $\Theta_2 = [\theta_{2,inf}, \theta_{2,sup}]$ are:

$$\Theta_1 + \Theta_2 = \left[\theta_{1,inf} + \theta_{2,inf}, \theta_{1,sup} + \theta_{2,sup}\right] \tag{2.10}$$

$$\Theta_1 - \Theta_2 = \left[\theta_{1,inf} - \theta_{2,sup}, \theta_{1,sup} + \theta_{2,inf}\right] \tag{2.11}$$

$$\Theta_1 \cdot \Theta_2 = [\min(\theta_{1,inf}\theta_{2,inf}, \theta_{1,inf}\theta_{2,sup}, \theta_{1,sup}\theta_{2,inf}, \theta_{1,sup}\theta_{2,sup}), \max(\theta_{1,inf}\theta_{2,inf}, \theta_{1,inf}\theta_{2,sup}, \theta_{1,sup}\theta_{2,inf}, \theta_{1,sup}\theta_{2,sup})] \quad (2.12)$$

$$\Theta_1 / \Theta_2 = [\theta_{1,inf}, \theta_{1,sup}] \cdot \left[\frac{1}{\theta_{2,sup}}, \frac{1}{\theta_{2,inf}}\right] \; 0 \notin [\theta_{2,inf}, \theta_{2,sup}] \quad (2.13)$$

Computation of interval extensions for more complex functions can be obtained by composing these primitive operators. Based on Theorem 2.1, it is possible to conclude that if a function is evaluated using these IA-based operators, the resulting interval is guaranteed to enclose the range of function values.

IA-based computing has been applied for solving mathematical problems under uncertainty such as linear systems of equations (e.g., [42]), non-linear systems of equations (e.g., [43]), and optimization problems (e.g., [44]). The application of these algorithms typically yields to approximate interval solutions, called outer solutions, that are guaranteed to contain the exact interval solution. However, in many cases, these outer solutions are quite large; thus, as shown in [45], the use of IA-based computing in iterative solution algorithms may easily yield useless solutions. This is due to the fact that the IA formalism is unable to correctly represent the interaction between the problem variables, due to what is known as the "wrapping problem," as illustrated in Figure 2.1 [45], which plots the state space evolution of the harmonic oscillator, $\dot{\theta}_1 = \theta_2$, $\dot{\theta}_2 = -\theta_1$, whose initial condition ($t = 0$) is represented by the rectangle ABCD with its sides parallel to the axes. In this case, since the initial rectangle does not evolve into another rectangle parallel to the coordinate axes, when one represents the uncertain state of the dynamical system at time t' using the interval notation, the IA solution (rotated rectangle $A'B'C'D'$) adds a set of "spurious states"

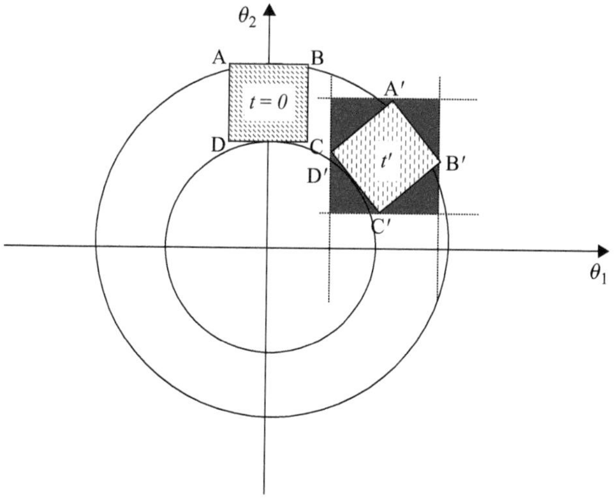

Figure 2.1 IA evolution of the external surface of the region of uncertainty for a second order oscillatory system ("wrapping" effect)

(black regions), which do not correspond to an evolution of points belonging to ABCD; thus, in a few iterations, the IA solution explodes, covering the entire phase space. As a consequence, the interval solutions produced by IA-based solvers are often much wider than the true range of the corresponding quantities, especially during long computational chains in which the interval width increases rapidly. This phenomenon is well known in the simulation of qualitative systems and requires the adoption of special techniques.

Another well-known issue which could limit the application of IA-based computing in real world application is the so-called "dependency problem," which derives directly from the definition of the interval difference operator (2.11):

$$\Theta - \Theta = [\theta_{inf}, \theta_{sup}] - [\theta_{inf}, \theta_{sup}] = [\theta_{inf} - \theta_{sup}, \theta_{sup} + \theta_{inf}] \neq 0 \tag{2.14}$$

This aberration is due to intrinsic inability of IA to discriminate the uncertainty sources, which are assumed to be independent for each interval variable. In particular, if the interval variables $\Theta_1 = [1, 2]$ and $\Theta_2 = [1, 2]$ describe two independent uncertain sources, then the results computed by applying the IA-based difference operator, $\Theta_1 - \Theta_2 = [-1, 3]$, is correct. On the other hand, if these interval variables describe the same uncertain source, then the corresponding result leads to a large overestimation error.

In [22], it is demonstrated that due to the introduction of spurious values in the result of IA-based PF analysis, there is excessive conservatism in the output intervals, especially when solving large scale problems. To address this limitation, the use of more advanced paradigms based on AA has been proposed in [38] to solve uncertain OPF problems.

2.3.3.2 Linear AA

AA is a self-validated computing method, which deploys range-based mathematical operators to process uncertain data. It is capable of handling correlated uncertainty sources, which leads to the mitigation of the dependency problem and wrapping effect in interval-based computing. An affine form $\hat{x}$ can be defined through the following expression:

$$\hat{x} = x_0 + x_1\varepsilon_1 + x_2\varepsilon_2 + \cdots + x_p\varepsilon_p \tag{2.15}$$

where x_0, $[x_1 \ldots x_n]$, and $[\varepsilon_1 \ldots \varepsilon_n]$ are the central value, the partial deviations, and the noise symbols of the affine form $\hat{x}$, respectively. Each noise symbol ε_i represents an independent uncertainty source affecting the uncertain variable, whose value ranges within the interval $[-1, 1]$. This means that if two uncertain variables share noise symbols, they are partially correlated, and their correlation depends on the value of the partial deviations associated with the common noise symbol.

Affine mathematical operators can be easily extended to affine forms as follows:

$$\alpha\hat{x} + \beta\hat{y} + \gamma = (\alpha x_0 + \beta y_0 + \gamma) + \sum_{k=1}^{P} \varepsilon_i(\alpha x_k + \beta y_k) \tag{2.16}$$

Nonlinear operations, on the other hand, require an approximation, which could assume different structures, depending on the desired degree of accuracy and the available computational resources [46]. The simplest solution is adding a new noise symbol, representing the approximation error. Thus, a multiplication between two affine forms yields the following nonlinear relationship:

$$\hat{x} * \hat{y} = x_0 y_0 + \sum_{k=1}^{P} (x_0 y_k + y_0 x_k)\varepsilon_k + \sum_{k=1}^{P} \sum_{h=1}^{P} (x_k y_h)\varepsilon_k \varepsilon_h \tag{2.17}$$

This non-affine operation is traditionally approximated by bounding the second-order terms as follows:

$$\sum_{k=1}^{P} \sum_{k=1}^{P} (x_k y_h)\varepsilon_k \varepsilon_h \leq \sum_{k=1}^{P} |x_k| \sum_{h=1}^{P} |y_h| = R(\hat{x})R(\hat{y}) \tag{2.18}$$

where the $R(.)$ is the radius of the affine form. In this case, a new noise symbol describing the approximation error can be defined as follows:

$$\hat{x} * \hat{y} = x_0 y_0 + \sum_{k=1}^{P} (x_0 y_k + y_0 x_k)\varepsilon_k + R(\hat{x})R(\hat{y})\varepsilon_{p+1} \tag{2.19}$$

2.3.4 Solving constrained optimization problems by linear AA

AA-based computing can be applied to reliably solve constrained optimization problems in the presence of data uncertainty. These kinds of problems can be formalized as follows:

$$\min_{(\hat{\mathbf{x}},\hat{\mathbf{u}})} \quad \hat{f}(\hat{\mathbf{x}}, \hat{\mathbf{u}}) \tag{2.20}$$

$$\text{s.t.} \quad \hat{g}_l(\hat{\mathbf{x}}, \hat{\mathbf{u}}) = 0 \qquad \forall l \in (1, \ldots, n) \tag{2.21}$$

$$\hat{h}_r(\hat{\mathbf{x}}, \hat{\mathbf{u}}) < 0 \qquad \forall r \in (1, \ldots, m) \tag{2.22}$$

where:

- $\hat{\mathbf{x}}$ and $\hat{\mathbf{u}}$ are the unknown affine forms describing the dependent and independent variables, respectively;
- $\hat{f}$ is a continuous and differentiable affine function describing the problem objectives;
- and $\hat{g}_l$ and $\hat{h}_r$ are continuous and differentiable affine functions describing the lth equality and rth inequality constraints, respectively.

To solve (2.20), novel AA-based mathematical operators are defined here aimed at extending to affine functions and affine forms the minimization operator and the main comparison operators $<$, $>$, $\leq$, $\geq$, and $==$, respectively. To accomplish this, starting from the definition of these novel operators and according to Theorem 2.1, it will be shown that (2.20) can be recasted as a dual deterministic problem, which can be solved employing a traditional numerical programming technique. In particular, the mathematical definitions introduced previously allow stating the following propriety of affine forms, which directly results from the definition of the difference operator:

Definition 2.1. (Equality operator for affine forms $\stackrel{A}{=}$) *Two affine forms* $\hat{x} = x_0 + \sum_{k=1}^{p_x} x_k \varepsilon_k^x$ *and* $\hat{y} = y_0 + \sum_{k=1}^{p_y} y_k \varepsilon_k^y$ *are equal, i.e.,* $\hat{x} \stackrel{A}{=} \hat{y}$, *if and only if:*

$$\hat{x} - \hat{y} = x_0 - y_0 + \sum_{k=1}^{p_x} x_k \varepsilon_k^x - \sum_{k=1}^{p_y} y_k \varepsilon_k^y = 0 \tag{2.23}$$

That is, two affine forms are equal if they have the same central value and share the same noise symbols with the same partial deviations, namely:

$$\hat{x} \stackrel{A}{=} \hat{y} \Leftrightarrow \begin{cases} x_0 = y_0 \\ \varepsilon_k^x = \varepsilon_k^y \;\; \forall k \in (1, \ldots, p) \\ x_k = y_k \;\; \forall k \in (1, \ldots, p) \\ p = P_x = p_y \end{cases} \tag{2.24}$$

These rigorous equality conditions cannot be satisfied when solving (2.20), by using the approach in (2.55) due to the presence of non-affine operations, as explained in Section 2.3.3.2 which introduce approximation and computational errors.

To address the aforementioned issue, a similarity criteria, which is based on the equality of the partial deviations of the "primitive" noise symbols, denoted here as ε_k $\forall k \in (1, \ldots, p)$, and on the definition of an approximation degree based on the radius of the uncertainties generated by the approximation of the non-affine operations, denoted here as ε_k $\forall k \in (p+1, \ldots, p+p_{na})$, is defined.

Definition 2.2. (Similarity operator for affine forms $\stackrel{A}{\approx}$) *Two affine forms* $\hat{x} = x_0 + \sum_{k=1}^{p+p_{na}} x_k \varepsilon_k$ *and* $\hat{y} = y_0 + \sum_{k=1}^{p+p_{na}} y_k \varepsilon_k$ *are similar with an approximation degree* $Ł_{x,y}$, *i.e.* $\hat{x} \stackrel{A}{\approx} \hat{y}$, *if and only if:*

$$\left(x_k = y_k \; \forall k \in (0, \ldots, p)\right) \wedge \left(Ł_{x,y} = \sum_{k=p+1}^{p+p_{na}} (|x_k| + |y_k|)\right) \tag{2.25}$$

By following the same approach, it is possible to define an inequality operator for affine forms as follows.

Definition 2.3. (Inequality operator for affine forms $\stackrel{A}{<}$) *Given two affine forms* $\hat{x} = x_0 + \sum_{k=1}^{p_x} x_k \varepsilon_k^x$ *and* $\hat{y} = y_0 + \sum_{k=1}^{p_y} y_k \varepsilon_k^y$, *then* $\hat{x} \stackrel{A}{<} \hat{y}$ *if and only if:*

$$x_0 + \sum_{k=1}^{p_x} |x_k| < y_0 - \sum_{k=1}^{p_y} |y_k| \tag{2.26}$$

This definition directly follows from the basic theory of interval analysis, since this states that the upper bound of $\hat{x}$ is less than the lower bound of $\hat{y}$.

Once the aforementioned relational operators are introduced, the problem of the minimization of a scalar and non-linear affine function could be effectively addressed by defining the following operator.

Definition 2.4. (Minimization operator for functions of affine forms) *Given a non-linear function* $f: \Re \rightarrow \Re$, *and the affine form* $\hat{x} = x_0 + \sum_{k=1}^{P} x_k \varepsilon_k$, *then the following AA-based minimization problem:*

$$\min_{\hat{x}}^{A} \hat{f}(\hat{x}) = f_0(\hat{x}) + \sum_{k=1}^{P} f_k(\hat{x})\varepsilon_k + \sum_{k=p+1}^{p+p_{na}} f_k(\hat{x})\varepsilon_k \tag{2.27}$$

is equivalent to the following deterministic multi-objective programming problem:

$$\min_{(x_0,x_1,\ldots,x_p)} \left\{ f_0(x_0, x_1, \ldots, x_p), \sum_{k=1}^{p+p_{na}} |f_k(x_0, x_1, \ldots, x_p)| \right\} \tag{2.28}$$

This definition follows from the AA-based robust circuit design approach proposed in [47], and the principles of risk-based programming theory, since the minimization of the affine central value aims at identifying the most effective solutions, without considering the uncertainty represented by the noise symbols, while the minimization of the affine radius aims at identifying the most reliable solutions that exhibit the lowest tolerance to data uncertainty. The tradeoff between these two conflicting objectives basically represents the decision maker's risk. Based on this, the minimization of an affine function should be equivalent to finding an affine form which minimizes both its central value and its radius.

From (2.28), problem (2.20) can be solved by means of the following deterministic multi-objective constrained optimization problem:

$$\min_{\hat{\mathbf{z}}} \left\{ f_0(\hat{\mathbf{z}}), \sum_{k=1}^{p+p_{na}} |f_k(\hat{\mathbf{z}})| \right\} \tag{2.29}$$

$$\text{s.t.} \quad \hat{g}_l(\hat{\mathbf{z}}) \overset{A}{\approx} 0 \qquad \forall l \in (1, \ldots, n) \tag{2.30}$$

$$\hat{h}_r(\hat{\mathbf{z}}) \overset{A}{<} 0 \qquad \forall r \in (1, \ldots, m) \tag{2.31}$$

To solve this problem, a two-stage solution algorithm is used here. In the first stage, the main idea is to identify the central values of the unknown state vector by first considering the system operating at its nominal condition, which defines these central values. In this case, which is referred here as the "nominal state," uncertainties are not considered and thus the corresponding solution can be computed by solving the following deterministic optimization problem:

$$\min_{(z_{1_0},\ldots,z_{N_x+N_{u0}})} f_0(z_{1_0}, \ldots, z_{N_x+N_{u0}}) \tag{2.32}$$

$$\text{s.t.} \quad g_l(z_{1_0}, \ldots, z_{N_x+N_{u0}}) = 0 \qquad \forall l \in (1, \ldots, n) \tag{2.33}$$

$$h_r(z_{1_0}, \ldots, z_{N_x+N_{u0}}) < 0 \qquad \forall r \in (1, \ldots, m) \tag{2.34}$$

In the second stage, referred here as the "perturbed state," the effect of data uncertainty is considered, computing the partial deviations of the unknown state vector by solving the following deterministic optimization problem:

$$\min_{(z_{1_1},\ldots,z_{N_x+N_{u1}},\ldots,z_{1_p},\ldots,z_{N_x+N_{up}})} \sum_{k=1}^{p+p_{na}} |f_k(z_{1_1},\ldots,z_{N_x+N_{u1}},\ldots,z_{1_p},\ldots,z_{N_x+N_{up}})| \tag{2.35}$$

$$\text{s.t.} \quad \hat{g}_l(z_{1_1},\ldots,z_{N_x+N_{u1}},\ldots,z_{1_p},\ldots,z_{N_x+N_{up}}) \overset{A}{\approx} 0 \qquad \forall l \in (1,\ldots,n) \tag{2.36}$$

$$\hat{h}_r(z_{1_1},\ldots,z_{N_x+N_{u1}},\ldots,z_{1_p},\ldots,z_{N_x+N_{up}}) \overset{A}{<} 0 \qquad \forall r \in (1,\ldots,m) \tag{2.37}$$

The following example should help understand the proposed method:

$$\min_{(\hat{x},\hat{y})} \hat{f}(\hat{x},\hat{y}) = \hat{x}^2 + 4\hat{y}^2 - (3 + 0.1\varepsilon_1 + 0.1\varepsilon_2) \tag{2.38}$$

$$\text{s.t.} \quad \hat{g}(\hat{x},\hat{y}) = 4\hat{x}^2 - 16\hat{x} + \hat{y}^2 \overset{A}{\approx} -12 + 0.2\varepsilon_1 \tag{2.39}$$

where the central values and the partial deviations of the unknown affine forms $\hat{x} = x_0 + x_1\varepsilon_1 + x_2\varepsilon_2$ and $\hat{y} = y_0 + y_1\varepsilon_1 + y_2\varepsilon_2$ are to be identified by solving the optimization problem in the "nominal" and "perturbed" state, namely:

$$\min_{(x_0,y_0)} x_0^2 + 4y_0^2 - 3 \tag{2.40}$$

$$\text{s.t.} \quad 4x_0^2 - 16x_0 + y_0^2 + 12 = 0 \tag{2.41}$$

$$\begin{aligned} \min_{(x_1,x_2,y_1,y_2)} \quad & |(2x_0x_1 + 8y_0y_1 - 0.1)| + |2x_0x_2 + 8y_0y_2 - 0.1| \\ & +(|x_1| + |x_2|)^2 + 4(|y_1| + |y_2|)^2 \\ \text{s.t.} \quad & 8x_0x_1 - 16x_1 + 2y_0y_1 = 0.2 \\ & 8x_0x_2 - 16x_2 + 2y_0y_2 = 0 \end{aligned} \tag{2.42}$$

The solution of these problems yields the following results:

$$\hat{x}_s = 1 - 0.026\varepsilon_1 \Rightarrow \Theta_{\hat{x}_s} = [0.9750, 1.0259] \tag{2.43}$$

$$\hat{y}_s = 0 \tag{2.44}$$

$$\hat{f}(\hat{x}_s,\hat{y}_s) = -2 - 0.15\varepsilon_1 - 0.1\varepsilon_2 + 0.00062\varepsilon_3 \Rightarrow \Theta_{\hat{f}} = [-2.25, -1.75] \tag{2.45}$$

$$\hat{g}(\hat{x}_s,\hat{y}_s) = -12 + 0.2\varepsilon_1 + 0.0025\varepsilon_3 \Rightarrow \Theta_{\hat{g}} = [-12.2025, -11.7975] \tag{2.46}$$

To check the consistency of results (2.43)–(2.46), the same problem has been solved by a Monte Carlo-based simulation, obtaining the following intervals:

$$\Theta_{\hat{x}_s} = [0.9753, 1.0253] \tag{2.47}$$

$$\Theta_{\hat{y}_s} = 0 \tag{2.48}$$

$$\Theta_{\hat{f}} = [-2.2443, -1.7535] \tag{2.49}$$

$$\Theta_{\hat{g}} = [-12.2124, -11.8035] \tag{2.50}$$

Observe that the AA-based computing paradigm allows obtaining accurate intervals. Hence, thanks to the definition of rigorous relational and minimization operators, it is possible to obtain more precise solution bounds compared to those obtained by applying others AA-based solution paradigms, as those proposed in [23, 24], which typically employ approximated minimization operators (i.e., domain contraction).

It should be noted that the solution of the AA-based optimization problem for the "perturbed state" requires the identification of a larger number of state variables, i.e., $p \times (N_x + N_u)$. Nevertheless, the number of the noise symbols describing the affine forms of the state vector, which sensible influences the problem cardinality, can be significantly reduced by exploring the statistical correlation between the uncertainty sources, as explained in [38].

2.4 AA-based linear methods for uncertain power system analysis

2.4.1 Uncertain PF problems

The uncertain PF problem can be effectively solved by applying the described AA-based linear framework, since, as proposed in [39], it can be stated as a particular instance of the OPF problem (2.20), as follows:

$$\begin{aligned}
&\min_{(\hat{V}_j,\hat{\delta}_b,\hat{V}_{Ag},\hat{V}_{Bg})} && \sum_{i\in\mathcal{N}_P} (\hat{P}_i^F - \hat{P}_i)^2 + \sum_{j\in\mathcal{N}_Q} (\hat{Q}_j^F - \hat{Q}_j)^2 \\
&\text{s.t.} && \hat{P}_i \overset{A}{\approx} \hat{V}_i \sum_{b=1}^{N} \hat{V}_b Y_{ib} \cos\left(\hat{\delta}_i - \hat{\delta}_b - \theta_{ib}\right) \quad \forall i \in \mathcal{N}_P \\
& && \hat{Q}_j \overset{A}{\approx} \hat{V}_j \sum_{b=1}^{N} \hat{V}_b Y_{jb} \sin\left(\hat{\delta}_j - \hat{\delta}_b - \theta_{jb}\right) \quad \forall j \in \mathcal{N}_Q, \\
& && \hat{V}_g \overset{A}{\approx} V_g + \hat{V}_{Ag} - \hat{V}_{Bg} \quad \forall g \in \mathcal{N}_G \\
& && 0 \overset{A}{\leq} (\hat{Q}_g - Q_{j,\min}) \perp \hat{V}_{Ag} \overset{A}{\geq} 0 \quad \forall g \in \mathcal{N}_G \\
& && 0 \overset{A}{\leq} (Q_{j,\max} - \hat{Q}_g) \perp \hat{V}_{Bg} \overset{A}{\geq} 0 \quad \forall g \in \mathcal{N}_G \\
& && \hat{V}_g, \hat{V}_{Ag}, \hat{V}_{Bg} \overset{A}{\geq} 0 \quad \forall g \in \mathcal{N}_G
\end{aligned} \tag{2.51}$$

where $\hat{V}_b = V_{b_0} + \sum_{k=1}^{p} V_{b_k}\varepsilon_k$ is the affine form describing the voltage magnitude of the bth bus to represent the source of uncertainties affecting the power system operation; $\hat{V}_{Ag}$, $\hat{V}_{Bg}$ are auxiliary variables for modeling the voltage regulation of generators; and $\hat{\delta}_b = \delta_{b_0} + \sum_{k=1}^{p} \delta_{b_k}\varepsilon_k$ is the affine form describing the voltage angle for the bth bus, except for the slack bus. Consequently, the dependent variables of the AA-based PF problem are:

- $(V_{j_0}, V_{j_1}, \ldots, V_{j_p})$ for each load bus, $\forall j \in \mathcal{N}_Q$;
- $(\delta_{i_0}, \delta_{i_1}, \ldots, \delta_{i_p})$ for each bus except the slack bus, $\forall i \in \mathcal{N}_P$.

2.4.2 Uncertain OPF problems

The application of the described AA-based framework for OPF analysis is straightforward, since the dependent variables are the affine forms of the voltage magnitudes at the load buses, and the voltage angles at all buses except the slack bus, as in the case of the PF problem. On the other hand, the control variables depend on the particular application domain; thus, for example, in optimal economic dispatch analysis, these variables include active power generated by the dispatchable generators, namely $\hat{P}_g \; \forall g \in \mathcal{N}_G$.

The equality constraints in the OPF problem are described by the PF equations, while the inequality constraints typically include in practice the maximum allowable apparent PF P_l on each line. Consequently, the AA-based formulation of the OPF problem is as follows:

$$
\begin{aligned}
\min_{(\hat{V}_j, \hat{\delta}_b, \hat{P}_j)} \quad & \hat{f}(\hat{V}_i, \hat{\delta}_b, \hat{P}_j) \\
\text{s.t.} \quad & \hat{P}_i^F - \hat{V}_i \sum_{b=1}^{N} \hat{V}_b Y_{ib} \cos(\hat{\delta}_i - \hat{\delta}_b - \theta_{ib}) \overset{A}{\approx} 0 && \forall i \in \mathcal{N}_P \\
& \hat{Q}_j^F - \hat{V}_j \sum_{b=1}^{N} \hat{V}_b Y_{jb} \sin(\hat{\delta}_j - \hat{\delta}_b - \theta_{jb}) \overset{A}{\approx} 0 && \forall j \in \mathcal{N}_Q \\
& \hat{Q}_g - \hat{V}_g \sum_{b=1}^{N} \hat{V}_b Y_{gb} \sin(\hat{\delta}_g - \hat{\delta}_b - \theta_{gb}) \overset{A}{\approx} 0 && \forall g \in \mathcal{N}_G \\
& V_j^{\min} \overset{A}{\leq} \hat{V}_j \overset{A}{\leq} V_j^{\max} && \forall j \in \mathcal{N}_Q \\
& P_g^{\min} \overset{A}{\leq} \hat{P}_g \overset{A}{\leq} P_g^{\max} && \forall g \in \mathcal{N}_G \\
& Q_g^{\min} \overset{A}{\leq} \hat{Q}_g \overset{A}{\leq} Q_g^{\max} && \forall g \in \mathcal{N}_G \\
& P_l^{\min} \overset{A}{\leq} \hat{P}_l(\hat{V}_1, \ldots, \hat{V}_N, \hat{\delta}_1, \ldots, \hat{\delta}_N) \overset{A}{\leq} P_l^{\max} && \forall l \in \mathcal{N}_L
\end{aligned}
\tag{2.52}
$$

where $\mathcal{N}_L$ is the set of constrained lines.

2.4.3 Numerical results

To assess the benefits of the AA-based framework, referred here as a unified AA method, for solving both uncertain PF and OPF problems, various case studies are analyzed in this section, comparing PF and OPF solution bounds to those calculated using the AA-based PF method proposed in [22], and the range arithmetic-based OPF proposed in [23]. In all the analyzed case studies, a Monte Carlo simulation with a uniform distribution is assumed to yield the "correct" solution intervals. For the latter, 5,000 different values of the input variables within the assumed input bounds were randomly selected, and a conventional solution was obtained for each one; this procedure yielded the desired interval solutions defined by the largest and the smallest values of the bus voltage magnitudes and angles as well as line flows. It should be noted that increasing the number of Monte Carlo simulations beyond 5,000 did not yield any significant changes to the solution intervals.

Without loss of generality, a ±20% (40%) tolerance on load and generator powers was assumed, since this would define an interval wide enough to represent, for example, uncertain wind and solar generation. Based on the assumed load and generator power bounds to represent input data uncertainty, the unified AA-based method was applied to estimate the bounds of the PF and OPF solution by assuming the independence of all the input uncertainties, i.e., by considering a different noise symbol for each injected active and reactive power. All the simulation studies were carried out on a PC workstation equipped with a 3 GHz Intel Core Duo CPU with 3 GB RAM. The obtained results are presented and discussed next.

2.4.3.1 PF Analysis

The uncertain PF problem for 3 IEEE test systems is solved here using the unified AA formulation. The obtained results are summarized in Figures 2.2 and 2.3 for the IEEE 30-bus test system; Figures 2.4 and 2.5 for the IEEE 57-bus test system; and Figure 2.6 for the IEEE 118-bus test system. In all these figures, observe that the unified AA method is characterized by an improved accuracy compared to the previous methods.

The benefits deriving by the application of the unified AA method are also confirmed by Tables 2.1 and 2.2, which report, for each case study, the average errors in the upper and lower bounds for the bus voltage magnitude and angle errors. computed as follows:

$$
\begin{aligned}
eV_b^{UB} &= \left(V_{b0} + \sum_{k=1}^{p} |V_{b_k}|\right) - \max\ V_b^{MC} \\
eV_b^{LB} &= \min\ V_b^{MC} - \left(V_{b0} - \sum_{k=1}^{p} |V_{b_k}|\right) \\
e\delta_b^{UB} &= \left(\delta_{b0} + \sum_{k=1}^{p} |\delta_{b_k}|\right) - \max\ \delta_b^{MC} \\
e\delta_b^{LB} &= \min\ \delta_k^{MC} - \left(\delta_{b0} - \sum_{k=1}^{p} |\delta_{b_k}|\right)
\end{aligned}
\tag{2.53}
$$

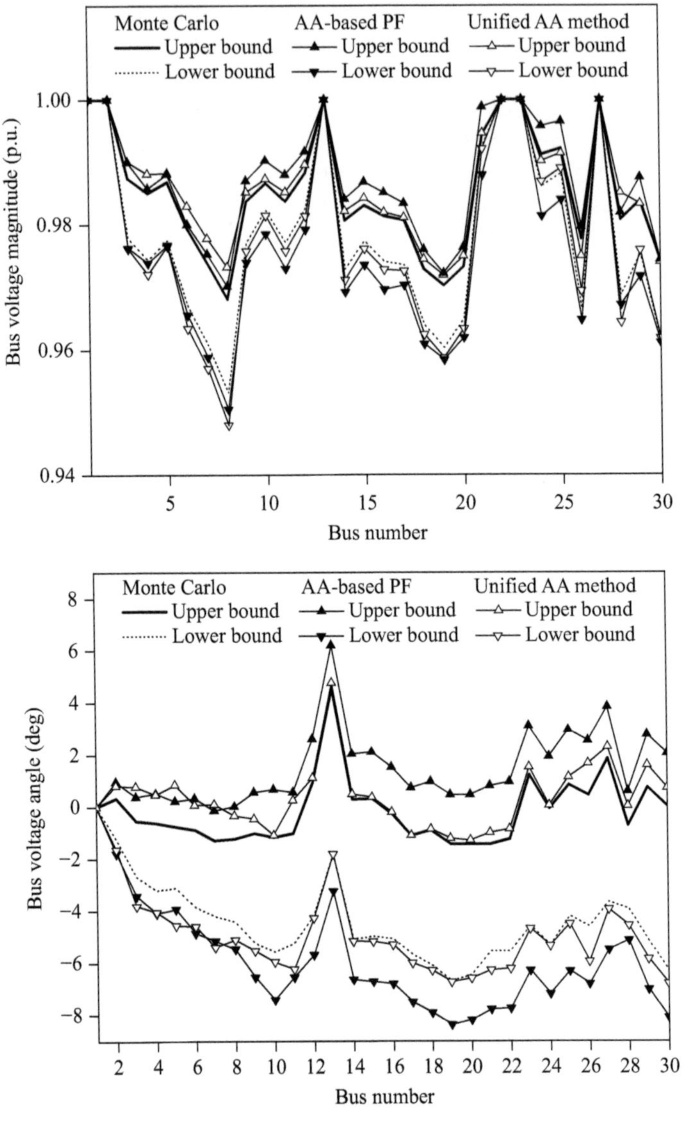

Figure 2.2 Bus voltage magnitude and angle bounds obtained for the IEEE 30-bus test system for the existing and presented linear AA-PF methods

where max and min of V_b^{MC} and δ_b^{MC} are the maximum and minimum voltage magnitudes and angles obtained from the Monte Carlo simulations, respectively. As expected, this accuracy improvement is obtained at the cost of increased computational burden, as confirmed in Table 2.3, which depicts the execution times registered for the simulations.

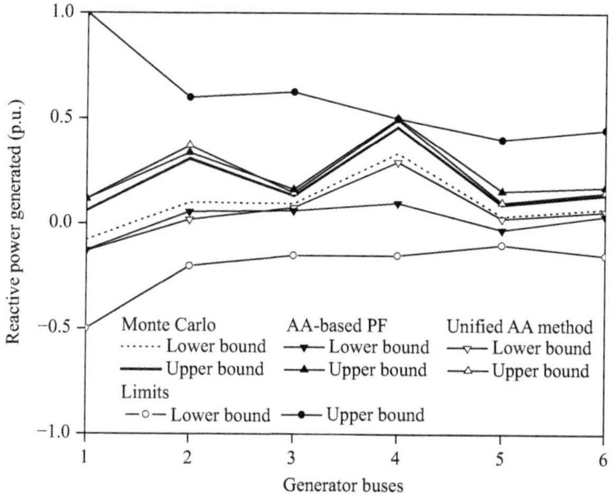

Figure 2.3 Reactive power bounds at the generation buses for the IEEE 30-bus test system for the existing and presented linear AA-PF methods

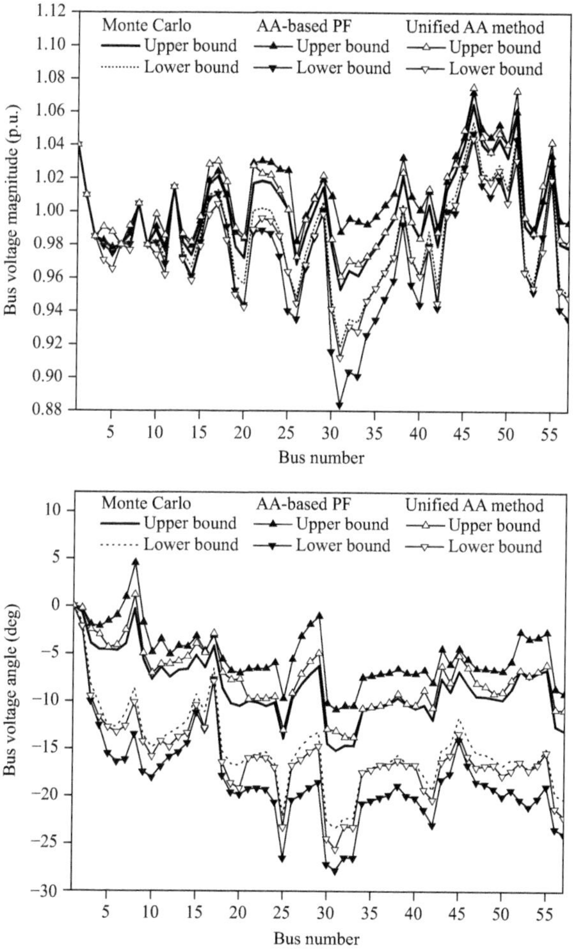

Figure 2.4 Bus voltage magnitude and angle bounds obtained for the IEEE 57-bus test system for the existing and presented linear AA-PF methods

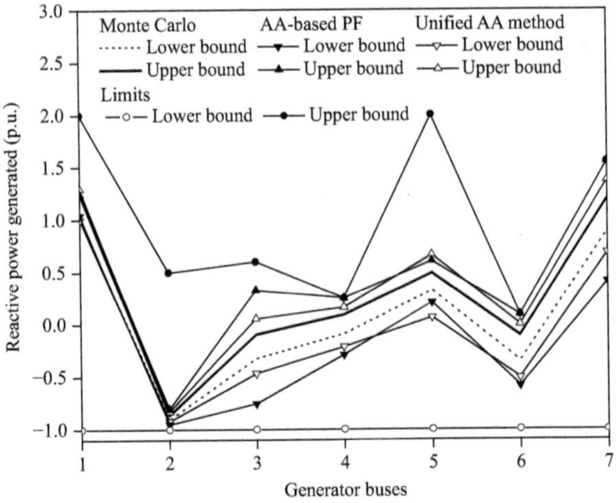

Figure 2.5 Reactive power bounds at the generation buses for the IEEE 57-bus test system for the existing and presented linear AA-PF methods

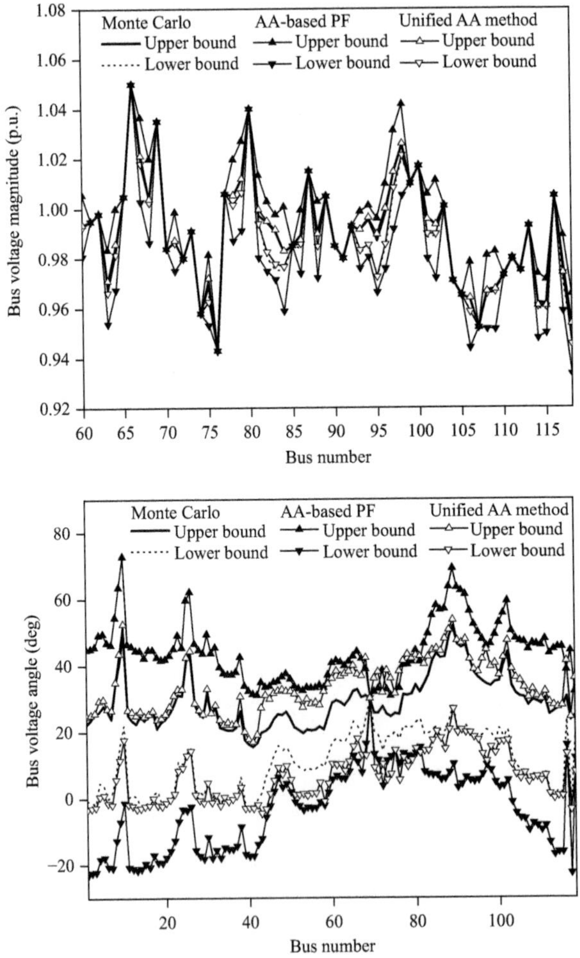

Figure 2.6 Bus voltage magnitude and angle bounds obtained for the IEEE 118-bus test system for the existing and presented linear AA-PF methods

Table 2.1 AA-PF average errors in bus voltage magnitude bounds

	30 bus		**57 bus**		**118 bus**	
	Up	**Low**	**Up**	**Low**	**Up**	**Low**
AA-based PF [10^{-4} p.u.]	55	46	90	88	102	101
Unified AA method [10^{-4} p.u.]	20	30	47	71	62	65

Table 2.2 AA-PF average errors in bus voltage angle bounds

	30 bus		**57 bus**		**118 bus**	
	Up	**Low**	**Up**	**Low**	**Up**	**Low**
AA-based PF (deg)	0.65	0.97	3.32	3.33	3.18	3.16
Unified AA method (deg)	0.26	0.10	0.96	0.98	0.99	0.01

Table 2.3 CPU times

	30 bus	**57 bus**	**118 bus**
Monte Carlo (5000 trials) (s)	149.9	211.8	603.1
AA-based PF (s)	1.7	2.5	5.7
Unified AA method (s)	110.72	167.8	406.5

2.4.3.2 Reactive power dispatch

To confirm the benefits deriving from the application of the unified AA method to solve a complex OPF problem, the uncertain optimal reactive power dispatch for the IEEE 118-bus test system is analyzed here. To solve this problem, the solution paradigm formalized in (2.52) is applied to solve the problem with the following objective function:

$$\hat{f}(\hat{V}_b) = \sum_{b=1}^{N} \left(\hat{V}_b^{\,2} - V_n^2 \right) \tag{2.54}$$

The solution of this problem, which required about 398 s, yields the results summarized in Figure 2.7, which depicts the bounds of the bus voltage magnitudes and

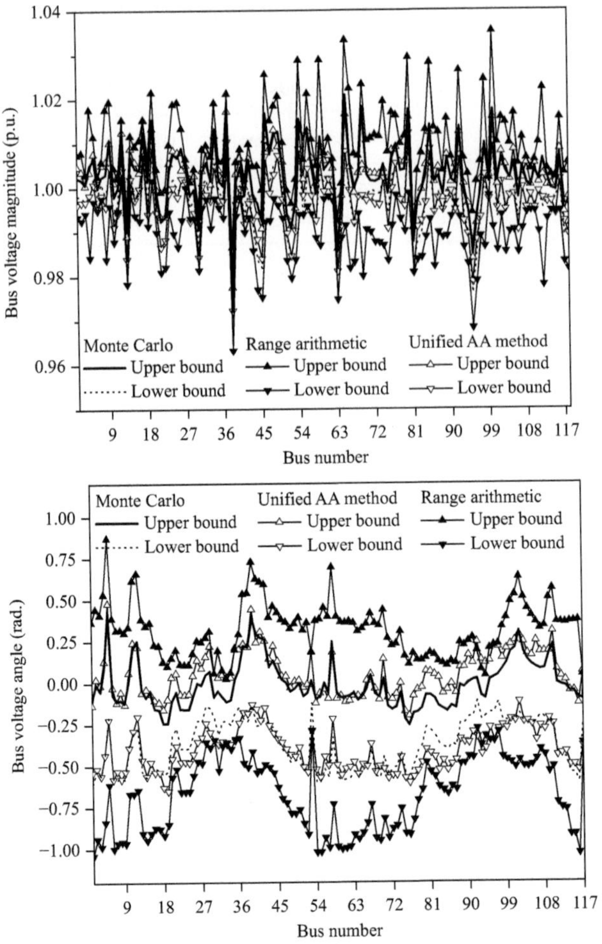

Figure 2.7 Voltage magnitude and angle bounds for all buses of the computed reactive power dispatch AA-OPF solutions for the 118-bus test system

angles, respectively. In this figure, the OPF solution bounds computed by a Monte Carlo model with 5,000 simulations, which required about 2,650 s, are also shown.

2.4.3.3 Computational requirements

In terms of computational requirements, the following observations can be made:

1. The AA-based PF and the range-arithmetic-based methods are the fastest techniques for uncertain PF and OPF analysis, respectively.
2. The computational cost of Monte Carlo and its accuracy is related to the number of required simulations.

3. Although the unified AA framework gives more accurate enclosures for the solution bounds, it is the costliest approach in terms of computational times, since this is mainly influenced by the large number of control variables, which depend on the number of noise symbols characterizing the parameter uncertainties. This poses computational difficulties for addressing uncertain PF and OPF analysis in large-scale power systems.

To overcome the latter limitation, it is necessary to design techniques aimed at identifying the optimal number of independent uncertainties, i.e., the optimal number of noise symbols, affecting the system variables. To address this issue, knowledge discovery paradigms from historical operation data can be used as explained in [38].

2.5 Elements of second-order AA

Ongoing research activities on uncertain power system analysis by AA-based methods are mainly oriented at developing new techniques for solving large-scale problems, and reducing the over-conservatism of the solutions. In this context, a promising research direction is based on the definition of new mathematical operators for the multiplication of affine forms, which aim at reducing the corresponding approximation errors [48, 49]. In particular, rather than representing the approximation errors by linearly introducing new noise symbols, these new operators properly distribute the corresponding uncertainties on the partial deviation of the primitive noise symbols, which are no longer numbers but affine forms. These new variables, which are referred to as second-order affine forms, can be managed by proper mathematical operators in order to reliably solve uncertain programming problems, considering all the possible combination of the exogenous uncertainty, and the worst case instance of the approximation errors. To clarify this concept, consider the following simple uncertain problem: Assume that $\hat{x} = x_0 + x_1\varepsilon_1 + x_2\varepsilon_2$ is an unknown affine form, thus to compute its parameters x_0, x_1 and x_2 in order to match the following nonlinear equation:

$$\underbrace{(x_0 + x_1\varepsilon_1 + x_2\varepsilon_2)}_{\hat{x}}{}^2 = \underbrace{4 + 0.8\varepsilon_1 + 0.4\varepsilon_2}_{\hat{z}^F} \tag{2.55}$$

one can solve this equation by computing the central value x_0 and the partial deviations x_1 and x_2, which share the same noise symbols of the known term, using (2.19), as follows:

$$x_0^2 + 2x_0x_1\varepsilon_1 + 2x_0x_2\varepsilon_2 + R(\hat{x})^2\varepsilon_3 = 4 + 0.8\varepsilon_1 + 0.4\varepsilon_2 \tag{2.56}$$

where $R(\hat{x})^2\varepsilon_3 = x_1^2\varepsilon_1^2 + x_2^2\varepsilon_2^2 + 2x_1x_2\varepsilon_1\varepsilon_2$. This problem can be approximately solved by using the similarity operator between affine forms defined in Section 2.3.4, as follows:

$$\begin{cases} x_0^2 = 4 \\ 2x_0x_1 = 0.8 \\ 2x_0x_2 = 0.4 \end{cases} \tag{2.57}$$

This allows to approximately computing the parameters of the unknown affine form, namely:

$$\hat{x} \approx \hat{x}_a = 2 + 0.2\varepsilon_1 + 0.1\varepsilon_2 \tag{2.58}$$

This affine form represents the solution to the uncertain problem (2.55); thus, by applying the multiplication operator between affine forms and considering that $\varepsilon_1 = [-1, 1]$ and $\varepsilon_2 = [-1, 1]$, it follows that:

$$E = \hat{x}_a^2 - \hat{z}^F = R(\hat{x})^2\varepsilon_3 = [-0.09, 0.09] \tag{2.59}$$

where the term E is the affine approximation error, which can be considered as the equivalent of the rounding error deriving from using floating-point numbers to approximate real numbers.

The aforementioned mathematical operators do not represent accurately the uncertain propagation process. In this case, a new method for approximating the multiplication between two affine forms is applied here. Thus, the main idea is to exploit the original formalization (2.18), rewritten as follows:

$$\hat{x} * \hat{y} = x_0 y_0 + \sum_{k=1}^{P}\left(x_0 y_k + y_0 x_k + \sum_{h=1}^{P} x_k y_h \varepsilon_h\right)\varepsilon_k \tag{2.60}$$

Instead of introducing a new noise symbol, this alternative formulation distributes the endogenous uncertainty on the partial deviations of the primitive noise symbols, where the resulting affine forms are called second-order affine forms. This new formulation allows reliably solving non-linear uncertain equations, containing unknown variables products. Thus, consider the following equation:

$$\hat{x}\hat{y} = z_0^F + \sum_{k=1}^{p} z_k^F \varepsilon_k \tag{2.61}$$

where $\hat{z}^F$ is a fixed or given affine form, while $\hat{x}$ and $\hat{y}$ are the unknown terms. To solve this problem, the original uncertain equation is recast by explicitly considering the effect of the "second-order" noise symbols introduced by the product operator as follows:

$$\hat{x} * \hat{y} = x_0 y_0 + \sum_{k=1}^{P}\left(x_0 y_k + y_0 x_k + \sum_{h=1}^{p} x_k y_h \varepsilon_h\right)\varepsilon_k = \hat{z}^F \tag{2.62}$$

Hence, by matching the central values and the partial deviations of the two affine forms, the following system of equations can be obtained:

$$\begin{cases} x_0 y_0 = z_0^F \\ x_0 y_k + y_0 x_k + \sum_{h=1}^{P} x_k y_h \varepsilon_h = z_k^F \;\; \forall k \in (1, p) \end{cases} \tag{2.63}$$

where the term $\sum_{h=1}^{p} x_k y_h \varepsilon_h$ allows to explicitly quantify the effect of the approximation error on each partial deviation.

The fundamental result in (2.63) allows quantifying the effect of approximation error on solution sets a priori, demonstrating that for some problems, such as the ones involving multiplications, the partial deviations are not deterministic numbers but uncertain variables that share the same primitive noise symbol as the initial affine form. Thus, through this approach, it is possible to solve the uncertain equations described in (2.63) by considering the worst-case bounds of the approximation error, which can be computed as follows:

$$E_k = \pm \sum_{h=1}^{P} |x_k y_h| \quad \forall k \in (1,p) \tag{2.64}$$

Robust enclosures of the real solution set can then be obtained by considering the worst value of the uncertainty in (2.63), which is described by the following set of deterministic equations:

$$\begin{cases} x_0 y_0 = z_0^F \\ x_0 y_k + y_0 x_k - \sum_{h=1}^{P} |x_k y_h| = z_k^F \ \forall k \in (1,p) \end{cases} \tag{2.65}$$

To assess the benefits deriving from the application of (2.65), consider again the example formalized in (2.55), which can be recast as follows:

$$\begin{cases} x_0^2 = 4 \\ 2x_0 x_1 + x_1^2 \varepsilon_1 + x_1 x_2 \varepsilon_2 = 0.8 \\ 2x_0 x_2 + x_1 x_2 \varepsilon_1 + x_2^2 \varepsilon_2 = 0.4 \end{cases} \tag{2.66}$$

According to (2.65), a robust solution to this uncertain problem can be obtained by solving the following set of deterministic equations:

$$\begin{cases} x_0^2 = 4 \\ 2x_0 x_1 - x_1^2 - |x_1 x_2| = 0.8 \\ 2x_0 x_2 - |x_1 x_2| - x_2^2 = 0.4 \end{cases} \tag{2.67}$$

which results in the following affine form:

$$\hat{x} = 2 + 0.2178\varepsilon_1 + 0.1089\varepsilon_2 \tag{2.68}$$

In order to assess the robustness of this solution, one can compute the corresponding bounds using (2.19) as follows:

$$\begin{aligned} \hat{x}^2 &= 4 + (0.8712 + 0.0474\varepsilon_1 + 0.0237\varepsilon_2)\varepsilon_1 \\ &\quad + (0.4356 + 0.0237\varepsilon_1 + 0.0474\varepsilon_2)\varepsilon_2 \end{aligned} \tag{2.69}$$

Hence, by considering the worst case, the solution bounds in this case are [2.8, 5.2], which, as expected, match the bounds of the fixed affine form (2.55). On the contrary, if one computes the approximate bounds of the approximate solution using (2.58), the following is obtained:

$$\hat{x}^2 \approx 4 + 0.8\varepsilon_1 + 0.4\varepsilon_2 + 0.09\varepsilon_3 \tag{2.70}$$

with the corresponding worst-case bounds of [2.89, 5.11], which are an underestimation of the real solution bounds.

The main advantages associated with this formulation are the following:

- The computed solutions are robust, as they guarantee the rigorous inclusion of the real solution set.
- The resulting uncertain problems can be formalized through equivalent deterministic ones, which can be solved by using state-of-the-art solvers.
- The uncertain equations and the corresponding Jacobian matrix can be explicitly expressed, allowing the employment of solution algorithms like Newton-based methods.

2.5.1 Uncertain power system analysis by second-order AA

Second-order AA can be adopted to solve the uncertain PF equations expressed in rectangular coordinates. To this aim, define the following "fixed" affine forms:

$$\hat{V}_g^F = V_{g0}^F \quad \forall g \in \mathcal{N}_G \tag{2.71}$$

$$\hat{P}_i^F = P_{i0}^F + \sum_{k=1}^{P} P_{i_k}^F \varepsilon_k \quad \forall i \in \mathcal{N}_P \tag{2.72}$$

$$\hat{Q}_j^F = Q_{j0}^F + \sum_{k=1}^{P} Q_{j_k}^F \varepsilon_k \quad \forall j \in \mathcal{N}_Q \tag{2.73}$$

These affine forms represent the voltage magnitude at the $g \in \mathcal{N}_G$ generator buses, the active power injected at the $i \in \mathcal{N}_P$ buses, and the reactive power injected at the $j \in \mathcal{N}_Q$ buses, respectively. Observe that the voltage magnitude at the generator buses is fixed and thus independent of the p noise symbols.

The voltage at the load buses can then be defined in a rectangular form as follows: $\forall\, b \in \mathcal{N}$:

$$\hat{V}_{Rj} = V_{Rj0} + \sum_{k=1}^{P} V_{Rj_k} \varepsilon_k \tag{2.74}$$

$$\hat{V}_{Ij} = V_{Ij0} + \sum_{k=1}^{P} V_{Ij_k} \varepsilon_k \quad \forall\, j \in \mathcal{N}_Q \tag{2.75}$$

The central value and the partial deviations of these affine forms can be computed by solving the PF equations (2.2) in the AA domain, which are described by the following set of nonlinear affine equations $\forall\, g \in \mathcal{N}_G,\ i \in \mathcal{N}_P,\ j \in \mathcal{N}_Q$:

$$(V_{g0}^F)^2 = V_{Rg0}^2 + V_{Ig0}^2 \tag{2.76}$$

$$\hat{P}_i^F = \sum_{b=1}^{N} G_{ib}(\hat{V}_{Ri}\hat{V}_{Rb} + \hat{V}_{Ii}\hat{V}_{Ib}) + B_{ib}(\hat{V}_{Ii}\hat{V}_{Rb} - \hat{V}_{Ri}\hat{V}_{Ib}) \tag{2.77}$$

$$\hat{Q}_j^F = \sum_{b=1}^{N} G_{jb}(\hat{V}_{Ij}\hat{V}_{Rb} - \hat{V}_{Rj}\hat{V}_{Ib}) - B_{jb}(\hat{V}_{Ij}\hat{V}_{Ib} + \hat{V}_{Rj}\hat{V}_{Rb}) \tag{2.78}$$

According to the second-order AA-based formulation, this problem can be recast as the following set of non-linear deterministic equations:

$$(V_{g0}^F)^2 = V_{Rg0}^2 + V_{Ig0}^2 \quad \forall\, g \in \mathcal{N}_G \tag{2.79}$$

$$\hat{P}_{i0}^F = \sum_{b=1}^{N} G_{ib}(V_{Ri0}V_{Rb0} + V_{Ii0}V_{Ib0}) + B_{ib}(V_{Ii0}V_{Rb0} - V_{Ri0}V_{Ib0}) \quad \forall\, i \in \mathcal{N}_P \tag{2.80}$$

$$\begin{aligned}\hat{P}_{ik}^F =& \sum_{b=1}^{N} G_{ib}(V_{Ri0}V_{Rbk} + V_{Rik}V_{Rb0} + V_{Ii0}V_{Ibk} + V_{Iik}V_{Ib0}) + \\ &+ B_{ib}(V_{Ii0}V_{Rbk} - V_{Iik}V_{Rb0} - V_{Ri0}V_{Ibk} - V_{Rik}V_{Ib0}) - EP_{ik} \\ &\forall\, i \in \mathcal{N}_P,\; k = [1,p]\end{aligned} \tag{2.81}$$

$$Q_{j0}^F = \sum_{b=1}^{N} G_{jb}(V_{Ij0}V_{Rb0} - V_{Rj0}V_{Ib0}) - B_{jb}(V_{Ij0}V_{Ib0} + V_{Rj0}V_{Rb0}) \quad \forall\, j \in \mathcal{N}_Q \tag{2.82}$$

$$\begin{aligned}\hat{Q}_{jk}^F =& \sum_{b=1}^{N} G_{jb}(V_{Ij0}V_{Rbk} + V_{Ijk}V_{Rb0} - V_{Ij0}V_{Ibk} - V_{Ijk}V_{Rb0}) + \\ &- B_{jb}(V_{Ij0}V_{Ibk} - V_{Ijk}V_{Ib0} - V_{Rj0}V_{Rbk} - V_{Rjk}V_{Rb0}) - EQ_{jk} \\ &\forall\, j \in \mathcal{N}_Q,\; k = [1,p]\end{aligned} \tag{2.83}$$

where the worst-case instances of the endogenous uncertainties are explicitly integrated in these equations by considering the maximum approximation errors, as follows:

$$EP_{ik} = \sum_{b=1}^{N} G_{ib}\Big(\sum_{h=1}^{P} |V_{Rih}V_{Rbh}| + |V_{Iih}V_{Ibh}|\Big) + B_{ib}\Big(\sum_{h=1}^{P} |V_{Rih}V_{Ibh}| + |V_{Iih}V_{Rbh}|\Big) \tag{2.84}$$

$$EQ_{jk} = \sum_{b=1}^{N} G_{jb}\Big(\sum_{h=1}^{P} |V_{Ijh}V_{Rbh}| - |V_{Rjh}V_{Ibh}|\Big) + B_{jb}\Big(\sum_{h=1}^{P} |V_{Ijh}V_{Ibh}| + |V_{Rjh}V_{Rbh}|\Big) \tag{2.85}$$

The deterministic problem formalized in (2.79)–(2.83) is well determined, since it is composed by a set of $2N - 1 + (2N - 1 - N_G)p$ equations with as many unknowns and can be effectively solved by traditional Newton-type iterative methods, whose convergence could be sensibly improved by computing the sparse Jacobian matrix of the PF equations. Note that this characteristic makes this uncertain PF formulation more accurate and faster to solve than the linear AA-based PF approach described in Section 2.4.1, which is based on an optimization model, and thus it is more challenging to solve. Furthermore, reactive power limits can be readily introduced in the presented model using the same techniques used in deterministic PF problems.

The described formulation can used for solving uncertain security-constrained OPF problems, where the uncertain equations are represented by the multiplication of affine forms, since the PF equations, the system constraints and the cost functions can be properly expressed by quadratic functions. The solution of these problems is extremely useful in robust power system analysis, as discussed in Section 2.3.2.

2.5.2 Numerical results

In order to show the effectiveness of second-order AA in solving uncertain PF problems, the IEEE 118 test network is considered here, assuming the uncertain active and reactive power demands vary $\pm 20\%$ with respect to their nominal values and ignoring reactive power limits. The results obtained by the second-order AA-based method can then be compared to those obtained by applying a Monte Carlo-based solution technique with 10,000 samples and the linear AA-based approach presented in Section 2.4.1.

Figure 2.8 depicts the bounds of the bus voltages computed by the second-order AA-method benchmarked against those obtained by the Monte Carlo technique. These

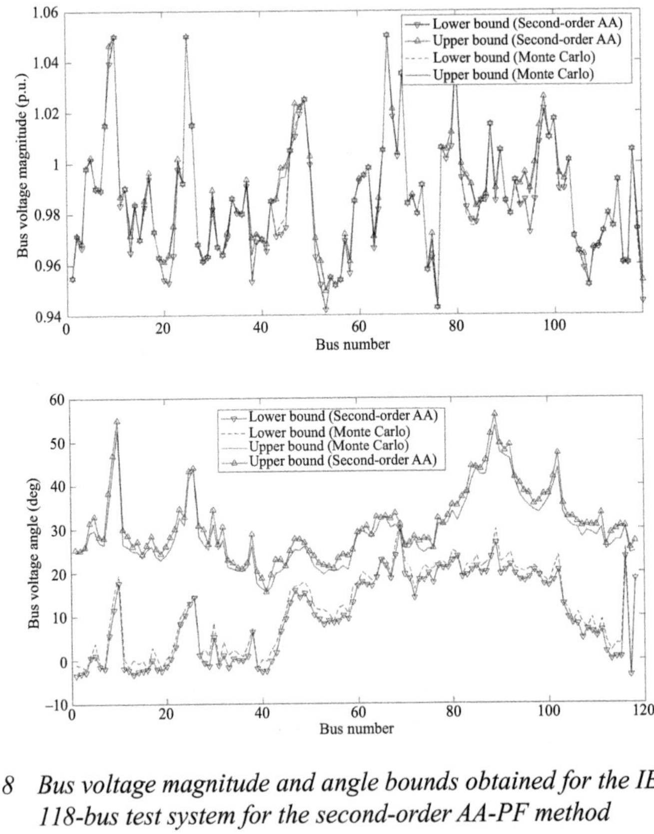

Figure 2.8 Bus voltage magnitude and angle bounds obtained for the IEEE 118-bus test system for the second-order AA-PF method

Table 2.4 PF simulation times for the 118-bus test system

Linear AA	**Monte Carlo**	**Second-order AA**
5.7 s	192.81 s	0.076 s

Table 2.5 PF simulation time comparison

Case	**Monte Carlo**	**Second-order AA**
1,354 buses	35.91 min	3.37 s
1,951 buses	1.14 h	6.77 s

results confirm the better level of approximation of the solution enclosures computed by the second-order AA-based technique with respect to the sampling-based approach, especially when compared with the same bounds illustrated in Figure 2.6, since reactive power limits are not active in that case.

Table 2.4 shows a drastic reduction of the computation burden of the proposed approach with respect to Monte Carlo as well as to the linear AA-based results discussed in Section 2.4.3.1, as expected, since the former only requires the solution of a well-determined set of non-linear equations without the need for an optimization approach. These computational benefits are more obvious in case studies based on large-scale test networks as presented in Table 2.5, which reports the computational times for the second-order AA-based technique to solve the uncertain PF problems viz-a-viz a Monte Carlo approach. The results confirm the good scalability of the second-order AA-based solution algorithm for solving realistic PF problems.

2.6 Conclusion

In this chapter, AA-based computing paradigms aimed at achieving better enclosures for AA PF and OPF solution sets were presented. Compared to existing AA-based solution paradigms for uncertain PF and OPF analysis, this formulation allows to drastically reduce the approximation errors by obtaining a better estimation of the OPF solution sets. However, compared to the previous proposed AA-based computing paradigms, the linear AA-based approach results in higher computational costs, mainly due to the large number of control variables required to solve the "perturbed state" problem, which poses computational difficulties for large scale power system applications. To address this problem, principal component analysis (PCA)-based paradigms for knowledge discovery from historical operation data-sets have been proposed to identify the optimal affine forms describing the uncertain variables in AA-based PF and OPF analysis.

On the other hand, the described second-order AA approach for solving uncertain PF problems is more precise and faster, since it allows modeling the affine approximation errors by distributing the corresponding endogenous uncertainties on the partial deviation of the primitive noise symbols, which are no longer numbers but affine forms, hence mitigating the over-conservatism of the conventional AA-based non-linear operators. Thus, the application of this approach to solve uncertain security-constrained OPF problems is being investigated.

Bibliography

[1] A. Rezaee Jordehi. How to deal with uncertainties in electric power systems? A review. *Renewable and Sustainable Energy Reviews*, 96:145–155, 2018.

[2] P. Chen, Z. Chen, and B. Bak-Jensen. Probabilistic load flow: a review. In *Proceedings of the 3rd International Conference on Deregulation and Restructuring and Power Technologies, DRPT 2008*, pp. 1586–1591, 2008.

[3] B. Zou and Q. Xiao. Solving probabilistic optimal power flow problem using quasi Monte Carlo method and ninth-order polynomial normal transformation. *IEEE Transactions on Power Systems*, 29(1):300–306, 2014.

[4] M. Hajian, W. Rosehart, and H. Zareipour. Probabilistic power flow by Monte Carlo simulation with Latin supercube sampling. *IEEE Transactions on Power Systems*, 28(2):1550–1559, 2013.

[5] C.L. Su. Probabilistic load-flow computation using point estimate method. *IEEE Transactions on Power Systems*, 20(4):1843–1851, 2005.

[6] H. Zhang and P. Li. Probabilistic analysis for optimal power flow under uncertainty. *IET Generation, Transmission and Distribution*, 4(5):553–561, 2010.

[7] S.R. Kasim, M.M. Othman, N.F.A. Ghani, and I. Musirin. Application of bootstrap technique in power system risk assessment. In *Proceedings of the 2010 4th International Power Engineering and Optimization Conference*, pp. 82–88, 2010.

[8] A. Schellenberg, W. Rosehart, and J. Aguado. Cumulant-based probabilistic optimal power flow (p-opf) with Gaussian and gamma distributions. *IEEE Transactions on Power Systems*, 20(2):773–781, 2005.

[9] P. Zhang and S.T. Lee. Probabilistic load flow computation using the method of combined cumulants and Gram–Charlier expansion. *IEEE Transactions on Power Systems*, 19(1):676–682, 2004.

[10] L.A. Sanabria and T.S. Dillon. Stochastic power flow using cumulants and Von Mises functions. *International Journal of Electrical Power & Energy Systems*, 8(1):47–60, 1986.

[11] H. Zhang and P. Li. Chance constrained programming for optimal power flow under uncertainty. *IEEE Transactions on Power Systems*, 26(4):2417–2424, 2011.

[12] A. Vaccaro and D. Villacci. Radial power flow tolerance analysis by interval constraint propagation. *IEEE Transactions on Power Systems*, 24(1):28–39, 2009.

[13] M. Madrigal, K. Ponnambalam, and V.H. Quintana. Probabilistic optimal power flow. In *Proceedings of the IEEE Canadian Conference on Electrical and Computer Engineering*, Vol. 1, pp. 385–388, 1998.

[14] J.M. Morales and J. Perez-Ruiz. Point estimate schemes to solve the probabilistic power flow. *IEEE Transactions on Power Systems*, 22(4):1594–1601, 2007.

[15] G. Verbic and C.A Canizares. Probabilistic optimal power flow in electricity markets based on a two-point estimate method. *IEEE Transactions on Power Systems*, 21(4):1883–1893, 2006.

[16] G. Verbic and C.A. Cañizares. Probabilistic optimal power flow in electricity markets based on a two-point estimate method. *IEEE Transactions on Power Systems*, 21(4):1883–1893, 2006.

[17] M. Mohammadi, A. Shayegani, and H. Adaminejad. A new approach of point estimate method for probabilistic load flow. *International Journal of Electrical Power and Energy Systems*, 51(1):54 – 60, 2013.

[18] P.R. Bijwe and G.K. Viswanadha Raju. Fuzzy distribution power flow for weakly meshed systems. *IEEE Transactions on Power Systems*, 21(4): 1645–1652, 2006.

[19] V. Miranda and J.T. Saraiva. Fuzzy modelling of power system optimal load flow. In *Proceedings of the Power Industry Computer Application Conference*, pp. 386–392, 1991.

[20] X. Guan, W.H. Edwin Liu, and A.D. Papalexopoulos. Application of a fuzzy set method in an optimal power flow. *Electric Power Systems Research*, 34(1): 11–18, 1995.

[21] F. Alvarado, Y. Hu, and R. Adapa. Uncertainty in power system modeling and computation. In *Proceedings of the IEEE International Conference on Systems, Man and Cybernetics*, pp. 754–760, 199

[22] A. Vaccaro, C.A. Cañizares, and D. Villacci. An affine arithmetic-based methodology for reliable power flow analysis in the presence of data uncertainty. *IEEE Transactions on Power Systems*, 25(2):624–632, 2010.

[23] A. Vaccaro, C.A. Cañizares, and K. Bhattacharya. A range arithmetic-based optimization model for power flow analysis under interval uncertainty. *IEEE Transactions on Power Systems*, 28(2):1179–1186, 2013.

[24] M. Pirnia, C.A. Cañizares, K. Bhattacharya, and A. Vaccaro. An affine arithmetic method to solve the stochastic power flow problem based on a mixed complementarity formulation. *IEEE Transactions on Power Systems*, 29(6):2775–2783, 2014.

[25] H. Liang, A.K. Tamang, W. Zhuang, and X.S. Shen. Stochastic information management in smart grid. *IEEE Communications Tutorials and Surveys*, 16(3):1746–1770, 2014.

[26] C. Rakpenthai, S. Uatrongjit, and S. Premrudeepreechacharn. State estimation of power system considering network parameter uncertainty based on

parametric interval linear systems. *IEEE Transactions on Power Systems*, 27(1):305–313, 2012.

[27] M. Pirnia, C.A. Cañizares, K. Bhattacharya, and A. Vaccaro. A novel affine arithmetic method to solve optimal power flow problems with uncertainties. In *Proceedings of the IEEE Power and Energy Society General Meeting*, pp. 1–7, 2012.

[28] R. Bo, Q. Guo, H. Sun, W. Wu, and B. Zhang. A non-iterative affine arithmetic methodology for interval power flow analysis of transmission network. *Proceedings of the Chinese Society for Electrical Engineering*, 33(19):76–83, 2013.

[29] A. Pepiciello, A. Vaccaro, and M. Mañana. Robust optimization of energy hubs operation based on extended affine arithmetic. *Energies*, 12(12):2420, 2019.

[30] S. Chen, Z. Wei, G. Sun, H. Zang, Y. Zhang, and Z. Ma. An interval state estimation for electricity-gas urban energy systems. In *2017 IEEE Conference on Energy Internet and Energy System Integration (EI2)*, pp. 1–5. IEEE, New York, NY, 2017.

[31] D. Romero-Quete and C.A. Cañizares. An affine arithmetic-based energy management system for isolated microgrids. *IEEE Transactions on Smart Grid*, 10(3):2989–2998, 2018.

[32] S. Wang, K. Wang, F. Teng, G. Strbac, and L. Wu. An affine arithmetic-based multi-objective optimization method for energy storage systems operating in active distribution networks with uncertainties. *Applied Energy*, 223:215–228, 2018.

[33] M.F. Zambroni de Souza, C.A. Cañizares, and K. Bhattacharya. Self-scheduling models of a caes facility under uncertainties. *IEEE Transactions on Power Systems*, 36(4):3607–3617, 2021.

[34] J. Luo, L. Shi, and Y. Ni. A solution of optimal power flow incorporating wind generation and power grid uncertainties. *IEEE Access*, 6:19681–19690, 2018.

[35] S. Wang, L. Han, and P. Zhang. Affine arithmetic-based dc power flow for automatic contingency selection with consideration of load and generation uncertainties. *Electric Power Components and Systems*, 42(8):852–860, 2014.

[36] G. Wei, L. Lizi, D. Tao, M. Xiaoli, and S. Wanxing. An affine arithmetic-based algorithm for radial distribution system power flow with uncertainties. *International Journal of Electrical Power & Energy Systems*, 58(0):242 – 245, 2014.

[37] T. Ding, H.Z. Cui, W. Gu, and Q.L. Wan. An uncertainty power flow algorithm based on interval and affine arithmetic. *Automation of Electric Power Systems*, 36(13):51–55, 2012.

[38] A. Vaccaro. *Affine arithmetic for power and optimal power flow analyses in the presence of uncertainties*. Ph.D. dissertation, Electrical and Computer Engineering, University of Waterloo, Waterloo, Ontario, Canada, 2015.

[39] M. Pirnia, C.A. Cañizares, and K. Bhattacharya. Revisiting the power flow problem based on a mixed complementarity formulation approach. *IET Generation, Transmission & Distribution*, 7(11):1194–1201, 2013.

[40] A. Gomez-Exposito, A. Conejo, and C.A. Canizares. *Electric Energy Systems: Analysis and Operation*. CRC Press, London, 2009.

[41] R. Moore. *Methods and Applications of Interval Analysis*, Vol. 2, SIAM, Philadelphia, PA, 1979.

[42] L.V. Kolev. A method for outer interval solution of linear parametric systems. *Reliable Computing*, 10(3):227–239, 2004.

[43] C Jiang, X. Han, and G.P. Liu. A sequential nonlinear interval number programming method for uncertain structures. *Computer Methods in Applied Mechanics and Engineering*, 197(49):4250–4265, 2008.

[44] C. Jiang, X. Han, G.R. Liu, and G.P. Liu. A nonlinear interval number programming method for uncertain optimization problems. *European Journal of Operational Research*, 188(1):1–13, 2008.

[45] G. Bontempi, A. Vaccaro, and D. Villacci. Power cables' thermal protection by interval simulation of imprecise dynamical systems. *IEE Proceedings-Generation, Transmission and Distribution*, 151(6):673–680, 2004.

[46] A. Vaccaro, C.A. Canizares, and D. Villacci. An affine arithmetic-based methodology for reliable power flow analysis in the presence of data uncertainty. *IEEE Transactions on Power Systems*, 25(2):624–632, 2010.

[47] X. Liu, W.-S. Luk, Y. Song, P. Tang, and X. Zeng. Robust analog circuit sizing using ellipsoid method and affine arithmetic. In *Proceedings of the 2007 Asia and South Pacific Design Automation Conference*, pp. 203–208, 2007.

[48] A. Vaccaro and A. Pepiciello. *Affine Arithmetic-Based Methods for Uncertain Power System Analysis*. Elsevier, New York, NY, 2022.

[49] A. Pepiciello, A. Vaccaro, and M. Mañana. Robust optimization of energy hubs operation based on extended affine arithmetic. *Energies*, 12(12):2420, 2019.

Chapter 3

DFT-based synchrophasor estimation processes for Phasor Measurement Units applications: algorithms definition and performance analysis

Mario Paolone[1] *and Paolo Romano*[1]

Among the various logical components of a phasor measurement unit (PMU), the synchrophasor estimation (SE) algorithm definitely represents the core one. Its choice is driven by two main factors: its accuracy in steady-state and dynamic conditions as well as its computational complexity.

Most of the SE algorithms proposed in the literature are based on the direct implementation of the Discrete Fourier Transform (DFT). This is due to the relatively low computational complexity of such technique and to the inherent DFT capability to isolate and identify the main tone of a discrete sinusoidal signal and to reject close-by harmonics. Nevertheless, these qualities come with non-negligible drawbacks and limitations that typically characterize the DFT: mainly they refer to the fact that the DFT theory assumes a periodic signal with time-invariant parameters, at least along the observation window. The latter, from one side should be as short as possible to be closer to the above-mentioned quasi-steady-state hypothesis also during power system transient; on the other hand, longer windows are needed when interested in rejecting and isolating harmonic and inter-harmonic signals that are quite frequent in power systems.

In this respect, this chapter first analyses the DFT with a particular focus on the origin of the well-known aliasing and spectral leakage effects. Then it formulates and validates in a simulation environment a novel SE algorithm, hereafter referred as iterative-Interpolated DFT (i-IpDFT), which considerably improves the accuracies of classical DFT- and IpDFT-based techniques and is capable of keeping the same static and dynamic performances independently of the adopted window length that can be reduced down to two cycles of signal at the nominal frequency of the power system.

This chapter is organized as follows: Section 3.2 introduces the nomenclature and some basic concepts in the field of synchrophasors. Section 3.3 presents the theoretical background of the DFT, with a specific focus on the detrimental effects of aliasing and spectral leakage. Next, Section 3.4 discusses advantages and drawbacks of DFT-based SE algorithms and derives the analytical formulation of the i-IpDFT method. Finally, Section 3.5, after illustrating the procedure presented in Reference 1 to assess the performances of a PMU, analyses the performances of the i-IpDFT

[1]School of Engineering, École Polytechnique Fédérale de Lausanne (EPFL), Switzerland

algorithm using two of the testing conditions presented in Reference 1 and compares them with those of the classical IpDFT technique.

3.1 Literature review

The scientific literature in the field of synchrophasor is quite recent but it already contains several contributions presenting novel algorithms for the calculation of phasors and/or local system frequency and rate of change of frequency (ROCOF).

In general, according to the adopted signal model, SE algorithms can be grouped into two categories [2]:

- algorithms based on a *static* signal model that assume that the waveform parameters are constant within the adopted window length;
- algorithms based on a *dynamic* signal model that assume a more sophisticated but generic signal model that includes also the possibility that the waveform parameters are time varying within the observation time window.

The main representatives of the latter category are based on the so-called Taylor–Fourier transform and were initiated with Reference 3. The advantages of these algorithms are evident and refer to the possibility of tracking and potentially estimating the power system dynamics (e.g., References 4,5). Nevertheless, they are characterized by a non-negligible drawback, represented by their computational complexity that generally tends to be higher and does not typically match the available computational resources of standard hardware platforms. Additionally, it is barely impossible to track any kind of dynamics. In particular sudden ones, like those happening during faults, still deteriorate the quality of the estimated synchrophasor for the duration of the window length, that is typically quite high to keep the estimation uncertainty within reasonable limits.

On the other hand, SE algorithms based on a static signal model are the most common one, due to their good their trade-off between computational complexity and estimation accuracies. The majority of these methods is based on the direct implementation of the DFT [6]. Non-DFT-based SE methods based on a static signal model have also been proposed along the years. These include, among others, *zero-crossing methods* [7, 8], *demodulation filters* [7, 9], *adaptive filters* [10, 11], *compressive sensing* algorithms [12, 13], *wavelet*-based algorithms (e.g., Reference 14), *resampling* methods [15], *Prony*'s estimation methods [16] and *Matrix Pencil* methods [17].

Based on the window length, DFT-based SE algorithms can be grouped into multi-cycle, one-cycle or fractional-cycle DFT estimators performing recursive and non-recursive updates (e.g., References 18–20). In order to improve their accuracy, DFT-based algorithms have been sometimes proposed in combination with weighted least squares (e.g., Reference 21) or Kalman filter-based methods (e.g., Reference 22). Within this category, in order to reduce the effects of leakage and of the so-called picket-fence effect, the use of time windows in combination with the well-known Interpolated-DFT (IpDFT) technique has been first proposed in References 23,24 and further developed in References 25–28. More in particular, contributions

[27, 28] have proven that the effects of long- and short-range leakage can be considerably minimized by adopting suitable windows functions and IpDFT schemes, respectively (see also References 29,30). The advantages of this kind of approaches refer to the relatively simple implementation and low computational complexity capable of achieving reasonable accuracy and response times (RTs) after a careful selection of the algorithm parameters.

3.2 Definitions

This section recalls the theoretical definition of a *phasor* and derives the concept of *synchrophasor* starting from a given signal model. Then, it illustrates the general architecture of a PMU together with a preliminary analysis of its components.

3.2.1 Signal model

Electrical power is traditionally delivered from the generators to the end users through an infrastructure that is mainly composed by components operating in alternating current (AC). As a consequence, during normal operating conditions of the power system, voltage and current waveforms are usually modelled as signals characterized by a single sinusoidal component with constant parameters:

$$x(t) = A_0 \cdot \cos(2\pi f_0 t + \varphi_0) \tag{3.1}$$

being A_0 the nominal peak amplitude, φ_0 the initial phase, i.e., for $t = 0$, and f_0 the nominal frequency of the power system, i.e., 50 or 60 Hz.

However, even in normal operating conditions, a power system is never in a steady state. As a consequence, the parameters of (3.1) are rarely time-invariant and typically exhibit various *dynamics*. Frequency fluctuations are definitely the most evident phenomena and are typically related to changes in load or imbalances in generation and to the interactions between power demand in the grid, inertia of large generators and the operation of governors equipping the majority of power generators [31]. Additionally, when faults or other switching events take place, those variations can involve even larger frequency fluctuations [32].

Similarly, the waveform amplitude and phase are also affected by transient phenomena. Those can be relatively slow like in the case of power swings (i.e., amplitude and phase oscillations typically characterized by frequencies below few hertz) or faster like in the case of switching events or faults that usually produce step changes in voltage and current waveforms with spectral components that can even reach several hundreds of kilohertz.

Additionally, the main tone is often corrupted by other superposed signals that can be of different nature. In the literature, the following *interfering signals* are usually considered:

- *Harmonics*, namely spectral components at frequencies that are multiple integers of the AC system instantaneous frequency that, as previously stated, can be different from the nominal one (f_0). These signals are typically produced by power

electronics devices: in transmission systems, these can be flexible AC transmission systems (FACTS) or high-voltage direct current (HVDC) connections. On the other hand, at distribution level most of the harmonics are generated by converters typically interfacing distributed generation units or non-linear loads [31]. The literature on PMUs usually considers the effects of harmonics up to the 50th, assuming that higher frequency components are either attenuated by the analogue front-end filters of the PMU or too far in the frequency spectrum to be considered relevant [1].
- *Inter-* and *sub-harmonics*, namely spectral components at frequencies that are not multiple integers of the system frequency: inter-harmonics are characterized by a frequency that is bigger than the nominal one f_0 and sub-harmonics by a frequency that is smaller than f_0. The causes of inter- and sub-harmonics are usually static frequency converters, cycloconverters, subsynchronous converter cascades, induction motors, arc furnaces and all loads not synchronous with the fundamental power system frequency [33].
- Aperiodic components like *decaying direct current (DC) offsets* that are likely to appear during power system transients (consider, for instance, the case of a decaying short-circuit current or inrush of transformers/induction motors). The involved time constants can vary in the range between 0.1 and 10 s.
- *Wide-band noise* that includes both the "measurement noise" (namely the noise added by any measurement equipment) and the so-called grid noise. Regarding the latter, the most well-known phenomena are the thermal noise (also known as Johnson/Nyquist noise), the corona effect, and partial discharges. Both sources are usually modelled as a zero-mean Gaussian noise processes.

As a consequence, a generalized and more complete signal model that takes into account each one of the above-mentioned *dynamics* and *interfering signals* can be formulated as follows:

$$x(t) = A(t)\cos(2\pi f(t)t + \varphi_0) + \sum_{h=1}^{H} A_h(t)\cos(2\pi f_h(t)t + \varphi_h) + A_{DC}(t)e^{-\frac{t}{\tau}} + \epsilon(t) \tag{3.2}$$

where

- The first term represents the main tone of the spectrum, characterized by an instantaneous frequency $f(t)$ that is typically very close to the nominal frequency of the power system f_0, an instantaneous peak amplitude $A(t)$, and an instantaneous phase $\psi(t) = 2\pi f(t)t + \varphi_0$.*
- The second term models the contribution of any superposed sinusoidal tone excluding the main one. In other words, it includes all the effects of harmonic,

*The assumption of having a constant initial phase φ_0 is both mathematically and physically correct. It implies that the instantaneous phase variations are only due to the power system frequency variations and it avoids the possibility of having multiple couples of frequency and initial phase that produce the same instantaneous phase $\psi(t)$.

inter- and sub-harmonic tones, characterized by an instantaneous frequency $f_h(t)$ (that does not necessarily need to be an integer multiple of the fundamental frequency $f(t)$), peak amplitude $A_h(t)$ and instantaneous phase $\psi_h = 2\pi f_h(t)t + \varphi_h$, being φ_h the initial phase. During normal operating conditions, the peak amplitude is at least one order of magnitude lower than the one of the main tone $A(t)$, but no hypothesis can be made with respect to the instantaneous phase that does not necessarily have to match the one of the main tone.

- The third term models a decaying DC component characterized by an initial amplitude A_{DC} and an arbitrary time constant τ.
- The last element models a superposed wide-band noise that includes any other contribution not included in the previous terms.

3.2.2 *Phasor*

The phasor transformation has been historically adopted in electrical engineering to simplify the analysis of electrical systems in sinusoidal steady state. It consists in a one-to-one mapping between time-harmonic functions and complex numbers that can be adopted if and only if the instantaneous frequency f, peak amplitude A and initial phase φ are *time-invariant*, namely, if and only if the sinusoidal signal $x(t)$ is *stationary* (see (3.1)). In particular, such a transformation allows to represent a sinusoidal function of time like the one expressed by (3.1) with a single complex constant and vice versa. In order to derive it, we can rewrite (3.1) as:

$$x(t) = A \cdot \cos(2\pi f t + \varphi) \tag{3.3}$$

$$= \mathrm{Re}\{A \cdot e^{j(2\pi f t + \varphi)}\} \tag{3.4}$$

$$= \mathrm{Re}\{A \cdot e^{j\psi(t)}\} \tag{3.5}$$

where the dependency of $\psi(t)$ on time can be disregarded due to the stationarity of $x(t)$. We can then associate to the sinusoid $x(t)$ the complex number X and call it *phasor*:

$$x(t) \rightleftarrows X \triangleq A \cdot e^{j\psi} \tag{3.6}$$

$$= A \cdot [\cos(\psi) + j\sin(\psi)] \tag{3.7}$$

$$= X_r + jX_i \tag{3.8}$$

where the subscripts r and i identify the real and imaginary parts of X. A phasor can be represented either in polar (3.6) or rectangular (3.8) coordinates and the transformation from one set of coordinates to the other can be done using the well-known Euler identity (3.7).

3.2.3 *Synchrophasor*

The phasor concept cannot be directly applied to the analysis of a real AC power system for the simple reason that the *stationarity hypothesis* does not hold in practice. Nevertheless, even when the power system is not in steady state (e.g., during electromechanical oscillations), the voltage and current variations can be assumed relatively slow and be treated as a series of steady-state conditions where the stationarity hypothesis holds (*quasi-steady-state* approximation).

With such an assumption the phasor analysis can be still applied but it requires that the measured values are referenced to a common time reference and taken at the same instant in time. For this reason the phasor concept has been extended to the *synchrophasor* one that has been first defined in the IEEE Std. 1344-1995 [34] as:

> [...] a phasor calculated from data samples using a standard time signal as the reference for the sampling process. In this case, the phasors from remote sites have a defined common phase relationship. [...]

Thanks to the explicit reference to a common time reference t, the synchrophasor concept allows to define a common phase relationship between phasors from remote sites. Additionally, it is not restricted to the analysis of stationary systems since a synchrophasor can be associated to any waveform characterized by a dominant sinusoidal component, also, during power system transients and with superposed interfering signals.[†]

The synchrophasor concept, and particularly the analytical definition of its phase, has been reviewed later in the IEEE Std. C37.118.1-2011 [1] that defined it as follows:

> [...] the instantaneous phase angle relative to a cosine function at the nominal system frequency synchronized to Coordinated Universal Time (UTC). [...] Under this definition ψ is the offset from a cosine function at the nominal system frequency synchronized to UTC. [...]

As shown in Figure 3.1, such a convention implies that, if a sinusoidal waveform representing a generic current or voltage, has its maximum at the UTC-second rollover (i.e., when the cosine has its maximum) the synchrophasor angle is 0°. It is 90° if the positive zero-crossing occurs at the UTC-second rollover.

The above IEEE Std. definition, which refers the phase to an hypothetical signal at the nominal frequency f_0 synchronized to the UTC, is restrictive and might generate ambiguities. A better and more generic definition of synchrophasor can be given by referencing both amplitude and phase estimations to the same time reference:

Definition. *The synchrophasor representation of the signal $x(t)$ in (3.2) is the complex function $X(t)$ characterized by an instantaneous amplitude and phase corresponding to the instantaneous amplitude $A(t)$ and phase $\psi(t)$ of the main tone of $x(t)$, respectively, being t the UTC time reference.*

[†]It is worth pointing out that the synchrophasor refers only to the main tone of $x(t)$ and does not take into account any other contribution that might be present in the original signal.

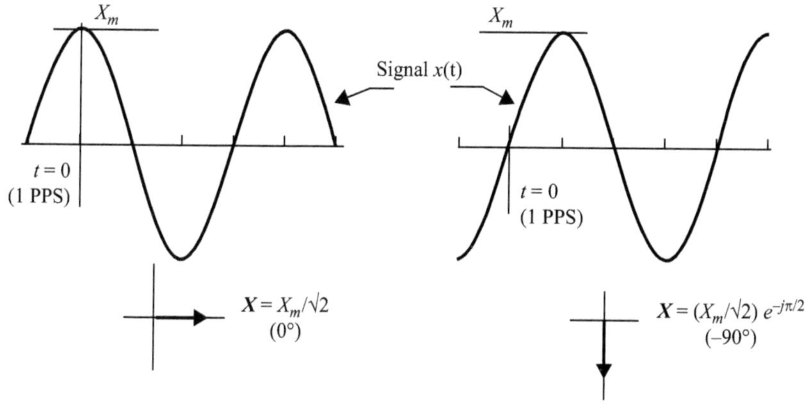

Figure 3.1 Phase convention for synchrophasors for a signal of amplitude X_m (adapted from Reference 1)

As a consequence, in the case of the signal $x(t)$ represented in (3.2), the associated synchrophasor is:

$$x(t) \rightleftarrows X(t) \triangleq A(t) \cdot e^{j\psi(t)} \tag{3.9}$$

$$= A(t) \cdot e^{j(2\pi f(t)t+\varphi)} \tag{3.10}$$

where $A(t)$ and $\psi(t)$ are the instantaneous peak amplitude and phase of the main tone of $x(t)$ (see (3.2)).

3.2.4 Frequency and rate of change of frequency

According to the IEEE Std. C37.118, frequency is defined starting from the first-order derivative of the instantaneous phase $\psi(t)$:

$$\frac{d\psi(t)}{dt} = \frac{d}{dt}[2\pi f(t)t + \varphi] \tag{3.11}$$

$$= 2\pi \left[f(t) + \frac{df}{dt}t \right] \tag{3.12}$$

With the quasi-steady-state assumption, the term $df/dt \approx 0$ and frequency can be defined as:

$$f(t) = \frac{1}{2\pi}\frac{d\psi(t)}{dt} \tag{3.13}$$

Similarly, the ROCOF is defined in the IEEE Std. C37.118 as the first-order derivative of the frequency:

$$ROCOF(t) = \frac{df(t)}{dt} \tag{3.14}$$

It is worth underlying that the definitions given in (3.13) and (3.14) disagree as (3.13) has been obtained with an approximation ($df/dt = 0$) that would lead to $ROCOF = 0$. However, for slow-varying frequencies (3.14) can be still considered as a correct approximation to compute the frequency time derivative.

3.2.5 *Phasor measurement unit*

According to the IEEE Std. C37.118.1-2011 [1], a PMU is a stand-alone physical unit or a functional unit within another physical unit

> [...] that produces synchronized phasor, frequency, and ROCOF estimates from voltage and/or current signals and a time synchronizing signal. [...]

The logic architecture of a generic PMU is shown in Figure 3.2. The PMU, in order to report synchronized measurements, needs to be equipped with a time-synchronization module capable of receiving the UTC absolute time from a reliable and accurate time source. The time-sync unit internally generates the "time base", namely a stable and accurate internal time reference, used by the signal conditioning and analog to digital (A/D) conversion unit to discipline‡ the sampling process of the input waveforms (as many as the number of connected input channels). The sampled waveforms are then transferred, sample by sample, to the SE algorithm, the core component of any PMU, that extracts the fundamental tone of a distorted sinusoidal waveform from a previously acquired set of samples and estimates its amplitude, phase and frequency, and ROCOF. The estimated values are then transferred to the data encapsulation and streaming unit that encapsulates and streams the data according, for instance, to the IEEE Std. C37.118.2-2011 [35] or IEC 61850-90-5 [36] data-transmission protocols.

It might be obvious, but still worth pointing out, that each one of the logical components highlighted in Figure 3.2 contributes to the global uncertainty that intrinsically characterizes the PMU estimations. In particular:

- The *time-sync unit*, depending on the adopted time source and dissemination technology, might deteriorate the synchronism of the sampling process (i.e., its alignment to the UTC-second rollover and the accuracy of the sampling time). Such an uncertainty can have non-negligible effects on the overall measurement accuracy, particularly in the estimation of the phase.§ According to the PMU accuracy requirements dictated by Reference 1, the maximum acceptable timing uncertainty is ± 26 μs for a 60 Hz system and ± 31 μs for a 50 Hz system. In practice, PMUs adopt much more accurate timing sources, to further reduce the timing uncertainty.

‡Such a functionality is quite common in PMUs but not mandatory. The synchronization of the sampling process to a common time reference can be also achieved by post-processing the acquired samples.
§The time-sync uncertainty linearly translates in phase uncertainty based on the instantaneous frequency values following the formula $\Delta\psi = 2\pi f \Delta t$, being $\Delta\psi$ and Δt the phase and time uncertainties.

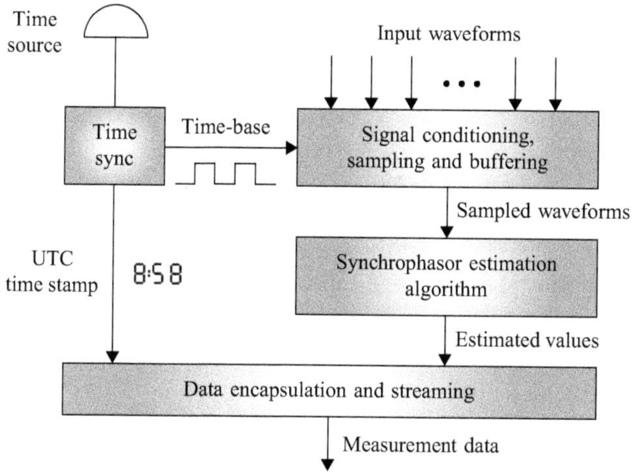

Figure 3.2 Scheme of a generic PMU

- The *signal conditioning and sampling unit* introduces an uncertainty that is mainly related to the intrinsic noise, non-linearities and gain errors introduced by the analogue filtering stage (anti-aliasing filters for instance) and to the quantization error introduced by the A/D converters.
- The *SE algorithm* is characterized by its own accuracy levels that are typically assessed in a simulation environment following the guidelines given by the IEEE Std. C37.118 [1] (see also Section 3.5). Nevertheless, this IEEE Std. does not take into account every possible power system operating condition and the SE algorithm estimations might be biased when processing non-contemplated waveforms. For instance, it has already been demonstrated that different SE algorithms might produce different estimations, particularly when exposed to power system transients [37].

Finally, it should be noted that, in order to transform the power system voltage and current waveforms to levels appropriate for the PMU analogue front-end, the PMU must be interfaced to the electrical grid using instrument transformers (refer to Reference 38 for a partial review of the available instrument transformer technologies).

Although this aspect will not be treated along this chapter, it is worth pointing out that the uncertainty introduced by this transformation stage, if not properly designed, might be dominant and exceed those introduced by the PMU components. For instance, in the case of standard magnetic core voltage and current transformers (VTs and CTs), their accuracy is generally limited to class 0.5. This, according to the definition of the standards [39, 40], translates, at full scale, to a maximum ratio error of 0.5% and a maximum phase error of 6 mrad for VTs and 9 mrad for CTs. Such an uncertainty, in most of the cases, exceeds the one of the PMU and deteriorates, therefore, the expected accuracies.

3.3 The discrete Fourier transform

The DFT is the most common analytical tool to extract the frequency content of a finite and discrete sequence of samples, obtained from the periodic sampling of a continuous waveform in the time domain. This also applies to the field of synchrophasors, where, as it will be shown later, the DFT represents the most common technique to extract the waveform parameters out of a sequence of samples.

In this respect, this section derives the analytical formulation of the DFT starting from the theoretical formulation of the Fourier transform of a continuous signal of infinite duration. Next, it highlights the most relevant DFT properties and investigates two well-known effects caused by the computation of the DFT such as *aliasing* and *spectral leakage*.

3.3.1 From the Fourier transform to the DFT

The Fourier transform of a continuous function of time $x(t)$, satisfying integrability constraints [41] is defined as:

$$X(f) = \int_{-\infty}^{+\infty} x(t)e^{-j2\pi ft}\,\mathrm{d}t \tag{3.15}$$

and it is used to transform a continuous time-domain function $x(t)$ to a continuous frequency-domain function $X(f)$.

In practice, continuous signals are sampled using A/D converters producing a sequence of samples $x(n)$ that can be easily treated by any digital hardware (samples are assumed to be equally spaced by $T_s = 1/F_s$, being F_s the PMU sampling rate). In this respect, the discrete-time Fourier transform (DTFT) has been defined to transform a time-domain sequence of infinite length $x(n),\ n \in \mathbb{N}$ into a continuous frequency-domain function $X(f)$ according to the following equation [6]:

$$X(f) = \sum_{n=-\infty}^{\infty} x(n)e^{-j2\pi fn} \tag{3.16}$$

Such a transformation cannot be applied to the analysis of real signals, as it assumes the possibility to calculate a continuum of functional values $X(f)$ by means of an infinite summation that is unfeasible in digital computers.

A more practical transformation is represented by the so-called discrete Fourier series (DFS), a frequency analysis tool conceived for periodic sequences of infinite length $\tilde{x}(n)$ characterized by period N. The DFS can be seen as a frequency-discretized version of the DTFT [6] and it involves a finite summation of N complex terms:

$$\widetilde{X}(k) = \sum_{n=0}^{N-1} \tilde{x}(n)e^{-j\frac{2\pi kn}{N}}, \qquad 0 \le k \le N-1 \tag{3.17}$$

In particular, the DFS differs from the DTFT in that its input and output sequences are both finite; it is therefore said to be the Fourier analysis of periodic discrete-time functions.

When the DFS is used to represent a generic (i.e., not necessarily periodic) finite-length sequence of samples $x(n)$, $n \in [0, N-1]$, it is called the DFT that is defined as follows:

$$X(k) \triangleq \frac{2}{B}\sum_{n=0}^{N-1} w(n)x(n)W_N^{kn}, \qquad 0 \le k \le N-1 \tag{3.18}$$

where $w(n)$ is a discrete windowing function used to extract a portion of the infinite length original sequence (see Section 3.3.4.3 for further details about windowing functions),

$$B \triangleq \sum_{n=0}^{N-1} w(n) \tag{3.19}$$

is the DFT normalization factor and

$$W_N \triangleq e^{-j2\pi/N} = \cos(2\pi/N) - j\sin(2\pi/N), \qquad W_N^{kN} = 1, k \in \mathbb{N} \tag{3.20}$$

is the so-called *twiddle factor*.

The DFT spectrum can be equivalently expressed in matrix form for a more intuitive understanding of its logic:

$$\begin{bmatrix} X(0) \\ X(1) \\ \vdots \\ X(k) \\ \vdots \\ X(N-1) \end{bmatrix} = \begin{bmatrix} 1 & 1 & 1 & \cdots & 1 \\ 1 & W_N & W_N^2 & \cdots & W_N^{N-1} \\ \vdots & \vdots & \vdots & & \vdots \\ 1 & W_N^{(k)_N} & W_N^{(2k)_N} & \cdots & W_N^{((N-1)k)_N} \\ \vdots & \vdots & \vdots & & \vdots \\ 1 & W_N^{N-1} & W_N^{N-2} & \cdots & W_N \end{bmatrix} \begin{bmatrix} x(0) \\ x(1) \\ \vdots \\ x(k) \\ \vdots \\ x(N-1) \end{bmatrix} \tag{3.21}$$

where $(\cdot)_N$ identifies the mod$N$ operator. It is easy to show that the columns of matrix $[W_N^{kN}]$ of (3.21) are linearly independent. Therefore, they are a base of $\mathbb{C}^N$.

3.3.2 DFT interpretation and relevant properties

The DFT is a sequence of complex values that are equally spaced in the frequency spectrum and represent, under specific conditions, a portion of the Fourier transform of the original continuous-time signal $x(t)$ (see Section 3.3.3 for further details about the original assumptions to guarantee an exact matching between the DFT and the Fourier transform). It is the result of a frequency decomposition of the finite-length discrete signal that is projected onto the sinusoidal basis set W_N^{kn}, $0 \le k \le N-1$.

In order to correctly interpret the DFT spectrum obtained by applying (3.18) to a generic real-valued finite sequence $x(n)$, some considerations must be made.

3.3.2.1 DFT periodicity

As shown in Section 3.3, the DFT of a finite sequence of real values $x(n)$ is a finite sequence of complex values $X(k)$ (also called "bins") defined in the interval $0 \leq k \leq N-1$. This does not mean that the DFT cannot be computed outside of the interval $0 \leq k \leq N-1$, but simply that this will result into a periodic extension of $X(k)$ due to the periodicity of the theoretical spectrum of the sampled signal.

Usually, when analysing signals through the DFT, the convention is to associate the DFT bins in the interval $0 \leq k \leq N/2-1$ to the "positive" frequencies $0 \leq f \leq F_s/2$ and the bins in the interval $N/2 \leq k \leq N-1$ to the "negative" frequency range $F_s/2 \leq f < 0$ (see also Figure 3.4(b)).

3.3.2.2 DFT symmetry

The DFT of a real-valued sequence is symmetric. In particular, the DFT bins in the interval $0 \leq k \leq N/2-1$ (positive frequency range) are related to those in the interval $N/2 \leq k \leq N-1$ (negative frequency range) based on the following equivalences:

$$\begin{cases} X(k) = X^*((k)_N) \\ \mathrm{Re}\,(X(k)) = \mathrm{Re}\,(X((-k)_N)) \\ \mathrm{Im}\,(X(k)) = -\,\mathrm{Im}\,(X((-k)_N)) \\ |X(k)| = |X((-k)_N)| \\ \angle X(k) = -\angle X((-k)_N) \end{cases} \tag{3.22}$$

As a consequence, when applying the DFT to real-valued signals, each frequency component will appear twice in the DFT spectrum: once in the positive frequency range (the so-called *positive image*) and once in the negative frequency range (the so-called *negative image*).

3.3.2.3 DFT frequency discretization

According to (3.18), the DFT of the sequence $x(n)$, $n \in [0, N-1]$, provides samples of the DTFT at N equally spaced discrete frequencies. In particular the kth DFT bin represents the frequency content of the original signal at the normalized frequency $2\pi k/N$ as it is the result of the projection of the finite-length sequence into the basis vector characterized by that frequency (see (3.18)). As a consequence, consecutive bins are separated by the normalized frequency interval $2\pi/N$ and the whole DFT spectrum will cover the normalized frequency interval $[0, 2\pi]$ or equivalently, due to the DFT periodicity, $[-\pi, \pi]$.

In order to derive a more practical scale for the DFT frequency axis, it should be noted that the finite-length sequence $x(n)$ is associated to specific instants in time according to the adopted sampling rate F_s. Therefore, based on the sampling rate, the window length can be expressed as a function of time as $T = N/F_s$ and, accordingly, each basis vector W_N^{kn} can be referred to an absolute frequency. In particular, the bin for $k = 0$ can be associated to the DC component ($f = 0$), the bin for $k = 1$ to a frequency $f = 1/T$ and a generic kth bin to a frequency $f = k/T$. It is then clear that the frequency separation between two consecutive bins is $\Delta f = 1/T$ and can only be increased at the price of enlarging the window length T. On the other hand, the frequency range that can be represented with the DFT of the finite-length sequence $x(n)$ obtained by sampling the original signal $x(t)$ with a sampling rate F_s is $[-F_s/2, F_s/2]$.

3.3.3 DFT effects

The DFT is often improperly considered as a "sampled" version of the continuous Fourier transform. This statement is valid only if both of the following assumptions are satisfied:

- the original signal $x(t)$ can be perfectly reconstructed from the discrete sequence of samples $x(n)$;
- the original signal $x(t)$ is periodic and characterized by a period that is contained an integer number of times in the chosen window length T.

In general, these hypotheses do not always hold and, as demonstrated in Section 3.3.1, the DFT can be correctly interpreted only from a precise knowledge of the theoretical background of the Fourier analysis. In this respect, Figure 3.3 shows the necessary steps required to analyse a continuous-time signal $x(t)$ with the DFT. The first step is represented by the continuous-to-discrete-time conversion that allows to transform the original continuous signal into an equivalent sequence of samples $x(n)$. This is typically performed by an A/D converter that, for the time being, is considered to be ideal; in other words, it is assumed to produce a sequence of equally spaced samples that are equal to the original signal evaluated at regular time intervals $x(n) = x(nT_s)$, being $T_s = 1/F_s$ the sampling time as introduced before. The A/D conversion process produces an infinite sequence of samples that, in order to be processed by the DFT, needs to be clustered in finite portions containing the same number of samples N. This is done by applying to the infinite sequence $x(n)$ a finite duration windowing function $w(n)$. After this steps the DFT can be applied to each portion of the sampled signal.

In what follows, each one of the above-mentioned steps is analytically modelled and analysed using the Fourier transform theory. This will help understanding the previously mentioned assumptions and derive the two main DFT error sources, namely, *aliasing* and *spectral leakage*.

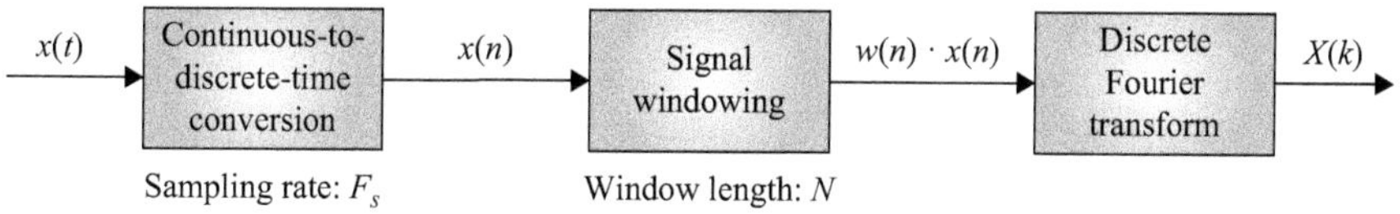

Figure 3.3 Main steps to apply a DFT-based digital signal processing technique to a continuous-time signal $x(t)$

3.3.3.1 Aliasing

As known, the sampling process can be modelled as the multiplication of the input continuous signal $x(t)$ with a periodic impulse train $s(t)$ [6]:

$$x_s(t) = x(n \cdot T_s) = x(t) \cdot s(t) \tag{3.23}$$

$$= x(t) \cdot \sum_{n=-\infty}^{\infty} \delta(t - nT_s) \tag{3.24}$$

$$= \sum_{n=-\infty}^{\infty} x(nT_s) \cdot \delta(t - nT_s) \tag{3.25}$$

being δ the unit *impulse function* or *Dirac delta function*.

The frequency-domain representation of this transformation can be given by applying the Fourier transform theory and properties to (3.24). In particular, by recalling that the Fourier transform of the product of two functions is the convolution of the Fourier transforms $\mathfrak{F}$ of the two functions, the Fourier transform of the sampled signal can be analytically derived as follows:

$$x_s(t) = x(t) \cdot \sum_{n=-\infty}^{\infty} \delta(t - nT_s) \xrightarrow{\mathfrak{F}} X_s(f) = X(f) * \frac{1}{T_s} \sum_{k=-\infty}^{\infty} \delta\left(f - \frac{k}{T_s}\right) \tag{3.26}$$

$$= \frac{1}{T_s} \sum_{n=-\infty}^{\infty} X\left(f - \frac{k}{T_s}\right) \tag{3.27}$$

where, to obtain (3.27), we have taken advantage of the property that the convolution between a Dirac δ function and any generic function is the value of the generic function evaluated at the location of the Dirac δ function.

As a consequence, the Fourier transform of the sampled signal is composed by infinite copies of the spectrum $X(f)$ (see Figure 3.4(a)) of the original continuous signal $x(t)$. These copies are shifted by integer multiples of the sampling frequency $F_s = 1/T_s$ and, then, superimposed to produce the periodic Fourier transform depicted in Figure 3.4 for a band-limited original spectrum characterized by a bandwidth F_m and a sampling rate F_s.

From Figure 3.4(b) it is evident that, if the signal is band limited with bandwidth $F_m < F_s/2$, the spectrum copies are not overlapping and the original spectrum $X(f)$ can be reconstructed by low-pass filtering the base-band copy of the spectrum $X_s(f)$. On the other hand, if this is not the case and the bandwidth F_m of the original signal is higher than half of the sampling rate F_s (i.e., $F_m > F_s/2$), the spectrum copies are overlapping so that when they add together, the original spectrum $X(f)$ is no longer recoverable by low-pass filtering (see Figure 3.4(c)).

The latter phenomenon is called *aliasing* and it results into a distortion of the original signal that cannot be any longer be reconstructed from the sampled signal. Such a phenomenon is the basis of the well-known Nyquist–Shannon sampling theorem [6]:

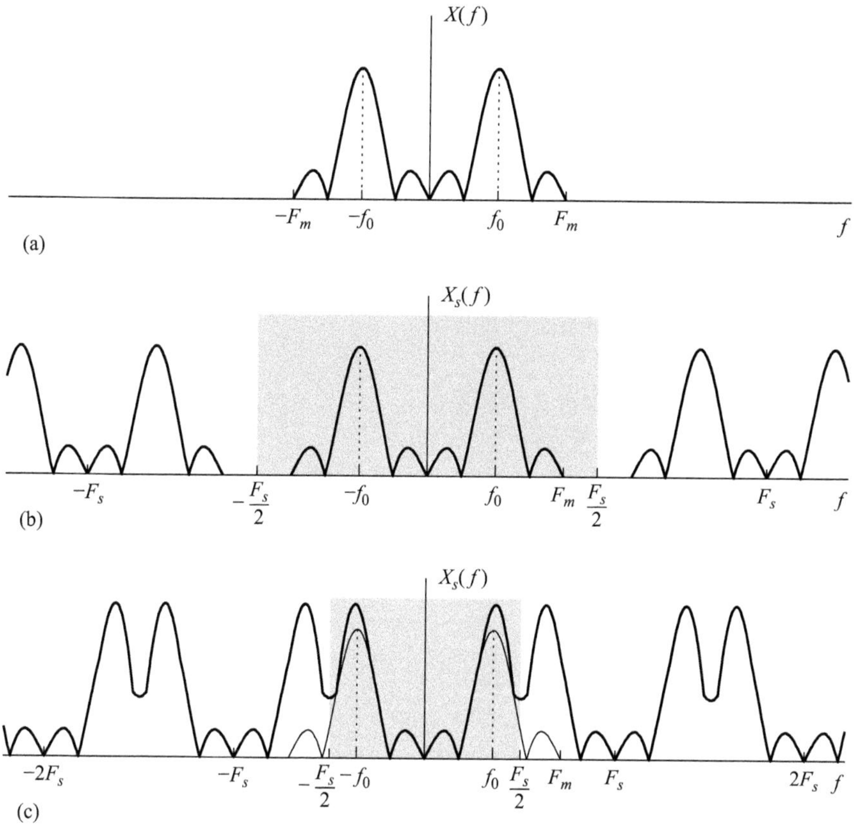

Figure 3.4 Effects of aliasing for a signal characterized by a main frequency component at f_0 and a bandwidth F_m, sampled with a sampling rate F_s. The figure shows, in order, the Fourier transforms of the continuous-time signal (a) and the Fourier transform of the sampled signal with a sampling rate respecting (b) or violating (c) the Nyquist–Shannon theorem

Theorem 3.1 (Nyquist–Shannon). *Let $x(t)$ be a band-limited signal and $X(f)$ its Fourier transform with*

$$X(f) = 0, \qquad f > F_m$$

Let $x(n) = x(nT_s)$ ($n \in \mathbb{N}$) be an infinite sequence of equally spaced samples obtained by sampling the continuous signal $x(t)$ with a sampling frequency $F_s = 1/T_s$. Then $x(t)$ is uniquely determined by the sequence of samples $x(n) = x(nT_s)$ if

$$F_s > 2F_m$$

In other words, in order to be able to correctly reconstruct the signal $x(t)$ from the infinite sequence of samples $x(n)$, the original signal must be sampled at a sampling rate F_s that must be at least two times higher than the maximum frequency component contained in the original spectrum $X(f)$.

3.3.3.2 Spectral leakage

Once the original continuous signal $x(t)$ has been sampled, it must be clustered in portions to be analysed by the DFT. This process is called *windowing* and consists in multiplying the infinite sequence of samples $x(n)$ by a specific windowing function $w(n)$ (see Figure 3.5).

In order to derive the effects of windowing, let us consider, without loss of generality, a signal $x(t)$ that is only composed by a sinusoidal component at the nominal frequency of the power system f_0:

$$x(t) = A\cos(2\pi f_0 t) \tag{3.28}$$

As known, its Fourier transform is simply:

$$X(f) = \frac{A}{2}[\delta(f - f_0) + \delta(f + f_0)] \tag{3.29}$$

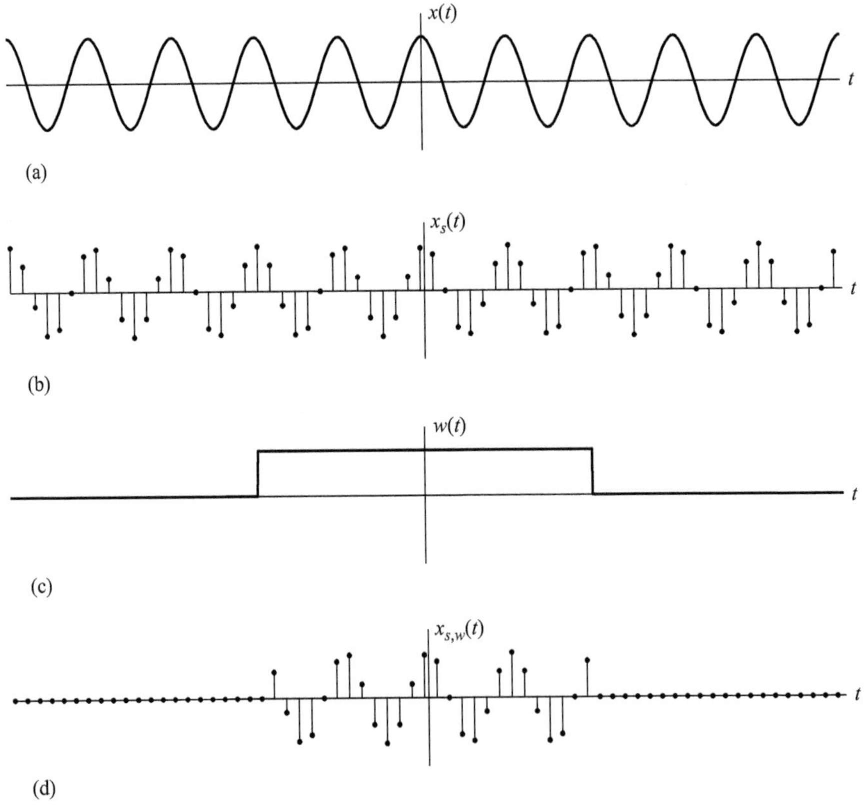

Figure 3.5 Successive applications of sampling and windowing to obtain a portion of samples of the original input signal $x(t)$. The figure shows, in order, (a) the continuous-time signal $x(t)$, (b) the sampled signal $x_s(t)$, (c) the rectangular windowing function $w(t)$ and (d) the sampled and windowed signal $x_{s,w}(t)$

The windowing operation can be modelled as the multiplication between the sampled sinusoidal signal $x_s(t)$ and the adopted window function $w(t)$ (see Figure 3.5):

$$x_{s,w}(t) = w(t) \cdot x_s(t) = w(t) \cdot [x(t) \cdot s(t)] \tag{3.30}$$

$$= [w(t) \cdot x(t)] \cdot s(t) \tag{3.31}$$

$$= [w(t) \cdot A\cos(2\pi f_0 t)] \cdot \sum_{n=-\infty}^{\infty} \delta(t - nT_s) \tag{3.32}$$

By recalling that (i) the Fourier transform of the product between two functions equals the convolution between the Fourier transform of each one of the two functions and (ii) the convolution between a Dirac δ function and any generic function is the value of the generic function evaluated at the location of the Dirac δ function, the Fourier transform of $x_{s,w}(t)$ can be computed as:

$$x_{s,w}(t) \xrightarrow{\mathfrak{F}} X_{s,w}(f) = [W(f) * X(f)] * S(f) \tag{3.33}$$

$$= \frac{A}{2}[W(f - f_0) + W(f + f_0)] * \frac{1}{T_s} \sum_{k=-\infty}^{\infty} \delta\left(f - \frac{k}{T_s}\right) \tag{3.34}$$

$$= \frac{A}{2T_s} \sum_{n=-\infty}^{\infty} \left[W\left(f - \frac{k}{T_s} - f_0\right) + W\left(f - \frac{k}{T_s} + f_0\right)\right] \tag{3.35}$$

In other words, the spectrum of the sampled and windowed signal $x_{s,w}(t)$ is composed by infinite copies of the spectrum of the windowed signal $W(f) * X(f)$ shifted by integer multiples of the sampling rate $F_s = 1/T_s$ and, then, superimposed.

In order to illustrate the effects of leakage on the DFT, let us consider just the base-band copy of the spectrum of $X_{s,w}(f)$ for $k = 0$. Additionally, let us adopt the simplest windowing function, namely the rectangular one.[‖] The Fourier transform of the rectangular window is the so-called *sinc function* (see Figure 3.6(a))

$$\text{sinc}\,(fT) = \frac{\sin(\pi fT)}{\pi fT} \tag{3.36}$$

that is characterized by the peculiar property:

$$\text{sinc}\,(fT)|_{f=\frac{1}{T}} = 0 \tag{3.37}$$

namely, the zero-crossing of the Fourier transform of the rectangular window is equally spaced and happens at integer multiples of $1/T$.

According to (3.35), the base-band copy of the Fourier transform of $X_{s,w}(f)$ will be composed by two sinc functions centred around $\pm f_0$. In case f_0 is a multiple of the DFT frequency resolution $\Delta f = 1/T$ (i.e., if the window contains an integer number

[‖]Nevertheless, such an analysis can be made with any other kind of window (see Section 3.3.4.3 for further details about the various type of windowing functions).

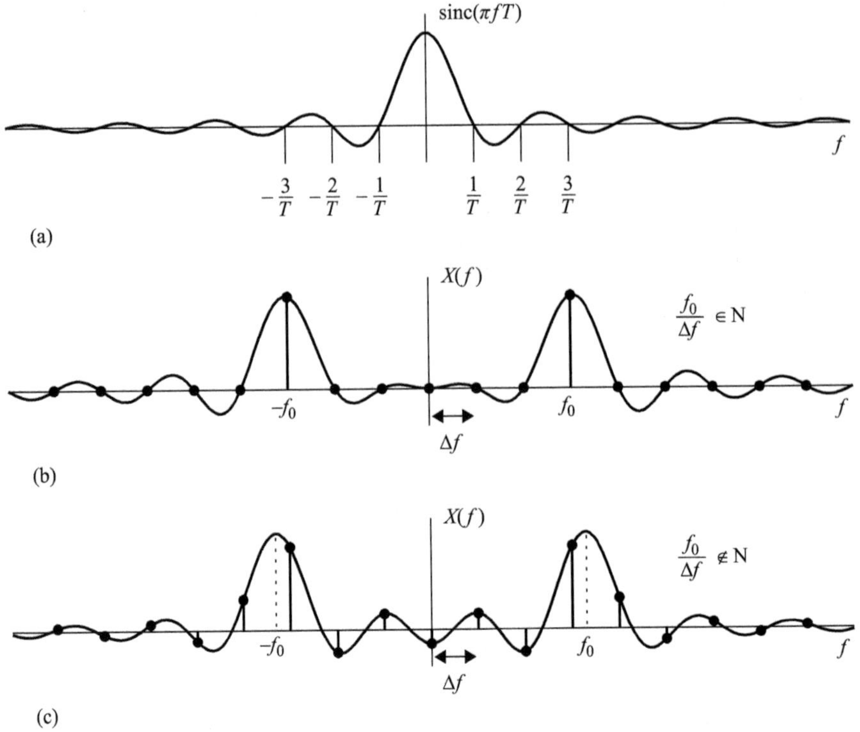

Figure 3.6 Graphical representation of the spectral leakage effects on the Fourier Transform (continuous line) and the DFT (black dots) when using a rectangular window of length T to analyze a sinusoidal signal characterized by a main tone at f_0. In (a) the sinc function is depicted. In (b) the effects of leakage are not visible since the adopted window contains an integer number of periods of the input signal; they appear in (c) where the adopted window length is not a multiple of the period of the input signal

of periods of the signal), the zero-crossings of the translated sinc functions happen exactly at multiples of $1/T$. The only frequency that will have a non-zero projection into the DFT basis set will be $f = \pm f_0$ (i.e., the DFT bin with index $k = f_0/\Delta f$) and the resulting DFT will be characterized by only two non-zero bins, at index $\pm k$ (see Figure 3.6(b)).

On the other hand, if f_0 is not a multiple of the frequency resolution Δf (i.e., if the window does not contain an integer number of periods of the signal), the zero-crossings of the translated sinc functions do not happen exactly at multiples of $1/T$. Therefore, all the discrete frequencies will exhibit non-zero projections on the DFT basis set even though the majority of the spectrum energy will still be concentrated around $f = f_0$ (see Figure 3.6(c)).

This effect is the so-called *spectral leakage* and it evidently arises when the sampling process is not synchronized with the fundamental tone of the signal under analysis and the DFT is computed over a non-integer number of cycles of the input signal. As it will be discussed later, spectral leakage can be separated into *short-term* and *long-term* spectral leakage: the first refers to the effects of the main lobe width of the Fourier transform of the adopted window that can cause difficulties in identifying the "true" maximum of a specific portion of the DFT spectrum. The latter, on the other hand, refers to the effect caused by the side lobes (i.e., the "tails") of the Fourier transform of the adopted window that can generate the so-called spectral interference between nearby tones.

3.3.3.3 Spectral sampling

As shown before, the DFT of the sequence $x(n)$, $n \in [0, N-1]$, provides samples of the DTFT of the equivalent windowed signal, at N equally spaced discrete frequencies $f_k = k \cdot \Delta f = k/T$, being $-N/2 \leq k \leq N/2 - 1$ (see Figure 3.6). In Section 3.3.3.2, we have shown how such a reduced frequency resolution can turn out to be beneficial when the input signal $x(n)$ is composed by a single tone and sampled coherently with respect to its frequency (i.e., when $F_s = k \cdot f_0$, being F_s and f_0 the sampling rate and the main-tone frequency, respectively). Nevertheless, coherent sampling is purely ideal and the DFT energy is spread along the whole spectrum. Consequently, the DFT bin values and the frequency axis discretization change based on the adopted number of samples N and window length T.

In general, when the target is identifying the parameters of a signal, the frequency resolution Δf plays a major role as it defines the accuracy in locating the correct position in the frequency spectrum of the tone under analysis. As we will see later, the IpDFT method will partially overcome such a limitation with a proper interpolation of the DFT bins.

3.3.4 DFT parameters

The DFT output can be modified by acting on three main parameters: the sampling rate F_s, the window length T, and the window profile $w(n)$. Together, they determine the amount of aliasing and spectral leakage and the frequency resolution of the DFT.

3.3.4.1 Sampling rate

The sampling rate F_s defines the frequency range that can be correctly analysed with the DFT. It is limited on one side by the Nyquist–Shannon sampling theorem (see Section 3.3.3.1) and on the other side by the hardware limitations of the platform where the DFT has to be implemented. In particular, in the real world, F_s is limited by the maximum and possible sampling rates of the adopted A/D conversion technology and by the processing and data storage capabilities of the processing unit (the higher the sampling rate the higher the amount of samples to be processed in real time by the DFT).

3.3.4.2 Window length

The window length is the most critical parameter when analysing the frequency content of a finite sequence of samples through the DFT. Its selection depends on the specific field of application and it is typically chosen as a trade-off between the required frequency resolution and the desired bandwidth.

According to what stated in the previous sections, the window length T defines the DFT frequency resolution $\Delta f = 1/T$ that determines the uncertainty to correctly identify the position of a tone and, therefore, the uncertainty in identifying its parameters. In Section 3.3.3.2, we have also shown how the window length T influences the frequency separation between the zero-crossings of the sinc function. This, in turn, determines the main lobe width, namely the DFT capability to detect nearby tones, and the side-lobes decaying rate, namely the amount of spectral interference produced by each tone of the spectrum. In general, the longer the window length T, the higher the possibility to detect nearby tones and the lower the effects of the spectral interference produced by near-by tones (see Figure 3.7). Nevertheless, the higher the window length the more probable the possibility that the waveform parameters within the window length are not constant at that therefore the DFT assumptions are not respected, introducing therefore an inherent error in the DFT spectrum.

In this respect, it is worth pointing out that the window length T is a crucial parameter when applying the DFT to SE. As defined in Reference 1 (see also Section 3.5.1.3), the RTs and measurement reporting latencies are two very important characteristics of a PMU that are mainly influenced by the adopted window length. In general, the lower the window length, the lower the RT and measurement reporting latencies and therefore the higher the possibility to use the PMU for specific applications that require fast response and reduced latencies (e.g., power system protections).

Last but not least, it should be observed that the window length T determines, together with the sampling rate F_s, the number of samples $N = T/F_s$ to be processed by the DFT. Even though this is usually forgotten, the number of samples is an important parameter when processing waveforms corrupted by white noise. As shown

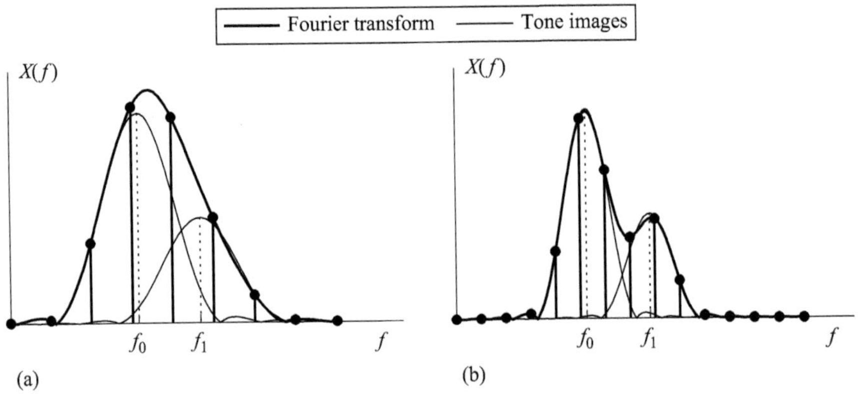

Figure 3.7 Effects of the window length on the detectability of a nearby tone characterized by a frequency f_1 and half of the amplitude of the main tone at f_0. In panel (a), a shorter window length does not allow to detect the tone at f_1 whereas this is possible in panel (b) where a longer window length is adopted

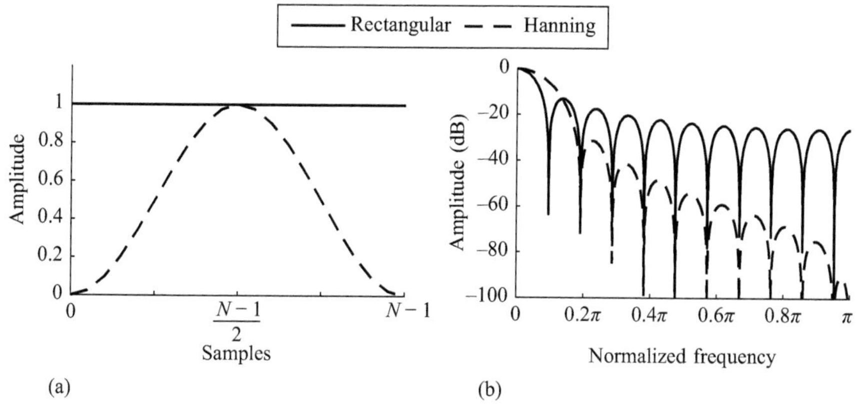

Figure 3.8 Comparison between the time and frequency profiles of the rectangular (continuous line) and the Hanning (dashed line) window

in Reference 31, the effects of white noise on the DFT uncertainty can be effectively reduced by increasing the number of samples to be processed.

3.3.4.3 Window profile

In Section 3.3.3.2, it has been shown that spectral leakage arises when the input signal is not periodic within the window length. This has been demonstrated in the case of a rectangular window function but the proposed analysis could be extended to any kind of windows. Nevertheless, the spectral leakage effects might be reduced by adopting specific windowing functions.

As shown in Reference 29, the characteristics of a window function can be expressed in terms of several metrics. For our purposes we can restrict the analysis to:

- the *main lobe width*, namely, the width of the highest lobe of the adopted window (see Figure 3.6 in the case of a rectangular window), that is typically measured by evaluating the −3 dB bandwidth of the Fourier transform of the adopted window;
- the *side-lobe levels* and *decaying rate*, namely, the relative height of the second, third, etc. lobes with respect to the height of the main one and their characteristic decaying rate.

Whereas both parameters impact the tone detectability, namely the capability to detect two nearby tones through DFT, the latter is also responsible of the aforementioned spectral interference, namely the detrimental effect that neighbouring tones produce on the main tone under analysis. This effect is quite noticeable when using a rectangular window (see Figure 3.8) that, besides being the windowing function with the minimum main lobe width, it is characterized by the highest side lobes (see Reference 29 for a detailed analysis of the window parameters). To improve this characteristic, Harris in Reference 29 has first derived a set of bell-shaped windowing functions that, by reducing the discontinuities at the edge of a window, reduce the side-lobe levels. Unfortunately, this comes at the price of enlarging the main lobe width and, therefore, reducing the tone detectability (see Figure 3.8).

3.3.5 DFT calculation in real time

If not properly designed and implemented, the DFT calculation might represent a considerable bottleneck when running a DFT-based SE algorithm in a real-time environment. As discussed in Section 3.5.1.3, it also affect both measurement reporting latencies and achievable reporting rates. Indeed, particularly when the DFT is calculated over a discrete sequence of samples obtained with a sampling rate of some tens of kilohertz, the high number of samples within a single observation window makes the DFT calculation according to (3.18) computationally intensive. This, combined with the maximum reporting rate and measurement latency requirements defined in Reference 1, makes the development of a PMU based on the DFT, a quite challenging task.

In this respect, in order to improve both latencies and throughput, several efficient techniques to compute the DFT spectrum have been proposed in the literature. They can be separated into two main categories: *recursive* and *non-recursive* algorithms.

Within the group of non-recursive algorithms the well-known *fast Fourier transform* (FFT) algorithm (e.g., Reference 42) is widely used. Typically, this implementation is adopted to perform harmonic analysis over an extended portion of the spectrum even though its deployment on embedded system is usually onerous. When, on the other hand, only a subset of the overall DFT spectrum is used to estimate the synchrophasor (see for instance Reference 24), the so-called *short-time Fourier transform* (STFT) turns out to be very effective [6]. In both cases, the measurement reporting latencies are proportional to the amount of samples to be processed. As a consequence, the algorithm throughput can only be improved at the cost of deteriorating the PMU accuracy levels. The first option is to reduce the sampling rate and eventually originate aliasing (see Section 3.3.3.1); the other one refers to the adoption of shorter window lengths and potentially increase the spectral leakage effects (see Section 3.3.3.2).

In order to increase the throughput without decreasing the precision of the adopted DFT-based SE algorithm, DFT can be calculated via recursive algorithms that are usually characterized by a lower number of operations to update the values of a single DFT bin (e.g., Reference 31). Despite this evident advantage with respect to the class of non-recursive DFT algorithms, the two categories do not generally have identical performances. In particular, the majority of the recursive algorithms suffers of errors due to either the approximations made to perform the recursive update, or the accumulation of the quantization errors produced by the finite word length of computers [43].

A very effective method for sample-by-sample DFT bins computation is represented by the so-called sliding-DFT (SDFT) technique presented in Reference 44. This reference demonstrates the efficiency of this method in comparison with the popular Goertzel algorithm and its computational advantages over the more traditional DFT and FFT, but also its drawbacks. Unfortunately, the approach proposed in Reference 44 is only marginally stable. In particular, if the truncation errors on the computation of the filter coefficients are not severe, the SDFT is bounded-input, bounded-output stable. Otherwise, the algorithm suffers from accumulated errors

Table 3.1 Characteristic computational complexity and numerical stability of the STFT, SDFT (stable and unstable version) and MSDFT methods (M is the number of samples within a window of length T)

Method	**Computational workload**		**Numerical stability**
	ADD	**MUL**	
Short-time Fourier transform (STFT)	M	M	✓
Sliding DFT (SDFT)	4	4	✗
Guaranteed-stable SDFT (rSDFT)	5	4	✓
Modulated sliding DFT (MSDFT)	10	7	✓

and is, consequently, potentially unstable. Whereas common approaches found in the literature [44, 45] face this problem compromising results accuracy for guaranteed stability, the method proposed in Reference 46 and called modulated sliding DFT (MSDFT) is guaranteed stable without sacrificing accuracy.

In what follows, three of the most efficient techniques to compute a portion of the DFT spectrum, namely, the STFT, SDFT and MSDFT, will be presented and analysed with respect to their precision and computational complexity (see also Table 3.1).

3.3.5.1 Short-time Fourier transform

Starting from what formulated in (3.18), the DFT can be potentially updated every time-step n, based on the most recent set of samples $\{x(n-N+1), x(n-N+2), \ldots, x(n)\}$, according to the following time-dependent equation:

$$X_k(n) = \sum_{m=0}^{N-1} x(q+m) \cdot W_N^{-km} \tag{3.38}$$

being N the number of samples within the window of length T, n the time-step index, k the DFT-bin index, $q = n - N + 1$, and $W_N^{-km} = e^{-j2\pi km/N}$ the DFT complex twiddle factor.

The derived equation is the so-called STFT, namely the simplest technique to apply the DFT theory to a real signal. It assumes to split the signal into partially overlapping block of samples of equal length N and apply the DFT computation independently to each one of them. Consequently, such a calculation is extremely inefficient: although two consecutive estimations are derived from a partially overlapping windows, in order to update the DFT estimation, it assumes to re-process already analysed portions of the signal. Nevertheless, such a technique can still be applied to SE, as long as the PMU does not need to report estimated data too frequently and the adopted hardware platform has enough available computational resources to host a parallel computation of the DFT, according to (3.38), over the whole set of input channels.

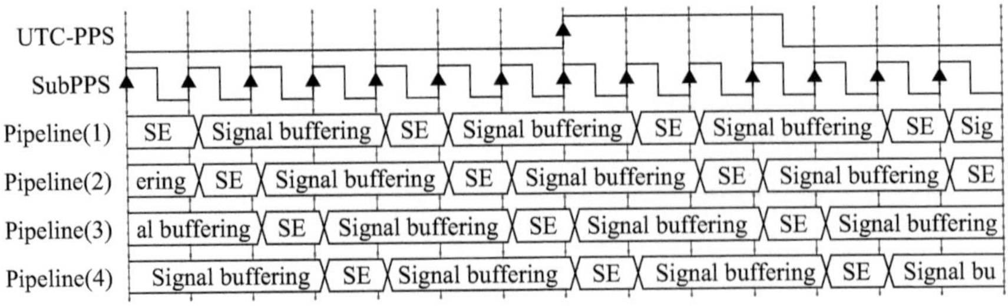

Figure 3.9 Digital timing diagram of a hypothetical PMU that estimates synchrophasors 50 times per second using the STFT technique. A four-stage pipeline is here adopted to achieve the desired reporting rate with a three-cycle window to calculate the DFT and estimate the synchrophasor (SE stands for synchrophasor estimation)

For instance, in order to achieve the highest reporting rates required by the IEEE Std. [1] (i.e., a new estimation every nominal period of the power system) with a three-cycle DFT-based SE algorithm, one solution would be to apply the STFT calculation expressed by (3.38) to partially overlapped portions of data. As shown in Figure 3.9, a four-stage pipeline architecture for each input channel can be adopted: based on the rising edges of a square waveform aligned to PPS and characterized by a frequency corresponding to the PMU reporting rate, each pipeline will alternately collects the required amount of data (N samples) in dedicated memories and, once they have been filled, activates a flag that triggers the SE on the previously acquired set of data.

3.3.5.2 Sliding DFT

The SDFT structure is depicted in Figure 3.10(a) and, as demonstrated in Reference 44, it can be derived from (3.18) as follows:

$$X_k(n) = \sum_{m=0}^{N-1} x(q+m) \cdot W_N^{-km} \tag{3.39}$$

$$= \sum_{m=0}^{N-1} x(q+m-1) \cdot W_N^{-k(m-1)} - x(q-1) \cdot W_N^k + x(q+N-1) \cdot W_N^{-k(N-1)} \tag{3.40}$$

$$= W_N^k \cdot \sum_{m=0}^{N-1} x(q+m-1) \cdot W_N^{-km} - x(q-1) \cdot W_N^k + x(q+N-1) \cdot W_N^{-k(N-1)} \tag{3.41}$$

$$= W_N^k \cdot (X_k(n-1) - x(q-1) + x(q+N-1)) \tag{3.42}$$

$$= W_N^k \cdot (X_k(n-1) - x(n-N) + x(n)) \tag{3.43}$$

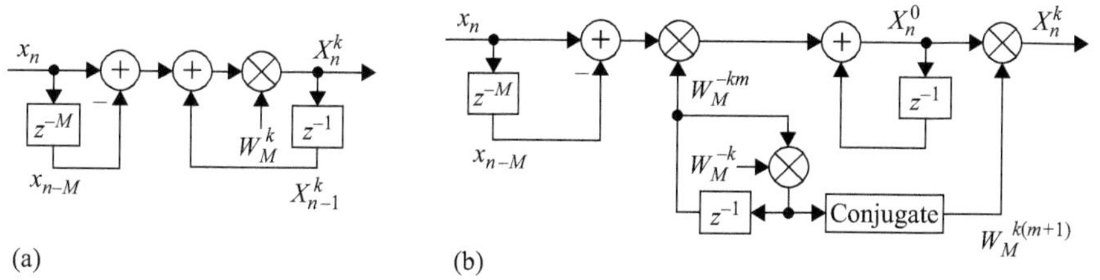

Figure 3.10 Structure of the sliding DFT (a) and of the modulated sliding DFT (b) (adapted from Reference 46)

As it can be noticed by looking at (3.43), the SDFT is a quite efficient method as, it allows to update the values of a single DFT bin every time a new sample is acquired with few multiplication and additions (see also Table 3.1).

As previously mentioned, the SDFT filter is not numerically stable and suffers from accumulated errors. Common approaches that can be found in the literature [44, 45] face this problem compromising results accuracy for guaranteed stability. This is the case of the method presented in Reference 44 where the stability is guaranteed by including in the DFT formula a damping factor r that force the SDFT pole to reside within the z-domain's unit circle (this method is here referenced as rSDFT):

$$X_k(n) = rW_N^k \cdot (X_k(n-1) - r^N x(n-N) + x(n)) \tag{3.44}$$

In this context, the next section will examine a sample-by-sample DFT update method, called MSDFT that is guaranteed stable without sacrificing accuracy [46].

3.3.5.3 Modulated sliding DFT

With reference to (3.43), it is easy to observe that the recursive formula for the computation of X_k when $k = 0$ does not involve the complex twiddle factor and is, therefore, by definition stable:

$$X_0(n) = X_0(n-1) - x(n-m) + x(n) \tag{3.45}$$

The MSDFT takes advantage of this SDFT property in order to derive a recursive formula for the DFT computation that is intrinsically stable.

In particular, by taking advantage of the so-called Fourier modulation property [6], the generic kth DFT bin can be shifted to the position $k = 0$ multiplying the input signal by the complex twiddle factor W_N^{-km}:

$$X_k(n) \Rightarrow W_N^{-km} X_0(n) = X_0(n-1) - x(n-N) \cdot W_N^{-k(m-N)} + x(n) \cdot W_N^{-km} \tag{3.46}$$

$$= X_0(n-1) + W_N^{-km} \cdot (-x(n-N) + x(n)) \tag{3.47}$$

where (3.47) is obtained thanks to the periodicity of the modulating sequence W_N^{-km}.

The twiddle factor modulation only introduces a phase shift that is changing with index m: $\angle W_N^{-km} = 0$ for $m = 0$, it increases by the W_N^{-k} factor at each iteration and is periodically reset to 0 every N samples. Indeed, at every iteration the modulating sequence can be recursively computed as:

$$W_N^{-km} = W_N^{-k(m-1)} \cdot W_N^{-k}, \qquad m = 0, 1, \ldots, N-1 \tag{3.48}$$

It is clear that, in order to prevent that accumulated errors corrupt our estimation, the modulating sequence must be reset to 1 every N samples.¶ In view of this, the kth bin of the DFT can be derived from (3.47), as:

$$X_k(n) = W_N^{k(m+1)} \cdot X_0(n) \tag{3.49}$$

$$= W_N^{k(m+1)} \cdot (X_0(n-1) + W_N^{-km} \cdot (-x(n-N) + x(n))) \tag{3.50}$$

where $W_N^{-k(m+1)}$ compensate for the phase-shift due to the modulating sequence.

Equation (3.50) defines the MSDFT method for the update of the value of a single bin of the entire DFT spectrum and the related block scheme is given in Figure 3.10(b).

3.3.5.4 Integrating the MSDFT with signal windowing

As already discussed in Section 3.3.4.3, signal windowing is a powerful technique that allows to reduce the effects of long-range spectral leakage. Windowing is applied as in the time domain by weighting a finite sequence of samples with a particular window profile like the Hanning one (see Section 3.3.4.3):

$$x_w(n) = x(n) \cdot w(n), \quad 0 \leq n \leq N-1 \tag{3.51}$$

However, windowing by time-domain multiplication would compromise the computational simplicity of the MSDFT or any other sample-by-sample DFT calculation technique. For this reason, when adopting this kind of methods, it is of common use to apply the signal windowing in the frequency domain, namely after the DFT has been computed. Indeed, by recalling that the multiplication between two functions in the time domain corresponds to the convolution between the Fourier transform of the two functions in the frequency domain, the time-domain multiplication could be replaced by a frequency-domain convolution and obtain equivalent results.

In particular, in the case of the Hanning window, this will result into the following linear combination of adjacent DFT bins $X_k(n)$:

$$X_k(n) = -0.25 \cdot X_{k-1}(n) + 0.5 \cdot X_k(n) - 0.25 \cdot X_{k+1}(n) \tag{3.52}$$

From (3.52), it is clear that, in order to compute three windowed DFT bins, we need to compute five MSDFT bins, namely, those associated to indices $k_m + \{-2, -1, 0, 1, 2\}$.

3.4 DFT-based SE algorithms

The SE algorithm is definitely the most relevant and challenging component of a PMU. The main task of an SE algorithm is to identify and assess the parameters of the fundamental tone of a signal by using a previously acquired set of samples representing

¶To be noticed that, for practical implementation, the modulating sequence can be either (i) precomputed and stored into memory or (ii) computed online based on (3.48).

a portion of an acquired waveform (i.e., node voltage and/or branch/nodal current). As a consequence, it mainly influences not only the PMU measurement uncertainty, but also the PMU measurement reporting latency and the maximum achievable reporting rate.

This section first reviews the literature on SE algorithms based on the DFT. This category is characterized by some inherent advantages when applied to the extraction of the main-tone parameter, namely: (i) the inherent DFT capability to isolate and identify the main tone of a discrete sinusoidal signal and (ii) reject close-by harmonics; (iii) the relatively low computational complexity, particularly when the DFT spectrum is computed through one of the well-known algorithms (e.g., FFT or sliding DFT). Nevertheless, these qualities come with non-negligible drawbacks and limitations characterizing the DFT theory that assumes a periodic signal with time-invariant parameters, at least along the observation window. The latter, on the one hand, should be as short as possible to be closer to the above-mentioned quasi-steady-state hypothesis also during power system transient; on the other hand, longer windows are needed when interested in identifying close-by tones and rejecting harmonic and inter-harmonic components. Indeed, as demonstrated in Section 3.3, the observation window T is inversely proportional to the DFT frequency resolution, as it defines both the frequency separation between consecutive DFT bins and the main lobe width.

In this respect, Section 3.4.1 presents the IpDFT, a well-established technique to estimate the parameters of a waveform out of a finite sequence of samples, which reduces the effects of spectral leakage and overcome the limitations introduced by adopting a relatively short window length. Finally, on the basis of Reference 47, Section 3.4.2 illustrates a computationally affordable method that is capable of improving the performances of standard IpDFT methods and to keep the same static and dynamic performances independently of the adopted window length.

3.4.1 The Interpolated-DFT technique

As reported in Section 3.3.3, DFT-based SE algorithms are characterized by three main sources of uncertainty: aliasing, spectral leakage and spectral sampling. Whereas the effects of aliasing can be disregarded by simply increasing the sampling frequency to values much larger than the highest spectrum component contained in the sampled signal, the combined effects of spectral leakage and spectral sampling can be detrimental if not properly treated.

As discussed in Section 3.3.3.2, spectral leakage arises when the sampling process is not synchronized with the fundamental tone of the signal under analysis and the DFT is computed over a non-integer number of cycles of the input signal [30]. Since the precise synchronization of the sampling process with the fundamental frequency component of the signal is purely theoretical (indeed such synchronization involves the a priori knowledge of the signal main tone, which is, by hypothesis, unknown), several approaches have been proposed to reduce this bias.

Among them the IpDFT technique has outperformed the others for its higher accuracies combined with a lower computational complexity. Such a method refers to the usage of:

- windowing functions aiming at mitigating the effect of long-range spectral leakage [29, 48];
- proper DFT interpolation schemes aiming at correcting the effects of the short-range leakage and reducing the inaccuracies introduced by the DFT spectral sampling (e.g., References 23,24).

The IpDFT problem has been originally defined for a discrete sequence of samples windowed using the rectangular window [23]. In order to reduce the effects of long-range spectral leakage, the input sequence can be windowed using one of the "special" windowing functions defined in the literature (see References 29,48). The first to combine such an approach with the IpDFT technique was Grandke in Reference 24 using the Hanning window. More recently, the IpDFT problem has been formulated using various windowing functions belonging to the Rife–Vincent class I (RVCI) [49] or parametric windowing function non-belonging to the cosine windows class, like Kaiser–Bessel or Dolph–Chebyshev windows [50].

In what follows, the IpDFT algorithm will be first formulated and solved for the case of Hanning window; then its performances are analysed, with a specific focus on the effect of the spectral interference on the accuracy of the IpDFT method.

3.4.1.1 Formulation of the IpDFT problem

The IpDFT is a technique that allows to estimate the parameters of a tone (i.e., its frequency, amplitude, and phase) by interpolating the DFT spectrum obtained from a finite sequence of N samples of a discrete signal $x(n)$ that includes the tone under analysis.

In this respect, let us consider the following finite sequence obtained by sampling with a sampling rate $F_s = 1/T_s$ a continuous waveform $x(t)$ characterized by a single frequency component at frequency f_0:

$$x(n) = A\cos(2\pi f_0 n T_s + \varphi), \quad 0 \le n \le N-1 \tag{3.53}$$

Let us assume to window the sequence $x(n)$ with a known windowing function $w(n)$ and compute the DFT of the obtained sequence using (3.18).

As demonstrated in Section 3.3.3.2, if the window does not contain an integer number of periods of the signal $x(n)$ leakage occurs and the main tone of the signal will be located between two consecutive DFT bins (see Figure 3.11). Its frequency can therefore be expressed as follows:

$$f_0 = (k_m + \delta)\Delta f \tag{3.54}$$

being k_m the index of the DFT bin characterized by the highest amplitude and $-0.5 \le \delta < 0.5$ a fractional correction term.

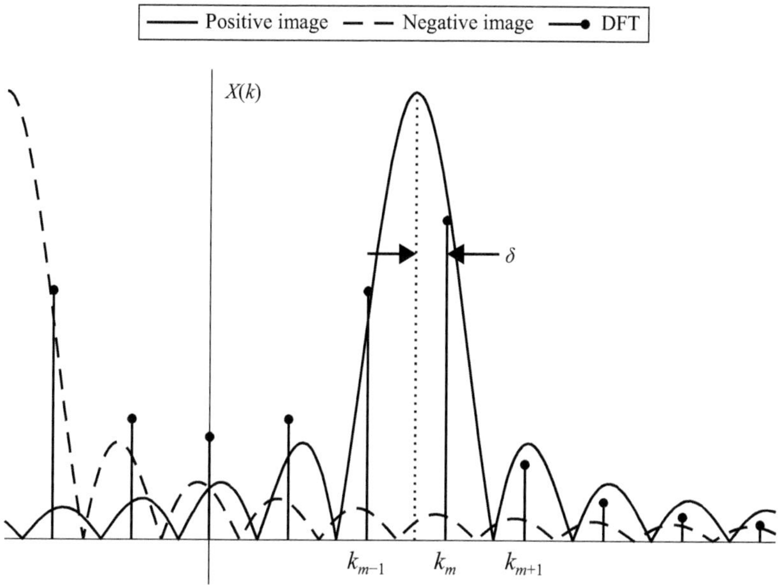

Figure 3.11 *Zoom on the portion of the DFT spectrum surrounding the bin $X(k_m)$ that highlights the correction term δ and the spectral contributions produced by the positive (continuous line) and negative (dashed line) images of the spectrum*

From (3.54), the IpDFT problem can be formulated as follows:

Based on the DFT spectrum $X(k)$ of the signal $x(n)$ analysed with the known windowing function $w(n)$, find the correction term δ that better approximates the exact location of the main spectrum tone.

According to the adopted window profile and number of DFT bins used to perform the interpolation, various analytical or approximated solutions to the problem can be given [49]. In what follows, the solution will be presented for a signal windowed using a Hanning window (see (3.55)–(3.57)) and using a two-point interpolation.

3.4.1.2 Solution of the IpDFT problem using the Hanning window

The Hanning window (also known as Hann window) is defined as:

$$w_H(n) = \frac{1 - \cos\left(\frac{2\pi n}{N}\right)}{2}, \quad n \in [0, N-1] \tag{3.55}$$

and its Fourier transform is

$$W_H(\omega) = -0.25 \cdot D_N\left(\omega - \frac{2\pi}{N}\right) + 0.5 \cdot D_N(\omega) - 0.25 \cdot D_N\left(\omega + \frac{2\pi}{N}\right) \tag{3.56}$$

being $D_N(\omega)$ the *Dirichlet kernel* defined as:

$$D_N(\omega) = e^{-j\omega(N-1)/2} \frac{\sin\left(\frac{\omega N}{2}\right)}{\sin\left(\frac{\omega}{2}\right)} \tag{3.57}$$

also known as the generalized Fourier transform of the rectangular window.

As demonstrated in Section 3.3, the spectrum of the sampled signal $x(n)$ in (3.53) can be expressed in terms of its positive and negative images:

$$X(f) = X^+(f) + X^-(f) \tag{3.58}$$

$$= \frac{A}{2} e^{j\psi} W_H(f - f_0) + \frac{A}{2} e^{-j\psi} W_H(f + f_0) \tag{3.59}$$

being $W_H(f)$ the Fourier transform of the Hanning window, A and ψ the amplitude and instantaneous phase of the signal $x(t)$, respectively.

Assuming that the effects of leakage are properly compensated by windowing, we can reasonably neglect the long-range spectral leakage produced by the negative spectrum image on the positive frequency range and assume that the DFT bins in the positive frequency range are only generated from the positive image of the spectrum, namely:

$$X(k) \approx X^+(k), \quad 0 \leq k \leq \frac{N}{2} \tag{3.60}$$

The fractional term δ can be estimated starting from the ratio between the two highest bins of the DFT $X(k_m)$ and $X(k_m + \varepsilon)$ that, for $N \gg 0$, can be approximated as follows [49]:

$$\frac{X(k_m + \varepsilon)}{X(k_m)} \approx \frac{W_H\left((\varepsilon - \delta) \cdot \frac{2\pi}{N}\right)}{W_H\left(-\delta \cdot \frac{2\pi}{N}\right)} \tag{3.61}$$

where $W_H(\cdot)$ is the Fourier transform of the Hanning window (see (3.56)) and

$$\varepsilon = \begin{cases} 1 & \text{if } |X(k_m + 1)| > |X(k_m - 1)| \\ -1 & \text{if } |X(k_m + 1)| < |X(k_m - 1)| \end{cases} \tag{3.62}$$

If $N \gg 0$, the following approximation holds:

$$e^{\pm j\pi(N-1/N)} \approx -1 \pm \frac{j\pi}{N} \tag{3.63}$$

and the absolute value of $W_H(\omega)$ can be approximated as:

$$|W_H(\omega)| \approx \left|\sin\left(\frac{\omega N}{2}\right)\right| \cdot \left| -\frac{0.25}{\sin\left(\frac{\omega}{2} - \frac{\pi}{N}\right)} + \frac{0.5}{\sin\left(\frac{\omega}{2}\right)} - \frac{0.25}{\sin\left(\frac{\omega}{2} + \frac{\pi}{N}\right)} \right| \tag{3.64}$$

By replacing (3.64) in (3.61) and recalling that $\lim_{x \to 0} \sin(x) = x$, we get:

$$\begin{aligned}
\frac{X(k_m+\varepsilon)}{X(k_m)} &= \frac{\left|W_H\left((\varepsilon-\delta)\cdot\frac{2\pi}{N}\right)\right|}{\left|W_H\left(-\delta\cdot\frac{2\pi}{N}\right)\right|} \\
&\approx \left|\frac{0.25}{\delta-\varepsilon+1}-\frac{0.5}{\delta-\varepsilon}+\frac{0.25}{\delta-\varepsilon-1}\right| \Big/ \left|\frac{0.25}{\delta+1}-\frac{0.5}{\delta}+\frac{0.25}{\delta-1}\right| \\
&= \left|\frac{0.5}{\delta(\delta-\varepsilon)(\delta-2\varepsilon)}\right|\left|\frac{\delta(\delta+1)(\delta-1)}{-0.5}\right| \\
&= \left|\frac{\delta+\varepsilon}{\delta-2\varepsilon}\right| \qquad (3.65)
\end{aligned}$$

that, solved for the frequency correction δ gives:

$$\widehat{\delta} = \varepsilon\,\frac{2\,|X(k_m+\varepsilon)| - |X(k_m)|}{|X(k_m)| + |X(k_m+\varepsilon)|} \qquad (3.66)$$

where the $\widehat{\ }$ symbol was used to indicate the estimation of the fractional correction term δ. The waveforms parameters (i.e., its frequency, amplitude and phase) can then be computed as follows:

$$\widehat{f} = (k_m + \widehat{\delta})\Delta f \qquad (3.67)$$

$$\widehat{A} = |X(k_m)|\left|\frac{\pi\widehat{\delta}}{\sin(\pi\widehat{\delta})}\right|\left|\widehat{\delta}^2 - 1\right| \qquad (3.68)$$

$$\widehat{\varphi} = \angle X(k_m) - \pi\widehat{\delta} \qquad (3.69)$$

3.4.1.3 Optimal selection of the IpDFT parameters for SE

The relevant parameters of the IpDFT technique should be properly chosen according to the characteristics of the input sequence $x(n)$ (i.e., the DFT spectrum $X(k)$).

In this respect, this section presents the optimal selection of the three main IpDFT parameters (i.e., the *sampling rate*, the *window length*, and the *window profile*) when applying such a technique to SE and argument their choice by making reference to the signal model presented in Section 3.2.1 (see (3.2)). In particular, in the following of this chapter, both the SE algorithm formulation and the discussion of the experimental results will make reference to a DFT calculated on a sequence of samples obtained by sampling the input signal $x(t)$ with a sampling rate F_s of 50 kHz over a window length T containing three periods of a waveform at the nominal frequency of a power system f_0 and windowed using a Hanning window.

Sampling rate

The IpDFT sampling rate F_s must be primarily selected following the considerations presented in Section 3.3.4.1 regarding the detrimental effects of aliasing. In this respect, it is worth pointing out that in the field of power system, harmonic and inter-harmonic components rarely exceed few kilohertz and consequently a PMU that does not adopt any anti-aliasing filter must sample the signal with a sampling rate F_s of

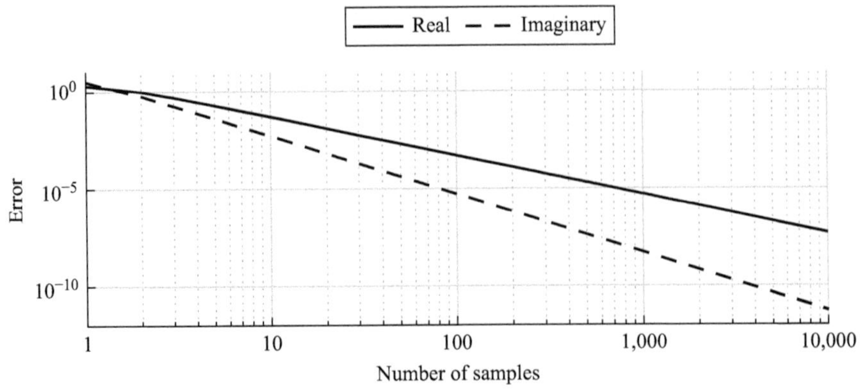

Figure 3.12 Behaviour of the real (continuous line) and imaginary parts (dashed line) of $e^{\pm j\pi(N-1/N)} - (-1 \pm j\pi/N)$ *(see 3.63)) as a function of the number of samples N*

at least 10 kHz. Additionally, a relatively high sampling rate allows to process a higher number of samples and consequently mitigate the effects of wideband noise.

Last but not least, it should not be forgotten that, the sampling rate F_s together with the window length T, defines the number of samples N to be processed by DFT. In this respect, it is worth pointing out that the IpDFT solution given in (3.66) for the case of the Hanning window (the same hypothesis is actually valid for any IpDFT estimator based on the class of RVCI windows) can only be obtained if the number of samples N is sufficiently large so that the approximation presented in (3.63) is valid. Figure 3.12 shows the residuals of (3.63). As it can be seen, by adopting a relatively high sampling rate F_s, the number of samples N can be increased up to values where the effects of such an approximation are not visible and it can be considered exact.

Window length

As reported in Section 3.3.4.2, the window length T is the most critical DFT parameter, and it must be selected first of all to guarantee the stationarity of the sequence of samples $x(n)$. Additionally, as reported in Section 3.5.1.3, window lengths directly affect the PMU RTs and measurement reporting latencies. In this respect, a shorter window length including, at most, four periods of a signal at the nominal frequency of the power system f_0, more easily ensures the stationarity hypothesis in most of the PMU operating conditions. At the same time, it guarantees to satisfy the most stringent PMU RT and measurement reporting latency requirements dictated by Reference 1, particularly with respect to the P-class measurement requirements (see also Section 3.5.1.2).

Nevertheless, this choice comes with at the price of a reduced DFT frequency resolution that, depending on the adopted window profile, might not allow to detect nearby tones like inter- or sub-harmonics. Furthermore, as it will be discussed in the next section, shorter window lengths amplify the effect of long-range spectral leakage and therefore the spectral interference between close-by tones.

Window profile

The solution of the IpDFT problem (Section 3.4.1.2) has been presented when using a Hanning window in order to reduce the effects of long-range spectral leakage. This windowing function has demonstrated to represent a good trade-off between the long-range spectral leakage suppression (i.e., the side-lobe levels) and the DFT frequency resolution (i.e., the main lobe width). In particular, compared to the rectangular window, the Hanning window is characterized by a wider (almost double) main lobe but a much higher side-lobes decaying rate (see Figure 3.8).

Nevertheless, a similar solution could be analytically derived for any other function belonging to the category of RVCI windows or based on a polynomial approximation for the class of parametric non-cosine windows (see Reference 49 for the IpDFT solution for both category of windows).

In general, the window profile must be chosen according to the field of application of the IpDFT technique. Whether the interest is reducing the effect of long-range spectral leakage produced by interfering tones or it is identifying and estimating the parameters of nearby frequency components, a window profile characterized by lower side lobes or a narrower main lobe must be preferred, respectively.

3.4.1.4 IpDFT sensitivity to spectral interference

The IpDFT is a powerful method to artificially increase the DFT frequency resolution, and accurately estimate the parameters of a sinusoidal waveform. Nevertheless, its accuracy is mainly limited by the *spectral interference* produced by nearby tones that might not allow to distinguish the portion of spectrum that has been generated by the tone under analysis from other frequency components.

Indeed, as reported in Sections 3.4.1.1 and 3.4.1.2, the main hypothesis behind the analytical solution of the IpDFT problem is that the DFT bins used to perform the interpolation are only generated by the tone under analysis. As a consequence, the IpDFT theory assumes that, independently of the adopted window profile, the DFT spectrum only contains a single component characterized by a frequency $f_0 \gg \Delta f$, so that the positive and negative images of the spectrum are sufficiently distant and the approximation expressed by (3.60) is satisfied. In case the input signal includes more frequency components (e.g., harmonics and/or inter- and sub-harmonics), they should be sufficiently separate so that the effects of long-range spectral leakage can be neglected. If these assumptions are not satisfied, the DFT bins used to perform the interpolation, and therefore the IpDFT estimations, might be partially biased by the tails of the nearby tones. Eventually, if the energy content of the spectral interference exceed the one of the tone under analysis, the tone detectability is not even guaranteed.

As reported in the previous section, the application of the IpDFT technique to the field of SE requires the adoption of quite short windows, which contain few periods of the signal under analysis, in order to assume the stationarity of the signal and, at the same time, reduce the PMU RTs and measurement reporting latencies. This choice causes the energy of the DFT spectrum to be concentrated in the lower frequency range and the positive and negative images of the spectrum to be relatively close. Independently of the adopted window profile, the side lobes of the negative image of the spectrum leak in the positive frequency range and bias the DFT bins used to perform the interpolation that are not only originated by the positive image of $X(k)$ but also influenced by the tails produced by the negative image of the spectrum (see Figure 3.11). The assumption expressed by (3.60) cannot be satisfied and the IpDFT estimations are consequently degraded.

This phenomenon can be slightly reduced by adopting windowing functions with good side-lobe behaviours [48], but most of the time, this is not sufficient to achieve the higher accuracies required by certain synchrophasor-based applications. In this respect, the following of this chapter will focus on the formulation of an SE algorithm that enhances the IpDFT performances by combining such a technique with an iterative approach for the compensation of the spectral interference produced by the negative image of the main tone of the spectrum.

3.4.2 The iterative-Interpolated DFT technique

The performances of any IpDFT method are definitely related to the accuracy in the estimation of the fractional term δ that, ideally, could be improved by processing a DFT spectrum that only contains the positive image of the tone under analysis. In this respect, in what follows, an iterative technique for the compensation of the spectral interference produced by the negative image of the main tone of the spectrum is proposed and combined with the two-point IpDFT estimator presented in Section 3.4.1.2, to derive a novel SE algorithm that hereafter will be called i-IpDFT.

3.4.2.1 Iterative compensation of the spectral interference

By neglecting the spectral interference produced by other tones, the DFT spectrum of the signal $x(n)$ in (3.53) can be expressed, as shown in (3.58), in function of the contribution of the positive and negative images of the main tone. As a consequence, the highest and second-highest DFT bins, which are used to estimate δ according to (3.66), can be expressed as:

$$X(k_m) = \frac{1}{B}\left[\frac{A}{2}e^{j\psi}\cdot W(-\delta) + \frac{A}{2}e^{-j\psi}\cdot W(2k_m+\delta)\right] \tag{3.70}$$

$$X(k_m+\varepsilon) = \frac{1}{B}\left[\frac{A}{2}e^{j\psi}\cdot W(\varepsilon-\delta) + \frac{A}{2}e^{-j\psi}\cdot W(2k_m+\varepsilon+\delta)\right] \tag{3.71}$$

where $W(\cdot)$ is the Fourier transform of the adopted windowing function and the spectral interference coming from the negative spectrum image is represented by the following terms:

$$X^-(k_m) \triangleq \frac{1}{B}\left[\frac{A}{2}e^{-j\psi}\cdot W(2k_m+\delta)\right] \tag{3.72}$$

$$X^-(k_m+\varepsilon) \triangleq \frac{1}{B}\left[\frac{A}{2}e^{-j\psi}\cdot W(2k_m+\varepsilon+\delta)\right] \tag{3.73}$$

Since $W(\cdot)$ is analytically known once the windowing function has been selected, the amount of spectral interference generated by the negative image of the spectrum on the above DFT bins can be estimated. In particular, (3.72) and (3.73) can be evaluated using an initial estimation of the waveform parameters $\widehat{A}$ and $\widehat{\psi}$ obtained using the classical IpDFT technique presented in Section 3.4.1.2. These estimations can then be subtracted from the DFT bins to reduce the spectral interference so that the "compensated" DFT bins

$$\widehat{X}(k_m) = X(k_m) - X^{-}(k_m) \tag{3.74}$$

$$\widehat{X}(k_m + \varepsilon) = X(k_m + \varepsilon) - X^{-}(k_m + \varepsilon) \tag{3.75}$$

are mostly generated by the positive image of the spectrum.

Then, the estimation of the fractional term δ and the related waveform parameters can be improved by re-using the IpDFT algorithm on the new set of DFT bins. This process can be either iterated a predefined number of times or performed until a given convergence criterion is achieved and bring to a more accurate estimation of the set of parameters $\{\widehat{f}, \widehat{A}, \widehat{\psi}\}$.

3.4.2.2 Formulation of the i-IpDFT method

The proposed i-IpDFT SE algorithm is presented, on the basis of Reference 47, in form of a pseudo-code (see next page) that explains the fundamental steps necessary to correctly estimate the synchrophasor according to the i-IpDFT technique.

First, the continuous input waveform (voltage or current) is sampled with a sampling rate F_s that is sufficiently high to neglect the effects of aliasing on the DFT spectrum. In this respect, in Section 3.4.1.3, we have adopted a sampling rate F_s of 50 kHz. The samples are then collected into a finite sequence of length N that should be sufficiently short so that the signal can be assumed stationary at least within its boundaries. In this respect, a window containing three periods of a signal at the nominal frequency f_0 (namely, $T = NT_s = 3/f_0$) has proven to be a good trade-off between the accuracy and RT requirements (see Section 3.4.1.3). The finite sequence of samples is then windowed using a Hanning window and the three highest bins of the DFT, namely those corresponding to DFT indices $k = k_m + \{-1, 0, +1\}$, computed according to what stated in (3.18).**

Next, a first estimation of the parameters of the main tone can be given according to the classical two-point IpDFT technique based on a Hanning window formulated in Section 3.4.1.2. Although such an estimation could be considerably affected by the effect of the spectral interference produced by the image component, it can be used to approximate the effects of its "tails" on the two highest DFT bins that are used to estimate δ according to what stated in Section 3.4.2.1. Such an amount can be subtracted from the original DFT spectrum to reduce the effect of spectral interference and the waveform parameters estimation refined. As shown in the pseudo-code, this approach is either iterated a predefined amount of time or until the spectral interference compensation does not bring any additional advantage.

As it can be noticed, the i-IpDFT technique has a clear advantage with respect to other SE methods: the algorithm is composed by few well-defined macro-functionalities that are often recalled along the execution of the pseudo-code through

**It is worth pointing out that by adopting a sufficiently short window length T, the index corresponding to the DFT maximum k_m can be fixed a priori and calculated as $k_m = \lceil f_0/\Delta f \rfloor = \lceil f_0 N/F_s \rfloor$. Consequently, there is no need to perform a maximum search on the DFT bins.

a for loop structure. The i-IpDFT SE algorithm can therefore exploit the speed of high-speed digital architectures like field programmable gate arrays (FPGAs) and reduce the required amount of hardware resources simply reusing already allocated portion of the hardware design.

3.5 Performance analysis of SE algorithm

This section first presents a procedure to assess the performances of SE algorithms that is based on the IEEE Std. C37.118. Next, it verifies the performances of the i-IpDFT SE algorithms by making reference to the most relevant tests defined in Reference 1.

The reader should consider that this section is only meant to quantify the performances of the adopted SE algorithm and does not deal with the other components of a PMU that might introduce a further and non-negligible uncertainty to the estimated synchrophasor (see Section 3.2.5).

3.5.1 The IEEE Std. C37.118

The performances of PMUs are constrained by a single international standard, the IEEE Std. C37.118. This standard derives from a preliminary version of an IEEE Standard for synchrophasors, the IEEE Std. 1344-1995 [34], and has been first issued in 2005 [51], reviewed in 2011 [1, 35], and amended in 2014 [52]. It is composed

---

Algorithm 1: The iterative-IpDFT synchrophasor estimation algorithm.

---

procedure ITERATIVE-INTERPOLATED DFT($x(t)$) ▷ $x(t)$ is the input signal
    Signal sampling: $x(n) = x(nT_s)$ ▷ Sampling rate: $F_s = 1/T_s = 50$ kHz
    Signal buffering: $x(n), \quad n = 0, \ldots, N-1$ ▷ Window length $T = NT_s = 3/f_0$
    Signal windowing (Hanning): $x_h(n) = x(n) \cdot w_H(n)$ ▷ see (3.55);
    DFT calculation: $X(k), \quad k = k_m + \{-1, 0, +1\}$ ▷ see (3.18);
    Two-point DFT interpolation: $\{\widehat{f}, \widehat{A}, \widehat{\psi}\}_0$ ▷ see (3.67)–(3.69);
    for $r = 1 \rightarrow R$ **do**
        Spectral interference estimation: $X^{-}(k)$ ▷ see (3.72) and (3.73) ;
        DFT enhancement: $\widehat{X}(k)$ ▷ see (3.74) and (3.75);
        Two-point DFT interpolation: $\{\widehat{f}, \widehat{A}, \widehat{\psi}\}_r$ ▷ see (3.67)–(3.69);
        if $\{\widehat{f}, \widehat{A}, \widehat{\psi}\}_r \approx \{\widehat{f}, \widehat{A}, \widehat{\psi}\}_{r-1}$ **then**
            break for
        end if
    end for
    return $\{\widehat{f}, \widehat{A}, \widehat{\psi}\}_r$
end procedure

---

by two parts: (i) Part I, the *IEEE Std. for Synchrophasor Measurements for Power Systems*, deals with the synchrophasor terminology and the synchrophasor measurement requirements and compliance verification; (ii) Part II, the *IEEE Standard for Synchrophasor Data Transfer for Power Systems*, mainly defines the synchrophasor message format, message types, and communications.

The PMU compliance to the IEEE Std. is achieved by satisfying both the measurement requirements specified in Reference 1 and amended in Reference 52 and the communication requirements as described in Reference 35. Whereas the latter are straightforward to be accomplished, the former is a more challenging achievement as it involves several engineering competencies necessary to build a compliant PMU prototype. To this purpose, Reference 1 has proposed a metrological procedure aimed at assessing the PMU accuracies during both steady state and dynamic conditions of a power system. This is done by defining specific tests and reference signals, aimed at artificially simulating various operating conditions of the electrical grid, together with the related compliance requirements. The PMU compliance is then assessed by comparing the PMU performances to the IEEE Std. requirements and verifying if they respect the previously defined limits.

In what follows, the PMU compliance verification procedure proposed in Reference 1 is summarized and commented. For further details, the reader should make reference to what originally stated in the IEEE Standard [1], the related IEEE Guide [38] and the latest Test Suite Specification (TSS) [53].

3.5.1.1 Reporting rates and reporting times

Synchrophasor measurements produced by PMUs must be reported regularly at a reporting rate F_r that is an integer number of frames per second [1]. The reporting times shall be evenly spaced through each second, with the time of the first frame within the second coincident with the UTC-second rollover (i.e., with a fractional second of 0).

According to Reference 1, a PMU shall support the reporting rates 10, 25, 50 or 10,12, 15, 20, 30, 60 frames-per-second (fps) depending on whether the power system frequency f_0 is 50 or 60 Hz, respectively. Higher reporting rates, such as 100 or 120 fps, are also encouraged in the Standard to support time-critical applications like power system protection and fault management [54]. Nevertheless, it is worth pointing out that only those PMU characterized by a shorter window length can afford to push their reporting rates up to these values, as the correlation between consecutive measurements can make such an increased information rate useless.

3.5.1.2 Performance classes

PMUs shall be tested against the expected *operating conditions* and the targeted *power system applications*. With respect to the latter, the IEEE Std. [1] has defined two performance classes, corresponding to two distinct PMU applications and different compliance requirements:

- *P-class* is intended for Protection applications or any application requiring fast RTs and reduced reporting latencies (see Section 3.5.1.3).

- *M-class* is intended for Measurement applications where more importance is given to the accuracy of the PMU estimations, and particularly to its capability to reject inter-harmonics, rather than their RTs and reporting latencies.

It is worth noting that the performance classes defined in the IEEE Std. [1] do not discriminate between the use of PMUs in transmission or distribution networks and the various test types and compliance requirements have been derived by assuming the operating conditions of transmission networks. This is because PMUs have been originally deployed at such a level, mainly for the higher investment costs and the critical reliability of this infrastructure. As a consequence, it is important to keep in mind that such a compliance verification might not be sufficient for a PMU that wants to operate at the power distribution level.

3.5.1.3 PMU performance evaluation

The evaluation of the PMU performances is typically performed by a dedicated PMU calibrator capable of characterizing the metrological performances of the PMU under test by comparing its estimations to a "true" value†† and assessing its conformity to some predefined accuracy limits.

In what follows, in the following sections, the metrics defined in Reference 1 to evaluate the PMU estimation errors together with those for the PMU RT and measurement reporting latency are presented.

Total vector error

The PMU accuracy in estimating the true synchrophasor can be expressed independently in terms of the amplitude and phase errors or the real- and imaginary-part errors if rectangular coordinates are adopted. This is, for instance, the approach used in the case of instrument transformers, where the acceptable errors are expressed separately in terms of the allowed phase angle and magnitude error (see Reference 55).

Nevertheless, sometimes it is useful to express the SE error with a single value that includes both components. This quantity has been defined in Reference 1 and is called *total vector error* (TVE). The TVE is a real number that expresses the Euclidean distance between the true and estimated synchrophasors, normalized with respect to the amplitude of the true synchrophasor (see Figure 3.13):

$$TVE \triangleq \frac{|\widehat{X} - X|}{|X|} \tag{3.76}$$

$$= \frac{|(\widehat{X}_r + j\widehat{X}_i) - (X_r + jX_i)|}{|X_r + jX_i|} \tag{3.77}$$

$$= \sqrt{\frac{(\widehat{X}_r - X_r)^2 + (\widehat{X}_i - X_i)^2}{X_r^2 + X_i^2}} \tag{3.78}$$

††Since the "true" value of a quantity is hidden by definition, in what follows, the term "true" will be used to define a reference quantity characterized by a variance that is known and can be a priori considered much smaller than the one of the PMU estimation.

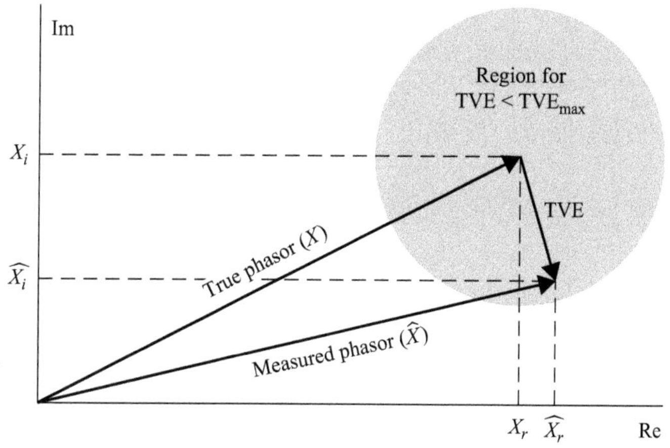

Figure 3.13 Complex plane plot showing the "true" phasor and the measured one that is, in this case, compliant with the TVE requirements (grey area centered around the "true" phasor) (adapted from Reference 1)

where $\widehat{X}$ identifies the synchrophasor estimated by the PMU, X the true one, and the subscripts r and i identify the real and imaginary parts of the synchrophasor, respectively.

The TVE can be used to set the PMU accuracy requirements by defining the maximum allowable limit like any other error quantity. It can be visualized as a circle centred around the true synchrophasor X characterized by a radius corresponding to the maximum allowable TVE (that also correspond to the maximum allowable magnitude error). If the estimated synchrophasor falls within the circle, the PMU is compliant else it is not (see Figure 3.13).

Treating a single number to characterize the PMU accuracy in estimating the synchrophasor has its advantages and disadvantages. Among these, the impossibility to decouple the contribution of amplitude and phase estimation error to understand eventual PMU asymmetries in the SE. For these reasons, in what follows, the TVE will be considered together with the amplitude and phase errors as additional metrics.

Frequency error

The frequency error (FE) is defined in the IEEE Std. C37.118 [1] as follows:

$$FE \triangleq \left| f - \widehat{f} \right| \tag{3.79}$$

being f the true frequency value and $\widehat{f}$ the one estimated by the PMU.

ROCOF error
The rate of change of frequency error (RFE) is defined in the IEEE Std. C37.118 [1] as follows:

$$RFE \triangleq \left| ROCOF - \widehat{ROCOF} \right| \tag{3.80}$$

being $ROCOF$ the true ROCOF value and $\widehat{ROCOF}$ the one estimated by the PMU.

Magnitude error
The magnitude error is defined as follows:

$$\varepsilon_A \triangleq A - \widehat{A} \tag{3.81}$$

being A the true magnitude of the signal and $\widehat{A}$ the one estimated by the PMU.

Phase error
The phase error is defined as follows:

$$\varepsilon_\psi \triangleq \psi - \widehat{\psi} \tag{3.82}$$

being ψ the true instantaneous phase of the signal and $\widehat{\psi}$ the one estimated by the PMU.

Response time
The RT is defined as the transition time between two consecutive steady-state measurements, before and after a step change is applied at one or more waveforms acquired by the PMU.

It is evaluated by applying a step change in the amplitude or phase of the input waveforms and measuring the time interval where the PMU errors exceed some predefined accuracy limits (see Figure 3.14). If t_{start} is the time of the first PMU estimation that exceed the limit and t_{end} is the time of the first PMU estimation that re-enters and stays within that limit, then the RT can be computed as:

$$RT \triangleq t_{end} - t_{start} \tag{3.83}$$

The evaluation of the RT is useful to understand the effect of the adopted window length during power system transients.

Delay time
The delay time is defined as the time interval between the instant that a step change is applied to the input of a PMU and measurement time that the stepped parameter achieves a value that is halfway between the initial and final steady-state values [1].

It is evaluated by applying a step change in the amplitude or phase of the input waveforms and measuring the time interval between the effective time the step change takes place and the moment the PMU estimation of the stepped parameter reached a value that is halfway between the initial and final steady-state values (see Figure 3.14).

The evaluation of the delay time is useful to verify that the timestamp has been properly compensated for any delay introduced by the PMU filtering system.

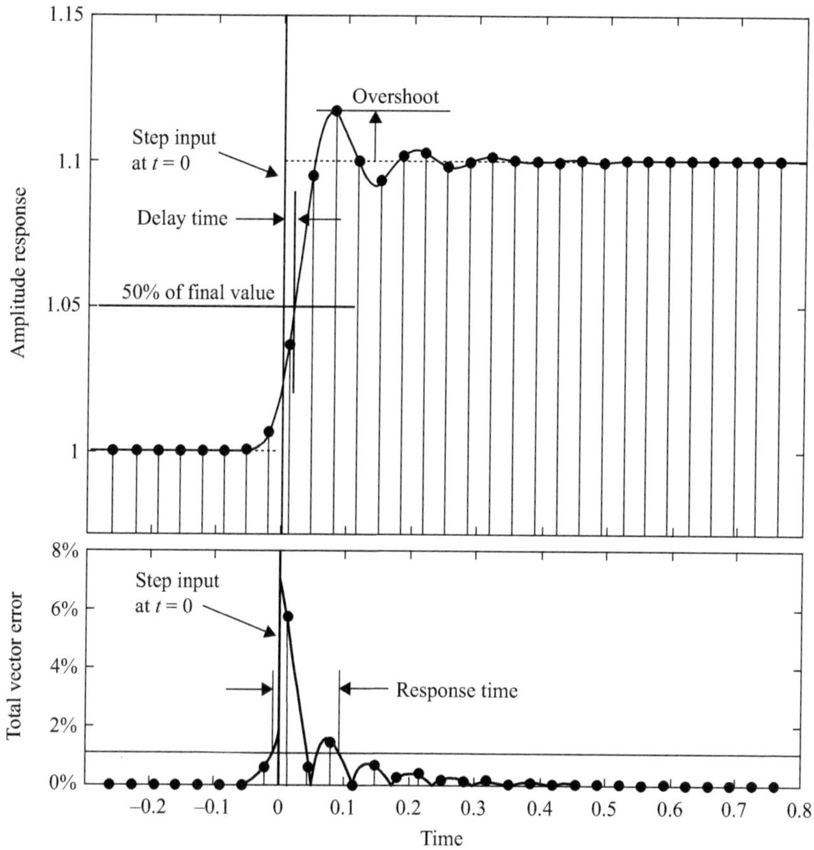

Figure 3.14 Representation of the PMU RT, delay time and maximum overshoot during an amplitude step change with a TVE limit of 1% (adapted from Reference 1)

Maximum overshoot

The maximum overshoot is defined as the measure of the maximum peak value of the estimated synchrophasor by the PMU during a step change of the instantaneous amplitude or phase (see Figure 3.14).

Reporting latency

The reporting latency is the time delay between the time a specific event has occurred in the power system and the time it is measured and reported by the PMU. The main contribution to this quantity are: (i) the adopted window length, (ii) the processing time needed by the SE algorithm and (iii) the processing time needed by the encapsulation and streaming process.

It is typically measured by connecting the PMU point-to-point with a PMU calibrator that has the possibility to time-tag the moment a PMU data frame is received at

Table 3.2 IEEE Std. C37.118 compliance requirements for steady-state conditions (in the table F_s is the PMU reporting rate). The symbol "–" means that either the specific performance class does not include that specific test or the specific requirements have been suspended by the IEEE Std. C37.118 amendment [52]

Influence quantity		**TVE (%)**		**FE (Hz)**		**RFE (Hz/s)**	
		P-class	**M-class**	**P-class**	**M-class**	**P-class**	**M-class**
Signal frequency		1	1	0.005	0.005	0.4	0.1
Signal magnitude		1	1	0.005	0.005	0.4	0.1
Harmonic distortion	$F_r > 20$	1	1	0.005	0.025	0.4	–
	$F_r \leq 20$	1	1	0.005	0.005	0.4	–
OOB interference		–	1.3	–	0.01	–	–

its network adapter. By assuming the network delays negligible, the reporting latency can be computed as follows:

$$RL \triangleq t_r - t_s \tag{3.84}$$

being t_s the PMU timestamp of a specific data frame and t_r the time that data frame has been received in the calibrator.

The maximum allowable reporting latencies are independent of the testing conditions and equal to $2/F_r$ and $7/F_r$ for P-class and M-class, respectively, being F_r the PMU reporting rate.

3.5.1.4 The steady-state compliance tests

Steady-state conditions are defined as conditions where the instantaneous parameters of the reference signals (including also the parameters of the interfering signals other than the main tone) are constant along the whole duration of each sub-test. According to Reference 1, the PMU steady-state compliance can be accomplished by satisfying the accuracy requirements for TVE, FE and RFE of the specific performance class, while varying the influence quantities that are listed below. Table 3.2 summarizes the IEEE Std. C37.118 [1] steady-state compliance limits (including the latest amendment reported in Reference 52) for TVE, FE and RFE for both performance classes P and M.

Signal frequency

During this test, the reference signals are composed by a single sinusoidal component characterized by a frequency f that is varied by 0.1 Hz between each sub-test, whereas the other tone parameters (A and φ) are kept constant. The frequency bandwidth to be tested depends on the targeted performance class and reporting rate but cannot be higher than 10 Hz, centred around the nominal frequency f_0 (e.g., in the case of a nominal frequency of 50 Hz, the widest frequency range to be tested is between 45 and 55 Hz).

Signal magnitude (voltage and current)
During this test, the reference signal is composed by a single sinusoidal component characterized by an amplitude A that is varied by 10% between each sub-test, whereas the other tone parameters (f and φ) are kept constant. The amplitude range to be tested depends on the targeted performance class and whether the measured quantity is a voltage or a current. In general, it cannot be lower than 10% of the nominal amplitude A_0 and higher than 120% of A_0 for voltage signals or 200% of A_0 for current signals.

Harmonic distortion
This test allows to assess the influence of harmonics on the quality of the estimated synchrophasor, frequency, and ROCOF. It assumes that a steady-state single-tone signal at the nominal frequency f_0 and characterized by a nominal amplitude A_0 is corrupted by a single superposed harmonic characterized by an amplitude that is either 1% or 10% of the nominal amplitude A_0 in case of P-class or M-class compliance, respectively. The harmonic order is varied between each test, starting from the second up to the 50th harmonic.

Out-of-band interference
This test allows to assess the influence of inter- and sub-harmonics on the quality of the estimated synchrophasor, frequency, and ROCOF and it is only defined for M-class compliance. Out-of-band (OOB) compliance must be tested with a main-tone signal characterized by a frequency within $f_0 - 0.1F_r/2$ and $f_0 + 0.1F_r/2$, being F_r the reporting rate and $F_r/2$ the associated Nyquist limit (e.g., between 47.5 and 52.5 Hz if f_0 is 50 Hz) and a nominal amplitude A_0. Single inter- or sub-harmonic signals must be superposed to such a main tone and be characterized by an amplitude equal to 10% of A_0 and a frequency that is either within 10 Hz and $f_0 - F_r/2$ or within $f_0 + F_r/2$ and $2f_0$.

3.5.1.5 The dynamic compliance tests

Dynamic conditions are defined in Reference 1 as conditions where the instantaneous parameters of the main tone of the reference signal (namely its frequency f, amplitude A, and initial phase φ) are not constant along the duration of each sub-test. Therefore, with the term dynamic conditions, the IEEE Std. [1] does not include the evaluation of the PMU performances during dynamic behaviour of any interfering signals. According to Reference 1, the PMU dynamic compliance can be accomplished by satisfying the accuracy requirements for TVE, FE and RFE together those for the RT for the specific performance class, during the following tests.

Measurement bandwidth
This test aims at testing the quality of the PMU estimations during increasing variation of the instantaneous amplitude $A(t)$ and phase $\psi(t)$ of the main tone of the reference signal. In particular, during this test, the reference signals are characterized

by independent sinusoidal amplitude or phase modulations according to the following formula:

$$x(t) = A_0 [1 + k_a \cos (2\pi f_m t)] \cdot \cos \left[2\pi f_0 t + k_p \cos (2\pi f_m t - \pi)\right] \tag{3.85}$$

being f_m the modulating frequency, k_a and k_p the amplitude and modulation factors, respectively. According to Reference 1, f_m must be varied by steps of 0.2 Hz or smaller between each sub-test, within a range that goes from 0.1 to $F_m \triangleq \min (F_r/10, 2)$ or from 0.1 to $F_m \triangleq \min (F_r/10, 5)$ in the case of P-class or M-class compliance, respectively. On the other hand, the modulation factors k_a and k_p are set constant to 0.1 and the two modulations applied separately.

System frequency ramp

This test is defined to verify the correct positioning of the timestamp within the time window and to test the linearity of the PMU "filter". During this test, the main-tone parameters are kept constant except the instantaneous frequency that is varied linearly with a constant rate of 1 Hz/s. Both positive and negative ramps must be tested within a frequency range that is between $f_0 - 2\,\text{Hz}$ and $f_0 + 2\,\text{Hz}$ for P-class compliance and between $f_0 - \min (F_r/5, 5)$ Hz and $f_0 + \min (F_r/5, 5)$ Hz in the case of M-class compliance.

Step changes in amplitude and phase

This test is designed to simulate sudden power system events like voltage and current variations during faults, short circuits, or the synchronization of islanded/separated networks. It is aimed at testing the PMU RT, that is mainly influenced by the adopted time-window length. According to Reference 1, the step change is applied simultaneously to all three phases and to both voltage and current inputs. The IEEE Std. requires to test 10% positive and negative amplitude steps and 10° positive and negative phase steps.

Tables 3.4 and 3.5 summarize the IEEE Std. C37.118 [1] dynamic compliance limits (including the latest amendment reported in Reference 52) for TVE, FE and RFE (see Table 3.3) together with those for the RT (see Table 3.4), delay time and maximum overshoot (see Table 3.5) for both performance classes P and M.

Table 3.3 IEEE Std. C37.118 limits for TVE, FE and RFE during dynamic conditions (in the table F_m is the maximum modulating frequency for a specific PMU reporting rate F_r)

Test	**TVE (%)**		**FE (Hz)**		**RFE (Hz/s)**	
	P-class	**M-class**	**P-class**	**M-class**	**P-class**	**M-class**
Measurement bandwidth	3	3	$0.03F_m$	$0.06F_m$	$0.18\pi F_m^2$	$0.18\pi F_m^2$
Frequency ramp	1	1	0.01	0.01	0.4	0.2
Amplitude/phase steps	1	1	0.005	0.005	0.4	0.1

Table 3.4 IEEE Std. C37.118 limits for RT during amplitude and phase steps

Test	Phasor RT (s)		Frequency RT (s)		ROCOF RT (s)	
	P-class	**M-class**	**P-class**	**M-class**	**P-class**	**M-class**
Amplitude/ phase steps	$2/f_0$	$7/F_r$	$4.5/f_0$	$\max(14/F_r, 14/f_0)$	$6/f_0$	$\max(14/F_r, 14/f_0)$

Table 3.5 IEEE Std. C37.118 limits for delay time and maximum overshoot during amplitude and phase steps

Test	Delay time (s)		Maximum overshoot	
	P-class	**M-class**	**P-class**	**M-class**
Amplitude/phase steps	$1/(4F_r)$	$1/(4F_r)$	$0.005A_0$	$0.01A_0$

3.5.2 Performance assessment of the i-IpDFT SE algorithm

This section presents a partial performance assessment of the i-IpDFT SE algorithm presented in Section 3.4.2, aimed at demonstrating the improvement introduced by the proposed iterative compensation of the spectral interference produced by the negative image component on the positive frequency range. In particular, this section will analyse the effects of the number of iterations on the estimated synchrophasor accuracy in both static and dynamic conditions by making reference to two test conditions defined in the IEEE Std. C37.118, namely, the *signal frequency* test (see Section 3.5.1.4) and the *system frequency ramp* test (see Section 3.5.1.5).

The reference signals have been synthesized in a software environment where the previously presented i-IpDFT technique has been also implemented. These tests were chosen because they are those in the IEEE Std. [1] that magnify the effects of spectral interference and therefore the tests where the i-IpDFT method should mostly improve the performances of classical IpDFT methods.

The performance of the SE is shown in function of the frequency, amplitude and phase estimation errors as stated in Section 3.5.1.3. In particular, in order to combine multiple plots in a single graph, the simulation results have been presented in a logarithmic scale and therefore the above errors are presented in terms of their absolute value.

3.5.2.1 Steady-state performances

The improvement introduced by the i-IpDFT in steady-state conditions is presented during the signal frequency test (see Section 3.5.1.4) with a nominal frequency of

50 Hz. Nevertheless, equivalent results can be obtained for 60 Hz. The i-IpDFT performances are shown in function of the nominal frequency of each sub-test that, in the case of a 50 Hz power system, must span the frequency interval between 45 and 55 Hz.

In particular, Figure 3.15 shows the maximum frequency, amplitude, and phase errors for various number of iterations (i.e., the parameter r in Algorithm 1), starting from $r = 0$ (i.e., the classical IpDFT approach without any compensation of the spectral interference) up to $r = 4$. The effects of the iterative compensation of the spectral interference produced by the negative image of the spectrum are evident and self-explanatory: the estimation accuracies of frequency, amplitude and phase are improved of almost two orders of magnitude every new iteration, up to the fourth. After, the effects of the spectral interference compensation are no longer visible.

Additionally, it is possible observing that, independently of the number of iterations, the best accuracies are obtained for $f \approx f_0$ (namely for values of frequency closer to 50 Hz) since the effects of spectral leakage interference are here minimized. They deteriorate as the nominal frequency of each sub-test deviates from the rated one f_0 and the effects of spectral interference increase. In particular, the i-IpDFT accuracies are asymmetric with respect to f_0; in other words, the frequency interval on the left of f_0 usually exhibits poorer performances than the frequency interval on the right of f_0. The cause of this behaviour is still related to the spectral interference produced by the negative image spectrum that is higher as the two images get closer, namely for smaller nominal frequencies. Nevertheless, such an asymmetry between the frequency intervals on the left and on the right of f_0 is relative and becomes less and less visible by increasing the number of iterations.

Finally, Figure 3.16 compares the performances of the i-IpDFT SE algorithm with those provided by the classical IpDFT technique that adopt the Hanning window during the signal frequency test. In particular, the five-, four-, three-, and two-period IpDFT SE techniques (dashed lines, various gray scales) are compared to the three- and two-period i-IpDFT SE techniques that adopt four and seven iterations, respectively (continuous lines).

As it can be noticed, the classical IpDFT accuracy is strongly related to the adopted window length and, due to the higher immunity to spectral leakage, the longer the window length the higher the steady-state accuracies. On the other hand, the i-IpDFT technique definitely improves such an accuracy and, at the same time, allows to reduce the window length down to two periods of a signal at the nominal frequency f_0. Such a possibility introduces non-negligible advantages as it allows to considerably reduce the PMU RTs and measurement reporting latencies (see Section 3.5.1.3) without deteriorating the PMU accuracies.

3.5.2.2 Dynamic performances

The improvement introduced by the i-IpDFT technique during dynamic conditions can be evaluated by making reference to the frequency ramp test, defined in the IEEE Std. C37.118 (see Section 3.5.1.5). During this test, the frequency of the reference signal is linearly increased from 45 to 55 Hz at a rate of 1 Hz/s.

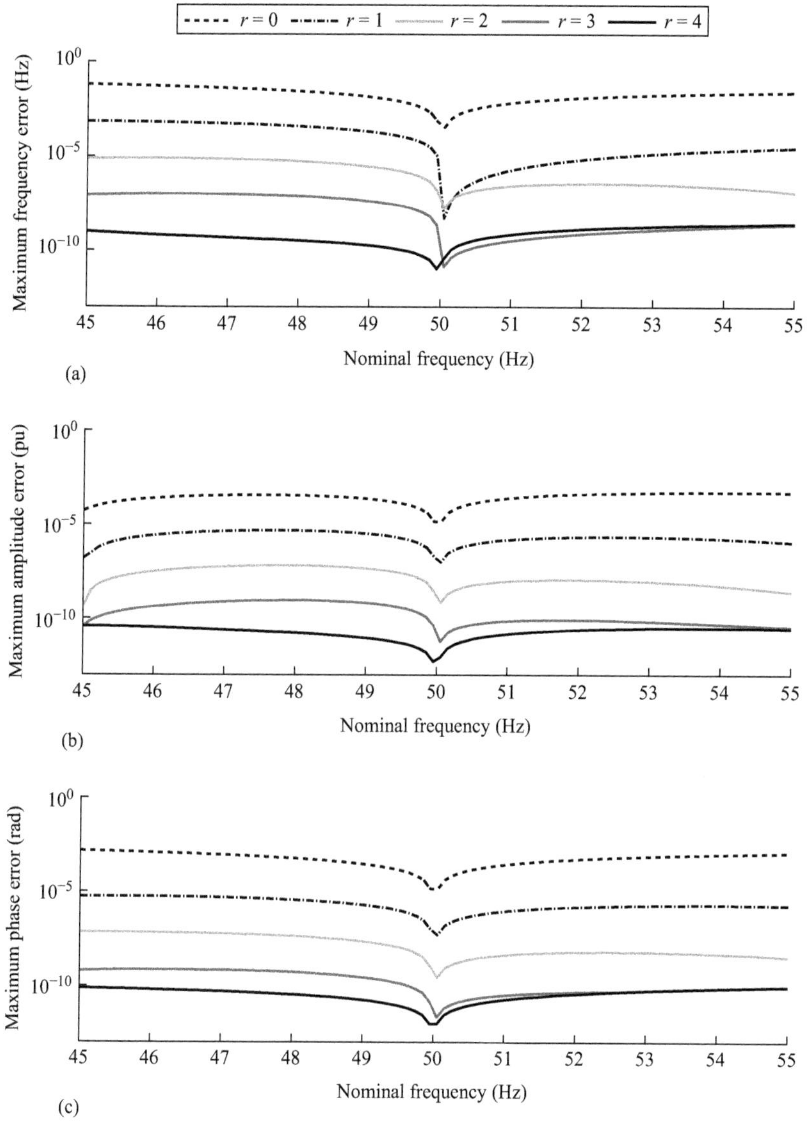

Figure 3.15 Maximum frequency (a), amplitude (b) and phase estimation errors (c) of the proposed i-IpDFT method as a function of the nominal frequency of a steady-state reference signal. The errors are plotted for various number of iterations ($r = 0,1,2,3,4$), being $r = 0$ the classical IpDFT method

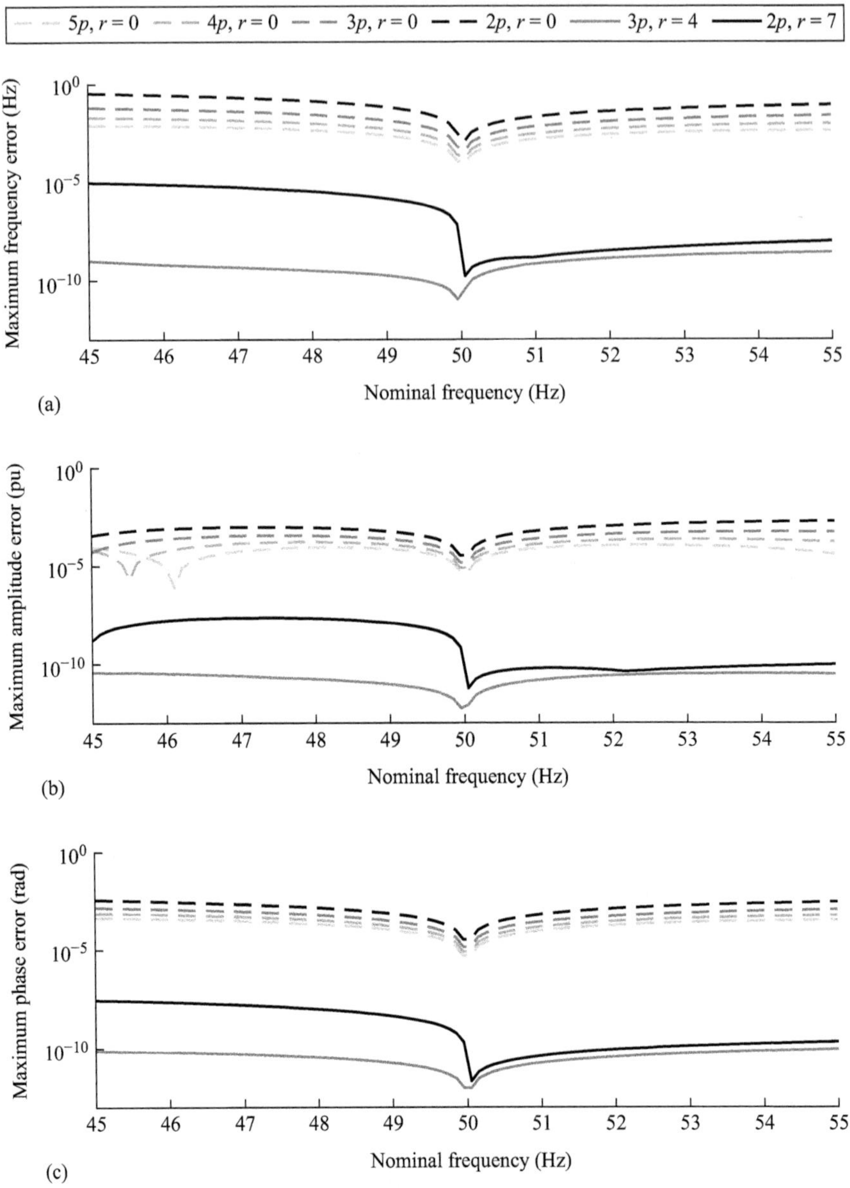

Figure 3.16 *Maximum frequency (a), amplitude (b), and phase estimation errors (c) of the proposed i-IpDFT technique versus the classical IpDFT approach. The results are plotted as a function of the nominal frequency of a steady-state reference signal and present, with different gray scales, the results for five-, four-, three- and two-period IpDFT estimators with (continuous) and without (dashed) the proposed iterative compensation of the spectral interference*

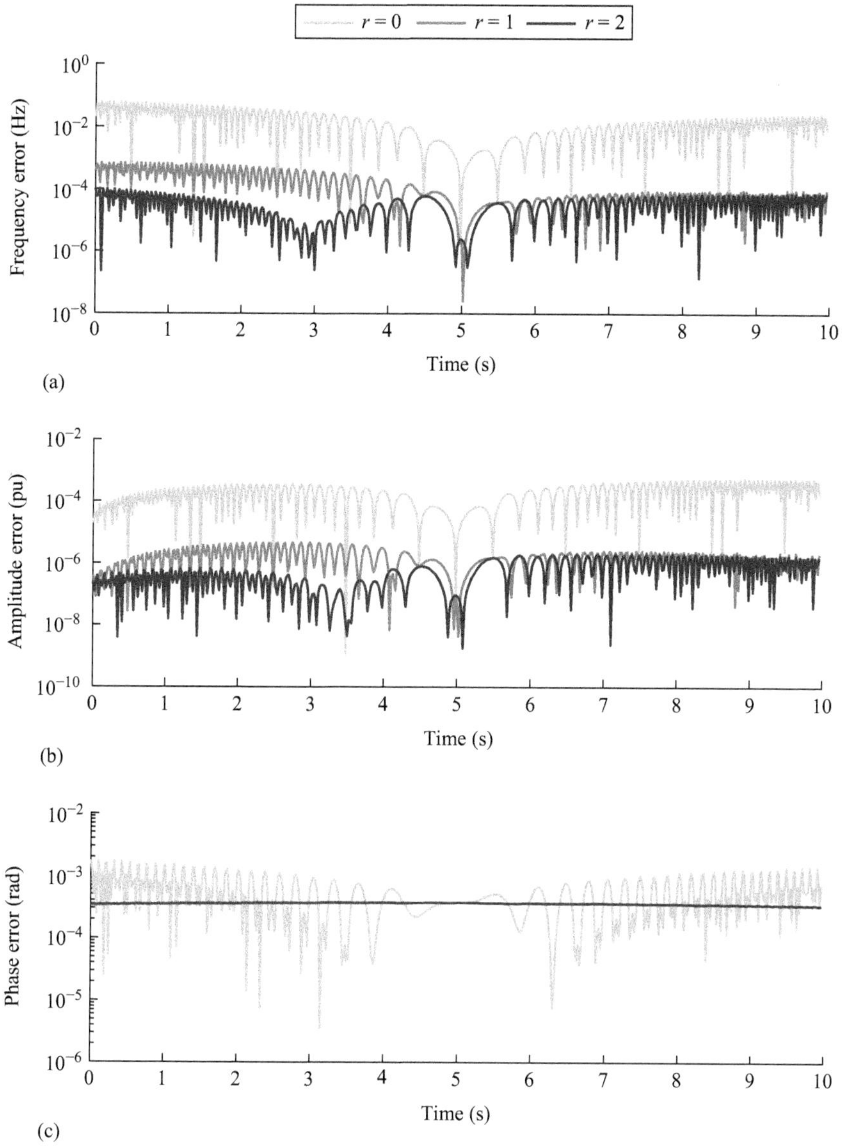

Figure 3.17 Instantaneous frequency (a), amplitude (b), and phase estimation errors (c) of the proposed i-IpDFT technique during a positive ramp of the system frequency. Errors are plotted for various number of iterations ($r = 1,2,0$), being $r = 0$ the classical IpDFT method

Figure 3.17 shows the instantaneous errors in the estimation of frequency amplitude and phase during a positive frequency ramp (the results for the negative ramp are here omitted but lead to equivalent conclusions) for various number of iterations. Similarly to the previous figure, the light gray line is obtained for $r = 0$ and makes reference to a classical IpDFT technique where the effects of the spectral interference are neglected. The other two lines are the result of the i-IpDFT with one and two iterations, respectively. Also, during this test, the improvements introduced by the iterative compensation of the spectral interference produced by the negative image of the tone are evident and lead to enhanced accuracies of the estimated synchrophasor that can reach accuracies similar to those obtained during steady state. In particular, the effects of the iterative compensation presented in Section 3.4.2.1 are visible up to the third iteration in the case of the estimated value of frequency and amplitude and up to the second iteration for the estimated phase.

3.6 Conclusions

This chapter has discussed the main elements related to the definition of DFT-based SE algorithms since they represent the most commonly adopted ones in real PMU devices. In particular, this chapter has focused on the analysis of spectral leakage as it represents the most relevant source of uncertainty when using the DFT to estimate the parameters of a sinusoidal signal. This aspect is of importance in SE processes since they usually adopt relatively short windows to reduce the PMU measurement reporting latencies and RTs.

This chapter has also presented state-of-the-art SE algorithms belonging to the family of IpDFT estimators. This chapter has discussed both the classical IpDFT as well as its iterative counterpart (i-IpDFT) capable of dealing with the compensation of the effects of the self-interaction between the positive and the negative images of the spectrum. Such a technique has demonstrated to improve the classical IpDFT performances during both static and dynamic conditions described in the IEEE Std. C37.118 and to be immune to the instantaneous frequency variations of a power system. Furthermore, it has been demonstrated that the i-IpDFT technique outperforms classical IpDFT methods, also by adopting shorter windows (up to two periods) that are usually worsening the estimation uncertainty of any SE algorithm.

Bibliography

[1] IEEE Standard for Synchrophasor Measurements for Power Systems. IEEE Std C37.118.1-2011 (Revision of IEEE Std C37118-2005). 2011; p. 1–61.

[2] Castello P, Lixia M, Muscas C, Pegoraro PA. "Impact of the model on the accuracy of synchrophasor measurement". *IEEE Transactions on Instrumentation and Measurement* 2012 Aug;61(8):2179–2188.

[3] de la Serna JAd. "Dynamic phasor estimates for power system oscillations". *IEEE Transactions on Instrumentation and Measurement* 2007 Oct; 56(5):1648–1657.

[4] Castello P, Liu J, Muscas C, Pegoraro PA, Ponci F, Monti A. "A fast and accurate PMU algorithm for P and M class measurement of synchrophasor and frequency". *IEEE Transactions on Instrumentation and Measurement* 2014 Dec;63(12):2837–2845.

[5] Bertocco M, Frigo G, Narduzzi C, Muscas C, Pegoraro PA. "Compressive sensing of a Taylor–Fourier multi-frequency model for synchrophasor estimation". *IEEE Transactions on Instrumentation and Measurement* 2015 Dec;64(12):3274–3283.

[6] Oppenheim AV, Schafer RW, Buck JR, *et al. Discrete-Time Signal Processing*, vol. 2. Englewood Cliffs, NJ: Prentice-Hall; 1989.

[7] Begovic MM, Djuric PM, Dunlap S, Phadke AG. "Frequency tracking in power networks in the presence of harmonics". *IEEE Transactions on Power Delivery* 1993 Apr;8(2):480–486.

[8] Duric MB, Durisic ZR. "Frequency measurement in power networks in the presence of harmonics using Fourier and zero crossing technique". In: *Power Tech, 2005*, IEEE, Russia; 2005. p. 1–6.

[9] Kamwa I, Leclerc M, McNabb D. "Performance of demodulation-based frequency measurement algorithms used in typical PMUs". *IEEE Transactions on Power Delivery* 2004 Apr;19(2):505–514.

[10] Roscoe AJ, Abdulhadi IF, Burt GM. "P and M class phasor measurement unit algorithms using adaptive cascaded filters". *IEEE Transactions on Power Delivery* 2013 Jul;28(3):1447–1459.

[11] Roscoe AJ. "Exploring the relative performance of frequency-tracking and fixed-filter phasor measurement unit algorithms under C37.118 test procedures, the effects of interharmonics, and initial attempts at merging P-class response with M-class filtering". *IEEE Transactions on Instrumentation and Measurement* 2013 Aug;62(8):2140–2153.

[12] Bertocco M, Frigo G, Narduzzi C. "On compressed sensing and super-resolution in DFT-based spectral analysis". *Proceedings 19th IMEKO TC-4 Symposium and 17th IWADC Workshop Advances in Instrumentation and Sensors Interoperability*. 2013; p. 615–620.

[13] Bertocco M, Frigo G, Narduzzi C, Tramarin F. "Resolution enhancement by compressive sensing in power quality and phasor measurement". *IEEE Transactions on Instrumentation and Measurement* 2014 Oct;63(10):2358–2367.

[14] Wong CK, Leong LT, Wu JT, *et al.* "A novel algorithm for phasor calculation based on wavelet analysis". In: *Power Engineering Society Summer Meeting*, 2001. vol. 3. IEEE; 2001. p. 1500–1503.

[15] Guo Q, Zhang C, Shi Y. "Resampling a signal to perform power quality and synchrophasor measurement". Google Patents; 2013. US Patent App. 13/326,676.

[16] de la OS, Antonio J. "Synchrophasor estimation using Prony's method". *IEEE Transactions on Instrumentation and Measurement* 2013;62(8): 2119–2128.

[17] Yang L, Jiao Z, Kang X, Wang X. "A novel matrix pencil method for real-time power frequency phasor estimation under power system transients". In: *12th IET International Conference on Developments in Power System Protection (DPSP 2014)*, Copenhagen, 2014. p. 1–5.

[18] Sidhu TS, Zhang X, Balamourougan V. "A new half-cycle phasor estimation algorithm". *IEEE Transactions on Power Delivery*. 2005;20(2): 1299–1305.

[19] Phadke A, Thorp J, Adamiak M. "A new measurement technique for tracking voltage phasors, local system frequency, and rate of change of frequency". *IEEE Transactions on Power Apparatus and Systems* 1983;(5): 1025–1038.

[20] Warichet J, Sezi T, Maun JC. "Considerations about synchrophasors measurement in dynamic system conditions". *International Journal of Electrical Power & Energy Systems* 2009;31(9):452–464.

[21] Sachdev M, Nagpal M. "A recursive least error squares algorithm for power system relaying and measurement applications". *IEEE Transactions on Power Delivery* 1991;6(3):1008–1015.

[22] Pradhan A, Routray A, Sethi D. "Voltage phasor estimation using complex linear Kalman filter". In: *Eighth IEE International Conference on Developments in Power System Protection, 2004*, vol. 1. IET; 2004. p. 24–27.

[23] Jain VK, Collins WL, Davis DC. "High-accuracy analog measurements via interpolated FFT". *IEEE Transactions on Instrumentation and Measurement*. 1979;28(2):113–122.

[24] Grandke T. "Interpolation algorithms for discrete Fourier transforms of weighted signals". *IEEE Transactions on Instrumentation and Measurement*. 1983;32(2):350–355.

[25] Belega D, Dallet D. "Multifrequency signal analysis by Interpolated DFT method with maximum sidelobe decay windows". *Measurement* 2009;42(3):420–426.

[26] Paolone M, Borghetti A, Nucci CA. "A synchrophasor estimation algorithm for the monitoring of active distribution networks in steady state and transient conditions". In: *Proceedings of the 17th Power Systems Computation Conference (PSCC 2011)*, Stockholm, Sweden, Aug; 2011.

[27] Belega D, Petri D. "Accuracy analysis of the multicycle synchrophasor estimator provided by the Interpolated DFT algorithm". *IEEE Transactions on Instrumentation and Measurement* 2013;62(5):942–953.

[28] Belega D, Macii D, Petri D. "Fast synchrophasor estimation by means of frequency-domain and time-domain algorithms". In: *IEEE Transactions on Instrumentation and Measurement*, vol. 63, no. 2, 2014, p. 388–401.

[29] Harris FJ. "On the use of windows for harmonic analysis with the discrete Fourier transform". *Proceedings of the IEEE*. 1978;66(1):51–83.

[30] Andria G, Savino M, Trotta A. "Windows and interpolation algorithms to improve electrical measurement accuracy". *Instrumentation and Measurement, IEEE Transactions on Instrumentation and Measurement*. 1989;38(4):856–863.

[31] Phadke AG, Thorp JS. "Synchronized phasor measurements and their applications". *Power Electronics and Power Systems*. New York, NY: Springer US; 2008.

[32] Borghetti A, Nucci CA, Paolone M, Ciappi G, Solari A. "Synchronized phasors monitoring during the islanding maneuver of an active distribution network". *IEEE Transactions on Smart Grid* 2011;2(1):82–91.

[33] Gunther EW. "Interharmonics in power systems". In: *Power Engineering Society Summer Meeting, 2001*, Vancouver, BC, Canada, vol. 2; 2001. p. 813–817.

[34] IEEE Standard for Synchrophasors for Power Systems. IEEE Std 1344-1995. 1995.

[35] IEEE Standard for Synchrophasor Data Transfer for Power Systems. IEEE Std. C37.118.2-2011 (Revision of IEEE Std C37118-2005). 2011; p. 1–53.

[36] Communication Networks and Systems for Power Utility Automation – Part 90-5: Use of IEC 61850 to Transmit Synchrophasor Information According to IEEE C37.118. IEC TR 61850-90-5:2012. 2012 May.

[37] Castello P, Muscas C, Pegoraro PA. "Performance comparison of algorithms for synchrophasors measurements under dynamic conditions". In: *2011 IEEE International Workshop on Applied Measurements for Power Systems (AMPS)*; Aachen, 2011. p. 25–30.

[38] IEEE Guide for Synchronization, Calibration, Testing, and Installation of Phasor Measurement Units (PMUs) for Power System Protection and Control. IEEE Std C37242-2013. 2013 Mar; p. 1–107.

[39] Instrument Transformers – Part l: Current Transformers. Int Std IEC 60044-1. 1996.

[40] Instrument Transformers – Part 2: Inductive Voltage Transformers. Int Std IEC 60044-2. 1997.

[41] Papoulis A. *The Fourier Integral and Its Applications. McGraw-Hill Electronic Sciences Series*. New York, NY: McGraw-Hill; 1962.

[42] Cooley JW, Lewis PA, Welch PD. "The fast Fourier transform and its applications". *IEEE Transactions on Education* 1969;12(1):27–34.

[43] Darwish HA, Fikri M. "Practical considerations for recursive DFT implementation in numerical relays". *IEEE Transactions on Power Delivery* 2007;22(1):42–49.

[44] Jacobsen E, Lyons R. "The sliding DFT". *Signal Processing Magazine, IEEE*. 2003;20(2):74–80.

[45] Douglas S, Soh J. "A numerically-stable sliding-window estimator and its application to adaptive filters". In: *Conference Record of the Thirty-First Asilomar Conference on Signals, Systems & Computers, 1997*, Pacific Grove, CA, USA, vol. 1. IEEE; 1997. p. 111–115.

[46] Duda K. "Accurate, guaranteed stable, sliding discrete Fourier transform [DSP tips & tricks]". *Signal Processing Magazine, IEEE*. 2010;27(6):124–127.

[47] Romano P, Paolone M. "Enhanced Interpolated-DFT for synchrophasor estimation in FPGAs: theory, implementation, and validation of a PMU prototype". *IEEE Transactions on Instrumentation and Measurement* 2014 Dec;63(12):2824–2836.

[48] Nuttall AH. "Some windows with very good sidelobe behavior". *IEEE Transactions on Acoustics, Speech and Signal Processing* 1981 Feb;29(1):84–91.

[49] Salih MS, ed. *Fourier Transform: Signal Processing*. InTech, 2012.

[50] Duda K. "DFT interpolation algorithm for Kaiser–Bessel and Dolph–Chebyshev windows". *IEEE Transactions on Instrumentation and Measurement* 2011 Mar;60(3):784–790.

[51] IEEE Standard for Synchrophasors for Power Systems. IEEE Std C37118-2005 (Revision of IEEE Std 1344-1995). 2006; p. 1–57.

[52] IEEE Standard for Synchrophasor Measurements for Power Systems – Amendment 1: Modification of Selected Performance Requirements. IEEE Std C371181a-2014 (Amendment to IEEE Std C371181-2011). 2014 Apr; p. 1–25.

[53] IEEE Synchrophasor Measurement Test Suite Specification. IEEE Synchrophasor Measurement Test Suite Specification. 2014 Dec; p. 1–44.

[54] Romano P, Paolone M. "An enhanced interpolated-modulated sliding DFT for high reporting rate PMUs". In: *2014 IEEE International Workshop on Applied Measurements for Power Systems Proceedings (AMPS)* Aachen, 2014. p. 1–6.

[55] IEEE Standard Requirements for Instrument Transformers. IEEE Std C5713-2008 (Revision of IEEE Std C5713-1993). 2008 Jul; p. 1–106.

Chapter 4
Modeling power systems with stochastic processes

Guðrún Margrét Jónsdóttir[1], Muhammad Adeen[2], Rafael Zárate-Miñano[3] and Federico Milano[2]

Any physical system, and, thus, also power systems, contains randomness and uncertainty. For example, load power consumption is not fully deterministic. Moreover, in recent years, the massive installation of non-dispatchable technologies, e.g., wind and solar parks, has increased the degree of randomness in power systems.

This chapter summarizes the state-of-the-art of the modeling of stochastic perturbations in power systems by means of stochastic differential equations (SDEs). The chapter begins with a brief theoretical introduction to SDEs and provides designing methods of SDEs to represent perturbations with given statistical properties. Perturbation models derived from the application of the presented methods are illustrated through numerical examples. The chapter also describes a general procedure to define stochastic dynamic models for power system components. Practical issues related to the numerical integration of the resulting power system model are discussed. Finally, the dynamic behaviour of power systems subjected to stochastic phenomena is illustrated through simulations of the 1,479 bus Irish power system model.

4.1 Literature review

The literature on SDEs is vast. Theoretical background on SDEs can be found in, e.g., [1,2]. SDEs are widely applied in finance to model stochastic fluctuations of stock prices and other financial assets and in several fields of science and engineering to study physical systems affected by different stochastic phenomena [3,4].

Explicit solutions of SDEs are only known in very particular cases. Thus, SDEs are generally solved through numerical methods. The computational burden of integrating SDEs is extremely high compared to the integration of deterministic dynamic

[1]Norconsult, Kópavogur, Iceland
[2]School of Electrical and Electronic Engineering, University College Dublin, Ireland
[3]Department of Electrical Engineering, Electronics, Automation and Communications, University of Castilla-La Mancha, Spain

systems. The two main issues are apart from theoretical complexity: (i) the need to solve *thousands* of time domain integrations for every simulation, as opposed to a single solution required in the deterministic case; and (ii) the need to consistently reduce the step length of the integration method to ensure that stochastic processes are properly accounted for. The latter issue is also due to the fact that implicit integration methods that works well for deterministic approaches are extremely involved when stochastic processes are considered. Another ingredient that complicates the numerical analysis of SDEs is that the literature on numerical methods for SDEs is not as huge as that on theoretical analysis, and no industry-grade open-source numerical library to integrate SDEs is available, as opposed to the several tools available for deterministic ODEs and differential algebraic equations (DAEs), e.g., DASSL [5].* Among the few relevant references on numerical analysis of SDEs, we cite the classical monograph [6] that provides detailed descriptions of the available fixed step size methods for the numerical solution of SDEs.

Power system models are typically formalized as a set of DAEs. This allows for the SDEs to be readily incorporated into the system model and the resulting model is a set of stochastic differential algebraic equations (SDAEs). There is no well-established theory on how SDAEs that include stochastic processes within algebraic constraints should be integrated. Then, due to the inherent stiffness of DAEs, implicit numerical methods are required to avoid numerical instability, but stochastic processes are thought to be integrated through explicit numerical schemes. Also in this case, the literature does not help much. For example, in [7], the authors discuss the adequacy of different implicit fixed step size numerical methods for SDAEs. As for applications of SDAEs to engineering problems, Ref. [8] shows that, in the field of electronic circuit simulation, implicit numerical methods with fixed step size used to solve SDEs are also suitable for being applied to SDAEs.

Uncertainty, introduced through diverse sources, such as loads, wind and solar, can negatively impact the reliability, safety, and economy of power systems. The uncertainties in the system can be modeled using SDEs. Several SDE-based approaches on the stochastic modeling of specific energy sources such as wind [9–12] and solar [13,14], as well as loads [9,15–17], have been proposed in the literature.

In [18], a systematic method to model power systems as SDAEs is presented. The method is demonstrated through a case study of the IEEE 145-bus 50-machine system. A few more studies on multi-bus systems including uncertainties, modeled as SDEs, have been presented in the literature. For example, in [19], the IEEE 145-bus test system is modified to include wind generation, formulated using SDEs and the system stability is studied. The Icelandic system with uncertain variations is modeled as a set of linearized SDEs and its frequency stability assessed in [20]. The Irish system SDAEs model has also been studied, including stochastic wind and tidal generation in [21] and including correlated loads and wind generation in [22].

*This situation might change in the future. For example, the interested reader can refer to the interesting project StochSS, available at `www.stochss.org`. StochSS provides an integrated development environment (IDE) for discrete stochastic simulations of biochemical networks and population systems.

4.2 Stochastic differential equations

SDEs can be used to model continuous-time stochastic processes. They are a prominent mathematical modeling technique employed in areas such as finance for modeling stock prices or interest rates and physics to model particles in fluids. In this section, a brief introduction to SDEs is provided. A more extensive presentation of SDEs can be found in [1,2,6].

A generic multi-dimensional SDE has the following form:

$$d\eta(t) = \boldsymbol{a}(t, \eta(t))dt + \boldsymbol{b}(t, \eta(t))d\boldsymbol{W}(t), \tag{4.1}$$

where $\boldsymbol{a}$ ($\boldsymbol{a} : \mathbb{R}^{n_\eta} \times \mathbb{R}^+ \mapsto \mathbb{R}^{n_\eta}$) and $\boldsymbol{b}$ ($\boldsymbol{b} : \mathbb{R}^{n_\eta} \times \mathbb{R}^+ \mapsto \mathbb{R}^{n_\eta} \times \mathbb{R}^{n_w}$) are continuous functions and are referred to as the drift and diffusion term of the SDE, respectively. $\boldsymbol{W}$ represents the n_w-dimensional vector of stochastic components driving the SDE. This vector is composed of n_w independent scalar Wiener processes, $\{W(t), t > 0\}$, which is a random function characterized by the following properties:

1. $W(0) = 0$, with probability 1.
2. The function $t \mapsto W(t)$ is continuous in t.
3. If $t_1 \neq t_2$, then $W(t_1)$ and $W(t_2)$ are independent.
4. For $\forall t_i \geq 0$, all increments, $\Delta W_i = W(t_{i+1}) - W(t_i)$, are normally distributed, with mean 0 and variance $h = t_{i+1} - t_i$, i.e., $\Delta W_i \sim \mathcal{N}(0, h)$.

Examples of Wiener processes are shown in Figure 4.1. Wiener processes cannot be integrated in the conventional Riemann–Stieltjes sense as the derivative is not defined, i.e., the limit $\lim_{\Delta t \to 0} (W(t + \Delta t) - W(t))/\Delta t$ does not exist. Therefore, the correct mathematical formulation of (4.1) is the integral form:

$$\eta(t) = \eta(t_0) + \int_{t_0}^{t} \boldsymbol{a}(s, \eta(s))ds + \int_{t_0}^{t} \boldsymbol{b}(s, \eta(s))d\boldsymbol{W}(s), \quad t \in [t_0, t_f], \tag{4.2}$$

where the first integral is an ordinary Riemann–Stieltjes integral and the second one is a stochastic integral. There are several different ways to interpret stochastic integrals. Mainly, Itô's approach, Stratonovich's approach, and the backward integral. The most widely used approach is applied here, namely, the Itô integral.

A stochastic process is said to be *stationary* if it is characterized by a *probability density function (PDF)* that does not change with time. An example of a non-stationary process is the Wiener processes. Its mean is constant but its standard deviation, $\sqrt{t}$, changes with time. In this chapter, only stationary stochastic processes are discussed, even though some non-stationary stochastic processes affect power systems. For example, wind speed with an hourly resolution or bigger over several days/weeks may be modeled more accurately with non-stationary models [23]. This is due to the daily, and seasonal variations in the wind speed. However, for the purpose of simulating wind speed within a 10 min time frame, non-stationary models do not perform better than stationary ones [23]. This is typically, the case for short-term analysis of stochastic processes within the seconds to minutes time frame, which is the time frame considered in dynamic analysis of power systems.

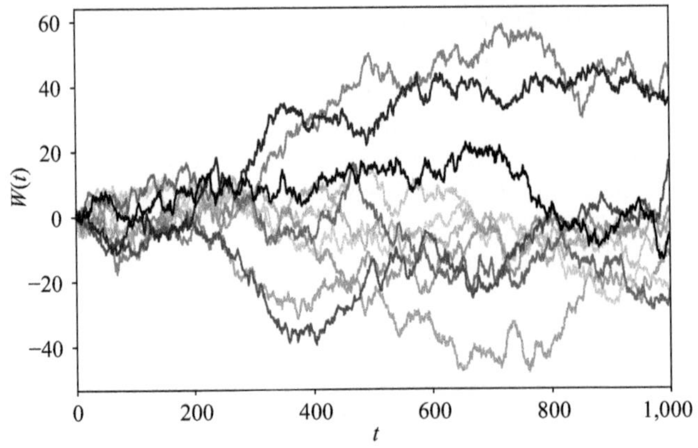

Figure 4.1 Ten simulated sample paths of Wiener processes, illustrating their great variability

A stationary stochastic process is characterized by its probability density function (PDF) and autocorrelation function (ACF). The *autocorrelation function (ACF)* is a measure of how the process changes over time. That is, the ACF gives a measure of the relationship between the current process value and past process values. Its mathematical definition for a stationary stochastic process η is:

$$R_\eta(\tau) = \frac{\mathrm{E}[(\eta(t) - \mu_\eta)(\eta(t+\tau) - \mu_\eta)]}{\sigma_\eta^2}, \tag{4.3}$$

where τ stands for the time lag. μ_η and σ_η^2 are the mean of and the variance of the stochastic process η. The *autocovariance function* of the stochastic process η is defined as $R_\eta(\tau)\sigma_\eta^2$.

The *Ornstein–Uhlenbeck (OU)* process is a popular building block in SDE modeling. It is stationary, with a Gaussian PDF which exhibits mean reversion, i.e., it drifts toward its mean value at an exponential rate. It is considered as a modification of the Wiener process, with a bounded standard deviation which makes it suitable to model physical processes. The general form of an OU SDE process is:

$$d\eta(t) = \alpha(\mu - \eta(t))dt + bdW(t), \tag{4.4}$$

where α, $b > 0$. α is the mean reversion speed of the process, $\eta(t)$, which defines the slope of its exponentially decaying ACF. The process $\eta(t)$ is Gaussian distributed with the mean μ and variance $\sigma^2 = b^2/(2\alpha)$. In Figure 4.2, examples of simulated OU process trajectories for the OU SDE $\eta_1(t)$ are shown, where $\alpha = 0.01$, $\mu = 0$, and $b = 1$. This shows that this modification to the Wiener process enables the modeling of bounded mean-reverting stochastic processes.

In Figure 4.3, examples of simulated OU process trajectories for another SDE η_2 are shown. However, in this case, $\alpha = 0.001$. α defines the ACF of the OU process. Thus, by comparing Figures 4.2 and 4.3, the difference that the ACF makes can be observed. That is, the smaller α, the slower the OU process varies over time. The ACFs of η_1 and η_2 are shown in Figure 4.4.

The OU process has been used in power systems for modeling both stochastic wind generation and load consumption volatility [11,15,16]. However, the stochastic perturbations affecting power systems may not always have a PDF that is Gaussian. For example, wind speed is typically considered to have a Weibull PDF. In the next sections, two methods for defining SDEs with non-Gaussian PDFs are presented.

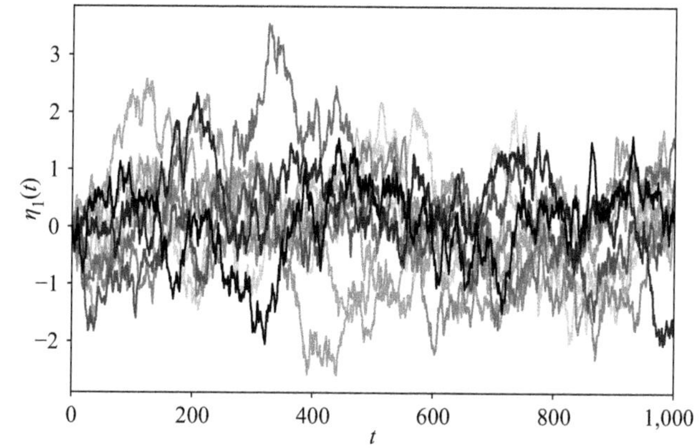

Figure 4.2 Ten simulated sample paths of an OU SDE where $\alpha = 0.01\ s^{-1}$, $\mu = 0$ and $\sigma = 1$ ($b = 0.14$)

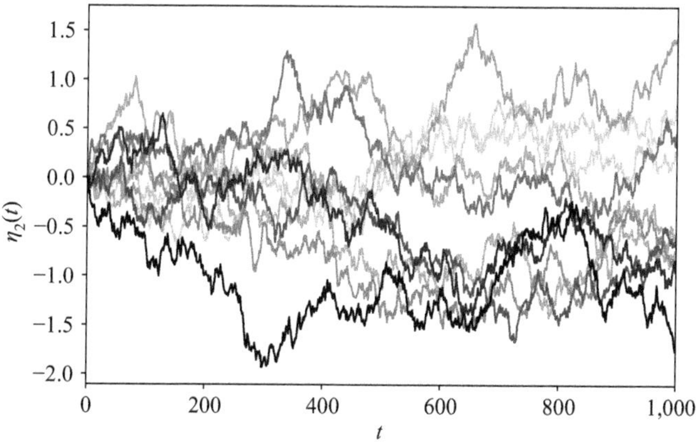

Figure 4.3 Ten simulated sample paths of an OU SDE, where $\alpha = 0.001\ s^{-1}$, $\mu = 0$ and $\sigma = 1$ ($b = 0.04$)

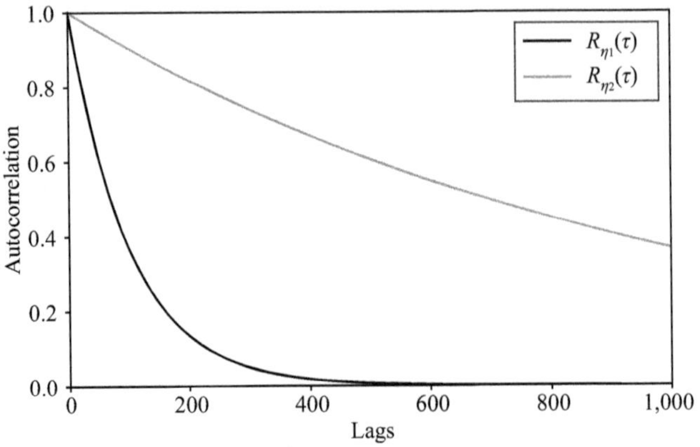

Figure 4.4 ACFs of OU processes η_1 and η_2

4.2.1 Method for modeling an SDE with an analytically defined PDF and an exponentially decaying ACF

This method, presented in [24], can be used to build a SDE that has an exponentially decaying ACF and an analytically defined PDF. The method is based on defining the relation between the drift and the diffusion terms of a stationary SDE $\eta(t)$ in order for the SDE to have the analytically defined PDF $p(\eta(t))$. This relation is obtained from the Fokker–Planck equation. For a stationary process, $a(\eta(t), t) = a(\eta(t))$, $b(\eta(t), t) = b(\eta(t))$, and $p(\eta(t), t) = p(\eta(t))$. To model a stochastic process with a given PDF $p(\eta(t))$ and an exponential decaying ACF by means of a SDE, it is a sufficient condition to define the drift term in the form:

$$a(\eta(t)) = -\alpha \cdot (\eta(t) - \mu), \tag{4.5}$$

where μ is the mean of the particular PDF $p(\eta(t))$, and the diffusion term computed by solving

$$b^2(\eta(t)) = \frac{2}{p(\eta(t))} \int_{-\infty}^{\eta(t)} -\alpha \cdot (\zeta(t) - \mu) \cdot p(\zeta(t)) \cdot d\zeta(t). \tag{4.6}$$

The details on how these expressions for the drift and diffusion term are derived as provided in [24].

To illustrate the ability of this method to produce processes with the statistical properties for which they are designed, two examples are presented here below. The first example is for the Gaussian distribution. The PDF $p_N(\eta(t))$ of the Gaussian distribution is:

$$p_N(\eta(t)) = \frac{1}{\sigma \cdot \sqrt{2 \cdot \pi}} \cdot e^{-(\eta(t)-\mu)^2/2 \cdot \sigma^2}, \tag{4.7}$$

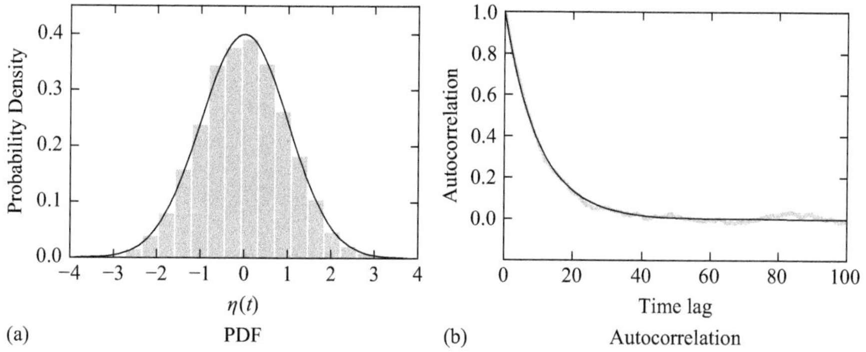

Figure 4.5 The Gaussian PDF. Model (4.5) and (4.8) with parameters: $\alpha = 0.1$, $\mu = 0$, and $\sigma = 1.0$. The values for the process generated by the SDE-based model are represented in gray, whereas theoretical values are represented in black

where μ is the mean and σ is the standard deviation. The drift term is defined as in (4.5) and, solving (4.6), the diffusion term is:

$$b(\eta(t)) = \sqrt{2 \cdot \alpha} \cdot \sigma. \tag{4.8}$$

The resulting SDE represents the OU process introduced in (4.4). To demonstrate the ability of the resulting model to capture the desired statistical properties the histogram and the ACF of the generated trajectories are compared to the Gaussian PDF and the imposed exponentially decaying ACF in Figure 4.5.

The second example is for the Weibull distribution. The PDF $p_W(\eta(t))$ of the two-parameter Weibull distribution is:

$$p_W(\eta(t)) = \begin{cases} \frac{\lambda_1}{\lambda_2} \cdot \left(\frac{\eta(t)}{\lambda_2}\right)^{\lambda_1 - 1} \cdot e^{-(\eta(t)/\lambda_2)^{\lambda_1}} & \text{if } \eta(t) \geq 0, \\ 0 & \text{if } \eta(t) < 0, \end{cases} \tag{4.9}$$

where λ_1 is a shape parameter and λ_2 is a scale parameter. The mean of the Weibull distribution is expressed as:

$$\mu = \lambda_2 \cdot \Gamma\left(1 + \frac{1}{\lambda_1}\right). \tag{4.10}$$

Therefore, according to (4.5), the drift term is:

$$a(\eta(t)) = -\alpha \cdot \left(\eta(t) - \lambda_2 \cdot \Gamma\left(1 + \frac{1}{\lambda_1}\right)\right), \tag{4.11}$$

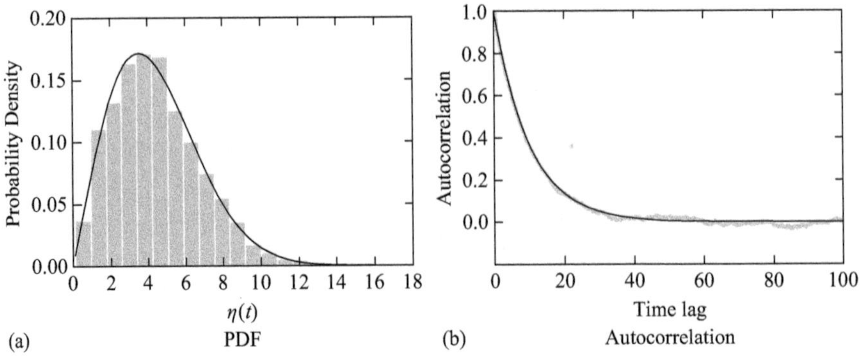

Figure 4.6 The two-parameter Weibull PDF. Model (4.11)–(4.14) with parameters: $\alpha = 0.1$, $\lambda_1 = 2.0$, and $\lambda_2 = 5.0$. The values for the process generated by SDE-based model are represented in gray, whereas theoretical values are represented in black

and solving (4.6), the diffusion term is:

$$b(\eta(t)) = \sqrt{b_1(\eta(t)) \cdot b_2(\eta(t))}, \tag{4.12}$$

with

$$b_1(\eta(t)) = 2 \cdot \alpha \cdot \frac{\lambda_2}{{\lambda_1}^2} \cdot \eta(t) \cdot \left(\frac{\lambda_2}{\eta(t)}\right)^{\lambda_1}, \tag{4.13}$$

and

$$b_2(\eta(t)) = \lambda_1 \cdot \Gamma\left(1 + \frac{1}{\lambda_1}, \left(\frac{\eta(t)}{\lambda_2}\right)^{\lambda_1}\right) \cdot e^{(\eta(t)/\lambda_2)^{\lambda_1}} - \Gamma\left(\frac{1}{\lambda_1}\right), \tag{4.14}$$

where $\Gamma(\cdot)$ is the Gamma function and $\Gamma(\cdot,\cdot)$ is the Incomplete Gamma function.

The SDE model built with the Weibull PDF is used to generate example trajectories. PDF and ACF computed from simulations are compared to the theoretical PDF and ACF in Figure 4.6. In [24], examples of SDEs with other given PDFs are presented.

4.2.2 *Method for modeling a SDE with an arbitrary PDF and ACF*

The procedure outlined in this subsection enables the transformation of a stochastic process with a given PDF into another stochastic process with the desired distribution while retaining the ACF of the original process. A relevant advantage of this approach, based on the memoryless transformation, is that it can be used with any analytical or numerical PDF that has been defined to model the probability distribution of measured data.

The memoryless transformation is used to impose the desired PDF to a Gaussian SDE. This is achieved by applying the Gaussian cumulative distribution function (CDF) to the inverse of the target CDF, as follows:

$$Z_F(t) = F^{-1}(\Phi(\eta(t))), \tag{4.15}$$

where $F^{-1}(\cdot)$ is the inverse CDF of the desired process and $\Phi(\cdot)$ is the Gaussian CDF of the SDE, $\eta(t)$[25]. The resulting process has the desired PDF that can be numerically or analytically defined [12].

In power systems, the stochastic perturbations maybe a combination of fast and slow dynamics. In that case, the ACF is better described as a weighted sum of exponentially decaying functions. The stochastic perturbations may also be subject to periodical variations which result in sinusoidal components in the ACF of the process. In this subsection, a modeling method to build stochastic processes with a wider range of ACFs is discussed.

The following 2-dimensional OU is utilized as the building block for this method:

$$\begin{pmatrix} d\eta_1(t) \\ d\eta_2(t) \end{pmatrix} = \begin{pmatrix} -\alpha & -\omega \\ \omega & -\alpha \end{pmatrix}\begin{pmatrix} \eta_1(t) \\ \eta_2(t) \end{pmatrix} dt + \begin{pmatrix} b \\ 0 \end{pmatrix} dW(t), \tag{4.16}$$

where $\alpha > 0, \sigma > 0, \omega \geq 0$ and $W(t)$ is a standard Wiener process. The autocorrelation matrix of the SDE in (4.16) is:

$$R(\tau) = \mathbb{E}\left(\begin{pmatrix} \eta_1(t+\tau) \\ \eta_2(t+\tau) \end{pmatrix}(\eta_1(t), \eta_2(t))\right) = \exp(-\alpha\tau)\begin{pmatrix} \cos(\omega\tau) & -\sin(\omega\tau) \\ \sin(\omega\tau) & \cos(\omega\tau) \end{pmatrix}. \tag{4.17}$$

The process η_1 has the ACF:

$$R_{\eta_1}(\tau) = \exp(-\alpha\tau)\cos(\omega\tau), \tag{4.18}$$

In stationary conditions, η_1 is Gaussian distributed with zero mean and variance $\sigma^2 = b^2/(2\alpha)$.

For $\omega = 0$, η_1 and η_2 are decoupled and η_1 becomes a conventional 1-dimensional Ornstein–Uhlenbeck (OU) process as presented in (4.4) with an exponentially decaying ACF:

$$R_{\eta_1}(\tau) = \exp(-\alpha\tau). \tag{4.19}$$

The main idea behind this method is to use a summation of a set of SDEs of the form (4.16) to impose the target ACF. This enables, for example, the modeling of ACFs that are a combination of exponentially decaying and periodical behaviors.

Let $\tilde{\eta}$ be a stochastic process obtained as the weighted sum of n SDE processes, as follows:

$$\tilde{\eta}(t) = \sum_{i=1}^{n} \sqrt{w_i}\,\eta_i(t), \tag{4.20}$$

where η_i, $i = 1, \ldots n$, are SDE processes with ACFs $R_{\eta_i}(\tau)$, $w_i > 0$ and

$$\sum_{i=1}^{n} w_i = 1. \tag{4.21}$$

If all n processes have an identical Gaussian PDF $\mathcal{N}(\mu_\eta, \sigma_\eta^2)$, the stochastic process $\tilde{\eta}$ has the same Gaussian PDF, $\mathcal{N}(\mu_\eta, \sigma_\eta^2)$, and an ACF which is a weighted sum of the ACFs of the n SDE processes, that is:

$$R_{\tilde{\eta}}(\tau) = \sum_{i=1}^{n} w_i R_{\eta_i}(\tau). \tag{4.22}$$

If the n SDE processes in (4.20) are processes as in (4.16), the resulting ACF of $\tilde{\eta}$ is a weighted sum of damped sinusoidal and decaying exponential functions and (4.22) can be rewritten as:

$$R_{\tilde{\eta}}(\tau) = \sum_{i=1}^{n} w_i \exp(-\alpha_i \tau) \cos(\omega_i \tau). \tag{4.23}$$

Hence, the superposition of SDE processes allows capturing any ACF that can be modeled as a weighted sum of exponential and/or sinusoidal functions. If the ACF does not show a periodic behavior then $\omega_i = 0, \forall i = 1, \ldots, n$.

4.2.2.1 Application to wind speed modeling

The fitting procedure for this SDE modeling method involves fitting the function in (4.23) to the ACF of the data as well as identifying a PDF that best captures the probability distribution of the data. It is presented in more detail in [12] and Appendix A. For illustration, the modelling method presented here is used to model wind speed. The details of the two example measured wind speed data sets used are provided in Table 4.1. In Figures 4.7 and 4.8, the histograms for the two data sets are shown. The memoryless transformation is used to build SDEs with both an analytically fitted PDF and a numerically fitted PDF. The analytically fitted PDFs are the 1-parameter Rayleigh distribution for Data set 1 and the 3-parameter Beta distribution for Data set 2. In the case of Data set 1 the numerically fitted PDF gives a better fit as there is a slight shift in the probability distribution of the measured wind

Table 4.1 The wind speed data sets modeled

#	Sampling rate	Averaged	Duration	Location
1	1 hour	Yes	3 years	Malin Head, Donegal, Ireland
2	1 second	No	1 month	Tracy, California

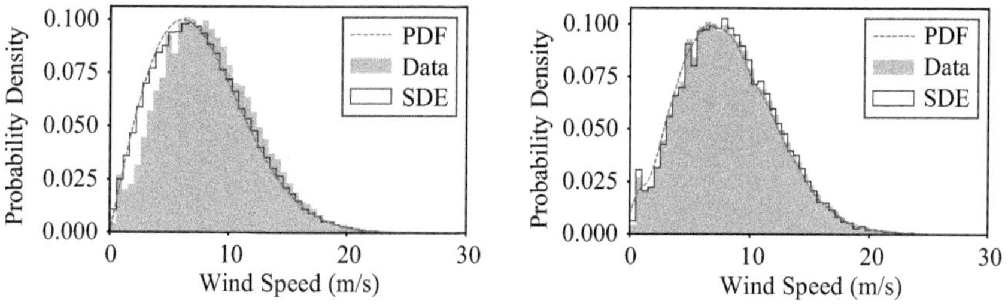

(a) Analytical fitted PDF (b) Numerical fitted PDF

Figure 4.7 Histogram for measured wind speed Data set 1, fitted analytical and numerical PDFs and histogram of the equivalent simulated SDEs

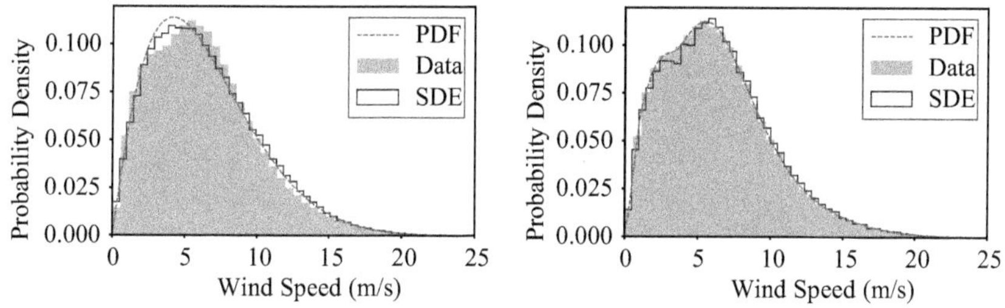

(a) Analytical fitted PDF (b) Numerical fitted PDF

Figure 4.8 Histogram for measured wind speed Data set 2, fitted analytical and numerical PDFs and histogram of the equivalent simulated SDEs

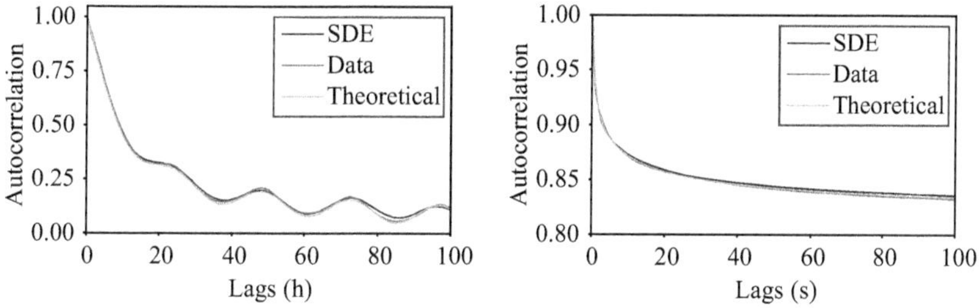

(a) Data set 1 (b) Data set 2

Figure 4.9 ACFs of the data, the fitted theoretical ACFs and the ACFs of the simulated SDE models

speed data. For Data set 2, both the analytically and numerically defined PDFs give a good fit.

In Figure 4.9, the ACFs of Data sets 1 and 2 are shown. The ACF is highly dependent on the sampling rate and has different shapes for different sampling rates.

The ACF of Data set 2 shows a poorly damped sinusoidal mode with a period of 24 h, while the ACF of Data set 2 drops rapidly in the first 20 sec before settling to a more gradual slope. The method outlined above is able to reproduce these different shapes of wind speed ACFs as a weighted sum of exponential and sinusoidal functions.

4.2.3 Method for modeling SDEs with jumps

Certain processes in power systems will not only be stochastic but also have faster variations, termed as jumps [26]. These are for example seen in load consumption and solar generation. SDE jump diffusion models can be used to model such stochastic processes with jumps. A general one-dimensional jump diffusion model, i.e. SDE with jumps is defined as:

$$d\eta_J(t) = a(t, \eta_J)dt + b(t, \eta_J)dW(t) + c(t, \eta_J)dJ(t). \tag{4.24}$$

Here $c(t, \eta_J)$ is the jump coefficient which determines the jump size and $\{J(t), t > 0\}$ is the Poisson process. The Poisson process with intensity $\lambda > 0$ is defined as a type of stochastic process called a counting process which is characterized by the following properties:

1. $J(0) = 0$, with probability 1.
2. It has stationary independent increments.
3. The number of events (or points) in any given time interval of length t is a Poisson random variable with the mean λt.
4. Its probability distribution is:

$$f(k, \lambda) = \frac{\lambda^k}{k!}\exp(-\lambda), \tag{4.25}$$

where $k \in \mathbb{N}$.

In Sections 4.3.3.1 and 4.3.3.3, jump diffusion SDEs are used to model loads and solar irradiance, respectively.

4.2.4 Method for modeling correlation

Stochastic processes observed in power systems are, in some cases, not fully independent. For example, the voltage on the one side of a line/transformer is correlated to the voltage on the other side of that line/transformer. Similarly, there exists a correlation between the active and reactive load power consumption at a given bus [17].

The set of multi-dimensional SDEs defined in (4.1) are uncorrelated if $\boldsymbol{W}$ is a vector of independent Wiener processes. In that case, the elements of the covariance matrix $\mathbf{P} \in \mathbb{R}^{n\times n}$ of the increments $d\boldsymbol{W}$ are defined as follows:

$$P_{i,j} = \text{cov}[dW_i, dW_j] = \begin{cases} \sigma_i^2, & \text{if } i = j, \\ 0, & \text{if } i \neq j, \end{cases} \tag{4.26}$$

To model the multi-dimensional SDEs as correlated, a vector of correlated Wiener processes is constructed. Suppose $\boldsymbol{V}$ is a vector of n elements that are obtained as the linear combination of the n uncorrelated Wiener processes $\boldsymbol{W}$. The correlation matrix $\mathbf{R} \in \mathbb{R}^{n\times n}$ for $\boldsymbol{V}$ has the form:

$$\mathbf{R} = \begin{bmatrix} 1 & r_{1,2} & r_{1,3} & \dots & r_{1,n} \\ r_{2,1} & 1 & r_{2,3} & \dots & r_{2,n} \\ r_{3,1} & r_{3,2} & 1 & \dots & r_{3,n} \\ \vdots & \vdots & \vdots & \ddots & \vdots \\ r_{n,1} & r_{n,2} & r_{n,3} & \cdots & 1 \end{bmatrix},$$

where $r_{i,j} = \text{corr}[dV_i, dV_j]$ represents the correlation between dV_i and dV_j. Naturally, $r_{i,j} = 1$ if $i = j$, since the correlation of any variable with itself is always 1.

The value of r can be calculated using the Pearson's correlation coefficient. The elements of covariance matrix $\mathbf{P} \in \mathbb{R}^{n\times n}$ of $d\boldsymbol{V}$ are written as:

$$P_{i,j} = \text{cov}[dV_i, dV_j] = \begin{cases} \sigma_i^2, & \text{if } i = j, \\ r_{i,j}\sigma_i\sigma_j, & \text{if } i \neq j, \end{cases} \tag{4.27}$$

The correlation between each set of SDEs can be modeled as either a constant or a stationary stochastic process. The technique presented here is equally applicable to both cases, as the value of correlation is just an entry in the correlation matrix $\mathbf{R}$ which can be updated at every time step in case the correlation is modeled as a stochastic process.

The procedure to write $d\boldsymbol{V}$ in terms of $d\boldsymbol{W}$ is involved and is thoroughly explained in [27]. The final expression is:

$$d\boldsymbol{V}(t) = \mathbf{C}\, d\boldsymbol{W}(t), \tag{4.28}$$

where $\mathbf{C} \in \mathbb{R}^{n\times n}$ is chosen such that:

$$\mathbf{R} = \mathbf{C}\,\mathbf{C}^T. \tag{4.29}$$

A family of $\mathbf{C}$ matrices satisfies (4.29) but the best choice of $\mathbf{C}$ is a lower triangular matrix as it reduces memory requirements and the computational burden of numerical implementations. A lower triangular matrix is obtained by performing Cholesky-decomposition of $\mathbf{R}$. Cholesky decomposition requires that the input matrix is positive semi-definite. Generally, this condition is satisfied for stochastic processes of power systems.

Based on the definitions above, a n-dimensional correlated SDE is constructed by substituting (4.28) into (4.1) as follows:

$$\begin{aligned} d\boldsymbol{\eta}(t) &= \boldsymbol{a}(t, \boldsymbol{\eta}(t))dt + \boldsymbol{b}(t, \boldsymbol{\eta}(t))d\boldsymbol{V}(t), \\ d\boldsymbol{V}(t) &= \mathbf{C}\, d\boldsymbol{W}(t), \end{aligned} \tag{4.30}$$

where $\boldsymbol{a}$, $\boldsymbol{b}$, and $\boldsymbol{W}$ have the same meaning as in (4.1); $\mathbf{C}$ satisfies (4.29); $\boldsymbol{\eta} \in \mathbb{R}^n$ is the vector of correlated stochastic processes; and $d\boldsymbol{V} \in \mathbb{R}^n$ is the vector of correlated white noises.

4.3 Modeling power systems as SDAEs

The transient behaviour of electric power systems is traditionally described through a set of differential algebraic equations (DAEs) as follows:

$$\begin{aligned} d\boldsymbol{x}(t) &= \boldsymbol{f}(\boldsymbol{x}(t), \boldsymbol{y}(t))dt\,, \\ \boldsymbol{0} &= \boldsymbol{g}(\boldsymbol{x}(t), \boldsymbol{y}(t))\,, \end{aligned} \tag{4.31}$$

where $\boldsymbol{f}$ ($\boldsymbol{f} : \mathbb{R}^n \times \mathbb{R}^m \mapsto \mathbb{R}^n$) are the differential equations; $\boldsymbol{g}$ ($\boldsymbol{g} : \mathbb{R}^n \times \mathbb{R}^m \mapsto \mathbb{R}^{n_y}$) are the algebraic equations; $\boldsymbol{x}$ ($\boldsymbol{x} \in \mathbb{R}^n$) are the state variables, e.g., rotor speeds and rotor angles of synchronous machines, the dynamic states of loads and system controllers; $\boldsymbol{y}$ ($\boldsymbol{y} \in \mathbb{R}^m$) are the algebraic variables, e.g., voltage magnitudes at system buses; and $t \in \mathbb{R}^+$ is the time.

Despite the fact that (4.31) are well accepted and are the common choice in power system software packages, some aspects of the reality are missing from this formulation, e.g., stochastic behaviour and variable functional relations. It is of interest to define the possible effects of stochastic perturbations on the transient behavior of (4.31). These kinds of perturbations can be originated by the stochastic variations of loads, transient rotor vibrations of synchronous machines, harmonics, EMT transients, measurement errors of control devices, etc. Additionally, there could be wind and solar power plants connected to the system whose production is stochastic due to the stochastic nature of the primary energy they use. The effect of such perturbations can lead to stochastic behaviors of the main system variables, e.g., frequency, voltages, and power flows.

By introducing stochastic phenomena, modeled as presented in Section 4.2, the power system DAEs are transformed into a set of SDAEs, as follows:

$$\begin{aligned} d\boldsymbol{x}(t) &= \boldsymbol{f}(\boldsymbol{x}(t), \boldsymbol{y}(t), \boldsymbol{\eta}(t))dt + \mathbf{H}\, d\boldsymbol{\eta}(t)\,, \\ \boldsymbol{0}_m &= \boldsymbol{g}(\boldsymbol{x}(t), \boldsymbol{y}(t), \boldsymbol{\eta}(t))\,, \\ d\boldsymbol{\eta}(t) &= \boldsymbol{a}(\boldsymbol{\eta}(t))dt + \boldsymbol{b}(\boldsymbol{\eta}(t))d\boldsymbol{W}(t) + \boldsymbol{c}(\boldsymbol{\eta}(t))d\boldsymbol{J}(t)\,, \end{aligned} \tag{4.32}$$

where the functions $\boldsymbol{f}$ ($\boldsymbol{f} : \mathbb{R}^n \times \mathbb{R}^m \times \mathbb{R}^p \mapsto \mathbb{R}^n$) and $\boldsymbol{g}$ ($\boldsymbol{g} : \mathbb{R}^m \times \mathbb{R}^m \times \mathbb{R}^p \mapsto \mathbb{R}^m$) are updated to include the effect of the stochastic term $\boldsymbol{\eta}$ ($\boldsymbol{\eta} \in \mathbb{R}^p$) and $d\boldsymbol{W}$ ($d\boldsymbol{W} \in \mathbb{R}^q$) and $d\boldsymbol{J}$ ($d\boldsymbol{J} \in \mathbb{R}^\nu$) is the vector of increments of the Wiener and Poisson processes, respectively. Each of the SDEs is characterized by its drift $\boldsymbol{a}$ ($\boldsymbol{a} : \mathbb{R}^p \mapsto \mathbb{R}^p$) and diffusion terms $\boldsymbol{b}$ ($\boldsymbol{b} : \mathbb{R}^p \mapsto \mathbb{R}^p \times \mathbb{R}^q$) and $\boldsymbol{c}$ ($\boldsymbol{c} : \mathbb{R}^p \mapsto \mathbb{R}^p \times \mathbb{R}^\nu$). The term $\mathbf{H}\, d\boldsymbol{\eta}$ (with $\mathbf{H} : \mathbb{R}^n \times \mathbb{R}^p$) expresses the possibility to have additive noise in the differential equations. Observe that, in (4.32), $\boldsymbol{g}$ does not explicitly depend on the Wiener and/or Poisson processes nor on $d\boldsymbol{\eta}$, which allows solving (4.32) by means of state-of-art integration techniques for SDAEs [6].

The model (4.32) includes both continuous and discrete stochastic processes. However, the latter are utilized in the remainder of the chapter only for modelling fluctuations of load power consumption and solar generation. To simplify the notation, thus, in the next sections, unless stated otherwise and without loss of generality, $\boldsymbol{c} = \mathbf{0}$ is assumed.

4.3.1 Initialization of stochastic power system models

Equations (4.32) can be used to define a Stochastic Initial Value Problem (SIVP), which is formally obtained by (4.32) and the initial conditions:

$$\begin{aligned} \boldsymbol{x}(t_0) &= \boldsymbol{x}_0\,, \\ \boldsymbol{y}(t_0) &= \boldsymbol{y}_0\,, \\ \boldsymbol{\eta}(t_0) &= \boldsymbol{\eta}_0\,, \end{aligned} \tag{4.33}$$

For the sake of unifying the notation, the SIVP (4.32)–(4.33) is rewritten in the following compact expression:

$$\begin{aligned} d\boldsymbol{x}(t) &= \boldsymbol{f}(\boldsymbol{x}(t), \boldsymbol{y}(t), \boldsymbol{\eta}(t))dt + \mathbf{H}d\boldsymbol{\eta}(t)\,, & \boldsymbol{x}(t_0) &= \boldsymbol{x}_0\,, \\ \mathbf{0} &= \boldsymbol{g}(\boldsymbol{x}(t), \boldsymbol{y}(t), \boldsymbol{\eta}(t))\,, & \boldsymbol{y}(t_0) &= \boldsymbol{y}_0\,, \\ d\boldsymbol{\eta}(t) &= \boldsymbol{a}(\boldsymbol{\eta}(t))dt + \boldsymbol{b}(\boldsymbol{\eta}(t))d\boldsymbol{W}(t)\,, & \boldsymbol{\eta}(t_0) &= \boldsymbol{\eta}_0\,. \end{aligned} \tag{4.34}$$

Preserving the structure of (4.34), the Deterministic Initial Value Problem (DIVP) is:

$$\begin{aligned} d\boldsymbol{x}(t) &= \boldsymbol{f}(\boldsymbol{x}(t), \boldsymbol{y}(t), \mathbf{0})dt\,, & \boldsymbol{x}(t_0) &= \boldsymbol{x}_0\,, \\ \mathbf{0} &= \boldsymbol{g}(\boldsymbol{x}(t), \boldsymbol{y}(t)\,, \boldsymbol{0})\,, & \boldsymbol{y}(t_0) &= \boldsymbol{y}_0\,, \\ \dot{\boldsymbol{\eta}}(t) &= \boldsymbol{0}\,, & \boldsymbol{\eta}(t_0) &= \mathbf{0}\,. \end{aligned} \tag{4.35}$$

Finally, the following probabilistic initial value problem (PIVP) has been largely used in power system analysis:

$$\begin{aligned} d\boldsymbol{x}(t) &= \boldsymbol{f}(\boldsymbol{x}(t), \boldsymbol{y}(t), \boldsymbol{\eta}_0)dt\,, & \boldsymbol{x}(t_0) &= \boldsymbol{x}_0\,, \\ \mathbf{0} &= \boldsymbol{g}(\boldsymbol{x}(t), \boldsymbol{y}(t)\,, \boldsymbol{\eta}_0), & \boldsymbol{y}(t_0) &= \boldsymbol{y}_0\,, \\ \dot{\boldsymbol{\eta}}(t) &= \mathbf{0}\,, & \boldsymbol{\eta}(t_0) &= \boldsymbol{\eta}_0\,, \end{aligned} \tag{4.36}$$

where the effect of the stochastic variables $\boldsymbol{\eta}(t)$ is just due to their initial values $\boldsymbol{\eta}_0$, but for $t > t_0$ the evolution of the system is purely deterministic. Hence the trajectories of (4.35) and (4.36) are regulated by the same differential equations and only differ because of the initial condition. It is relevant to note that (4.35) and (4.36) are particular cases of the general formulation given in (4.34).

The solution to the DIVP in (4.35) consists in finding a (stable) equilibrium point $(\boldsymbol{x}_d, \boldsymbol{y}_d)$ that satisfies the condition:

$$\begin{aligned} \mathbf{0}_n &= \boldsymbol{f}(\boldsymbol{x}_d, \boldsymbol{y}_d)\,, \\ \mathbf{0}_m &= \boldsymbol{g}(\boldsymbol{x}_d, \boldsymbol{y}_d)\,. \end{aligned} \tag{4.37}$$

The initial values of the deterministic state and algebraic variables are then set as $\boldsymbol{x}(t_0) = \boldsymbol{x}_d$ and $\boldsymbol{y}(t_0) = \boldsymbol{y}_d$, respectively.

Solving the SIVP, (4.34), on the other hand, is not as straightforward. Three methods to solve the SIVP are presented here. The first two methods are:

- **Method 1:** The SDAEs are initialized in a deterministic way. That is the deterministic state and the algebraic variables are initialized as shown in (4.37) and the initial values of the stochastic processes are set equal to their expectation, i.e. $\boldsymbol{\eta}(t_0) = \boldsymbol{\eta}_d$.
- **Method 2:** The initialization of the deterministic states and the algebraic variables is the same as in Method 1, thus leading to $\boldsymbol{x}(t_0) = \boldsymbol{x}_d$ and $\boldsymbol{y}(t_0) = \boldsymbol{y}_d$, which are obtained using $\boldsymbol{\eta}(t_0) = \boldsymbol{\eta}_d$. Once this step is completed, the stochastic processes $\boldsymbol{\eta}$ are initialized in a probabilistic way. That is, the initial value of the stochastic processes are selected at random based on their probability distributions, say $\boldsymbol{\eta}_s$. The complete initial point is thus represented by $\boldsymbol{x}(t_0) = \boldsymbol{x}_d$, $\boldsymbol{y}(t_0) = \boldsymbol{y}_d$ and $\boldsymbol{\eta}(t_0) = \boldsymbol{\eta}_s$.

Method 3, presented below, proposes an alternative and efficient way to initialize the SDAE system as a whole. Note also that, in general, the initial point obtained with Method 2 yields $\boldsymbol{g}(\boldsymbol{x}_d, \boldsymbol{y}_d, \boldsymbol{\eta}_s) \neq \boldsymbol{0}_m$. This, however, is not an issue as, at the first step of the time domain simulation, the condition $\boldsymbol{g}(\boldsymbol{x}_d, \boldsymbol{y}(t_0), \boldsymbol{\eta}_s) = \boldsymbol{0}_m$ is recovered by means of the internal loop of the numerical integration scheme.

- **Method 3:** The starting point of this initialization method is the set of SDAEs linearized at the equilibrium point $(\boldsymbol{x}_d, \boldsymbol{y}_d, \boldsymbol{\eta}_d)$ as per Method 1 above. The linearization of (4.32) gives:

$$\begin{bmatrix} d\tilde{x}(t) \\ \boldsymbol{0}_m \\ d\tilde{\eta}(t) \end{bmatrix} = \begin{bmatrix} \boldsymbol{f}_x & \boldsymbol{f}_y & \boldsymbol{f}_\eta \\ \boldsymbol{g}_x & \boldsymbol{g}_y & \boldsymbol{g}_\eta \\ \boldsymbol{0}_{p,n} & \boldsymbol{0}_{p,m} & \boldsymbol{a}_\eta \end{bmatrix} \begin{bmatrix} \tilde{x}(t) \\ \tilde{y}(t) \\ \tilde{\eta}(t) \end{bmatrix} dt + \begin{bmatrix} \boldsymbol{0}_{n,q} \\ \boldsymbol{0}_{m,q} \\ \boldsymbol{b}(\boldsymbol{\eta}_o) \end{bmatrix} d\boldsymbol{W}(t), \tag{4.38}$$

where $\boldsymbol{f}_x, \boldsymbol{f}_y, \boldsymbol{f}_\eta, \boldsymbol{g}_x, \boldsymbol{g}_y, \boldsymbol{g}_\eta, \boldsymbol{a}_\eta$ are the Jacobian matrices of the system calculated at $(\boldsymbol{x}_o, \boldsymbol{y}_o, \boldsymbol{\eta}_o)$. $\tilde{x}$ and $\tilde{\eta}$ represent the deterministic and the stochastic states of the linearized system. Eliminating the algebraic variables from (4.38) and defining $z = (\tilde{x}, \tilde{\eta})$ leads to a set of linear SDEs, as follows:

$$\begin{aligned} \begin{bmatrix} d\tilde{x}(t) \\ d\tilde{\eta}(t) \end{bmatrix} &= \begin{bmatrix} \boldsymbol{f}_x - \boldsymbol{f}_y \boldsymbol{g}_y^{-1} \boldsymbol{g}_x & \boldsymbol{f}_\eta - \boldsymbol{f}_y \boldsymbol{g}_y^{-1} \boldsymbol{g}_\eta \\ \boldsymbol{0}_{p,n} & \boldsymbol{a}_\eta \end{bmatrix} \begin{bmatrix} \tilde{x}(t) \\ \tilde{\eta}(t) \end{bmatrix} dt + \begin{bmatrix} \boldsymbol{0}_{n,q} \\ \boldsymbol{b}(\boldsymbol{\eta}_o) \end{bmatrix} d\boldsymbol{W}(t) \\ &= \mathbf{A}_o\, z(t) dt + \mathbf{B}_o\, d\boldsymbol{W}(t). \end{aligned} \tag{4.39}$$

Based on the Fokker–Planck equation the probability distribution of all state variables in stationary conditions satisfies:

$$\mathbf{P}(z) = (\det |\, 2\pi \mathbf{C}\, |)^{-1/2} \cdot \exp\left(-\frac{1}{2} z^T \mathbf{C}^{-1} z\right), \tag{4.40}$$

where $\mathbf{C}$ is the covariance matrix of the state variables in (4.39). This matrix is symmetric and satisfies the Lyapunov equation:

$$\mathbf{A}_o\mathbf{C} + \mathbf{C}\mathbf{A}_o^T = -\mathbf{B}_o\mathbf{B}_o^T \,, \tag{4.41}$$

which is a special case of the Riccati equation. This equation can be solved numerically for systems of arbitrary size.
The covariance matrix can be clustered as:

$$\mathbf{C} = \begin{bmatrix} \mathbf{C}_{xx} & \mathbf{C}_{x\eta} \\ \mathbf{C}_{\eta x} & \mathbf{C}_{\eta\eta} \end{bmatrix}, \tag{4.42}$$

where $\mathbf{C}_{xx}$ and $\mathbf{C}_{\eta\eta}$ are the covariance matrices of $\tilde{x}$ and $\tilde{\eta}$, respectively. The two remaining sub-matrices $\mathbf{C}_{x\eta}$ and $\mathbf{C}_{\eta x} = \mathbf{C}_{x\eta}^T$ represent the covariance between the deterministic $\tilde{x}$ and stochastic $\tilde{\eta}$ state variables.†

This method to initialize the SDAE power system model in (4.32) begins with the initialization of the stochastic state variables, $\boldsymbol{\eta}$. The initial stochastic state variables are set independently at random, based on their probability distribution, i.e. $\boldsymbol{\eta}(t_0) = \boldsymbol{\eta}_s$. That is, in the same way as in Method 2. Then, the state variables $\boldsymbol{x}$ are initialized using the mean of the conditional distribution of $\boldsymbol{x}$, given $\boldsymbol{\eta}$, that is:

$$\boldsymbol{x}_s = \boldsymbol{x}_0 + \mathbf{C}_{x\eta}\mathbf{C}_{\eta\eta}^{-1}\,\boldsymbol{\eta}_s \,. \tag{4.43}$$

Thus, in this way the initial value of $\boldsymbol{x}$, set using Method 1 is modified to consider the effect of the random stochastic state initial value $\boldsymbol{\eta}_s$, considering their correlation.

As discussed for Method 2, there is no need to update the algebraic variables as they will be determined by the time integration routine. Hence, the resulting initial point is $\boldsymbol{x}(t_0) = \boldsymbol{x}_s$, $\boldsymbol{y}(t_0) = \boldsymbol{y}_d$ and $\boldsymbol{\eta}(t_0) = \boldsymbol{\eta}_s$.

These three initialization methods are further discussed and compared through a case study of the Irish power system model, presented in Section 4.5.1.

4.3.2 Modeling stochastic perturbations in power systems

To complete the modeling of stochastic phenomena in power systems, it is necessary to derive expressions that couple such a phenomena with the variables and parameters of power systems. With this aim, stochastic inputs are considered *perturbations* and stochastic models are derived by introducing stochastic variations, as discussed above.

†Once the covariance matrix $\mathbf{C}$ is known, it is straightforward to determine the covariance matrix of algebraic variables, as follows [28]:

$$\mathbf{K} = \mathbf{G}_o\,\mathbf{C}\,\mathbf{G}_o^T \,,$$

where

$$\mathbf{G}_o = -\boldsymbol{g}_y^{-1}\left[\boldsymbol{g}_x \;\; \boldsymbol{g}_\eta\right].$$

The diagonal elements of $\mathbf{K}$ are the sought variances of the algebraic variables $\tilde{y}$.

For the sake of simplicity, but without lack of generality, the discussion is restricted to a one-dimensional model.

Let $z(t)$ be a system variable or parameter, $z_s(t)$ a vector with the rest of system variables and parameters, and $\psi(z(t), z_s(t))$ a function representing the deterministic trajectory of $z(t)$. A stochastic model for this trajectory can be derived by introducing a stochastic variation $\eta_z(t)$ as follows:

$$z(t) = \psi(z(t), z_s(t)) + \eta_z(t). \tag{4.44}$$

Equation (4.44) can be particularized as follows:

1. If $z(t)$ represents a state variable ($z(t) \equiv x(t)$), the trajectory $\psi(x(t), z_s(t))$ is generally not known a priori, but it is the solution of a differential equation. Therefore, by taking derivatives with respect to time in both sides of equation (4.44) we have

$$dx(t) = f(x(t), z_s(t))dt + d\eta_x(t), \tag{4.45}$$

where

$$f(x(t), z_s(t))dt = d\psi(x(t), z_s(t)). \tag{4.46}$$

The dynamic behaviour of the stochastic variable $\eta_x(t)$ is a SDE, as follows:

$$d\eta_x(t) = a_x(\eta_x(t))dt + b_x(\eta_x(t))dW_x(t). \tag{4.47}$$

Therefore, (4.45) and (4.47) can be rewritten as the following set of SDEs:

$$\begin{aligned} dx(t) &= f(x(t), z_s(t))dt + d\eta_x(t), \\ d\eta_x(t) &= a_x(\eta_x(t))dt + b_x(\eta_x(t))dW_x(t). \end{aligned} \tag{4.48}$$

2. The variable $z(t)$ can represent an algebraic variable, e.g., $z(t) \equiv y(t)$. Taking into account that this kind of variable is constrained by an algebraic equation, e.g., $g(y(t), z_s(t)) = 0$, its deterministic trajectory $\psi(y(t), z_s(t))$ can be obtained by formally imposing that:

$$y(t) = g^{-1}(z_s(t)), \tag{4.49}$$

where $g^{-1}(\cdot)$ is the inverse of $g(\cdot)$. By combining (4.49) in (4.44), one obtains:

$$y(t) = g^{-1}(z_s(t)) + \eta_y(t). \tag{4.50}$$

Therefore, the stochastic model for an algebraic variable $y(t)$ is a SDAE where the stochastic perturbation $\eta_y(t)$ imposes that $y(t)$ varies according to the constraint $g(\cdot)$, as follows:

$$\begin{aligned} 0 &= g(y(t) - \eta_y(t), z_s(t)), \\ d\eta_y(t) &= a_y(\eta_y(t))dt + b_y(\eta_y(t))dW_y(t). \end{aligned} \tag{4.51}$$

3. In case $z(t)$ represents a system parameter, say $p(t)$ ($z(t) \equiv p(t)$), its deterministic trajectory is a constant, e.g., $\psi(p(t), z_s(t)) = p_0$. Therefore, according to (4.44), the stochastic model for a parameter is as follows:

$$
\begin{aligned}
p(t) &= p_0 + \eta_p(t)\,, \\
d\eta_p(t) &= a_p(\eta_p(t))dt + b_p(\eta_p(t))dW_p(t)\,.
\end{aligned}
\tag{4.52}
$$

4. Models (4.48), (4.51), and (4.52) have been obtained by adding a stochastic perturbation to a deterministic value or trajectory. If required, pure stochastic state and algebraic variables and parameters can be modeled by setting $dx(t) = d\eta_x(t)$, $y(t) = \eta_y(t)$, and $p(t) = \eta_p(t)$, where $d\eta_x(t)$ is given in (4.47) and $\eta_y(t)$, $\eta_p(t)$ are the solution of the SDEs in (4.51) and (4.52), respectively.
5. The multidimensional case can be derived by simply defining as many sets of (4.48), (4.51), and (4.52), as needed. Correlations between stochastic processes can be taken into account as presented in Section 4.2.4.

The formulation given in this section enables deriving SDAEs from DAEs by simply including additional (stochastic) differential equations to the original DAE model.

4.3.3 Examples

This subsection presents three examples of stochastic perturbation models. In particular, stochastic load power consumption, wind, and solar generation are modeled. For modeling the stochastic perturbations, the modeling techniques presented in Section 4.2 are used.

4.3.3.1 Load modeling

The stochastic load model presented here is developed based on the well-known voltage-dependent load model. It models the load variations using an OU SDE model with jumps where the correlation in the active and reactive power consumption of the load can be modeled. The SDE-based load model discussed is:

$$
\begin{aligned}
p_L(t) &= (p_{L0} + \eta_p(t))(v(t)/v_0)^k\,, \\
q_L(t) &= (q_{L0} + \eta_q(t))(v(t)/v_0)^k\,, \\
d\eta_p(t) &= \alpha_p(\mu_p - \eta_p(t))dt + b_p dW_p(t) + \varsigma_p(t)dJ_p(t)\,, \\
d\eta_q(t) &= \alpha_q(\mu_q - \eta_q(t))dt + b_q dW_q(t) + \varsigma_q(t)dJ_q(t)\,,
\end{aligned}
\tag{4.53}
$$

where $p_L(t)$ and $q_L(t)$ are the active and reactive power of the load, respectively, and p_{L0} and q_{L0} are parameters representing active and reactive load powers at $t = 0$. $v(t)$ is the voltage magnitude at the bus where the load is connected and v_0 is the value of this voltage magnitude at $t = 0$.

The model in (4.53) can, through the exponent k, define whether the load is a constant power load ($k = 0$), a constant current load ($k = 1$) or a constant impedance

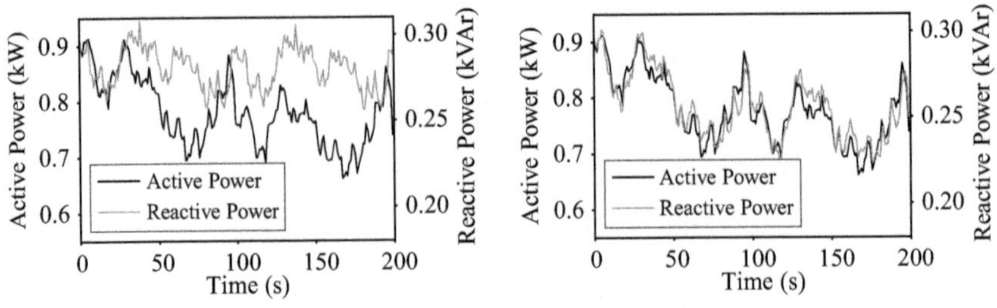

(a) No correlation (b) Correlated with $\rho W = 0.8$.

Figure 4.10 Active and reactive power of a load modeled without jumps

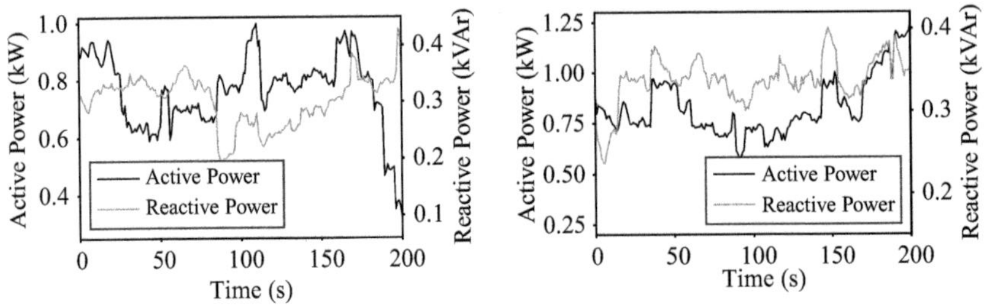

(a) No correlation (b) Correlated with $\rho J = 0.8$ and $\rho J_s = 0.8$.

Figure 4.11 Active and reactive power of a load modeled with jumps

load ($k = 2$). The stochastic variability of the load is modeled through the stochastic processes $\eta_p(t)$ and $\eta_q(t)$ which are formulated as SDE jump diffusion processes, where α is the mean-reversion speed, μ is the mean and σ is the diffusion component of the OU process (4.4).

The jump amplitudes $\varsigma_p(t)$ and $\varsigma_q(t)$ are normally distributed random numbers, namely $\varsigma_p(t) \sim \mathcal{N}(0, \sigma^2_{\varsigma_p})$ and $\varsigma_q(t) \sim \mathcal{N}(0, \sigma^2_{\varsigma_q})$. They are modeled to have a correlation ρ_{J_s} in the same way as the Wiener process. Details on the modeling of the correlation in the Wiener ($W_p(t)$ and $W_q(t)$), ρ_W, and Poisson processes ($J_p(t)$ and $J_q(t)$), ρ_J, are provided in [17] and Appendix B.

Examples of the load consumption of a load modeled using (4.53) are shown in Figures 4.10 and 4.11. In Figure 4.10, the loads are modeled without jumps, that is $\varsigma_p(t) = \varsigma_q(t) = 0$. Figure 4.10 allows us to compare two scenarios, that is what the load consumption looks like without correlation and with a correlation of $\rho_W = 0.8$. The load model parameters used to generate these example trajectories are provided in Table 4.2.

Figure 4.11(a) shows load consumption modeled with jumps where there is no correlation in the active and reactive power consumption. This results in the active and

Table 4.2 The SDE parameters for the load model. p_{L0} and q_{L0} are the active and reactive powers of the respective loads at time $t = 0$

Parameters		
$\alpha_p = \alpha_q = 0.02,$	$\sigma_p = 0.05p_{L0},$	$\sigma_q = 0.05q_{L0},$
$\sigma_{\varsigma_p} = 0.1p_{L0},$	$\sigma_{\varsigma_q} = 0.1q_{L0},$	$\mu_p = \mu_q = 0.$

reactive power trajectories being very different from one and other and for example at the end of the simulation it can be seen that while the active power is dropping the reactive power is increasing. In Figure 4.11(b), an example is shown where both the jump times and jump amplitudes are correlated, that is $\rho_J = \rho_{J_s} = 0.8$. In this case we can see that the two trajectories are following the same trend and in many cases the jumps are proportionally similar in the active and reactive power consumed.

4.3.3.2 Wind generation

The power generated by a wind turbine depends on the weather conditions which makes it a highly volatile power source. In order to ensure a reliable and secure operation of the grid, it is essential to model the source of such a volatility, i.e., the wind speed.

The aggregated wind speed model presented here consists of two parts: a constant-mean wind speed, v_c, and a Gaussian stochastic process, $\eta_w(t)$. The wind is modeled in this way since the wind speed variations within a 10 minute time frame can be assumed to be normally distributed around a certain mean v_c. Thus, the wind speed model for an aggregated wind farm is:

$$v_{\text{wind}}(t) = v_c + \eta_w(t), \tag{4.54}$$

where $v_{\text{wind}}(t)$ is the modeled wind speed time-series and $\eta_w(t)$ is a stochastic process modeled as a SDE. The aggregated wind speed model considers both the filtering across the blade area of a single turbine and the aggregation effect across the whole wind farm.

Wind speed is not uniform across the rotor blade area. For example, wind speed at the tip, center and hub can differ. However, these variations even out over the blade area. This damping effect by the rotor blades is modeled as a low-pass filter, shown in Figure 4.12. The input is the model in (4.54) generating wind speed at hub height of the wind turbine. The output is the equivalent wind speed that produces the same torque as the actual wind field.

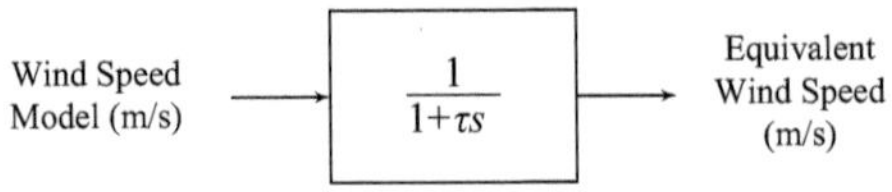

Figure 4.12 Low-pass filter that represents the damping effect of the blades of a wind turbine

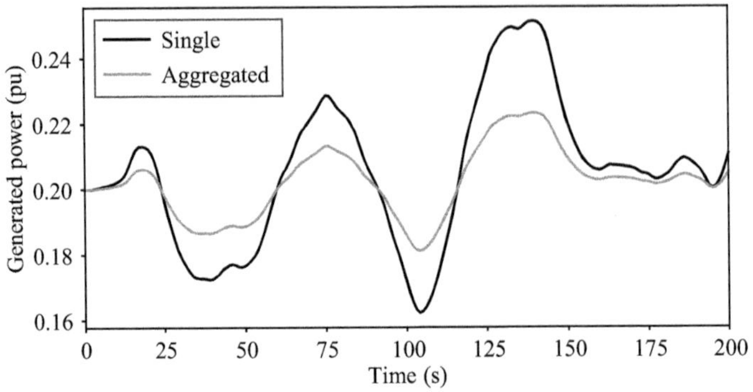

Figure 4.13 Generated power of a single wind turbine and an aggregated wind farm of 20 turbines

The time constant, τ [s], of the low-pass filter is proportional to the rotor radius, R [m], and the average wind speed at the hub height, v_c [m/s]. The time constant of the low-pass filter is:

$$\tau = \frac{\gamma R}{v_c}, \tag{4.55}$$

where γ is the decay factor over the disc [29].

The stochastic process $\eta_w(t)$ is defined using the method presented in Section 4.2.2. Thus, $\eta_w(t)$ has the probability distribution $\mathcal{N}(0, \sigma^2_{\eta w})$ and an ACF as in (4.23). The parameters of the stochastic process will differ depending on the hub heights of the wind turbines and their locations. The standard deviation of the process can be set as $10 - 20$ % of the mean wind speed v_c. The aggregated wind speed is defined through the $\sigma_{\eta w}$ parameter:

$$\sigma_{\eta w} = 0.2 \cdot v_c / \sqrt{2 n_{\text{turb}}}, \tag{4.56}$$

where n_{turb} is the number of turbines in the wind farm. Thus, the standard deviation of the modeled wind speed decreases in proportion to the number of turbines, as the variability of wind speed averages out over a spread wind farm.

The power output of an example wind farm is shown in Figure 4.13. It shows two cases. The first case is when the wind speed is modeled as for one turbine. The

Table 4.3 The SDE parameters for the wind speed model

Parameters	
$w_1 = 0.35$	$\alpha_1 = -0.3$
$w_2 = 0.5$	$\alpha_2 = -0.03$
$w_3 = 0.15$	$\alpha_3 = -0.0001$

second case is where the aggregated wind speed model presented in this section is used. The number of turbines for the aggregated case is 20. The parameters used for the wind speed model are presented in Table 4.3.

4.3.3.3 Solar generation

Solar generation is renewable and eco-friendly but also highly volatile due to the position of the sun and clouds changing. The solar irradiance measured at a single photovoltaic (PV) panel is known to be very jumpy during mixed clouding conditions. Solar irradiance of a single PV can be modeled through the combination of a deterministic model and a stochastic model:

$$G(t) = G_C \cdot k(t), \tag{4.57}$$

where G_C is the deterministic clear-sky model, which represents the highly predictable daily variations in the solar irradiance due to the apparent movement of the sun, and k is the clear-sky index, which models the stochastic variations effecting the solar irradiance due to cloud coverage [13].

The deterministic clear-sky model used is:

$$G_C = a \cdot (\cos(z))^b, \tag{4.58}$$

where z is the zenith angle, which is estimated based on the location and time of day. The parameters a and b are determined by fitting (4.58) to the measured data for clear-sky days. These coefficients change day by day and are thus found for each clear day of the data set used.

The clear-sky stochastic variations are modeled using the OU process:

$$d\eta(t) = -\alpha\, \eta(t)dt + bdW(t). \tag{4.59}$$

The jumps are modeled as:

$$h_i(t) = mP_i(t), \tag{4.60}$$

where m is the jump amplitude assumed to be a normally distributed random number, namely, $m \sim \mathcal{N}(\mu_m, \sigma_m^2)$. $P_i(t)$ is a step function that can get only 0/1 values, where the number of transitions per period are determined with a Poisson distribution with the parameter λ. The duration of each jump is determined with a normal distribution $\delta \sim |\mathcal{N}(0, \sigma_\delta^2)|$. In turn, each time $P_i(t)$ switches from 0 to 1 (or to 1 to 0), it remains constant for a time δ.

Visual inspection of the measured clear-sky index data allows identifying two types of jumps of the clear-sky index:

- **Jump model 1 (JM1):** Big clouds passing over the PV that block most of the solar irradiance.
- **Jump model 2 (JM2):** Small clouds that typically pass by more frequently and only partially reduce the solar irradiance.

The resulting model of the clear-sky index is:

$$k(t) = \eta(t) + u(t)\,h(t), \tag{4.61}$$

where $u(t)$ is a function that defines the duration of a clouding event:

$$u(t) = \begin{cases} 1 & \text{if } u_{\text{start}} \le t \le u_{\text{stop}} \\ 0 & \text{otherwise,} \end{cases} \tag{4.62}$$

where u_{start} and u_{stop} are the starting and ending times of the clouding event and

$$h(t) = \begin{cases} -h_1(t) + h_2(t) & \text{if } h_1(t) > 0 \\ -h_2(t) & \text{otherwise,} \end{cases} \tag{4.63}$$

where h_1 and h_2 are JM1 and JM2, respectively.

The aggregation of a whole plant of PV panels smooths the effect of the most rapid jumps. This smoothing effect is dependent on the area the plant covers. The aggregated irradiance for a PV plant can be modeled as a low pass filter, as shown in Figure 4.14 [30]. The cut-off frequency for the filter is directly dependent on the square root of the plant area S, measured in Ha.

In Figure 4.15, the generated power of a farm of PV panels is shown. There, the output power is compared for when the irradiance is modeled for a single point to a case where the aggregated solar irradiance is considered, i.e., using the low-pass filter presented in Figure 4.14. In this case $S = 2$ Ha. The parameters used for the solar model are presented in Table 4.4.

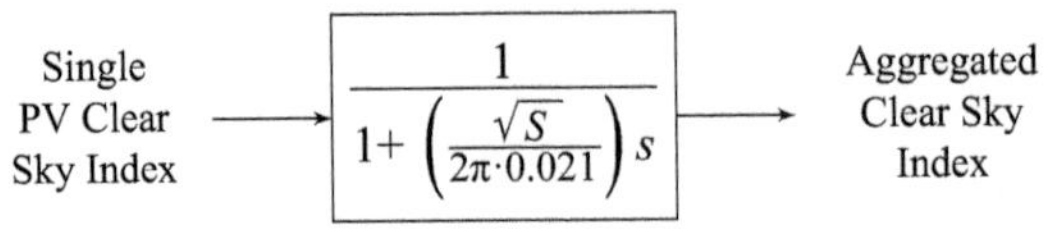

Figure 4.14 Low-pass filter that represents the smoothing effect of a PV plant

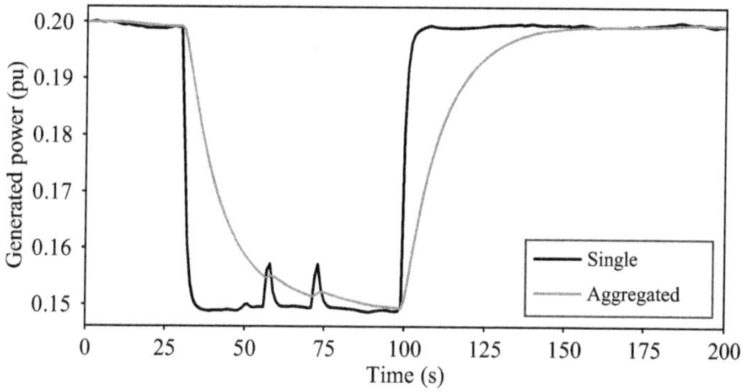

Figure 4.15 Generated power by a single PV panel and aggregated power output of a PV park covering an area of 2 Ha

Table 4.4 Parameters used for the SDE solar clear-sky index model

Parameters	**Jump model 1**	**Jump model 2**	**Clear sky OU**
λ	0.007	0.05	
μ_m	0.7	0	
σ_δ^2	30	3	
σ_m^2	0.05	0.001	
α			0.0001
σ			0.007

4.4 Time-domain integration of SDAEs

Very few SDEs can be explicitly solved, thus, numerical methods have to be applied. Reference [6] provides a detailed description of the available methods for the numerical solution of SDEs. Numerical methods for SDEs can show two types of convergence: *strong* and *weak*.

Strong convergence refers to the goodness of the approximation when the focus is on the process trajectories themselves, and is a straightforward generalization of the usual convergence criterion applied to the numerical schemes for DAEs. Formally, an approximation $\boldsymbol{\eta}_N$ converges strongly with order $\beta_S > 0$ to the solution $\boldsymbol{\eta}$ at time t_N if there exist a positive constant c, independent of Δt, such that

$$||\boldsymbol{\eta}(t_N) - \boldsymbol{\eta}_N|| \leq c(\Delta t)^{\beta_S}, \tag{4.64}$$

with $\Delta t \in (0, \overline{\Delta t})$, and $\overline{\Delta t} > 0$ is a given step length.

On the other hand, weak convergence refers to the goodness of the approximation of the statistical properties of the solutions to the statistical properties of the process.

Formally, an approximation $\boldsymbol{\eta}_N$ converges weakly with order $\beta_W > 0$ to the solution $\boldsymbol{\eta}$ at time t_N if there exists a positive constant c, independent of Δt, such that

$$||E[\mathcal{M}(\boldsymbol{\eta}(t_N)] - E[\mathcal{M}(\boldsymbol{\eta}_N)]|| \leq c(\Delta t)^{\beta_W}, \tag{4.65}$$

with $\Delta t \in (0, \overline{\Delta t})$, where $\mathcal{M}$ is a smooth function satisfying certain polynomial growth conditions [6], that usually represents a moment. A weak convergence is sufficient when Monte Carlo simulation is used. In power systems simulations, we are often more concerned with the weak convergence as it enables us to study the statistical properties of the stochastic system.

The most common methods used for integrating SDEs and SDAEs are explicit, being the most popular ones the Euler–Maruyama and Milstein explicit schemes [6]. These methods have been used for studying power system transients, e.g., [31]. However, the stiffness of power system equations makes explicit integration schemes particularly prone to numerical issues and errors.

A way to deal with the aforementioned difficulties is to consider an implicit integration scheme for the deterministic functions $\boldsymbol{f}(\cdot)$ and the drift terms $\boldsymbol{a}(\cdot)$, and an explicit scheme for the diffusion terms $\boldsymbol{b}(\cdot)$. To illustrate the resulting integration scheme, let us suppose a power system where all variables and parameters are affected by stochastic perturbations according to models (4.48), (4.51), and (4.52). The set of SDAEs is as follows:

$$\begin{aligned} d\boldsymbol{x}(t) &= \boldsymbol{f}(\boldsymbol{x}(t), \boldsymbol{y}(t) - \boldsymbol{\eta}_y(t), \boldsymbol{p}(t))dt + d\boldsymbol{\eta}_x(t), \\ \boldsymbol{0} &= \boldsymbol{g}(\boldsymbol{x}(t), \boldsymbol{y}(t) - \boldsymbol{\eta}_y(t), \boldsymbol{p}(t)), \\ \boldsymbol{p}(t) &= \boldsymbol{p}_0 + \boldsymbol{\eta}_p(t), \\ d\boldsymbol{\eta}(t) &= \boldsymbol{a}(\boldsymbol{\eta}(t))dt + \boldsymbol{b}(\boldsymbol{\eta}(t))\, d\boldsymbol{W}(t), \end{aligned} \tag{4.66}$$

where the last equation encompasses all SDEs in (4.48), (4.51), and (4.52), i.e.:

$$d\boldsymbol{\eta}(t) = \begin{bmatrix} d\boldsymbol{\eta}_x(t) \\ d\boldsymbol{\eta}_y(t) \\ d\boldsymbol{\eta}_p(t) \end{bmatrix} = \begin{bmatrix} \boldsymbol{a}_x(\boldsymbol{\eta}_x(t)) \\ \boldsymbol{a}_y(\boldsymbol{\eta}_y(t)) \\ \boldsymbol{a}_p(\boldsymbol{\eta}_p(t)) \end{bmatrix} dt + \begin{bmatrix} \boldsymbol{b}_x(\boldsymbol{\eta}_x(t)) \\ \boldsymbol{b}_y(\boldsymbol{\eta}_y(t)) \\ \boldsymbol{b}_p(\boldsymbol{\eta}_p(t)) \end{bmatrix} \begin{bmatrix} d\boldsymbol{W}_x(t) \\ d\boldsymbol{W}_y(t) \\ d\boldsymbol{W}_p(t) \end{bmatrix}. \tag{4.67}$$

For example, if the implicit scheme is the trapezoidal method and the explicit scheme is the Milstein scheme, the jth step of the resulting integration scheme for model (4.66) is as follows:

$$\begin{aligned} \boldsymbol{x}(t_j) &= \boldsymbol{x}(t_{j-1}) + \frac{1}{2} \cdot \big(\boldsymbol{f}(\boldsymbol{x}(t_j), \hat{\boldsymbol{y}}(t_j), \boldsymbol{p}(t_j)) + \boldsymbol{f}(\boldsymbol{x}(t_{j-1}), \hat{\boldsymbol{y}}(t_{j-1}), \boldsymbol{p}(t_{j-1}))\big) \cdot \Delta t \\ &\quad + \boldsymbol{\eta}_x(t_j) - \boldsymbol{\eta}_x(t_{j-1}), \\ \boldsymbol{0} &= \boldsymbol{g}(\boldsymbol{x}(t_j), \hat{\boldsymbol{y}}(t_j), \boldsymbol{p}(t_j)), \\ \boldsymbol{p}(t_j) &= \boldsymbol{p}_0 + \boldsymbol{\eta}_p(t_j), \end{aligned} \tag{4.68}$$

$$\eta(t_j) = \eta(t_{j-1}) + \frac{1}{2} \cdot \left(a(\eta(t_j)) + a(\eta(t_{j-1}))\right) \cdot \Delta t$$
$$+ b(\eta(t_{j-1})) \cdot \Delta W$$
$$+ \frac{1}{2} \cdot b(\eta(t_{j-1})) \cdot \frac{\partial b(\eta(t_{j-1}))}{\partial \eta(t_{j-1})} \cdot \left((\Delta W)^2 - \Delta t\right),$$

where $\hat{y}(t) = y(t) - \eta_y(t)$, $\Delta t = t_j - t_{j-1}$ is the time step length of the integration scheme, and $\Delta W = W(t_j) - W(t_{j-1})$ is the vector of increments of the Wiener process. This scheme provides a strong convergence order of 1.0 ($O(\Delta t)$) and a weak convergence order of 1.0 ($O(\Delta t)$).

When considering the application of a numerical scheme like (4.68), one has to restrict the attention to a finite sub-interval $[t_0, t_f]$. Moreover, it is necessary to choose an appropriate discretization $t_0 < t_1 < \cdots < t_j < \cdots < t_N = t_f$ of $[t_0, t_f]$. The other crucial problem is simulating sample paths of the Wiener process over the discretization of $[t_0, t_f]$. Considering an equally-spaced discretization $t_j - t_{j-1} = (t_f - t_0)/N = \Delta t, j = 1, \ldots, N$, the random independent increments are

$$W(t_j) - W(t_{j-1}) \sim \mathcal{N}(0, \Delta t), \qquad n = 1, \ldots, N \tag{4.69}$$

of the Wiener process $\{W(t), t_0 \leq t \leq t_f\}$. From (4.69), it is clear that the values of the Wiener process increments depend on the size of Δt and, therefore, the use of a different step size Δt will lead to a different path of the Wiener process.

If the analysis is on the trajectories themselves, that is in regards to strong convergence, it may be of interest to simulate the Wiener process with a time step $\Delta t^{\mathrm{w}} \leq \Delta t$. For example if trajectories of the solutions of SDAEs with different discretization steps Δt are being compared. Then, the step size Δt^{w} can be set equal to the smallest discretization step Δt used in the integration scheme. When a larger step size is required for the integration scheme, a new path of the Wiener process is constructed by adding intermediate values of the path that was generated with the smallest discretization step.

Figure 4.16 shows an example of this procedure. The solid gray line represents the original sample path of the Wiener process which was generated by using a step size $\Delta t^{\mathrm{w}} = 0.001$ for a $\Delta t = 0.001$ in the integration scheme, whereas the solid black line is the approximation to the Wiener process generated by using $\Delta t^{\mathrm{w}} = 0.001$ and constructed by adding intermediate values of this path in order to accommodate a new step size of $\Delta t = 0.01$ in the integration scheme. The dashed gray line in Figure 4.16 represents a sample path of the Wiener process directly generated by using a step size of $\Delta t^{\mathrm{w}} = 0.01$. It is apparent that both paths differ substantially.

For stochastic simulations of power systems where the focus is on the statistical properties of the system variables, we are concerned with the weak convergence criteria. This is for example for Monte Carlo simulations. When the weak convergence of the integration method is enough the computation time of the simulations can be

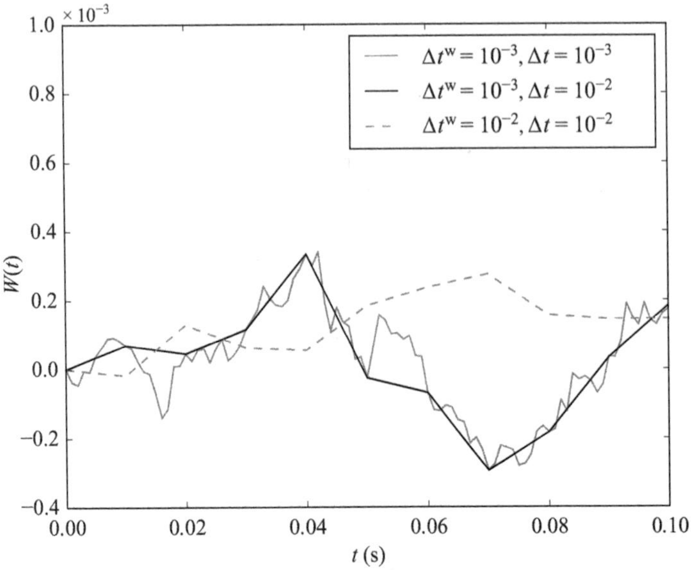

Figure 4.16 Sample paths of the Wiener process with different step sizes

reduced by replacing the Gaussian Wiener increments ΔW with another two-point random distributed variable $\Delta \hat{W}$, with

$$P(\Delta \hat{W} = \pm\sqrt{\Delta}) = \frac{1}{2}. \tag{4.70}$$

4.5 Stochastic power system case studies

The case studies presented in this section are based on the all-island Irish transmission system shown in Figure 4.17. The Irish transmission system model utilized consists of 1,479 buses, 1,851 transmission lines and transformers, 245 loads, 22 conventional synchronous power plants, modeled with sixth-order synchronous machine models, with AVRs and turbine governors, 6 PSSs, and 169 wind power plants, of which 159 are doubly-fed induction generators (DFIGs) and 10 are constant speed wind turbines (CSWTs). In the base system used for the case studies presented here the total load of the system is 2,215 MW and 25% of the total generated power is supplied by wind.

All simulations are carried out using Dome, a Python-based software tool for power system analysis [33]. Dome solves the SDEs using the Itô integral. It supports solving the SDEs using either the Euler–Maruyama or Milstein integration method. For this work, the Euler–Maruyama is used [34].

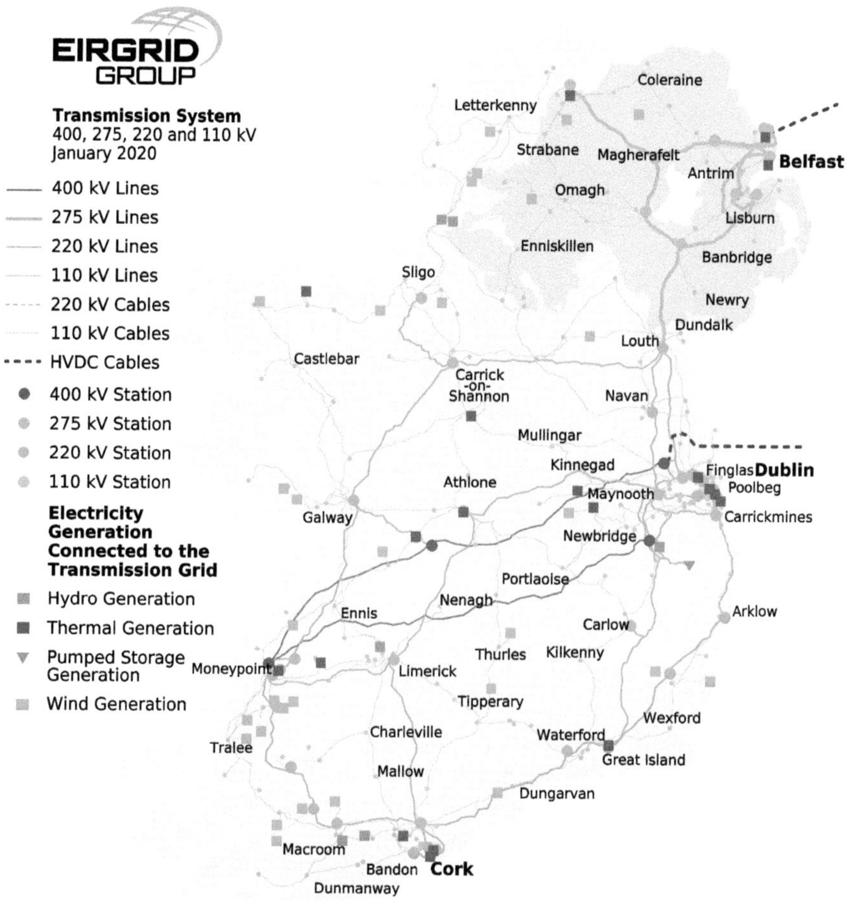

Figure 4.17 *Map of the all-island Irish transmission system (courtesy of EirGrid Group [32])*

4.5.1 Initialization of Irish power system models

The initialization methods presented in Section 4.3.1 are compared in the case study outlined in this section. In this case the only source of uncertainty in the Irish system model is the wind. The wind speed inputs for the wind power plants are modeled as presented in Section 4.3.3.2 where $\eta_w(t)$ are OU processes as presented in (4.4) with the standard deviation σ_{η_w} set as 20% of the mean wind speed and $\alpha = 0.01$. The Irish test system is simulated using the Monte Carlo method. The system is simulated 1, 000 times for 60 s with a time step of 0.1 s, using the three different initialization methods discussed in Section 4.3.1.

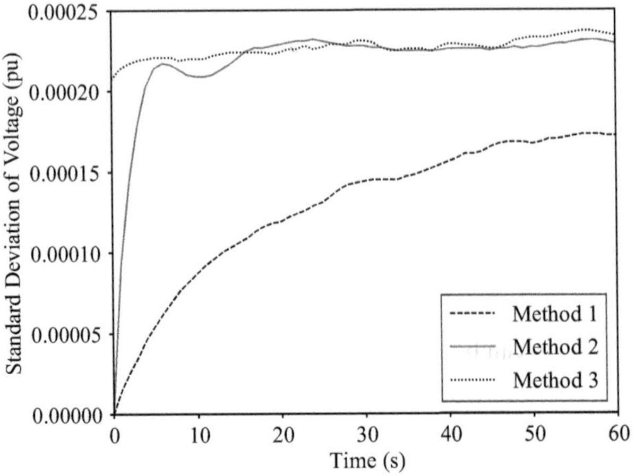

Figure 4.18 Standard deviation of voltage at a centrally located bus in the Irish transmission system model

Figure 4.18 shows the standard deviation of the voltage at a centrally located bus (Athlone in county Westmeath) in the Irish system. This figure shows that the standard deviation of the voltage is still growing at the end simulation time. Thus, it has not reached stationarity. The time it takes the system variables to reach stationarity for Method I is dictated by the ACF of the stochastic processes in the system. In this case the ACF of the single stochastic process in (4.4) is $\exp(-\alpha t)$. Therefore, this particular system needs to be simulated for at least $2/\alpha = 200$ s to reach stationarity, which requires a relatively large computing time.

To reduce the computing time required to reach stationarity, an option is to increase the time step. However, in general, this is not a feasible solution for power systems. The transient stability model, in fact, is stiff, i.e., it combines fast and slow dynamics. Fast dynamics quickly reach stationarity but need a small time step. On the other hand, the slow dynamics dictate when the system trajectories reach stationarity.

In this method, the slowest dynamics of the differential equation $\boldsymbol{f}$ (in this case, wind power plant dynamics) determine the time that the system takes to reach stationarity. This occurs, in this case, in approximately 10 s of simulated time. Thus, if the stochastic processes have a slow autocorrelation, Method II requires less computing time than Method I to reach stationarity. Note that, for systems with slow dynamics, Methods I and II have an equivalent computational burden.

As expected, Method 3 allows starting the time domain simulations with points that are in stationary conditions. Only a very short and, effectively, negligible transient of the standard deviation can be observed due to the approximations introduced by the linearization and the solution of (4.41) obtained with the SLICOT library [35].

The speedup of using Method 3 as apposed to Methods 1 and 2 depends on a trade-off between the number of trajectories simulated with the Monte Carlo approach and the size of the system. If the number of trajectories is sufficiently high, the proposed method will always save time compared to the conventional methods.

Another benefit of Method 3, besides saving time is that less disc space is required. This is because the initial non-stationary part of the Monte-Carlo simulations for Methods 1 and 2 are typically discarded as the full probability distribution of the variables is not simulated. With the proposed approach, Method 3, the whole trajectories are meaningful.

Finally, when using Methods 1 and 2, the slowest processes in the system need to be identified to be able to know before-hand how long a simulation time is needed to reach stationarity. This can be involved and time consuming for large systems. With the proposed method, on the other hand, this analysis is not required.

4.5.2 Irish system with inclusion of wind and solar generation

This section presents a case study of the Irish system with solar and wind generation. The Irish system model is divided up into 10 areas. In seven of those areas, there is wind generation. For this case study, a part of the wind generation in the system is replaced with solar PV generation. The capacity of the installed solar PV is distributed equally between the seven areas. Figure 4.19 indicates in which areas of the Irish system solar generation is installed and how much of the total generation of each area is produced with solar PV generation.

4.5.2.1 Scenarios

Three different scenarios for the Irish system model are studied:

- **Scenario a:** The onshore wind, 25% of generation, is stochastic.
- **Scenario b:** Stochastic onshore wind, 15% of generation, and stochastic solar, 10% of generation, for a mostly clear sky.
- **Scenario c:** Stochastic onshore wind, 15% of generation, and stochastic solar, 10% of generation, for a mixed clouding condition.

The stochastic models for loads, and the wind and solar generation are the SDE-based models presented in Sections 4.3.3.1, 4.3.3.2, and 4.3.3.3, respectively. The parameters used for the stochastic models are presented in Tables 4.2, 4.3, and 4.4, respectively. For Scenario b, the jump parameters for the solar model are all set to zero. The plant area parameter S for the aggregated solar model is set based on the capacity of the individual PV farms. All PV farms are assumed to produce 150 W/m^2.

4.5.2.2 Wind versus solar generation

All scenarios are simulated for 10,000 s with a time step of 0.1 s. As a metric to compare the different cases, the ramp rates of the the center-of-inertia frequency of the system are used, defined as:

$$\Delta_h \omega_t = \omega_t - \omega_{t-h}, \tag{4.71}$$

where ω_t is the system angular frequency at time t, and $\Delta_h \omega_t$ gives the probability of the sizes of ramps in a time step h. The ramp rates are computed for time lags $h = 0 - 100$ s. Then, the standard deviation of the ramp rates, $\Delta_h \omega_t$, for each time step h is computed. In this way, the ramp rates give information on how the ramps in the system frequency evolve with time.

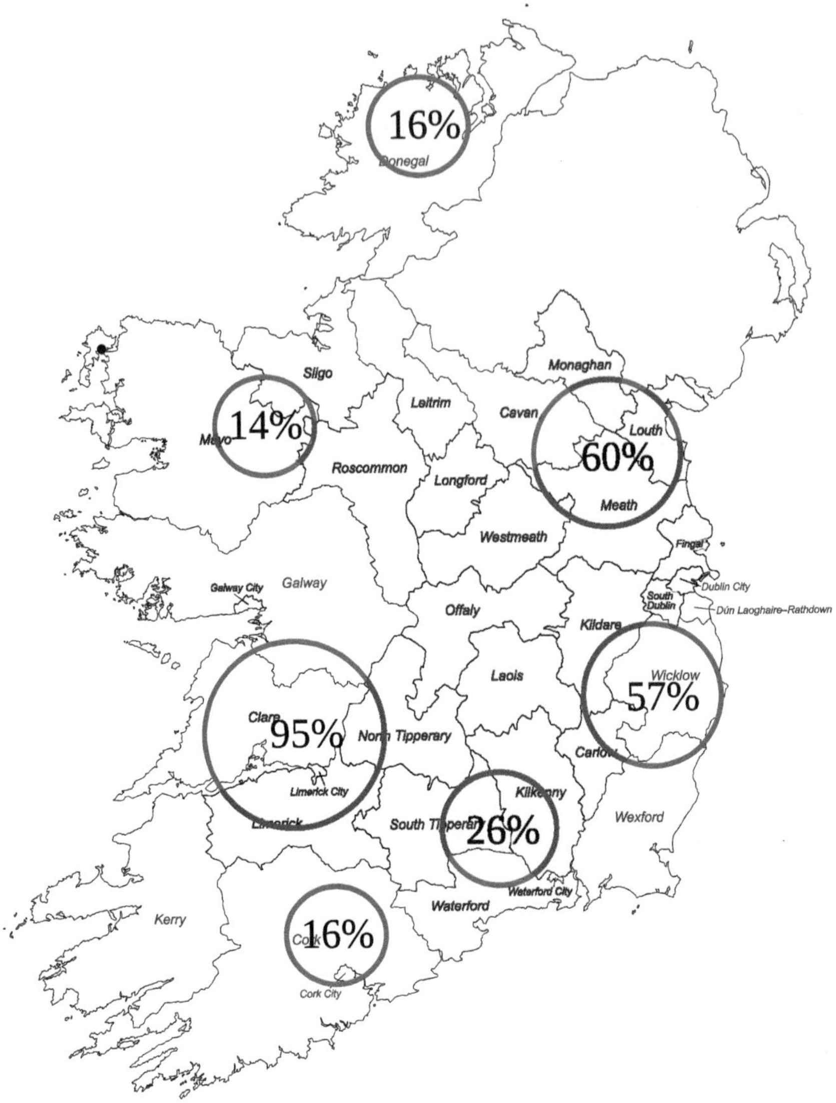

Figure 4.19 The areas of the Irish system where solar generation is included. It is indicated how much of the areas total generation is replaced with solar PV generation

In Figure 4.20, the standard deviation of the ramps in frequency of the system is shown for scenarios a, b, and c. It can be seen that for the first 10 s the standard deviation is approximately the same for all three scenarios. In the time frame from 10 to 40 s, scenario a, with only wind, has a slightly higher standard deviation. Scenario b, which represents a mostly clear sky, has a lower standard deviation than scenario a in time frames up to approximately 60 s, after that the two scenarios are similar.

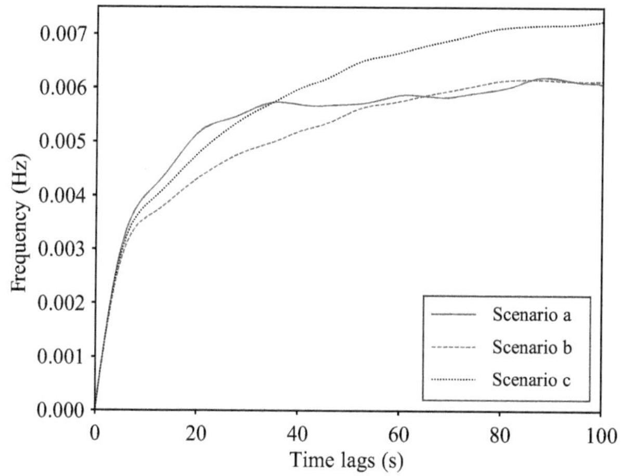

Figure 4.20 Standard deviation of the ramps in the frequency of the center of inertia for scenarios a, b, and c

Scenario c shows bigger ramps in the frequency for time steps above 40 s. This is as expected because of the jumps in the solar generation. However, the filtering effect of aggregating multiple PV generators results in the jumps ramping up over longer time frames.

It is worth noting that the stochastic solar model used in scenario c represents the worst case scenario in terms of short-term variability of solar PV. That is the scenario where there are intermittent clouds blocking the solar PVs. In all other weather conditions it is likely that the solar generation will not add more than wind to the system frequency uncertainty as can be seen for scenario b.

4.5.3 Irish system with correlated load and wind generation

In this case study, the effect of correlation between the stochastic perturbations to the system is considered. Correlated stochastic load consumption, wind generation and bus voltage phasors are modeled as outlined in Appendix B. The parameters used in this case study for the correlated stochastic models are provided in Table 4.5.

Table 4.5 The SDE parameters for the correlated models in the case study in Subsection 4.5.3.

Parameters
$\alpha_p = \alpha_q = \alpha_\theta = \alpha_v = \alpha_w = 1,$ $\sigma_p = 0.005p_{L0},\ \ \sigma_q = 0.005q_{L0},$ $\sigma_v = 0.003v_0,\ \ \sigma_\theta = 0.003\theta_0.$

Different levels of correlation are compared through these three scenarios S1, S2, and S3, defined as follows:

- S1 represents the fully uncorrelated SDAE model, i.e. the correlation between any two stochastic processes i and j is $r_{i,j} = 0$.
- S2 considers a low level of correlation among processes, i.e. the correlation between any two stochastic processes i and j is set to $r_{i,j} = 0.4$ if they belong to the same area, 0 otherwise.
- S3 considers a high level of correlation among processes, i.e., the value of correlation between any two stochastic processes i and j is set to $r_{i,j} = 0.8$ if they belong to the same area, 0 otherwise.

These scenarios are used through out the three different cases studied: Case a, including stochastic correlated load consumption; Case b, including correlated stochastic load consumption and wind generation; Case c, including correlated stochastic load consumption, wind generation and bus voltage phasors.

4.5.3.1 Case a

To begin this case study the impact of correlated load consumption is considered. An effective way to evaluate the effect of correlation between the loads is through observing the statistical properties of the relevant quantities. The statistical property, and the quantity chosen in this case are the standard deviation of the active and reactive power generation of synchronous generators, namely, σ_{p_g} and σ_{q_g}, respectively. Table 4.6 shows σ_{p_g} and σ_{q_g} of selected synchronous generators calculated for the three scenarios S1, S2, and S3. The correlation among the stochastic loads has a

Table 4.6 Standard deviation of active and reactive powers of synchronous generators for the Irish power system model with correlated stochastic loads

Standard deviation	S1 absolute	S2 % increase	S3 % increase
$\sigma_{p_{g\text{HUNT CT}}}$	0.0025	44	76
$\sigma_{p_{g\text{DUBLIN B}}}$	0.0037	56.76	94.59
$\sigma_{p_{g\text{PBEGG4}}}$	0.0013	53.85	92.31
$\sigma_{p_{g\text{PBEGG5}}}$	0.0012	58.33	100
$\sigma_{p_{g\text{PBEGG6}}}$	0.002	55	90
$\sigma_{q_{g\text{HUNT CT}}}$	0.0004	25	50
$\sigma_{q_{g\text{DUBLIN B}}}$	0.001	50	80
$\sigma_{q_{g\text{PBEGG4}}}$	0.0003	33.33	66.67
$\sigma_{q_{g\text{PBEGG5}}}$	0.0004	50	75
$\sigma_{q_{g\text{PBEGG6}}}$	0.0006	50	83.33

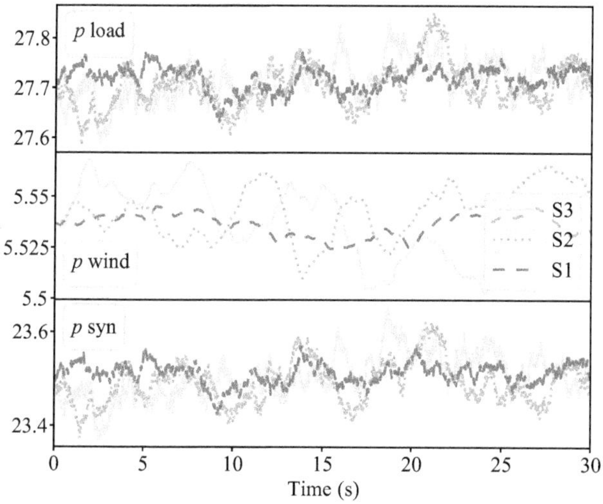

Figure 4.21 *Total active power load consumption (p_{load}); total active power generation (p_{wind}) by wind power plants; and total active power generation (p_{syn}) by conventional power plants for the three scenarios of correlation, i.e., S1, S2, and S3. All the values are in [pu], with a base of 100 MVA*

direct impact on σ_{p_g} and σ_{q_g} of the generators. The value of σ_{p_g} and σ_{q_g} almost doubles when the correlation among stochastic loads is doubled. This is a noteworthy result as the standard deviation of the loads remains the same in all three scenarios.

4.5.3.2 Case b

In this case, we model the Irish system incorporating correlated stochastic loads and correlated stochastic wind generation. To observe the effect of correlation between different stochastic perturbations on the system dynamics, we consider the sum of the trajectories of the relevant quantities such as the active power consumption and generation of all devices connected in the same area. Figure 4.21 illustrates the sum of the active powers p_{load} consumed by all loads; the sum of the active powers p_{wind} generated by all wind power plants; and the sum of the active powers p_{syn} generated by all synchronous generators for the three scenarios of correlation, i.e., S1, S2, and S3. Despite the fact that the standard deviation of the individual stochastic processes remains the same regardless of the level of correlation being used, Figure 4.21 shows that the spread, in terms of standard deviation, of the sum of the quantities above increases as the correlation between the stochastic process is increased.

4.5.3.3 Case c

This last example considers the Irish system model with inclusion of correlated stochastic loads, correlated stochastic bus voltage phasors, and correlated stochastic

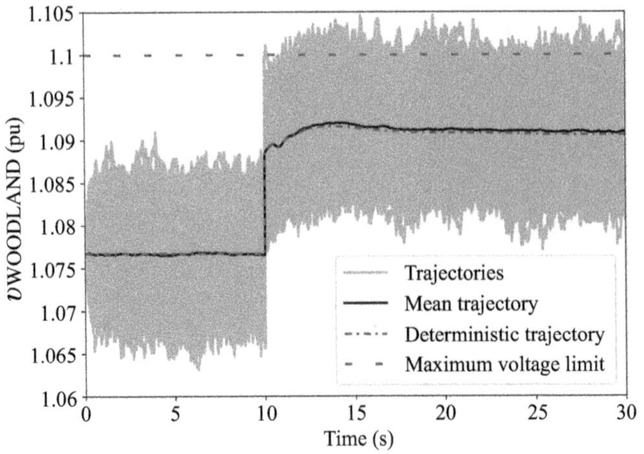

Figure 4.22 Bus voltage magnitude at bus a centrally located bus in the Irish transmission system model for S1

wind generation. In addition to the stochastic perturbations, the system undergoes a disconnection of a load connected to the East–West interconnector at $t = 10$ s.

Figure 4.22 shows the time domain profile of voltage magnitude at a centrally located bus in the system, for the 1,000 simulations, for S1, i.e., for the fully uncorrelated SDAE model. The black solid line shows the mean value of the 1,000 trajectories, which reflects the voltage profile of a deterministic solution, since all Wiener processes have zero average. The mean trajectory coincides with the deterministic trajectory. The deterministic trajectory is obtained for simulating the system for the same fault conditions without including the stochastic processes. Figure 4.22 indicates that the voltage profile for the deterministic solution is below the maximum voltage limit, which is shown by a dashed line. It is also relevant to note that 24.4% of the trajectories exceed the maximum voltage limit at least once in the period 10 s $< t <$ 30 s. This result can be relevant for TSOs as grid codes do not allow voltage variations above or below 10%. Moreover, overvoltage protections are implemented in some systems and these protections can be triggered if voltage limits are violated in transient conditions.

Figures 4.23 and 4.24 illustrate the 1,000 trajectories of voltage magnitude at a centrally located bus in the system, for S2 and S3, respectively. Results indicate that the higher the correlation among the processes the lower the standard deviation of the trajectories. For S3, i.e., for the maximum correlation considered in this case study, no trajectory crosses the maximum voltage limit. These results are summarized in Table 4.7. In this example, the uncorrelated stochastic model shows more conservative results than the scenarios that take into account correlation.

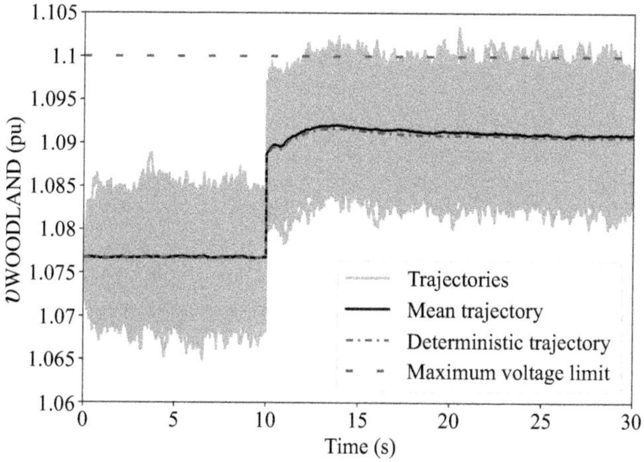

Figure 4.23 Bus voltage magnitude at a centrally located bus in the Irish transmission system model for S2

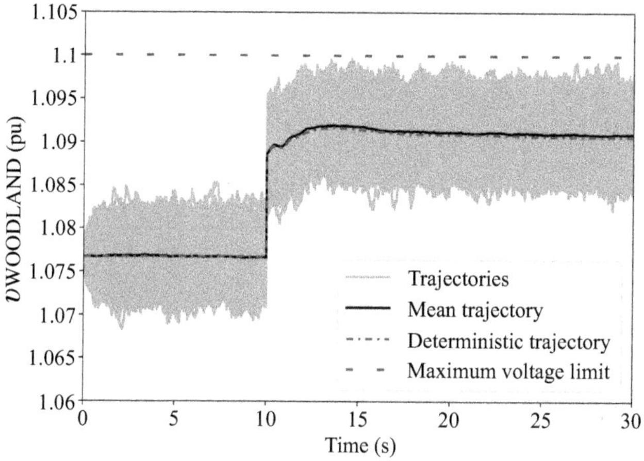

Figure 4.24 Bus voltage magnitude at a centrally located bus in the Irish transmission system model for S3

Table 4.7 Trajectories with over-voltages for the Irish system model in Case c.

Scenarios	Trajectories with over-voltages
S1	244 (24.4%)
S2	70 (7%)
S3	0

4.6 Conclusions

This chapter illustrates a general formulation for modeling power systems as stochastic differential algebraic equations (SDAEs), as well as a systematic method to model the stochastic perturbations present in these systems. One of the main advantages of the presented approach is that it can be applied systematically to both system variables and parameters. Methods to model random perturbations as stochastic differential equations (SDEs) are presented. These methods are general and allow to model a variety of stochastic properties encountered in perturbations to power systems, including jumps and correlation. The convenience of the stochastic modeling approach is illustrated through simulations on the Irish power system model. These case studies highlight the importance of considering the stochastic perturbations that affect power systems in dynamic system studies. The perturbations introduced by the various sources of random variations affect the system in different ways. Results indicate that a precise representation of power system noise can be a useful tool for system security analysis.

The methodology presented in this chapter can leverage several applications. For example, the tools provided in this chapter can be used to study transient stability as well as long-term voltage stability of power systems affected by stochastic inputs. We believe that stability analysis and control of power systems modeled as SDAEs is just at its beginning and is one of the most challenging promising tools in the field of power system dynamics.

Bibliography

[1] T. C. Gard. *Introduction to Stochastic Differential Equations*. Marcel Dekkler, New York, NY, 1987.

[2] B. Øksendal. *Stochastic Differential Equations: An Introduction with Applications*, 6th edition. Springer, New York, NY, 2003.

[3] M. Grigoriu. *Stochastic Calculus. Applications in Science and Engineering*. Birkhäuser, Boston, MA, 2002.

[4] C. Gardiner. *Stochastic Methods: A Handbook for the Natural and Social Sciences*, 4th edition. Springer-Verlag, New York, NY, 2009.

[5] K. E. Brenan, S. L. Campbell, and L. R. Petzold. *Numerical Solution of Initial-Value Problems in Differential-Algebraic Equations*. North-Holland, New York, NY, 1989.

[6] E. Kloeden and E. Platen. *Numerical Solution of Stochastic Differential Equations*, 3rd edition. Springer, New York, NY, 1999.

[7] T. Tian and K. Burrage. Implicit Taylor methods for stiff stochastic differential equations. *Applied Numerical Mathematics*, 38(1–2):167–187, 2001.

[8] R. Winkler. Stochastic differential algebraic equations of index 1 and applications in circuit simulation. *Journal of Computational and Applied Mathematics*, 157(2):477–505, 2003.

[9] Humberto Verdejo, Almendra Awerkin, Wolfgang Kliemann, and Cristhian Becker. Modelling uncertainties in electrical power systems with stochastic differential equations. *International Journal of Electrical Power & Energy Systems*, 113:322–332, 2019.

[10] R. Zárate-Miñano, M. Anghel, and F. Milano. Continuous wind speed models based on stochastic differential equations. *Applied Energy*, 104:42–49, 2013.

[11] M. Olsson, M. Perninge, and L. Soder. Modeling real-time balancing power demands in wind power systems using stochastic differential equations. *Electric Power Systems Research*, 80(8):966 – 974, 2010.

[12] G. M. Jónsdóttir and F. Milano. Data-based continuous wind speed models with arbitrary probability distribution and autocorrelation. *Renewable Energy*, 143:368–376, 2019.

[13] G. M. Jónsdóttir and F. Milano. Modeling solar irradiance for short-term dynamic analysis of power systems. *IEEE PES General Meeting*, 2019.

[14] M. Anvari, B. Werther, G. Lohmann, M. Wächter, J. Peinke, and H. -P. Beck. Suppressing power output fluctuations of photovoltaic power plants. *Solar Energy*, 157:735–743, 2017.

[15] C. Roberts, E. M. Stewart, and F. Milano. Validation of the Ornstein–Uhlenbeck process for load modeling based on μPMU measurements. In *Power Systems Computation Conference (PSCC)*, 2016.

[16] M. Perninge, M. Amelin, and V. Knyazkins. Load modeling using the Ornstein–Uhlenbeck process. In *IEEE 2nd International Power and Energy Conference (PECon)*, pp. 819–821, Johor Bahru, Malaysia, 2008.

[17] G. M. Jónsdóttir and F. Milano. Modeling correlation of active and reactive power of loads for short-term analysis of power systems. In *IEEE 20th International Conference on Environment and Electrical Engineering (EEEIC)*, 2020.

[18] F. Milano and R. Zárate-Miñano. A systematic method to model power systems as stochastic differential algebraic equations. *IEEE Transactions on Power Systems*, 28(4):4537–4544, 2013.

[19] B. Yuan, M. Zhou, G. Li, and X.-P. Zhang. Stochastic small-signal stability of power systems with wind power generation. *IEEE Transactions on Power Systems*, 30(4):1680–1689, 2014.

[20] H. Li, P. Ju, C. Gan, S. You, F. Wu, and Y. Liu. Analytic analysis for dynamic system frequency in power systems under uncertain variability. *IEEE Transactions on Power Systems*, 34(2):982–993, 2018.

[21] G. M. Jónsdóttir and F. Milano. Stochastic modeling of tidal generation for transient stability analysis: a case study based on the all-island Irish transmission system. In *Power Systems Computation Conference (PSCC)*, 2020.

[22] M. Adeen and F. Milano. Modeling of correlated stochastic processes for the transient stability analysis of power systems. *IEEE Transactions on Power Systems*, 2021.

[23] J. Chen, M. C. H. Hui, and Y. L. Xu. A comparative study of stationary and non-stationary wind models using field measurements. *Boundary-Layer Meteorology*, 122(1):105–121, 2007.

[24] R. Zárate-Miñano and F. Milano. Construction of sde-based wind speed models with exponentially decaying autocorrelation. *Renewable Energy*, 94:186–196, 2016.

[25] M. Grigoriu. *Applied Non-Gaussian Processes: Examples, Theory, Simulation, Linear Random Vibration, and MATLAB Solutions*. Prentice-Hall, Prentice, NJ, 1995.

[26] E. Platen and N. Bruti-Liberati. *Numerical Solution of Stochastic Differential Equations with Jumps in Finance*. Springer-Verlag, Berlin Heidelberg, 2010.

[27] S. Dipple, A. Choudhary, J. Flamino, B. K. Szymanski, and G. Korniss. Using correlated stochastic differential equations to forecast cryptocurrency rates and social media activities. *Applied Network Science*, 5(1):1–30, 2020.

[28] S. B. Provost and A. M. Mathai. *Quadratic Forms in Random Variables: Theory and Applications/A.M. Mathai, Serge B. Provost*. Statistics: Textbooks and Monographs. Marcel Dekker, New York, NY, 1992.

[29] T. Petru and T. Thiringer. Modeling of wind turbines for power system studies. *IEEE Transactions on Power Systems*, 17(4):1132–1139, 2002.

[30] J. Marcos, Í. de la Parra, M. García, and L. Marroyo. Simulating the variability of dispersed large PV plants. *Progress in Photovoltaics: Research and Applications*, 24(5):680–691, 2016.

[31] Z. Y. Dong, J. H. Zhao, and D. J. Hill. Numerical simulation for stochastic transient stability assessment. *IEEE Transactions on Power Systems*, 27(4):1741–1749, November 2012.

[32] EirGrid Group. EirGrid Group Transmission System in January 2020, 2020.

[33] F. Milano. A Python-based software tool for power system analysis. *IEEE PES General Meeting*, 2013.

[34] P. E. Kloeden, E. Platen, and H. Schurz. *Numerical Solution of SDE Through Computer Experiments*. Springer Science & Business Media, New York, NY, 2012.

[35] P. Benner, V. Mehrmann, V. Sima, S. Van Huffel, and A. Varga. SLICOT – a subroutine library in systems and control theory. In Biswa N. Datta, editor, *Applied and Computational Control, Signal and Circuits*, Vol. 1, Chapter 10, pp. 499–539. Birkauser, Switzerland, 1999.

Chapter 5

Detailed modeling of inverter-based resources

Younes Seyedi[1], Ulas Karaagac[2], Jean Mahseredjian[3], Aboutaleb Haddadi[4], Keijo Jacobs[5] and Houshang Karimi[1]

Inverter-based resources (IBRs) are increasingly deployed in modern power systems. The response of a bulk-power system connected IBR to a change in system state is substantially different from that of a traditional synchronous generator (SG). The reasons are the fundamental differences in the physical equipment between an IBR and a traditional synchronous generator. A key difference is a power electronic converter which interfaces an IBR with the grid. The output current of this converter is tightly controlled through fast switching of power electronics devices dependent upon manufacturer-specific and often proprietary IBR control schemes. Hence, the response of an IBR to a change in system state becomes dependent on non-universal IBR control schemes. Due to these differences, classical generator models do not apply to IBRs, and there is a need for sufficiently detailed IBR models for system level studies.

With increasing integration of IBRs in the power system, the power industry is faced with emerging issues related to safe operation of the bulk power system which require expanded applications of detailed IBR models at planning and interconnection stage. Interconnection queues across the world are inundated with IBRs, which are expected to further drive potential grid issues such as sub-synchronous control interactions, controls instability, and low short-circuit strength condition. Detailed IBR models can capture such abnormalities during the planning and/or interconnection study process, thus allowing for the issue to be addressed prior to the resource interconnection. In the absence of such models, these performance issues may go unnoticed until after the time of IBR interconnection, at which point there is limited capability to address them. Having detailed and accurate IBR models early in the process is critical to ensure adequate performance assessment and mitigation. IBR models can be classified into manufacturer-specific and generic models. Manufacturer-specific models are developed by IBR manufacturer to represent specific hardware/control,

[1]Department of Electrical Engineering, Polytechnique Montreal, Canada
[2]Department of Electrical Engineering, The Hong Kong Polytechnic University, China
[3]Faculty of Electrical Engineering, Polytechnique Montréal, Canada
[4]Electric Power Research Institute, New York City, USA
[5]Division of Electrical Power and Energy Systems, KTH Royal Institute of Technology, Sweden

are often proprietary and black-boxed, and are used for site-specific studies. By contrast, generic models are developed independent of any specific vendor's equipment or control structure, are transparent white-box models, and are intended to provide a reasonable representation of the trend of dynamic behavior of an IBR without accurately representing the minute details. These models can be used for long-term planning studies, wherein an actual IBR installation does not exist, for exploratory studies of long-term future conditions or other research-focused studies, and for informing forward looking IBR interconnection standards and technical interconnection requirements. The focus of this chapter is on generic IBR models.

This chapter aims to discuss important aspects of modeling and simulation of IBRs so that their behavior can be accurately analyzed, and their impacts on power system operations can be adequately studied. The focus is on accurate and generic time-domain models. Wind turbine (WT) modeling is presented with corresponding modeling blocks which include control functions, protection, and power hardware.

The chapter will provide an overview on modeling and allow the reader to better understand underlying problems. It will contribute to researchers and engineers dealing with IBRs. It will also provide didactic material for WT models.

5.1 Introduction

With recent advances in wind turbine (WT) and solar photovoltaic (PV) technologies, the solar and wind power penetration levels increase as well as the capacity of PV and wind parks (WPs) [1]. Large-scale WPs employ variable speed WTs (VSWTs) in order to increase the energy harvest, reduce the drive train stresses and comply with grid code requirements [2]. Full size converter (FSC) and doubly-fed induction generator (DFIG) WTs belong to this category.

Connecting a large-scale IBR installation into a bulk power system has become a more important issue due to potential impacts on safe operation of the bulk power system [3]. Increased integration of IBRs is expected to further drive potential grid issues such as sub-synchronous control interactions, controls instability, and low short-circuit strength condition. Failure to perform proper interconnection studies using sufficiently accurate models not only can lead to non-optimal design and operation of WPs but also may result in unexpected contingencies and even stability issues [4]. Manufacturer-specific models of WPs are required for interconnection studies due to their fidelity. However, such models have limited applicability in long-term planning studies, wherein an actual IBR installation does not exist, or for exploratory and research focused studies of long-term future conditions, which often require knowledge of IBR internal control and protection schemes [5]. Further, utilities and project developers require accurate and detailed WP models to carry out preliminary grid integration studies before an actual design is selected [6]. Generic IBR models enable such studies, thereby enabling researchers to better understand IBR-related grid issues and develop countermeasures ensuring reliability. A key aspect of generic models is parameterization, i.e., the capability to reproduce a specific performance by tuning model parameters based on experimental or theoretical input data. Motivated by the above considerations, this chapter elaborates on the detailed modeling and simulation of WPs based on the generic mechanical, electrical, control and protection subsystems.

5.2 Variable speed WT models

This section presents generic EMT models for VSWT-based WPs that can be used for a wide range of integration studies. The collector grid and the WTs are represented with their aggregated models; however the overall control structure of the WP is maintained. The WT and the WP control system models account for the non-linearities and include necessary transient and protection functions. In addition, they are crucial for the simulation of the accurate transient behavior of WPs subjected to external power system disturbances. The presented models are developed by the IEEE Task Force for EMT-type Modeling of WT Generators and Parks [7].

5.2.1 WT aerodynamics

WTs extract kinetic energy from the swept area of the blades. The mechanical power extracted from the wind is given by [8]:

$$P_t = (1/2)\rho A \upsilon^3 C_p(\lambda, \beta) \tag{5.1}$$

where ρ is the air density (approximately 1.225 kg/m^3), A is the swept area of the rotor (m^2), υ is the upwind free wind speed (m/s), and C_p is the power coefficient. The parameter β in (5.1) represents the blade pitch angle. The parameter C_p is usually provided as a set of curves that relate the power coefficient to the tip-speed-ratio, denoted by λ, as shown in Figure 5.1 with $\beta = \theta$ [9]. The tip-speed-ratio is defined as

$$\lambda = (\omega_t R) / \upsilon \tag{5.2}$$

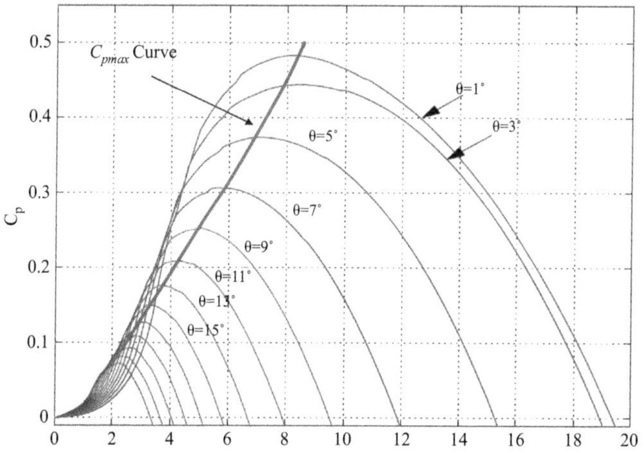

Figure 5.1 Wind power C_p curves

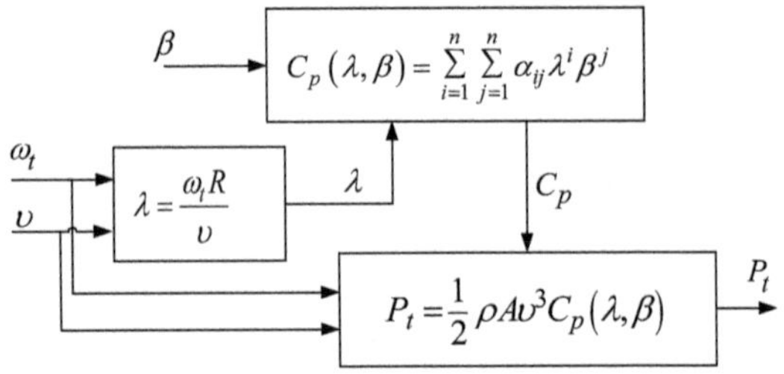

Figure 5.2 WT aerodynamics model

where ω_t is the WT rotational speed (rad/s) and R is the blade radius (m). At a specific wind speed and pitch angle, there is a unique WT rotational speed that achieves the maximum power coefficient $C_{P\max}$, shown as the red line in Figure 5.1. The mathematical model of the WT aerodynamics is shown in Figure 5.2. In this modeling approach, the C_p curves of the WT are fitted with high order polynomials as a function of λ and β as follows:

$$C_p(\lambda,\beta)=\sum_{i=1}^{n}\sum_{j=1}^{n}\alpha_{ij}\lambda^i\beta^j \tag{5.3}$$

5.2.1.1 Mechanical system

The mechanical system is constituted by the blades linked to the hub coupled to the slow shaft. The shaft is linked to the gearbox which multiplies the rotational speed of the fast shaft connected to the generator. Although the mechanical representation of the entire WT is complex, representing the fundamental resonance frequency of the drive train using its two-mass model is sufficient as the other resonance frequencies are much higher and their magnitudes are significantly lower [10].

5.2.2 Control of variable speed WTs

The controller of VSWT calculates the generator output power and the pitch angle of the rotor blades in order to extract the maximum energy from the wind while keeping the WT in a safe operating mode. There are four operation areas:

- Shut down: for too low wind speeds
- MPPT control: when operating below rated speed
- Pitch control: when operating above rated speed
- Shut down: for too high wind speeds

The WT remains shut down when the wind speed is too low for energy production, i.e., below the cut-in speed denoted by υ_{cut-in}. When the wind speed is above υ_{cut-in} and below the rated speed υ_{rated}, the pitch angle is kept at zero ($\beta=0°$) and the power reference of the WT generator is determined by the maximum power point tracking

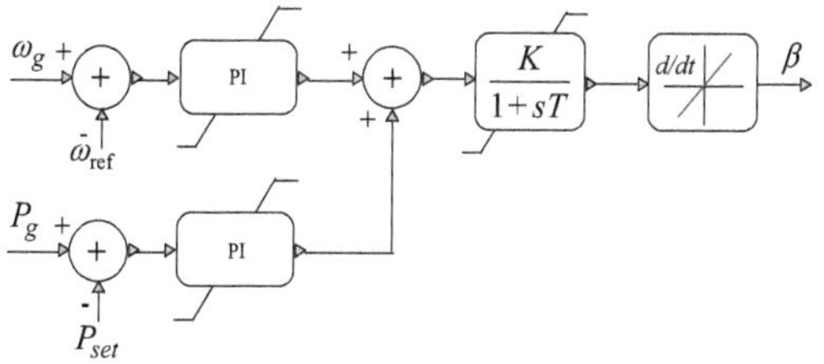

Figure 5.3 Schematic diagram of the pitch controller

(MPPT) scheme to achieve the optimal operation. The conventional method is to calculate the power reference using a cubic function of the turbine angular speed:

$$P_{ref} = K_{opt}\,\omega_t^3 \tag{5.4}$$

where

$$K_{opt} = (1/2)\,C_{p-\max}\,\rho\,A\,(R/\lambda_{opt})^3 \tag{5.5}$$

When the wind speed is above υ_{rated}, the pitch angle is increased by the pitch controller (see Figure 5.3) in order to limit the mechanical power produced by the wind and to reduce the mechanical loads on the drive train. It should be noted that the pitch controller should ensure zèro pitch angle ($\beta = 0°$) for the wind speeds below υ_{rated} [11]. The WT is shut down if the wind speed exceeds the cut-off threshold $\upsilon_{cut-off}$.

5.2.3 Wind parks with variable speed WTs

The power produced by the WTs is transmitted to the high-voltage (HV) transmission grid through the medium-voltage (MV) collector grid and the WP transformer as shown in Figure 5.4. Usually, the WP transformer has an on-load-tap-changer to keep the MV collector bus voltage around its nominal value. The active power at the point of interconnection (POI) depends on the wind conditions at each WT inside the WP and is determined by the MPPT function when the wind speed is between υ_{cut-in} and υ_{rated}. On the other hand, the voltage/reactive power at the POI is controlled by a central wind park controller (WPC) which is located at the WP substation.

The voltage at the POI, denoted by V_{POI}, is controlled by a proportional controller (V-control). Generally, the proportional gain of this controller and the reference value for the voltage at the POI, denoted by V'_{POI}, are defined by the transmission system operator [12]. The proportional voltage regulator gain of the WPC can be defined as

$$K_{Vpoi} = \Delta Q_{POI} / \Delta V_{POI} \tag{5.6}$$

When the WT control (WTC) uses an automatic reactive power regulator (AQR) for actuating the WT reactive currents, the reactive power reference values calculated by the WPC voltage regulator are sent to the WTs. On the other hand, when the WTC uses an automatic voltage regulator (AVR) for actuating the reactive currents,

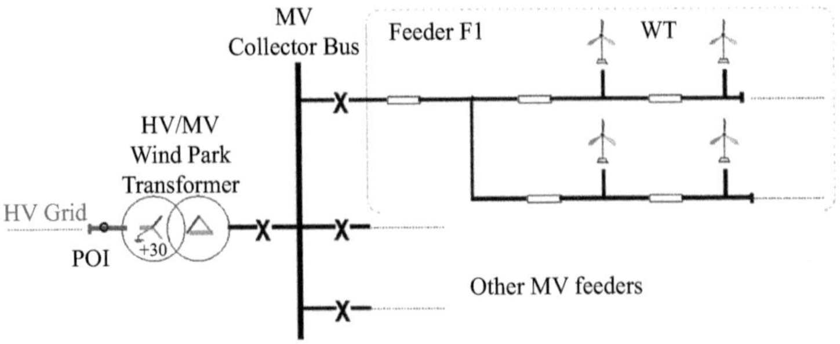

Figure 5.4 Simplified single-line diagram of a typical wind park

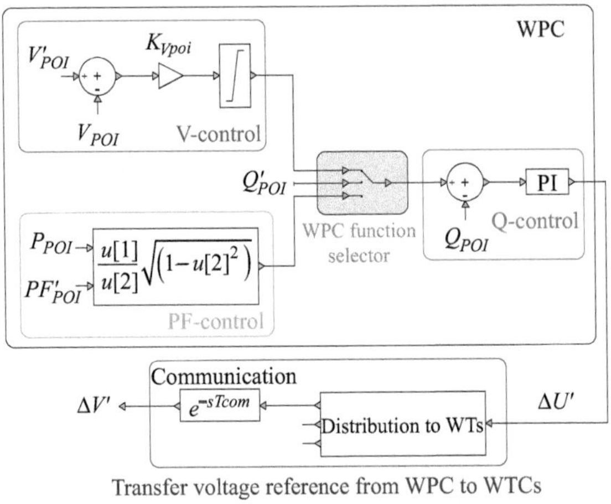

Figure 5.5 Structure of the wind park controller

the WPC also contains a proportional-integral (PI) reactive power regulator which modifies the WTC reference voltage values, denoted by V', as shown in Figure 5.5. Figure 5.5 also illustrates the WPC options that regulate the reactive power at the POI (represented as Q-control) and the power factor (represented as PF-control). The parameter T_{com} in Figure 5.5 is the communication delay.

5.2.4 FSC WTs

A full size converter (FSC) WT may or may not have a gearbox. A wide range of electrical generators such as asynchronous, synchronous, and permanent magnet can be used. The WT power is transferred through an AC–DC–AC converter system, and the dynamics of the electrical generator are isolated from the grid [13].

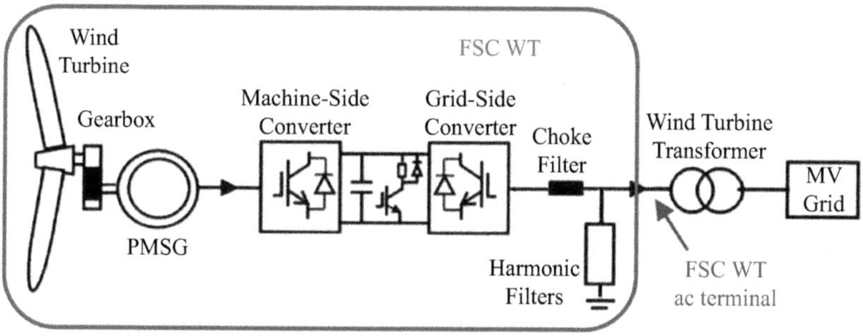

Figure 5.6 Main components of a typical FSC WT

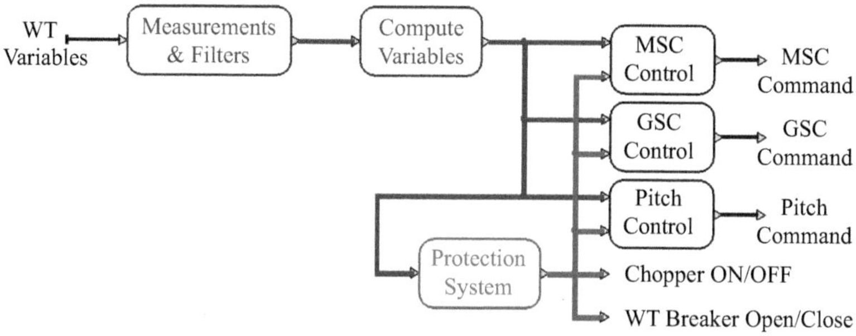

Figure 5.7 Functional block diagram of a FSC WT control and protection systems

The main components of a FSC WT are shown in Figure 5.6. The structure in Figure 5.6 uses a permanent magnet synchronous generator (PMSG) and an AC–DC–AC converter system. The converter consists of two pulse-width modulated (PWM) voltage source converters (VSCs) as:

- Machine-side converter (MSC)
- Grid-side converter (GSC)

The DC resistive chopper is used for the DC bus overvoltage protection. The line inductor (choke filter) and AC harmonic filters are used at the GSC to improve the power quality.

Figure 5.7 depicts the simplified diagram of FSC WT control and protection systems. The functionalities of the constituent block diagrams are explained below:

- Measurements and filters block: samples and converts signals to per-unit. Next, the signals are filtered by low-pass filters.
- Compute variables block: calculates different variables used by the FSC WT control and protection systems.

- Protection system block: contains DC resistive chopper control, MSC and GSC overcurrent protections, low voltage and overvoltage relays, and the cut-in and cut-off speed relays.
- Pitch control block: limits the mechanical power extracted from wind by increasing the pitch angle when the wind speed is above its rated value.

As demonstrated in Figure 5.8, the WT converters are controlled using a vector control scheme. The MSC and GSC signals are transformed to the flux and voltage reference frames, respectively. Both converters are controlled by a two-level controller. The slow outer control determines the reference dq-frame currents, and the fast inner control generates the AC voltage reference signals for the converter.

5.2.4.1 Machine side converter control

The MSC outer control loop controls the references for the machine currents. The q- and d-axis currents of the MSC are denoted as and i_{dm} in Figure 5.8, respectively. i_{qm} and i_{dm} are used to control the active and reactive power outputs of the PMSG, respectively. The q-axis current reference ($f(T')$ in Figure 5.8) is given by

$$i'_{qm} = T'/\lambda_m \tag{5.7}$$

where λ_m is the constant flux generated by the permanent magnet and $T' = K_{opt}\,\omega_t^2$ is the reference for PMSG electromagnetic torque given by the MPPT scheme. The d-axis current reference is set to zero, i.e., $i'_{dm} = 0$, to achieve a unity power factor.

The MSC inner control loop is designed based on the internal model control (IMC) method [14]. This method enables calculation of dq-frame PI controller parameters (gain and integration time constant) using certain machine parameters and the desired

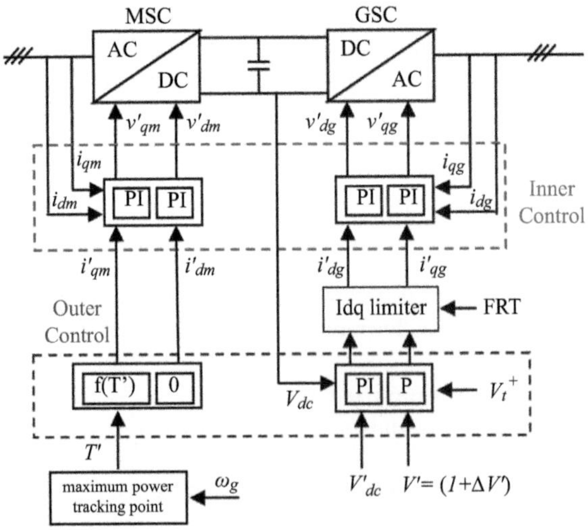

Figure 5.8 Schematic diagram of the FSC WT control

closed-loop bandwidth. This method simplifies the controller design procedure and eliminates or reduces the need for trial-and-error controller tuning.

The PMSG stator voltages are defined as

$$v_{dm} = -R_s i_{dm} - L_d \left(d\, i_{dm}/dt\right) + \omega_g L_q i_{qm} \tag{5.8}$$

$$v_{qm} = -R_s i_{qm} - L_q \left(d\, i_{qm}/dt\right) + \omega_g \left(L_d i_{dm} + \lambda_m\right) \tag{5.9}$$

where R_s is the armature resistance, L_d and L_q are the d- and q-axis inductances of the PMSG, respectively.

The instantaneous errors in i_{dm} and i_{qm} are processed by the PI controller to yield v_{dm} and v_{qm}, respectively. To ensure a good reference tracking, feed-forward compensating terms, $\omega_g L_q i_{qm}$ in (5.8) and $\omega_g\left(L_d i_{dm} + \lambda_m\right)$ in (5.9), are added. The converter reference voltages are therefore given by

$$v'_{dm} = -\left(k_p^d + k_i^d/s\right)\left(i'_{dm} - i_{dm}\right) + \omega_g L_q i_{qm} \tag{5.10}$$

$$v'_{qm} = -\left(k_p^q + k_i^q/s\right)\left(i'_{qm} - i_{qm}\right) + \omega_g \left(L_d i_{dm} + \lambda_m\right) \tag{5.11}$$

Based on the IMC, the PI controller parameters are found as

$$k_p^d = \alpha_c L_d, \quad k_p^q = \alpha_c L_q, \quad k_i^d = k_i^q = \alpha_c R_s \tag{5.12}$$

where α_c is the bandwidth. The relationship between the bandwidth and the rise time (10–90%) is $\alpha_c = \ln(9)/t_{rise}$. As the considered converter control and protection systems use per-unit quantities, the PI control parameters are calculated for per-unit quantities including angular frequency and time. With ω_b and $t_b = 1/\omega_b$ being the base quantities for angular frequency and time, and τ be the per-unit time. It can be written:

$$\frac{d}{d\tau} = \frac{d}{d(t/t_b)} = \frac{1}{\omega_b}\frac{d}{dt} \tag{5.13}$$

Hence, the calculated integral constants k_i^d and k_i^q should be multiplied by ω_b for real-time applications.

5.2.4.2 GSC control

The GSC function is to maintain the DC bus voltage V_{dc} at its nominal value, and control the positive-sequence AC terminal voltage, denoted by V_t^+ when equipped with AVR, as can be seen in Figure 5.8. The outer control consists of the proportional AC voltage control and the PI DC voltage control. The q-axis reference current is calculated by the proportional outer voltage control as

$$i'_{qg} = K_V\left(V' - V_t^+\right) \tag{5.14}$$

where K_V is the voltage regulator gain. The reference signal for the FSC positive sequence voltage, i.e., $V' = 1 + \Delta V'$, is calculated by the WPC (refer to Figure 5.5).

When AQR is employed to control the GSC reactive current output, the q-axis reference current is calculated by a PI reactive power regulator as

$$i'_{qg} = \left(K_p^Q + K_i^Q/s\right)\left(Q'_{WT} - Q_{WT}\right) \tag{5.15}$$

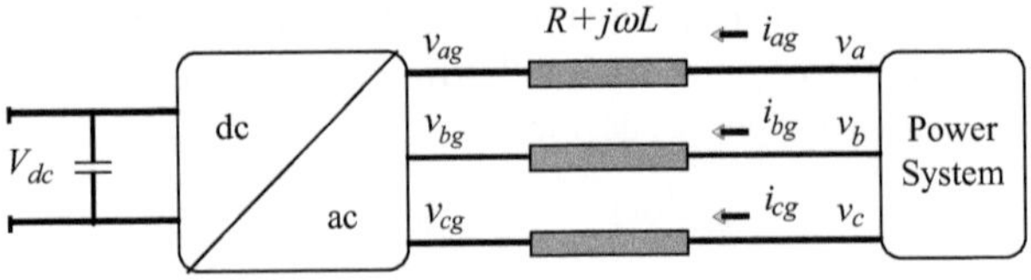

Figure 5.9 The GSC connection to the power system

where K_p^Q and K_i^Q are the reactive power regulator parameters, Q_{WT} is the reactive power output of the FSC WT (including harmonic filters), and Q'_{WT} is the reference calculated by the WPC. It should be noted that the use of the AVR for actuating the reactive current provides faster response. Moreover, the WPC tuning becomes easier as the system dependency on the short circuit ratio (SCR) is reduced compared to a WP with an AQR at the WTs [12].

The d-axis reference current is calculated by the outer DC voltage controller. It is a PI controller which can be tuned based on the inertia emulation:

$$k_p = \omega_0^2 (2H_{Cdc}) \ , \qquad k_i = 2\xi\omega_0 (2H_{Cdc}) \tag{5.16}$$

where ω_0 is the natural frequency of the closed loop system and ξ is the damping factor. $H_{Cdc} = E_{Cdc}/S_{wt}$ is the static moment of inertia, E_{Cdc} is the stored energy in the DC bus capacitor (in Joules), and S_{wt} is the WT WT-rated power (in VA).

The schematic of the GSC connected to the power system is shown in Figure 5.9. The impedance $Z = R + j\omega L$ represents the total impedance between the aggregated GSC and the external Thevenin source of the HV system including the transformers as well as the choke filter of the aggregated GSC. The aggregated WT model per-unit parameters are the same with the single WT per-unit parameters in aggregation when $S_{agg} = N\ S_{WT}$ (N is the number of WTs in aggregation and $S_{agg}$is the base power for aggregated WTs.

The voltage is given by the following equation where the small and capital letters in bold are, respectively, used for vectors and matrices

$$\mathbf{v_{abc}} = \mathbf{R\ i_{gabc}} + \mathbf{L}\left(d\ \mathbf{i_{gabc}}/dt\right) + \mathbf{v_{gabc}} \tag{5.17}$$

The link between the GSC output current and voltage can be described by the transfer function

$$G(s) = 1/\left(R + sL\right) \tag{5.18}$$

Based on the IMC method, the PI controller parameters of the inner current control loop are found as

$$k_p = \alpha_c L\ , \qquad k_i = \alpha_c R \tag{5.19}$$

Similar to the MSC, the feed-forward compensating terms, i.e., $(\omega L_{choke} i_{qg} + v_{dt})$ and $(-\omega L_{choke} i_{dg} + v_{qt})$, are added to the d- and q-axis voltages which are given by the PI regulators. L_{choke} is the inductance of the aggregated WT choke filter, v_{dt} and v_{qt}

are the FSC terminal voltages in dq reference frame. The FSC terminal is illustrated in Figure 5.6.

As the PI control parameters are calculated for per-unit quantities including angular frequency and time, the calculated integral constant k_i in (5.19) should be multiplied by ω_b for real-time applications.

5.2.4.3 Fault-ride-through (FRT) function

During normal operation, the controller gives priority to the active currents, i.e.,

$$i'_{dg} < I_{dg}^{\lim}$$

$$i'_{qg} < I_{qg}^{\lim} = \sqrt{\left(I_g^{\lim}\right)^2 - \left(i'_{dg}\right)^2} \tag{5.20}$$

where $I_{dg}^{\lim}$, $I_{qg}^{\lim}$, and $I_g^{\lim}$ are the limits for d-axis, q-axis, and total GSC currents, respectively.

The grid code requirements, as described in [15] and demonstrated in Figure 5.10, deal with the WT transient response upon severe voltage disturbances. To comply with these requirements, a FRT function is traditionally added to the WTC. The FRT function is activated when the voltage $|1 - V_{MV}^{+}|$ exceeds the pre-defined value V_{FRT-ON}

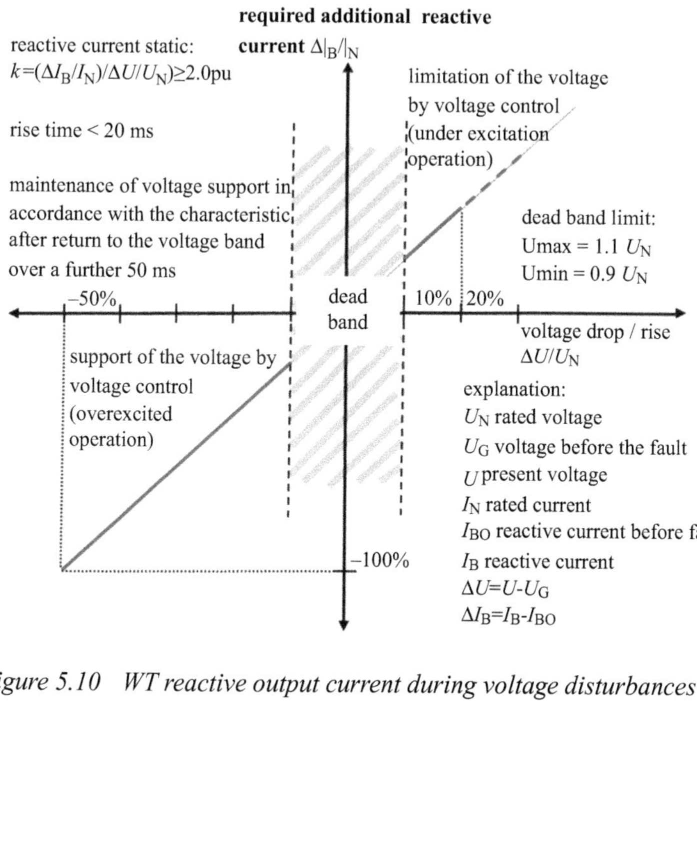

Figure 5.10 WT reactive output current during voltage disturbances [15]

and deactivated when $|1 - V_{MV}^{+}|$ reduces below the pre-defined value $V_{FRT-OFF}$ after a pre-specified release time t_{FRT}. V_{MV}^{+} is the WT transformer MV terminal positive sequence voltage and it is estimated in "Compute Variables" block (refer to Figure 5.7) using the WT transformer parameters and measured FSC WT AC terminal voltages and currents. During the FRT operation, the GSC controller gives priority to the reactive current by reversing the d- and q-axis current limits given in (5.20). The limits for d-axis, q-axis, and total GSC currents might be also different during FRT operation. Further details are discussed in [16].

Due to the AVR, the voltage control is continuous even inside the dead band region which is shown in Figure 5.10. On the other hand, the reactive current output is limited with the available reserve on the GSC since the priority is given to the active currents according to (5.20).

When the AQR is used for controlling reactive current output of the GSC, it is switched to the AVR during FRT operation to achieve the desired reactive current injection from the GSC. The voltage reference of the AVR is set to the pre-disturbance voltage value and the AQR input is blocked. Further details are given in [17].

5.2.4.4 Decoupled sequence control (DSC) of GSC

Ideally, the GSC with traditional coupled sequence control (CSC) scheme is not expected to inject any negative sequence currents into the grid during unbalanced loading conditions or faults. In practice, it injects a very small amount due to the phase shift in low-pass measurement filters [18]. Unlike its output currents, the GSC terminal voltages contain a negative sequence component during unbalanced loading conditions or faults, and this brings about second harmonic oscillations in the GSC active power output as well as the DC bus capacitor voltage. Such second harmonic oscillations can be eliminated by adopting a DSC scheme [19].

The implementation in [7] employs the outer control and Idq limiter shown in Figure 5.11 for calculating i'_{dg}, i'_{qg}, $I_{dg}^{\lim}$, and $I_{qg}^{\lim}$. These values are used to calculate the GSC current references i_{dg}^{+}, i_{qg}^{+}, i_{dg}^{-}, and i_{qg}^{-} for the decoupled sequence current controller. As i_{dg}^{+}, i_{qg}^{+}, i_{dg}^{-}, and i_{qg}^{-} are controlled, the DSC contains four PI regulators and requires sequence extraction for the GSC currents and voltages. Different methods have been proposed for sequence extraction in the literature, for example, the implementation in [7] uses the sequence decoupling method [10].

During normal operation, the priority is given to the average value of the instantaneous active power. The second harmonic oscillating terms of the instantaneous active power are eliminated at the expense of a reduction in reactive power generation capacity of the GSC. On the other hand, priority is given to the positive sequence reactive currents during FRT operation. In that case, the second harmonic oscillating terms are eliminated at the expense of a reduction in active power generation capacity of the GSC. Further details are found in [7,16].

In the recent VDE-AR-N 4120 technical connection rules [20], there is also a required additional negative sequence reactive current as a function of the voltage change in the negative sequence domain, as shown in Figure 5.12. The objective is

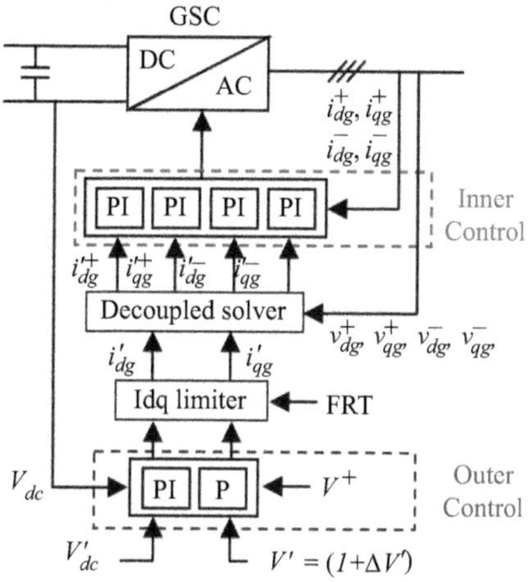

Figure 5.11 The GSC decoupled sequence control scheme

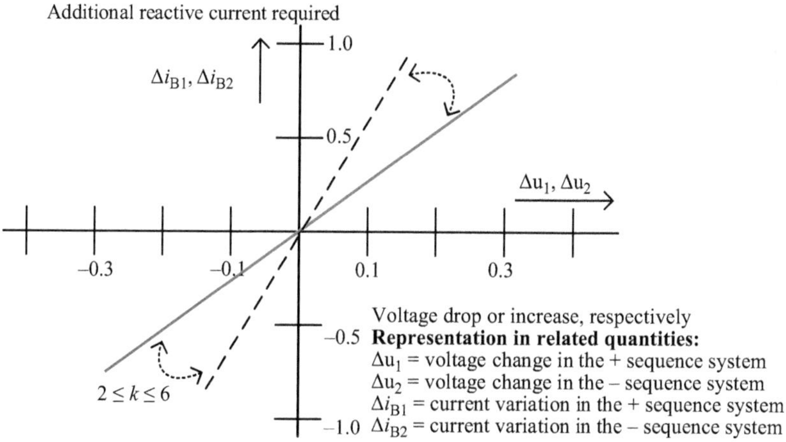

Figure 5.12 WT reactive output current during voltage disturbances [20]

to reduce the negative sequence voltage by consuming negative sequence reactive power. The negative sequence reactive current is proportional to the voltage

$$i_{qg}^{-} = K_{neg} V^{-} \tag{5.21}$$

where K_{neg} is the proportional gain between negative sequence voltage and reactive current with recommended values between 2 and 6, as shown in Figure 5.12.

The positive sequence reactive current reference is calculated by the outer control proportional voltage regulator as given in (5.14) for $i_{qg}^{+\prime}$. The reactive current references must be revised according to (5.22) when $\left(i_{qg}^{+\prime} + i_{qg}^{-\prime}\right) > I_{qg}^{\lim}$

$$i_{qg}^{-\prime\prime} = i_{qg}^{-\prime}\left[I_{qg}^{\lim} \Big/ \left(i_{qg}^{+\prime} + i_{qg}^{-\prime}\right)\right] \tag{5.22}$$

The positive sequence active current reference is generated by the DC bus voltage regulator ($i_{dg}^{+\prime} = i_{dg}^{\prime}$) and $i_{dg}^{-\prime} = 0$ as there is no active power exchange in the negative sequence domain. During the normal operation, the controller gives priority to the positive sequence active current, as given in (5.20). On the other hand, during the FRT operation, the GSC controller gives priority to the reactive current by reversing the d- and q-axis current limits given in (5.20). As there is no dead-band region in VDE-AR-N 4120, AVR usage is essential to control the GSC reactive current output.

5.2.5 DFIG WTs

In WTs with DFIGs, the stator of the induction generator (IG) is directly connected to the grid whereas the wound rotor is connected to the grid through an AC–DC–AC converter system as shown in Figure 5.13. The AC–DC–AC converter consists of two VSCs:

- Rotor side converter (RSC)
- Grid-side converter (GSC)

A line inductor (current choke) along with shunt harmonic AC filters is used at the GSC terminal to improve the power quality (not shown in Figure 5.13). A crowbar is used to protect the RSC against overcurrent and the DC capacitor against overvoltage. During crowbar ignition, the RSC is blocked, and the IG consumes reactive power. To avoid the crowbar ignition during faults, the DC resistive chopper is widely used to limit the DC voltage. The DFIG WT protection system also contains RSC and GSC overcurrent protections, low-voltage and overvoltage relays, cut-in and cut-off speed relays, and the pitch control as shown in Figure 5.3.

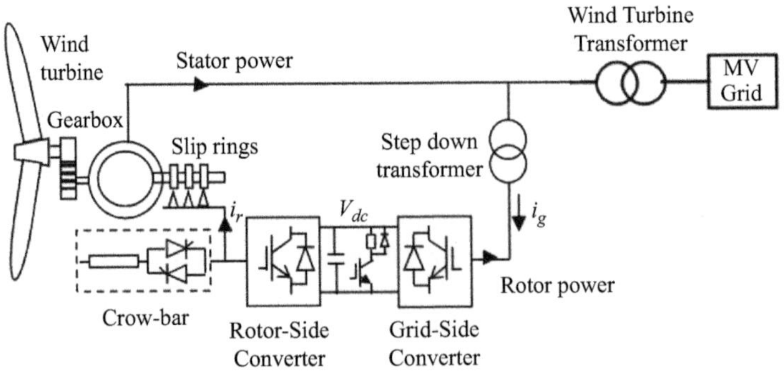

Figure 5.13 Generic structure of a DFIG WT

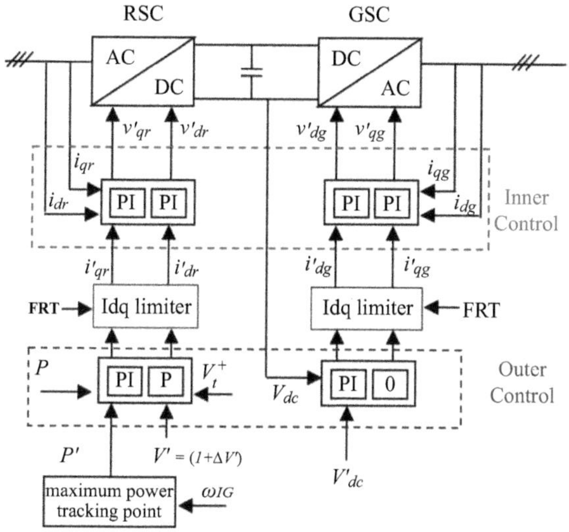

Figure 5.14 DFIG WT control scheme

The control scheme is illustrated in Figure 5.14. In this figure, i_{qr} and i_{dr} are the q- and d-axis currents of the RSC, i_{qg} and i_{dg} are the q- and d-axis currents of the GSC, V_{dc} is the DC bus voltage, P is the active power output of the DFIG, and V^+ is the DFIG positive sequence terminal voltage. The RSC operates in flux reference frame and the GSC operates in voltage reference frame. Variables i_{qr} and i_{dr} are used to control P and V^+, respectively. On the other hand, i_{dg} is used to maintain the dc bus voltage (V_{dc}) and i_{qg} is used to support the grid with reactive power severe voltage sags and swells.

The RSC and the GSC are both controlled by a two-level controller. The slow outer control calculates the reference dq-frame currents, i.e., i'_{dr}, i'_{qr}, i'_{dg}, and i'_{qg}, while the fast inner control deals with the converter AC voltage reference. The reference for DFIG active power output P' is given by MPPT control as given in (5.4). The reference for DFIG positive sequence voltage v' is calculated by the WPC as can be seen in Figure 5.5. Similar to the FSC WTs, the DFIG WTs are also equipped with an FRT function. When the FRT function is enabled, the DFIG injects reactive current proportionally to the voltage deviation from 1 p.u. as demonstrated in Figure 5.10.

5.2.5.1 RSC control

The RSC inner control loop is designed based on the IMC method [14] considering the Γ representation of the IG [21]. Γ representation is obtained by adjusting the rotor/stator turn ratio for eliminating the stator leakage inductance and eliminates the complexity of the well-known T representation without loss of information or accuracy [22]. Due to symmetry of the IG, the PI parameters are the same for both d- and q-axis channels and are given by

$$k_p = \alpha_c L_\sigma \,, \qquad k_i = \alpha_c R_R \tag{5.23}$$

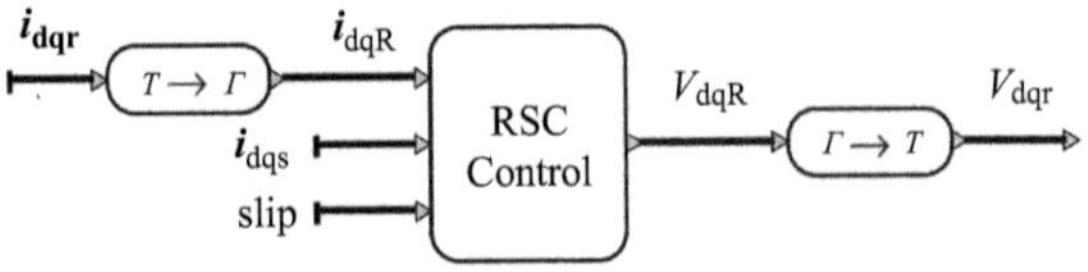

Figure 5.15 Conversion at RSC input and output variables

where L_σ is the equivalent leakage inductance and R_R is the equivalent rotor resistance in Γ representation of the IG. References [16,21] provide further calculation details in this regard. Similar to the MSC in the FSC WTs, the controller includes the feed-forward compensating terms to ensure good reference tracking. It should be noted that the RSC inner current control needs conversion for the input RSC currents and the output RSC voltages, as shown in Figure 5.15.

The d-axis reference current is determined by the proportional outer voltage control

$$i'_{dr} = K_V \left(V' - V^+\right) + i_{dr-m} \tag{5.24}$$

In (5.24), i_{dr-m} is the compensating term for the reactive current that is absorbed by the IG and is approximated by

$$i_{dr-m} = V_{wt}^+ / \left(\omega_s L_m\right) \tag{5.25}$$

where V_{wt}^+ is the positive sequence terminal voltage at the DFIG terminal and L_m is the IG magnetizing inductance.

The q-axis reference current is calculated by the power controller

$$i'_{qr} = \left(K_{PP} + K_{IP}/s\right)\left(P' - P\right) \tag{5.26}$$

During normal operation, the controller gives the priority to the active currents. However, when FRT function is activated, the priority is given to the reactive currents to achieve the desired response given in Figure 5.10.

5.2.5.2 GSC control

Except the q-axis reference current calculation, the GSC is similar to the GSC in FSC WTs. In DFIG WTs, the GSC operates at a unity power factor, hence the q-axis reference current is set to zero, i.e., $i'_{qg} = 0$. However, the GSC starts injecting reactive currents during faults when the RSC reactive current contribution is not sufficient to satisfy the grid code requirement due to the reactive current absorbed by the IG. Under such circumstances, the GSC q-axis reference current is given by [16]

$$i'_{qg} = K_V \left(V' - V^+\right) - \left(I_{dr}^{\text{lim}} - i_{dr-m}\right) \tag{5.27}$$

Similar to the RSC, the priority is given to the GSC reactive currents when the FRT function is activated. In order to improve the high-voltage ride through (HVRT) capability of the DFIG WT, reactive current contribution of the GSC can be also used.

5.2.5.3 DSC

The DFG is exposed to high stator and rotor currents as well as oscillating torque during unbalanced steady-state operation and fault conditions [23]. The mitigation techniques include RSC compensation by supplying negative sequence voltages to the rotor circuits [24–26]. Suppression of the DC-link voltage ripples can be achieved in addition to the rotor current and torque pulsations with coordinated control of RSC and GSC. In this scheme, the RSC is used to limit the torque and rotor current pulsations while the GSC is used to suppress the DC-link voltage ripples [27–32].

The effectiveness of such mitigation methods is limited to the RSC voltage limits. They are particularly effective for small imbalances since the limitations with regard to the positive sequence control are not to be expected. It should be noted that, the priority is given to the positive sequence control. In case of large asymmetry, the torque and rotor current pulsations can be limited effectively when the priority is given to negative sequence control. However, this would significantly reduce the operational range for the positive sequence control and leads to an increased load current. Moreover, providing voltage support as per the grid code requirements would normally be no longer possible [32]. It should be also noted that the negative sequence fault current contribution of the DFIG reduces when operating under this control scheme especially when the GSC is operating under CSC. The mitigation techniques also include the GSC compensation by compensating negative sequence currents to the grid [33] to keep the stator currents free of negative sequence components and thus eliminate the negative sequence components in the rotor currents. It is thus expected that the effectiveness of this method is limited to the GSC current limit.

The traditional CSC scheme shown in Figure .5.14 is expected to comply with the recent grid code VDE-AR-N 4120 as the IG rotor circuits provide a low inductive impedance path to the negative sequence currents. However, depending on severity of the asymmetry, the circulating negative sequence currents on rotor circuits may cause overvoltage at rotor side converter and the crowbar operation. During crowbar operation, RSC is blocked, and IG starts consuming reactive power. The crowbar operation is not desirable for the faults at transmission system, as the DFIG would fail to comply with the grid code. However, Ref. [34] demonstrated that crowbar operation is possible when the unbalanced transmission system fault takes place electrically close to the DFIG-based WP. Recent research such as [34,35] focus on coordinated DSC of RSC and GSC to develop strategies from the perspective of grid code compliance, limiting the shaft torque pulsations and suppression of the DC-link voltage ripples.

5.3 Software implementation

Given that the developed models provide full details of IBR architecture, control loops, and parameters, they can be readily implemented in a typical EMT software package. With proper parameterization, the models can produce accurate simulation results for planning and operation of WPs. For instance, the complete models of the

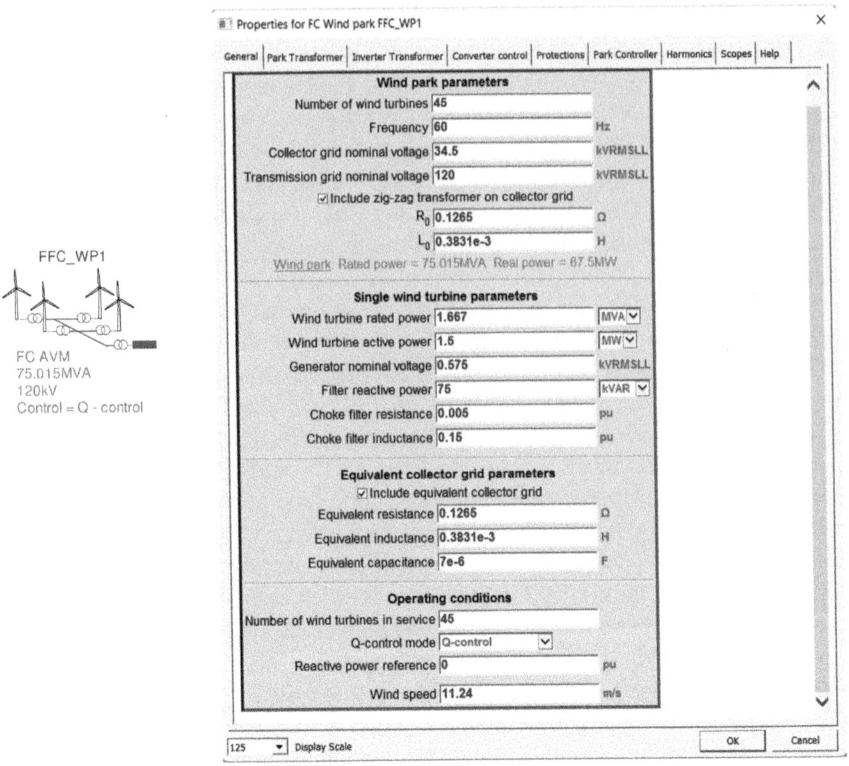

Figure 5.16 FSC WP device and its mask in EMTP

FSC and DFIG WTs have been implemented in Electromagnetic Transients Program (EMTP®) [36] based on a hierarchical and versatile approach. In this approach, the WPs consist of different interconnected modules and subnetworks (e.g., building blocks) with masking capability and tunable parameters.

The WPs contain the aggregated WT, the aggregated LV/MV WT transformer, and equivalent PI circuit of the MV collector grid and the MV/HV WP transformer. A snapshot of the FSC WP device and its mask (data input interface) is shown in Figure 5.16. In this implementation approach, the average wind speed is used under the assumption of aggregated modeling which provides acceptable accuracy for different scenarios including SSO studies [37].

The first tab of the wind park mask enables the user to modify the general parameters of the system:

- Wind park parameters: number of WTs in the WP, POI and collector grid voltage levels, collector grid equivalent and zig zag transformer parameters
- WT parameters: WT rated power, voltage, and frequency

- Wind park operating conditions: number of WTs in service, wind speed, WPC operating mode, reactive power or power factor

The second and the third tabs are used for MV/HV WP transformer and LV/MV WT transformer parameters, respectively.

The fourth tab is used to modify the parameters of the converter control system as given below:

- Sampling rate and PWM frequency at WT converters
- WT input measuring filter parameters
- MSC (or RSC) control parameters
- GSC control parameters
- Coupled/DSC options

The fifth tab is used to modify the parameters of protection system, such as chopper, crowbar (for DFIG only), overcurrent protection, low voltage and overvoltage relays.

The sixth tab is used to modify the WPC parameters.

The internal model parameters are calculated automatically by EMTP. The WP device also handles the data that is not accessible from the mask, such as the data for WT aerodynamics, mechanical system, and pitch control.

The aggregated and single unit model per unit parameters are the same for both the WT and the WT transformer if the base power for the aggregated unit is selected as

$$S_{agg} = N\ S_{single} \tag{5.28}$$

where S_{single} is the single unit base power and N is the number of units in aggregation.

The WT control offers both AVR and AQR schemes. When the AQR scheme is selected as WT control, the reactive power reference for the aggregated WT model is produced by the proportional voltage control of the WPC. When the AVR scheme is selected, the reactive power regulator of the WPC adjusts the voltage reference value of the aggregated WT model, as demonstrated in Figure 5.5. For the AVR scheme, the WPC offers POI reactive power and POI power factor control options (Q-control and PF-control, respectively) in addition to the POI voltage control option as shown in Figure 5.5. In addition, the user can deactivate the WPC. In that case, the AVR option uses the user defined voltage reference and the AQR option uses either user defined reactive power or power factor as suggested in IEC 61400-27 [17].

The FSC WT control offers both presented DSC options in [7] in addition to the traditional CSC option when AVR is selected to control the WT reactive currents. The DSC option compliant with VDE-AR-N 4120 technical connection rules [20] is not available for the AQR scheme. The DFIG WT control offers the DSC scheme in [32] in which the RSC is used to suppress the negative sequence rotor currents, and the DSC scheme in [33] in which the GSC compensates the negative sequence current required in the network during any unbalanced operation.

The developed generic models (both FSC and DFIG) feature two versions with different converter models: detailed model (DM) and average value model (AVM). In the DM, the WT converters are represented based on the circuit of Figure 5.17(a) in

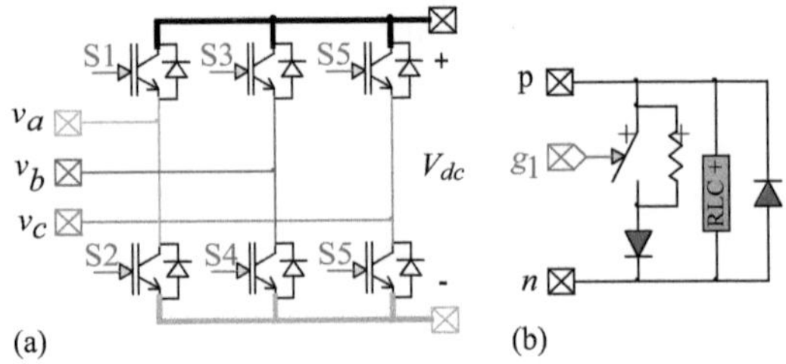

Figure 5.17 DM for WT converters: (a) two level VSC circuit and (b) IGBT valve model

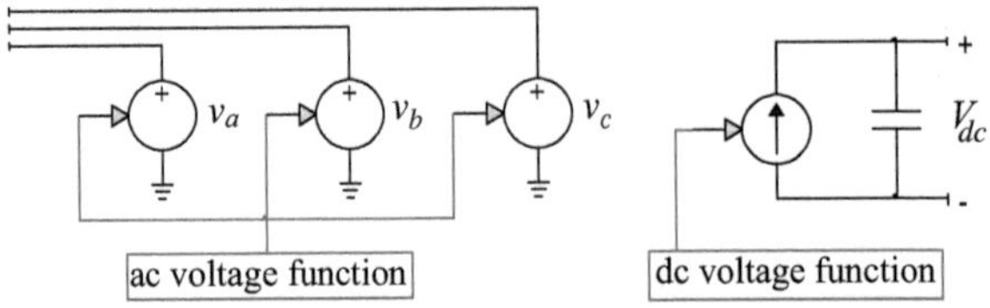

Figure 5.18 AVM for WTs employing two level VSC

which the insulated-gate bipolar transistor (IGBT)/diode is modeled by an ideal switch and nonlinear resistors shown in Figure 5.17(b) to mimic the actual behavior more accurately. The simulation of such switching circuits with variable topology requires many time-consuming mathematical operations and the high frequency PWM signals forces small simulation time-step usage. These computational inefficiencies can be eliminated by using the AVM, which replicates the average response of converters through simplified functions and controlled sources [38]. The AVMs are widely used for wind generation technologies [39,40]. The AVM is obtained by replacing the DMs of converters with controlled voltage sources on the AC side and controlled current sources on the DC side as can be seen in Figure 5.18 [41]. Interested readers may refer to [16] for further details on the modeling, implementation and utilization of the presented generic models.

5.4 Simulation results

5.4.1 FSC-based WP

In the first case study, the WP consists of 45 FSC WTs where each WT has a rated power of 1.5 MW. The single-line diagram of the 120 kV, 60 Hz test system is shown in Figure 5.19. The WP operates at the full load (under nominal wind speed) and under Q-control function of the WPC with the reference power of $Q'_{POI} = 0$. References [16,42,43] explain the details of the wind park and the test transmission system.

Several simulations are performed for different fault types and locations using the simulation models (M1 to M4) presented in Table 5.1. However, only the 250 ms

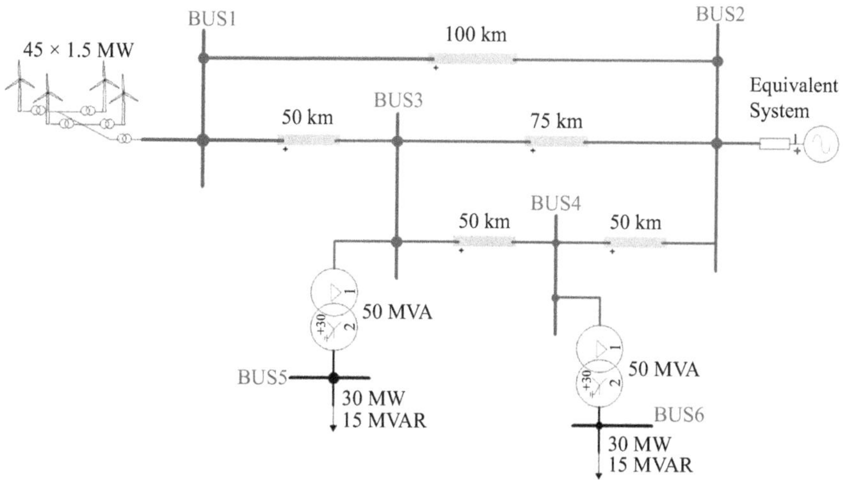

Figure 5.19 The test system for simulation of FSC-based WP

Table 5.1 Different models for simulations of the test FSC WP

Model	**M1**	**M2**	**M3**	**M4**
GSC control	CSC	DSC1	DSC2	DSC1
Converter model	DM	DM	DM	AVM

DSC1: DSC eliminates second harmonic oscillations in the GSC active power output; ***DSC2***: DSC complies with VDE-AR-N 4120.

double line-to-ground (DLG) fault at BUS4 scenario is presented in this section. A long duration fault is applied for testing purposes. The simulation time-step is 10 μs (a typical value in DM usage) and the total simulation time is 2 s.

As shown in Figure 5.20(a), the simulated unbalanced fault results in second harmonic pulsations in the active power output of M1. These second harmonic pulsations are eliminated in M2 at the expense of a reduction in the active power output of the WT, as can be seen in Figure 5.20(b). M3 achieves injection of desired negative sequence reactive current at the expense of a reduction in the active power output of the FSC WT, as shown in Figure 5.20(b), and an increase in the second harmonic oscillations in the GSC active power output which is evident in Figure 5.20(a). On the other hand, the reactive power output of the FSC WT is similar in M1, M2, and M3. This is due to the same FRT requirement on positive sequence reactive currents.

The performance of M2 and M3 is limited to the GSC rating. The DSC objectives cannot be achieved in both M2 and M3 when the required GSC current output exceeds its rating. It should be noted that, when the electrical distance between the WP and

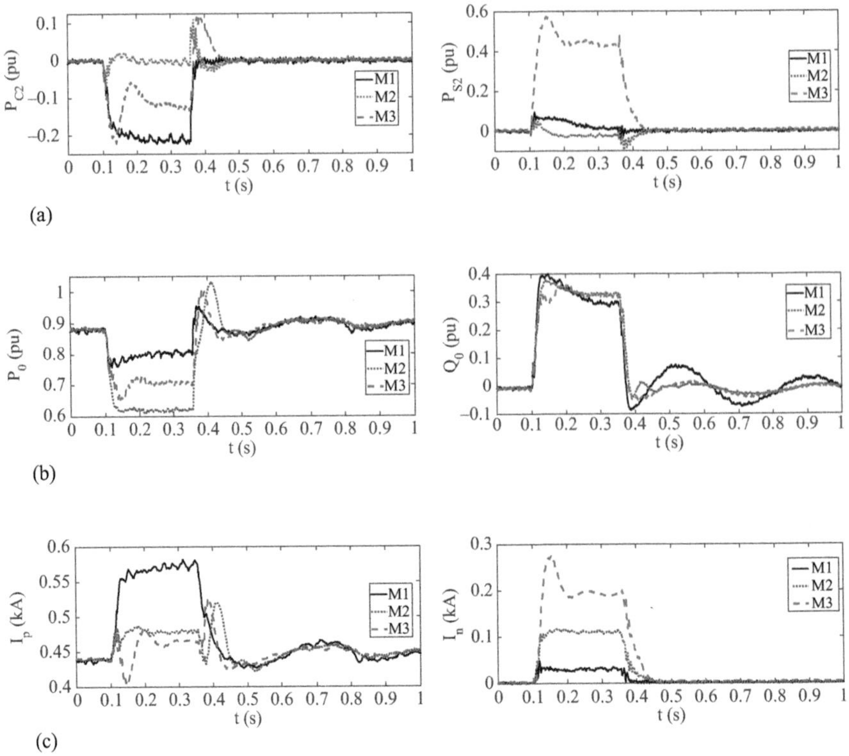

Figure 5.20 FSC WT behavior during DLG fault at BUS4 for different GSC control schemes. (a) Second harmonic pulsations in the active power output (P_{C2}, P_{S2}) of aggregated FSC WT in M1, M2, and M3. (b) Average values of instantaneous active and reactive powers (P_0, Q_0) of aggregated FSC WT in M1, M2, and M3. (c) The negative and positive sequence fault currents (I_n and I_p) of the WP in M1, M2, and M3

unbalanced fault decreases, larger GSC currents are required to achieve the DSC objectives under M2 and M3.

The negative and positive sequence fault currents of the WP in M1, M2, and M3 are illustrated in Figure 5.20(c). The small negative sequence current injection in M1 is due to a phase shift in low pass measuring filters [18]. M2 injects a considerable amount of negative sequence current to achieve mitigation of second harmonic power oscillations, but still quite low compared to M3. It should be noted that, this difference strongly depends on the unbalanced fault type, its electrical distance to the WP and GSC rating. It becomes less noticeable especially for the electrically distant faults such as an unbalanced fault at BUS6.

To illustrate the differences in computation time between models with different level of detail (DM and AVM converter models), a comparison is given in Figure 5.21. The Model M4* represents a copy of M1 with a timestep of 50 μs,

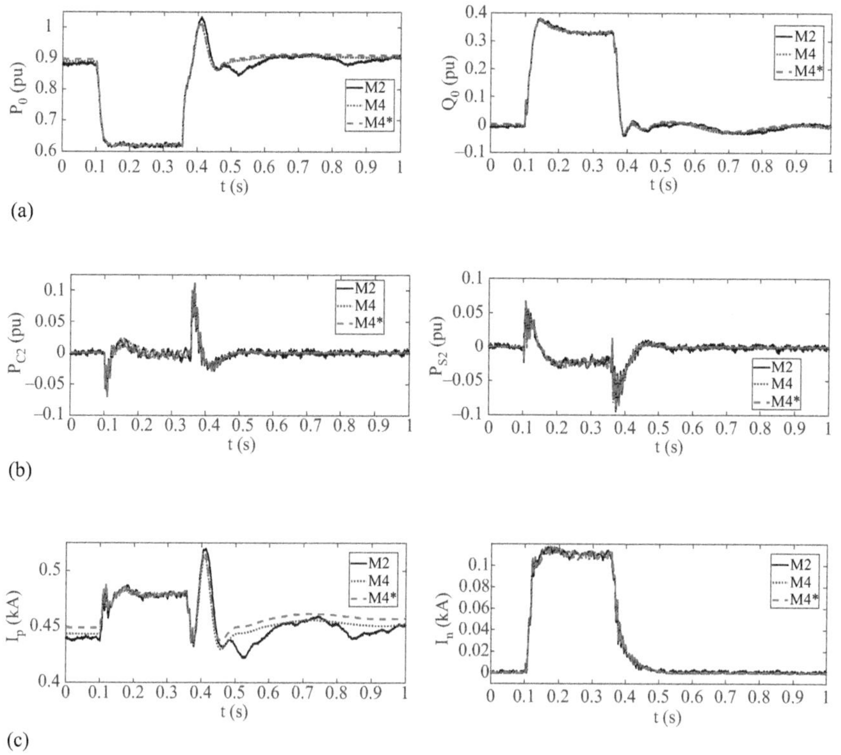

Figure 5.21 Impact of AVM usage for FSC WT converters. (a) Average values of instantaneous active and reactive powers (P_0, Q_0) of aggregated FSC WT in M2 and M4. (b) Second harmonic pulsations in the active power output (P_{C2}, P_{S2}) of aggregated FSC WT in M2 and M4. (c) The negative and positive sequence fault currents (I_n and I_p) of the WP in M2 and M4

Table 5.2 CPU timings (Intel i7-4900MQ CPU @ 2.8 GHz)

Model	**M2**	**M4**	**M4***
CPU time	144.7 s	99.5 s	28.8 s

for faster computation time. As shown in Figure 5.21, the AVM (M4 and M4*) provides acceptable accuracy with 10 μs, and even with 50 μs time step. The gain in computation time is evident when comparing to the DM (M2) as shown in Table 5.2. For this simulation case, the use of M4* yields a reduction of computation time by 80%. A higher computational gain can be expected in the simulations of a large-scale power system. In a large-scale system, the wind parks can be simulated on separate CPUs to accelerate simulations. More powerful computers will also contribute to much better computational performance. It is noted here that all control block diagrams are solved simultaneously without artificial numerical delays.

5.4.2 DFIG-based WP

The simulations in the previous part of this section are repeated using the developed DFIG WP model. Similar to the FSC WP model, acceptable accuracy is obtained with the AVM even with 50 μs time step. As shown in Figure 5.22, the DSC control scheme does not eliminate the torque pulsations effectively when the unbalanced fault occurs close to the WP. This is an expected result due to the prioritization of the positive sequence control and the RSC voltage limits.

As mentioned, a key aspect of a generic IBR model is parameterization, i.e., the capability to reproduce a specific performance by tuning model parameters. Reference [45] has provided a parameter tuning example in which the developed generic DFIG WP model operating under CSC was parameterized to replicate the fault response of two actual WP installations subjected to different types of unbalanced faults. One of those WPs is connected to a 230-kV transmission system as shown in Figure 5.23. In this case, the WP consists of 66 DFIG WTs each with the rated power of 1.5 MW. The WTs are connected to a collector substation through 34.5-kV collector grid and a wye-delta-wye 34.5/230 kV step-up transformer. An 18.7 km tie-line connects the collector substation to a 230-kV POI substation. The fault is phase-to-phase BC occurring on the tie-line 3.5 km from the POI substation. Prior to the fault, all 66 WT generators were connected to the system, the WP was delivering 25.69 MW and absorbing 1.35 MVAR from the 230-kV system at the collector substation. The fault event was recorded by the line relays on the 230-kV tie line. The current differential relay systems that are applied to this line recorded the currents at both terminals in each relay. Further details on the simulated system and the pertinent models can be found in [45].

The simulation results and the field measurements are compared in Figure 5.24. The fault occurs at 3.8 cycles after the simulation start time and is cleared within 3 cycles. Although the information about the WTs were limited and simulations were performed

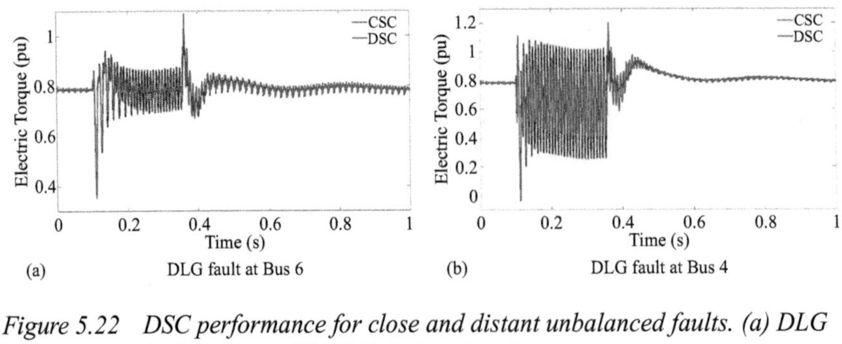

Figure 5.22 DSC performance for close and distant unbalanced faults. (a) DLG fault at bus 6. (b) DLG fault at bus 4

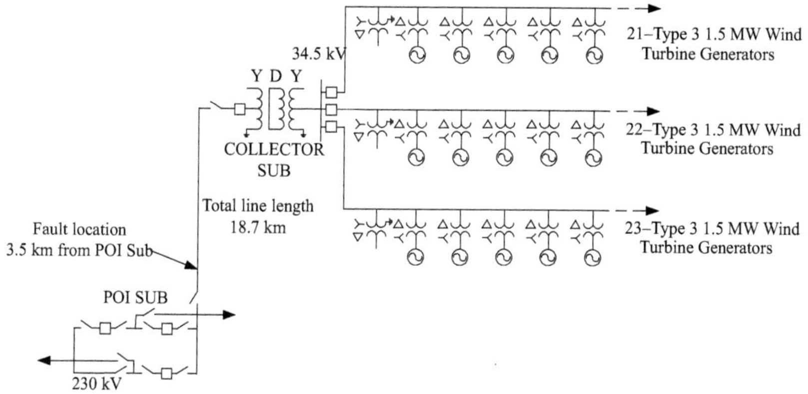

Figure 5.23 The test system used in simulation and validation of the DFIG-based WP

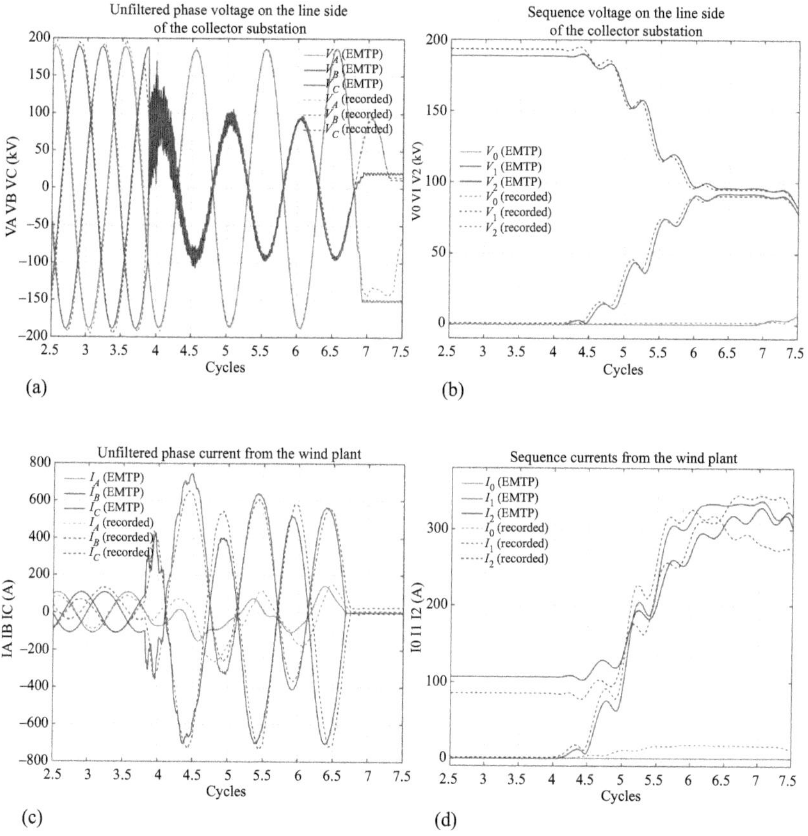

Figure 5.24 Actual response recorded by the relays (dashed line) and the EMT simulation results (solid line). (a) Phase voltage on the line side of the collector substation. (b) Sequence voltage on the line side of the collector substation. (c) Phase current from the wind park measured at the collector substation. (d) Sequence current from the wind park measured at the collector substation

by keeping the default (typical) values of the generic model, acceptable accuracy was obtained in EMT simulations.

5.5 Conclusion

WPs consist of a large variety of interconnected components including mechanical parts, power electronic devices, control and protection systems, etc. Accurate and generic models for different types of WTs are crucial for reliable design and planning of modern power systems that incorporate WPs. This chapter introduces an effective approach towards detailed modeling and simulation of WPs that employ variable speed WTs. Specifically, converters and their control schemes for DFIG and FSC WTs are thoroughly discussed, and their main parameters are explained. Moreover, software implementation, average value and detailed models, and controller design are addressed. EMTP® is used to verify the accuracy of the generic models under different test cases. Time-domain simulation results are analyzed and compared with the real-life measurements of post-fault transients in the test scenarios. The results confirm that the developed models in conjunction with the EMT simulations can accurately predict the response of DFIG and FSC WTs under both steady-state and transient conditions.

Bibliography

[1] M. Shafiul Alam, F. S. Al-Ismail, A. Salem and M. A. Abido, "High-Level Penetration of Renewable Energy Sources into Grid Utility: Challenges and Solutions," *IEEE Access*, vol. 8, pp. 190277–190299, 2020.

[2] S. V. Bozhko, R. Blasco-Gimenez, R. Li, J. C. Clare and G. M. Asher, "Control of Offshore DFIG-Based Wind Farm Grid with Line-Commutated HVDC Connection," *IEEE Transactions on Energy Conversion*, vol. 22, no. 1, pp. 71–78, 2007.

[3] M. El Moursi, G. Joos and C. Abbey, "A Secondary Voltage Control Strategy for Transmission Level Interconnection of Wind Generation," *IEEE Transactions on Power Electronics*, vol. 23, no. 3, pp. 1178–1190, 2008.

[4] P. E. Sutherland, "Ensuring Stable Operation with Grid Codes: A Look at Canadian Wind Farm Interconnections," *IEEE Industry Applications Magazine*, vol. 22, no. 1, pp. 60–67, 2016.

[5] U. Karaagac, J. Mahseredjian, S. Jensen, R. Gagnon, M. Fecteau and I. Kocar, "Safe Operation of DFIG based Wind Parks in Series Compensated Systems," *IEEE Transactions on Power Delivery*, vol. 33, no. 2, pp. 709–718, 2018.

[6] B. Novakovic, Y. Duan, M. G. Solveson, A. Nasiri and D. M. Ionel, "From Wind to the Electric Grid: Comprehensive Modeling of Wind Turbine Systems," *IEEE Industry Applications Magazine*, vol. 22, no. 5, pp. 73–84, 2016.

[7] U. Karaagac, J. Mahseredjian, R. Gagnon, *et al.*, "A Generic EMT-type Simulation Model for Wind Parks with Permanent Magnet Synchronous Generator

Full Size Converter Wind Turbines," *IEEE Power and Energy Technology Systems Journal*, vol. 6, no. 3, pp. 131–141, 2019.

[8] O. Anaya-Lara, N. Jenkins, J. B. Ekanayake, P. Cartwright and M. Hughes, *Wind Energy Generation: Modelling and Control*, John Wiley & Sons, New York, NY, 2009.

[9] N. Miller, J. J. Sanchez-Gasca, W. W. Price and R. W. Delmerico, *Dynamic Modeling of GE 1.5 and 3.6 Wind Turbine-Generators*, GE-Power System Energy Consulting, Schenectady, NY, 2003.

[10] G. Abad, J. Lopez, M. Rodriguez, L. Marroyo and G. Iwanski, *Doubly Fed Induction Machine: Modeling and Control for Wind Energy Generation*, Wiley, New York, NY, 2011.

[11] M. Singh and S. Santoso, *Dynamic Models for Wind Turbines and Wind Power Plants*, University of Texas at Austin and National Renewable Energy, 2011.

[12] J. M. Garcia, *Voltage Control in Wind Power Plants with Doubly Fed Generators*, Ph.D. dissertation, Aalborg University, 2010.

[13] V. Akhmatov, A. H. Nielsen, J. K. Pedersen and O. Nymann, "Variable-Speed Wind Turbines with Multi-Pole Synchronous Permanent Magnet Generators. Part I: Modelling in Dynamic Simulation Tools," *Wind Engineering*, vol. 27, no. 6, pp. 531–548, 2003.

[14] L. Harnefors and H.-P. Nee, "Model-Based Current Control of AC Machines Using the Internal Model Control Method," *IEEE Transactions on Industry Applications*, vol. 34, no. 1, pp. 133–141, 1998.

[15] *Grid Code – High and Extra High Voltage*, E.ON Netz GmbH, Bayreuth, Germany, 2006.

[16] U. Karaagac, J. Mahseredjian, H. Gras, H. Saad, J. Peralta and L. D. Bellomo, *Simulation Models for Wind Parks with Variable Speed Wind Turbines in EMTP-RV*, Polytechnique Montréal, Montréal, 2017.

[17] *Wind Turbines – Part 27-1: Electrical Simulation Models – Wind Turbines*, IEC Standard 61400-27-1, 2015.

[18] A. Haddadi, I. Kocar, U. Karaagac, J. Mahseredjian and E. Farantatos, *Impact of Renewables on System Protection: Wind/PV Short-Circuit Phasor Model Library and Guidelines for System Protection Studies*, EPRI Technical Report, Report number: 3002008367, Palo Alto, CA, 2016.

[19] R. Teodorescu, M. Liserre and P. Rodriguez, *Grid Converters for Photovoltaic and Wind Power Systems*, John Wiley & Sons, New York, NY, 2011.

[20] *Technische Regeln für den Anschluss von Kundenanlagen an das Hochspannungsnetz und deren Betrieb* (TAR Hochspannung), VDE-AR-N 4120 Anwendungsregel, 2018.

[21] R. Pena, J. C. Clare and G. M. Asher, "Doubly Fed Induction Generator Using Back-to-Back PWM Converters and Its Application to Variable-Speed Wind-Energy Generation," *EE Proceedings-Electric Power Applications*, vol. 143, no. 3, pp. 231–241, 1996.

[22] G. R. Slemon, "Modelling of Induction Machines for Electric Drives," *IEEE Transactions on Industry Applications*, vol. 25, no. 6, pp. 1126–1131, 1989.

[23] E. Muljadi, D. Yildirim, T. Batan and C. P. Butterfield, "Understanding the Unbalanced-Voltage Problem in Wind Turbine Generation," *Conference Record of the 1999 IEEE Industry Applications Conference. Thirty-Forth IAS Annual Meeting*, 1999.

[24] T. K. A. Brekken and N. Mohan, "Control of a Doubly Fed Induction Wind Generator Under Unbalanced Grid Voltage Conditions," *IEEE Transactions on Energy Conversion*, vol. 22, no. 1, pp. 129–135, 2007.

[25] L. Xu and Y. Wang, "Dynamic Modeling and Control of DFIG-Based Wind Turbines Under Unbalanced Network Conditions," *IEEE Transactions on Power Systems*, vol. 22, no. 1, pp. 314–323, 2007.

[26] J. Hu and Y. He, "Modeling and Enhanced Control of DFIG Under Unbalanced Grid Voltage Conditions," *Electric Power Systems Research*, vol. 79, no. 2, pp. 273–281, 2009.

[27] L. Xu, "Enhanced Control and Operation of DFIG-Based Wind Farms During Network Unbalance," *IEEE Transactions on Energy Conversion,* vol. 23, no. 4, pp. 1073–1081, 2008.

[28] O. Gomis-Bellmunt, A. Junyent-Ferré, A. Sumper and J. Bergas-Jané, "Ride-Through Control of a Doubly Fed Induction Generator under Unbalanced Voltage Sags," *IEEE Transactions on Energy Conversion*, vol. 23, no. 4, pp. 1036–1045, 2008.

[29] Y. Zhou, P. Bauer, J. A. Ferreira and J. Pierik, "Operation of Grid-Connected DFIG Under Unbalanced Grid Voltage Condition," *IEEE Transactions on Energy Conversion*, vol. 24, no. 1, pp. 240–246, 2009.

[30] J. Hu and Y. He, "Modeling and Control of Grid-Connected Voltage-Sourced Converters Under Generalized Unbalanced Operation Conditions," *IEEE Transactions on Energy Conversion*, vol. 23, no. 3, pp. 903–913, 2008.

[31] L. Fan, H. Yin and Z. Miao, "A Novel Control Scheme for DFIG-Based Wind Energy Systems Under Unbalanced Grid Conditions," *Electric Power Systems Research*, vol. 81, no. 2, pp. 254–262, 2011.

[32] S. Engelhardt, J. Kretschmann, J. Fortmann, F. Shewarega, I. Erlich and C. Feltes, "Negative Sequence Control of DFG Based Wind Turbines," in *IEEE Power and Energy Society General Meeting*, 2011.

[33] R. Pena, R. Cardenas, E. Escobar, J. Clare and P. Wheeler, "Control System for Unbalanced Operation of Stand-Alone Doubly Fed Induction Generators," *EEE Transactions on Energy Conversion*, vol. 22, no. 2, pp. 544–545, 2007.

[34] Y. Chang, I. Kocar, J. Hu, U. Karaagac, K. W. Chan and J. Mahseredjian, "Coordinated Control of DFIG Converters to Comply with Reactive Current Requirements in Emerging Grid Codes," *Journal of Modern Power Systems and Clean Energy*, vol. 10, no. 2, pp. 502–514, 2022.

[35] H. Xu, Y. Zhang, Z. Li, R. Zhao and J. Hu, "Reactive Current Constraints and Coordinated Control of DFIG's RSC and GSC During Asymmetric Grid Condition," *IEEE Access*, vol. 8, pp. 184339–184349, 2020.

[36] J. Mahseredjian, S. Dennetière, L. Dubé, B. Khodabakhchian and L. Gérin-Lajoie, "On a New Approach for the Simulation of Transients in Power

Systems," *Electric Power Systems Research*, vol. 77, no. 1, pp. 1514–1520, 2007.

[37] M. Ghafouri, U. Karaagac, J. Mahseredjian and H. Karimi, "SSCI Damping Controller Design for Series Compensated DFIG based Wind Parks Considering Implementation Challenges," *IEEE Transactions on Power Systems*, vol. 34, no. 4, pp. 2644–2653, 2019.

[38] S. R. Sanders, J. M. Noworolski, X. Liu and G. C. Verghese, "Generalized Averaging Method for Power Conversion Circuits," *IEEE Transactions on Power Electronics*, vol. 6, no. 2, pp. 251–259, 1991.

[39] J. Morren, S. W. H. de Haan, P. Bauer, J. T. G. Pierik and J. Bozelie, "Comparison of Complete and Reduced Models of a Wind Turbine with Doubly-Fed Induction Generator," *Tenth European Conference on Power Electronics and Applications*, Toulouse, France, 2003.

[40] J. G. Slootweg, H. Polinder and W. L. Kling, "Representing Wind Turbine Electrical Generating Systems in Fundamental Frequency Simulations," *EEE Transactions on Energy Conversion*, vol. 18, no. 4, pp. 516–524, 2003.

[41] J. Peralta, H. Saad, U. Karaagac, J. Mahseredjian, S. Dennetière and X. Legrand, "Dynamic Modeling of MMC-based MTDC Systems for the Integration of Offshore Wind Generation," in *Procedings of the CIGRE Canada Conference on Power Systems*, Montreal, Canada QC, 2012.

[42] "Impact of Renewables on System Protection: Short-Circuit Phasor Models of Renewables and Impact of Renewables on Power Swing Detection and Distance Protection," EPRI, Palo Alto, CA, 2016.

[43] T. Kauffmann, U. Karaagac, I. Kocar, S. Jensen, J. Mahseredjian and E. Farantatos, "An Accurate Type III Wind Turbine Generator Short Circuit Model for Protection Applications," *IEEE Transactions on Power Delivery*, vol. 32, no. 6, pp. 2370–2379, 2016.

[44] A. Haddadi, I. Kocar, T. Kauffmann, U. Karaagac, E. Farantatos and J. Mahseredjian, "Field Validation of Generic Wind Park Models using Fault Records," *Journal of Modern Power Systems and Clean Energy*, vol. 7, no. 4, p. 826–836, 2019.

Chapter 6

Isomorphism-based simulation of modular multilevel converters

Federico Bizzarri[1]*, Angelo Brambilla*[1]*, Daniele Linaro*[1] *and Davide del Giudice*[1]

The modular multilevel converter (MMC) has become an increasingly recurrent element in high-voltage direct current systems. The simplest structure of this converter includes three phase legs, each of which consists of two arms. In turn, every arm comprises a filter and a cascading stack of up to several hundreds of identical submodules (SMs) that consist of a capacitor and semiconductor devices.

This modular structure is both a blessing and a curse. On the one hand, it grants MMCs low switching losses, minimum filter requirements, and scalability to high voltage and power ratings. On the other hand, it poses a significant challenge in several fields of power system simulation, including electromagnetic transient (EMT) simulation. Indeed, the multitude of cascaded SMs in each arm leads to a high computational burden if conventional EMT simulation tools are adopted. To address this issue, scholars proposed several MMC simulation approaches based on simplified SM models. Despite paving the way towards compact equivalent MMC arm representations and higher simulation speed, these models neglect the switching dynamics within each SM. Thus, they are not suitable for thorough simulations of AC/DC networks, which may require the implementation of detailed transistor-level SM models.

In this chapter, we first give an overview of some MMC simulation models that lower the computational burden and then we describe a different MMC simulation paradigm based on sub-circuit isomorphism. This technique has been originally adopted to analyze electronic circuits made up of many structurally identical cells, such as RAMs. If applied to MMCs, this approach exploits the common behavior shown by the SMs of each arm by dynamically grouping them based on their current working conditions. While simulating, this paradigm selects only a single element for each cluster, and its evolution over time is replicated to all the SMs of the same group. By doing so, the number of nodes and semiconductor devices to be evaluated reduces significantly, thereby leading to simulations that are much faster than those obtained with conventional EMT solvers but just as accurate.

[1]Department of Electronics, Information Technology and Bioengineering, Polytechnic of Milan, Italy

Since this approach has general validity, it can be applied to any SM model. Thus, depending on the degree of detail required, the user can flexibly choose a simplified SM model or a more detailed one to analyze the switching phenomena inside it.

6.1 Introduction

The high-voltage direct current (HVDC) systems are an ideal solution to connect possibly asynchronous AC grids and efficiently integrate the growing amount of concentrated energy resources typically located in remote areas, such as offshore wind farms [1]. Among the solutions proposed in the literature, the modular multilevel converter (MMC) – first proposed in [2,3] – gained increasing popularity over time in the industry and research community, thereby evolving as the technology of choice for this kind of systems.

Figure 6.1(a) depicts the classical schematic of an MMC. It comprises three legs, each of which consists of an upper and lower arm. Every arm includes a cascading stack of up to several hundreds of identical submodules (SMs) (i.e., a SM string) that can implement several topologies [4,5]. This chapter focuses on the half-bridge SM, shown in Figure 6.1(b): it comprises two IGBTs, two free-wheeling diodes, and a capacitor C_{sm}. Based on the gate signals of its IGBTs, each SM is:

- *Inserted*, if only S_1 is switched on. In this case, the SM impresses at its pins the voltage of its capacitor, which is either charged or discharged based on the sign of its corresponding arm current.
- *Bypassed*, when only S_2 is switched on. In this working mode, most of the arm current circulates through the lower valve and only a negligible amount (i.e., leakage current of the IGBT) flows through C_{sm}, which is therefore bypassed. Thus, the SM voltage v_{SM} is basically null, regardless of arm current.
- *Blocked*, if S_1 and S_2 are both switched off. In this case, the SM behaves as an uncontrolled diode bridge. In particular, this configuration is such that v_{SM} depends on the current direction and C_{sm} can only charge but not discharge.

The SMs are blocked either during converter start-up or when faults occur. The former case is exploited to charge, if needed, the SM capacitors up to their rated value at the beginning of MMC operation. A start-up resistor is often added in the AC side to limit inrush currents when the start-up sequence begins and then shorted to minimize losses once the standard system operation starts. On the contrary, the latter case prevents the discharge of SM capacitors during DC-side faults, which would otherwise further aggravate fault currents. It is worth highlighting that half-bridge SMs cannot limit DC fault currents. Thus, after blocking the SMs, the DC-side fault must be managed either by opening *ad hoc* DC breakers or those at the AC-side of the converter (and then isolating the fault using off-load isolators in the DC grid).

On the contrary, during the normal operation of the MMC, the IGBT gate signals are regulated so that each SM alternates time intervals when they are inserted and bypassed. So doing, the SM capacitor voltages are profitably exploited to synthesize a given AC-side voltage required by the MMC control scheme to fulfill specific desiderata. In any

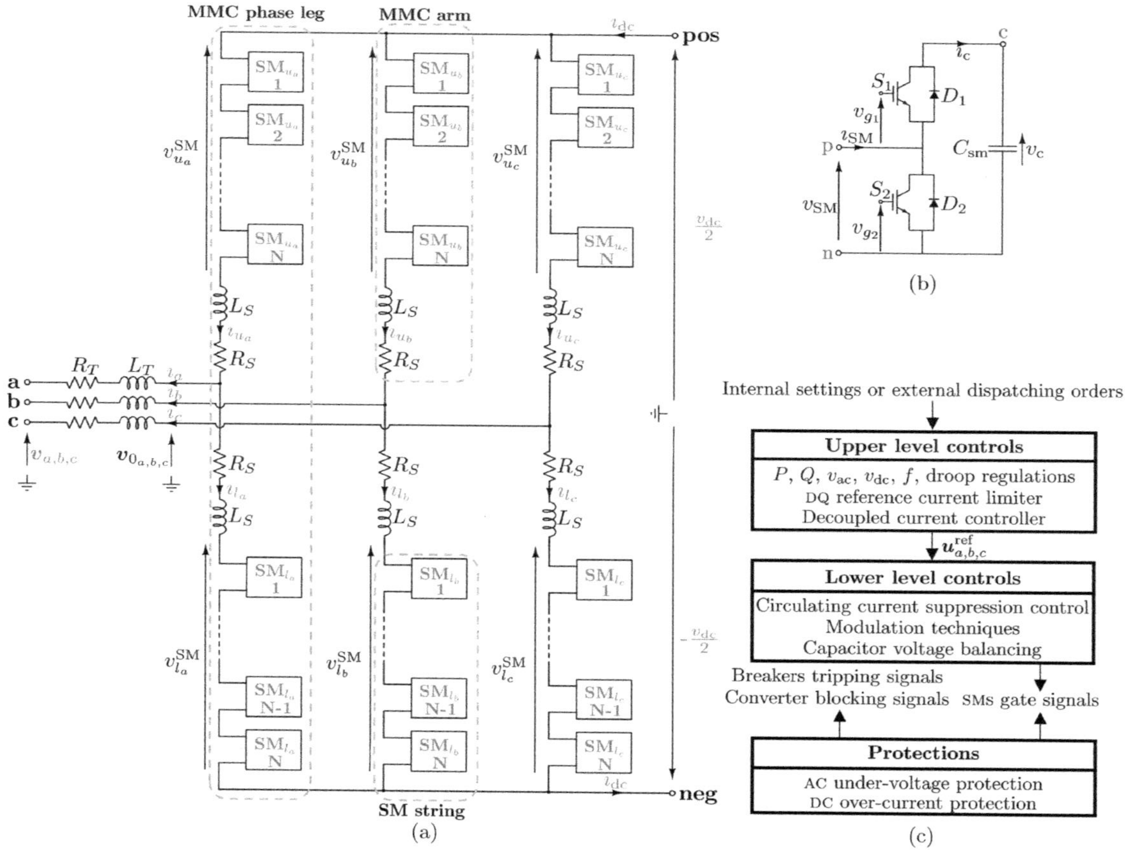

Figure 6.1 *The schematic of an* MMC *that comprises a generic number N* SMs *per arm (a), the schematic of a half-bridge* SM *(b), and a synthetic representation of the* MMC *control scheme (c). In (a), the electrical variables labeled in gray represent those measured and used as input for the control scheme. For each* SM*, the measured variable is the capacitor voltage* v_c

case, the actual transition of each SM from one operating mode to another requires all gate signals to be equipped with rising, falling, and dead-times to prevent cross conduction (i.e., the creation of dangerous low-impedance paths involving the SM capacitors).

The MMC control strategy, outlined in Figure 6.1(c), is more complex than that of other converter architectures, since it needs to process a large number of gate signals and regulate different aspects concurrently. *Upper level controls* determine the phase voltage reference values needed at the AC-side point of connection of the converter to fulfil specific objectives, determined by *internal settings or external dispatching orders*, such as a given active and reactive power exchange (or a desired AC and DC side voltage) [6]. The *lower level controls* translate the previously obtained reference phase voltages into a specific gate signal sequence for the SMs. To do so, the controls rely on a specific modulation technique (e.g., nearest-level modulation) [7,8], a *capacitor voltage balancing algorithm* [9–11] and a *circulating current suppression strategy* [12–15]. The gate signals can be modified at any time by the MMC *protections*, which could order the tripping of the converter breakers and block all SMs if abnormal operating conditions are detected (i.e., AC-side undervoltages or DC-side overcurrents following faults) [16]. To implement the previously described tasks, the variables labeled in gray in Figure 6.1(a) are measured and exploited. The interested reader can refer to [16–19] to an in-depth discussion about all the previously mentioned controls, as well as some MMC control scheme implementation examples compatible with balanced and unbalanced operating conditions, which are not reported here for brevity.

As its name suggests, the distinctive feature of the MMC is its modular structure. Contrary to other converter technologies (e.g., diode clamped and flying capacitor converters), this feature allows MMCs to increase the number of SMs per arm (and, thus, number of levels) without leading to a complex converter design. Therefore, MMCs are easily scalable to high voltage and power applications employing semiconductor devices of relatively low voltage rating. In addition, thanks to the high number of converter levels, they are characterized by minimum filtering requirements and switching losses. While on the one hand these characteristics have made MMCs the technology of choice in HVDC systems [8], on the other hand they pose significant challenges from a simulation perspective. Indeed, the presence of MMCs in modern networks leads to difficulties in performing several tasks typical of many power system simulators, such as power flow [20,21], initialization [22,23], small-signal analysis [24–26], and EMT simulation [8,27,28].

This chapter focuses on the last issue. It is worth pointing out that many power system simulators (e.g., PSCAD, EMTP) rely on the nodal admittance method and adopt a sufficiently short and often fixed integration time step to perform EMT simulations [29,30]. If MMCs need to be simulated, this paradigm leads to high CPU times for three main reasons. First, thousands of IGBTs and diodes must be considered, which results in the MMC being described by a big nodal admittance matrix. Provided that techniques exploiting circuit sparsity are employed, the computational burden required to LU-factorize this matrix varies almost quadratically with its size. This constitutes the main bottleneck for the EMT simulation of MMCs. Second, due to the non-linear nature of semiconductor devices inside the SMs, repetitive re-triangularizations of the matrix are required to determine the power system solution at each time step [8,31]. Third, frequent interpolations are necessary to accurately track the switching instants of the semiconductor devices (i.e., the exact moment when they shift from conduction to interdiction operating mode) [32].

To reduce the computational burden, scholars proposed several MMC models based on more simplified SM representations that implement different trade-offs between simulation speed and accuracy [8,27]. Albeit successful in boosting CPU time, these models are not suitable for thorough analyses of AC/DC networks, which may require the adoption of detailed transistor-level SM representations. In this framework, the main contribution of this chapter is the description of a simulation approach based on isomorphism. This technique, which has its roots in the field of electronics, can be profitably exploited to efficiently simulate MMCs regardless of the model used to describe its SMs, including the most accurate ones.

The chapter is organized as follows. Section 6.2 presents the most popular MMC models in the literature. After illustrating it from a general perspective, Section 6.3 describes how the isomorphism-based approach can be specifically applied to efficiently simulate MMCs and its features of interest. These characteristics are validated in Section 6.4 by considering two MMCs in a benchmark test system during different operating conditions and comparing the results of the conventional and isomorphism-based simulation approaches.

6.2 MMC models for EMT simulations: a review

The increasing presence of MMCs in HVDC systems requires ensuring that their performance is satisfactory in a wide range of operating conditions by performing EMT simulations. In this framework, depending on the degree of detail needed, three main analyses could be carried out [8,33].

- Component level studies, which are particularly useful during the early design phase of the MMC, focus on analyzing the performance of the semiconductor devices inside each SM. As a result, they require the adoption of detailed semiconductor models and integration time steps in the order of nanoseconds. Variables of interest include current stresses and switching losses of the diodes and IGBTs.
- System level studies evaluate the interactions between the MMC and a power system of limited extension. In this case, the main goal is to examine the correct operation of the converter filters, controls, and protections. Since the focus is no longer on analyzing the switching transients inside the SMs, simpler representations of the semiconductor devices and higher integration time steps (e.g., in the order of microseconds) can be used.
- Network level studies investigate how the presence of the MMC in a large AC grid affects its electromechanical transients and steady-state operation. In this case, integration time steps in the order of milliseconds are typically employed, together with simpler representations of the semiconductor devices (or even of the MMC as a whole).

In principle, the most detailed MMC and SM models would suffice to perform all of the studies above. However, as mentioned in the Introduction, conventional simulation approaches, which consider each SM individually, would lead to prohibitively high simulation times, especially if detailed SM models are adopted. The most straightforward solution to this issue would be employing simplified models of the SMs and, thus, of the whole MMC. Over the years, this option has been pursued intensively by academia and industry, and spurred the development of several MMC models. Some of the most popular ones are grouped and sketched from left to right in Figure 6.2 in decreasing order of complexity. In a nutshell, the full physics (FP), full detailed (FD), and bi-value resistor (BVR) models consist of different representations of the semiconductor devices inside each SM that preserve the overall MMC topology. On the contrary, the Thévenin Equivalent Model (TEM) and switching function model (SFM) lead to a more compact representation of the SM strings in each MMC arm. Lastly, by using the average value model (AVM), the entire MMC topology is further simplified in favor of a much lower computational burden.

Each of the models above implements a different trade-off between simulation speed and accuracy. For instance, the TEM (or the SFM) and the AVM are ideal solutions to efficiently perform system and network level studies, respectively. However, to the best of the authors' knowledge, a still unresolved issue is the absence of efficient MMC models to perform component level studies. Since these analyses require the implementation of detailed transistor-level representations of the converter SMs, the simplification of the SM model is not a viable option to reduce CPU times. Rather, as shown in the following for the case of the isomorphism-based method, the exploitation of advanced simulation approaches is essential.

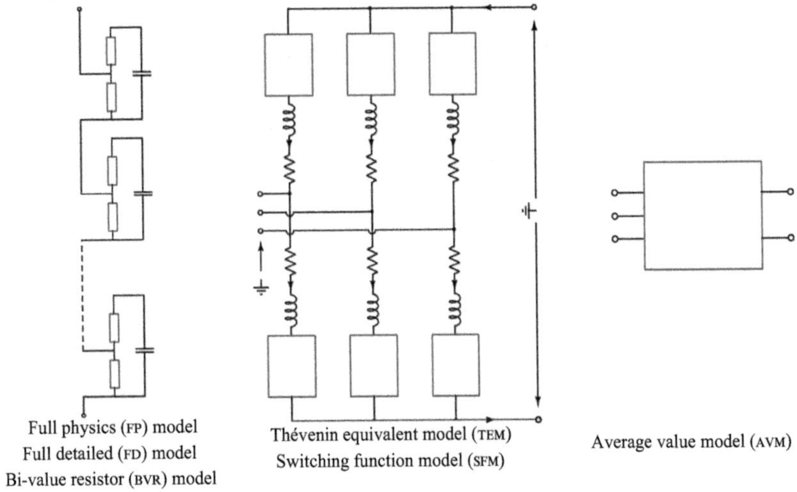

Figure 6.2 The MMC models grouped and listed in decreasing order of complexity. For each model, the red boxes indicate the sections of the MMC that are simplified

The next subsections describe the MMC models presented above by highlighting their key properties, strengths, and weaknesses. For the sake of brevity, the following overview does not dwell on the details of each model. Indeed, its only purpose is to provide a logical path aimed at better showcasing the potentialities of the isomorphism-based approach. The readers interested to an in-depth literature review of all the models can refer to [8,34] and their related references.

6.2.1 Full physics, full detailed, and bi-value resistor models

All the MMC representations described in this section stem from the full physics (FP) model, which represents SMs as shown in Figure 6.3(a). It retains the highest accuracy because it describes the semiconductor devices in each SM through detailed, non-linear dynamic representations. Other than allowing accurate analyses of the switching transients of the semiconductor devices (e.g., turn-on and turn-off characteristic, transient recovery voltage) and their parasitics effects, the FP model is the only one capable of assessing both conduction and switching losses. These features, however, come at the cost of a high computational burden due to the multitude of SMs in each arm, which generally leads to prohibitive simulation times.

This kind of detail is not available in commercial power system simulators, such as PSCAD and EMTP-RV. The second best representation in terms of accuracy that is commonly offered in these programs is the full detailed (FD) model shown in Figure 6.3(b). Compared to the FP representation, in this model, each SM valve is replaced with an ideal switch controlled by its corresponding gate signal and two non-ideal diodes (one in series, the other in anti-parallel). The schematic might include a snubber circuit that acts as a surrogate of the dynamic behavior exhibited by IGBTs and diodes during switching, which otherwise would be lost. This representation can be used as a benchmark to validate simpler MMC models and for system level studies. For instance, it could be adopted to analyze converter start-up, normal and abnormal operating conditions (e.g., AC and DC-side faults), and to test the effectiveness of MMC protection and control systems. However, due to the simplifications put forward, this

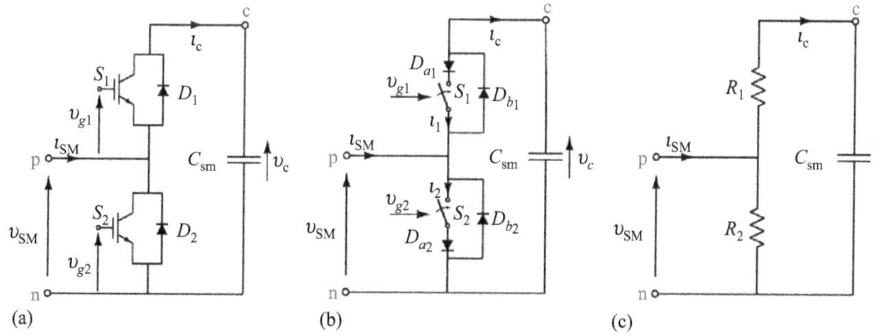

Figure 6.3 The SM schematic employed in the FP (a), FD (b), and BVR (c) models. The voltages v_{g_1} and v_{g_2} denote the gate signals of the IGBTs

Values of R_1 and R_2 of a generic SM at time t

if S_1 is ON && S_2 is OFF (i.e., SM inserted state)
    $R_1 = R_{ON}, R_2 = R_{OFF}$
else if S_1 is OFF && S_2 is ON (i.e., SM bypassed state)
    $R_1 = R_{OFF}, R_2 = R_{ON}$
else if S_1 && S_2 are OFF (i.e., SM blocked state)
    if $\imath_{SM}(t) > 0$ && $\upsilon_{SM}(t - \Delta T) > \upsilon_c(t - \Delta T)$ (i,e., D_1 conducts)
        $R_1 = R_{ON}, R_2 = R_{OFF}$
    else if $\imath_{SM}(t) < 0$ && $\upsilon_{SM}(t - \Delta T) < 0$ (i,e., D_2 conducts)
        $R_1 = R_{OFF}, R_2 = R_{ON}$
    else (i,e., no diode is conducting)
        $R_1 = R_{OFF}, R_2 = R_{OFF}$
    end
end

Figure 6.4 A pseudo-code that determines the values of R_1 and R_2 of a generic SM *at a given time instant using the* BVR *representation. In the code, ΔT corresponds to the (fixed) time step adopted*

model cannot be used for component level studies. This drawback is common to all the other SM representations described in the following.

A further simplification of the semiconductor models inside the SMs is offered by the BVR representation shown in Figure 6.3(c). Indeed, the couples of diodes and IGBTs are replaced with two time-varying resistors R_1 and R_2, whose resistance is either small ($R_{ON} \sim$ mΩ) or high ($R_{OFF} \sim$ MΩ) if one of the elements in each pair is conducting or not, respectively. These values, which depend on the gate signals and the SM current and voltage, can be determined at each time step through the pseudo-code in Figure 6.4. When the SM is inserted or bypassed, R_1 and R_2 are easy to determine. Indeed, based on the sign of the current $\imath_{SM}$, either the IGBT in the ON-state or the free-wheeling diode in parallel to it conducts. On the contrary, in the blocked state, the derivation of these values is more complex because the SM behaves as an uncontrolled diode bridge. As a result, the diode D_1, D_2, or none of them conducts depending on the current and the voltages of the SM and its capacitor. Except for a slightly lower computational burden, the BVR and FD share similar traits. However, note that due to the adoption of mere resistors to synthesize the SM valves, any dynamic of the semiconductors is lost.

It is worth pointing out that, despite employing a simplified representation of the semiconductor devices, neither the FD nor the BVR SM models lead to a satisfactory boost in simulation efficiency. Indeed, since they consider each SM individually, both models still lead to a big nodal admittance matrix* and, thus, a computational burden that grows almost quadratically with the number of SMs in each arm. To address this

*Consider for example an MMC comprising 200 SMs per arm, each of which is described with the BVR model (see Figure 6.3(c)). In this case, simulating one MMC requires taking into account a number of nodes equal to $200 \times 6 \times 2 = 2,400$. Indeed, there are 200 SMs per arm, 6 arms, and 2 nodes per SM (note that each consecutive pair of series-connected SMs shares one of its three internal nodes). This counting includes neither the nodes introduced by the filters in each arm nor by MMC controllers and drivers.

issue, the TEM and the SFM have been developed. The former and the latter exploit different simplified SM models to represent the SM strings in each arm through a compact equivalent circuit that boosts simulation speed while preserving the overall behavior of the MMC in a wide range of operating conditions.

6.2.2 Thévenin equivalent model

The TEM proposed in [31] relies on the BVR SM representation. In this model, the constitutive equation of the SM capacitor is discretized so that it can be substituted with an (algebraic) Thévenin equivalent, also known in the literature as *companion model*. As shown in Figure 6.5(a), this resulting circuit comprises a resistor $R_{\mathrm{c_{eq}}}$ in series with a voltage source $v_{\mathrm{c_{eq}}}$. In the case of [31], which adopts an integration scheme based on a fixed step ΔT and the trapezoidal rule, the value of these parameters at a given time t amounts to

$$\begin{aligned} R_{\mathrm{c_{eq}}}(t) &= \frac{\Delta T}{2C_{\mathrm{sm}}} \\ v_{\mathrm{c_{eq}}}(t) &= v_{\mathrm{c}}(t-\Delta T) + \frac{\Delta T}{2C_{\mathrm{sm}}} i_{\mathrm{c}}(t-\Delta T)\,. \end{aligned} \tag{6.1}$$

By deriving the Thévenin equivalent circuit at the terminals of each SM, this circuit can be further simplified, thus leading to that in Figure 6.5(b). Simple calculations demonstrate that the equivalent resistance $R_{\mathrm{eq}}^{\mathrm{SM}}$ and voltage source $v_{\mathrm{eq}}^{\mathrm{SM}}$ equal

$$\begin{aligned} R_{\mathrm{eq}}^{\mathrm{SM}} &= \frac{R_2\left(R_1 + R_{\mathrm{c_{eq}}}\right)}{R_1 + R_2 + R_{\mathrm{c_{eq}}}} \\ v_{\mathrm{eq}}^{\mathrm{SM}} &= v_{\mathrm{c_{eq}}} \frac{R_2}{R_1 + R_2 + R_{\mathrm{c_{eq}}}}, \end{aligned} \tag{6.2}$$

where the time dependency of $R_{\mathrm{c_{eq}}}(t)$ and $R_{\mathrm{c_{eq}}}(t)$ has been omitted for simplicity. The steps carried out so far were aimed at simplifying the structure of a single SM. To reduce that of an entire cascaded SM string (referred to in [31] as MV), the model exploits the fact that the N SMs in each arm are connected in series and, thus, flown

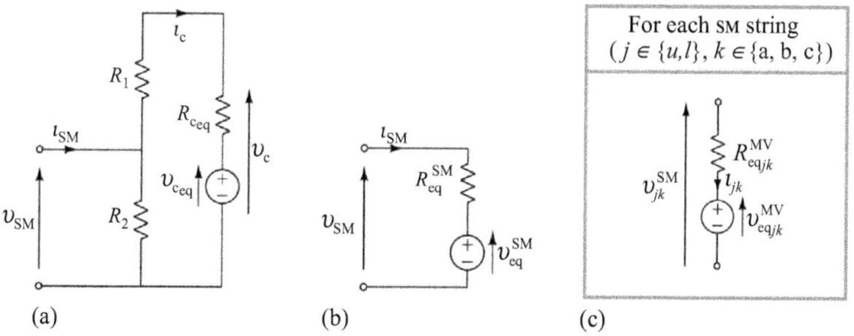

Figure 6.5 Simplification process of the SM schematic (a and b) and equivalent SM string schematic based on the TEM (c)

by the same current. Thus, the terms $R_{\text{eq}}^{\text{SM}}$ and $v_{\text{eq}}^{\text{SM}}$ of each SM in a given arm can be summed, thereby obtaining

$$\begin{aligned} R_{\text{eq}}^{\text{MV}} &= \sum_{n=1}^{N} R_{\text{eq}_n}^{\text{SM}} \\ v_{\text{eq}}^{\text{MV}} &= \sum_{n=1}^{N} v_{\text{eq}_n}^{\text{SM}}. \end{aligned} \tag{6.3}$$

As a result, the SM string in each of the six arms of the MMC can be represented with the equivalent circuit in Figure 6.5(c). In particular, at each time step, the derivation of this equivalent circuit requires executing the pseudo-code of Figure 6.4 to determine the values of R_1 and R_2 of each SM and then computing the values of $R_{\text{eq}}^{\text{MV}}$ and $v_{\text{eq}}^{\text{MV}}$ of every arm with (6.1)–(6.3).

The TEM presents three key features. First, regardless of the number of SMs in each arm, the equivalent circuit in Figure 6.5(c) consists only of three nodes. Hence, instead of dealing with a big nodal admittance matrix as in the BVR case, the TEM handles a smaller matrix, thus significantly reducing the computational burden. Second, despite replacing them with an equivalent circuit, the TEM retains the individual behavior of the SMs in one string, including their capacitor voltages. Indeed, these variables, which are essential to perform the CBA in the *lower level controls* of Figure 6.1(c), can still be recovered over time with (6.4). Such equations, which hold for a generic SM and are hard-coded in the simulator, can be easily derived from Figure 6.5(a)

$$\begin{aligned} i_{\text{c}}(t) &= \frac{R_2 i_{\text{SM}}(t) - v_{\text{c}_{\text{eq}}}(t - \Delta T)}{R_1 + R_2 + R_{\text{c}_{\text{eq}}}} \\ v_{\text{c}}(t) &= R_{\text{c}_{\text{eq}}} i_{\text{c}}(t) + v_{\text{c}_{\text{eq}}}. \end{aligned} \tag{6.4}$$

The third feature is that, due to the simplification process, the CPU time needed to simulate MMCs with the TEM is not quadratically dependent on the number of SMs per arm N. Rather, it is almost linearly dependent. Such a trend is due to different reasons. First, as N increases, so does the switching activity of the MMC (i.e., the number of times that the *modulation technique* adopted in Figure 6.1(c) requires one or more SMs to change their operating state). In turn, to better detect these occurrences (hereafter referred to as *threshold crossings*), the time step ΔT must be suitably reduced, thereby leading to proportionally higher simulation times. Nonetheless, even if a fixed and sufficiently low ΔT was used during a set of simulations given by different values of N, a linear increase in CPU time would still be incurred because of the presence of the CBA, which requires sorting the SM capacitor voltages [9]. Since these algorithms are run at every threshold crossing, the number of times they are executed (and, thus, their computational overhead) rises linearly with N.

To sum up, in comparison with the BVR representation, the TEM enhances simulation speed while maintaining the same accuracy. Like the BVR model, the TEM allows studying both normal operating conditions and contingencies (e.g., AC and DC-side faults). In addition, this model lends itself to parallel computation [16,35]: at every time step, (6.1)–(6.4) could be solved for each arm in separate threads. In the light of the above, the TEM is an excellent candidate for real-time simulation of MMCs through CPUs and Field-Programmable Gate Arrays.

6.2.3 *Switching function model*

Contrary to the TEM, the SFM relies on a more simplified SM representation than the BVR: instead of bi-value resistors, ideal switches are used to describe each SM valve. In this case, considering Figure 6.6(a), the voltage and current of a SM and its capacitor are related to one another as

$$\begin{aligned} i_c &= S\, i_{SM} \\ v_{SM} &= S\, v_c\,, \end{aligned} \tag{6.5}$$

where S is the SM *binary switching function*, which gives the name to this model. This variable, which is regulated by the modulation technique in the MMC control scheme, is equal to 1 when a given SM is inserted and 0 when bypassed. On the contrary, as explained in the sequel, the behavior of the SMs during the blocked condition is mimicked through the implementation of *ad hoc* equivalent circuits.

Another variable adopted in the SFM is the *arm* switching function s, defined as

$$s = \frac{1}{N}\sum_{n=1}^{N} S_n\,, \tag{6.6}$$

with S_n and N being respectively the binary function associated with the nth SM and N the overall number of SMs in one arm. During normal operating conditions, s ranges between 0 and 1 (i.e., all SMs are bypassed and inserted, respectively). In particular, s changes at each threshold crossing of the modulation scheme.

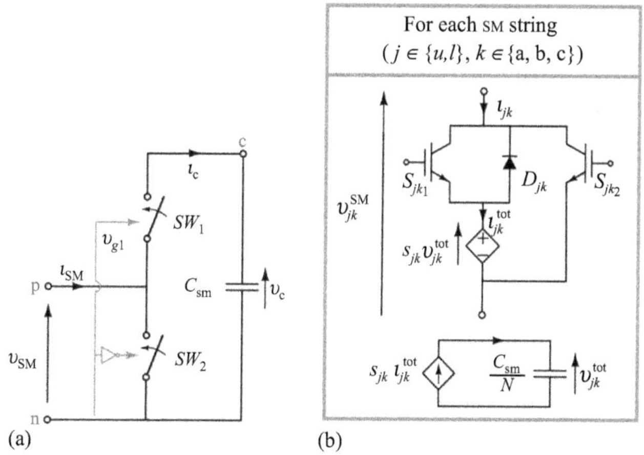

Figure 6.6 The SM representation used in the SFM (a) and the resulting equivalent circuit of one SM string (b)

There are several versions of the SFM [34]. The simplest one, described hereafter, assumes all SM capacitor voltages to be always perfectly balanced [36]†, that is

$$v_{c_1} = v_{c_2} = \ldots = \frac{1}{N}\sum_{n=1}^{N} v_{c_n} = \frac{v_c^{\mathrm{tot}}}{N}. \tag{6.7}$$

In this case, each SM string is replaced with the equivalent circuit depicted in Figure 6.6(b). As done in standard power system simulation programs (e.g., PSCAD), the diodes and IGBTs are described by a simple representation based on an ON-state and OFF-state resistance. In particular, their ON-state resistance could be suitably tuned to cope with the fact that the original resistors in the BVR model are lost because they are replaced by ideal switches. So doing, the SFM would still allow computing the conduction losses of the MMC correctly.

The behavior of the equivalent circuit in Figure 6.6(b) can be explained as follows.‡ In normal operating conditions, the IGBTs S_{jk_1} and S_{jk_2} are, respectively, ON and OFF. Thus, the current $\imath_{jk}$ flows through a voltage-controlled voltage source (i.e., $\imath_{jk}^{\mathrm{tot}} = \imath_{jk}$) and, depending on its sign, through either S_{jk_1} or D_{jk}. The voltage across this source amounts to $s_{jk}v_{jk}^{\mathrm{tot}}$, with s_{jk} and v_{jk}^{tot} being, respectively, the arm switching function and the voltage of a capacitor $\frac{C_{\mathrm{sm}}}{N}$. In turn, this latter component is connected in series with a current-controlled current source $s_{jk}\imath_{jk}^{\mathrm{tot}}$.

The controlled sources are a ploy to simulate the behavior of all SMs in one arm by merging all their capacitors into a single one. On the one hand, the controlled current source mirrors the charge and discharge of the SM capacitors. On the other hand, the controlled voltage source mimics the fact that, as the insertion indices vary, so does the voltage $s_{jk}v_{jk}^{\mathrm{tot}}$ across the SM strings because the number of SMs inserted into the arms (i.e., the only ones impressing a non-null voltage at their terminals) changes.

For what concerns the blocked state of the MMC, the following holds. As already stated, this condition is such that all the capacitors inside the SMs can charge but not discharge. To replicate this behavior, the variable s_{jk} is set to 1 and the IGBTs S_{jk_1} and S_{jk_2} are turned off and on, respectively. So doing, if $\imath_{jk}$ is positive, the diode D_{jk} conducts and every SM capacitor charges because $\imath_{jk}^{\mathrm{tot}} = \imath_{jk} > 0$. Otherwise, the IGBT S_{jk_2} conducts and all capacitors neither charge nor discharge because $\imath_{jk}^{\mathrm{tot}} = 0$.

In short, this model can be used for system level studies to analyze all the MMC protections and controls except for the Capacitor voltage Balancing Algorithm (CBA), which is in this case absent because a perfect voltage balance in the SMs is assumed. In addition, just like the TEM, normal operating conditions and a wide range of contingencies can still be analyzed with the SFM. It is important to point out that, in comparison with the TEM, the computational burden of this SFM is basically constant

†This assumption also holds for the AVM presented in the next section. However, note that this may not be the case for other implementations of the SFM described in the literature. For instance, the model proposed in [37] retains the individual evolution of each SM capacitor. However, this comes at the cost of a higher computational burden.

‡Hereafter, $j \in \{u, l\}$ denotes either the upper or the lower arm of one MMC phase leg, where as $k \in \{a, b, c\}$ indicates one of the phase legs of the converter.

regardless of the number N of SMs in one arm.§ This feature stems from the fact that (i) the equivalent circuit is the same independently from the number of SMs in each arm (i.e., the MMC nodal admittance matrix has a constant size) and (ii) the linear overhead resulting from capacitor voltage sorting—incurred when adopting the TEM—is absent because no CBA is adopted in this version of the SFM.

6.2.4 Average value model

The AVM owes its name to the fact that its disregards the voltage and current ripples associated to the commutations in the SMs, thereby leading to waveforms that are in a sense an *average* of those obtained with more detailed models (e.g., the FD one). In particular, this representation significantly simplifies the MMC topology by resorting to controlled voltage and current sources and relying on two assumptions. First, the SM capacitor voltages are always perfectly balanced. Second, considering Figure 6.1(a), the voltages of the SM strings in each phase leg are such that $v_{l_k}^{\mathrm{SM}} + v_{u_k}^{\mathrm{SM}} = v_{\mathrm{dc}}$. As explained in [16], these assumptions imply that the so-called circulating current of the MMC is suppressed without implementing an *ad hoc* control scheme (see Figure 6.1(c)).

Figure 6.7 depicts the schematic of one of the possible implementations of the AVM [16]. Compared to the original circuit in Figure 6.1, the MMC is divided into two parts, one in the AC and the other in the DC side. In the AC network, the upper and lower arms of each phase leg are replaced by a resistor $\frac{1}{2}R_S$, an inductor $\frac{1}{2}L_S$, and a controlled voltage source, which mirrors the output of the upper level controls in Figure 6.1(c). For what concerns the DC network, the couple of capacitors C_{dc}

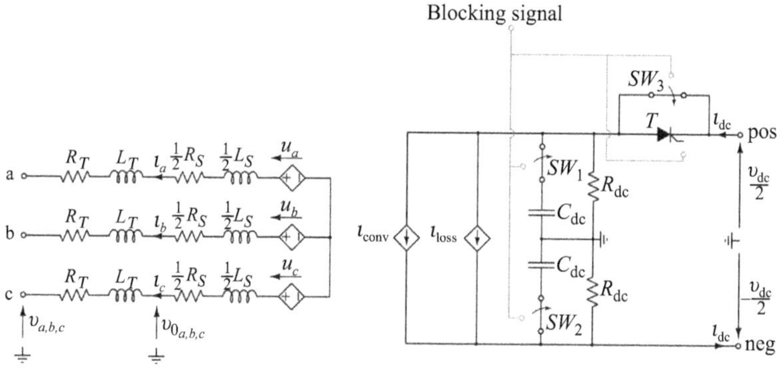

Figure 6.7 The AVM of the MMC

§This holds provided that a sufficiently small time step is adopted to correctly detect threshold crossings for any value of N.

approximates those in the SM strings of each arm, while the controlled current sources implement the following power balance:

$$\sum_{k\in\{a,b,c\}} u_k \imath_k = v_{dc}\imath_{dc} - P_{loss}\,, \tag{6.8}$$

where P_{loss} are the estimated power losses of the converter. The current controlled sources $\imath_{conv}$ and $\imath_{loss}$ in Figure 6.7 can be obtained as follows by isolating the $\imath_{dc}$ term:

$$\imath_{dc} = \underbrace{\frac{\sum_{k\in\{a,b,c\}} u_k \imath_k}{v_{dc}}}_{\imath_{conv}} + \underbrace{\frac{P_{loss}}{v_{dc}}}_{\imath_{loss}}\,. \tag{6.9}$$

The literature proposes several approaches to model converter losses. In [16], for instance, $P_{loss} = R\imath_{conv}^2$, with R being associated with switching and conduction losses. By doing so, the losses considered are only those dependent on the DC-side current. As shown in Figure 6.7, two resistors R_{dc} could be added in parallel to the controlled current sources to take into account the losses depending on the DC-side pole-to-pole voltage. Lastly, the switches SW_1, SW_2, SW_3 and the thyristor T could be included to mirror the behavior of the MMC during the blocked state [16].

On the one hand, the AVM described here is the fastest model among those proposed in the literature. For reasons analogous to the SFM presented in Section 6.2.3, its computational burden does not change with the number of SMs in each arm. In particular, it is the ideal solution when analyzing MMCs in large-scale AC grids. On the other hand, due to the simplifications introduced, it cannot be used to analyze any of the *lower level controls* in Figure 6.1(c). In addition, compared to other representations, this model is less accurate when DC-side faults are simulated [27].

6.3 EMT simulation of MMCs based on isomorphism

The previous section showed several simplified MMC representations. Based on the required trade-off between simulation speed and accuracy, each of them might be more suitable for system or network level studies. However, none of them is compatible with component level studies for two main reasons. First, some MMC models (such as the AVM and some versions of the SFM) do not retain the individual behavior of the SMs. Second, even if the evolution of each SM over time was accessible (as in the case of the TEM), switching transients inside the SMs could not still be analyzed because simplified semiconductor devices representations are adopted.

According to the previous section, only the MMC representation based on the FP SM model is capable of performing component level studies. However, other than being typically unavailable in commercial power system simulation programs, the adoption of this SM model would lead to prohibitive simulation times because conventional simulation approaches consider each SM model individually. This ultimately leads to a big nodal admittance matrix and, thus, a prohibitive computational burden. This issue is further aggravated by the adoption of time steps in the order of nanoseconds, which are necessary to examine the fast dynamics of IGBTs and diodes.

This problem could be solved by applying the so-called isomorphism-based approach to simulate MMCs. This method was originally developed in [38,39] to efficiently simulate RAMs. Before describing its principle of operation in detail, it is worth pointing out that, in the specific case of RAMs, this approach exploits two main features, namely (i) the presence of a multitude of structurally identical memory cells and (ii) their limited number of working states (i.e., cells in "read", "write", or "idle" mode) [40]. By comparing RAMs and MMCs, one can notice that the latter shares similar traits. Indeed, each arm of an MMC has a high number of topologically identical SMs. In turn, at a given time, each SM only has three main operating states: inserted, bypassed, and blocked. As also proven by the simulation results shown in the following, this comparison suggests that, regardless of the SM model adopted, MMCs lend themselves to isomorphism-based simulation, too. To the best of the authors' knowledge, such an approach has never been used before in this framework (except for [34,41], written by the same authors of this chapter).

6.3.1 Operating principle of the isomorphism-based approach

The following paragraphs summarize the key features of the isomorphism-based simulation strategies presented in [38,39]. Although these works applied the technique to efficiently simulate the memory cells inside RAMs, this approach has general validity and applicability. Thus, it can be applied to any SC, including the SMs in one arm of an MMC.

Contrary to conventional approaches, the isomorphism-based method actually simulates only a limited number of SMs in each string while preserving the time evolution of each of them at the cost of a minor sacrifice in terms of accuracy. To this aim, the method resorts to two logics, known as *static* and *dynamic* partitioning.

Static partitioning, carried out at the beginning of the simulation, regards as stand-alone modules the SCs characterized by the same topology (i.e., the way internal components are connected) and nature of components (i.e., constitutive equations). Provided that they fulfil these requirements, all the series-connected SMs in each string are identified as static isomorphic SCs.‖

On the other hand, dynamic partitioning runs continuously while simulating and groups all the static isomorphic SCs with common dynamics in a single equivalent entity. Instead of considering each SC individually, only these entities are simulated. Nevertheless, the approach still retains the individual behavior of each SC. Indeed, the evolution of the state and algebraic variables of the simulated entities is replicated to every SC of the same group. So doing, the intensive computations that would be otherwise required by conventional simulation approaches are avoided. Provided that all the SCs grouped in a class really have sufficiently similar dynamics, this approach leads to a minor sacrifice in simulation accuracy.

‖If in one arm the SMs present different internal parameters (e.g., several IGBTs are employed) or implementations (e.g., one arm comprises both half-bridge and full-bridge SMs), so are their topology and constitutive equations. Thus, the SMs in one stack are not isomorphic from a "*static*" point of view. Therefore, multiple static isomorphic SMs must be inevitably considered as stand-alone modules. Hereafter it is assumed for simplicity that every SM is of the half-bridge type and comprises the same components.

Over time, the behavior of each SC may change. For instance, in the case of RAMS, one memory cell that was previously in "read" mode could become "idle" or vice-versa. Therefore, in specific time intervals, some SCS can be merged because their dynamics are similar, while in others previously merged SCS must be split because their behaviors have diverged so much that they cannot be represented by a common equivalent entity any longer. In particular, when a SC leaves a given class (e.g., $\mathcal{C}^k$), it can be simulated alone or it may enter another class (e.g., $\mathcal{C}^l$) of dynamic isomorphic SCS. While simulating, the possible variables monitored to keep track of the behavior of each SC include its input electrical quantities (e.g., IGBT gate signals) and state variables (e.g., inductor currents and capacitor voltages).

To assess whether two or more SCS share similar dynamics or not (i.e., they should be merged or split), the works in [38,39] rely during simulation on proper *isomorphism indicators*. To explain their usage at a high level, consider for example a generic $\mathcal{S}_j$ sub-circuit (SC) that belongs to the $\mathcal{C}^k$ class of equivalent entities described by the following set of differential algebraic equation (DAES):

$$\begin{cases} \dot{x}_{k,j} = f_k(x_{k,j}, y_{k,j}, t) \\ 0 \;\; = g_k(x_{k,j}, y_{k,j}, t)\,, \end{cases} \tag{6.10}$$

where $x_{k,j} \in \mathbb{R}^{N_k}$ and $y_{k,j} \in \mathbb{R}^{M_k}$ are, respectively, the vectors of the state and algebraic variables of $\mathcal{S}_j$. Each class of isomorphic modules can be associated with a $(N_k + M_k)$-dimensional hyper-sphere of a given radius. A SC enters (leaves) the $\mathcal{C}^k$ class if its $N_k + M_k$ state and algebraic variables are within (outside) this hyper-sphere. A generic $\mathcal{S}_j$ SC belongs to the $\mathcal{C}^k$ class at a given time t only if the following constraints hold [39]:

$$\begin{cases} \left|x_{k,j}(t) - \hat{x}_k(t)\right| \le \boldsymbol{\alpha}_{\mathrm{R}_k} \left|\hat{x}_k(t)\right| + \boldsymbol{\alpha}_{\mathrm{A}_k} \\ \left|y_{k,j}(t) - \hat{y}_k(t)\right| \le \boldsymbol{\beta}_{\mathrm{R}_k} \left|\hat{y}_k(t)\right| + \boldsymbol{\beta}_{\mathrm{A}_k}\,. \end{cases} \tag{6.11}$$

In (6.11), $\hat{x}_k(t)$ and $\hat{y}_k(t)$ are the reference state and algebraic variables vectors of the equivalent module of the $\mathcal{C}^k$ class at time t. For instance, these vectors could be computed as the average value of the state and algebraic variables of all the SCS belonging to the same $\mathcal{C}^k$ class at each time instant. In addition, $\boldsymbol{\alpha}_{\mathrm{A}_k}$, $\boldsymbol{\alpha}_{\mathrm{R}_k}$, $\boldsymbol{\beta}_{\mathrm{A}_k}$, and $\boldsymbol{\beta}_{\mathrm{R}_k}$ are the isomorphism indicators of the $\mathcal{C}^k$ class. Such parameters – whose role is analogs to that of absolute and relative tolerances in DAES variable-step size solvers – determine the radius of the hyper-sphere of each class [29,42]. On the one hand, employing larger hyper-spheres makes an SC more likely to be simulated by an equivalent class with other SCS rather than being simulated alone, thereby speeding up the simulation process. On the other hand, the merging and splitting procedure of the SCS becomes less strict and simulation accuracy may decrease.

6.3.2 *Dynamic partitioning in the* MMC

What is stated so far about dynamic partitioning holds for any kind of SCS, be it a memory cell in a RAM or a SM in an MMC. In particular, when focusing on the latter case, four main classes of dynamic isomorphic SMS can be identified: *inserted*, *bypassed*, *blocked*, and *switching* (denoted, respectively, with $\mathcal{C}^{\mathrm{in}}$, $\mathcal{C}^{\mathrm{by}}$, $\mathcal{C}^{\mathrm{bl}}$, and $\mathcal{C}^{\mathrm{sw}}$).

The $\mathcal{C}^{\text{in}}$, $\mathcal{C}^{\text{by}}$, and $\mathcal{C}^{\text{bl}}$ classes retrace the operating states of the half-bridge SMs mentioned in Section 6.1. In all of these classes, the gate signals are *constant*. Considering the FP SM model in Figure 6.3(a), in the $\mathcal{C}^{\text{in}}$ class, the gate signals v_{g_1} and v_{g2} are, respectively, in the ON and the OFF state, whereas the opposite holds in the $\mathcal{C}^{\text{by}}$ group. On the contrary, in the $\mathcal{C}^{\text{bl}}$ class, both gate signals are OFF.

The $\mathcal{C}^{\text{sw}}$ class constitutes an additional isomorphic group that takes into account the fact that, depending on the SM model used, the transition from the inserted to the bypassed/blocked condition and vice versa may not be instantaneous. This is the case of the FP representation, which does not neglect the rising, falling, and dead-times of the gate signals. In the light of the above, the $\mathcal{C}^{\text{sw}}$ class is the only isomorphic group whose gate signals are *time-varying*. During normal operating conditions, this group comprises two major sub-classes: one refers to the SMs switching from the inserted to the bypassed class, whereas another refers to the SMs switching in the opposite direction. For example, in the former case, the gate signal v_{g_1} goes from ON to OFF, while the opposite holds for v_{g_2}.

Several factors influence the number of SMs in the $\mathcal{C}^{\text{sw}}$ group. To begin with, this class is empty when the FD and BVR representations are employed because they neglect the switching transients inside the SMs. On the contrary, if the FP model is used, the number of SMs in this class depends on the algorithm controlling the SM gate signals and the presence of contingencies. If faults are absent (i.e., the $\mathcal{C}^{\text{bl}}$ class is empty) and the *lower-level controls* of the MMC switch just one SM per arm at a time, the $\mathcal{C}^{\text{sw}}$ class consists of no more than just one SM, which leaves this class and enters the $\mathcal{C}^{\text{in}}$ or $\mathcal{C}^{\text{by}}$ classes as soon as the switching transient is exhausted [11].

The number of SMs populating each class changes in time. For example, during normal operating conditions, each SM alternates moments when it is inserted and bypassed. If faults occur, all SMs move to the $\mathcal{C}^{\text{bl}}$ class. Thus, dynamic partitioning must be continuously performed while simulating. Moreover, based on the constraints in (6.11), these classes can be further split in sub-classes. Assuming that this is not the case, the SM string in each arm can be replaced with the equivalent circuit in Figure 6.8. As in the TEM and SFM, there are six circuits of this kind, one for every MMC arm. In all classes, the FP SM model is adopted as an example: in any case, the method is compatible with any SM representation.

The schematic comprises four controlled voltage sources, each of which mirrors the voltage at the pins of a SM associated to one of the four isomorphic classes described earlier. For instance, the SM of the $\mathcal{C}^{\text{in}}$ class is driven by the $\imath_{\text{m}}$ arm current, which is injected by its current controlled current source. The $v_{\text{pn}}^{\text{in}}$ voltage at the SM pins is multiplied by the N^{in} (time-varying) number of SMs in the class and added to build the v_{arm} voltage using one of the previously mentioned controlled voltage sources. Similar electrical considerations hold for the SMs of the other classes.

It is important to underline once again that the v_{g_1} and v_{g_2} gate signals of the IGBTS in the $\mathcal{C}^{\text{in}}$, $\mathcal{C}^{\text{by}}$, and $\mathcal{C}^{\text{bl}}$ classes are constant during the entire simulation, whereas they are time-varying in the $\mathcal{C}^{\text{sw}}$ class. Thus, switching phenomena, power dissipation, and current conduction by IGBTs or diodes are correctly simulated (i.e., provided that the FP model and a sufficiently low integration time step are used).

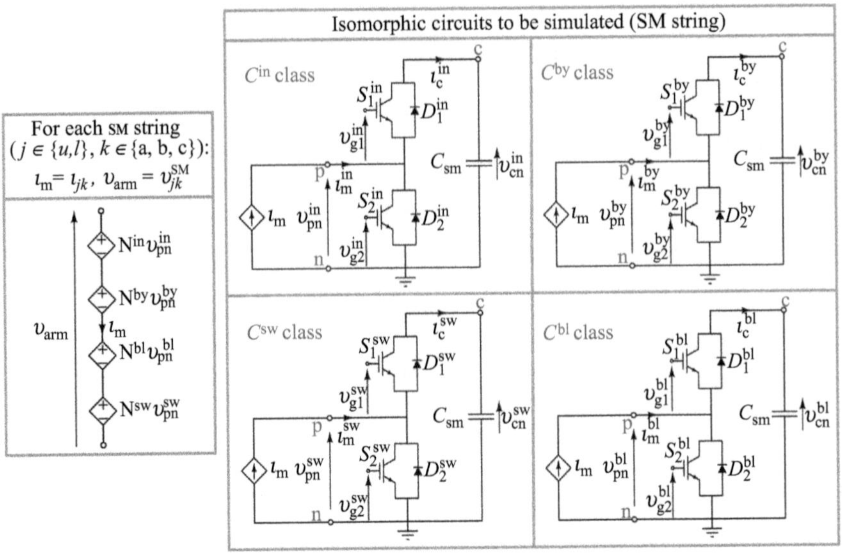

Figure 6.8 The equivalent circuit of a SM *string in one* MMC *arm obtained by adopting the* FP SM *model and the isomorphism-based approach*

In the light of the above, the constitutive equation of the equivalent circuit representing the stack of SMs in one MMC arm can be formalized as

$$
\begin{aligned}
v_{\text{arm}} &= v^{\text{in}} + v^{\text{by}} + v^{\text{bl}} + v^{\text{sw}} \\
v^{\text{in}} &= N^{\text{in}} v^{\text{in}}_{\text{pn}} = N^{\text{in}} v^{\text{in}}_{\text{pc}} + \sum_{j=1}^{N^{\text{in}}} v^{\text{in}}_{\text{cn}_j} \\
v^{\text{by}} &= N^{\text{by}} v^{\text{by}}_{\text{pn}} \\
v^{\text{bl}} &= N^{\text{bl}} (v^{\text{bl}}_{\text{pc}} + v^{\text{bl}}_{\text{cn}}) \\
v^{\text{sw}} &= N^{\text{sw}} (v^{\text{sw}}_{\text{pc}} + v^{\text{sw}}_{\text{cn}}) \\
i_{\text{m}} &= i^{\text{in}}_{\text{m}} = i^{\text{by}}_{\text{m}} = i^{\text{bl}}_{\text{m}} = i^{\text{sw}}_{\text{m}} \\
f^{\text{in}} &\left(\dot{v}^{\text{in}}_{\text{pc}}, v^{\text{in}}_{\text{pc}}, v^{\text{in}}_{\text{g1}}, v^{\text{in}}_{\text{g2}}, \dot{v}^{\text{in}}_{\text{cn}}, v^{\text{in}}_{\text{cn}}, i^{\text{in}}_{\text{m}}\right) = 0 \\
f^{\text{by}} &\left(\dot{v}^{\text{by}}_{\text{pc}}, v^{\text{by}}_{\text{pc}}, v^{\text{by}}_{\text{g1}}, v^{\text{by}}_{\text{g2}}, \dot{v}^{\text{by}}_{\text{cn}}, v^{\text{by}}_{\text{cn}}, i^{\text{by}}_{\text{m}}\right) = 0 \\
f^{\text{bl}} &\left(\dot{v}^{\text{bl}}_{\text{pc}}, v^{\text{bl}}_{\text{pc}}, v^{\text{bl}}_{\text{g1}}, v^{\text{bl}}_{\text{g2}}, \dot{v}^{\text{bl}}_{\text{cn}}, v^{\text{bl}}_{\text{cn}}, i^{\text{bl}}_{\text{m}}\right) = 0 \\
f^{\text{sw}} &\left(\dot{v}^{\text{sw}}_{\text{pc}}, v^{\text{sw}}_{\text{pc}}, v^{\text{sw}}_{\text{g1}}, v^{\text{sw}}_{\text{g2}}, \dot{v}^{\text{sw}}_{\text{cn}}, v^{\text{sw}}_{\text{cn}}, i^{\text{sw}}_{\text{m}}\right) = 0\,,
\end{aligned}
\tag{6.12}
$$

where the $^{\text{in, by, sw, bl}}$ superscripts refer to a given class of isomorphic SMs.

The $f^{\cdot}(\cdot)$ vector-valued multivariate functions are a compact way of describing the model of the diodes and IGBTs of each class of isomorphic SMs and their interconnections. Note that time derivatives, denoted by dots above the variables, appear among the arguments of these functions (i.e., the dynamical models of the diodes and IGBTs are employed [43–45]). Each one of the last four equations of (6.12) is a specialization of (6.10). To give a handy example of what happens by further specializing one of these equations, we expand $f^{\mathrm{sw}}(\cdot)$ when the full detailed (FD) model is adopted (see Figure 6.3(b)). In this case, by letting $\imath_{\mathrm{SM}} = \imath_{\mathrm{m}}$, we have

$$C_{\mathrm{sm}}\dot{v}_{\mathrm{c}}^{\mathrm{sw}} = I_o\left(e^{\kappa_d v_{\mathrm{pc}}^{\mathrm{sw}}} - 1\right) - \imath_1^{\mathrm{sw}} \tag{6.13}$$

$$\imath_{\mathrm{m}}^{\mathrm{sw}} - I_o\left(e^{\kappa_d v_{\mathrm{pc}}^{\mathrm{sw}}} - e^{-\kappa_d\left(v_{\mathrm{pc}}^{\mathrm{sw}}+v_{\mathrm{cn}}^{\mathrm{sw}}\right)}\right) + \imath_1^{\mathrm{sw}} - \imath_2^{\mathrm{sw}} = 0 \tag{6.14}$$

$$\imath_2^{\mathrm{sw}} - \chi(v_{\mathrm{g2}}^{\mathrm{sw}})I_o\left(e^{\kappa_d\left(v_{\mathrm{pc}}^{\mathrm{sw}}+v_{\mathrm{cn}}^{\mathrm{sw}}\right)} - 1\right) = 0 \tag{6.15}$$

$$\imath_1^{\mathrm{sw}} - \chi(v_{\mathrm{g1}}^{\mathrm{sw}})I_o\left(e^{-\kappa_d v_{\mathrm{pc}}^{\mathrm{sw}}} - 1\right) = 0\,, \tag{6.16}$$

where $\kappa_d = \frac{q}{\eta k_B T}$, being q the electron charge, η the emission coefficient, k_B the Boltzmann constant, and T the junction temperature. In this example, to keep notation simple, the I_o saturation current is assumed identical in all diodes and charge equations were dropped [46]. The $\chi(\cdot)$ function assumes values from the two-element set $\{0, 1\}$ according to its argument (i.e., the gate driving voltages, in this case). The charging of the SM capacitor is governed by (6.13). Equation (6.14) expresses the $\imath_{\mathrm{m}}^{\mathrm{sw}}$ submodule current as the sum of the currents flowing in the two switches and the anti-parallel diodes. Equations (6.15) and (6.16) implement the characteristic of the two switches. These equations are more complex in the FP case.

Having in mind (6.10), which collects the DAEs governing the generic class $\mathcal{C}^k$, for the jth SM belonging to the $\mathcal{C}^{\mathrm{sw}}$ class, the state variable $v_{\mathrm{c}}^{\mathrm{sw}}$ in (6.13) matches with $x_{k,j}$ ($N_k = 1$), while the $(v_{\mathrm{sm}}^{\mathrm{sw}}, v_{\mathrm{cn}}^{\mathrm{sw}}, v_{\mathrm{g2}}^{\mathrm{sw}}, v_{\mathrm{g1}}^{\mathrm{sw}}, \imath_1^{\mathrm{sw}}, \imath_2^{\mathrm{sw}}, \imath_{\mathrm{m}}^{\mathrm{sw}})$ algebraic variables in (6.14)–(6.16) match with $y_{k,j}$ ($M_k = 7$). Note that $f_k(\cdot)$ reduces to the right hand side of (6.13), whereas $g_k(\cdot)$ is a vector-valued function collecting the left hand side of (6.14)–(6.16).

When it comes to isomorphism-based MMC simulation, (6.11), which governs clustering and class building, becomes in the case of a SM of the $\mathcal{C}^{\mathrm{sw}}$ class

$$\begin{aligned}
\left|v_{\mathrm{cn}}^{\mathrm{sw}}(t) - \hat{v}_{\mathrm{cn}}^{\mathrm{sw}}(t)\right| &\le \alpha_{\mathrm{R}_{\mathrm{cn}}^{\mathrm{sw}}}\left|\hat{v}_{\mathrm{cn}}^{\mathrm{sw}}(t)\right| + \alpha_{\mathrm{A}_{\mathrm{cn}}^{\mathrm{sw}}}\\
\left|v_{\mathrm{g1}}^{\mathrm{sw}}(t) - \hat{v}_{\mathrm{g1}}^{\mathrm{sw}}(t)\right| &\le \beta_{\mathrm{R}_{\mathrm{g1}}^{\mathrm{sw}}}\left|\hat{v}_{\mathrm{g1}}^{\mathrm{sw}}(t)\right| + \beta_{\mathrm{A}_{\mathrm{g1}}^{\mathrm{sw}}}\\
\left|v_{\mathrm{g2}}^{\mathrm{sw}}(t) - \hat{v}_{\mathrm{g2}}^{\mathrm{sw}}(t)\right| &\le \beta_{\mathrm{R}_{\mathrm{g2}}^{\mathrm{sw}}}\left|\hat{v}_{\mathrm{g2}}^{\mathrm{sw}}(t)\right| + \beta_{\mathrm{A}_{\mathrm{g2}}^{\mathrm{sw}}}\\
\left|\imath_{\mathrm{m}}^{\mathrm{sw}}(t) - \hat{\imath}_{\mathrm{m}}^{\mathrm{sw}}(t)\right| &\le \beta_{\mathrm{R}_{\mathrm{m}}^{\mathrm{sw}}}\left|\hat{\imath}_{\mathrm{m}}^{\mathrm{sw}}(t)\right| + \beta_{\mathrm{A}_{\mathrm{m}}^{\mathrm{sw}}}\,,
\end{aligned}$$

where being $\zeta \in \{v_{\mathrm{cn}}^{\mathrm{sw}}, v_{\mathrm{g1}}^{\mathrm{sw}}, v_{\mathrm{g2}}^{\mathrm{sw}}, \imath_{\mathrm{m}}^{\mathrm{sw}}\}$, $\hat{\zeta}(t) = \dfrac{1}{N^{\mathrm{sw}}}\displaystyle\sum_{j=0}^{N^{\mathrm{sw}}} \zeta_j(t)$ is the average value of ζ, $\alpha_{\mathrm{R}_{\mathrm{cn}}^{\mathrm{sw}}}$ $(\beta_{\mathrm{R}_{\mathrm{g1}}^{\mathrm{sw}}}, \beta_{\mathrm{R}_{\mathrm{g2}}^{\mathrm{sw}}}\ \beta_{\mathrm{R}_{m}^{\mathrm{sw}}})$ and $\alpha_{\mathrm{A}_{\mathrm{cn}}^{\mathrm{sw}}}$ $(\beta_{\mathrm{A}_{\mathrm{g1}}^{\mathrm{sw}}}, \beta_{\mathrm{A}_{\mathrm{g2}}^{\mathrm{sw}}}\ \beta_{\mathrm{R}_{\mathrm{m}}^{\mathrm{sw}}})$ define the relative and the absolute error, respectively, with respect to $\hat{v}_{\mathrm{cn}}^{\mathrm{sw}}$ $(\hat{v}_{\mathrm{g1}}^{\mathrm{sw}}, \hat{v}_{\mathrm{g2}}^{\mathrm{sw}}\ \hat{\imath}_{\mathrm{m}}^{\mathrm{sw}})$. The $\beta_{\mathrm{R}_{\mathrm{m}}^{\mathrm{sw}}}$ and $\beta_{\mathrm{A}_{\mathrm{m}}^{\mathrm{sw}}}$ thresholds can be chosen very close to similar constants (10^{-6}) governing the accuracy of the DAE solver that computes the solution of the circuit since the modules in $\mathcal{C}^{\mathrm{sw}}$ share by construction the same $\imath_{\mathrm{m}} \equiv \imath_{\mathrm{m}}^{\mathrm{sw}}$ current. This leads to a good clustering level of $\mathcal{C}^{\mathrm{sw}}$.

To give more insight into the use of isomorphic indicators consider for instance the SMS populating the C^{in} class. It is reasonable to accept that a negligible current (μA/mA saturation current against kA arm current) flows through the S_2 IGBT and the D_2 diode. This assumption is justified by the fact that, once the switching transient is completed, in the C^{in} SMS the current $\imath_{\text{m}} \equiv \imath_{\text{m}}^{\text{in}}$ mostly flows through the capacitor C_{sm} and the S_1 IGBT (or the D_1 diode, based on the sign of the current). If this negligible current satisfies the constraint in (6.11) independently from the $v_{\text{pc}}^{\text{in}}$ and $v_{\text{cn}}^{\text{in}}$ voltages, the C^{in} class collects all the N^{in} arm SMS that are in this *inserted* working condition. Thus, the voltage *variation* of the capacitor C_{sm} and the voltage $v_{\text{pc}}^{\text{in}}$ (across the p and c nodes) are the same for all the SMS of such class. Note that the voltage of each capacitor can be significantly different but this marginally affects the current flowing through the SMS of each arm (fraction of mA versus kA). Thus, the isomorphic behavior of each SM in the arm is not affected. Consequently, one can relax the $\alpha_{\text{R}_{\text{cn}}}$ and $\alpha_{\text{A}_{\text{cn}}}$ tolerances of the C^{in} class.

6.3.3 Key features of the isomorphism-based approach

The isomorphism-based simulation approach has three main properties of interest.

The first signature feature of the approach is its versatility: based on the degree of accuracy required while simulating, one could adopt the detailed FP model, the simpler BVR representation, or even a brand new one. This is a marked difference with the accelerated MMC representations in the literature, which are exclusively compatible with a specific SM configuration. For example, the TEM relies on the BVR model of SMS, while the SFM approximates the semiconductor devices in each valve with ideal switches.

Second, the proposed method is suitable for network level studies because it can correctly simulate MMCS during every operating condition (including contingencies) and validate every item of the list in Figure 6.1(c). Indeed, contrary to the AVM and some versions of the SFM, the isomorphism-based approach requires no simplification whatsoever (e.g., perfectly balanced SM voltage, no circulating current). The MMC control and protections can be analyzed also because, just like the TEM, the proposed method retains the behavior of each SM. Thus, all of its internal variables are accessible and available as input for the MMC control architecture.¶ In addition, if the FP SM model is adopted, the approach allows evaluating both conduction and switching losses of semiconductor devices with a high degree of accuracy. As a consequence, the method constitutes also a viable option for component levels studies, thereby filling the gap left by the state-of-the-art MMC representations.

Third, the isomorphism-based approach allows increasing the simulation efficiency of MMCS with a minor sacrifice in simulation accuracy. The TEM attains a similar result by grouping the stack of SMS in each arm (described by the BVR model) in a

¶For instance, this is the case of the SM capacitor voltages used in the CBA. When employing the isomorphism-based method, the capacitor voltage variation of a given class is replicated to all the capacitors of the SMS belonging to the same class. In other words, the voltage variation is summed to the voltage that the capacitor of each SM of the same class had at the beginning of a new dynamic clustering.

compact linear, time-varying Thévenin equivalent circuit. An analogous circuit, however, could not be obtained with the FP and FD models because of the non-linearities introduced by diodes and IGBTs. In fact, the Thévenin equivalent exists only for linear time-varying circuits. On the contrary, the adoption of the proposed method results in a compact representation of the SM strings regardless of the SM adopted, be it linear or not. Indeed, for each SM string, only a single element of a given group of isomorphic SMs is simulated and its electrical evolution is replicated to all the SMs in the same class. In the best-case scenario depicted in Figure 6.8, each SM string comprises only four classes. Thus, the number of nodes and model equations to be considered reduces significantly, and simulation speed increases. In the light of the above, this approach can also simulate the highly detailed FP model of SMs within reasonable CPU times, which would not be possible with conventional simulation approaches.

In addition, similar to the TEM, with the proposed methodology the MMC computational burden increases linearly with the number of SMs per arm N, instead of (almost) quadratically. As stated in Section 6.2.2, this happens because, as N grows, so does the switching activity of the MMC and the number of times CBAS are executed. The slope associated with this linear trend depends on the integration time step adopted (fixed or variable) and the SM model chosen (i.e., in general, the more complex the model, the steeper the linear trend).

6.4 Validation

6.4.1 Simulation setting

To demonstrate the features of the isomorphism-based approach listed in the previous section, the DCS1 HVDC CIGRE test system in Figure 6.9 was selected as a benchmark for simulations. It comprises two MMCs connected to two separate AC grids through just as many submarine cables in a symmetrical monopole configuration. The data required to implement every component of this test system is described in [16].

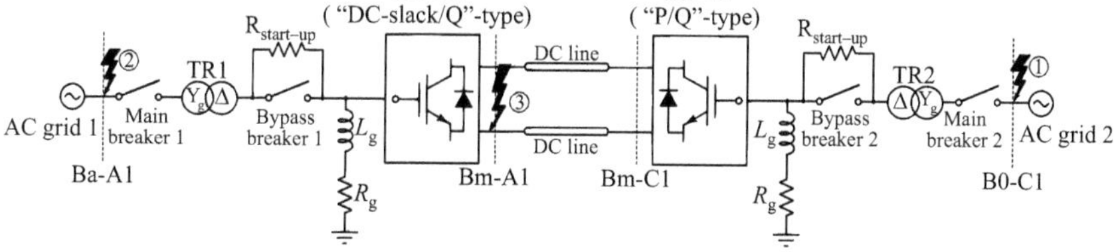

Figure 6.9 The block schematic of the CIGRE DCS1 HVDC test system [16]. Dashed lines indicate busses. Lightning bolts and circled numbers next to those, respectively, indicate faults and the order in which they are considered in this chapter.

The AC networks are represented by a three-phase balanced voltage sources connected to an impedance. The DC lines associated with the submarine cables are described by the frequency-dependent model developed in [47], whose parameters can be retrieved from the detailed geometrical data of the submarine cables in [16,48].

Due to their scalability to high voltage and power applications, MMCs do not strictly require to be connected to their respective AC grids with the Y_g/Δ transformers TR1 and TR2. Nonetheless, they are employed because they ensure that in most of the operating conditions the zero sequence component of the current flowing through the MMCs is null, which simplifies their control scheme. The resistance R_g and inductance L_g are added to provide a voltage reference at the Δ-side of the transformers [49]. Moreover, the leakage impedance of the transformers is exploited to satisfy the minimum filter requirements of the MMCs.

The Cm-A1 and Cm-C1 MMCs are the cornerstones of the simulation results shown in this section. From a control respective, the former and the latter are defined respectively as of "DC-slack/Q" an "P/Q" type. By referring to the *upper level controls* in Figure 6.1(b), this means that the Cm-A1 MMC regulates the DC-side voltage and reactive power, while the Cm-C1 MMC controls active and reactive power. Except for this difference, both MMCs share the same topology, *lower level controls*, *protections*, and design parameters. Some of these are listed in Table 6.1: the interested reader can find more detailed information about these converters in [16].

The simulation results shown in the following vary in terms of the SM model adopted. Indeed, throughout the whole section, the FP, FD, and BVR models of the SMS are used to highlight different properties of the isomorphism-based approach.

For what concerns specifically the FP model in Figure 6.3(a), the D2700U45X122 diode by Infineon and the ST3000GXH31A IGBT by Toshiba were adopted in this presentation because they are compatible with the MMC characteristics in Table 6.1. In particular, the IGBT is represented by the macro model depicted in Figure 6.10. In brief, the collector–emitter current is conducted by the PNP-type BI BJT. This BJT is controlled by the MI MOSFET, which can conduct a fraction of the total $\imath_{EB}$ current. The modeling equations of the BJT and MOSFET (LEVEL2) can be found in [46]. The circuit in the dashed-box and the f_{12},f_{11} controlled sources implement the non-linear

Table 6.1 Parameters of the Cm-A1 and Cm-C1 MMCs in Figure 6.9

Parameter	Value	Comment
f_{nom}	50 Hz	Nominal frequency
S_{nom}	800 MVA	Nominal power
$V_{\text{ac}_{\text{nom}}}$	245 kV	Nominal line-to-line RMS AC-side voltage
$V_{\text{dc}_{\text{nom}}}$	400 kV	Nominal pole-to-pole DC-side voltage
N	200	Number of SMS per arm

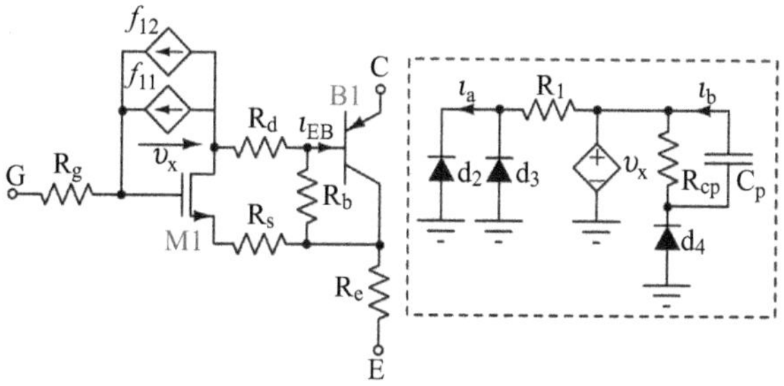

Figure 6.10 The schematic of the sub-circuit implementing the model of the ST3000GXH31A IGBT *adopted in the* FP SM *representation. The* B1 BJT *is a* PNP *type, so its emitter corresponds to the collector terminal of the* IGBT *model*

charge characteristic of the gate, whose behavior is governed by the D2, D3, and D4 diodes [46]. This non-linear charge characteristic is the main actor in determining the switching losses of the SM and the current flowing through the gate driver. The gate driver is *ideal* (i.e., it does not introduce any limitation in its branch current). The gate signal of each IGBT in every SM is given by a voltage-dependent source connected directly to terminal G in Figure 6.10 (note that the gate includes an internal resistance R_g). The value of this source depends on the *Capacitor voltage Balancing Algorithm* (CBA) and *protections* of the MMC. In addition, the rising, falling, and dead-times of the gate signal amount to 100 ns.

When the FD model in Figure 6.3(b) is used, the switches S_1 and S_2 are modeled as bi-value resistors, whose ON and OFF states are associated with 1 μΩ and 1 MΩ resistors, respectively. The diode model is analogous to the static one used in [46].

Lastly, the R_1 and R_2 resistances of the BVR SM model in Figure 6.3(c) in the ON and OFF states amount to 1.36 mΩ or 1 MΩ, respectively [16]. These values are determined with the pseudo-code in Figure 6.4.

Regardless of the model used, the capacitance of each SM equals $C_{sm} = 10$ mF.

6.4.2 Simulated scenarios

All the simulations reported here were carried out with PAN simulator [50,51] on a 1.8 *GHz* Intel i7, 14 *MByte* processor, whose RAM allocation never exceeded 100 MByte. Considering Figure 6.9, each simulation is structured as follows:

- The simulation starts with a start-up sequence to bring the system to steady-state periodic operation through an initialization transient. The SM capacitors are pre-charged to the value $v_{dc_{nom}}/N$, the main breakers 1 and 2 are closed, and the start-up resistors are inserted (i.e., the bypass breakers 1 and 2 are opened). The DC-side voltage and active power setpoints of the Cm-A1 and Cm-C1 MMCs are, respectively, 0 and 400 kV. For both MMCs, the reactive power setpoint is null.

- At $t = 0.06$ s, bypass breakers are closed (i.e., start-up resistors are short-circuited).
- At $t = 0.15$ s, the active power setpoint of Cm-C1 is set to −400 MW (i.e., power is injected by AC grid 2 toward the HVDC link). The other setpoints are unaltered.
- AC or DC faults (denoted by the lightning bolts in Figure 6.9) occur at $t = 0.6$ s.

We want to stress that our goal here is neither validating the MMC controls, nor investigating the HVDC system in use. Rather, the focus is on showcasing the features of the isomorphism-based approach with some case studies. In particular, all the scenarios considered in the following compare two simulation approaches, hereafter referred to as **Iso** and **Conv**. The former relies on the isomorphism-based method to simulate the SM strings in each arm. On the contrary, the latter is a *conventional* approach, in the sense that it considers each SM individually. Thus, as already stated, it is characterized by a high computational burden.

6.4.3 Analysis of simulation accuracy

This subsection analyses a three-phase fault at AC grid 2 (i.e., the ① mark in Figure 6.9). The left and right column panels of Figure 6.11, respectively, show the simulation results obtained with the **Iso** and **Conv** approaches. Both cases use the FD model and a fixed integration time step $\Delta T = 4\ \mu\text{s}$.

When the fault occurs, the AC-side currents of the Cm-C1 MMC increase and drop to zero after about 20 *ms*, which is the time required by the MMC protection system to open the main breaker 2 in Figure 6.9 when under-voltages at the AC side are detected. Simultaneously, the Cm-C1 MMC enters the blocked state, and its active power setpoint is set to zero. To achieve this, as shown in Figure 6.11, the currents at its AC-side and, thus, also that of the HVDC link decay to zero.

The comparison between the **Conv** and **Iso** approaches confirms the accuracy of the latter. To further validate this statement, the last panel of Figure 6.11 (left column) includes a number, which corresponds to the maximum RSE obtained among the variables shown in all the panels. While computing the RSE, the results of the **Conv** approach are taken as reference, as it introduces no simplification. This indicator is computed in the same way in the next two scenarios.

6.4.4 Analysis of submodules variables

This subsection considers a three-phase to ground fault at AC grid 1 (i.e., the ② mark in Figure 6.9). Specifically, this scenario adopts the FP SM model and a variable time step integration scheme based on the implicit linear multi-step method by Gear up to order six or the trapezoidal method, with absolute and relative error tolerances on the electrical variables of 10^{-3} and 10^{-6}, respectively.

Figure 6.12 reports the simulation results. Mirroring the case of Figure 6.11, the AC-side currents of converter Cm-A1 start increasing right after the fault occurs. After 20 ms, they drop to zero because the main breaker 1 is opened by the MMC under-voltage protection. At the same time, the Cm-A1 MMC enters the blocked state. The AC-side currents of the Cm-C1 MMC drop smoothly with some magnitude oscillations

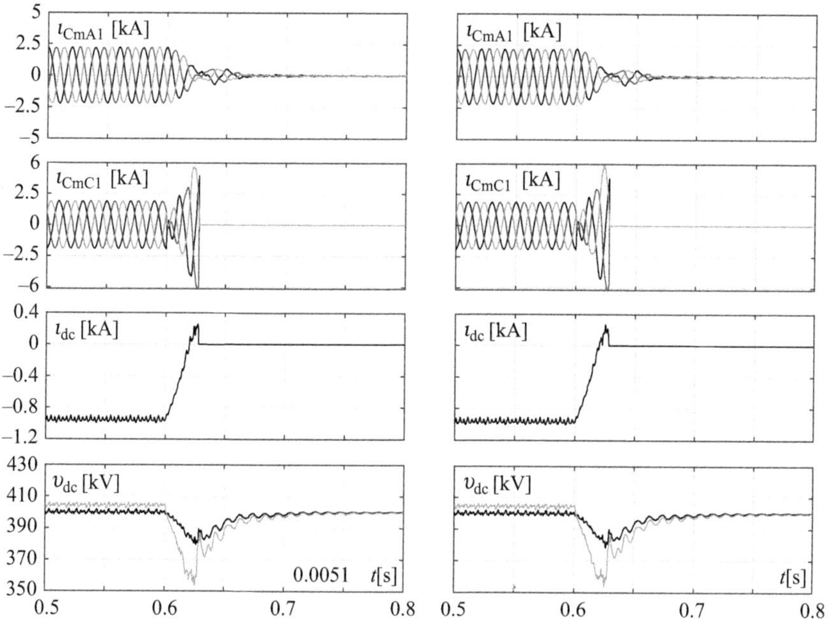

Figure 6.11 Simulation results of a three-phase fault occurring in AC *grid 2 at 0.6 s, obtained with the* ***Iso*** *(left column panels) and* ***Conv*** *(right column panels) approaches. From top: Panel 1 (ι_{CmA1}): three-phase* AC *currents of Cm-A1* MMC*. Panel 2 (ι_{CmC1}): three-phase* AC *currents of Cm-C1* MMC*. Panel 3 (ι_{dc}):* DC *current of Cm-C1* MMC*. Panel 4 (v_{dc}): pole-to-pole* DC *voltages at Bm-A1 (black) and Bm-C1 (gray). The number in the fourth panel of the left column corresponds to* RSE *of the variable plotted in the same panel. Section 6.4.3 explains how this parameter is computed*

and so does the HVDC link current. This behavior originates from the presence of a regulation in the *upper level controls* in Figure 6.1(c), referred to in [16] as *deadband control*, which progressively lowers the power exchange of Cm-C1 MMC as the pole-to-pole voltage v_{dc} reaches 470 kV [28].

Other than accuracy, an additional advantage of the isomorphism-based method is its compatibility with any SM model. For instance, when the FP model is adopted, this paves the way to the analysis of specific features like switching losses and IGBT currents during gate signal commutations. As already stated, this is possible because the proposed approach retains the individual behavior of each SM.

For example, Figure 6.13 reports some electrical variables of a single SM in the time interval between a single switching-off of the S_1 IGBT and the subsequent switching-on of the S_2 IGBT in the previous case study. Panel B of Figure 6.13 reports the v_{g_1} gate-emitter voltage of the S_1 IGBT at turn-off. The fall time amounts to 100 ns.

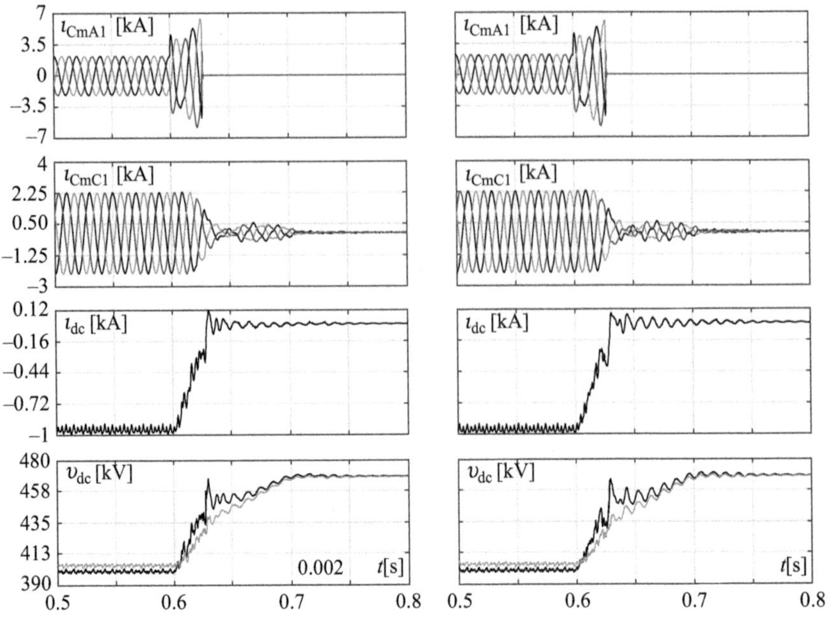

Figure 6.12 Simulation results of a three-phase fault occurring in AC *grid 1 at 0.6 s, obtained with the **Iso** (left column panels) and **Conv** (right column panels) approaches. Labels in panels have the same meaning as those in Figure 6.11*

The corresponding gate voltage of the M1 MOSFET of the IGBT model in Figure 6.10 is reported in Panel D. The inset of the same panel highlights the so-called Miller plateau, caused by the homonymous effect. The IGBT collector–emitter voltage starts to increase only when the Miller effect takes place (see Panel E, which depicts the SM voltage drop). The collector–emitter current of the IGBT is 1,566 A (Panel G), which leads to a peak in the instantaneous switching losses of 2.77 MW (Panel F). The fall of the collector–emitter current of the IGBT is evident, and so is its current switching toward the D_2 diode (Panel H). Panel J shows the current of the S_1 IGBT ideal gate driver. After 100 ns from the falling edge of the gate driving waveform of the S_1 IGBT, there is the rising edge of the gate driving waveform of the S_2 IGBT, which also lasts 100 ns (Panel A). The inset of Panel B (i.e., the gate voltage of the M2 MOSFET) shows again the Miller effect.

6.4.5 Analysis of simulation speed

The last scenario examined is a DC pole-to-pole fault at bus Bm-A1 (i.e., the ③ mark in Figure 6.9). Analogously to Section 6.4.3, this case study employs a FD SM model and a fixed integration time step of $\Delta T = 4$ μs. Figure 6.14 depicts some simulation

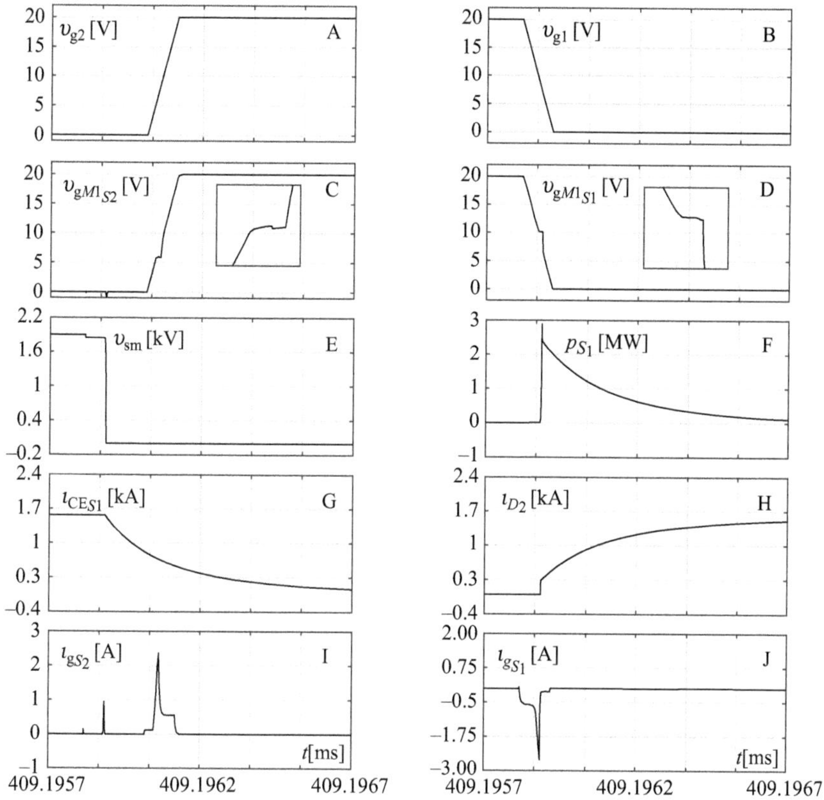

*Figure 6.13 Main waveforms during switching of a submodule described by a full physics model. Simulation results are obtained by adopting the isomorphism-based approach (**Iso**). Devices and circuit nodes to which waveforms refer to are those in Figures 6.3(a) and 6.10. Panel A: the v_{g_2} gate-emitter voltage (turn-on) of the S_2* IGBT. *Panel B: the v_{g_1} gate-emitter voltage (turn-off) of the S_1* IGBT. *Panel C: voltage at the gate of the* MI MOSFET *in Figure 6.10 of the S_2* IGBT *model. Panel D: voltage at the gate of the* MI MOSFET *in Figure 6.10 of the S_1* IGBT *model. The Miller Plateau, given by the homonymous effect and highlighted in the insets of Panel C and D, is clearly visible. Panel E: the v_{sm}^{sw} voltage. Panel F: the instantaneous power dissipated by the S_1* IGBT. *Panel G: the collector–emitter current of the S_1* IGBT. *Panel H: the current through the D_2 diode. Panel I: the gate current of the S_2* IGBT. *Panel J: the gate current of the S_1* IGBT. *x-axis: time (ms).*

results. Immediately after the fault, the DC overcurrent protection blocks both MMCs. Then, at about 0.65 s, the main breakers 1 and 2 are opened.

Aside from confirming once again the accuracy of the proposed method, this scenario is exploited to show that, through isomorphism, the computational burden

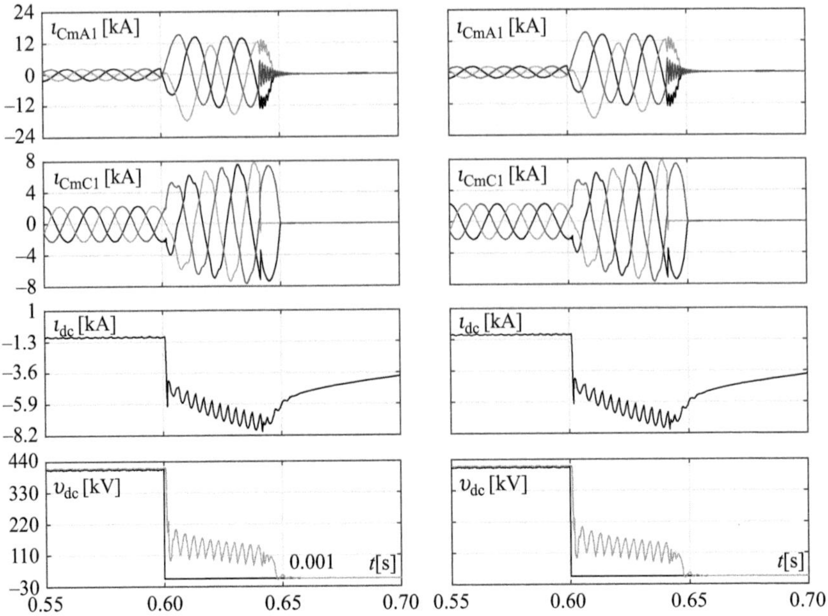

Figure 6.14 Simulation results of a DC *pole-to-pole fault at bus Bm-A1 occurring at 0.6 s, obtained with the* ***Iso*** *(left column panels) and* ***Conv*** *(right column panels) approaches. Labels in panels have the same meaning as those in Figure 6.11*

grows almost linearly with the number of SMs N in each MMC arm instead of almost quadratically. To this aim, the previous simulation was repeated by varying N from 20 to 200 in both MMCs and using an integration scheme based on a fixed or variable time step ΔT (in the latter case, the same one used in Section 6.4.4 is adopted). To further prove that the isomorphism-based approach allows choosing any SM model, different representations (i.e., FP, FD, and BVR) were considered.**

Table 6.2 reports the CPU times required to perform the simulations. The entries in the last row correspond to the values of ΔT used in the simulations based on a fixed time step. As mentioned in Section 6.2.2, as N increases, so does the number of threshold crossings during each period associated with the *modulation technique* implemented in the MMC control scheme. To attain good simulation efficiency, ΔT must be set as large as possible to minimize CPU times but small enough to correctly identify these crossings, which would otherwise lead to inaccurate results.

**As the number of SMs per arm changes, so does the average working voltage across each SM capacitor (see Section 6.4.1). To allow them to withstand the correct voltage value, the semiconductor models of the IGBTS used in the FP SM were suitably adapted.

Table 6.2 CPU *time in [s] (shaded grey area) required to simulate the grid in Figure 6.9 for 1 s (with a* DC *fault at 0.6 s) with the isomorphism-based method and different numbers of* MMC SMs *N*

SM model	Integration time step ΔT	Number of SMs per arm N						
		20	**50**	**80**	**110**	**140**	**170**	**200**
FD	variable	1206	2568	3623	4086	5095	5870	7097
FP	fixed	103	167	226	316	430	484	548
FD	fixed	116	176	262	333	450	508	578
BVR	fixed	50	86	132	180	246	278	312
Fixed time step ΔT		60 μs	30 μs	18 μs	13 μs	9 μs	8 μs	7 μs

The comparison among the first and the other rows of the table highlights how adopting a fixed ΔT instead of a variable one reduces simulation times significantly, thus boosting the simulation efficiency of the isomorphism-based method. Indeed, with a fixed ΔT, the simulator computes the solution at fewer time points than with a variable ΔT. However, it is worth highlighting that the latter is the most suitable option for detailed MMC simulations based on the FP model. Indeed, the analysis of switching transients requires using ΔT in the order of ns to accurately simulate gate driving signals (this is not the case in Table 6.2). If a fixed time step integration scheme is used, numerical inefficiency and a high computational burden are incurred [27]. On the contrary, with a variable time-step integration scheme, the simulation algorithm adjusts ΔT by estimating the local truncation error and following an optimal trade-off between accuracy and efficiency [29,42,52]. For example, ΔT is automatically shortened to accurately identify threshold crossings and then possibly increased between subsequent ones to boost CPU times. This is the reason why this integration scheme was used in Section 6.4.4.

As expected and also shown in Figure 6.15, the trend in CPU time shown in each row of Table 6.2 increases almost linearly with the number of SMs per arm N, with different slopes based on the SM model and the kind of integration time step employed. By comparing Table 6.2 and Table II of [27] (which simulates an HVDC system similar to that considered in this section), it is possible to better grasp the potentialities of the isomorphism-based method with respect to some MMC models proposed in the literature. As N grows, regardless of the SM model used, the CPU times based on a fixed ΔT in Table 6.2 are one or two orders of magnitude lower than those shown in the "Model 1" row of Table II in [27], corresponding to the **Conv** approach based on the FD SM model (i.e., the most accurate representation available in commercial power system simulators). Rather, simulation times obtained with the isomorphism-based method are comparable to those shown in the "Model 2" row of Table II in [27], based on the TEM described in Section 6.2.2 (which is only compatible, thus, with the BVR model).

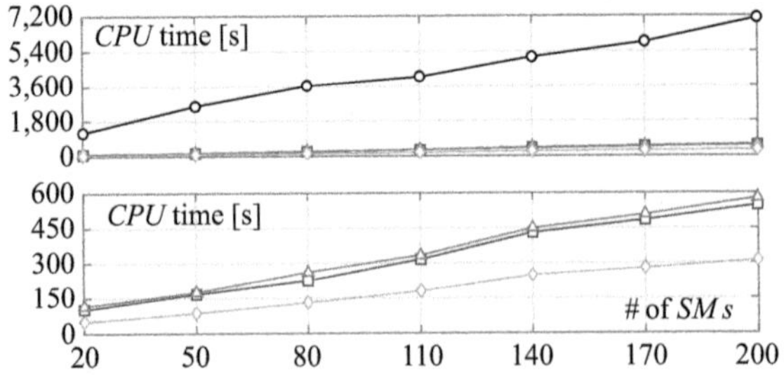

Figure 6.15 Plot of the simulation times shown in Table 6.2. The lines with circle, square, triangle and diamond-shaped markers refer, respectively, to the cases "FD-variable ΔT", "FP-fixed ΔT", "FD-fixed ΔT", and "BVR-variable ΔT".

Lastly, as mentioned in Section 6.2.1, simulations based on FP models are typically not considered in power system simulators because they lead to prohibitively high simulation times. However, when the isomorphism-based approach is adopted, the increase in simulation time incurred when using this SM representation becomes tolerable. Indeed, Table 6.2 shows that the CPU times associated with a fixed ΔT and the FP (or FD) model are only about twice those of the BVR one.[††] This stems from the fact that, with isomorphism, a limited number of SMs is actually simulated. Thus, since the number of nodes is reduced, so is the number of equations to be considered and the size of the nodal admittance matrix to be inverted at each time step, which ultimately leads to a boost in simulation speed.

6.5 Conclusions

This chapter shows how the isomorphism-based approach can be applied to perform efficient EMT simulations of MMCs. This approach has three main features, which have been validated by adopting it to simulate a benchmark system comprising two MMCs. First, the approach is compatible with any MMC operating condition (ranging from ordinary operation to faults in the AC and DC networks) and SM model. Indeed, depending on the degree of detail required, the user can flexibly choose the complex FP, the simple BVR representation, or even a brand new custom SM representation.

[††]Despite being relatively simpler, the CPU times of the "FD-fixed ΔT" case are higher than that of the "FP-fixed ΔT" one (i.e., second and third rows of Table 6.2). The reason for this lies in the ideal commutations of switches in the FD model, which introduce more "time-discontinuities" in the solution than in the FP one (i.e., a much more dynamic variation of the solution from a time point to the next one is observed). This leads to more iterations with the Newton method to compute the solution at each time point and, thus, a higher CPU time. On the contrary, with a variable ΔT, if too many Newton iterations are performed, the method is stopped, and ΔT is shortened. In summary, the total CPU time depends on both ΔT and the number of Newton iterations performed every time step.

Second, regardless of the SM model adopted, the CPU time required to simulate MMCs increases only almost linearly with the number of SMs in each MMC arm. Lastly, the proposed simulation approach paves the way for component level studies that require the adoption of the FP model of SMs. This kind of analysis would be possible with neither conventional simulation approaches nor the MMC representations outlined in Section 6.2. Indeed, in the former case, the adoption of the FP representation would lead to prohibitive simulation times. On the contrary, in the latter case, the use of accelerated MMC models prevents the analysis of switching transients within the SMs either because they adopt too simplified SM models or they do not retain their individual behavior.

Bibliography

[1] Barnes M, Van Hertem D, Teeuwsen SP, Callavik M. HVDC systems in smart grids. *Proceedings of the IEEE* 2017;105(11):2082–2098.

[2] Lesnicar A, Marquardt R. An innovative modular multilevel converter topology suitable for a wide power range. In: 2003 IEEE Bologna PowerTech Conference Proceedings, vol. 3, 2003, 6 pp.

[3] Glinka M, Marquardt R. A new AC/AC-multilevel converter family applied to a single-phase converter. In: The Fifth International Conference on Power Electronics and Drive Systems, 2003, PEDS 2003, vol. 1, 2003, pp. 16–23.

[4] Nami A, Liang J, Dijkhuizen F, Demetriades GD. Modular multilevel converters for HVDC applications: review on converter cells and functionalities. *IEEE Transactions on Power Electronics* 2015;30(1):18–36.

[5] Dekka A, Wu B, Fuentes RL, Perez M, Zargari NR. Evolution of topologies, modeling, control schemes, and applications of modular multilevel converters. *IEEE Journal of Emerging and Selected Topics in Power Electronics* 2017;5(4):1631–1656.

[6] Yazdani A, Iravani R. *Voltage-Sourced Converters in Power Systems: Modeling, Control, and Applications*, John Wiley & Sons, New York, NY, 2010.

[7] Debnath S, Qin J, Bahrani B, Saeedifard M, Barbosa P. Operation, control, and applications of the modular multilevel converter: a review. *IEEE Transactions on Power Electronics* 2015;30(1):37–53.

[8] Khan S, Tedeschi E. Modeling of MMC for fast and accurate simulation of electromagnetic transients: a review. *Energies* 2017;10(8):1161. Available from: http://dx.doi.org/10.3390/en10081161.

[9] Saeedifard M, Iravani R. Dynamic performance of a modular multilevel back-to-back HVDC system. *IEEE Transactions on Power Delivery* 2010;25(4):2903–2912.

[10] Ghazanfari A, Mohamed YARI. A hierarchical permutation cyclic coding strategy for sensorless capacitor voltage balancing in modular multilevel converters. *IEEE Journal of Emerging and Selected Topics in Power Electronics* 2016;4(2):576–588.

[11] Mathe L. Performance comparison of the modulators with balancing capability used in MMC applications. In: 2017 IEEE 26th International Symposium on Industrial Electronics (ISIE), 2017, pp. 815–820.

[12] Tu Q, Xu Z, Xu L. Reduced switching-frequency modulation and circulating current suppression for modular multilevel converters. *IEEE Transactions on Power Delivery* 2011;26(3):2009–2017.

[13] Li S, Wang X, Yao Z, Li T, Peng Z. Circulating current suppressing strategy for MMC-HVDC based on nonideal proportional resonant controllers under unbalanced grid conditions. *IEEE Transactions on Power Electronics* 2015;30(1):387–397.

[14] Bahrani B, Debnath S, Saeedifard M. Circulating current suppression of the modular multilevel converter in a double-frequency rotating reference frame. *IEEE Transactions on Power Electronics* 2016;31(1):783–792.

[15] Li B, Xu Z, Shi S, Xu D, Wang W. Comparative study of the active and passive circulating current suppression methods for modular multilevel converters. *IEEE Transactions on Power Electronics* 2018;33(3):1878–1883.

[16] B4-57 CW. Guide for the Development of Models for HVDC Converters in a HVDC Grid. CIGRÉ (WG Brochure), 2014.

[17] Aji H, Ndreko M, Popov M, van der Meijden MAMM. Investigation on different negative sequence current control options for MMC-HVDC during single line to ground AC faults. In: 2016 IEEE PES Innovative Smart Grid Technologies Conference Europe (ISGT-Europe), 2016, pp. 1–6.

[18] Sixing D, Apparao D, Bin W, Navid Z. Modular multilevel converters: analysis, control, and applications. In: *IEEE Press Series on Power Engineering*, Wiley, New York, NY, 2018.

[19] Li J, Konstantinou G, Wickramasinghe HR, Pou J. Operation and control methods of modular multilevel converters in unbalanced AC grids: a review. *IEEE Journal of Emerging and Selected Topics in Power Electronics* 2019;7(2):1258–1271.

[20] Nguyen Q, Todeschini G, Santoso S. Power flow in a multi-frequency HVac and HVdc system: formulation, solution, and validation. *IEEE Transactions on Power Systems* 2019;34(4):2487–2497.

[21] Bizzarri F, del Giudice D, Linaro D, Brambilla A. Partitioning-based unified power flow algorithm for mixed MTDC/AC power systems. *IEEE Transactions on Power Systems* 2021;36(4):3406–3415.

[22] Stepanov A, Saad H, Karaagac U, Mahseredjian J. Initialization of modular multilevel converter models for the simulation of electromagnetic transients. *IEEE Transactions on Power Delivery* 2019;34(1):290–300.

[23] Liu Y, Song Y, Zhao L, Chen Y, Shen C. A general initialization scheme for electromagnetic transient simulation: towards large-scale hybrid AC–DC grids. In: 2020 IEEE Power Energy Society General Meeting (PESGM), 2020, pp. 1–5.

[24] Lyu J, Zhang X, Cai X, Molinas M. Harmonic state-space based small-signal impedance modeling of a modular multilevel converter with consideration

of internal harmonic dynamics. *IEEE Transactions on Power Electronics* 2019;34(3):2134–2148.

[25] Sakinci ÖC, Beerten J. Equivalent multiple *dq*-Frame model of the MMC using dynamic phasor theory in the $\alpha\beta z$-frame. *IEEE Transactions on Power Delivery* 2020;35(6):2916–2927. doi: 10.1109/TPWRD.2020.3025388.

[26] del Giudice D, Brambilla A, Linaro D, Bizzarri F. Modular multilevel converter impedance computation based on periodic small-signal analysis and vector fitting. *IEEE Transactions on Circuits and Systems I: Regular Papers* 2022;69(4):1832–1842. doi: 10.1109/TCSI.2021.3138515.

[27] Saad H, Dennetière S, Mahseredjian J, *et al.* Modular multilevel converter models for electromagnetic transients. *IEEE Transactions on Power Delivery* 2014;29(3):1481–1489.

[28] del Giudice D, Bizzarri F, Linaro D, Brambilla A. Stability analysis of MMC/MTDC systems considering DC-link dynamics. In: 2021 IEEE International Symposium on Circuits and Systems (ISCAS), 2021, pp. 1–5.

[29] Vlach J, Singhal K. *Computer Methods for Circuit Analysis and Design*, Van Nostrand Reinhold Company, 1983.

[30] Chua LO, Desoer CA, Kuh ES. *Linear and Nonlinear Circuits*, McGraw-Hill Editions, New York, NY, 1987.

[31] Gnanarathna UN, Gole AM, Jayasinghe RP. Efficient modeling of modular multilevel HVDC converters (MMC) on electromagnetic transient simulation programs. *IEEE Transactions on Power Delivery* 2011;26(1):316–324.

[32] Watson N, Arillaga J. Power systems electromagnetic transients simulation. In: IEEE Press Series on Power Engineering. IET Power and Energy Series, 2003.

[33] Teeuwsen SP. Simplified dynamic model of a voltage-sourced converter with modular multilevel converter design. In: 2009 IEEE/PES Power Systems Conference and Exposition, 2009, pp. 1–6.

[34] del Giudice D, Brambilla A, Linaro D, Bizzarri F. Isomorphic circuit clustering for fast and accurate electromagnetic transient simulations of MMCs. *IEEE Transactions on Energy Conversion* 2022;37(2):800–810. doi: 10.1109/TEC.2021.3113719.

[35] Ashourloo M, Mirzahosseini R, Iravani R. Enhanced model and real-time simulation architecture for modular multilevel converter. *IEEE Transactions on Power Delivery* 2018;33(1):466–476.

[36] Guo D, Rahman MH, Ased GP, *et al.* Detailed quantitative comparison of half-bridge modular multilevel converter modelling methods. *The Journal of Engineering* 2019;2019(16):1292–1298.

[37] Adam GP, Li P, Gowaid IA, Williams BW. Generalized switching function model of modular multilevel converter. In: 2015 IEEE International Conference on Industrial Technology (ICIT), 2015, pp. 2702–2707.

[38] Inc. Synopsys, Assignee. SPICE Optimized for Arrays. US20070133245A, 2007.

[39] McGaughy BW, Kong J. Method and system for partitioning integrated circuits. Google Patents, 2010. US Patent 7,836,419. Available from: http://www.google.com/patents/US7836419.

[40] Rewieński M. In: *A Perspective on Fast-SPICE Simulation Technology*, Springer Netherlands, Dordrecht, 2011, pp. 23–42.

[41] del Giudice D, Bizzarri F, Linaro D, Brambilla A. Efficient isomorphism based simulation of modular multilevel converters. In: 2019 IEEE Milan PowerTech, 2019, pp. 1–5, doi: 10.1109/PTC.2019.8810733.

[42] Gear CW. *Numerical Initial Value Problems in Ordinary Differential Equations*, Prentice-Hall, Hoboken, NJ, 1971.

[43] Hefner AR, Diebolt DM. An experimentally verified IGBT model implemented in the Saber circuit simulator. In: PESC '91 Record 22nd Annual IEEE PESC, 1991, pp. 10–19.

[44] Azar R, Udrea F, De Silva M, *et al.* Advanced SPICE modeling of large power IGBT modules. In: Conference Record of the 2002 IEEE Industry Applications Conference. 37th IAS Annual Meeting (Cat. No. 02CH37344), Vol. 4, 2002, pp. 2433–2436.

[45] Chen H, Wakeman F, Pitman J, Li G. Design, analysis, and testing of PP-IGBT-based submodule stack for the MMC VSC HVDC with 3000 A DC bus current. *The Journal of Engineering* 2019;2019(16):917–923.

[46] Antognetti P, editor. *Semiconductor Device Modeling with SPICE*, McGraw-Hill, Inc., New York, NY, 1988.

[47] Beerten J, D'Arco S, Suul JA. Frequency-dependent cable modelling for small-signal stability analysis of VSC-HVDC systems. *IET Generation, Transmission Distribution* 2016;10(6):1370–1381.

[48] Leterme W, Ahmed N, Beerten J, *et al.* A new HVDC grid test system for HVDC grid dynamics and protection studies in EMT-type software. In: 11th IET International Conference on AC and DC Power Transmission, 2015, pp. 1–7.

[49] Ajaei FB, Iravani R. Enhanced equivalent model of the modular multilevel converter. *IEEE Transactions on Power Delivery* 2015;30(2):666–673.

[50] Bizzarri F, Brambilla A, Gajani GS, Banerjee S. Simulation of real world circuits: extending conventional analysis methods to circuits described by heterogeneous languages. *IEEE Circuits and Systems Magazine* 2014 Fourthquarter;14(4):51–70.

[51] Bizzarri F, Brambilla A. PAN and MPanSuite: simulation vehicles towards the analysis and design of heterogeneous mixed electrical systems. In: NGCAS, IEEE, New York, NY, 2017, pp. 1–4.

[52] Lin N, Dinavahi V. Variable time-stepping modular multilevel converter model for fast and parallel transient simulation of multiterminal DC grid. *IEEE Transactions on Industrial Electronics*, 2019;66(9):6661–6670. doi: 10.1109/TIE.2018.2880671.

Part II

Control

Chapter 7

Optimization methods for preventive/corrective control in transmission systems

Massimo La Scala[1] *and Sergio Bruno*[1]

This chapter discusses a methodology for the assessment of corrective control actions to be implemented in power system after the occurrence of a severe contingency. This pre-calculation can be intended as one of the contingency screening functions to be developed within each Supervisory Control and Data Acquisition (SCADA)/Energy Management Systems (EMS) control cycle and in the framework of online Dynamic Security Assessment (DSA). It is assumed that, based on latest system state information and calculations carried out for a selected number of severe fault events, pre-calculated control actions can be applied on actuators during the power system transient that follows such contingencies. The methodology can be adopted, for example, for the arming of Remedial Actions Schemes based on load shedding, generation shedding, Flexible AC Transmission Systems (FACTS) or any other fast actuator. The goal is preserving power system stability and its integrity even after that a specific severe contingency had occurred.

The proposed methodology is based on the conversion of a dynamic optimization problem in the continuous time domain, into a static optimization problem in the discrete time domain. In order to show the key working principle of the proposed approach, a dynamic optimization problem is first formulated in his general form and then solved with the classical method and through discretization. Finally, the methodology is explicitly adapted to the set of equations and constraints that represent the dynamic behaviour of power systems. The same general methodology is then also applied to the solution of a preventive control optimization problem.

7.1 Formulation of a time-continuous dynamic optimization problem for corrective control

The problem that is hereby formulated is aimed to find an optimal time-varying control rule that, if applied during the power system transient following a specific contingency, can preserve system stability and allow to reach a stable equilibrium point.

[1]Department of Electrical Engineering and Information, Politecnico di Bari, Italy

This problem can always be formulated as a generic dynamic optimization problem. The dynamic of power systems, not differently from many other physical systems, can be described through its generic state space equations:

$$\begin{aligned} \dot{\boldsymbol{x}}(t) &= f(\boldsymbol{x}(t), \boldsymbol{u}(t), t) \\ \boldsymbol{x}(t_0) &= \boldsymbol{x}_0 \end{aligned} \tag{7.1}$$

where $\boldsymbol{x}$ is an n-dimensional state vector, $\boldsymbol{x}_0$ is the initial state reached after the perturbation of the system, $\boldsymbol{u}$ is an m-dimensional control vector and f is a set of continuous doubly differentiable analytical vector function (i.e., C^2 class function). The control variable vector $\boldsymbol{u}$ represents the state that the selected actuators must assume during the transient in order to preserve stability.

Having supposed that t_f represents the final time, the control problem to be solved is to choose the optimal trajectory $\boldsymbol{u}(t)$ along the interval $[t_0, t_f]$ so that the system described by (7.1) has a suitable dynamical behaviour. If the power system operates normally in steady-state conditions when it receives an unknown large disturbance which changes its state to some known (or computable) state $\boldsymbol{x}_0$, the desired dynamic behaviour could be represented by the control vector $\boldsymbol{u}$ which minimizes some function of the states and controls all along the trajectory.

A possible approach is to search the time-varying control $\boldsymbol{u}$ that minimizes a generic function

$$J = h(\boldsymbol{x}(t_f), t_f) + \int_{t_0}^{t_f} g(\boldsymbol{x}(t), \boldsymbol{u}(t), t)\,\mathrm{d}t \tag{7.2}$$

with h and g being, in general, scalar non-linear functions of class C^2.

The function h is the objective function and allows to ensure that at the final time t_f the system state $\boldsymbol{x}$ will approach a specific target state, whereas the integral function g ensures to avoid, over the optimization interval, excessive control effort or significant deviations from any given "desired" trajectories that the controlled system is expected to follow.

In addition, since our model equation (7.1) represents a physical system and since a physical system does not permit infinite controls or states, states and controls must be constrained so that each one is physically realizable. It is defined an admissible state $\boldsymbol{x} \in \boldsymbol{X}$ when it satisfies the additional state constraints, aside from the dynamic equation (7.1), and an admissible control $\boldsymbol{u} \in \boldsymbol{U}$ when it satisfies the additional control constraints. $\boldsymbol{X}$ and $\boldsymbol{U}$ are, respectively, the sets of admissible states and controls over the period $[t_0, t_f]$.

The dynamical optimization problem is to find an admissible control $\boldsymbol{u}^*$ which forces the dynamic system in (7.1) to follow an admissible trajectory $\boldsymbol{x}^*$ that minimizes the performance measure J given in (7.2).

In practice, except for the special case of linear-quadratic problems, it is difficult to obtain the global optimal control numerically. However, obtaining local optima is always quite easy, and local optimum solutions can always be accepted when system security is at stake.

The most common approach for tackling such problems is to develop the necessary conditions for the solution of the dynamical optimization problem using the calculus of variations. Satisfaction of these conditions will yield a local optimum. In order to write the necessary conditions for optimality in a more compact form, the function H, or Hamiltonian, is defined:

$$\boldsymbol{H}(\boldsymbol{x}(t), \boldsymbol{u}(t), \boldsymbol{\lambda}(t), t) = \boldsymbol{g}(\boldsymbol{x}, \boldsymbol{u}, t) + \boldsymbol{\lambda}^{\mathrm{T}} \boldsymbol{f}(\boldsymbol{x}, \boldsymbol{u}, t) \tag{7.3}$$

where $\boldsymbol{\lambda}$ are Lagrangian multipliers. If the terminal time t_f is fixed and the final state $\boldsymbol{x}(t_f)$ is free the first-order conditions are given by [1,2]:

$$\dot{\boldsymbol{x}}(t) = \frac{\partial \boldsymbol{H}}{\partial \boldsymbol{\lambda}} = \boldsymbol{f}(\boldsymbol{x}, \boldsymbol{u}, t) \tag{7.4}$$

$$\dot{\boldsymbol{\lambda}}(t) = -\frac{\partial \boldsymbol{H}}{\partial \boldsymbol{x}} = -\left[\frac{\partial \boldsymbol{f}}{\partial \boldsymbol{x}}\right]^{\mathrm{T}} \boldsymbol{\lambda}(t) - \frac{\partial \boldsymbol{g}}{\partial \boldsymbol{x}} \tag{7.5}$$

$$\frac{\partial H}{\partial \boldsymbol{u}} = \boldsymbol{0} = \left[\frac{\partial \boldsymbol{f}}{\partial \boldsymbol{u}}\right]^{\mathrm{T}} \boldsymbol{\lambda}(t) + \frac{\partial \boldsymbol{g}}{\partial \boldsymbol{u}} \tag{7.6}$$

$$\boldsymbol{x}(t_0) = \boldsymbol{x}_0 \tag{7.7}$$

$$\boldsymbol{\lambda}(t_f) = \frac{\partial h}{\partial \boldsymbol{x}}(\boldsymbol{x}(t_f)) \tag{7.8}$$

The problem given in (7.4)–(7.8) is a very complex problem that, in the case of the thousands of non-linear equations which represent a full-scale power system, cannot be solved any another way than numerically and, therefore, through discretization of system trajectories. The typical approach based on a steepest descent algorithm starts with a guess of certain variables or trajectories, obtaining a solution of the equations in such way that just one or more of the necessary conditions for optimality is not satisfied whilst all others are satisfied. This first solution is used to adjust the initial guess in the attempt of getting closer to a (sub)-optimal solution. By iterating the steps, if the procedure converges, the necessary conditions will eventually be satisfied. An iterative algorithm to solve this problem can be implemented as follows.

STEP 1. An initial nominal control trajectory is chosen. Having fixed a time step Δt for discretization of all variables, the interval $[t_0, t_f]$ is then divided into N equally large subintervals. The iteration index i is set to zero. The function $\boldsymbol{u}(t)$ assumes a piecewise-constant shape $\boldsymbol{u}^0(t)$ that follows the following formulation:

$$\boldsymbol{u}^0(t) = \boldsymbol{u}^0(t_k) \tag{7.9}$$

with $t \in [t_k, t_{k+1}]$ $\quad t_N = t_f \quad$ and $\quad k = 0, 1, \ldots, N-1$.

STEP 2. Using the piecewise-constant $\boldsymbol{u}^i(t)$, the state equations (7.1) are integrated from t_0 to t_f, starting from the initial conditions in (7.7), obtaining a time-varying trajectory of state variables. This trajectory is also discretized following the same criteria given in (7.9), obtaining a piecewise-constant state trajectory $\boldsymbol{x}^i(t)$.

STEP 3. By substituting $\boldsymbol{x}^i(t_f)$ in (7.8), $\boldsymbol{\lambda}^i(t_f)$ is obtained. This value is the terminal condition used for integrating the costate equation (7.5) backwards from t_f to t_0, so that a time-varying costate trajectory is obtained. This trajectory is again discretized so that it assumes a piecewise-constant shape, and is then used for evaluating $\partial H^i(\boldsymbol{t})/\partial \boldsymbol{u}$ from (7.6).

STEP 4. Having chosen a small positive constant number ε as tolerance of the iterative procedure and having calculated

$$\left\| \frac{\partial H^i}{\partial \boldsymbol{u}} \right\| = \int_{t_0}^{t_f} \left[\frac{\partial H^i}{\partial \boldsymbol{u}}(t) \right]^{\mathrm{T}} \cdot \left[\frac{\partial H^i}{\partial \boldsymbol{u}}(t) \right] \cdot dt \tag{7.10}$$

$$\text{if} \quad \left\| \frac{\partial H^i}{\partial \boldsymbol{u}} \right\| < \varepsilon \tag{7.11}$$

the optimality conditions are considered satisfied and $\boldsymbol{u}^i(t)$ represents the solution of the optimal control problems (7.4)–(7.8).

If condition (7.11) is not respected a new piecewise-constant control rule is defined

$$\boldsymbol{u}^{i+1}(t_k) = \boldsymbol{u}^i(t_k) + \alpha \cdot \frac{\partial H^i}{\partial \boldsymbol{u}}(t_k) \quad \text{for} \quad k = 0, 1, \ldots, N-1 \tag{7.12}$$

the iteration index i is incremented by one and the algorithm goes back to STEP 2 until a solution is found.

The step size α can be a fixed number or can be determined by a single variable search. For example, having assumed an arbitrary initial value of α, a linear search sub-algorithm can be used in order to select the control rule $\boldsymbol{u}^{i+1}(t)$ that minimizes equation (7.2). The obtained control and state trajectories are then used in the next iteration of the main solving algorithm.

The value adopted for ε depends mainly on the required accuracy. Several trial runs can be needed before fixing ε.

7.2 Formulation of a time-discrete static optimization problem for corrective control

The method showed in the previous section, based on applying the calculus of variations to obtain necessary conditions for optimization and solving differential equations in $\boldsymbol{x}$ and $\boldsymbol{\lambda}$ through classical numerical integration methods, is a standard procedure suitable for small-scale systems [2]. In this section, a different approach to solve dynamic optimization problems is showed. It is based on the idea of discretizing first the differential equations, and then treating the problem as a static optimization problem to be solved through non-linear programming techniques.

Having chosen a fixed a time step Δt so that the interval $[t_0, t_f]$ is divided into N subintervals, the state equation (7.1) is discretized with an Euler integration rule obtaining

$$F_{t_k}(\boldsymbol{x}(t_{k-1}), \boldsymbol{x}(t_k), \boldsymbol{u}(t_{k-1})) = 0, \qquad k = 1, \ldots, N \tag{7.13}$$

Equation (7.1) can be written in a compact form as a unique large set of discretized equations and discretized variables

$$\hat{\boldsymbol{F}}(\hat{\boldsymbol{x}}, \hat{\boldsymbol{u}}) = 0 \tag{7.14}$$

with

$$\hat{F} = [F_{t_1}^{\mathrm{T}}, F_{t_2}^{\mathrm{T}}, \dots, F_{t_N}^{\mathrm{T}}]^{\mathrm{T}}$$
$$\hat{x} = [x_{t_1}^{\mathrm{T}}, x_{t_2}^{\mathrm{T}}, \dots, x_{t_N}^{\mathrm{T}}]^{\mathrm{T}}$$
$$\hat{u} = [u_{t_0}^{\mathrm{T}}, u_{t_1}^{\mathrm{T}}, \dots, u_{t_{N-1}}^{\mathrm{T}}]^{\mathrm{T}}$$
$$x_{t_0} = x_0$$

and it represents the dynamic constraints of the problem that are now reduced to algebraic equations instead of differential equations.

Assuming the simplest rule for discretization, the cost function J becomes

$$\hat{J}(\hat{x}, \hat{u}) = h(x_{t_N}) + \sum_{k=0}^{N-1} \left[g_k(x_{t_k}, u_{t_k}) \cdot \Delta t \right] \tag{7.15}$$

where h and g_k are scalar non-linear C^2 functions discretized at each time step. Also, in this case, the function $h(x_{t_N})$ allows to ensure that at the final time $t_N = t_f$ the state x will approach a target state.

The time-discrete static optimization problem is therefore a minimization of equation (7.15), subject to the equality constraints (7.14). Further, inequality constraints should be introduced in order to that take into account the feasibility of the solution

$$\hat{x}_{t_k} \in X \quad \forall k = 1, 2, \cdots, N \tag{7.16}$$

$$\hat{u}_{t_k} \in U \quad \forall k = 0, 1, \cdots, N-1 \tag{7.17}$$

An approach to treat inequality constraints (7.16) is usually based on treating them as soft inequality constraints and introducing in (7.15) penalty functions depending on state. Constraints on control variables (7.17) can also be treated introducing a penalty function but, more easily, are usually considered as hard constraints which do not require particular mathematical formulations. The presence of constraints (7.16) and (7.17) is neglected at this stage, but will be treated in detail in the next section.

The static optimization problem just formulated can be solved straightforwardly with the Lagrangian multipliers method. Having defined the vector of Lagrangian multipliers $\hat{\lambda} = [\lambda_{t_1}^{\mathrm{T}}, \dots, \lambda_{t_N}^{\mathrm{T}}]^{\mathrm{T}}$, the overall cost function to be minimized is

$$L(\hat{x}, \hat{u}, \hat{\lambda}) = J(\hat{x}, \hat{u}) - \hat{\lambda}^{\mathrm{T}} \hat{F}(\hat{x}, \hat{u}) \tag{7.18}$$

and the first-order conditions are obtained as follows:

$$\frac{\partial L}{\partial \hat{\lambda}} = \hat{F}(\hat{x}, \hat{u}) = 0 \tag{7.19}$$

$$\frac{\partial L}{\partial \hat{x}} = \frac{\partial J}{\partial \hat{x}} - \hat{\lambda}^{\mathrm{T}} \frac{\partial \hat{F}}{\partial \hat{x}} = 0 \tag{7.20}$$

$$\frac{\partial L}{\partial \hat{u}} = \frac{\partial J}{\partial \hat{u}} - \hat{\lambda}^{\mathrm{T}} \frac{\partial \hat{F}}{\partial \hat{u}} = 0 \tag{7.21}$$

By adopting a gradient method, for static optimization, it is possible to compare the solution of the time-continuous dynamic optimization problem with the time-discrete static approach. The algorithm to be used for solving (7.19)–(7.21) can be implemented as follows.

STEP 1. The iteration index i is first set to zero and an initial discretized nominal control trajectory $\hat{\boldsymbol{u}}^0$ is assumed.
STEP 2. Using the control vector $\hat{\boldsymbol{u}}^i$ and having considered that $\boldsymbol{x}^i(t_0) = \boldsymbol{x}_0$, (7.13) are solved so that the state variable vector $\hat{\boldsymbol{x}}^i$ is known.
STEP 3. The vector $\hat{\boldsymbol{\lambda}}^i$ is calculated solving (7.20). Note that assuming an explicit Euler integration rule, (7.13) can be written as

$$F^i_{t_k} = \boldsymbol{x}_{t_k} - \boldsymbol{x}_{t_{k-1}} - \Delta t \cdot f\left(\boldsymbol{x}_{t_{k-1}}, \boldsymbol{u}_{t_{k-1}}\right) = 0, \qquad k = 1, \ldots, N \tag{7.22}$$

and the matrix $\partial F^i/\partial \boldsymbol{x}$ assumes a very helpful block bi-diagonal form with

$$\frac{\partial F^i_{t_k}}{\partial \boldsymbol{x}_{t_k}} = \boldsymbol{I}, \quad k = 1, \ldots, N \tag{7.23}$$

$$\frac{\partial F^i_{t_k}}{\partial \boldsymbol{x}_{t_{k-1}}} = -\left[\boldsymbol{I} + \Delta t \cdot \frac{\partial f}{\partial \boldsymbol{x}_{t_{k-1}}}\right], \quad k = 2, \ldots, N \tag{7.24}$$

The system of equations (7.20) can be solved easily by backward substitutions considering that such equations can be written as

$$\frac{\partial F^i_{t_k}}{\partial \boldsymbol{x}_{t_k}}^{\mathrm{T}} \cdot \boldsymbol{\lambda}^i_{t_k} + \frac{\partial F^i_{t_{k+1}}}{\partial \boldsymbol{x}_{t_k}}^{\mathrm{T}} \cdot \boldsymbol{\lambda}^i_{t_{k+1}} - \Delta t \cdot \frac{\partial g_k}{\partial \boldsymbol{x}_{t_k}} = 0, \quad k = 1, \ldots, N-1 \tag{7.25}$$

and

$$\frac{\partial F^i_{t_N}}{\partial \boldsymbol{x}_{t_N}}^{\mathrm{T}} \cdot \boldsymbol{\lambda}^i_{t_N} - \frac{\partial h}{\partial \boldsymbol{x}_{t_N}} = 0 \tag{7.26}$$

Considering the relationship (7.21), (7.24) gives the condition

$$\boldsymbol{\lambda}^i_{t_N} = \frac{\partial h}{\partial \boldsymbol{x}_{t_N}} \tag{7.27}$$

which is equivalent to (7.8) in the time-continuous formulation. Knowing this value, all other Lagrangians can be evaluated with a backwards substitution considering that from (7.23) to (7.25)

$$\boldsymbol{\lambda}^i_{t_k} = \boldsymbol{\lambda}^i_{t_{k+1}} + \boldsymbol{\lambda}^i_{t_{k+1}} \cdot \Delta t \cdot \frac{\partial f}{\partial \boldsymbol{x}_{t_k}} + \Delta t \cdot \frac{\partial g_k}{\partial \boldsymbol{x}_{t_k}} \quad \text{with} \quad k = N-1, N-2, \ldots, 1 \tag{7.28}$$

Once the entire vector $\hat{\boldsymbol{\lambda}}^i$ is known, (7.21) gives the value of $\partial L^i/\partial \hat{\boldsymbol{u}}$.

STEP 4. Having chosen a small positive constant number ε, if

$$\left\| \frac{\partial L^i}{\partial \hat{\boldsymbol{u}}} \right\| < \varepsilon \tag{7.29}$$

the control vector $\hat{\boldsymbol{u}}^i$ is considered a solution of the problem.

Otherwise, a new control rule is calculated

$$\hat{\boldsymbol{u}}^{i+1} = \hat{\boldsymbol{u}}^i + \alpha \frac{\partial L^i}{\partial \hat{\boldsymbol{u}}} \tag{7.30}$$

the iteration number i is incremented by one and the iterative process goes back to STEP 2.

If the initial control rule is the same and if an Euler integration rule is adopted, $\partial L^i / \partial \hat{\boldsymbol{u}}$ is equal to $\partial H^i(\boldsymbol{t}) / \partial \boldsymbol{u}$ evaluated in the previous case, neglecting second-order errors. Consequently, the same results are expected by the two approaches. However, it should be kept in mind that the equivalence of the two approaches is valid as long as an Euler integration scheme is adopted. Other integration rules will bring different results.

The use of this approach has several advantages with respect to the classical approach. For example, the possibility of solving (7.1) simultaneously instead of using step-by-step procedures allows to adopt more complex integration rules that work on the whole trajectory, such as Boundary Value Methods applied to initial value problems [3]. Moreover, the static optimization approach permits to exploit any non-linear programming techniques, granting the availability a wider choice of solving tools with respect to the classical approach where only gradient methods can be applied. The last and most important advantage is that the static approach permits to integrate easily in the procedure the differential-algebraic equations (DAEs) that describe the behaviour of power systems and permits also to include in the formulation inequality constraints based on technical constraints as shown in the next section.

7.3 Application to power system DAEs

The following passages are aimed at showing how power system dynamic equations and physical constraints are treated within the formulation of an optimal control problem aimed at ensuring transient stability after the occurrence of a severe fault event.

The dynamic behaviour of a power system having n_b nodes and n_g generating units can be represented by a set of non-linear DAEs:

$$\begin{aligned} \dot{\boldsymbol{x}}(t) &= f_d\left(\boldsymbol{x}(t), \boldsymbol{V}(t), \boldsymbol{u}(t), t\right) \\ f_s\left(\boldsymbol{x}(t), \boldsymbol{V}(t), \boldsymbol{u}(t), t\right) &= 0 \end{aligned} \tag{7.31}$$

where f_d represents the dynamic model of generators and their controllers, f_s are the static load-flow equations, $\boldsymbol{x}$ is a state p-dimensional vector whose dimension depends on the number of equations used for representing the dynamic of the generators and their controllers, $\boldsymbol{V}$ is the $2n_b$-dimensional vector of nodal voltages and $\boldsymbol{u}$ is the m-dimensional vector of control variables. The nature and dimension of $\boldsymbol{u}$ depends of course on the actuators that have been selected for controlling transient security.

Having supposed that t_f represents the time at which the main electromechanical modes are supposed to be damped out, and having chosen a fixed integration step

Δt, N is defined as the total number of time steps relative to the integration interval $[t_0, t_f]$. By adopting any explicit integration rule, (7.30) can be discretized at each kth time step as follows:

$$F_k(\boldsymbol{y}_k, \boldsymbol{y}_{k,j}, \boldsymbol{u}_k) = 0 \quad \forall k = 1, 2, \ldots, N \tag{7.32}$$

where

$$\begin{aligned} \boldsymbol{y}_k &= \begin{bmatrix} \boldsymbol{x}_k^{\mathrm{T}} & \boldsymbol{V}_k^{\mathrm{T}} \end{bmatrix}^{\mathrm{T}} \\ \boldsymbol{y}_{k,j} &= \begin{bmatrix} \boldsymbol{y}_{k-1}^{\mathrm{T}} & \boldsymbol{y}_{k-2}^{\mathrm{T}} & \cdots & \boldsymbol{y}_{k-j}^{\mathrm{T}} \end{bmatrix}^{\mathrm{T}} \\ \boldsymbol{y}_0 &= \begin{bmatrix} \boldsymbol{x}(t_0)^{\mathrm{T}} & \boldsymbol{V}(t_0)^{\mathrm{T}} \end{bmatrix}^{\mathrm{T}} \end{aligned} \tag{7.33}$$

with $j \leq k$ and j representing the number of steps of the adopted implicit discretization rule. If an explicit Euler integration rule is chosen then $j = 1$. Function F_k represents both discretized differential equations and load-flow algebraic equations at the kth time step.

Note that (7.31) can always be written in a more compact form so that is equivalent to (7.13) formulated beforehand:

$$\hat{\boldsymbol{F}}(\hat{\boldsymbol{y}}, \hat{\boldsymbol{u}}) = 0 \tag{7.34}$$

with

$$\begin{aligned} \hat{\boldsymbol{F}} &= [F_1^{\mathrm{T}}, F_2^{\mathrm{T}}, \ldots, F_N^{\mathrm{T}}]^{\mathrm{T}} \\ \hat{\boldsymbol{y}} &= [\boldsymbol{y}_1^{\mathrm{T}}, \boldsymbol{y}_2^{\mathrm{T}}, \ldots, \boldsymbol{y}_N^{\mathrm{T}}]^{\mathrm{T}} \\ \hat{\boldsymbol{u}} &= [\boldsymbol{u}_0^{\mathrm{T}}, \boldsymbol{u}_1^{\mathrm{T}}, \ldots, \boldsymbol{u}_{N-1}^{\mathrm{T}}]^{\mathrm{T}} \end{aligned}$$

The function to be minimized (see (7.15)) is composed by two terms. The first term is related to the final point to be reached at t_f, whereas the second term represents a cost function. Very often, the term depending on the final state is not helpful for practical applications. In fact, controlling the final state at the end of the trajectory requires to represent much longer time intervals (oscillation modes should be more or less damped out) hence slowing and degrading the overall convergence properties of the solving algorithm. Moreover, the control effort required to ensure a specific condition on the final state may result too burdensome if not unfeasible. It should be reminded that, whenever system security is at risk, the essential goal is to maintain stability and preserve system integrity, and that all calculations should always be performed in the shortest time possible. In this context, the fine research of a specific final target can be inappropriate if not harmful.

According to such premises, the cost function can assume the general formulation:

$$J = g_1(\hat{\boldsymbol{u}}) + \sum_{k=1}^{N} g_2(\boldsymbol{y}_k) \cdot \Delta \mathrm{t} \tag{7.35}$$

where g_1 is a term depending on the sole control variable vector, taking into account the cost of control (for example, the cost for thermal units [4,5], or the cost of load interruption [6]), and g_2 is an objective function aimed at ensuring system stability. Its formulation can be based, for example, on the concept of energy signal associated to voltages during the post-fault transient [5] or on a function oriented to improve transient stability by the minimization of the transient kinetic energy [7].

Clearly (7.35) can be written as

$$J = J_1(\hat{\boldsymbol{u}}) + J_2(\hat{\boldsymbol{y}}) \tag{7.36}$$

It should be reminded that the system just represented by means of equations (7.31) is indeed composed by physical components that cannot withstand any possible state. Along the transient, certain variables (states or voltages) that represent the behaviour of power system can experience large excursions and assume states incompatible with power quality requirements. Moreover, such fluctuations can cause undesirable effects such as instability or incorrect switching of protections. For example, frequency dropping below a certain level might cause the automatic switching of underfrequency relays, or certain degraded power system trajectories might cross several times the tripping areas of distance relays causing an improper activation of distance protections as it had been observed during some large blackout events experienced in the recent past [8]. These rationales are at the basis of the concept of "practical stability" which has been introduced in References 4–7 in order to achieve higher qualifying stability performances rather the simple asymptotic stability.

In order to consider such criteria, time-varying inequality constraints can be introduced so that a domain where the trajectories of the system should be contained in order to satisfy practical requirements can be defined. In general, such constraints can be formulated as a set of equations:

$$q(\boldsymbol{x}(t), \boldsymbol{V}(t), \boldsymbol{u}(t)) \le 0 \tag{7.37}$$

Thanks to the discretization of all system variable, the constraints (7.37) can be taken into account by introducing any type of penalty function $J_P(\hat{\boldsymbol{y}}, \hat{\boldsymbol{u}})$, generically function of both $\hat{\boldsymbol{y}}$ and $\hat{\boldsymbol{u}}$ so that the overall problem becomes

$$\min_{\hat{\boldsymbol{u}}} J_1(\hat{\boldsymbol{u}}) + J_2(\hat{\boldsymbol{y}}) + J_P(\hat{\boldsymbol{y}}, \hat{\boldsymbol{u}}) \tag{7.38}$$

subject to

$$\hat{\boldsymbol{F}}(\hat{\boldsymbol{y}}, \hat{\boldsymbol{u}}) = 0$$

and

$$\boldsymbol{y}_0 = [\boldsymbol{x}(t_0), \boldsymbol{V}(t_0)]$$

$$\boldsymbol{u}_k^{\min} \le \boldsymbol{u}_k \le \boldsymbol{u}_k^{\max}, \quad k = 0, 1, \ldots, N-1$$

This result is possible on the basis of the static approach to non-linear dynamic optimization described above for power system applications. The basic idea is to conceive dynamic optimization as a minimization problem of some objective function constrained by time-varying inequality constraints and algebraic equations obtained from the discretization of DAEs representing the power system. The methodology permits application for both corrective and preventive control of power system dynamic security.

The problem given by (7.38), being a mere minimization in the presence of sole equality constraints, can be solved easily through the use of Lagrangian multipliers. The auxiliary variables to be introduced are given by the number of equality constraints to be respected. The problem is solved minimizing the unconstrained Lagrangian function:

$$L(\hat{\boldsymbol{y}},\hat{\boldsymbol{u}},\hat{\boldsymbol{\lambda}}) = J_1(\hat{\boldsymbol{u}}) + J_2(\hat{\boldsymbol{y}}) + J_P(\hat{\boldsymbol{y}},\hat{\boldsymbol{u}}) - \hat{\boldsymbol{\lambda}}^{\mathrm{T}}\hat{\boldsymbol{F}}(\hat{\boldsymbol{y}},\hat{\boldsymbol{u}}) \tag{7.39}$$

and considering the first-order conditions given by

$$\frac{\partial L}{\partial \hat{\boldsymbol{\lambda}}} = \hat{\boldsymbol{F}}(\hat{\boldsymbol{y}},\hat{\boldsymbol{u}}) = 0 \tag{7.40}$$

$$\frac{\partial L}{\partial \hat{\boldsymbol{y}}} = \frac{\partial J_2}{\partial \hat{\boldsymbol{y}}} + \frac{\partial J_P}{\partial \hat{\boldsymbol{y}}} - \hat{\boldsymbol{\lambda}}^{\mathrm{T}}\frac{\partial \hat{\boldsymbol{F}}}{\partial \hat{\boldsymbol{y}}} = 0 \tag{7.41}$$

$$\frac{\partial L}{\partial \hat{\boldsymbol{u}}} = \frac{\partial J_1}{\partial \hat{\boldsymbol{u}}} + \frac{\partial J_P}{\partial \hat{\boldsymbol{u}}} - \hat{\boldsymbol{\lambda}}^{\mathrm{T}}\frac{\partial \hat{\boldsymbol{F}}}{\partial \hat{\boldsymbol{u}}} = 0 \tag{7.42}$$

Equations (7.40)–(7.42) can be solved with a gradient method, such as in the procedure explained in the previous section, and according to the algorithm shown in Figure 7.1. The choice of a gradient method for solving the non-linear programming problem is not obligatory since, as a matter of fact, different algorithms can be used such as the interior point method or Newton-like methods. However, employing this method permits to simplify calculations and avoid the heavy computational burden necessary to evaluate Hessian matrices (for Newton-like methods). Since computational time constraints might be relevant for the feasibility of the proposed approach

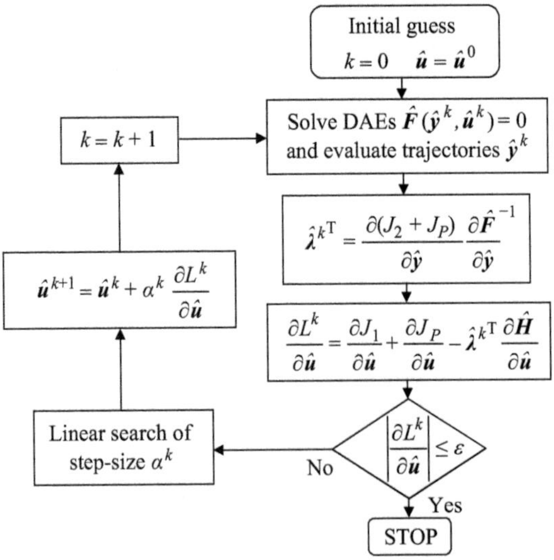

Figure 7.1 Solving algorithm for the solution of a time-discrete static optimization problem applied to power system DAEs

in an online DSA framework, decreasing significantly the algorithmic complexity of the problem is always a good choice provided that the overall convergence behaviour is not considerably worsened.

The methodology presented so far is general and can be adapted to the solution of any problem. Clearly, the actual formulation of the optimization problem and the programming of the solving algorithm depends on the specific problem that is under study. In the following paragraphs, a (non-exhaustive) description of possible formulations is given.

7.3.1 Control variables

One of the main choices that has to be made regards the formulation of control variable. Each control variable can have different dependence with respect to states, static equations, dynamic equations and objective functions. Clearly, the set of control variables to be adopted depends on the actuators that will be employed for implementing the corrective control actions. Several kind of control actions (both for preventive and corrective control) can be considered for preserving system stability.

A common approach to ensure dynamic security is based on shedding loads through special protection schemes that are activated after the onset of certain specific pre-postulated severe contingencies. This type of control action is very robust and reliable and, together with the automatic load shedding schemes that are operated when frequency drops under a minimum frequency threshold, is part of the defence plans of most power systems. However, the enforcement of such control actions comes with a price, namely the interruption of energy supply to entire portions of the network. Even the availability of long-term demand response (DR) contracts with large industrial customers, that are willing to be interrupted without notice, does not solve the problem since it is not known beforehand if these control resources will be located where load shedding is actually needed.

For such reason other corrective control actions and actuators can be studied and proposed. In general, "cost-free control actions" can always be preferred to load shedding schemes, provided that they ensure the same level of robustness and efficacy. The proposed methodology has been tested on several different possible set of cost-free control variables, for example, generated reactive power, phase-shifters and under-load tap changers or FACTS controllers [9]. However, recent studies have been oriented at exploring the possibility to conceive real-time protection schemes where the set of control variables is given by series impedance parameters of transmission lines [10]. It is imagined that the grid is a flexible net of links that can be stretched or shrunk depending on transient conditions in order to control power flows in the network and between areas. For illustration purposes, this chapter presents some test results on the application of the proposed approach to the control of series impedances by means of series-connected FACTS devices (thyristor-controlled series capacitor – TCSC).

7.3.2 Control effort minimization

The function J_1 in (7.36) is usually adopted for minimizing the control effort. This function can take into account the cost of implementing control actions (for example, the cost of load interruption [6]) and it is usually represented as a quadratic function depending on the difference between the uncontrolled initial value $\boldsymbol{u}^0$ and the value assumed during iterations. When "cost-free control actions" are adopted the formulation of this function can be skipped (there are no costs to be minimized). In any case, the introduction of such function is suggested since it improves algorithm's convergence behaviour. A typical discretized formulation for such function when time varying control variables $\boldsymbol{u}(t)$ are adopted is:

$$J_1 = \alpha_1(i) \cdot \sum_{k=0}^{N-1} \sum_{j=1}^{m} \frac{(u_{j,k} - u_j^0)^2}{(u_j^0)^2} \tag{7.43}$$

where $u_{j,k}$ is the value assumed by the jth control variable at the kth time step, u_j^0 is the value of the jth control variable in the uncontrolled case. The weighting factor $\alpha_1(i)$, just like all the other weighting factors that will be introduced in the followings, may vary with the number of iteration i depending on the strategy chosen for convergence. However, in most cases, it can be assumed as constant.

7.3.3 Kinetic energy cost function

Function J_2 can assume different formulation. J_2 is usually a function of system states that contains information about the stability of the system. Several formulations have been tried, for example, involving a dot product associated to the potential energy boundary surface (PEBS) [4] or the signal energy index [5]. However, in our experience the simplest, and still most efficient, function to be minimized is a function that measures the kinetic energy of the electrical masses with respect to the centre of inertia (COI) of the system.

This function can be formulated as:

$$J_2 = \alpha_2(i) \cdot \sum_{k=1}^{N} \sum_{j=1}^{n_g} M_j \cdot \left(\omega_{j,k} - \omega_{COI,k}\right)^2 \tag{7.44}$$

where $\omega_{j,k}$ is the angular velocity of the jth machine at the kth time step, $\omega_{COI,k}$ is the angular velocity of the COI at the kth time step, and M_j is the inertia coefficient of the jth machine; $\alpha_2(i)$ is a coefficient that takes into account the relative weight of this objective function with respect to other functions.

This function is always very sensitive in the cases of angle instability, since the separating generators tend to acquire kinetic energy with respect to the rest of the synchronous masses. Clearly, its minimization does not ensure that stability has been reached, but usually this condition can be verified with other classical tests included in the transient stability code (angle and voltage deviations, threshold on specific variables, damping assessment, hybrid methods, direct method-based tests and so on). An advantage of this formulation is that, being function of a state variable (angular velocity), the calculation of derivative $\partial J_2 / \partial \hat{\boldsymbol{y}}$ is straightforward. Given its limited ability in treating voltage instability transient behaviours, this function should always be accompanied by another cost function or penalty function aimed at constraining transient voltage trajectories.

7.3.4 Voltage penalty functions

One of the most efficient ways to obtain optimal solutions where both stability and technical feasibility are achieved is to control voltage magnitude along the transient at every bus. Poor voltage behaviour is always experienced in most unstable cases and in both cases of angle and voltage stability. Having a transient behaviour where all voltage trajectories stay contained within a confidence interval allows to improve the stability and, at the same time, to avoid the improper triggering of several protection schemes.

For example, distance relays, being sensitive to the ratio V/I, will always measure low impedances if voltage magnitude is too low, and improper activations of third or even second zone might be experienced. Keeping voltages always above a certain level also allows to ensure that undervoltage relays at substation level will not be activated. This condition might be risky, for example, during faults that include the loss of generation if a wide number of small generating units are installed at medium and low voltage level since the improper activation of undervoltage relays at substations might disconnect even more power right when it is really needed.

The function proposed is aimed at constraining all voltages within time-varying limits [5]. Assumed that the voltage magnitude at each a generic jth bus and kth time step must satisfy the condition

$$V_{j,k}^{\min} \leq V_{j,k} \leq V_{j,k}^{\max} \tag{7.45}$$

with $V_{j,k}^{\min}$ and $V_{j,k}^{\max}$ being time-varying thresholds fixed to ensure a desirable voltage transient and acceptable steady-state voltage levels, the following penalty function can be defined:

$$J_{PV}(\hat{y}) = \alpha_{PV}(i) \cdot \sum_{k=1}^{N} \sum_{j=1}^{n_b} \left(\frac{V_{j,k} - V_{\lim}}{\sigma_{j,k}} \right)^2 \tag{7.46}$$

where

$$V_{\lim} = \begin{cases} V_{j,k}^{\max} & \text{if } V_{j,k} \geq V_{j,k}^{\max} \\ V_{j,k}^{\min} & \text{if } V_{j,k} \leq V_{j,k}^{\min} \\ V_{j,k} & \text{otherwise} \end{cases} \tag{7.47}$$

The weighting factor $\sigma_{j,k}$ can be adopted in order to avoid dangerous low-voltage conditions such as the ones associated to transient voltage instability. A possible formulation is:

$$\sigma_{j,k} = \begin{cases} V_{j,k} & \text{if } V_{j,k} \leq V_{j,k}^{\min} \\ 1 & \text{otherwise} \end{cases} \tag{7.48}$$

One of the main advantages of adopting this formulation is that it is based on practical and perceptible indices that are also well known to the operators which can bring their own experience with the system and its operating devices and can set voltage constraints with confidence. A generally acceptable setting is to constrain voltages in the interval [0.8, 1.2] pu during the transient activity and in the range [0.9, 1.1] pu in steady-state conditions. Since steady-state condition cannot be considered reached within the time interval used for simulating the transient, steady-state conditions have to be ensured, conservatively, at the end of the simulation interval (i.e., at the latest time steps).

7.3.5 Distance relays penalty function

The proposed time-discrete formulation permits to include further inequality constraints that are aimed at ensuring that system trajectories will not cause the improper activation of automatic protections. For example, it is possible to force system trajectories to stay out of the tripping areas of distance protections [11]. This security requirement might be needed whenever the power system is characterized by line protection systems where distance relays are not equipped with power swing blocking relays. However, even when blocking relays are installed, keeping trajectories out of the distance relays tripping areas can be suggested in order to avoid the possibility of improper operation of power swing blocking relays or particularly unlucky events where trajectories linger too long in the third zone.

A typical quadrilateral-tripping characteristic of distance relays is illustrated in Figure 7.2. This model is characterized by separate settings for resistance and reactance directions, ensuring reliable tripping even in case of faults with arcing resistance.

For example, in the Italian high-voltage network, a common criterion is to use the first zone for detecting faults over 80% of the line next to the relay, with no time delay. The second zone protects the entire line and slightly beyond its end (120%) with a short time delay (300–500 ms) allowing the selective tripping of relays protecting

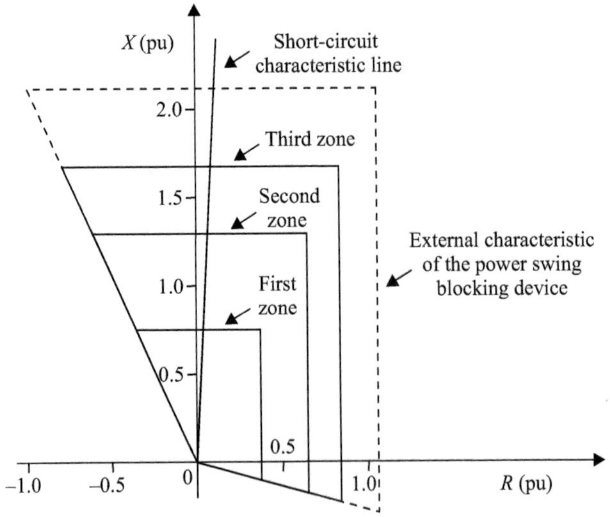

Figure 7.2 Typical quadrilateral distance relay characteristic

the neighbouring lines. The third zone that looks for faults well beyond the length of the line is set in order to provide a remote relay (or breaker) backup.

The presence of the third zone may become critical under typical emergency conditions and overloaded scenarios when large swings occur during post-fault transients. An improper activation of the third zone can cause a weakening of the network in cases where security is already comprised, leading to the triggering of cascading events and blackouts [12].

With the introduction of the following penalty function, power system transient trajectory can be forced out of the tripping areas of the system, thus preventing unwanted operation of distance relays.

A possible formulation is the following:

$$J_{PZ}(\hat{\boldsymbol{y}}, \hat{\boldsymbol{u}}) = \sum_{a=1}^{3} \alpha_{PZ}^{a}(i) \cdot J_{PZ}^{a}(\hat{\boldsymbol{y}}, \hat{\boldsymbol{u}}) \tag{7.49}$$

with

$$J_{PZ}^{a} = \sum_{k=1}^{N} \sum_{j=1}^{b} \alpha_{Zj,k} \left[\left(X_{j,k} - \overline{X}_{a,j} \right)^2 + \left(R_{j,k} - \overline{R}_{a,j} \right)^2 \right] \tag{7.50}$$

$$\alpha_{Zj,k} = \begin{cases} 1 & \text{if} \begin{cases} \left(R_{j,k}, X_{j,k} \right) \in \Omega_{a,j} \\ n_{a,j,k} > \overline{n}_{a,j} \end{cases} \\ 0 & \text{otherwise} \end{cases} \tag{7.51}$$

where

- b is the number of transmission lines equipped with distance relays;
- $a = 1, 2, 3$ is an index that represents, respectively, first, second and third zone;
- α_{PZ}^{a} is a weighting factor;
- $\overline{R}_{a,j}$ and $\overline{X}_{a,j}$ are the resistance and reactance thresholds for the jth line and the ath zone, expressed in per unit with respect to the line impedance;
- $(R_{j,k}, X_{j,k})$ is the generic point in the $R - X$ plane, assumed by the jth line at the kth time step;
- $\Omega_{a,j}$ is the tripping area associated to the ath zone of the distance relay installed on the jth line;
- $n_{a,j,k}$ is the number of consecutive steps during which it has been $(R_{j,k'}, X_{j,k'}) \in \Omega_{a,j}$ with $k' < k$;
- $\overline{n}_{a,j}$ is the relay time threshold for the ath zone, expressed in time steps.

The effect of the blocking scheme can be taken into account through a timer which switches J_{PZ}^{a} off. However, ignoring the power swing blocking device will always guarantee that conservative solutions are obtained. In the experienced accumulated so far, keeping all trajectories out of the third zone at any instant (meaning that $\overline{n}_{3,j}$ is set to zero) will always work towards the reaching of conservative and robust solutions.

An important remark is that the realistic simulation of distance relays tripping is possible thanks to the structure of the time-discrete static optimization problem that treat discretized dynamic trajectories altogether. Other approaches to transient stability analysis based, for example, on direct methods [13] will not be able to provide this kind of control because DAEs are not solved and $R-X$ trajectories are not known.

7.4 Application of the proposed methodology for the corrective control of a realistically sized power system (test results)

This paragraph contains an explicative example that will show how the proposed methodology is able to provide an optimal control rule for correcting the unstable behaviour of a realistic-sized power system.

The network adopted for tests is a detailed model of the Italian national grid and of a part of the interconnected ENTSO-E network. The overall model is characterized by 1 333 buses, 1 762 lines, 769 transformers and 294 generators and reproduces the state of the Italian power system (plus the ENTSO-E network) during the morning peak of a working day.

It has been supposed that two TCSCs have been installed on two 380 kV transmission lines in the Central Italy, namely Montalto-Villanova and Villanova-Larino (see Figure 7.3), and that these two devices are integrated within the DSA and SCADA/EMS functions devoted to the control of static and dynamic security of the system. The two devices are therefore treated as fast actuators of corrective control actions that can be triggered following the occurring of certain fault events that were selected through a contingency screening phase. The corrective control actions can be calculated, for each selected contingency, through the proposed methodology. The list of corrective actions, each one related to a specific contingency, is communicated to the TCSCs, arming such devices to apply remedial actions if any of the selected

Figure 7.3 Portion of the Italian 380 kV transmission grid (Central Italy)

contingencies should ever occur. A more detailed description of this architecture is given in Reference 14.

On the basis of such premises, the control variable vector is given by the degree of series compensation [15] the two transmission lines where the TCSC are installed:

$$u_{j,k} = \frac{X_{TSCS,j,k}}{X_{LINE,j}} \tag{7.52}$$

where $X_{TSCS,j,k}$ represents the equivalent series reactance provided at the kth time step by the jth TCSC and $X_{LINE,j}$ is the reactance value of the controlled line before compensation.

The formulation provided in the previous paragraphs is able to define an optimal time-varying control rule to be applied to TCSC. However, for many reasons such as the problems with synchronization of time-varying responses and the necessity of implementing robust and simple control actions, it is supposed that the control rule to be applied is a step-size variation in the degree of series compensation of each TCSC.

Considering that the control actions are applied at the time step t_{corr}, this condition can be represented in the overall problem adding the constraints

$$u_{j,k} = u_{j,t_{corr}} \quad \forall k > t_{corr} \tag{7.53}$$

$$u_{j,k} = u_j^0 \quad \forall k < t_{corr} \tag{7.54}$$

In practice there is no need to consider explicitly the two constraints (7.53) and (7.54) since these conditions can be treated easily reducing the dimension of the control variable vector and changing few operations in the calculation of $\partial\hat{\boldsymbol{F}}/\partial\hat{\boldsymbol{u}}$.

The optimization problem was tested on a test case characterized by a 20% uniformly distributed overload with respect to the base case. A three-phase fault applied to a line in Central Italy and cleared after 0.14 s produced an unstable behaviour typical of angle instability.

In Figure 7.4, the trajectory of all rotor angles is shown, each one calculated with respect to the reference machine (an equivalent generator in France). Figure 7.5 shows the (theoretical) trajectory of voltage magnitude assumed by each bus with respect to its initial value. It can be seen how power swings determine voltages close to zero in several nodes of the network. Clearly, this trajectory is purely theoretical since it does not take into account the triggering of protections during the transient.

Assuming that this fault is in the set of the selected contingency to be monitored, a corrective control rule, to be applied within 300 ms from the onset of the fault, was calculated through the proposed methodology. The solving algorithm was run enabling all the objective and penalty functions that have been introduced so far.

In Table 7.1, it is shown how the algorithm converges towards a solution that minimizes all objective and penalty functions. It can be observed how, if the evaluated corrective control action is applied during transient, no machines are accelerating with respect to the reference one (Figure 7.6) and voltages have an acceptable behaviour (Figure 7.7), even if few violations are still experienced in the first seconds of the transient.

Voltage-related inequality constraints have been considered soft constraints. Such constraints can be hardened increasing the weight of α_{PV} in (7.46). However, in this

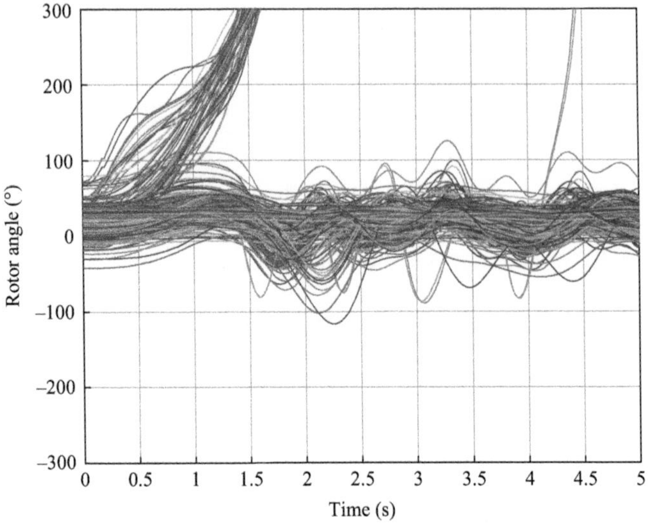

Figure 7.4 Rotor angle trajectories (unstable uncontrolled case)

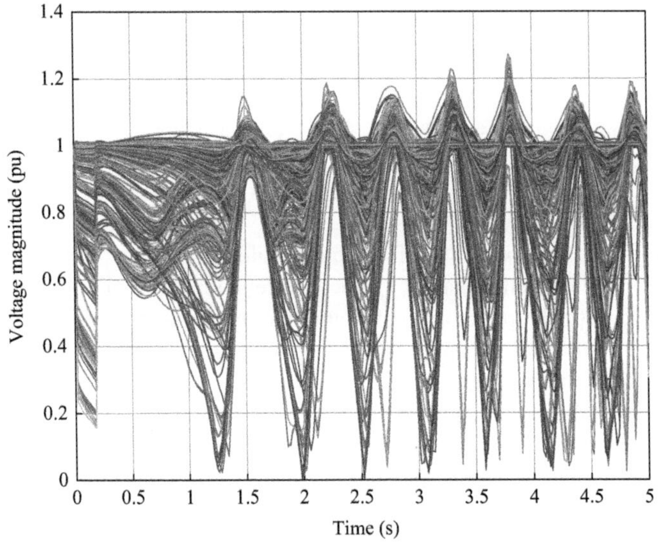

Figure 7.5 Voltage magnitude trajectories (unstable uncontrolled case)

specific case increasing such value does not yield different results, showing that more control resources are needed (i.e., more TCSCs) in order to obtain a null value of J_{PV} at final iteration.

Figure 7.8 shows in the plane $R - X$, the trajectories of impedances seen by distance protections. Assuming that all protections are set like the model proposed in

Table 7.1 Convergence behaviour of the corrective control algorithm

Iteration *i*	J_1	J_2	J_{PV}	J_{PZ}^1	J_{PZ}^2	J_{PZ}^3
0	0.000	210.85	579 062.00	41.12	548.15	2 366.34
1	0.008	21.31	27.54	0.00	0.00	0.00
2	0.008	18.35	19.62	0.00	0.00	0.00
3	0.008	16.81	16.48	0.00	0.00	0.00
4	0.008	15.92	15.35	0.00	0.00	0.00
5	0.009	15.48	15.04	0.00	0.00	0.00
6	0.009	15.31	14.98	0.00	0.00	0.00
7	0.009	15.25	14.96	0.00	0.00	0.00
8	0.010	15.22	14.94	0.00	0.00	0.00
9	0.010	15.20	14.94	0.00	0.00	0.00

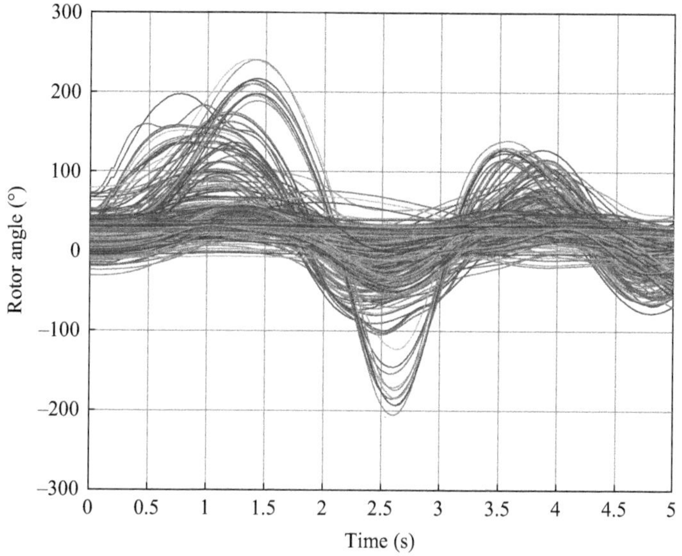

Figure 7.6 Rotor angle trajectories (stable case with corrective control)

Figure 7.2, it can be noted how such trajectories approach the tripping area but never cross.

The convergence was reached in about 100 s running the algorithm on a common desktop PC (Intel Core i7-4770, 3.40 GHz, 8 GB RAM, 64 bit). Such timing shows how this approach is definitely compatible with real-time requirements of online DSA functions. It should also be remarked that suboptimal stable solutions, good enough to preserve system stability, were already available at the first iterations and required just few tens of seconds to be computed.

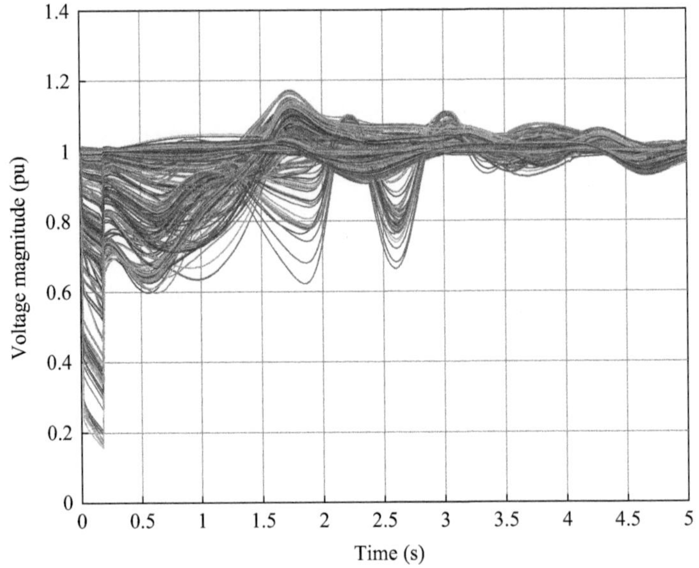

Figure 7.7 Voltage magnitude trajectories (stable case with corrective control)

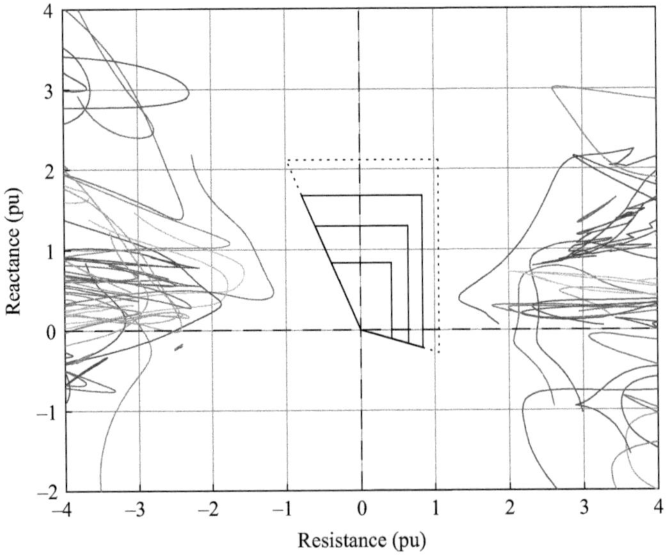

Figure 7.8 R–X trajectories at distance protections (stable case with corrective control)

7.5 Application to preventive control problems

The mathematical formulation previously described can be easily adapted to the solution of a slightly different optimization problem. If corrective control actions are supposed to be applied only "after" the occurrence of a contingency, preventive control actions can be used to change "in advance" system operating point so that transient stability is ensured if a selected contingency should ever occur.

The formulation so far provided is nearly equivalent to the one required for preventive control, but important differences should be outlined.

In the previous formulation, the time (step) 0 refers to the system's perturbed state before any corrective action could be applied. In the preventive control formulation, the time (step) 0 refers to the instant when the contingency occurs. In the discretized representation, two steps (a 0^- and a 0^+) must be considered, and different sets of DAEs must be formulated for the two instants. This condition does not bring changes in the system representation since (7.31) are already modelled as functions of time. Please note that any discrete event that causes a sudden modification of system parameters will always introduce a discontinuity and a replication of steps (i.e., a step$^-$ and a step$^+$ are needed).

Under such assumptions, the set of discretized equality constraints is now

$$\hat{\boldsymbol{F}} = [\boldsymbol{F}_{0+}^{\mathrm{T}}, \boldsymbol{F}_1^{\mathrm{T}}, \boldsymbol{F}_2^{\mathrm{T}}, \ldots, \boldsymbol{F}_N^{\mathrm{T}}]^{\mathrm{T}} \tag{7.55}$$

with state trajectories that will include all steps from 0^+ to N

$$\hat{\boldsymbol{y}} = [\boldsymbol{y}_{0+}^{\mathrm{T}}, \boldsymbol{y}_1^{\mathrm{T}}, \boldsymbol{y}_2^{\mathrm{T}}, \ldots, \boldsymbol{y}_N^{\mathrm{T}}]^{\mathrm{T}} \tag{7.56}$$

The control variable vector does not depend on time and can be represented as a $\boldsymbol{u}_{0^-}$.

A further equality constraint must be added for taking into account initial pre-fault steady-state conditions (at time 0^-). Analogously at what made before, all these equations can be represented as

$$\boldsymbol{F}_{0^-}(\boldsymbol{y}_{0^-}, \boldsymbol{u}_{0^-}) = 0 \tag{7.57}$$

where $\boldsymbol{y}_{0^-} = \left[\boldsymbol{x}_{0^-}^{\mathrm{T}}\ \boldsymbol{V}_{0^-}^{\mathrm{T}}\right]^{\mathrm{T}}$

Since equality constraints (7.57) have been added to the formulation, the function to be minimized is now

$$\begin{aligned} L(\hat{y}, y_{0^-}, u_{0^-}, \hat{\boldsymbol{\lambda}}, \boldsymbol{\mu}) &= J_1(u_{0^-}) + J_2(\hat{y}) + J_P(\hat{y}, u_{0^-}) - \hat{\boldsymbol{\lambda}}^{\mathrm{T}} \hat{F}(\hat{y}, y_{0^-}, \boldsymbol{u}_{0^-}) \\ &\quad - \boldsymbol{\mu}^{\mathrm{T}} F_{0^-}(\boldsymbol{y}_{0^-}, \boldsymbol{u}_{0^-}) \end{aligned} \tag{7.58}$$

and the first-order conditions are

$$\frac{\partial L}{\partial \hat{\boldsymbol{\lambda}}} = \hat{\boldsymbol{F}}(\hat{\boldsymbol{y}}, \boldsymbol{y}_{0^-}, \boldsymbol{u}_{0^-}) = 0 \tag{7.59}$$

$$\frac{\partial L}{\partial \boldsymbol{\mu}} = F_{0^-}(\boldsymbol{y}_{0^-}, \boldsymbol{u}_{0^-}) = 0 \tag{7.60}$$

$$\frac{\partial L}{\partial \hat{\boldsymbol{y}}} = \frac{\partial J_2}{\partial \hat{\boldsymbol{y}}} + \frac{\partial J_P}{\partial \hat{\boldsymbol{y}}} - \hat{\boldsymbol{\lambda}}^{\mathrm{T}} \frac{\partial \hat{\boldsymbol{F}}}{\partial \hat{\boldsymbol{y}}} = 0 \tag{7.61}$$

$$\frac{\partial L}{\partial \boldsymbol{y}_{0^-}} = \hat{\boldsymbol{\lambda}}^{\mathrm{T}} \frac{\partial L}{\partial \boldsymbol{y}_{0^-}} - \boldsymbol{\mu}^{\mathrm{T}} \frac{\partial L}{\partial \boldsymbol{y}_{0^-}} = 0 \tag{7.62}$$

$$\frac{\partial L}{\partial \boldsymbol{u}_{0^-}} = \frac{\partial J_1}{\partial \boldsymbol{u}_{0^-}} + \frac{\partial J_P}{\partial \boldsymbol{u}_{0^-}} - \hat{\boldsymbol{\lambda}}^{\mathrm{T}} \frac{\partial \hat{\boldsymbol{F}}}{\partial \boldsymbol{u}_{0^-}} - \boldsymbol{\mu}^{\mathrm{T}} \frac{\partial F_{0^-}}{\partial \boldsymbol{u}_{0^-}} = 0 \tag{7.63}$$

The solution of the system of equations (7.59)–(7.63) can be obtained iteratively like in the previous approach. With respect to corrective control algorithm only a single step is added since it is necessary to first calculate $\hat{\boldsymbol{\lambda}}$ through (7.61). Having substituted $\hat{\boldsymbol{\lambda}}$ in (7.62), $\boldsymbol{\mu}$ can be evaluated allowing to eventually calculate $\partial L/\partial \boldsymbol{u}_{0^-}$ through (7.63).

The procedure described in this paragraph has been tested on the same test case used for corrective control. Results are similar to the one obtained before as shown by the transient behaviour shown in Figures 7.9–7.11. In Table 7.2, the convergence

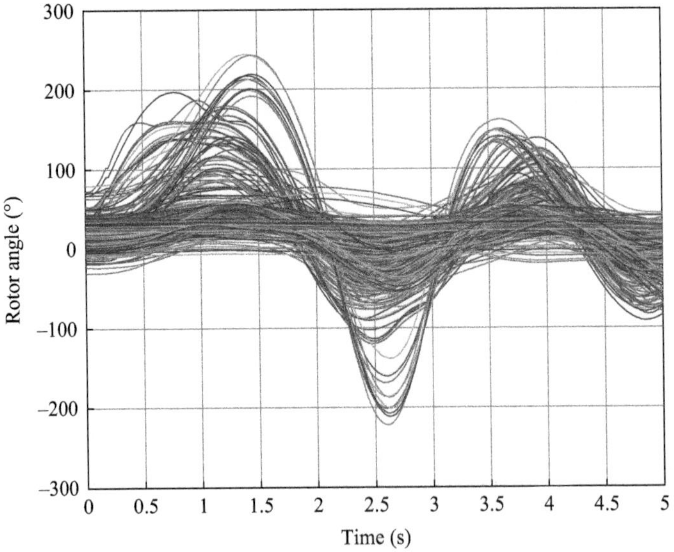

Figure 7.9 Rotor angle trajectories (stable case with preventive control)

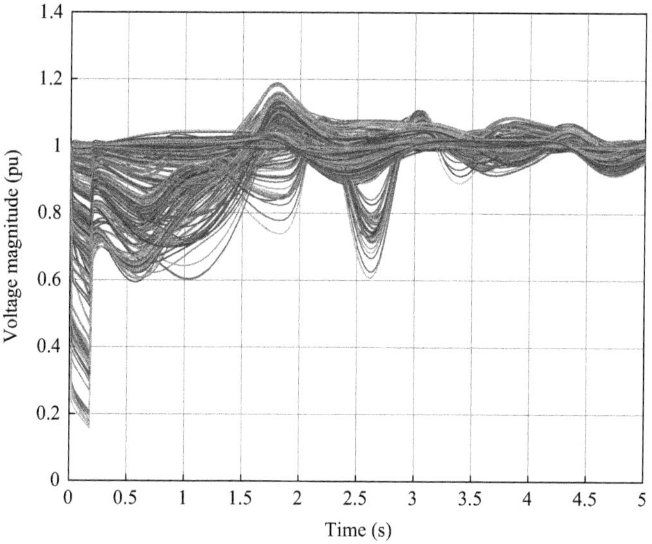

Figure 7.10 Voltage magnitude trajectories (stable case with preventive control)

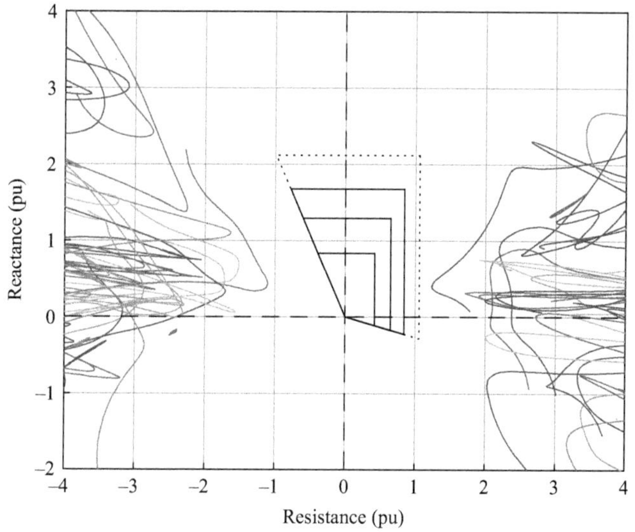

Figure 7.11 $R - X$ trajectories at distance protections (stable case with preventive control)

behaviour of the algorithm for preventive control is shown. Please note that all cost functions, even in the corrective control case, are calculated starting at the time 0^+. This means that the two solutions can be comparable and that the initial overall cost at iteration 0 is the same. Clearly, in the corrective control case, only the sensitivities associated to the piece of trajectory that can be controlled are used.

Table 7.2 Convergence behaviour of the preventive control algorithm

Iteration i	J_1	J_2	J_{PV}	J_{PZ}^1	J_{PZ}^2	J_{PZ}^3
0	0.000	210.85	579 062.00	41.12	548.15	2 366.34
1	0.007	20.45	24.12	0.00	0.00	0.00
2	0.007	19.12	19.34	0.00	0.00	0.00
3	0.008	17.31	16.68	0.00	0.00	0.00
4	0.008	15.32	15.49	0.00	0.00	0.00
5	0.009	15.01	14.73	0.00	0.00	0.00
6	0.009	14.36	14.61	0.00	0.00	0.00
7	0.009	14.33	14.58	0.00	0.00	0.00

It should be remarked that preventive control can often provide results more conservative with regard to system security, since it allows to control the whole trajectory starting from the onset of the contingency. However, in the presence of a set of multiple possible unstable contingencies, a corrective control strategy can provide a specific control action for each contingency, whereas preventive control must be able to find a new operating point where the system is able to withstand any possible contingency (possibly even some "stable" contingencies that having moved the operating point might have become "unstable"). Clearly, this condition can never be granted, but a preventive control methodology, derived from this same formulation, treating concurrently a set of possible contingencies can be developed as proposed in Reference 10.

Bibliography

[1] D. E. Kirk, *Optimal Control Theory: An Introduction*, Prentice-Hall, Englewood Cliffs, NJ, 1970.

[2] M. G. Singh, A. Titli, *Systems Decomposition, Optimization and Control*, Pergamon Press, Oxford, 1978.

[3] F. Iavernaro, M. La Scala, F. Mazzia, "Boundary values methods for time domain simulation of power system dynamic behaviour", *IEEE Transaction on Circuits & Systems Part I – Fundamental Theory & Applications*, vol. 45, no. 1, pp. 50–63, Jan. 1998.

[4] M. La Scala, M. Trovato, C. Antonelli, "On-line dynamic preventive control: an algorithm for transient security dispatch", *IEEE Transaction on Power Systems*, vol. 13, no. 2, pp. 601–610, May 1998.

[5] E. De Tuglie, M. La Scala, P. Scarpellini, "Real-time preventive actions for the enhancement of voltage-degraded trajectories", *IEEE Transaction on Power Systems*, vol. 14, no. 2, pp. 561–568, May 1999.

[6] S. Bruno, M. La Scala, P. Scarpellini, G. Vimercati, "A dynamic approach for transmission management through a contract curtailment strategy", *Proceedings of the 14th Power Systems Computation Conference (PSCC)*, Sevilla, Spain, June 24–28, 2002.

[7] W. Li, A. Bose, E. De Tuglie, M. La Scala, "On-line contingency screening and remedial action for dynamic security analysis", *International Conference on Large High Voltage Electric Systems – CIGRE*, Paris, France, August 29–September 6, 1998.

[8] G. Andersson, P. Donalek, R. Farmer, *et al.*, "Causes of the 2003 major grid blackouts in North America and Europe, and recommended means to improve system dynamic performance", *IEEE Transaction on Power Systems*, vol. 20, no. 4, pp. 1922–1928, Nov. 2005.

[9] S. Bruno, M. La Scala, "Unified power flow controllers for security constrained transmission management", *IEEE Transactions on Power Systems*, vol. 19, no. 1, pp. 418–426, Feb. 2004.

[10] S. Bruno, E. De Tuglie, M. La Scala, "Transient security dispatch for the concurrent optimization of plural postulated contingencies", *IEEE Transactions on Power Systems*, vol. 17, no. 3, pp. 707–714, Aug. 2002.

[11] S. Bruno, M. De Benedictis, M. Delfanti, M. La Scala, "Preventing blackouts through reactive rescheduling under dynamical and protection system constraints", *Proceedings of PowerTech*, St. Petersburg, Russia, June 27–30, 2005.

[12] S. H. Horowitz, A. G. Phadke, "Third zone revisited", *IEEE Transactions on Power Delivery*, vol. 21, no. 1, pp. 23–29, Jan. 2006.

[13] P. Kundur, *Power System Stability and Control*, McGraw-Hill, New York, NYY, 1994.

[14] S. Bruno, G. De Carne, M. La Scala, "Transmission grid control through TCSC dynamic series compensation", *IEEE Transactions on Power System*, vol. 31, no. 31, pp. 3202–3211, Jul. 2016.

[15] N.G. Hingorani, L. Gyugyi, *Understanding FACTS: Concepts and Technology of Flexible AC Transmission Systems*, IEEE Press, New York, NY, 2000.

Chapter 8

Static and recursive PMU-based state estimation processes for transmission and distribution power grids

Mario Paolone[1], Jean-Yves Le Boudec[2], Styliani Sarri[1] and Lorenzo Zanni[1]

In the operation of power systems, the knowledge of the system state is required by several fundamental functions, such as security assessment, voltage control and stability analysis. By making reference to the *static* state of the system represented by the voltage phasors at all the network buses, it is possible to infer the system operating conditions. Until the late 1970s, conventional load flow calculations provided the system state by directly using the raw measurements of voltage magnitudes and power injections. The loss of one measurement made the calculation impossible and the presence of measurement errors affected dramatically the computed state. To overcome these limitations, load flow theory has been combined with statistical estimation constituting the so-called *state estimation* (SE). The latter consists in the solution of an optimization problem that processes the measurements together with the network model to determine the optimal estimate of the system state. The outputs of load flow and SE are composed of the same quantities, typically the voltage magnitude and phase at all the network buses, but SE uses all the types of measurements (e.g., voltage and current magnitudes, nodal power injections and flows, synchrophasors) and evaluates their consistency using the network model. The measurement redundancy is key to tolerate measurement losses, identify measurement and network parameter errors, and filter out the measurement noise. The foregoing properties of SE allow the system operator to obtain an accurate and reliable estimate of the system state that consequently improves the performance of the functions relying on it.

Traditionally, SE has been performed at a relatively low refresh rate of a few minutes, dictated by the time requirements of the related functions together with the low measurement acquisition rate of remote terminal units (RTUs).* Nowadays,

[1]School of Engineering, École Polytechnique Fédérale de Lausanne (EPFL), Switzerland
[2]School of Computer and Communication Sciences, École Polytechnique Fédérale de Lausanne (EPFL), Switzerland

*RTUs are devices installed at the network substations that regularly send measurements, usually unsynchronized, to the network operator control centre. Typically, these measurements are composed of voltage and current magnitudes as well as active and reactive powers.

the emerging availability of phasor measurement units (PMUs) allows to acquire accurate and time-aligned phasors, called *synchrophasors*, with typical streaming rates in the order of some tens of measurements per second [1,2]. This technology is experiencing a fast evolution, which is triggered by an increasing number of power system applications that can benefit from the use of synchrophasors. SE processes can exploit the availability of synchrophasor measurements to achieve better accuracy performance and higher refresh rate (sub-second).

PMUs already compose the backbone of wide area monitoring systems in the context of transmission networks to which several real-time functionalities are connected, such as inter-area oscillations, relaying, fault location and real-time SE [3,4]. However, PMUs might represent fundamental monitoring tools even in the context of distribution networks for applications such as: SE [5,6], loss of main [7], fault event monitoring [8], synchronous islanded operation [9] and power quality monitoring [10]. The recent literature has discussed the use of PMUs for SE in distribution networks both from the methodological point of view [11] and also via dedicated real-scale experimental setups [12,13].

Since the pioneering works of Schweppe on power system SE in 1970 [14–16], most of the research on the subject has investigated *static* SE methods based on weighted least squares (WLS) [17–19]. Static SE computes the system state performing a "best fit" of the measurements belonging only to the current time-step. Another category of state estimators are the *recursive* methods, such as the Kalman filter (KF). In addition to the use of the measurements and their statistical properties, they also predict the system state by modelling its time evolution. In general, recursive estimators are characterized by higher complexity and the prediction introduces an additional source of uncertainty that, if not properly quantified, might worsen the accuracy of the estimated state. Besides, their ability to filter out measurement noise could not be exploited due to the low SE refresh rate: even in quasi-steady state conditions, the measurement noise was smaller than the state variations between two consecutive time-steps. However, the effectiveness of power system SE based on KF has been recently reconsidered thanks to the possibility to largely increase the SE refresh rate by using synchrophasor measurements.

The chapter starts by providing the measurement and process model of WLS and KF SE algorithms and continues with the analytical formulation of the two families of state estimators, including their linear and non-linear versions as a function of the type of available measurements. Finally, two case studies targeting IEEE transmission and distribution reference networks are given.

8.1 State estimation measurement and process model

In general, any SE algorithm relies on a *measurement model*, which expresses how the system state variables are related to the measurements and the measurement noise. The link between the state variables and the measurements can be either linear or non-linear, depending on the kind of measurements used. For the case of power systems, the link is represented by the network model that is composed of the network topology

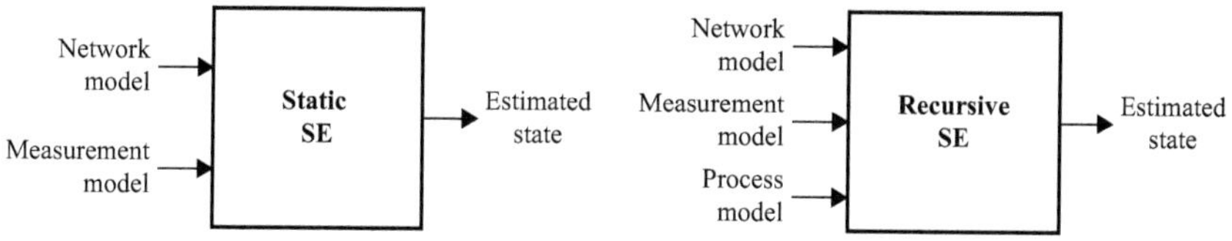

Figure 8.1 Inputs and outputs of static and recursive state estimators

and the electrical parameters of the various components, such as transmission lines and transformers. Additionally to the measurement model, recursive state estimators, such as KF, use a *process model* to represent the time evolution of the system state as a function of the previous system states, the controllable inputs and the process noise. Figure 8.1 shows the inputs and outputs of static and recursive state estimators.

8.1.1 Measurement model

The measurement model of the SE can be formulated in a common way regardless of the estimator type (static or recursive). In particular, if the measurements come only from PMUs, the measurement model is linear, whereas if the measurements come from RTUs or a hybrid set of RTUs and PMUs, the measurement model is non-linear. In what follows, examples of formulations of these measurement models are given.

8.1.1.1 Linear measurement model

In case the SE uses only measurements coming from PMUs composed of voltage and/or current phasors, the SE can be formulated in a linear way. The state variables are represented by the phase-to-ground nodal voltages, or the branch voltages, or nodal current injections or current flows (or even a mix of them, provided that they are independent state variables).

By defining the set of network buses $\mathcal{S}$, the number of network buses is equal to $s = |\mathcal{S}|$, where the operator $|\ |$ denotes the cardinality of a set. Then, the state of a three-phase (3-ph) network with s buses is denoted by $\mathbf{x} \in \mathbb{R}^n$ (where $n = 3 \cdot 2s$ is the number of state variables that compose the set of state variables $\mathcal{N}$) that, in most of the literature on the subject, is represented by the phase-to-ground nodal voltages. The set of three phases a, b, c is denoted by $\mathcal{P}$. Also, the set of network branches is denoted by $\mathcal{B}$. To obtain an exact linear measurement model, measurements and state variables are expressed in rectangular coordinates. Hence, the state is composed of the real and imaginary parts of the voltage phasors at every bus:

$$\mathbf{x} = [\mathbf{V}_{1,re}^{a,b,c}, \ldots, \mathbf{V}_{i,re}^{a,b,c}, \ldots, \mathbf{V}_{s,re}^{a,b,c}, \mathbf{V}_{1,im}^{a,b,c}, \ldots, \mathbf{V}_{i,im}^{a,b,c}, \ldots, \mathbf{V}_{s,im}^{a,b,c}]^T \tag{8.1}$$

where

$$\begin{aligned} \mathbf{V}_{i,re}^{a,b,c} &= [\mathbf{V}_{i,re}^{a}, \mathbf{V}_{i,re}^{b}, \mathbf{V}_{i,re}^{c}] \\ \mathbf{V}_{i,im}^{a,b,c} &= [\mathbf{V}_{i,im}^{a}, \mathbf{V}_{i,im}^{b}, \mathbf{V}_{i,im}^{c}] \end{aligned} \tag{8.2}$$

are, respectively, the real and imaginary parts of the voltage phasor at bus i in the three phases a, b and c. It is worth mentioning that, as explained in Reference 20, the existence of PMU measurements can eliminate the need to choose a reference bus, i.e., a bus where the phase or the imaginary part of the voltage is not included in the state and is assumed to have a certain value (usually equal to zero). Therefore, unlike the conventional SE formulation, in this case the phase or the imaginary part of the voltage is estimated at every bus.

To provide an example of linear measurement model, it is assumed that measurements come only from PMUs that measure nodal voltage phasors, current injection phasors and current flow phasors. The set of buses where PMUs measure nodal voltage phasors is $\mathcal{D}_1$ ($d_1 = |\mathcal{D}_1|$). Similarly, the set of buses where current injection phasors are measured is $\mathcal{D}_2$ ($d_2 = |\mathcal{D}_2|$) and the set of branches where current flow phasors are measured is $\mathcal{D}_3$ ($d_3 = |\mathcal{D}_3|$). Hence, the set of measurements $\mathcal{M}$ is composed of:

- $3d_1$ phase-to-ground voltage phasors;
- $3d_2$ current injection phasors;
- $3d_3$ current flow phasors;

and its cardinality is equal to $m = 2 \cdot (3d_1 + 3d_2 + 3d_3)$. Therefore, the measurement array $\mathbf{z} \in \mathbb{R}^m$ is equal to:

$$\mathbf{z} = [\mathbf{z}_V, \mathbf{z}_{I_{\text{inj}}}, \mathbf{z}_{I_{\text{flow}}}]^T \tag{8.3}$$

where

$$\begin{aligned}
\mathbf{z}_V &= [\mathbf{V}^{a,b,c}_{1,re}, \ldots, \mathbf{V}^{a,b,c}_{d_1,re}, \mathbf{V}^{a,b,c}_{1,im}, \ldots, \mathbf{V}^{a,b,c}_{d_1,im}] \\
\mathbf{z}_{I_{\text{inj}}} &= [\mathbf{I}^{a,b,c}_{\text{inj},1,re}, \ldots, \mathbf{I}^{a,b,c}_{\text{inj},d_2,re}, \mathbf{I}^{a,b,c}_{\text{inj},1,im}, \ldots, \mathbf{I}^{a,b,c}_{\text{inj},d_2,im}] \\
\mathbf{z}_{I_{\text{flow}}} &= [\mathbf{I}^{a,b,c}_{\text{flow},1,re}, \ldots, \mathbf{I}^{a,b,c}_{\text{flow},d_3,re}, \mathbf{I}^{a,b,c}_{\text{flow},1,im}, \ldots, \mathbf{I}^{a,b,c}_{\text{flow},d_3,im}].
\end{aligned} \tag{8.4}$$

The convention employed in this chapter for the currents injections, and also the most common, is to consider with positive sign the current flowing from the loads/generators to the respective bus.

The equation that relates the measurements with the system state variables is:

$$\mathbf{z} = \mathbf{H}\mathbf{x} + \boldsymbol{\varepsilon} \tag{8.5}$$

where $\mathbf{H}$ is an $m \times n$ matrix that represents the link between measurements and state variables and $\boldsymbol{\varepsilon}$ is the measurement noise. It is important to point out that, in the case of linear SE, $\mathbf{H}$ does not represent a linear approximation of the measurement model, since it corresponds to the exact link between measurements and state variables. The measurement noise is assumed to be white and Gaussian:

$$p(\boldsymbol{\varepsilon}) \sim \mathcal{N}(0, \mathbf{R}) \tag{8.6}$$

where $\mathbf{R}$ is the measurement noise covariance matrix. Note that the normality of PMU errors is based on experimental evidences of error distributions of actual PMUs (e.g., Reference 21). Therefore, the diagonal entries of $\mathbf{R}$ represent the variances

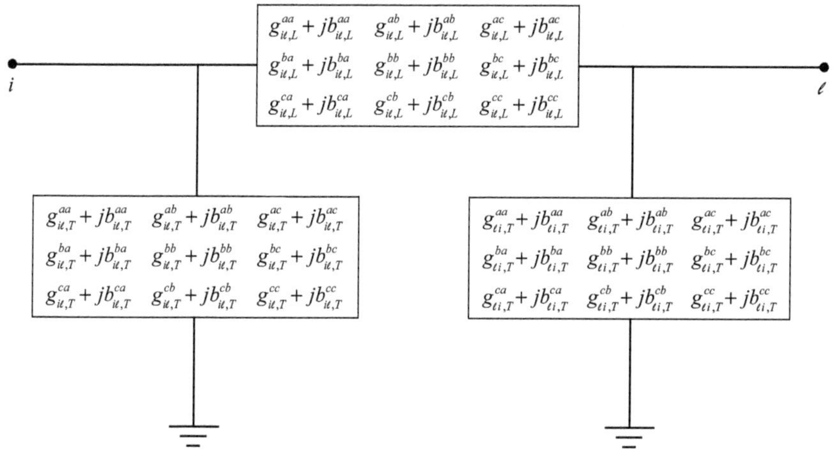

Figure 8.2 3-ph two-port π-model of a generic network branch

of the measurements, which correspond to the cumulative uncertainty of sensors[†] and meters. The off-diagonal entries account for eventual correlation between the measurements that occurs if mutual influence among meters is present.

For the case of power networks, matrix $\mathbf{H}$ of (8.5) is derived from the network topology and the electrical parameters of the various network components. To this end, the so-called *nodal analysis* linearly links the nodal current injections with the considered system state, i.e., the phase-to-ground nodal voltages. As it is described in what follows, such link consists in the so-called *compound network admittance matrix* $\mathbf{Y}$, whose derivation is given below.

A generic 3-ph network branch between buses i and ℓ can be represented by the two-port π-model shown in Figure 8.2. Its parameters are 3×3 matrices of complex numbers, i.e., the longitudinal admittance $\mathbf{y}_{i\ell,L}$ and the two transverse admittances $\mathbf{y}_{i\ell,T}$ and $\mathbf{y}_{\ell i,T}$:

$$\mathbf{y}_{i\ell,L} = \mathbf{g}_{i\ell,L} + j\mathbf{b}_{i\ell,L} \tag{8.7}$$

$$\mathbf{y}_{i\ell,T} = \mathbf{g}_{i\ell,T} + j\mathbf{b}_{i\ell,T} \tag{8.8}$$

$$\mathbf{y}_{\ell i,T} = \mathbf{g}_{\ell i,T} + j\mathbf{b}_{\ell i,T} \tag{8.9}$$

where

- j is the imaginary unit;
- $\mathbf{g}_{i\ell,L}$ and $\mathbf{b}_{i\ell,L}$ are, respectively, the longitudinal conductance and susceptance;

[†]In this chapter, the sensors refer to the transducers (such as voltage or current instrument transformers) that scale down the input voltage and current signals in order to interface the meters (represented, in this case, by PMUs and/or RTUs) with the electrical network.

- $\mathbf{g}_{i\ell,T}$ and $\mathbf{b}_{i\ell,T}$ are, respectively, the transverse conductance and susceptance from the side of bus i.
- $\mathbf{g}_{\ell i,T}$ and $\mathbf{b}_{\ell i,T}$ are, respectively, the transverse conductance and susceptance from the side of bus ℓ;

The 3×3 matrix representing the longitudinal admittance is given by (the same matrix form applies also to the transverse admittances as shown in Figure 8.2):

$$\mathbf{y}_{i\ell,L} = \begin{bmatrix} g_{i\ell,L}^{aa} + jb_{i\ell,L}^{aa} & g_{i\ell,L}^{ab} + jb_{i\ell,L}^{ab} & g_{i\ell,L}^{ac} + jb_{i\ell,L}^{ac} \\ g_{i\ell,L}^{ba} + jb_{i\ell,L}^{ba} & g_{i\ell,L}^{bb} + jb_{i\ell,L}^{bb} & g_{i\ell,L}^{bc} + jb_{i\ell,L}^{bc} \\ g_{i\ell,L}^{ca} + jb_{i\ell,L}^{ca} & g_{i\ell,L}^{cb} + jb_{i\ell,L}^{cb} & g_{i\ell,L}^{cc} + jb_{i\ell,L}^{cc} \end{bmatrix} \tag{8.10}$$

The expression of these parameters depends on the type of network branch. For instance, a 3-ph transmission line characterized by a longitudinal impedance $\mathbf{z}_{\text{line}}$ and a transverse admittance $\mathbf{y}_{\text{line}}$ can be represented by the following two-port π-model parameters:

$$\mathbf{y}_{i\ell,L} = \mathbf{z}_{\text{line}}^{-1} \tag{8.11}$$

$$\mathbf{y}_{i\ell,T} = \mathbf{y}_{\ell i,T} = \mathbf{y}_{\text{line},T}/2 \tag{8.12}$$

Note that $\mathbf{y}_{i\ell,T}$ can be different from $\mathbf{y}_{\ell i,T}$ for some network components, such as tap changing or phase-shifting transformers. For the particular case of phase-shifting transformers, the two-port network is not reciprocal.

A detailed description of the procedure to construct the network admittance matrix is given in Reference 17. Considering a 3-ph network of s buses, $\mathbf{Y}$ is a $3s \times 3s$ matrix with the following form:

$$\mathbf{Y} = \begin{bmatrix} \mathbf{Y}_{11} & \mathbf{Y}_{12} & \cdots & \mathbf{Y}_{1s} \\ \mathbf{Y}_{21} & \mathbf{Y}_{22} & \cdots & \mathbf{Y}_{2s} \\ \vdots & \vdots & \ddots & \vdots \\ \mathbf{Y}_{s1} & \mathbf{Y}_{s2} & \cdots & \mathbf{Y}_{ss} \end{bmatrix} \tag{8.13}$$

where

- the off-diagonal element $\mathbf{Y}_{i\ell}$ ($i \neq \ell$) is a 3×3 matrix equal to the opposite of the longitudinal admittance of the branch between buses i and ℓ:

$$\mathbf{Y}_{i\ell} = -\mathbf{y}_{i\ell,L} \tag{8.14}$$

  and $\mathbf{Y}_{i\ell} = \mathbf{Y}_{\ell i}$, so that $\mathbf{Y}$ is symmetrical;
- the diagonal element $\mathbf{Y}_{ii}$ is a 3×3 matrix equal to the sum of the longitudinal admittances of the branches connected to bus i and the transverse admittances of the branches connected between bus i and the neutral:

$$\mathbf{Y}_{ii} = \sum_{\ell=1}^{s} (\mathbf{y}_{i\ell,L} + \mathbf{y}_{i\ell,T}) \tag{8.15}$$

  Note that $\mathbf{y}_{i\ell,L}$ and $\mathbf{y}_{i\ell,T}$ are null matrices if bus i is not connected to bus ℓ.

The expression of the generic element $\mathbf{Y}_{i\ell}$ of the compound admittance matrix is:

$$\mathbf{Y}_{i\ell} = \begin{bmatrix} G_{i\ell}^{aa} + jB_{i\ell}^{aa} & G_{i\ell}^{ab} + jB_{i\ell}^{ab} & G_{i\ell}^{ac} + jB_{i\ell}^{ac} \\ G_{i\ell}^{ba} + jB_{i\ell}^{ba} & G_{i\ell}^{bb} + jB_{i\ell}^{bb} & G_{i\ell}^{bc} + jB_{i\ell}^{bc} \\ G_{i\ell}^{ca} + jB_{i\ell}^{ca} & G_{i\ell}^{cb} + jB_{i\ell}^{cb} & G_{i\ell}^{cc} + jB_{i\ell}^{cc} \end{bmatrix} \tag{8.16}$$

where $\mathbf{G}$ and $\mathbf{B}$ are the real and imaginary parts of the admittance matrix $\mathbf{Y}$, respectively.

The expressions of the real and imaginary parts of the 3-ph nodal current injection phasors at bus $i \in \mathcal{S}$ and phase $p \in \mathcal{P}$ are:

$$I_{i,re}^{p} = \sum_{\ell=1}^{s} \sum_{l \in \mathcal{P}} \left[G_{i\ell}^{pl} V_{\ell,re}^{l} - B_{i\ell}^{pl} V_{\ell,im}^{l} \right] \tag{8.17}$$

$$I_{i,im}^{p} = \sum_{\ell=1}^{s} \sum_{l \in \mathcal{P}} \left[G_{i\ell}^{pl} V_{\ell,im}^{l} + B_{i\ell}^{pl} V_{\ell,re}^{l} \right] \tag{8.18}$$

where the subscripts i and ℓ refer to the bus indices whereas the superscripts p and l refer to the phase indices.

The current flow phasor $I_{\text{flow},u}^{p}$ at branch $u \in \mathcal{B}$ and phase $p \in \mathcal{P}$ can be also indicated as $I_{i\ell}^{p}$ with respect to the two terminal buses i and ℓ of this branch. Note that $I_{i\ell}^{p}$ is measured from the side of bus i, while $I_{\ell i}^{p}$ is measured from the side of bus ℓ. The expressions of the real and imaginary parts of the 3-ph current flow phasors at the branch between buses i and ℓ and phase $p \in \mathcal{P}$ are:

$$I_{i\ell,re}^{p} = \sum_{l \in \mathcal{P}} [g_{i\ell,L}^{pl}(V_{i,re}^{l} - V_{\ell,re}^{l}) - b_{i\ell,L}^{pl}(V_{i,im}^{l} - V_{\ell,im}^{l}) + g_{i\ell,T}^{pl} V_{i,re}^{l} - b_{i\ell,T}^{pl} V_{i,im}^{l}] \tag{8.19}$$

$$I_{i\ell,im}^{p} = \sum_{l \in \mathcal{P}} [g_{i\ell,L}^{pl}(V_{i,im}^{l} - V_{\ell,im}^{l}) + b_{i\ell,L}^{pl}(V_{i,re}^{l} - V_{\ell,re}^{l}) + g_{i\ell,T}^{pl} V_{i,im}^{l} + b_{i\ell,T}^{pl} V_{i,re}^{l}] \tag{8.20}$$

The structure of the matrix $\mathbf{H}$ of (8.5) is:

$$\mathbf{H} = \begin{bmatrix} \mathbf{H}_V \\ \mathbf{H}_{I_{\text{inj}}} \\ \mathbf{H}_{I_{\text{flow}}} \end{bmatrix} \tag{8.21}$$

The part related to the nodal voltage phasor measurements $\mathbf{H}_V$ is:

$$\mathbf{H}_V = \begin{bmatrix} \boldsymbol{\beta} & \boldsymbol{\upsilon} \\ \boldsymbol{\zeta} & \boldsymbol{\eta} \end{bmatrix} \tag{8.22}$$

where

$$\beta_{\ell l,re}^{ip,re} = \begin{cases} 1, \text{ if } i = \ell \text{ and } p = l \\ 0, \text{ if } i \neq \ell \text{ or } p \neq l \end{cases} \tag{8.23}$$

$$\upsilon_{\ell l,im}^{ip,re} = 0 \tag{8.24}$$

$$\zeta_{\ell l,re}^{ip,im} = 0 \tag{8.25}$$

$$\eta_{\ell l,im}^{ip,im} = \begin{cases} 1, \text{ if } i = \ell \text{ and } p = l \\ 0, \text{ if } i \neq \ell \text{ or } p \neq l \end{cases} \tag{8.26}$$

In (8.23)–(8.26), the superscripts refer, respectively, to the bus, the phase and the real or imaginary part of the measurements, while the subscripts refer to the state variables. For instance, $\beta_{\ell l,re}^{ip,re}$ is the scalar that links the measurement $V_{i,re}^{p}$ with the state variable $V_{\ell,re}^{l}$ (in this specific case, the scalar is simply zero or one).

The part related to the current injection phasor measurements $\mathbf{H}_{I_{\text{inj}}}$ can be derived in a straightforward way from (8.17) and (8.18):

$$\mathbf{H}_{I_{\text{inj}}} = \begin{bmatrix} \boldsymbol{\Theta} & \boldsymbol{\Lambda} \\ \boldsymbol{\Xi} & \boldsymbol{\Upsilon} \end{bmatrix} \tag{8.27}$$

where

$$\Theta_{\ell l,re}^{ip,re} = G_{i\ell}^{pl} \tag{8.28}$$

$$\Lambda_{\ell l,im}^{ip,re} = -B_{i\ell}^{pl} \tag{8.29}$$

$$\Xi_{\ell l,re}^{ip,im} = B_{i\ell}^{pl} \tag{8.30}$$

$$\Upsilon_{\ell l,im}^{ip,im} = G_{i\ell}^{pl} \tag{8.31}$$

In (8.28)–(8.31), the superscripts and subscripts have the same meaning as in (8.23)–(8.26).

The part related to the current flow phasor measurements $\mathbf{H}_{I_{\text{flow}}}$ can be derived in a straightforward way from (8.19) and (8.20):

$$\mathbf{H}_{I_{\text{flow}}} = \begin{bmatrix} \boldsymbol{\theta} & \boldsymbol{\vartheta} & \boldsymbol{\iota} & \boldsymbol{\kappa} \\ \boldsymbol{\lambda} & \boldsymbol{\nu} & \boldsymbol{\xi} & \boldsymbol{\varpi} \end{bmatrix} \tag{8.32}$$

where

$$\theta_{il,re}^{i\ell p,re} = g_{i\ell,L}^{pl} + g_{i\ell,T}^{pl} \tag{8.33}$$

$$\vartheta_{\ell l,re}^{i\ell p,re} = -g_{i\ell,L}^{pl} \tag{8.34}$$

$$\iota_{il,im}^{i\ell p,re} = -(b_{i\ell,L}^{pl} + b_{i\ell,T}^{pl}) \tag{8.35}$$

$$\kappa_{\ell l,im}^{i\ell p,re} = b_{i\ell,L}^{pl} \tag{8.36}$$

$$\lambda_{il,re}^{i\ell p,im} = b_{i\ell,L}^{pl} + b_{i\ell,T}^{pl} \tag{8.37}$$

$$\nu_{\ell l,re}^{i\ell p,im} = -b_{i\ell,L}^{pl} \tag{8.38}$$

$$\xi_{il,im}^{i\ell p,im} = g_{i\ell,L}^{pl} + g_{i\ell,T}^{pl} \tag{8.39}$$

$$\varpi_{\ell l,im}^{i\ell p,im} = -g_{i\ell,L}^{pl} \tag{8.40}$$

In (8.33)–(8.40), the superscripts refer, respectively, to the two terminal buses of the branch, the phase and the real or imaginary part of the measurements, while the subscripts refer to the state variables. For instance, $\theta_{il,re}^{i\ell p,re}$ is the scalar that links the measurement $I_{i\ell,re}^{p}$ with the state variable $V_{i,re}^{l}$.

8.1.1.2 Non-linear measurement model

In the case of a mixed set of measurements that includes PMUs and conventional power and magnitude[‡] measurements, the SE becomes non-linear due to the non-linear equations that link the non-phasor measurements with the system state. In this case, the system state for a network with s buses is usually expressed in polar coordinates:

$$\mathbf{x} = [\boldsymbol{\delta}_1^{a,b,c}, \ldots, \boldsymbol{\delta}_i^{a,b,c}, \ldots, \boldsymbol{\delta}_s^{a,b,c}, \mathbf{V}_1^{a,b,c}, \ldots, \mathbf{V}_i^{a,b,c}, \ldots, \mathbf{V}_s^{a,b,c}]^T \tag{8.41}$$

where

$$\begin{aligned} \boldsymbol{\delta}_i^{a,b,c} &= [\delta_i^a, \delta_i^b, \delta_i^c] \\ \mathbf{V}_i^{a,b,c} &= [V_i^a, V_i^b, V_i^c] \end{aligned} \tag{8.42}$$

are, respectively, the phase and magnitude of the voltage phasor at bus i in the three phases a, b and c.

The measurements are assumed to come from PMUs that measure voltage phasors and from conventional power meters.[§] The set of network buses where active and reactive power injections are measured is $\mathcal{U}_1$ ($u_1 = |\mathcal{U}_1|$). Then, the set of network branches where active and reactive power flows are measured is $\mathcal{U}_2$ ($u_2 = |\mathcal{U}_2|$). Hence, it is assumed that the set of measurements $\mathcal{M}$ is composed of:

- $3d_1$ phase-to-ground voltage phasors;
- $3 \cdot (2u_1)$ active and reactive power injections;
- $3 \cdot (2u_2)$ active and reactive power flows.

‡The *magnitude* indicates the amplitude of voltage and current signals.
§The magnitude and phase of current injections and flows could be added to the measurement set. The equations related to current magnitude measurements can be found in Reference 19.

and its cardinality is equal to $m = 2 \cdot (3d_1 + 3u_1 + 3u_2)$. Therefore, the measurement array $\mathbf{z} \in \mathbb{R}^m$ is equal to:[‖]

$$\mathbf{z} = [\mathbf{z}_V, \mathbf{z}_{PQ_{\text{inj}}}, \mathbf{z}_{PQ_{\text{flow}}}]^T \tag{8.43}$$

where

$$\begin{aligned} \mathbf{z}_V^T &= [\boldsymbol{\delta}_1^{a,b,c}, \ldots, \boldsymbol{\delta}_{d_1}^{a,b,c}, \mathbf{V}_1^{a,b,c}, \ldots, \mathbf{V}_{d_1}^{a,b,c}] \\ \mathbf{z}_{PQ_{\text{inj}}}^T &= [\mathbf{P}_{\text{inj},1}^{a,b,c}, \ldots, \mathbf{P}_{\text{inj},u_1}^{a,b,c}, \mathbf{Q}_{\text{inj},1}^{a,b,c}, \ldots, \mathbf{Q}_{\text{inj},u_1}^{a,b,c}] \\ \mathbf{z}_{PQ_{\text{flow}}}^T &= [\mathbf{P}_{\text{flow},1}^{a,b,c}, \ldots, \mathbf{P}_{\text{flow},u_2}^{a,b,c}, \mathbf{Q}_{\text{flow},1}^{a,b,c}, \ldots, \mathbf{Q}_{\text{flow},u_2}^{a,b,c}] \end{aligned} \tag{8.44}$$

As stated before, in this case, the equation that links the measurements with the system state variables is:

$$\mathbf{z} = h(\mathbf{x}) + \boldsymbol{\varepsilon} \tag{8.45}$$

where the vector $h(\mathbf{x}) \in \mathbb{R}^m$ represents the non-linear function relating the system state variables to the measurements, i.e., the so-called *measurement function*. As a consequence, matrix $\mathbf{H}$ used in the SE process is a linearization of $h(\mathbf{x})$ and does not represent the exact link between measurements and state variables.

By using the same notation of (8.17) and (8.18), the expressions of the active and reactive power injections with respect to the state variables are given by:

$$P_i^p = V_i^p \sum_{\ell=1}^{s} \sum_{l \in \mathcal{P}} V_\ell^l (G_{i\ell}^{pl} \cos \delta_{i\ell}^{pl} + B_{i\ell}^{pl} \sin \delta_{i\ell}^{pl}) \tag{8.46}$$

$$Q_i^p = V_i^p \sum_{\ell=1}^{s} \sum_{l \in \mathcal{P}} V_\ell^l (G_{i\ell}^{pl} \sin \delta_{i\ell}^{pl} - B_{i\ell}^{pl} \cos \delta_{i\ell}^{pl}) \tag{8.47}$$

where $\delta_{i\ell}^{pl} = \delta_i^p - \delta_\ell^l$.

By using the same notation of (8.19) and (8.20), the expressions of the active and reactive power flows with respect to the state variables are given by:

$$\begin{aligned} P_{i\ell}^p = {} & V_i^p \sum_{l \in \mathcal{P}} \{V_i^l [(g_{i\ell,L}^{pl} + g_{i\ell,T}^{pl}) \cos \delta_{ii}^{pl} + (b_{i\ell,L}^{pl} + b_{i\ell,T}^{pl}) \sin \delta_{ii}^{pl}]\} \\ & - V_i^p \sum_{l \in \mathcal{P}} [V_\ell^l (g_{i\ell,L}^{pl} \cos \delta_{i\ell}^{pl} + b_{i\ell,L}^{pl} \sin \delta_{i\ell}^{pl})] \end{aligned} \tag{8.48}$$

[‖]In general, the voltage phasor measurements from PMUs and the power measurements from RTUs are not obtained at the same time. In particular, the RTU measurements are characterized by lack of global positioning system (GPS) synchronization, lower refresh rates and lower accuracy compared to the PMU measurements. However, here it is assumed that at time t a full set of measurements composed of synchronized voltage phasor and power measurements is available, as shown in (8.43).

$$Q_{i\ell}^{p} = V_i^p \sum_{l \in \mathcal{P}} \{V_i^l [(g_{i\ell,L}^{pl} + g_{i\ell,T}^{pl}) \sin \delta_{ii}^{pl} - (b_{i\ell,L}^{pl} + b_{i\ell,T}^{pl}) \cos \delta_{ii}^{pl}]\}$$
$$- V_i^p \sum_{l \in \mathcal{P}} [V_\ell^l (g_{i\ell,L}^{pl} \sin \delta_{i\ell}^{pl} - b_{i\ell,L}^{pl} \cos \delta_{i\ell}^{pl})] \quad (8.49)$$

As it will be clarified later, both static and recursive state estimators require the measurement model to be linear. Therefore, in the case of the non-linear measurement model, $\mathbf{H}$ is the Jacobian matrix of the measurement function $h(\mathbf{x})$:

$$\mathbf{H}(\mathbf{x}) = \frac{\partial h(\mathbf{x})}{\partial \mathbf{x}} \quad (8.50)$$

It is called *measurement Jacobian* and its sub-matrices for the considered available measurements are:

$$\mathbf{H} = \begin{bmatrix} \mathbf{H}_V \\ \mathbf{H}_{PQ_{\text{inj}}} \\ \mathbf{H}_{PQ_{\text{flow}}} \end{bmatrix} \quad (8.51)$$

where $\mathbf{H}_V$ is the part of the Jacobian that is related to the partial derivatives of the phase and magnitude of the voltages as a function of the state, $\mathbf{H}_{PQ_{\text{inj}}}$ is related to the partial derivatives of the active and the reactive power injections as a function of the state, and $\mathbf{H}_{PQ_{\text{flow}}}$ is related to the partial derivatives of the active and the reactive power flows as a function of the state.

The first part $\mathbf{H}_V$ is given by:

$$\mathbf{H}_V = \begin{bmatrix} \mathbf{\Psi} & \mathbf{\Gamma} \\ \mathbf{\Lambda} & \mathbf{\Phi} \end{bmatrix} \quad (8.52)$$

where

$$\Psi_{\ell l,re}^{ip,re} = \begin{cases} 1, & \text{if } i = \ell \text{ and } p = l \\ 0, & \text{if } i \neq \ell \text{ or } p \neq l \end{cases} \quad (8.53)$$

$$\Gamma_{\ell l,im}^{ip,re} = 0 \quad (8.54)$$

$$\Lambda_{\ell l,re}^{ip,im} = 0 \quad (8.55)$$

$$\Phi_{\ell l,im}^{ip,im} = \begin{cases} 1, & \text{if } i = \ell \text{ and } p = l \\ 0, & \text{if } i \neq \ell \text{ or } p \neq l \end{cases} \quad (8.56)$$

In (8.53)–(8.56), the superscripts and subscripts have the same meaning as in (8.23)–(8.26).

The second part $\mathbf{H}_{PQ_{\text{inj}}}$ is equal to:

$$\mathbf{H}_{PQ_{\text{inj}}} = \begin{bmatrix} \frac{\partial P_i^p}{\partial \delta_\ell^l} & \frac{\partial P_i^p}{\partial V_\ell^l} \\ \frac{\partial Q_i^p}{\partial \delta_\ell^l} & \frac{\partial Q_i^p}{\partial V_\ell^l} \end{bmatrix} \quad (8.57)$$

The partial derivatives that correspond to the active power injections are:

$$\frac{\partial P_i^p}{\partial \delta_i^p} = -(V_i^p)^2 B_{ii}^{pp} + V_i^p \sum_{\ell=1}^{s} \sum_{l \in \mathcal{P}} V_\ell^l(-G_{i\ell}^{pl} \sin \delta_{i\ell}^{pl} + B_{i\ell}^{pl} \cos \delta_{i\ell}^{pl}) \tag{8.58}$$

$$\frac{\partial P_i^p}{\partial \delta_\ell^l} = V_i^p V_\ell^l (G_{i\ell}^{pl} \sin \delta_{i\ell}^{pl} - B_{i\ell}^{pl} \cos \delta_{i\ell}^{pl}) \tag{8.59}$$

$$\frac{\partial P_i^p}{\partial V_i^p} = V_i^p G_{ii}^{pp} + \sum_{\ell=1}^{s} \sum_{l \in \mathcal{P}} V_\ell^l (G_{i\ell}^{pl} \cos \delta_{i\ell}^{pl} + B_{i\ell}^{pl} \sin \delta_{i\ell}^{pl}) \tag{8.60}$$

$$\frac{\partial P_i^p}{\partial V_\ell^l} = V_i^p (G_{i\ell}^{pl} \cos \delta_{i\ell}^{pl} + B_{i\ell}^{pl} \sin \delta_{i\ell}^{pl}) \tag{8.61}$$

The partial derivatives that correspond to the reactive power injections are:

$$\frac{\partial Q_i^p}{\partial \delta_i^p} = -(V_i^p)^2 G_{ii}^{pp} + V_i^p \sum_{\ell=1}^{s} \sum_{l \in \mathcal{P}} V_\ell^l (G_{i\ell}^{pl} \cos \delta_{i\ell}^{pl} + B_{i\ell}^{pl} \sin \delta_{i\ell}^{pl}) \tag{8.62}$$

$$\frac{\partial Q_i^p}{\partial \delta_\ell^l} = V_i^p V_\ell^l (-G_{i\ell}^{pl} \cos \delta_{i\ell}^{pl} - B_{i\ell}^{pl} \sin \delta_{i\ell}^{pl}) \tag{8.63}$$

$$\frac{\partial Q_i^p}{\partial V_i^p} = -V_i^p B_{ii}^{pp} + \sum_{\ell=1}^{s} \sum_{l \in \mathcal{P}} V_\ell^l (G_{i\ell}^{pl} \sin \delta_{i\ell}^{pl} - B_{i\ell}^{pl} \cos \delta_{i\ell}^{pl}) \tag{8.64}$$

$$\frac{\partial Q_i^p}{\partial V_\ell^l} = V_i^p (G_{i\ell}^{pl} \sin \delta_{i\ell}^{pl} - B_{i\ell}^{pl} \cos \delta_{i\ell}^{pl}) \tag{8.65}$$

The third part of the Jacobian $\mathbf{H}_{PQ_{\text{flow}}}$ is equal to:

$$\mathbf{H}_{PQ_{\text{flow}}} = \begin{bmatrix} \dfrac{\partial P_{i\ell}^p}{\partial \delta_\ell^l} & \dfrac{\partial P_{i\ell}^p}{\partial V_\ell^l} \\[2ex] \dfrac{\partial Q_{i\ell}^p}{\partial \delta_\ell^l} & \dfrac{\partial Q_{i\ell}^p}{\partial V_\ell^l} \end{bmatrix} \tag{8.66}$$

The partial derivatives that correspond to the active power flows are:

$$\begin{aligned} \frac{\partial P_{i\ell}^p}{\partial \delta_i^p} &= -(V_i^p)^2 (b_{i\ell,L}^{pl} + b_{i\ell,T}^{pl}) \\ &\quad + V_i^p \sum_{l \in \mathcal{P}} \{V_i^l[-(g_{i\ell,L}^{pl} + g_{i\ell,T}^{pl}) \sin \delta_{ii}^{pl} + (b_{i\ell,L}^{pl} + b_{i\ell,T}^{pl}) \cos \delta_{ii}^{pl}]\} \\ &\quad + V_i^p \sum_{l \in \mathcal{P}} [V_\ell^l (g_{i\ell,L}^{pl} \sin \delta_{i\ell}^{pl} - b_{i\ell,L}^{pl} \cos \delta_{i\ell}^{pl})] \end{aligned} \tag{8.67}$$

$$\frac{\partial P_{i\ell}^p}{\partial \delta_\ell^l} = -V_i^p V_\ell^l (g_{i\ell,L}^{pl} \sin \delta_{i\ell}^{pl} - b_{i\ell,L}^{pl} \cos \delta_{i\ell}^{pl}) \tag{8.68}$$

$$\frac{\partial P_{i\ell}^{p}}{\partial V_{i}^{p}} = V_{i}^{p}(g_{i\ell,L}^{pl} + g_{i\ell,T}^{pl}) + \sum_{l \in \mathcal{P}} \{V_{i}^{l}[(g_{i\ell,L}^{pl} + g_{i\ell,T}^{pl}) \cos \delta_{ii}^{pl} + (b_{i\ell,L}^{pl} + b_{i\ell,T}^{pl}) \sin \delta_{ii}^{pl}]\}$$

$$- \sum_{l \in \mathcal{P}} [V_{\ell}^{l}(g_{i\ell,L}^{pl} \cos \delta_{i\ell}^{pl} + b_{i\ell,L}^{pl} \sin \delta_{i\ell}^{pl})] \tag{8.69}$$

$$\frac{\partial P_{i\ell}^{p}}{\partial V_{\ell}^{l}} = -V_{i}^{p}(g_{i\ell,L}^{pl} \cos \delta_{i\ell}^{pl} + b_{i\ell,L}^{pl} \sin \delta_{i\ell}^{pl}) \tag{8.70}$$

The partial derivatives that correspond to the reactive power flows are:

$$\frac{\partial Q_{i\ell}^{p}}{\partial \delta_{i}^{p}} = -(V_{i}^{p})^{2}(g_{i\ell,L}^{pl} + g_{i\ell,T}^{pl})$$

$$+ V_{i}^{p} \sum_{l \in \mathcal{P}} \{V_{i}^{l}[(g_{i\ell,L}^{pl} + g_{i\ell,T}^{pl}) \cos \delta_{ii}^{pl} + (b_{i\ell,L}^{pl} + b_{i\ell,T}^{pl}) \sin \delta_{ii}^{pl}]\}$$

$$- V_{i}^{p} \sum_{l \in \mathcal{P}} [V_{\ell}^{l}(g_{i\ell,L}^{pl} \cos \delta_{i\ell}^{pl} + b_{i\ell,L}^{pl} \sin \delta_{i\ell}^{pl})] \tag{8.71}$$

$$\frac{\partial Q_{i\ell}^{p}}{\partial \delta_{\ell}^{l}} = V_{i}^{p} V_{\ell}^{l}(g_{i\ell,L}^{pl} \cos \delta_{i\ell}^{pl} + b_{i\ell,L}^{pl} \sin \delta_{i\ell}^{pl}) \tag{8.72}$$

$$\frac{\partial Q_{i\ell}^{p}}{\partial V_{i}^{p}} = -V_{i}^{p}(b_{i\ell,L}^{pl} + b_{i\ell,T}^{pl})$$

$$+ \sum_{l \in \mathcal{P}} \{V_{i}^{l}[(g_{i\ell,L}^{pl} + g_{i\ell,T}^{pl}) \sin \delta_{ii}^{pl} - (b_{i\ell,L}^{pl} + b_{i\ell,T}^{pl}) \cos \delta_{ii}^{pl}]\}$$

$$- \sum_{l \in \mathcal{P}} [V_{\ell}^{l}(g_{i\ell,L}^{pl} \sin \delta_{i\ell}^{pl} - b_{i\ell,L}^{pl} \cos \delta_{i\ell}^{pl})] \tag{8.73}$$

$$\frac{\partial Q_{i\ell}^{p}}{\partial V_{\ell}^{l}} = -V_{i}^{p}(g_{i\ell,L}^{pl} \sin \delta_{i\ell}^{pl} - b_{i\ell,L}^{pl} \cos \delta_{i\ell}^{pl}) \tag{8.74}$$

8.1.2 Network observability

A power grid is fully observable if it is possible to calculate all the system state variables using a given set of measurements. For a given network model, the network observability is influenced by the type of available measurements (e.g., nodal voltages, current/power injections and flows) and their locations.

A necessary (but not sufficient) condition for the network observability is that the total number of measurements should be equal or larger than the total number of state variables, i.e., $m \geq n$. However, the criterion that needs to be always satisfied for the network to be fully observable is that matrix **H** must be of full rank.

At the design stage of the network measurement infrastructure, the type and location of the measurements are chosen in order to satisfy the network observability criteria. However, in case of data losses or topology changes, a new observability study must be conducted in real-time before the SE computation. If the network becomes unobservable, it can be split in multiple observable sub-networks called *observable islands*, and a separate SE is performed for every island. Additionally, both at the design stage and in real-time, the observability criteria can be met by adding pseudo-measurements.

There are several methods to perform the observability analysis, using graph theory or mathematical techniques, e.g., References 19, 22, 23. The minimum set of n measurements that can guarantee the full network observability is called *critical set* and if one measurement is removed from this set, the system state cannot be calculated. Note that the critical set of measurements is not unique since it can be composed of measurements of different type and location. Adding measurements to the critical set is very beneficial for the SE process since it results in higher estimation accuracy, improved robustness against data loss, and enhanced bad data identification capability.

8.1.3 Process model

As shown in Figure 8.1, recursive SE exploits the statistical properties of the system state by modelling its time evolution via a process model. In particular, the considered linear discrete-time process model can be formulated as [24]:

$$\mathbf{x}_t = \mathbf{A}\mathbf{x}_{t-1} + \mathbf{B}\mathbf{u}_{t-1} + \mathbf{w}_{t-1} \tag{8.75}$$

where

- t is the time-step index;
- $\mathbf{x} \in \mathbb{R}^n$ represents the system state;
- $\mathbf{u} \in \mathbb{R}^{u_c}$ represents a set $\mathcal{U}_c$ ($u_c = |\mathcal{U}_c|$) of known controllable variables;
- $\mathbf{w} \in \mathbb{R}^n$ represents the process noise;
- $\mathbf{A}$ is an $n \times n$ matrix that links the system state $\mathbf{x}$ at time-step $t-1$ with the one at the time-step t, for the case of null controllable variables and null process noise;
- $\mathbf{B}$ is an $n \times u_c$ matrix that links the system state $\mathbf{x}$ at time-step t with the controllable variables $\mathbf{u}$ at time-step $t-1$, for the case of null process noise.

In general, also the matrices $\mathbf{A}$ and $\mathbf{B}$ might change at each time-step. The process noise $\mathbf{w}_{t-1}$ is assumed to be a Gaussian white sequence:

$$p(\mathbf{w}_{t-1}) \sim \mathcal{N}(0, \mathbf{Q}_{t-1}) \tag{8.76}$$

where $\mathbf{Q}_{t-1}$ is the *process noise covariance matrix*.

As it has been already clarified, power system SE is facilitated by the use of synchrophasor measurements streamed by PMUs at a high frame rate, e.g., 50 frames-per-second – fps. The advantage of this working condition is that the state exhibits small variations between two consecutive time-steps, so that a good approximation of matrix $\mathbf{A}$ can be the identity matrix $\mathbf{I}$. In addition, the power system inputs are typically not controllable, i.e. $\mathbf{B}$ is set equal to the null matrix $\mathbf{0}$. Therefore, a suitable process model for the case of power system SE is the autoregressive integrated moving average – ARIMA (0,1,0) model given by:

$$\mathbf{x}_t = \mathbf{x}_{t-1} + \mathbf{w}_{t-1} \tag{8.77}$$

The model of (8.77) can be derived from (8.75) by imposing $\mathbf{A} = \mathbf{I}$ and $\mathbf{B} = \mathbf{0}$ for the aforementioned reasons. The application of this process model to power system SE using KF was firstly proposed in 1970 by Debs and Larson [25] and then it was adopted by other authors, e.g., References 26,27. An advantage of using this process model is that only $\mathbf{Q}$ has to be assessed. An heuristic method for the assessment of $\mathbf{Q}$ in the context of power system SE is proposed in Reference 28 and recalled in Section 8.3.4. The numerical validation of the correctness of this process model is provided in Section 8.7 by using the procedure presented in Reference 5.

Observation. In most of the cases, including power system SE, it is reasonable to assume that the process and measurement noises are uncorrelated:

$$\mathbb{E}[\mathbf{w}\boldsymbol{\varepsilon}^T] = \mathbf{0} \tag{8.78}$$

8.2 Static state estimation: the weighted least squares

The WLS problem is generally formulated as an unconstrained optimization problem. While the least squares requires the measurement noises to have the same variances, the WLS is used when measurements are characterized by different accuracies (*heteroscedasticity*). Indeed, the WLS method is able to weight the measurements according to their accuracies. The WLS relies on the following assumptions:

1. The measurement noises are Gaussian-distributed with null mean value.
2. The measurement noises are uncorrelated; therefore, the measurement noise covariance matrix $\mathbf{R}$ is diagonal.
3. The measurement matrix $\mathbf{H}$ that links the measurements with the state variables is of full rank, so that the network is observable.

Observation. Assumption 2 could be relaxed if significant mutual correlation among measurement errors is present. For such purpose, the problem becomes the so-called *generalized least squares*, where $\mathbf{R}$ is a full matrix [29,30]. However, this case is unlikely to occur in power systems due to the following reasons [5]:

- Measurements provided by different meters can be reasonably considered independent [30], and it is assumed to use no 3-ph multi-function meters [29].
- The sensors are typically installed separately in each of the three phases and the cross-talk interferences are assumed to be negligible.
- The voltage and current magnitudes measured by the same PMU can be usually considered uncorrelated [30].
- As demonstrated in Reference 30, neglecting PMU correlations (both in magnitude and phase) in the estimator model does not lead to a significant decrease of the SE accuracy.

The WLS method can be derived using the maximum likelihood estimation concept as in Reference 31. Indeed, the goal is to compute the *most likely* system state given a set of measured quantities. The measurement noises are assumed to have the same and known Gaussian probability distribution with zero mean. The variances of the measurement noises compose the diagonal elements of **R**. The *measurement residual* vector **r** is defined as:

$$\mathbf{r} = \mathbf{z} - h(\mathbf{x}) \tag{8.79}$$

where the measurement function $h(\mathbf{x})$ can be either linear or non-linear, as in (8.5) and (8.45), respectively. Then, the objective of the WLS optimization problem is to minimize the weighted sum of the squares of the measurement residuals:¶

$$J(\mathbf{x}) = \mathbf{r}^T\mathbf{R}^{-1}\mathbf{r} \tag{8.80}$$

Equation (8.80), since **R** is diagonal (assumption 2 of the WLS), becomes:

$$J(\mathbf{x}) = \sum_{i=1}^{m} \frac{r_i^2}{R_{ii}} \tag{8.81}$$

It can be seen that the reciprocal of the measurement noise variances represents the weights assigned to each measurement, so that the higher the accuracy, the higher the weight.

At this point, the WLS algorithm has two different formulations depending on the types of measurements available. In particular, as it is explained in Sections 8.1.1.1 and 8.1.1.2, the choice of the types of measurements makes the measurement function linear or non-linear leading to the formulation of linear or non-linear SE, respectively.

8.2.1 Linear weighted least squares state estimator

In the case of linear weighted least squares (LWLS), the state is defined by using (8.1) and the exact linear measurement model is (8.5). The measurement residual vector is given by:

$$\mathbf{r}_t = \mathbf{z}_t - \mathbf{H}\mathbf{x}_t \tag{8.82}$$

where **H** is the exact matrix linking measurements and state variables, which contains no approximations and it is constant in time for a given network model. In this case, J is a quadratic function of the unknown state **x** with a minimum that can be computed analytically by imposing the derivative of J equal to zero evaluated in correspondence of the *estimated state* $\widehat{\mathbf{x}}$, as follows:

$$\left.\frac{\partial J(\mathbf{x}_t)}{\partial \mathbf{x}_t}\right|_{\widehat{\mathbf{x}}} = \left.\frac{\partial(\mathbf{r}_t^T\mathbf{R}_t^{-1}\mathbf{r}_t)}{\partial \mathbf{x}_t}\right|_{\widehat{\mathbf{x}}} = 0 \tag{8.83}$$

that yields:

$$\mathbf{H}^T\mathbf{R}_t^{-1}(\mathbf{z}_t - \mathbf{H}\widehat{\mathbf{x}}_t) = 0 \tag{8.84}$$

¶See Reference 19 for the formal derivation of (8.80) and (8.81).

Solving for $\widehat{\mathbf{x}}_t$ yields:

$$\widehat{\mathbf{x}}_t = \mathbf{G}_t^{-1}\mathbf{H}^T\mathbf{R}_t^{-1}\mathbf{z}_t \tag{8.85}$$

where $\mathbf{G}$ is the so-called *Gain matrix* that is defined as:

$$\mathbf{G}_t = \mathbf{H}^T\mathbf{R}_t^{-1}\mathbf{H} \tag{8.86}$$

The covariance matrix of $\widehat{\mathbf{x}}_t$ is:

$$\text{cov}(\widehat{\mathbf{x}}_t) = \mathbf{G}_t^{-1} \tag{8.87}$$

Note that while $\mathbf{H}$ is constant in time for a given network model, $\mathbf{R}$ may change at each time-step for the reasons explained in Section 8.4.

As discussed in detail later in Section 8.5, the presence of erroneous measurements can be detected by analysing the vector of the *normalized measurement estimation residual* vector $\widehat{\mathbf{r}}_t^N$. The measurement estimation residual vector and its covariance matrix are:

$$\widehat{\mathbf{r}}_t = \mathbf{z}_t - \mathbf{H}\widehat{\mathbf{x}}_t \tag{8.88}$$

$$\text{cov}(\widehat{\mathbf{r}}_t) = \mathbf{C}_t = \mathbf{R}_t - \mathbf{H}\mathbf{G}_t^{-1}\mathbf{H}^T \tag{8.89}$$

then, the ith element of $\widehat{\mathbf{r}}_t^N$ is computed as:

$$\widehat{r}_{t,i}^N = \frac{\left|\widehat{r}_{t,i}\right|}{\sqrt{C_{t,ii}}} \tag{8.90}$$

8.2.2 Non-linear weighted least squares

In the case of non-linear weighted least squares (NLWLS), the state is defined by (8.41) and the measurement model is non-linear and given by (8.45). The measurement residual vector is given by:

$$\mathbf{r}_t = \mathbf{z}_t - h(\mathbf{x}_t) \tag{8.91}$$

where $h(\mathbf{x})$ is the non-linear measurement function linking measurements and state variables. Thus, unlike the linear case, substituting (8.91) into (8.83) yields:

$$\mathbf{H}^T(\widehat{\mathbf{x}}_t)\mathbf{R}_t^{-1}[\mathbf{z}_t - h(\widehat{\mathbf{x}}_t)] = 0 \tag{8.92}$$

where $\mathbf{H}(\widehat{\mathbf{x}}_t)$ is the *measurement Jacobian* evaluated at $\widehat{\mathbf{x}}_t$:

$$\mathbf{H}(\widehat{\mathbf{x}}_t) = \left.\frac{\partial h(\mathbf{x}_t)}{\partial \mathbf{x}_t}\right|_{\widehat{\mathbf{x}}_t} \tag{8.93}$$

To solve (8.92), the non-linear function $h(\mathbf{x}_t)$ can be linearized around an initial vector $\mathbf{x}_{t,k}$ as:

$$h(\mathbf{x}_t) = h(\mathbf{x}_{t,k}) + \mathbf{H}(\mathbf{x}_{t,k})(\mathbf{x}_t - \mathbf{x}_{t,k}) \tag{8.94}$$

that substituted in (8.92) yields:

$$\mathbf{H}^T(\widehat{\mathbf{x}}_{t,k})\mathbf{R}_t^{-1}[\mathbf{z}_t - h(\widehat{\mathbf{x}}_{t,k})] - \mathbf{G}(\widehat{\mathbf{x}}_{t,k})(\widehat{\mathbf{x}}_{t,k+1} - \widehat{\mathbf{x}}_{t,k}) = 0 \tag{8.95}$$

where **G** is the Gain matrix already defined in (8.86). It can be seen that the linearization leads to an iterative algorithm where k is the iteration index. The matrices $\mathbf{H}(\widehat{\mathbf{x}}_{t,k})$ and $\mathbf{G}(\widehat{\mathbf{x}}_{t,k})$ can be abbreviated as $\mathbf{H}_{t,k}$ and $\mathbf{G}_{t,k}$, respectively. Rearranging (8.95) leads to the so-called *normal equation*:

$$\mathbf{G}_{t,k}(\widehat{\mathbf{x}}_{t,k+1} - \widehat{\mathbf{x}}_{t,k}) = \mathbf{H}_{t,k}^T \mathbf{R}_t^{-1}[\mathbf{z}_t - h(\widehat{\mathbf{x}}_{t,k})] \tag{8.96}$$

Note that for the case of NLWLS, **H** is a function of the network state; consequently, it is not constant in time and needs to be recomputed at every time-step. Equation (8.96) can be solved with various techniques (e.g., forward/backward substitution as proposed in Reference 19) or by directly inverting the Gain matrix as:

$$\widehat{\mathbf{x}}_{t,k+1} = \widehat{\mathbf{x}}_{t,k} + \mathbf{G}_{t,k}^{-1}\mathbf{H}_{t,k}^T \mathbf{R}_t^{-1}[\mathbf{z}_t - h(\widehat{\mathbf{x}}_{t,k})] \tag{8.97}$$

Equations (8.96) and (8.97) are iterated until convergence. Some possible stopping criteria are:

1. $\max|\widehat{\mathbf{x}}_{t,k+1} - \widehat{\mathbf{x}}_{t,k}| \leq \epsilon_1$
2. $|J(\widehat{\mathbf{x}}_{t,k+1}) - J(\widehat{\mathbf{x}}_{t,k})| < \epsilon_2$
3. $J(\widehat{\mathbf{x}}_{t,k+1}) < \epsilon_3$

where ϵ_1, ϵ_2 and ϵ_3 are a priori selected thresholds.

8.3 Recursive state estimation: the Kalman filter

In 1960, R.E. Kalman published his most famous paper, i.e., Reference 32, describing a recursive solution to the discrete data linear filtering problem. Since then, the KF has been used in a large number of different fields including power systems SE.

The KF relies on the following assumptions:

1. The process and measurement noises are Gaussian white sequences.
2. The process and measurement noises are uncorrelated, as indicated by (8.78).
3. The measurement matrix **H** that links the measurements with the state variables is of full rank, so that the network is observable.

There are several versions of KF. The Discrete Kalman Filter (DKF) is used for linear formulations of the SE problem, whereas the Extended Kalman Filter (EKF) is used when the process and/or the measurement model are non-linear. These two versions of KF are described, respectively, in Sections 8.3.1 and 8.3.2 with reference to the ARIMA (0,1,0) process model of (8.77).

8.3.1 Discrete Kalman filter

As known (e.g., Reference 24), DKF consists of two different parts, the so-called *time-update* (*prediction*) and the *measurement-update* (*estimation*). In what follows, the derivation of the KF equations is presented. The goal is to find an unbiased and minimum variance estimator.

Recursivity: The first assumption is that at time $t-1$ there is already an estimate $\widehat{\mathbf{x}}_{t-1}$ which includes all information up to – and including – time $t-1$. At the next time-step t, this estimate is used to compute the predicted state $\widetilde{\mathbf{x}}_t$. The goal is to find an estimate $\widehat{\mathbf{x}}_t$ for the state at time t that incorporates the new set of measurements $\mathbf{z}_t$ and the predicted state $\widetilde{\mathbf{x}}_t$, e.g., Reference 33:

$$\widehat{\mathbf{x}}_t = \mathbf{K}\prime_t \widetilde{\mathbf{x}}_t + \mathbf{K}_t \mathbf{z}_t \tag{8.98}$$

where $\mathbf{K}\prime_t$ and $\mathbf{K}_t$ are two weighting matrices whose values are determined as reported below.

The prediction and estimation errors are, respectively, defined as:

$$\widetilde{\mathbf{e}}_t \triangleq \mathbf{x}_t - \widetilde{\mathbf{x}}_t \tag{8.99}$$

$$\widehat{\mathbf{e}}_t \triangleq \mathbf{x}_t - \widehat{\mathbf{x}}_t \tag{8.100}$$

and the related prediction and estimation error covariance matrices are, respectively:

$$\widetilde{\mathbf{P}}_t = \mathbb{E}[\widetilde{\mathbf{e}}_t \widetilde{\mathbf{e}}_t^T] \tag{8.101}$$

$$\widehat{\mathbf{P}}_t = \mathbb{E}[\widehat{\mathbf{e}}_t \widehat{\mathbf{e}}_t^T] \tag{8.102}$$

where $\mathbb{E}$ is the expected value operator.

By subtracting $\mathbf{x}_t$ in both parts, (8.98) becomes:

$$\widehat{\mathbf{x}}_t - \mathbf{x}_t = \mathbf{K}\prime_t \widetilde{\mathbf{x}}_t + \mathbf{K}_t \mathbf{z}_t - \mathbf{x}_t \tag{8.103}$$

Then, by substituting (8.5) in (8.103) for the current time-step t, (8.103) becomes:

$$\widehat{\mathbf{x}}_t - \mathbf{x}_t = \mathbf{K}\prime_t \widetilde{\mathbf{x}}_t + \mathbf{K}_t(\mathbf{H}\mathbf{x}_t + \boldsymbol{\varepsilon}_t) - \mathbf{x}_t = \mathbf{K}\prime_t \widetilde{\mathbf{x}}_t + (\mathbf{K}_t\mathbf{H} - \mathbf{I})\mathbf{x}_t + \mathbf{K}_t\boldsymbol{\varepsilon}_t \tag{8.104}$$

By adding and subtracting $\mathbf{K}\prime_t \mathbf{x}_t$ in the right-hand side of (8.104), the latter becomes:

$$\begin{aligned}\widehat{\mathbf{x}}_t - \mathbf{x}_t &= \mathbf{K}\prime_t \widetilde{\mathbf{x}}_t + (\mathbf{K}_t\mathbf{H} - \mathbf{I})\mathbf{x}_t + \mathbf{K}_t\boldsymbol{\varepsilon}_t + \mathbf{K}\prime_t\mathbf{x}_t - \mathbf{K}\prime_t\mathbf{x}_t \\ &= \mathbf{K}\prime_t(\widetilde{\mathbf{x}}_t - \mathbf{x}_t) + (\mathbf{K}\prime_t + \mathbf{K}_t\mathbf{H} - \mathbf{I})\mathbf{x}_t + \mathbf{K}_t\boldsymbol{\varepsilon}_t\end{aligned} \tag{8.105}$$

In order for the estimation to be *unbiased* the expected value of the estimation error must be zero:

$$\mathbb{E}[\widehat{\mathbf{x}}_t - \mathbf{x}_t] = \mathbf{K}\prime_t\mathbb{E}[\widetilde{\mathbf{x}}_t - \mathbf{x}_t] - (\mathbf{I} - \mathbf{K}_t\mathbf{H} - \mathbf{K}\prime_t)\mathbb{E}[\mathbf{x}_t] + \mathbf{K}_t\mathbb{E}[\boldsymbol{\varepsilon}_t] = \mathbf{0} \tag{8.106}$$

or

$$\mathbb{E}[\widehat{\mathbf{e}}_t] = \mathbf{K}\prime_t\mathbb{E}[\widetilde{\mathbf{e}}_t] + (\mathbf{I} - \mathbf{K}_t\mathbf{H} - \mathbf{K}\prime_t)\mathbb{E}[\mathbf{x}_t] - \mathbf{K}_t\mathbb{E}[\boldsymbol{\varepsilon}_t] = \mathbf{0} \tag{8.107}$$

Since $\mathbb{E}[\boldsymbol{\varepsilon}_t] = \mathbf{0}$ (assumption 1 of the KF) and $\mathbb{E}[\widetilde{\mathbf{e}}_t] = \mathbf{0}$, the estimator is unbiased only if:

$$\mathbf{K}\prime_t = \mathbf{I} - \mathbf{K}_t\mathbf{H} \tag{8.108}$$

By substituting (8.108) in (8.98), the KF estimation equation is obtained:

$$\widehat{\mathbf{x}}_t = \widetilde{\mathbf{x}}_t + \mathbf{K}_t(\mathbf{z}_t - \mathbf{H}\widetilde{\mathbf{x}}_t) \tag{8.109}$$

where $\mathbf{K}_t$ is the so-called *Kalman Gain*.** The term $\boldsymbol{\gamma}_t = \mathbf{z}_t - \mathbf{H}\widetilde{\mathbf{x}}_t$ in (8.109) is known as *innovation*, which is a Gaussian white sequence with covariance matrix:

$$\mathbf{S}_t = \mathbf{R}_t + \mathbf{H}\widetilde{\mathbf{P}}_t\mathbf{H}^T \tag{8.110}$$

Some methods for erroneous measurement detection are based on the analysis of the normalized innovation, as explained in Section 8.5. The ith element of the normalized innovation vector is computed as:

$$\gamma_{t,i}^N = \frac{|\gamma_{t,i}|}{\sqrt{S_{t,ii}}} \tag{8.111}$$

Note that $\mathbf{H}\widetilde{\mathbf{x}}_t$ is essentially the vector of the predicted measurements, so that (8.109) can be interpreted as the sum of the prediction, plus the weighted difference between actual and predicted measurements.

Calculation of the estimation error covariance matrix: by definition, (8.102) can be also written as:

$$\widehat{\mathbf{P}}_t = \text{cov}(\widehat{\mathbf{e}}_t) = \text{cov}(\mathbf{x}_t - \widehat{\mathbf{x}}_t) \tag{8.112}$$

By using (8.109), (8.112) becomes:

$$\begin{aligned}\widehat{\mathbf{P}}_t &= \text{cov}(\mathbf{x}_t - (\widetilde{\mathbf{x}}_t + \mathbf{K}_t(\mathbf{z}_t - \mathbf{H}\widetilde{\mathbf{x}}_t))) \\ &= \text{cov}(\mathbf{x}_t - (\widetilde{\mathbf{x}}_t + \mathbf{K}_t(\mathbf{H}\mathbf{x}_t + \boldsymbol{\varepsilon}_t - \mathbf{H}\widetilde{\mathbf{x}}_t))) \\ &= \text{cov}((\mathbf{I} - \mathbf{K}_t\mathbf{H})(\mathbf{x}_t - \widetilde{\mathbf{x}}_t) - \mathbf{K}_t\boldsymbol{\varepsilon}_t)\end{aligned} \tag{8.113}$$

Since $\boldsymbol{\varepsilon}_t$ is not correlated with the other terms, (8.113) becomes:

$$\widehat{\mathbf{P}}_t = \text{cov}((\mathbf{I} - \mathbf{K}_t\mathbf{H})(\mathbf{x}_t - \widetilde{\mathbf{x}}_t)) + \text{cov}(\mathbf{K}_t\boldsymbol{\varepsilon}_t) \tag{8.114}$$

and thus:

$$\widehat{\mathbf{P}}_t = (\mathbf{I} - \mathbf{K}_t\mathbf{H})\text{cov}(\mathbf{x}_t - \widetilde{\mathbf{x}}_t)(\mathbf{I} - \mathbf{K}_t\mathbf{H})^T + \mathbf{K}_t\text{cov}(\boldsymbol{\varepsilon}_t)\mathbf{K}_t^T \tag{8.115}$$

By definition, $\text{cov}(\boldsymbol{\varepsilon}_t) = \mathbf{R}_t$ and $\text{cov}(\mathbf{x}_t - \widetilde{\mathbf{x}}_t) = \widetilde{\mathbf{P}}_t$, therefore:

$$\widehat{\mathbf{P}}_t = (\mathbf{I} - \mathbf{K}_t\mathbf{H})\widetilde{\mathbf{P}}_t(\mathbf{I} - \mathbf{K}_t\mathbf{H})^T + \mathbf{K}_t\mathbf{R}_t\mathbf{K}_t^T \tag{8.116}$$

The formula in (8.116) is also known as the *Joseph form* of the covariance update equation and is valid for any value of the Kalman Gain $\mathbf{K}_t$. However, if $\mathbf{K}_t$ is the optimal gain, (8.116) can be further simplified.

Optimal Kalman Gain derivation: KF is a *minimum variance* estimator. The goal is to minimize the expected value of the square of the magnitude of the estimation

**Note that the Kalman Gain $\mathbf{K}_t$ needs to be determined since, at this stage, it is still unknown.

error $\mathbb{E}[\|\mathbf{x}_t - \widehat{\mathbf{x}}_t\|^2]$. This is equivalent to minimizing the *trace* of $\widehat{\mathbf{P}}_t$. By expanding out the terms in (8.116):

$$\widehat{\mathbf{P}}_t = \widetilde{\mathbf{P}}_t - \mathbf{K}_t\mathbf{H}\widetilde{\mathbf{P}}_t - \widetilde{\mathbf{P}}_t\mathbf{H}^T\mathbf{K}_t^T + \mathbf{K}_t\mathbf{S}_t\mathbf{K}_t^T \tag{8.117}$$

The trace of $\widehat{\mathbf{P}}_t$ is minimized when its matrix derivative with respect to the Kalman Gain matrix is zero:

$$\frac{\partial \mathrm{tr}(\widehat{\mathbf{P}}_t)}{\partial \mathbf{K}_t} = 0 \tag{8.118}$$

Using the following identities:

$$\frac{\partial \mathrm{tr}(\mathbf{AB})}{\partial \mathbf{A}} = \mathbf{B}^T \tag{8.119}$$

$$\frac{\partial \mathrm{tr}(\mathbf{BA}^T)}{\partial \mathbf{A}} = \mathbf{B} \tag{8.120}$$

$$\frac{\partial \mathrm{tr}(\mathbf{ABA}^T)}{\partial \mathbf{A}} = 2\mathbf{AB}, \qquad \text{if } \mathbf{B} \text{ is a symmetric matrix} \tag{8.121}$$

and the fact that $\widetilde{\mathbf{P}}_t$ is a symmetric matrix, (8.118) becomes:

$$\frac{\partial \mathrm{tr}(\widehat{\mathbf{P}}_t)}{\partial \mathbf{K}_t} = -2\widetilde{\mathbf{P}}_t\mathbf{H}^T + 2\mathbf{K}_t\mathbf{S}_t = 0 \tag{8.122}$$

Solving (8.122) for $\mathbf{K}_t$ yields the optimal Kalman Gain:

$$\mathbf{K}_t = \widetilde{\mathbf{P}}_t\mathbf{H}^T\mathbf{S}_t^{-1} = \widetilde{\mathbf{P}}_t\mathbf{H}^T(\mathbf{H}\widetilde{\mathbf{P}}_t\mathbf{H}^T + \mathbf{R}_t)^{-1} \tag{8.123}$$

Hence, by multiplying both sides of (8.123) on the right by $\mathbf{S}_t\mathbf{K}_t^T$:

$$\mathbf{K}_t\mathbf{S}_t\mathbf{K}_t^T = \widetilde{\mathbf{P}}_t\mathbf{H}^T\mathbf{K}_t^T \tag{8.124}$$

Going back to (8.117), the last two terms are cancelled out, therefore:

$$\widehat{\mathbf{P}}_t = (\mathbf{I} - \mathbf{K}_t\mathbf{H})\widetilde{\mathbf{P}}_t \tag{8.125}$$

Calculation of the prediction error covariance matrix: the ARIMA (0,1,0) process model of (8.77) is recalled here below:

$$\mathbf{x}_t = \mathbf{x}_{t-1} + \mathbf{w}_{t-1} \tag{8.126}$$

Thus, the KF state prediction equation is:

$$\widetilde{\mathbf{x}}_t = \widetilde{\mathbf{x}}_{t-1} \tag{8.127}$$

Adding the true state $\mathbf{x}_t$ at both sides of (8.127) and using (8.126) yields:

$$\mathbf{x}_t - \widetilde{\mathbf{x}}_t = \mathbf{x}_{t-1} - \widehat{\mathbf{x}}_{t-1} + \mathbf{w}_{t-1} \tag{8.128}$$

and it follows that:

$$\widetilde{\mathbf{P}}_t = \mathrm{cov}(\widehat{\mathbf{e}}_{t-1} + \mathbf{w}_{t-1}) \tag{8.129}$$

Since $\mathbf{w}_{t-1}$ is the process noise between time $t-1$ and t, whereas $\widehat{\mathbf{e}}_{t-1}$ is the estimation error up to time $t-1$, they are uncorrelated; therefore:

$$\widetilde{\mathbf{P}}_t = \widehat{\mathbf{P}}_t + \mathbf{Q}_{t-1} \tag{8.130}$$

To sum up, the formulation of the DKF-SE for the optimal Kalman Gain is the following:

1. Time-update/prediction:

$$\widetilde{\mathbf{x}}_t = \widehat{\mathbf{x}}_{t-1} \tag{8.131}$$

$$\widetilde{\mathbf{P}}_t = \widehat{\mathbf{P}}_{t-1} + \mathbf{Q}_{t-1} \tag{8.132}$$

2. Measurement-update/estimation:

$$\mathbf{K}_t = \widetilde{\mathbf{P}}_t\mathbf{H}^T(\mathbf{H}\widetilde{\mathbf{P}}_t\mathbf{H}^T + \mathbf{R}_t)^{-1} \tag{8.133}$$

$$\widehat{\mathbf{x}}_t = \widetilde{\mathbf{x}}_t + \mathbf{K}_t(\mathbf{z}_t - \mathbf{H}\widetilde{\mathbf{x}}_t) \tag{8.134}$$

$$\widehat{\mathbf{P}}_t = (\mathbf{I} - \mathbf{K}_t\mathbf{H})\widetilde{\mathbf{P}}_t \tag{8.135}$$

8.3.2 Extended Kalman filter

The EKF is used when the process to be estimated and/or the relationship between measurements and state variables are non-linear, e.g., References 34, 35. In this case, the equations that describe the EKF are similar to the ones of the DKF except for the following differences:

- Similarly to (8.45), the equation that relates measurements and state variables is:

$$\mathbf{z}_t = h(\mathbf{x}_t) + \boldsymbol{\varepsilon}_t \tag{8.136}$$

- Matrix $\mathbf{H}$ is the measurement Jacobian defined by (8.50). It is not constant in time for a given network model, but it changes at every time-step as a function of $\mathbf{x}$.

The equations that describe the EKF-SE are:

1. Time-update/prediction:

$$\widetilde{\mathbf{x}}_t = \widehat{\mathbf{x}}_{t-1} \tag{8.137}$$

$$\widetilde{\mathbf{P}}_t = \mathbf{P}_{t-1} + \mathbf{Q}_{t-1} \tag{8.138}$$

2. Measurement-update/estimation:

$$\mathbf{K}_t = \widetilde{\mathbf{P}}_t\mathbf{H}_t^T(\mathbf{H}_t\widetilde{\mathbf{P}}_t\mathbf{H}_t^T + \mathbf{R}_t)^{-1} \tag{8.139}$$

$$\widehat{\mathbf{x}}_t = \widetilde{\mathbf{x}}_t + \mathbf{K}_t(\mathbf{z}_t - h(\widetilde{\mathbf{x}}_t)) \tag{8.140}$$

$$\widehat{\mathbf{P}}_t = (\mathbf{I} - \mathbf{K}_t\mathbf{H}_t)\widetilde{\mathbf{P}}_t \tag{8.141}$$

8.3.3 Kalman Filter sensitivity with respect to the measurement and process noise covariance matrices

As it has been already explained above, the KF equations involve two covariance matrices, the measurement noise covariance matrix $\mathbf{R}$ and the process noise covariance matrix $\mathbf{Q}$. The values of these two matrices influence significantly the performance of the KF. If they are not properly assessed, the quality of the estimated state given by the KF-SE is not guaranteed. Therefore, it is important to perform a sensitivity analysis of the KF with respect to $\mathbf{R}$ and $\mathbf{Q}$ (e.g., Reference 36).

The measurement noise covariance matrix represents the accuracies of the measurement devices and it weights how much the KF trusts the measurements. Its value can be easily inferred provided the knowledge of the accuracy of the measurement infrastructure (sensors plus PMUs and/or conventional metering devices) and pseudo-measurements (i.e., historical data and/or zero-injection buses). The process noise covariance matrix represents the uncertainties introduced by the process model to predict the system state. It is worth pointing out that in the literature dealing with power systems SE using KF, the values of $\mathbf{Q}$ are, usually, arbitrarily selected although they should be assessed in order to improve the estimation accuracy (e.g., References 25,37,38).

The influence of $\mathbf{R}_t$ and $\mathbf{Q}_t$ on the Kalman gain $\mathbf{K}_t$ can be explained by looking at (8.139). As $\mathbf{R}_t$ decreases, i.e., increasing the confidence on the measurement model, $\mathbf{K}_t$ increases. As a consequence, this leads to an increase of the contribution of the innovation $\boldsymbol{\gamma}_t = \mathbf{z}_t - \mathbf{H}\widetilde{\mathbf{x}}_t$ in the estimation equation (8.134), so that the KF estimates approach the WLS ones. In particular, as $\mathbf{R}_t \to 0$, also $\widehat{\mathbf{P}}_t \to 0$, so that $\widetilde{\mathbf{P}}_t \to \mathbf{Q}_{t-1}$. Therefore, the limit of $\mathbf{K}_t$ as $\mathbf{R}_t \to 0$ is:

$$\lim_{\mathbf{R}_t \to 0} \mathbf{K}_t = \mathbf{Q}_{t-1}\mathbf{H}^T(\mathbf{H}\mathbf{Q}_{t-1}\mathbf{H}^T)^{-1} \tag{8.142}$$

On the other hand, as $\mathbf{Q}_{t-1}$ decreases, i.e., the process model is trusted more, $\mathbf{K}_t$ decreases leading to an increasing weight of the predicted state with respect to the innovation. As $\mathbf{Q}_{t-1} \to 0$, both $\widehat{\mathbf{P}}_t$ and $\widetilde{\mathbf{P}}_t$ approach zero, so that also $\mathbf{K}_t$ tends to zero:

$$\lim_{\mathbf{Q}_{t-1} \to 0} \mathbf{K}_t = \mathbf{0} \tag{8.143}$$

Based on the above analysis, it becomes evident that a proper assessment of the matrix **Q** is of fundamental importance to maximize the performance of the KF-SE. The following section discusses this aspect.

8.3.4 Assessment of the process noise covariance matrix

The process noise covariance matrix can be assessed by using the heuristic method firstly proposed in Reference 28 and then further investigated in Reference 5. This method is effective in terms of estimation accuracy and suitable for real-time applications, although it needs a pre-tuning stage due to its heuristic nature. It assesses the diagonal elements of **Q** by using a moving window composed of the previous N estimated states. The method requires high-resolution state estimates to be effective. In this respect, reference is made to SEs based on PMU measurements (exclusively or through a mixed measurement set), characterized by high refresh rates in the order of tens of fps. The application of this method to the case of power system SE is given in Section 8.7.

In what follows, the formulation of the aforementioned method is briefly recalled. At time-step t, the estimation of $\mathbf{Q}_{t-1}$ is performed by using the last N estimated states. The procedure is the following:

1. Compute the cth element of the vector $\mathbf{g} \in \mathbb{R}^n$ as the sample variance of the cth element of the last N estimated states:

$$g_c = \text{var}[\widehat{x}_{t-1,c}, \ldots, \widehat{x}_{t-N,c}] \tag{8.144}$$

2. Then, the diagonal of $\mathbf{Q}_{t-1}$ is composed of the elements of **g** calculated in step 1:

$$\mathbf{Q}_{t-1} = \text{diag}(\mathbf{g}) \tag{8.145}$$

In general, the diagonal elements of **Q** are not all equal to each other.

8.4 Assessment of the measurement noise covariance matrix

The measurement noise covariance matrix **R** is a matrix that may change from one time-step to another. It can be diagonal or full, but it has to be *positive definite*. The elements of **R** include the accuracy of both the measurement electrical sensors and the PMUs/RTUs. The reasons making the measurement noise covariance matrix **R** changing at every time-step are:

- The measurement errors are calculated with respect to the measured values.
- When the state is expressed in rectangular coordinates, e.g., linear SE, whereas the measurement errors are given in polar coordinates, a transformation of coordinates (projection from polar to rectangular coordinates) is performed at each time-step. Another factor that generates time-dependent accuracies is that the

true system frequency is different from the rated one, since phasors rotate.[††] It is important to mention that, in general, the transformed measurement errors in rectangular coordinates are not Gaussian-distributed, unless the standard deviations of the original magnitude and phase errors are small enough.

Here below are described in detail the steps that lead to the derivation of **R**. In Section 8.7, it is given a numerical example of how matrix **R** is derived.

1. The first step is the calculation of the *maximum errors* that consist in the cumulative errors of the sensors and the PMUs/RTUs. The maximum errors are expressed in percentage for the magnitude and in radians for the phase.
2. The second step consists in the expression of the maximum magnitude errors with respect to the measured value.[‡‡]
3. The third step includes the calculation of the cumulative standard deviations, which are equal to one-third of the maximum errors, assuming that the 99.73% of the values lie within three standard deviations from the mean.
4. This step, in case the SE is linear, includes the transformation of the measurement errors and the associated standard deviations from polar coordinates (magnitudes and phases) to rectangular coordinates (real and imaginary parts).

In what follows, the polar-to-rectangular projection of the measurement uncertainties of phasors is described. Indeed, phasor uncertainties are usually expressed in polar coordinates and their transformation to rectangular coordinates requires the knowledge of the true value of the measured quantity that is, obviously, a hidden value. For this reason, it is interesting to derive the analytical relationship that allows to express the phasors uncertainties in rectangular coordinates as a function of the measurements. The process here illustrated is based on the work presented in Reference 39. In particular, compared to Reference 39, it is illustrated a more detailed derivation of the variances of the projected errors as well as a more detailed description of the mathematical justification of the individual steps of the process.

The measured voltage magnitude and phase are denoted with V_z and δ_z.[§§] Then:

$$V_z = V_x + \widetilde{V} \tag{8.146}$$

$$\delta_z = \delta_x + \widetilde{\delta} \tag{8.147}$$

where V_x and δ_x are the true voltage magnitude and phase, respectively, whereas $\widetilde{V}$ and $\widetilde{\delta}$ are the voltage magnitude and phase measurement errors. The latter are assumed to be independent white Gaussian sequences with standard deviations equal to σ_V and σ_δ, respectively.

[††]Note that even if all the phasor phases are referred to the phase of one of the phasors, i.e., they do not rotate, their phases change with time, making the projection from polar to rectangular coordinates time dependent.

[‡‡]For instance, if the maximum magnitude error is equal to 5% and the measurement is 1.1 pu, then the maximum error is equal to $0.05 \times 1.1 = 0.055$ pu.

[§§]The same process applies to the case of the current measurements.

The measurements in polar coordinates given in (8.146) and (8.147) are transformed to measurements in rectangular coordinates:

$$V_{re,z} = V_z \cos \delta_z \tag{8.148}$$

$$V_{im,z} = V_z \sin \delta_z \tag{8.149}$$

where $V_{re,z}$ and $V_{im,z}$ are the measured voltage real and imaginary parts, respectively.

The measured real and imaginary voltages ($V_{re,z}$ and $V_{im,z}$) can be expressed as a function of the corresponding true quantities ($V_{re,x}$ and $V_{im,x}$) and errors ($\widetilde{V}_{re}$ and $\widetilde{V}_{im}$) in the following way:

$$V_{re,z} = V_{re,x} + \widetilde{V}_{re} = (V_x + \widetilde{V}) \cos(\delta_x + \widetilde{\delta}) \tag{8.150}$$

$$V_{im,z} = V_{im,x} + \widetilde{V}_{im} = (V_x + \widetilde{V}) \sin(\delta_x + \widetilde{\delta}) \tag{8.151}$$

By making use of the trigonometric identities:

$$\cos(a + b) = \cos a \cos b - \sin a \sin b \tag{8.152}$$

$$\sin(a + b) = \sin a \cos b + \cos a \sin b \tag{8.153}$$

equations (8.150) and (8.151) become:

$$\begin{aligned} V_{re,z} &= (V_x + \widetilde{V})(\cos \delta_x \cos \widetilde{\delta} - \sin \delta_x \sin \widetilde{\delta}) \\ &= V_x \cos \delta_x \cos \widetilde{\delta} + \widetilde{V} \cos \delta_x \cos \widetilde{\delta} - V_x \sin \delta_x \sin \widetilde{\delta} - \widetilde{V} \sin \delta_x \sin \widetilde{\delta} \end{aligned} \tag{8.154}$$

and

$$\begin{aligned} V_{im,z} &= (V_x + \widetilde{V})(\sin \delta_x \cos \widetilde{\delta} + \cos \delta_x \sin \widetilde{\delta}) \\ &= V_x \sin \delta_x \cos \widetilde{\delta} + \widetilde{V} \sin \delta_x \cos \widetilde{\delta} + V_x \cos \delta_x \sin \widetilde{\delta} + \widetilde{V} \cos \delta_x \sin \widetilde{\delta} \end{aligned} \tag{8.155}$$

The measurement error of the real part of the voltage, using (8.150) and (8.154) is equal to:

$$\begin{aligned} \widetilde{V}_{re} &= V_{re,z} - V_{re,x} \\ &= V_x \cos \delta_x \cos \widetilde{\delta} + \widetilde{V} \cos \delta_x \cos \widetilde{\delta} - V_x \sin \delta_x \sin \widetilde{\delta} - \widetilde{V} \sin \delta_x \sin \widetilde{\delta} - V_x \cos \delta_x \\ &= V_x \cos \delta_x (\cos \widetilde{\delta} - 1) + \widetilde{V} \cos \delta_x \cos \widetilde{\delta} - V_x \sin \delta_x \sin \widetilde{\delta} - \widetilde{V} \sin \delta_x \sin \widetilde{\delta} \end{aligned} \tag{8.156}$$

and the measurement error of the imaginary part of the voltage, using (8.151) and (8.155) is equal to:

$$\begin{aligned}\widetilde{V}_{im} &= V_{im,z} - V_{im,x}\\ &= V_x \sin\delta_x \cos\widetilde{\delta} + \widetilde{V}\sin\delta_x\cos\widetilde{\delta} + V_x\cos\delta_x\sin\widetilde{\delta} + \widetilde{V}\cos\delta_x\sin\widetilde{\delta} - V_x\sin\delta_x\\ &= V_x\sin\delta_x(\cos\widetilde{\delta}-1) + \widetilde{V}\sin\delta_x\cos\widetilde{\delta} + V_x\cos\delta_x\sin\widetilde{\delta} + \widetilde{V}\cos\delta_x\sin\widetilde{\delta}\end{aligned} \tag{8.157}$$

As it can be observed from (8.156) and (8.157):

- The errors in rectangular coordinates depend on both the true quantities (magnitude and phase) as well as the errors in polar coordinates.
- The link between the errors in rectangular coordinates and the above-mentioned quantities is non-linear through trigonometric functions.

The mean value of the error of the real part of the voltage, starting from (8.156), is calculated as:

$$\begin{aligned}\mu_{\widetilde{V}re,x} &= \mathbb{E}[\widetilde{V}_{re}] = \mathbb{E}[V_x\cos\delta_x\cos\widetilde{\delta}] - \mathbb{E}[V_x\cos\delta_x]\\ &\quad + \mathbb{E}[\widetilde{V}\cos\delta_x\cos\widetilde{\delta}] - \mathbb{E}[V_x\sin\delta_x\sin\widetilde{\delta}] - \mathbb{E}[\widetilde{V}\sin\delta_x\sin\widetilde{\delta}]\end{aligned} \tag{8.158}$$

Since the errors in polar coordinates are assumed to be *zero-mean Gaussian and independent*:

$$\begin{aligned}\mathbb{E}[\widetilde{V}\cos\delta_x\cos\widetilde{\delta}] &= \cos\delta_x\mathbb{E}[\widetilde{V}]\mathbb{E}[\cos\widetilde{\delta}] = 0, \quad \text{since } \mathbb{E}[\widetilde{V}] = 0 \text{ and similarly}\\ \mathbb{E}[\widetilde{V}\sin\delta_x\sin\widetilde{\delta}] &= 0\end{aligned} \tag{8.159}$$

Then (8.158) becomes:

$$\mu_{\widetilde{V}re,x} = V_x\cos\delta_x\mathbb{E}[\cos\widetilde{\delta}] - V_x\cos\delta_x - V_x\sin\delta_x\mathbb{E}[\sin\widetilde{\delta}] \tag{8.160}$$

To calculate the value of $\mathbb{E}[\cos\widetilde{\delta}]$ and $\mathbb{E}[\sin\widetilde{\delta}]$, where $\widetilde{\delta} \sim \mathcal{N}(0, \sigma_{\widetilde{\delta}}^2)$, Euler's formula is used:

$$e^{j\delta} = \cos\delta + j\sin\delta \tag{8.161}$$

Then, the *characteristic function* $\Phi_X(t)$ for a Gaussian distributed variable $X \sim \mathcal{N}(\mu, \sigma^2)$ can be defined as:

$$\Phi_X(t) = \mathbb{E}[e^{jtX}] = e^{jt\mu - \frac{1}{2}\sigma^2 t^2} \tag{8.162}$$

where $t \in \mathbb{R}$ is the argument of the characteristic function. For $t = 1$ and since $\mu_\delta = 0$, (8.162) becomes:

$$\mathbb{E}[e^{j\widetilde{\delta}}] = e^{0-\frac{1}{2}\sigma_\delta^2} = e^{-\frac{1}{2}\sigma_\delta^2} \tag{8.163}$$

Going back to (8.161), by taking the expectations in both parts of the equation and by using (8.163), the imaginary part $\mathbb{E}[\sin(\widetilde{\delta})]$ is equal to zero and:

$$\mathbb{E}[\cos\widetilde{\delta}] = \mathbb{E}[e^{j\widetilde{\delta}}] = e^{-\frac{1}{2}\sigma_\delta^2} \tag{8.164}$$

Hence, (8.160) becomes:

$$\mu_{\widetilde{V}re,x} = V_x \cos\delta_x \left(e^{-\frac{1}{2}\sigma_\delta^2} - 1\right) \tag{8.165}$$

Similarly, the mean error of the imaginary part of the voltage is equal to:

$$\mu_{\widetilde{V}im,x} = V_x \sin\delta_x \left(e^{-\frac{1}{2}\sigma_\delta^2} - 1\right) \tag{8.166}$$

The bias becomes significant only for large values of magnitude and/or large phase errors.

The measurement variance of the real part of the voltage, without doing any approximations, is calculated as:

$$\mathrm{Var}_{\widetilde{V}re,x} = \mathrm{Var}[\widetilde{V}_{re}] \tag{8.167}$$

According to the properties of the variance, if X, Y and Z are independent random variables, then the covariance terms are zero and:

$$\mathrm{Var}[aX + bY - cZ] = a^2\mathrm{Var}[X] + b^2\mathrm{Var}[Y] + c^2\mathrm{Var}[Z] \tag{8.168}$$

In this respect, (8.167) becomes:

$$\begin{aligned}\mathrm{Var}_{\widetilde{V}re,x} = {} & V_x^2\cos^2\delta_x\mathrm{Var}[\cos\widetilde{\delta}] \\ & + \cos^2\delta_x\mathrm{Var}[\widetilde{V}\cos\widetilde{\delta}] + V_x^2\sin^2\delta_x\mathrm{Var}[\sin\widetilde{\delta}] + \sin^2\delta_x\mathrm{Var}[\widetilde{V}\sin\widetilde{\delta}]\end{aligned} \tag{8.169}$$

By using the definition of the variance

$$\mathrm{Var}[X] = \mathbb{E}[X^2] - (\mathbb{E}[X])^2 \tag{8.170}$$

and its following property for the product of two independent variables X, Y:

$$\mathrm{Var}[XY] = (\mathbb{E}[X])^2\mathrm{Var}[Y] + (\mathbb{E}[Y])^2\mathrm{Var}[X] + \mathrm{Var}[X]\mathrm{Var}[Y] \tag{8.171}$$

Equation (8.169) becomes:

$$\begin{aligned}\mathrm{Var}_{\widetilde{V}re,x} &= V_x^2\cos^2\delta_x(\mathbb{E}[\cos^2\widetilde{\delta}] - (\mathbb{E}[\cos\widetilde{\delta}])^2) \\ &\quad + \cos^2\delta_x((\mathbb{E}[\widetilde{V}])^2\mathrm{Var}[\cos\widetilde{\delta}] + (\mathbb{E}[\cos\widetilde{\delta}])^2\mathrm{Var}[\widetilde{V}] + \mathrm{Var}[\widetilde{V}]\mathrm{Var}[\cos\widetilde{\delta}]) \\ &\quad + V_x^2\sin^2\delta_x(\mathbb{E}[\sin^2\widetilde{\delta}] - (\mathbb{E}[\sin\widetilde{\delta}])^2) \\ &\quad + \sin^2\delta_x((\mathbb{E}[\widetilde{V}])^2\mathrm{Var}[\sin\widetilde{\delta}] + (\mathbb{E}[\sin\widetilde{\delta}])^2\mathrm{Var}[\widetilde{V}] + \mathrm{Var}[\widetilde{V}]\mathrm{Var}[\sin\widetilde{\delta}])\end{aligned} \tag{8.172}$$

The expectation $\mathbb{E}[\cos^2\widetilde{\delta}]$ is obtained in the following way:

$$\mathbb{E}[\cos^2\widetilde{\delta}] = \mathbb{E}\left[\frac{1}{2} + \frac{\cos(2\widetilde{\delta})}{2}\right] = \frac{1}{2}(1 + e^{-2\sigma_\delta^2}) \tag{8.173}$$

The expectation $\mathbb{E}[\sin^2\widetilde{\delta}]$ is computed in a similar way and is equal to:

$$\mathbb{E}[\sin^2\widetilde{\delta}] = \frac{1}{2}(1 - e^{-2\sigma_\delta^2}) \tag{8.174}$$

By using the fact that $\mathbb{E}[\sin(\widetilde{\delta})] = 0$, $\mathrm{Var}[\widetilde{V}] = \sigma_V^2$, $\mathbb{E}[\widetilde{V}] = 0$, as well as (8.164), (8.173) and (8.174), equation (8.172) becomes:

$$\begin{aligned}\mathrm{Var}_{\widetilde{V}re,x} &= V_x^2\cos^2\delta_x\frac{1}{2}(1 + e^{-2\sigma_\delta^2}) - V_x^2\cos^2\delta_x e^{-\sigma_\delta^2} + V_x^2\sin^2\delta_x\frac{1}{2}(1 - e^{-2\sigma_\delta^2}) \\ &\quad + \sigma_V^2\sin^2\delta_x\frac{1}{2}(1 - e^{-2\sigma_\delta^2}) + \sigma_V^2\cos^2\delta_x\frac{1}{2}(1 + e^{-2\sigma_\delta^2})\end{aligned} \tag{8.175}$$

Equation (8.175) can be re-written as:

$$\begin{aligned}\mathrm{Var}_{\widetilde{V}re,x} &= \frac{V_x^2\cos^2\delta_x}{2} + \frac{V_x^2\cos^2\delta_x e^{-2\sigma_\delta^2}}{2} - V_x^2\cos^2\delta_x e^{-\sigma_\delta^2} \\ &\quad + \frac{V_x^2\sin^2\delta_x}{2} - \frac{V_x^2\sin^2\delta_x e^{-2\sigma_\delta^2}}{2} + \frac{\sigma_V^2\sin^2\delta_x}{2} \\ &\quad - \frac{\sigma_V^2\sin^2\delta_x e^{-2\sigma_\delta^2}}{2} + \frac{\sigma_V^2\cos^2\delta_x}{2} + \frac{\sigma_V^2\cos^2\delta_x e^{-2\sigma_\delta^2}}{2}\end{aligned} \tag{8.176}$$

and

$$\mathrm{Var}_{\widetilde{V}re,x} = V_x^2\cos^2\delta_x e^{-\sigma_\delta^2}\left(\frac{e^{\sigma_\delta^2} + e^{-\sigma_\delta^2}}{2}\right) \tag{8.177}$$

$$\begin{aligned}&- V_x^2\cos^2\delta_x e^{-\sigma_\delta^2} + V_x^2\sin^2\delta_x e^{-\sigma_\delta^2}\left(\frac{e^{\sigma_\delta^2} - e^{-\sigma_\delta^2}}{2}\right) \\ &= \sigma_V^2\cos^2\delta_x e^{-\sigma_\delta^2}\left(\frac{e^{\sigma_\delta^2} + e^{-\sigma_\delta^2}}{2}\right) + \sigma_V^2\sin^2\delta_x e^{-\sigma_\delta^2}\left(\frac{e^{\sigma_\delta^2} - e^{-\sigma_\delta^2}}{2}\right)\end{aligned} \tag{8.178}$$

By using the trigonometric identities for the hyperbolic sine and cosine:

$$\sinh x = \frac{e^x - e^{-x}}{2} \tag{8.179}$$

$$\cosh x = \frac{e^x + e^{-x}}{2} \tag{8.180}$$

the measurement variance of the real part of the voltage can be finally written as:

$$\begin{aligned}\mathrm{Var}_{\widetilde{V}re,x} &= V_x^2 e^{-\sigma_\delta^2}\left[\cos^2\delta_x(\cosh(\sigma_\delta^2) - 1) + \sin^2\delta_x \sinh(\sigma_\delta^2)\right] \\ &\quad + \sigma_{\widetilde{V}}^2 e^{-\sigma_\delta^2}\left[\cos^2\delta_x \cosh(\sigma_\delta^2) + \sin^2\delta_x \sinh(\sigma_\delta^2)\right]\end{aligned} \tag{8.181}$$

The measurement variance of the imaginary part of the voltage is calculated in the same way and is equal to:

$$\begin{aligned}\mathrm{Var}_{\widetilde{V}im,x} &= V_x^2 e^{-\sigma_\delta^2}\left[\sin^2\delta_x(\cosh(\sigma_\delta^2) - 1) + \cos^2\delta_x \sinh(\sigma_\delta^2)\right] \\ &\quad + \sigma_{\widetilde{V}}^2 e^{-\sigma_\delta^2}\left[\sin^2\delta_x \cosh(\sigma_\delta^2) + \cos^2\delta_x \sinh(\sigma_\delta^2)\right]\end{aligned} \tag{8.182}$$

So far, the mean values and the variances of the errors in rectangular coordinates have been derived as a function of both the true quantities and the measurement errors expressed in polar coordinates. The mean values are given in (8.165) and (8.166), whereas the variances are given in (8.181) and (8.182). However, the true system state is *not* known. Therefore, the error statistics have to be re-expressed as a function of the measurements by introducing *secondary errors* [39].

Starting from (8.165) and using (8.146) and (8.147), the mean value of the error of the real part of the voltage can be expressed as:

$$\mu_{\widetilde{V}re,x} = (V_z - \widetilde{V})\cos(\delta_z - \widetilde{\delta})\left(e^{-\frac{1}{2}\sigma_\delta^2} - 1\right) \tag{8.183}$$

Then, by using the trigonometric identities (8.152) and (8.153) and after some algebraic manipulations, (8.183) becomes:

$$\begin{aligned}\mu_{\widetilde{V}re,x} &= V_z\cos\delta_z\cos\widetilde{\delta}\left(e^{-\frac{1}{2}\sigma_\delta^2} - 1\right) - \widetilde{V}\cos\delta_z\cos\widetilde{\delta}\left(e^{-\frac{1}{2}\sigma_\delta^2} - 1\right) \\ &\quad + V_z\sin\delta_z\sin\widetilde{\delta}\left(e^{-\frac{1}{2}\sigma_\delta^2} - 1\right) - \widetilde{V}\sin\delta_z\cos\widetilde{\delta}\left(e^{-\frac{1}{2}\sigma_\delta^2} - 1\right)\end{aligned} \tag{8.184}$$

As mentioned in Reference 39, the new mean value of the error of the real part of the voltage, conditioned on the measured state, is:

$$\mu_{\widetilde{V}re,z} = \mathbb{E}[\mu_{\widetilde{V}re,x} | V_z, \delta_z] \tag{8.185}$$

Since $\widetilde{V}$ and $\widetilde{\delta}$ are assumed to be independent and $\mathbb{E}[\sin(\widetilde{\delta})] = 0$, $\mathbb{E}[\widetilde{V}] = 0$, (8.185) becomes:

$$\mu_{\widetilde{V}re,z} = V_z\cos\delta_z\left(e^{-\sigma_\delta^2} - e^{-\frac{1}{2}\sigma_\delta^2}\right) \tag{8.186}$$

The mean value of the error of the imaginary part of the voltage, conditioned on the measured state, is equal to:

$$\mu_{\widetilde{V}im,z} = V_z\sin\delta_z\left(e^{-\sigma_\delta^2} - e^{-\frac{1}{2}\sigma_\delta^2}\right) \tag{8.187}$$

The new variances, associated to the new mean values, e.g., Reference 39, are given by:

$$\begin{aligned}\mathrm{Var}_{\tilde{V}re,z} = & V_z^2 e^{-2\sigma_\delta^2}[\cos^2\delta_z(\cosh(2\sigma_\delta^2) - \cosh(\sigma_\delta^2)) \\ & + \sin^2\delta_z(\sinh(2\sigma_\delta^2) - \sinh(\sigma_\delta^2))] \\ & + \sigma_V^2 e^{-2\sigma_\delta^2}[\cos^2\delta_z(2\cosh(2\sigma_\delta^2) - \cosh(\sigma_\delta^2)) \\ & + \sin^2\delta_z(2\sinh(2\sigma_\delta^2) - \sinh(\sigma_\delta^2))]\end{aligned} \tag{8.188}$$

and

$$\begin{aligned}\mathrm{Var}_{\tilde{V}im,z} = & V_z^2 e^{-2\sigma_\delta^2}[\sin^2\delta_z(\cosh(2\sigma_\delta^2) - \cosh(\sigma_\delta^2)) \\ & + \cos^2\delta_z(\sinh(2\sigma_\delta^2) - \sinh(\sigma_\delta^2))] \\ & + \sigma_V^2 e^{-2\sigma_\delta^2}[\sin^2\delta_z(2\cosh(2\sigma_\delta^2) - \cosh(\sigma_\delta^2)) \\ & + \cos^2\delta_z(2\sinh(2\sigma_\delta^2) - \sinh(\sigma_\delta^2))]\end{aligned} \tag{8.189}$$

Note that the variances conditioned on the measured values are larger than the ones conditioned on the true state. This is normal, since they account for the additional errors due to the evaluation at the measured position.

8.5 Data conditioning and bad data processing in PMU-based state estimators

One of the main features of SE is the ability to detect and identify errors in the measurements, the so-called *bad data*. Detection means recognizing the presence of bad data in the measurement set; identification means determining which measurements are bad data. In general, the inclusion of bad data in the estimation process significantly deteriorates the accuracy of the estimate, although the same bad data can have different influence on the estimate depending on the employed SE method and measurement redundancy. Besides, the presence of bad data is common in power systems, since the amount of collected measurements is usually very large and the error sources are several. Therefore, the coupling of a bad data processor to state estimators that are vulnerable to bad data, such as the WLS, is of fundamental importance to maintain the reliability of the estimates and, consequently, the confidence of the system operator on the state estimator. Instead, some state estimators belonging to the category of the robust state estimators, are designed to have an intrinsic bad data rejection capability, without the need of being equipped with a separate bad data processor [19]. One of the most common robust estimators is the least absolute value (LAV) estimator [40]. The main drawback of LAV is the high computational time that can be significantly decreased if the measurement model is linear. Thus, the use of synchrophasor-only measurements allows LAV to become computationally competitive with the LWLS.

Bad data consisting of gross errors are measurements very distant from their expected value, such as negative magnitude values, and can be identified with simple plausibility checks. Examples of possible causes of gross errors are telecommunication network or meter failure, erroneous wire connections, software bugs and loss of GPS signal (if GPS-synchronized meters are used). Thus, an algorithm that analyses the incoming measurements seeking for gross errors and missing data is suggested. This process can also include data conditioning algorithms that refine the raw measurements, such as procedures for the replacement of missing data or for the filtering of undesired disturbances. Data conditioning is particularly important for PMU measurements since they are vulnerable to additional sources of errors due to their high accuracy and time resolution [41,42].

Some other bad data consist of measurement errors with magnitude a few times larger than their expected standard deviation that still may significantly affect the estimate accuracy, but require more advanced methods to be detected. These errors can be due, for instance, to the degradation of the meter accuracy or to electromagnetic interferences. Static state estimators dealing with this kind of bad data are coupled with bad data processors placed after the estimation, called *post-estimation* methods. The availability of the estimate allows post-estimation methods to exploit the statistical properties of SE. Typically, the analysis is carried out on the objective function and the measurement estimation residuals, since they quantify how well the set of measurements fits the network model. The common assumption of post-estimation methods is the exact knowledge of the network model. The consequence of such hypothesis is that any inconsistency between measurements and network model is attributed to the former as a bad data. However, network parameter errors can have a similar effect of bad data on the state estimate, so that a bad data processor may identify persistent bad data in presence of network parameter errors.[‖‖] In most of the cases, post-estimation methods are reliable and precise, but this usually comes at the expense of computational time, because after the identification of bad data the state has to be re-estimated iteratively until no more bad data are diagnosed.

A well-known bad data detection method is the χ^2-test, but it usually fails to detect bad data if the error is lower than 20 standard deviations [19,44]. It is based on the assumption that each measurement residual r_i is a randomly distributed Gaussian variable with zero mean and variance R_{ii}. Thus, it can be easily demonstrated that the objective function $J(\widehat{\mathbf{x}})$ has a χ^2-distribution with $(m-n)$ degrees of freedom. The test reveals the presence of bad data if $J(\widehat{\mathbf{x}}) > \chi^2_{(m-n),\zeta}$, where ζ is the chosen detection confidence probability, e.g., 95% or 99%. The values of $\chi^2_{(m-n),\zeta}$ can be found in dedicated tables or using the MATLAB® function *chi2inv*($\zeta, m-n$).

Another widely used post-estimation method for bad data detection and identification is the largest normalized residual (LNR) test, thanks to its accuracy and

[‖‖]An effective method to distinguish between bad data and network parameter errors is presented in Reference 43.

straightforward implementation [19,44]. It requires the calculation of the normalized measurement estimation residual vector $\widehat{\mathbf{r}}^N$ as in (8.90). Each normalized residual should be distributed as $\sim \mathcal{N}(0, 1)$; therefore, the LNR test detects bad data if at least one element of $\widehat{\mathbf{r}}^N$ exceeds a certain threshold, e.g., 3 or 4. Assuming that $\widehat{\mathbf{r}}_i^N$ is the largest residual exceeding the threshold, the ith measurement is flagged as bad data and the SE is re-computed without using this measurement. This procedure is iterated until no bad data are detected. The LNR test limits its identification capability to a single bad data or multiple non-interacting bad data, namely multiple bad data appearing simultaneously whose residuals are not correlated. In case of multiple interacting bad data (whose residuals are correlated), the hypothesis testing identification method has shown better performance [19,45]. Moreover, critical measurements or measurements belonging to a critical pair cannot be identified as bad data by post-estimation algorithms, since their residuals are always equal to zero. A critical measurement is a measurement that makes the network unobservable if removed, while a critical pair is composed of two measurements that makes the network unobservable if removed simultaneously. It is evident that an abundant measurement redundancy helps substantially the detection and identification of bad data.

Concerning estimators that include a process model, such as the recursive ones, bad data processing can be performed before and/or after the estimation using, respectively, pre- and/or post-estimation methods. Pre-estimation methods consist in statistical procedures or simply logical checks that typically involve the examination of the normalized measurement innovation vector defined in (8.111) [46,47,42]. Therefore, they cannot be optimal and their reliability is enhanced if combined with a post-estimation method accounting also for the measurement estimation residuals [47]. High values of some elements of the innovation vector can be due to bad data or to sudden variation of the system state caused, for instance, by inrushes, faults and disconnection of loads or generators. Before performing bad data identification, pre-estimation methods need an algorithm that distinguishes between the two.

Historically, static state estimators, and therefore post-estimation methods, have been the main subject of the literature research and have been employed in real applications. The reasons are multiple and the main ones are listed hereafter. Static state estimators are characterized by lower computational complexity with respect to recursive ones. Furthermore, they are, in general, more reliable since they use only one measurement set without the need to consider a process model that may lead to estimation errors in case of sudden and unexpected changes of the power system state. The meters providing measurements for SE are usually voltage, current and power meters that send the measurement packet at intervals of several seconds or a few minutes. Hence, there is sufficient time to perform post-estimation analysis. However, the recent inclusion of PMU data in SE have made possible to dramatically increase the SE refresh rate to a few tens of milliseconds and to use effectively the process model of (8.77). Although the bad data processing theory proposed by the literature remains valid, PMUs have raised new interest on recursive state estimators, and consequently on pre-estimation methods, as well as on data conditioning of raw measurements.

8.6 Kalman filter vs. weighted least squares

This section compares theoretically the accuracy of KF vs. WLS. To formally quantify this difference, it is useful to recall that the KF process makes use of all the available measurements, past and present, whereas the WLS algorithm uses only measurements of the current time-step. The former should intuitively perform better, provided that the process model hypotheses that underlie the KF are correct. The following theorem formalizes this aspect. It states that the estimation error with the KF algorithm is always less than the estimation error with the WLS algorithm, the difference being given equal to the mean square difference between the two methods:

Theorem 8.1. *Assume that the true (unobserved) state $\mathbf{x}_t$ satisfies the process model in* (8.77). *Assume that the system parameters are known. Let $\widehat{\mathbf{x}}_{t,WLS}$ and $\widehat{\mathbf{x}}_{t,KF}$ be the state estimates obtained at time-step t with the WLS and KF algorithms, respectively. Then*

$$\mathbb{E}[\|\mathbf{x}_t - \widehat{\mathbf{x}}_{t,WLS}\|^2] = \mathbb{E}[\|\mathbf{x}_t - \widehat{\mathbf{x}}_{t,KF}\|^2] + \mathbb{E}[\|\widehat{\mathbf{x}}_{t,WLS} - \widehat{\mathbf{x}}_{t,KF}\|^2] \tag{8.190}$$

Proof. First of all, by standard KF theory, e.g., Reference 48, the estimation of the non-observable state is equal to its conditional expectation, given the sequence of measurements, i.e.,

$$\widehat{\mathbf{x}}_{t,KF} = \mathbb{E}[\mathbf{x}_t|\mathcal{F}_t] \tag{8.191}$$

where $\mathcal{F}_t$ the σ-field generated by all measurements up to and including time-step t.

Second, consider the Hilbert space of random vectors (with values in $\mathbb{R}^n$) equipped with the inner product $\langle \mathbf{X}, \mathbf{Y}\rangle_H \triangleq \mathbb{E}[\sum_{c=1}^{n} X_c Y_c]$. It is requested to show that the random vector $\mathbf{x}_t - \widehat{\mathbf{x}}_{t,KF}$ is orthogonal,[¶¶] in the sense of this Hilbert space, to all random vectors $\mathbf{Y}$ that are $\mathcal{F}_t$-measurable (i.e., that are a function of the measurements up to time t; in this context, the initial conditions of the estimation algorithms are assumed to be known and non-random). Note that what needs to be shown is that:

$$\mathbb{E}[\langle \mathbf{Y}, \mathbf{x}_t\rangle] = \mathbb{E}[\langle \mathbf{Y}, \widehat{\mathbf{x}}_{t,KF}\rangle] \tag{8.192}$$

with $\langle \mathbf{Y}, \mathbf{x}_t\rangle = \sum_{c=1}^{n} Y_c x_{t,c}$.

To prove (8.192), observe that, for any (real-valued) random variable U that is measurable with respect to $\mathcal{F}_t$, it can be written that $\mathbb{E}[U|\mathcal{F}_t] = U$ and further [48]:

$$\mathbb{E}[U x_{t,c}|\mathcal{F}_t] = U\mathbb{E}[x_{t,c}|\mathcal{F}_t] = U\widehat{x}_{t,c,KF}$$

Note that such a U can be any non-linear real-valued function of $(\mathbf{z}_1, \ldots, \mathbf{z}_t)$. Take expectations on both sides and use the fact that the expectation of the conditional expectation is the same as the original expectation (*law of total expectation*, e.g., Reference 48) and obtain

$$\mathbb{E}[U x_{t,c}] = \mathbb{E}[U\widehat{x}_{t,c,KF}] \tag{8.193}$$

¶¶We remind here that when two vectors are orthogonal, their inner product is equal to zero.

Consider now any $\mathcal{F}_t$-measurable random vector $\mathbf{Y}$, apply (8.193) to $U = Y_c$ for all coordinates c and sum over c; it comes:

$$\mathbb{E}\left[\sum_{c=1}^{n} Y_c x_{t,c}\right] = \mathbb{E}\left[\sum_{c=1}^{n} Y_c \widehat{x}_{t,c,KF}\right] \tag{8.194}$$

which shows (8.192) as required and it means that:

$$\mathbb{E}\left[\sum_{c=1}^{n} Y_c(\widehat{x}_{t,c} - \widehat{x}_{t,c,KF})\right] = \mathbb{E}[\langle \mathbf{Y}, \mathbf{x}_t - \widehat{\mathbf{x}}_{t,KF}\rangle] = 0 \tag{8.195}$$

namely $\mathbf{Y}$ and $\mathbf{x}_t - \widehat{\mathbf{x}}_{t,KF}$ are *orthogonal.*

Now observe that both $\widehat{\mathbf{x}}_{t,WLS}$ and $\widehat{\mathbf{x}}_{t,KF}$ are $\mathcal{F}_t$-measurable because they are derived from the measurements. Therefore, the previous result can be applied to $\mathbf{Y} = \widehat{\mathbf{x}}_{t,WLS} - \widehat{\mathbf{x}}_{t,KF}$. Equation (8.190) then follows from Pythagoras's equality. □

Final remark: The theorem applies as long as the process model in (8.77) holds. This explains why it is important to verify the adequacy of the process model.

8.7 Numerical validation and performance assessment of the state estimation

8.7.1 Linear state estimation case studies

8.7.1.1 Distribution network case study: IEEE 13-bus distribution test feeder

The adopted IEEE 13-bus distribution test feeder is shown in Figure 8.3 and is based on Reference 49. It is a 3-ph feeder of 13 buses where Bus 1 represents the connection to the sub-transmission network with a short-circuit power $S_{sc} = 300$ MVA and a ratio between the real and imaginary parts of the short-circuit impedance $R_{sc}/X_{sc} = 0.1$. The network has a rated voltage equal to 15 kV line-to-line root mean square (RMS) and the voltage base $V_b = 15$ kV (line-to-line RMS). The power base $S_b = 10$ MVA. The lines are unbalanced and the used line configuration is the #602 of Reference 49. The values of the resistance, reactance and susceptance and the line lengths are given in Appendix B. The loads/distributed energy resources (DERs) are also characterized by unbalanced power absorptions/injections, respectively. Figure 8.4 shows the aggregated active and reactive power consumption of the loads in three phases, whereas Figure 8.5 shows the power injected by DERs. The data comes from an experimental campaign in real distribution grids in the South-West region of Switzerland, in the EPFL campus and in a feeder in the Netherlands (BML 2.10 distribution feeder operated by Alliander). The DERs (one Mini-Hydro power plant at Bus 4 and one photovoltaic unit at Bus 12) inject only active power.

Table 8.1 shows the location of PMUs for the IEEE 13-bus distribution test feeder. In this case, the number of PMUs is equal to 7. It is assumed that each PMU measures the bus phase-to-ground voltage phasors and the nodal current injection phasors; therefore in this case, d_1 and d_2 of (8.4) are equal to 7 and d_3 is equal to 0.

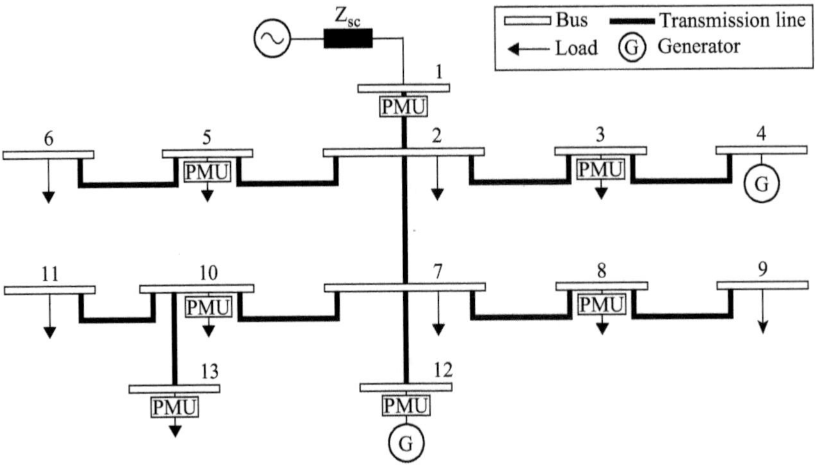

Figure 8.3 Network topology of the IEEE 13-bus distribution test feeder [49], together with the adopted PMU placement. We assume that Bus 1 is the connection point of the system to an external network that is represented by a voltage source in series with the short-circuit impedance Z_{sc}

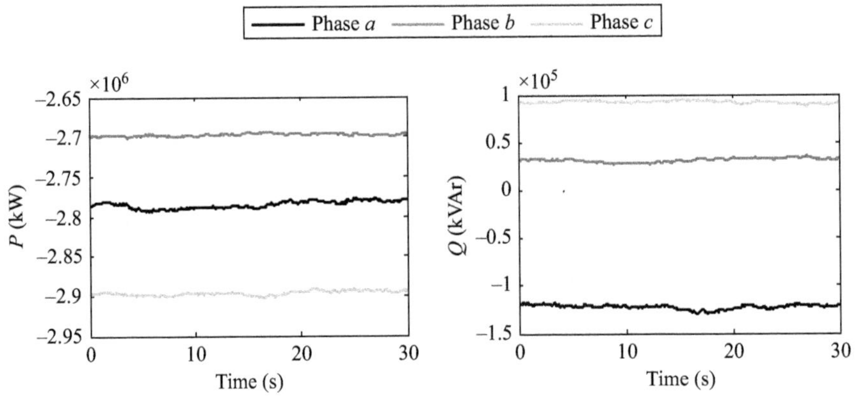

Figure 8.4 Time evolution of the active and reactive powers of the loads, per phase, used in the IEEE 13-bus distribution test feeder

The SE accuracy assessment of the IEEE 13-bus distribution test feeder is performed by using the LWLS-SE and the DKF-SE. This is justified by the fact that the measurements come only from PMUs; therefore, the SE is linear. The value of $\mathbf{Q}$ used by the DKF-SE is assessed using the method presented in Section 8.3.4, where the parameter $N = 20$. Current and voltage sensors used to perform the tests are assumed to be of 0.1-class. Limits of ratio error and phase displacement imposed by References 50,51 are shown in Tables 8.2 and 8.3, respectively.

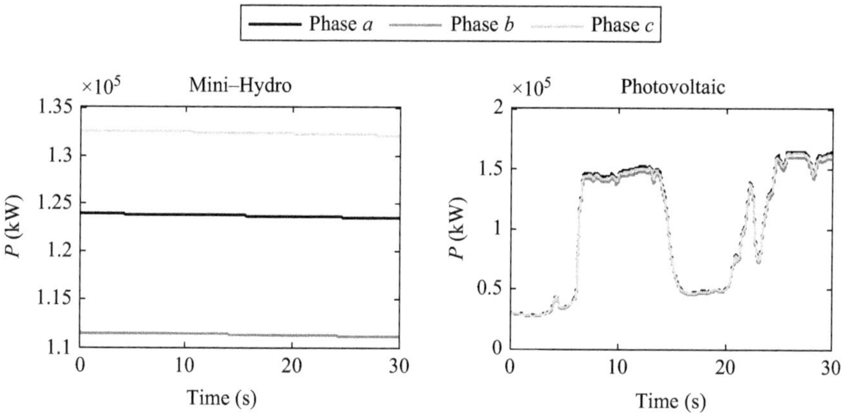

Figure 8.5 Time evolution of the active powers of the two DERs (one Mini-Hydro power plant at Bus 4 and one photovoltaic unit at Bus 12), per phase, used in the IEEE 13-bus distribution test feeder

Table 8.1 PMUs location in the IEEE 13-bus distribution test feeder

	Number	**Buses**
PMUs	7	1 3 5 8 10 12 13

Table 8.2 Limits of ratio error and phase displacement of the used current sensors in the IEEE 13-bus distribution test feeder, according to Reference 50

Class	**Ratio error (%)**	**Phase displacement (rad)**
0.1	0.1	1.5×10^{-3}

Table 8.3 Limits of ratio error and phase displacement of the used voltage sensors in the IEEE 13-bus distribution test feeder, according to Reference 51

Class	**Ratio error (%)**	**Phase displacement (rad)**
0.1	0.1	1.5×10^{-3}

Table 8.4 Limits of magnitude and phase errors for the used PMUs in the IEEE 13-bus distribution test feeder

TVE (in %)	Magnitude error (%)	Phase error (rad)
0.14	0.1	10^{-3}

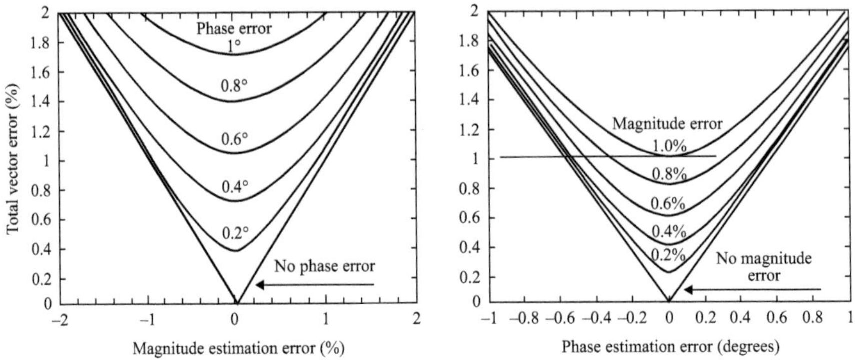

Figure 8.6 Equivalence of magnitude and phase error with PMUs TVE values (adapted from Reference 1)

The limits of magnitude and phase errors for the adopted PMUs are shown in Table 8.4 and correspond to a total vector error (TVE) equal to 0.14%. Such a TVE value results from assuming PMU class-P devices characterized by typical maximum errors in magnitude and phase of 0.1% and 10^{-3} rad, respectively.

Here below are given the steps for the numerical derivation of the measurement noise covariance matrix **R**, using the theory in Section 8.4.

1. The values of the maximum errors for the 0.1-class electrical sensors, in terms of magnitude error and phase displacement, are given in Tables 8.2 and 8.3. Hence, the limit for the magnitude error is 0.1%, whereas the maximum phase displacement is 1.5×10^{-3} rad.
   The maximum magnitude error of PMUs is obtained from Figure 8.6 by assuming a TVE equal to 0.1% and a null phase error. Then, the maximum phase error is obtained from Figure 8.6 by assuming a TVE equal to 0.1% and a null magnitude error. The maximum magnitude and phase errors of PMUs are, respectively, 0.1% and 1×10^{-3} rad, which corresponds to a TVE of 0.14%.
   Then, the cumulative maximum errors of sensors and PMUs are $(0.1 + 0.1)\%$ for the voltage magnitude and $(1.5 \times 10^{-3} + 1 \times 10^{-3})$ rad for the voltage phase.

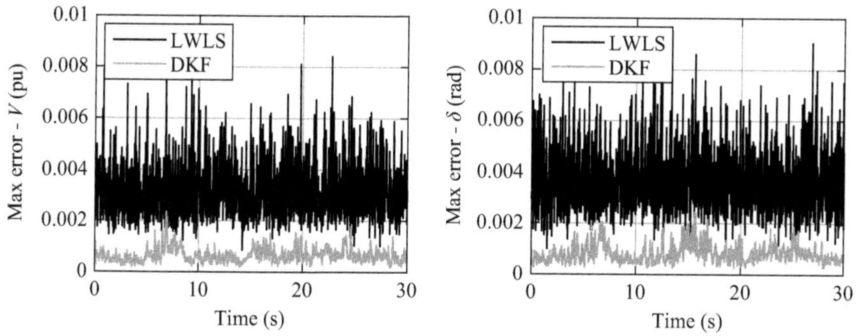

Figure 8.7 Time evolution of the maximum estimation errors calculated by using (8.196), for the LWLS-SE and DKF-SE, case of the IEEE 13-bus test feeder

2. The second step consists in the expression of the maximum magnitude errors with respect to the measured values.
3. The third step includes the calculation of the cumulative standard deviations, which are equal to one-third of the maximum errors, assuming that the 99.73% of the values lie within three standard deviations from the mean.
4. By using the cumulative standard deviations from the previous step and the measurements, the variances in rectangular coordinates are calculated by using (8.188) and (8.189).
5. The diagonal of the measurement noise covariance matrix is composed of the aforementioned variances.

Figure 8.7 shows the time evolution of the maximum errors of the estimated state vs. the true one for the LWLS-SE and DKF-SE, for a time window of 30 s and a resolution of 20 ms.*** Errors refer to both voltage magnitude V and phase δ. The magnitude error is expressed in pu and the phase error is expressed in radians. At time-step t, the maximum estimation error (MEE) is calculated considering the estimation errors in all the buses and the three phases as:

$$\begin{aligned} \mathrm{MEE}(V_t) &= \max[\mathrm{err}V^a_{1,t}, \mathrm{err}V^b_{1,t}, \mathrm{err}V^c_{1,t}, \ldots, \mathrm{err}V^a_{n,t}, \mathrm{err}V^b_{n,t}, \mathrm{err}V^c_{n,t}] \\ \mathrm{MEE}(\delta_t) &= \max[\mathrm{err}\delta^a_{1,t}, \mathrm{err}\delta^b_{1,t}, \mathrm{err}\delta^c_{1,t}, \ldots, \mathrm{err}\delta^a_{n,t}, \mathrm{err}\delta^b_{n,t}, \mathrm{err}\delta^c_{n,t}] \end{aligned} \tag{8.196}$$

Both the magnitude and the phase maximum errors of the LWLS-SE are larger than the ones of the DKF-SE along the overall simulation time.

To clarify further this aspect, Figure 8.8 shows the root mean squared errors (RMSEs) for the LWLS-SE and the DKF-SE and considering the same time window of Figure 8.7. Only the maximum RMSE among the three phases for each bus is

***The KF estimates have always an initial phase in which they converge from an arbitrary initial value (e.g., a flat-start initialization) towards the true state. For brevity, in all the figures that show the KF accuracy performance, this initial phase is not shown.

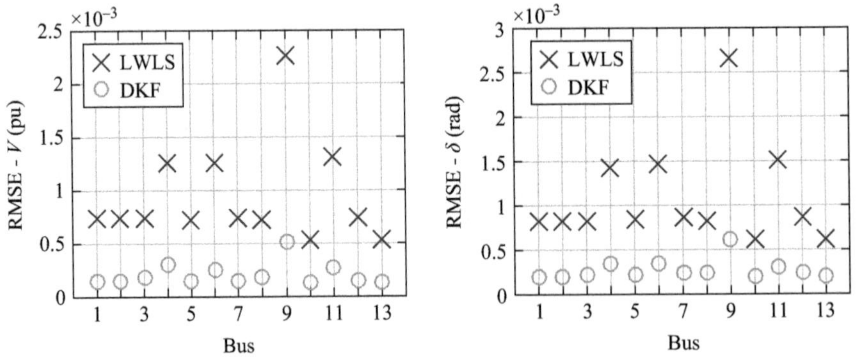

Figure 8.8 RMSEs calculated by using (8.197)*, for the LWLS-SE and DKF-SE, case of the IEEE 13-bus test feeder. Only the maximum RMSE among the three phases for each bus is shown*

shown. The errors refer to both voltage magnitude V and phase δ. The magnitude error is expressed in pu and the phase error is expressed in radians. As known, the RMSE for the voltage magnitude and phase at bus i and phase p is calculated as:

$$\text{RMSE}(V_i^p) = \sqrt{\frac{1}{N_s}\sum_{n_s=1}^{N_s}(\widehat{V}_{i,n_s}^p - V_{i,n_s}^p)^2}$$
$$\text{RMSE}(\delta_i^p) = \sqrt{\frac{1}{N_s}\sum_{n_s=1}^{N_s}(\widehat{\delta}_{i,n_s}^p - \delta_{i,n_s}^p)^2} \tag{8.197}$$

As it can be observed, the RMSEs of LWLS-SE are larger than the ones of the DKF-SE for all the 13 network buses.

8.7.1.2 Transmission system case study: IEEE 39-bus transmission test system

The IEEE 39-bus test system [52], shown in Figure 8.9, is balanced; therefore, only the direct sequence has been considered. Bus 31 in Figure 8.9 is the connection point of the system to an external grid characterized by a short-circuit power $S_{sc} = 50$ GVA and a ratio between real and imaginary parts of the short-circuit impedance $R_{sc}/X_{sc} = 0$, which is a standard assumption for transmission power systems. The IEEE 39-bus system is assumed to have four different voltage levels, i.e., 380 kV, 230 kV, 125 kV and 15 kV. The transformer ratios are given in Appendix B. The chosen value for the base power is $S_b = 100$ MVA. The values of the resistance, reactance and susceptance are given in Appendix B.

Table 8.5 gives the values of the active and reactive power injections at the respective buses, which are in accordance with Reference 52. The convention is that the absorbed powers are marked with a minus, whereas the generated powers have a positive sign.

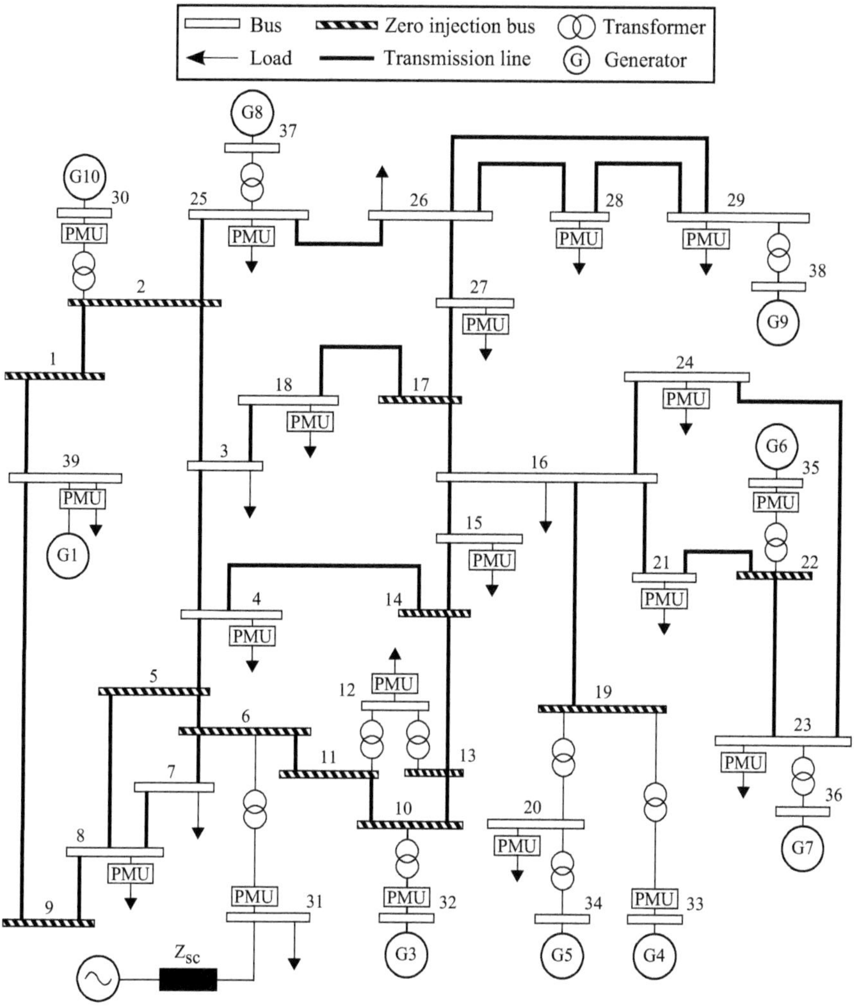

Figure 8.9 Network topology of the IEEE 39-bus test system [52], together with the adopted PMU placement. We assume that Bus 31 is the connection point of the system to an external network that is represented by a voltage source in series with the short-circuit impedance Z_{sc}

The SE accuracy assessment of the IEEE 39-bus transmission test system is performed by using the LWLS-SE and the DKF-SE. The current and voltage sensors used to perform the tests are assumed to be of 0.5-class. The PMU and sensor accuracies are reported in Tables 8.6–8.8. The assumed PMU locations and the zero-injection buses are given in Table 8.9. The variance assigned to zero-injection buses is lower than the ones of the other measurements since they are not affected by error. It is assumed that each PMU measures the bus phase-to-ground voltage phasors and the

Table 8.5 Active and reactive power injections for the IEEE 39-bus transmission test system

Bus	Type	*P* (MW)	*Q* (MW)
3	Load	−322	−2.4
4	Load	−500	−184
7	Load	−233.8	−84
8	Load	−522	−176
12	Load	−7.5	−88
15	Load	−320	−153
16	Load	−329	−32.3
18	Load	−158	−30
20	Load	−628	−103
21	Load	−274	−115
23	Load	−247.5	−84.6
24	Load	−308.6	92
25	Load	−224	−47.2
26	Load	−139	−17
27	Load	−281	−75.5
28	Load	−206	−27.6
29	Load	−283.5	−26.9
30	Generator	250	189.9
31	Load	−9.2	−4.6
32	Generator	650	204.8
33	Generator	632	72.3
34	Generator	508	149.8
35	Generator	650	258.7
36	Generator	560	212.8
37	Generator	540	30.8
38	Generator	830	63.0
39	Load	−104	−250.0

Table 8.6 Limits of magnitude and phase errors for the used PMUs of the IEEE 39-bus transmission test system

TVE (in %)	Magnitude error (%)	Phase error (rad)
0.14	0.1	10^{-3}

nodal current injection phasors; therefore in this case, d_1 and d_2 of (8.4) are equal to 19 and d_3 is equal to 0.

Figure 8.10 shows the time evolution of the maximum errors of the estimated state vs. the true one for the LWLS-SE and DKF-SE. The MEEs are calculated by using (8.196). As it can be observed, both the magnitude and the phase maximum errors of the LWLS-SE are significantly larger than the ones of the DKF-SE along the overall simulation time.

Table 8.7 Limits of ratio error and phase displacements of the used current sensors of the IEEE 39-bus transmission test system, according to Reference 50

Class	**Ratio error (%)**	**Phase displacement (rad)**
0.5	0.5	9×10^{-3}

Table 8.8 Limits of ratio error and phase displacement of the used voltage sensors of the IEEE 39-bus transmission test system, according to Reference 51

Class	**Ratio error (%)**	**Phase displacement (rad)**
0.5	0.5	6×10^{-3}

Table 8.9 PMU locations and zero-injection buses in the IEEE 39-bus test system

Case	**Number**	**Buses**
Zero-injection buses	12	1 2 5 6 9 10 11 13 14 17 19 22
PMUs	19	4 8 12 15 18 20 21 23 24 25 27 28 29 30 31 32 33 35 39

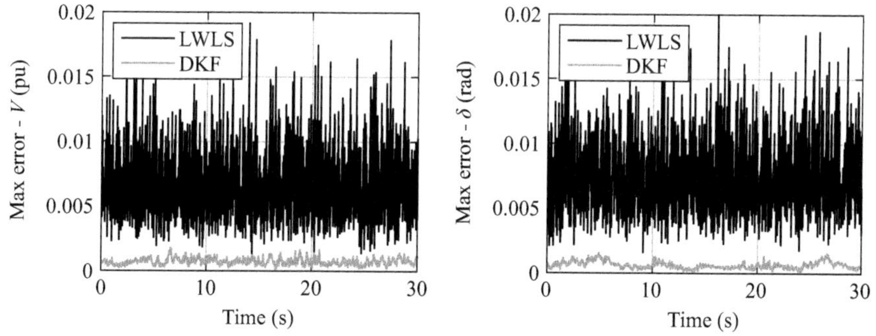

Figure 8.10 Time evolution of the maximum estimation errors calculated by using (8.196)*, for the LWLS-SE and DKF-SE, case of the IEEE 39-bus test system*

8.7.2 Non-linear SE case studies

The network used as a case study to show the performance of the non-linear SE is the 13-bus distribution test case. The network data is the same as the one reported in Section 8.7.1.1. Table 8.10 shows the location of PMUs and RTUs for the IEEE 13-bus distribution test feeder. It is assumed that each PMU measures the bus phase-to-ground voltage phasors and each RTU measures the nodal active and reactive power injections; therefore, in this case, d_1 and u_1 of (8.44) are equal to 7 and u_2 is equal to 0.

The accuracy of the PMUs is the one reported in Table 8.4. The cumulative error of sensors and RTUs is assumed to be equal to 3%. The SE accuracy assessment of the IEEE 13-bus distribution test feeder is in this case performed by using the NLWLS-SE and the EKF-SE.

Figure 8.11 shows the time evolution of the maximum errors of the estimated state vs. the true one for the NLWLS-SE and EKF-SE, for a time window of 30 s and a resolution of 20 ms. The errors refer to both voltage magnitude V and phase δ. The magnitude error is expressed in pu and the phase error is expressed in radians. At time-step t, the MEE is calculated by using (8.196). Both the magnitude and the

Table 8.10 PMUs and RTUs location in the IEEE 13-bus distribution test feeder

	Number	**Buses**
PMUs	7	1 3 5 8 10 12 13
RTUs	7	1 3 5 8 10 12 13

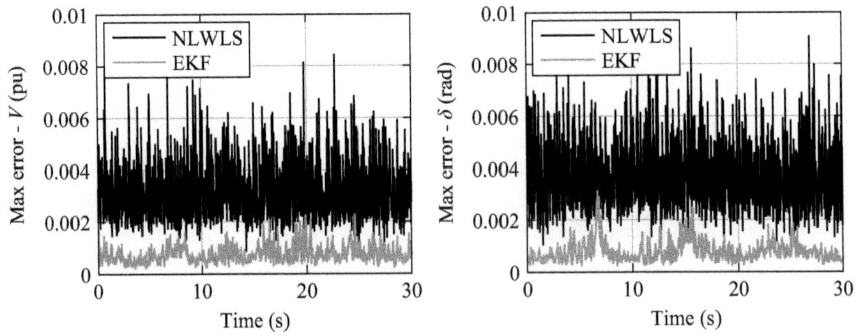

Figure 8.11 Time evolution of the maximum estimation errors calculated by using (8.196), for the NLWLS-SE and EKF-SE, case of the IEEE 13-bus test feeder

phase maximum errors of the NLWLS-SE are larger than the ones of the EKF-SE along the overall simulation time.

Figure 8.12 shows the RMSEs calculated by using (8.197), for the NLWLS-SE and the EKF-SE and considering the same time window of Figure 8.11. Only the maximum RMSE among the three phases for each bus is shown. The errors refer to both voltage magnitude V and phase δ. The magnitude error is expressed in pu and the phase error is expressed in radians. As it can be observed, the RMSEs of NLWLS-SE are larger than the ones of the EKF-SE for all the 13 network buses.

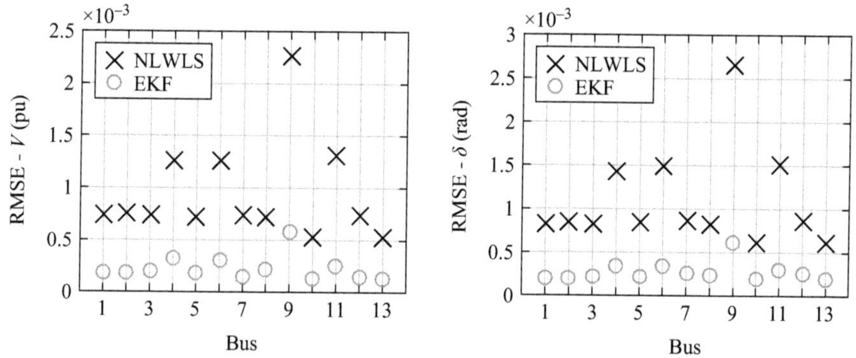

Figure 8.12 RMSEs calculated by using (8.197), *for the NLWLS-SE and EKF-SE, case of the IEEE 13-bus test feeder. Only the maximum RMSE among the three phases for each bus is shown*

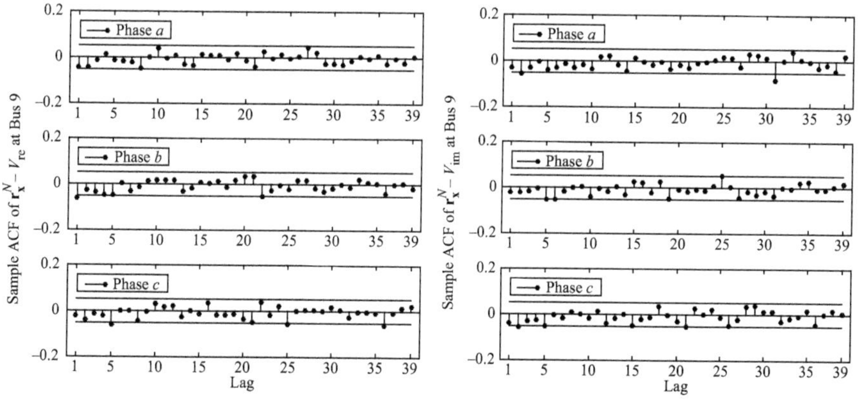

Figure 8.13 Process model validation for the case of linear SE (DKF) of the IEEE 13-bus distribution test feeder: sample ACFs of the normalized residuals of the state estimates referred to the real and imaginary parts of the voltage at Bus 9

8.8 Kalman filter process model validation

The validity of the KF algorithm depends on how accurate the underlying process model is. More precisely, since the KF equations use only second-order properties, it is sufficient to verify that the covariance properties of the model do hold. It is thus needed to verify whether the normalized residuals of the state estimates $\mathbf{r}_{\mathbf{x}}^{N}$ are uncorrelated by analyzing the sample auto correlation functions (ACFs). The ith element of $\mathbf{r}_{\mathbf{x}}^{N}$ is calculated as:

$$r_{\mathbf{x},i}^{N} = \frac{\widehat{x}_{t,i} - \widehat{x}_{t-1,i}}{\sqrt{Q_{t-1,ii}}} \tag{8.198}$$

The sample ACFs are computed by considering the first $\sim\sqrt{N_S}$ lags, where $N_S =$ 1 500 is the number of simulation time-steps. If the residuals are uncorrelated, the ACFs should be within the noise margins $\pm 1.96/\sqrt{N_S}$ with 95% of probability [53].

The above-mentioned condition is fulfilled for the case of the linear SE of the IEEE 13-bus distribution test feeder, as shown in Figure 8.13. The results refer to the real and imaginary parts of the voltage at Bus 9, which is the bus where the largest estimation error is observed. Since the normalized residuals of the state estimates are uncorrelated, the KF process model is accurate. Note that the sample ACF is, as a matter of fact, a statistically distributed quantity. The fact that, in few cases, the ACFs are slightly beyond the noise margins do not violate the validity of the result and the numerical proof of the statistical correctness of the process model. The same analysis for the case of the non-linear SE of the IEEE 13-bus distribution test feeder is shown in Figure 8.14 with respect to the magnitude and phase of the voltage at Bus 9.

It is worth pointing out that the validation of the KF process model together with the assessment of the process noise covariance matrix **Q** require further investigation. Future research shall focus on the analysis of the innovation and measurement residual vectors.

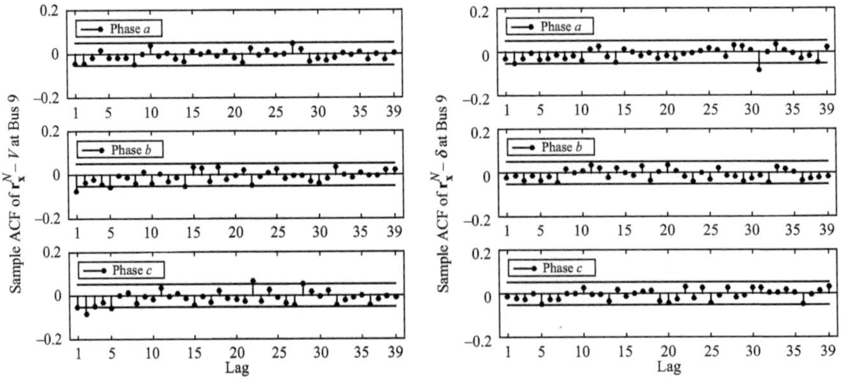

Figure 8.14 Process model validation for the case of non-linear SE (EKF) of the IEEE 13-bus distribution test feeder: sample ACFs of the normalized residuals of the state estimates referred to the magnitude and phase of the voltage at Bus 9

8.9 Numerical validation of Theorem 8.1

In this section, it is verified whether the quantitative conclusion of Theorem 8.1 of Section 8.6, namely (8.190), numerically holds. Figure 8.15 shows the left- and right-hand sides (LHS and RHS, respectively) of (8.190) for the case of the linear SE of the IEEE 13-bus distribution test feeder. The LHS and RHS of (8.190) are close to each other, which means that the equality in Theorem 8.1 is in expectation. In Figure 8.15, expectations are estimated by empirical averages; therefore, a small discrepancy is expected. Figure 8.15 also shows the contributions of the two terms of the RHS of (8.190). It can be observed that the contribution of the second one is predominant with respect to the first one, proving that the DKF is applied correctly.

Figure 8.16 shows the verification of (8.190) for the case of the non-linear SE of the IEEE 13-bus distribution test feeder. As it can be observed, the quantitative conclusion of Theorem 8.1 holds also in the case of the non-linear SE, proving that EKF is applied correctly.

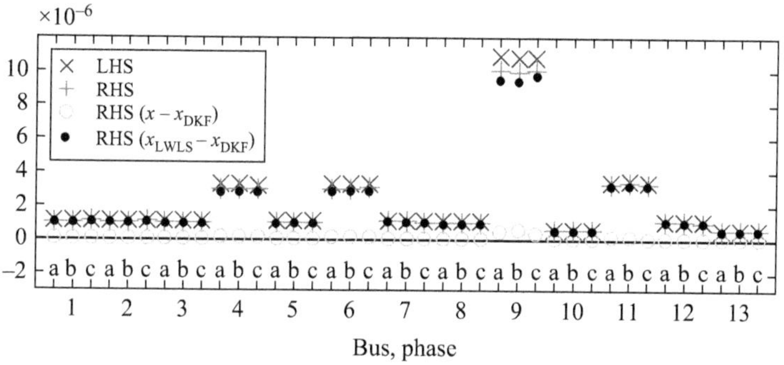

Figure 8.15 Numerical validation of (8.190) for the case of linear SE (DKF) of the IEEE 13-bus distribution test feeder: LHS vs. RHS of the equation. The separate contribution of the two terms of the RHS is also shown

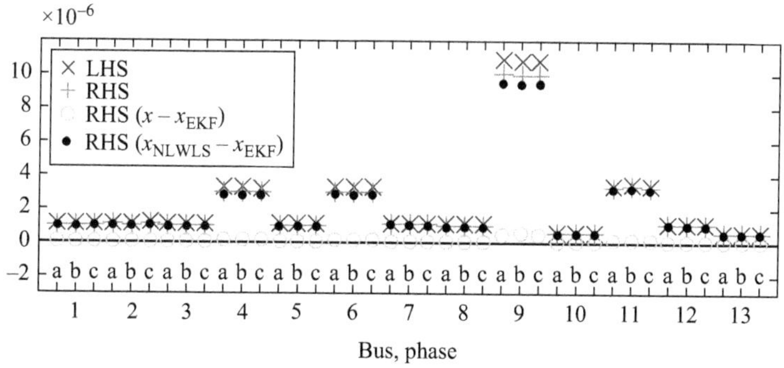

Figure 8.16 Numerical validation of (8.190) for the case of non-linear SE (EKF) of the IEEE 13-bus distribution test feeder: LHS vs. RHS of the equation. The separate contribution of the two terms of the RHS is also shown

Bibliography

[1] *IEEE Standard for Synchrophasor Measurements for Power Systems*. Standard IEEE C37.118.1; 2011.

[2] *IEEE Standard for Synchrophasor Data Transfer for Power Systems*. Standard IEEE C37.118.2; 2011.

[3] Jones KD, Thorp JS, Gardner RM. "Three-phase linear state estimation using phasor measurements". In: *IEEE PES Gen. Meet.* Vancouver, Canada; 2013.

[4] Zhang L, Bose A, Jampala A, Madani V, Giri J. "Design, testing, and implementation of a linear state estimator in a real power system". *IEEE Transactions on Smart Grid* (in press). 2016.

[5] Sarri S, Zanni L, Popovic M, Boudec JYL, Paolone M. "Performance assessment of linear state estimators using synchrophasor measurements". *IEEE Transactions on Instrumentation and Measurement*. 2016 Mar;65(3): 535–548.

[6] Powalko M, Rudion K, Komarnicki P, Blumschein J. "Observability of the distribution system". In: *Proc. 20th Int. Conf. and Exhib. on Electr. Distr., CIRED*. Prague, Czech Republic; 2009.

[7] Laverty DM, Morrow DJ, Best RJ, Crossley PA. "Differential ROCOF relay for loss-of-mains protection of renewable generation using phasor measurement over Internet protocol". In: *CIGRE/IEEE Power Energy Soc. Joint Symp. Integr. of Wide-Scale Ren. Res. into the Power Del. Syst.* Calgary, Canada; 2009.

[8] Samuelsson O, Hemmingsson M, Nielsen AH, Pedersen KOH, Rasmussen J. "Monitoring of power system events at transmission and distribution level". *IEEE Transactions on Power Systems*. 2006;21(2):1007–1008.

[9] Best RJ, Morrow DJ, Laverty DM, Crossley PA. "Synchrophasor broadcast over internet protocol for distributed generator synchronization". *IEEE Transactions on Power Delivery*. 2010 Oct;25(4):2835–2841.

[10] Carta A, Locci N, Muscas C. "GPS-based system for the measurement of synchronized harmonic phasors". *IEEE Transactions on Instrumentation and Measurement*. 2009 Mar;58(3):586–593.

[11] Liu J, Tang J, Ponci F, Monti A, Muscas C, Pegoraro PA. "Trade-offs in PMU deployment for state estimation in active distribution grids". *IEEE Transactions on Smart Grid*. 2012 Jun;3(2):915–924.

[12] Pignati M, Popovic M, Andrade SB, *et al.* "Real-time state estimation of the EPFL campus medium-voltage grid by using PMUs". In: *Sixth Conf. on Innov. Smart Grid Techn. (ISGT 2015)*. Washington, DC; 2015.

[13] Meier AV, Culler D, McEachern A, Arghandeh R. "Microsynchrophasors for distribution systems". In: *IEEE PES Innov. Smart Grid Techn. (ISGT) Conf.* Washington, DC; 2014.

[14] Schweppe FC, Wildes J. "Power system static-state estimation, part I: exact model". *IEEE Transactions on Power Apparatus and Systems*. 1970 Jan;PAS-89(1):120–125.

[15] Schweppe FC, Rom DB. "Power system static-state estimation, part II: approximate model". *IEEE Transactions on Power Apparatus and Systems*. 1970 Jan;PAS-89(1):125–130.
[16] Schweppe FC. "Power system static-state estimation, part III: implementation". *IEEE Transactions on Power Apparatus and Systems*. 1970 Jan;PAS-89(1):130–135.
[17] Grainger JJ, Stevenson WD. *Power System Analysis*. New York, NY: McGraw-Hill International Editions; 1994.
[18] Monticelli A. *State Estimation in Electric Power Systems: A Generalized Approach*. Berlin: Springer; 1999.
[19] Abur A, Expósito AG. *Power System State Estimation: Theory and Implementation*. New York, NY: CRC Press, Marcel Dekker; 2004.
[20] Zhu J, Abur A. "Effect of phasor measurements on the choice of reference bus for state estimation". In: *IEEE PES Gen. Meet.* Tampa, FL; 2007.
[21] Paolone M, Borghetti A, Nucci CA. "A synchrophasor estimation algorithm for the monitoring of active distribution networks in steady state and transient conditions". In: *Power Syst. Comp. Conf. (PSCC)*. Wrocław, Poland; 2014.
[22] Manousakis NM, Korres GN, Georgilakis PS. "Taxonomy of PMU placement methodologies". *IEEE Transactions on Power Systems*. 2012 May;27(2): 1070–1077.
[23] Xu B, Abur A. "Observability analysis and measurement placement for systems with PMUs". In: *Proc. of the IEEE PES Power System Conf. and Exp. 2004*, vol. 2. New York, NY; 2004. p. 943–946.
[24] Welch G, Bishop G. "An introduction to the Kalman filter". Department of Computer Science, University of North Carolina, Chapel Hill, NC; 2006. TR 95-041.
[25] Debs AS, Larson RE. "A dynamic estimator for tracking the state of a power system". *IEEE Transactions on Power Apparatus and Systems*. 1970 Sep;PAS-89(7):1670–1678.
[26] Nishiya K, Hasegawa J, Koike T. "Dynamic state estimation including anomaly detection and identification for power systems". In: *IEE Proc. on Gen., Transm. and Distr.*, vol. 129, no. 5. IET; 1982. p. 192–198.
[27] Zhang J, Welch F, Bishop G, Huang Z. "A two-stage Kalman filter approach for robust and real-time power system state estimation". *IEEE Transactions on Sustainable Energy*. 2014 Apr;5(2):629–636.
[28] Zanni L, Sarri S, Pignati M, Cherkaoui R, Paolone M. "Probabilistic assessment of the process-noise covariance matrix of discrete Kalman filter state estimation of active distribution networks". In: *Intern. Conf. Probab. Meth. Appl. to Power Syst. (PMAPS)*. Durham, UK; 2014.
[29] Caro E, Conejo AJ, Mínguez R. "Power system state estimation considering measurement dependencies". *IEEE Transactions on Power Systems*. 2009;24(4):1875–1885.
[30] Muscas C, Pau M, Pegoraro PA, Sulis S. "Effects of measurements and pseudo-measurements correlation in distribution system state estimation". *IEEE Transactions on Instrumentation and Measurement*. 2014;63(12):2813–2823.

[31] Boyd S, Vandenberghe L. *Convex Optimization*. Cambridge: Cambridge University Press; 2004.
[32] Kalman RE. "A new approach to linear filtering and prediction problems". *Transactions of the ASME-Journal of Basic Engineering*. 1960;82(1):33–45.
[33] Gelb A, Kasper JF, Nash RA, Price CF, Sutherland AA. *Applied Optimal Estimation*. Gelbe A, editor. Cambridge, MA: The MIT Press; 2001.
[34] Ghahremani E, Kamwa I. "Dynamic state estimation in power system by applying the extended Kalman filter with unknown inputs to phasor measurements". *IEEE Transactions on Power Systems*. 2011 Nov;26(4):2556–2566.
[35] Huang Z, Schneider K, Nieplocha J. "Feasibility studies of applying Kalman filter techniques to power system dynamic state estimation". In: *Inter. Power Eng. Conf. 2007 (IPEC 2007)*, Singapore; 2007. p. 376–382.
[36] Sarri S, Paolone M, Cherkaoui R, Borghetti A, Napolitano F, Nucci CA. "State estimation of active distribution networks: comparison between WLS and iterated Kalman-filter algorithm integrating PMUs". In: *Third Conf. on Innov. Smart Grid Techn. (ISGT Europe 2012)*. Berlin, Germany. IEEE; 2012.
[37] Jain A, Shivakumar NR. "Power system tracking and dynamic state estimation". In: *Power Syst. Conf. and Expos. (PSCE), 2009*. Seattle, WA. IEEE; 2009.
[38] Zhang J, Welch G, Bishop G, Huang Z. "Reduced measurement–space dynamic state estimation (ReMeDySE) for power systems". In: *2011 IEEE PowerTech*. Trondheim, Norway; 2011.
[39] Lerro D, Bar-Shalom Y. "Tracking with debiased consistent converted measurements versus EKF". *IEEE Transactions on Aerospace and Electronics Systems*. 1993 Jul;29(3):1015–1022.
[40] Göl M, Abur A. "LAV based robust state estimation for systems measured by PMUs". *IEEE Transactions on Smart Grid*. 2014 Jul;5(4):1808–1814.
[41] Jones KD, Pal A, Thorp JS. "Methodology for performing synchrophasor data conditioning and validation". *IEEE Transactions on Power Systems*. 2015 May;30(3):1121–1130.
[42] Pignati M, Zanni L, Sarri S, Cherkaoui R, Boudec JYL, Paolone M. "A pre-estimation filtering process of bad data for linear power systems state estimators using PMUs". In: *Power Syst. Comp. Conf. (PSCC)*. Wrocław, Poland; 2014.
[43] Zhu J, Abur A. "Identification of network parameter errors". *IEEE Transactions on Power Systems*. 2006 May;21(2):586–592.
[44] Monticelli A, Garcia A. "Reliable bad data processing for real-time state estimation". *IEEE Transactions on Power Apparatus and Systems*. 1983 May;PAS-102(5):1126–1139.
[45] Van Cutsem T, Ribbens-Pavella M, Mili L. "Hypothesis testing identification: a new method for bad data analysis in power system state estimation". *IEEE Transactions on Power Apparatus and Systems*. 1984 Nov;PAS-103(11):3239–3252.
[46] Falcao DM, Cooke PA, Brameller A. "Power system tracking state estimation and bad data processing". *IEEE Transactions on Power Apparatus and Systems*. 1982 Feb;PAS-101(2):325–333.

[47] Leite da Silva AM, Filho MBDC, Cantera JMC. "An efficient dynamic state estimation algorithm including bad data processing". *IEEE Power Engineering Review*. 1987 Nov;PER-7(11):49–49.
[48] Krishnan V. *Nonlinear Filtering and Smoothing: An Introduction to Martingales, Stochastic Integrals, and Estimation*. Mineola, NY: Courier Dover Publications; 2005.
[49] Group IDPW. "Radial distribution test feeders". *IEEE Transactions on Power Systems*. 1991;6:975–985.
[50] *Instrument Transformers: Additional Requirements for Electronic Current Transformers*. Standard IEC 61869-7; 2014.
[51] *Instrument Transformers: Additional Requirements for Electronic Voltage Transformers*. Standard IEC 61869-8; 2011.
[52] Athay T, Podmore R, Virmani S. "A practical method for the direct analysis of transient stability". *IEEE Transactions on Power Apparatus and Systems*. 1979 Mar;PAS-98(2):573–584.
[53] Le Boudec JY. *Performance Evaluation of Computer and Communication Systems*. Lausanne: EPFL Press; 2010.

Chapter 9
Real-time applications for electric power generation and voltage control

Massimiliano Chiandone[1], *Giorgio Sulligoi*[1] *and Vittorio Arcidiacono*[2]

This chapter deals with real-time applications of digital devices aimed at voltage control of transmission and distribution systems. The focus is on the implementation of the aforementioned devices as digital control systems using microprocessors and general-purpose computing systems, fitted out with operating systems for general use but with real-time characteristics. With this aim, high level simulation tools designed to model the plant process, and automatically generate a control code appear to be a promising approach to quickly prototype control systems. This chapter discusses in details the application and simulation results of such real-time operating systems (RTOSs) and hardware (HW)–software (SW) platforms for voltage control applications in some real-world distribution networks with inclusion of distributed generation.

9.1 Introduction

Voltage regulation and specifically some practical implementation issues related to voltage control strategies are the subject of this chapter. The main question is which real-time constraints have voltage regulators and therefore to present suitable HW and SW platforms that have been positively applied for the development of such devices.

The following sections aim at studying most relevant implementation aspects of the voltage regulation systems controlling the reactive power absorbed and injected into the network by generators. Voltage regulation in electrical systems is implemented acting on excitation fields of synchronous generators and controlling the reactive power of power stations. Ultimately, these regulations are both carried out by acting either (traditionally) on the excitation control systems (ECSs) of synchronous generators or on the voltage control systems of interface converters of the generators (modern distributed static generators).

A relevant aspect of the development of such regulators is their integration in distributed control systems (DCSs) through the use of computer networks. An emerging practice is the use of transmission control protocol (TCP) and the Internet

[1]Department of Engineering and Architecture, University of Trieste, Italy
[2]Retired Consultant, Milano, Italy

protocol (IP): some considerations and experimental measurements are presented. Two applications in power systems of the presented platform and methods are discussed in Section 9.3. Finally, possible applications of these platforms to voltage control with distributed generators are discussed.

This chapter is organized as follows. In Section 9.2, basic definitions related to real-time systems, with particular emphasis on RTOSs and real-time digital communications, are provided. These definitions are utilized to clarify the chosen HW/SW platform. Section 9.3 details the problem of voltage control and the real-time constraints involved. Two applications are presented: the development of an automatic voltage regulator (AVR) and of a reactive power control for secondary voltage regulator (SVR). The same HW and SW platforms are utilized for both AVRs and secondary voltage regulations. This chapter is completed by a case study that illustrates the possible application of the presented tools to the distributed generation. With this aim, simulation results showing the transient behaviour of a real-world distribution system are discussed.

9.2 Outlines of real-time system concepts

Real-time requirements for a specific control problem and real-time performance of a specific platform have to be compared to decide whether such a platform is suitable to implement the control. Other specifications such as the overall computing capabilities, eventually together with practical and economical considerations (e.g., cost of maintenance of the final device), finally lead to the decision on the platform to use.

This section provides some basic definitions for real-time systems. The interested reader can find a comprehensive description of real-time systems in, among others, Reference 1 and the IEEE standard IEEE 610.12 1990. The definitions provided below are the minimal necessary to follow the description of HW and SW platforms given in Section 9.3 and to compare the real-time performance with the requirements of the practical implementations, such as those illustrated in the case studies included in this chapter.

We start with defining *real-time systems*. Actually, several definitions can be found in the literature, depending on the application. In Reference 2, a real-time system is described as “any processing activity or system which has to respond to externally generated input stimuli within a finite and specifiable time delay”. In computer systems, the external generated stimulus is often called *interrupt* and the response of the system is the activation of a software routine that is known as *interrupt handler* or *interrupt service routine*. As for the definition above, concerning real-time computer systems, it should be possible to give an upper limit for the time elapsed between the interrupt occurrence and the execution by the processor of the software routine dedicated to that specific interrupt. It is thus relevant to define the *latency*, which is the measure of the delay passed from the occurrence of an external event and the execution of the related action; and the *worst-case execution time* (WCET), which is the finite and specifiable time for the execution of a specific task. The latency can include also the measure of the *scheduling latency* defined as the time elapsed between the scheduled time of a task and the time at which the task actually

```
while (true)
{
  read input values;
  calculate control values;
  write output data;
  wait for the next sampling period;
}
```

Figure 9.1 General scheme of a controller

starts [3]. The WCET is often a synonym of the maximum time to respond to external stimuli, e.g., an interrupt.

In some applications, it is theoretically possible to calculate the WCET [4]. However, in modern HW and SW platforms, the analytical calculation of the WCET is very difficult [5,6] if not even impossible. Due to the complexity resolving the actual WCET, estimates normally suffice. It should be also pointed out that theoretical analysis normally overestimates the WCET. Therefore, even when it is possible to calculate the WCET, theoretical appraisals are of limited interest. Thus, industrial engineers involved in real-time control development generally cannot count on a reliable estimation of the WCET provided by software developers.

The latency albeit limited in real-time systems is normally not constant. The variation of the latency is therefore called *jitter* (of the latency). The jitter is thus a measure of the predictability of the delay and, given a sufficient number of samples, allows evaluating statistical properties of such delay, e.g., its standard deviation.

In the following, we provide a couple of examples to explain the role and the significance of the latency and the jitter in practical real-time systems.

Let us consider the case of control algorithms first. A control law is designed according to discrete time control system theory and implemented as a real-time periodic task. The control algorithm is often implemented in such a way that the control task is in sleeping mode waiting for a signal from the scheduler. Then, once such a signal is received, the rescheduling has to occur exactly at a given time through an interrupt sent by an internal timer. A minimal structure of a control algorithm consists of the pseudo-code items shown in Figure 9.1. Given the constraint above on the execution of the rescheduling, it follows that the control action is a typical real-time task. It is also usual to name this case as time-driven real time in opposition with event-driven real time. Thus, we can apply the concept of latency. Clearly, from the designer's point of view, the variation of the scheduling time is a very important characteristic to take into account to determine if specific HW and SW platforms are suitable for a certain control task. As a rule of thumb, given a sampling frequency for the control task, a jitter of 10% of the scheduling period is an acceptable performance.

Another widely used application of the concept of latency is in the communication of data between two processing units. For real-time purposes, the delay in the communication channel has to be bounded by an upper limit. This can be a trivial requirement if the data are transmitted directly, e.g., through a wire, but can be somewhat difficult to calculate if a digitalized signal is transmitted by complex protocols not specifically

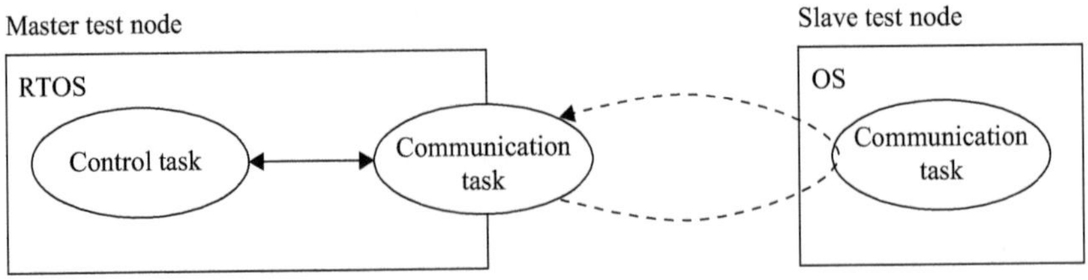

Figure 9.2 Round trip measurement of application level transmission delay

designed for real-time data communication. In such a case, it is necessary to measure the total transmission delay between the transmitting and receiving tasks. A common example of such delay and of how it can be measured is the so-called *round trip delay*, which is illustrated in Figure 9.2.

Likewise, the definition of WCET can be extended to communication delay. Unfortunately, some of the most common protocols used for data communications are inherently non-deterministic and therefore also the worst-case transmission time is not bounded. A relevant example is the standard protocol for Ethernet transmission described in IEEE 802.3 that, by construction, includes a random waiting period. This fact is sufficient to prevent real-time applications. Given their relevance to the matter discussed in this chapter, we dedicate the following subsection to briefly outline real-time operating systems (RTOSs).

9.2.1 Real-time operating systems

An operating system (OS) is identified as the software that controls and manages the hardware of a computing system; one of its main purposes is to implement the interface used by all the other users' programs to have access to the hardware itself.

Most OSs have been designed to have good average performance, but not to execute their jobs with strict temporal constraints. That is why normal OSs cannot be considered real time: there is not a WCET for any task; furthermore, the scheduling time is not upper limited. Moreover, OS runs on general purpose processors (GPPs) that introduces additional elements whose behaviour is not deterministic.

Real-time performance in modern OS is strictly related to one of the most common characteristics of general purpose OS, namely, *multitasking*. If a single task is running on a central processing unit (CPU), there are only minimal problems to control interrupts coming from the hardware. On the other hand, in a multitasking system, there are several active processes. Only one, however – in single processor systems – is running at a specific time. When an interrupt arrives from the hardware,* the running task has to be suspended to start the routine software dedicated to that interrupt. The time that elapses from the suspension of the task (including the operations to ensure a soundness computation) and the start of the handler of the interrupt causes the latency.

*For the interested reader, there are several ways in which interrupts are managed and the description of a complete implementation of an OS kernel can be found in Reference 7.

Together with multitasking, the other two major "enemies" for real-time applications in GPP are two hardware enhancements that can be found in all modern computers, namely *cache* (the mechanism to load data and instructions on a faster but smaller memory) and *pipelining* (the mechanism devoted to pre-fetch instructions to be executed in advance and in parallel). These two features improve the average performance, but worsen the efficiency of real-time applications.

In RTOS evaluation, these are the commonly considered latency types:

- *Interrupt latency*, which is the delay between an interrupt rise and the activation of the appropriate routine and the jitter.
- *Scheduling latency*, which is the delay between the time when a process is scheduled and the time when the process is actually started by the scheduler.

These two characteristics are similar. Based on the interrupt latency, in fact, it is possible to derive the scheduling latency. However, in practice, these quantities and their jitters are measured separately. For time-driven algorithms as in Figure 9.1, the scheduling latency is of major importance. A very simple method to measure it is to schedule a task to produce a square wave signal changing one digital output signal from zero to one and monitoring this signal through an oscilloscope.

To define the adequacy of a computing system for a defined control task, in addition to the real-time performance, the computational power shall be assessed. By inspecting the implementation code of an algorithm, it is possible to exactly determine how many floating-point operations have to be executed and to calculate the execution time of the control loop when executed on the target HW platform. In modern GPPs, the computational power is a constraint easily satisfied. Moreover, the progress in semiconductor manufacturing has made possible the industrial utilization of 32- and 64-bit GPPs with an extremely high computational power. This fact has led to the increasing interest in using such computing systems also for real-time tasks. Due to their high computing performance, relatively low price, and low maintenance cost, considerable efforts in extending standard OS to obtain RTOS have been carried out during the last three decades.

Several RTOSs running on modern GPPs have been proposed. Some of the RTOS have been designed from the ground up starting from a real-time kernel adding new features to become a complete OS, sometimes commercially supported. Others have been built as an extension of existing non-real-time open-source OS. In both cases, the main concern is to measure real-time performance to determine the applicability to real-time tasks.

In Reference 8 the latency and the jitter of four RTOSs for IBM-PC compatible computers (being IBM the leader in personal computers manufacturing at those time) are reported. A more recent comparison among different RTOSs on different hardware is presented in References 9,10. Comparing the results presented in those publications it is evident that in Reference 8 – back in 1993 – all times are expressed in milliseconds while in modern measurements delays are expressed in terms of tens of microseconds: the efforts to get real-time performance from those platform have been rewarded by satisfactory results.

In the literature, real-time performance is often considered in particular test conditions. It is important to note that real-time performance, being only an experimental proof of real time, has to be measured considering all tasks that will be performed by the CPU. The scheduling jitter is strongly dependent on the input and output (I/O) activity and with the general overall CPU activity and therefore it has been measured with the all I/O activity supposed to be necessary for the particular control system developed. Extended measures of the scheduling jitter in several different conditions have been carried out for the described HW and SW platforms.

Statistical analysis of the latency has to be performed in real-world conditions along with the recording of maximum measured jitter over a long-time range, given that many anomalies in RTOS on GPP in real-time scheduling are related to very low probability, sporadic events (with average periods of the order of minutes). Since an advantage of these real-time architectures is to support the execution of both real-time and best-effort tasks on the same hardware, a very interesting aspect of performance analysis is to assess the amount of adverse effect that best-effort execution may have on the real-time part of the system. Therefore, latency has to be measured with intense I/O activity and CPU load.

The measures discussed below refer to a system made with an industrial PC with a standard ATX form factor motherboard equipped with Intel Pentium 4 CPU running at 3 GHz, 512 MB of RAM, a low-cost IDE disk of 10 GB. The RTOS chosen is an RealTime Application Interface (RTAI) (version 3.6) over Linux kernel 2.6.23. Software has been compiled and installed on the specified hardware. References 11–13 provide more details on the hardware and software utilized in this experiment.

The scheduling jitter of a simple task scheduled with a sample time of 1.6667 ms has been monitored as described in Reference 14 and is shown in Figure 9.3. The probability density distribution of the jitter together with the calculated standard deviation is shown in Figure 9.4.

No other intense activity (neither real-time nor non-real-time tasks) is executed on the CPU in the examples in Figures 9.3 and 9.4. The probability density function of a similar task with the addition of several I/O channels (real-time tasks) and best-effort tasks is shown in Figure 9.5.

Testing an equipment for several hours it is possible to attest a sort of experimental WCET. For the application of such a platform to power system control devices, a test aimed to measure a three-phase system of currents and voltages is shown. It has been decided to read 12 samples in a period at industrial frequency, i.e., 50 Hz: a sample time of 1.666 ms is obtained.

Two sinusoids synchronized and shifted by 120° from each other are read (calculating the third one), and a mean value of the maximum amplitude is calculated with the following expression:

$$\begin{aligned} V &= \frac{2}{3}\left[V_M \sin^2(\omega t) + V_M \sin^2\left(\omega t - \frac{2\pi}{3}\right) + V_M \sin^2\left(\omega t + \frac{2\pi}{3}\right)\right] \\ &= \frac{2}{3}V_M + \frac{2}{3}V_M\left[\cos(2\omega t) + \cos\left(2\omega t - \frac{2\pi}{3}\right) + \cos\left(2\omega t + \frac{2\pi}{3}\right)\right] \end{aligned} \tag{9.1}$$

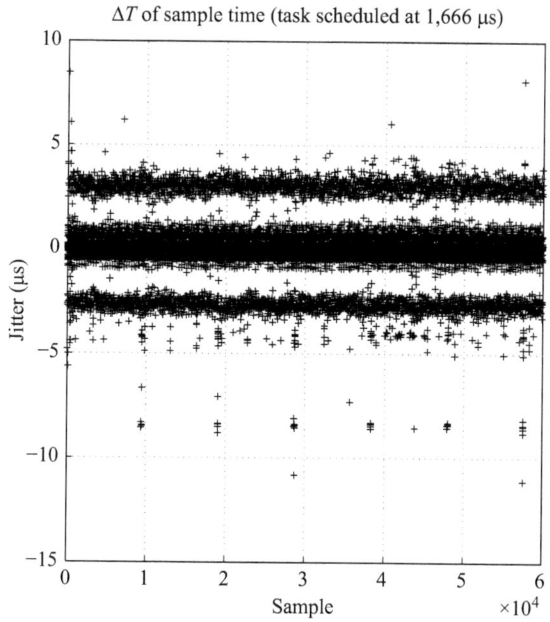

Figure 9.3 Latency of scheduling time with no I/O and no intense best-effort CPU activity

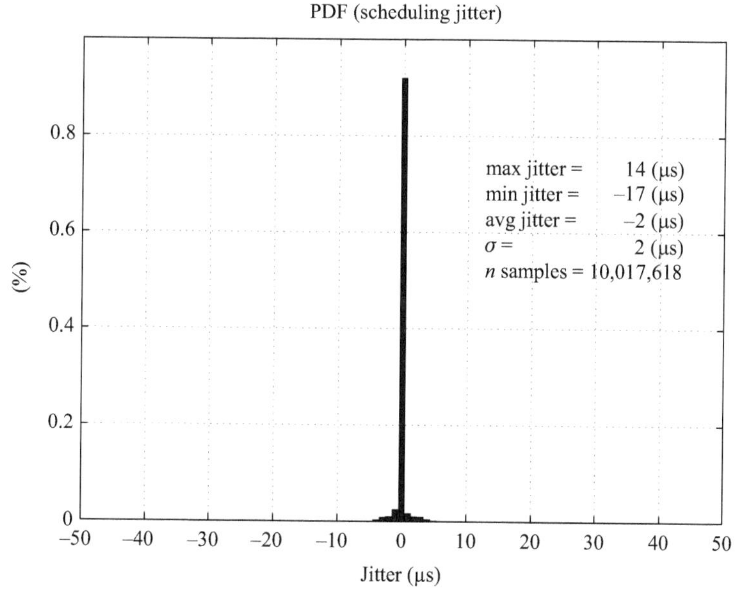

Figure 9.4 Probability density function of latency with no I/O and no intense best-effort CPU activity

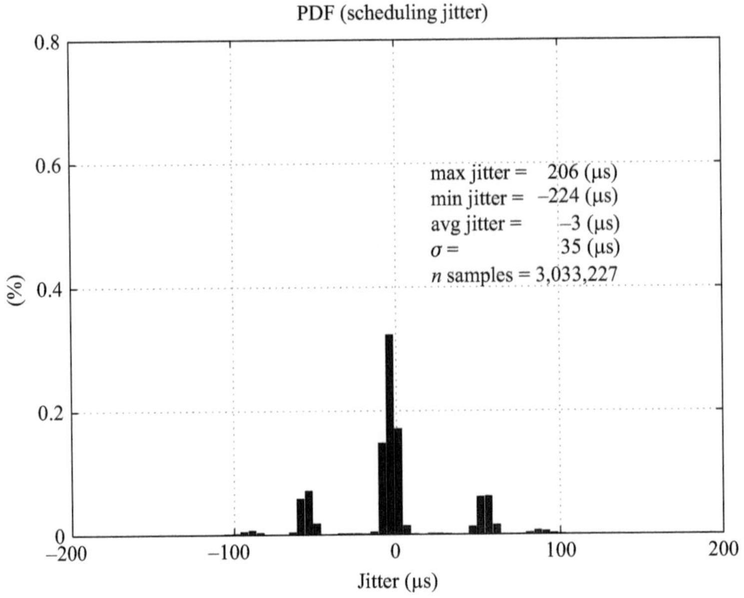

Figure 9.5 Scheduling latency PDF with intense CPU activity

Therefore, one obtains an average value of the maximum value and a harmonic with pulsation 2ω, which is not null if the three sinusoids are not balanced. Such a harmonic can be filtered through a second-order low-pass filter with a cut-off frequency of 40 rad/s. The chosen cut-off frequency allows attenuating the noises, but not limiting the control bandwidth. Note that, for example, for typical AVRs is far beyond 40 rad/s. Figure 9.6 shows a sampling activity test for a misconfigured PC. The ripple on the filtered voltage is higher than 0.4%. Figure 9.7 shows a sampling activity test on a well-configured PC where the jitter of the scheduling time is very low. The ripple on the filtered voltage is lower than 0.02%.

The scheduled time is shown for three periods, i.e., 0.06 s, along with the calculated v and the filtered v. The panel on the right shows the probability density function (PDF) of the sample time for a longer period while in the bottom the v calculated over a time long 9 s. These measures are the output of a basic block that has been coded and used in both devices described in the following section. This example shows that the voltage measurement is affected by the real-time performance of the system. Hence, systems with a WCET experimentally tested can be inappropriate if the jitter is too high.

The presented HW and SW platforms with the real-time performance shown in Figure 9.7 can be successfully used for real-time control tasks. Our experimental activity has been focused on the application of such real-time computing systems to several different fields. In References 11,12, we explored the possibility of designing and developing motion controls using high-level languages. In particular, a feedback acceleration control loop has been developed along with acceleration signal

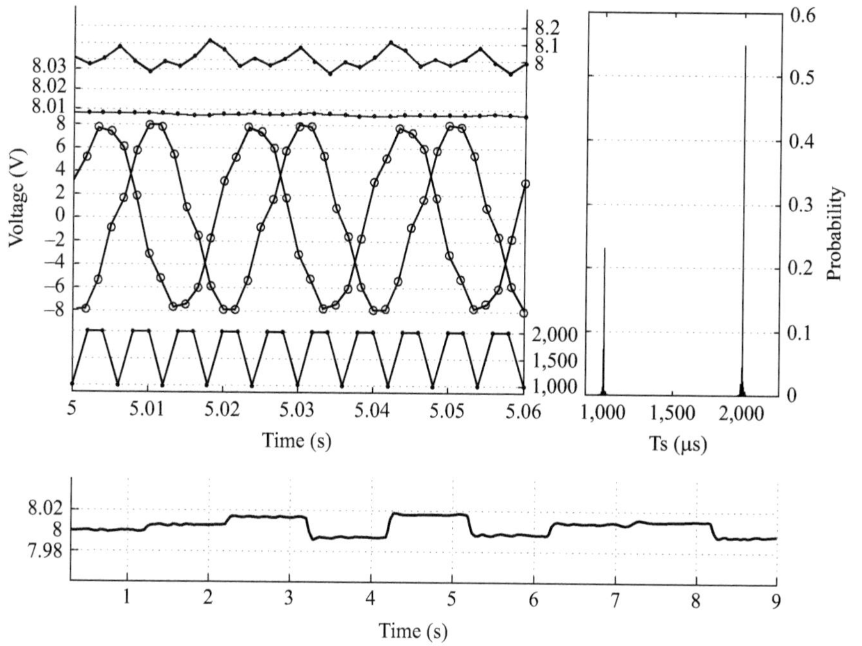

Figure 9.6 Three-phase voltage sampled with bad real-time performance

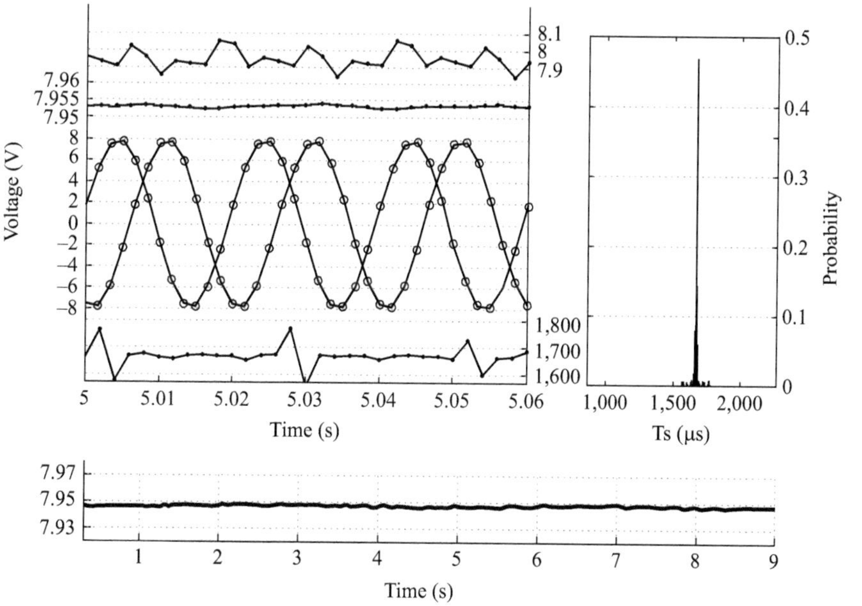

Figure 9.7 Three-phase voltage sampled with good real-time performance

processing tasks. An industrial PC-based acceleration control is presented to show that such a system offers adequate digital programming and signal processing capabilities for real-time motion control applications.

Another application is the monitoring of electrical and climate data for photovoltaics (PV) systems [13]. Data are taken both from the PV modules and from the power conditioning system of a given PV field, through appropriate sensors, to monitor: climate data (irradiance, temperature), electrical quantities (voltage, current, power). The system was designed to exchange data through a communication network, i.e., Ethernet/TCP-IP to download data, visualize the quantities and operative states. The aim is to collect data to be utilized in sensitivity studies of PV generators with respect to plant layout, devices and climate parameters.

The use of a control system based on an RTOS running on a GPP has been also suggested to improve the stability of an electromechanical shipboard integrated power systems in Reference 15 and for the particular case of islanded grid operation of the so-called microgrids.

In Section 9.3, two applications in power systems are presented, namely primary and secondary voltage controls. Such applications have been designed, developed and finally deployed in several real-world power systems.

9.2.2 Real-time communications

The need for communication functionalities for a real-time device comes mainly from the necessity to integrate the control device in Supervisory Control and Data Acquisition (SCADA) systems or a DCS or an energy management system (EMS). Not all the above require to be implemented in real time. For example, the connection to a SCADA system does not normally require hard real time, being the SCADA system mainly used as interface for human interactions. With the aim of developing real-time devices for power systems, three different types of communications are relevant:

- *Control*: real-time communications for the control itself with peripheral devices;
- *Protection*: real-time communications for auxiliary functions, i.e., redundancy;
- *Monitoring*: human–machine interface (HMI) communications, interfacing with SCADA systems, logging activities.

Several different technologies have been applied in DCSs: starting from wired protocols, e.g., the serial RS-232, to radio wireless protocols, thousands of different protocols have been developed and used so far. The convergence of computer and communication systems together with the digitalization of control devices and communications has led to a progressive use of standardized protocols. In the last decade, there has been a clear increase of the utilization in industrial automation of protocols that come from the ARPANET and have evolved into the Internet Protocols. The reference model of this architecture is often called TCP/IP. Its roots can be found in Reference 16, from where it has evolved into a complex suite of protocols.

The TCP/IP family of protocols has not been designed for specific control tasks, but is nowadays widely used also in the industrial automation context. One of the most common data link and physical implementation used with TCP/IP is Ethernet. Others exist, e.g., Token Ring, but the Ethernet is the one that leads in local area network (LAN) applications. Ethernet is a commonly used term that indicates several different protocols and physical media that can be used and have been standardized as IEEE 802.3: it comprises communication over copper wires or coaxial cables as well as fibre-optic links.

Two of the first media used for Ethernet were the thick and thin coaxial cables (formally known as 10Base5 and 10Base2). The media were shared among all the transmitters using a protocol named CSMA/CD (carrier sense multiple access with collision detection). The principle of CSMA/CD technique is that all transmitting stations access simultaneously the media (the cable), each one can try to transmit data and can detect if another station is transmitting at the same time (a collision occurs). If a collision is detected, the transmitting station waits for a random interval before trying to resend the data. Without going into the details of IEEE 802 standards, it is of paramount importance to highlight that the presence of a random waiting time leads to a theoretically unlimited delay in transmission. Thus, it is not possible to calculate a worst-case time for transmission and therefore CSMA/CD is not suited for real-time communications.

IEEE 802.3 protocols have evolved from the early 1970s to comprise different media as twisted-pair copper cables and fibre-optic cables. Also the protocols have changed to include different speeds in the transmission (from initial 10 Mbit/s to 100 Mbit/s and 10 Gbit/s), including full-duplex data flow, extending the frame (the unitary set of data) size. The major improvement towards an industrial use of Ethernet was the introduction of the switched Ethernet. When twisted copper wires are used as media, the stations are connected through a central device called hub. This is a passive device that repeats whatever it receives from an input port to all output ports connected to the stations. In the case of switched Ethernet, the hub is no more a passive device but an active one, which identifies the destination ports and relays the data only to these. This avoids collisions involving traffic from different ports. Moreover, in 1997, the full duplex was introduced: each interface was now capable of simultaneously transmitting and receiving remote traffic, eliminating the possibility of local collisions. Ethernet switch and full-duplex transmission mean that data delivery among multiple stations happens without collisions as long as the destinations are different.

Switched Ethernet with full-duplex communication and quality of service (QoS) at layer two render the Ethernet standard practically real time. Several industry solutions, e.g., PROFINET, have been developed to add real-time characteristics to Ethernet (building transmission algorithms with proved limited transmission time) but most of industrial proposals for real-time Ethernet require specific hardware. Efforts have been made to guarantee determinism over off-the-shelf Ethernet, i.e., to work with any Ethernet adapter. An example is RETHER [17] which is based on a token passing scheme that regulates the access to the network by passing a control token among the nodes on an Ethernet segment. Also RTnet [18] is a software framework to provide deterministic Ethernet that is already implemented on RTAI.

Performance analysis of Ethernet is not an easy task. In Reference 19, a low-cost distributed measuring instrument to measure timing characteristics of a real-time Ethernet is presented. To analyse network performance, Network Calculus [20,21] has been applied to switched Ethernet [22]. Measurements of TCP and UDP data transfer throughput [23] test how fast data can be sent between two nodes. This is not of interest in real-time controls where often the data to transmit is limited (e.g., commands to an actuator) and the interest is in timeliness of the transmission rather than in throughput. Performance of both UDP and TCP protocols is measured in Reference 24. Although it is seldom a good idea to design the control on the basis of the worst case (bringing to a resource waste), it is important to exactly know the worst-case time communication. In Reference 25, a worst-case communication delay with a frame size of 144 bits and a switch connecting 24 stations has been calculated in 1.745 ms.

In Reference 26, transmission delay and jitter are measured for non-switched, switched and switched with priorities – e.g., with QoS – Ethernet. It also shows the influence of external traffic in all three cases. The switched Ethernet with priorities shows a constant delay (around 100 μs) despite the increment of traffic and a very low jitter (tens of microns). With low external traffic also switched Ethernet without priorities shows similar performance. The practical effects of both switched and non-switched Ethernet delays in a motion control system have been shown in Reference 27. A delay under 1 ms was observed regardless the amount of traffic in switched Ethernet.

Based on the above results, a test of communication using a standard switched Ethernet (non-real-time) was set up between two stations equipped with RTAI to measure end-to-end delay transmission time and jitter. "End-to-end", in this context, means between two real-time tasks developed from MATLAB®/Simulink® or Scilab/Scicos or any other high-level language with C code automatically generated. Programs are compiled on the RTOS and are executed in user space. According to Figure 9.8, when using non-real-time Ethernet, an interface between the real-time task and the TCP/IP stack of the OS is needed. It is necessary to remark that, unless special care is used, sharing data between real-time and non-real-time tasks leads to serious scheduling anomalies, concerning, for example, unbounded priority inversion, which can (and probably will) undermine any real-time capability the system may offer.

The network link can be made up of several different media: a cross Ethernet cable for two units at a distance of few metres can be adequate, two cables connected to a switch, copper cables and fibre cables with appropriate interfaces for long distance. In the tests carried out, a round trip time of about 400 μs was measured with a jitter around 100 μs. Having set a control cycle of 1.666 ms, a transmission was able to be completed for each cycle. When several units are connected (as it is shown in the example), the rate transmission for each channel was limited to 10 ms. The data pass every control cycle from the control task to the communication tasks, but from here, the data are forwarded by UDP packets with a rate that can be set at run time. The same communication task, set with a lower transmission rate, is also used for other non-real-time communications: the UDP packet is sent each second to the SCADA system.

In the communication between two units, 30 doubles are transmitted every 10 ms. It means a transmission rate of less than 0.2 Mbit/s for each channel. The

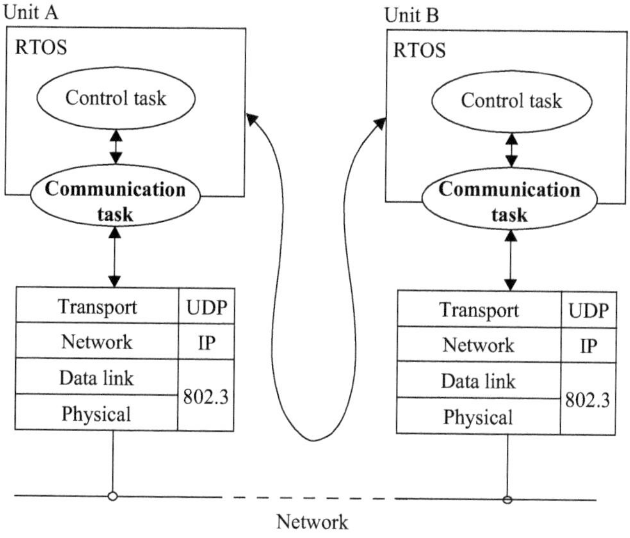

Figure 9.8 Communication between two real-time tasks

defragmentation of IP packets was avoided with consistent configuration of station's TCP/IP stacks, i.e., avoiding the transmission of data exceeding the standard packet size. A central unit with six peripheral units, each of them transmitting 30 doubles, means receiving 1.2 Mbit/s, thus quite lower than the Ethernet bandwidth.

Measurements concerning communication with commercial protocols commonly used in industrial automation (and specifically those very common in SCADA power systems as open platform communications (OPC) and Modbus) have been carried out. For example, Figure 9.9 shows a test bed for testing latency in Modbus communication.

A digital signal is sent to an oscilloscope through a digital output from a real-time task. The same signal is sent via Modbus protocol (using Ethernet over copper cables and optical fibre cable and appropriate converters) to a Modbus device in which a coil (the digital signal in the standard Modbus) is commanded. The output from the Modbus device is displayed on an oscilloscope. A transmission time of about 25 ms has been registered with a maximum jitter of ± 20 ms. This is considered adequate to transmit a pulse-width modulation (PWM) signal with a frequency of 1 Hz (this is the frequency normally adopted for the UP/DOWN pulses to the AVRs that are described in Section 9.3.1). With the above experimental measurements and the scientific literature proposed in the bibliography, the communication via TCP/IP over standard Ethernet has been chosen for the communication in the control devices that are described in the next section.

It is interesting to note that current communication via TCP/IP also over wireless physical media has been proposed and experimented in the literature for power systems [28].

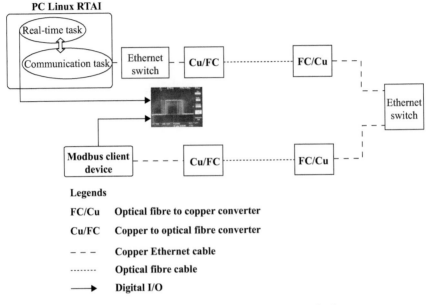

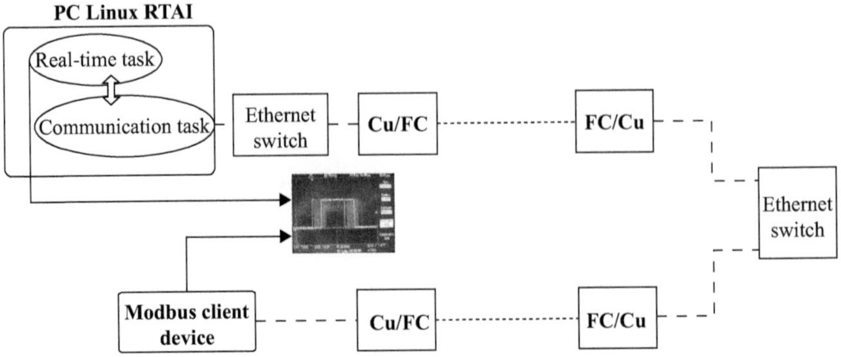

Legends

FC/Cu **Optical fibre to copper converter**

Cu/FC **Copper to optical fibre converter**

– – – **Copper Ethernet cable**

-------- **Optical fibre cable**

⟶ **Digital I/O**

Figure 9.9 Modbus communication test bed

9.3 Voltage control

An electric power system is essentially devoted to transforming the energy starting from several different primary sources into electrical energy and transfer the electrical energy to each of the several users with the required quality, e.g., a guaranteed frequency and voltage. The overall network is meshed and continuously subject to modifications in topology (due to faults, improvements, maintenance), production (insertion and disconnection of generators) and energy consumption (fluctuation of the loads); therefore, the operating point is continuously changing. A complete centralized control of the whole system has been always considered unfeasible.

Commonly, power system voltage control is divided into several levels. The first two levels of the hierarchical voltage control, as it is implemented in the Italian power system, are constituted by the primary voltage regulator actuated on generators (and this is the same in any power system with synchronous generators) and the secondary voltage control actuated by the reactive power controllers of reactive power sources (this can differ in different implementations of a hierarchical voltage control). The first level is implemented by the AVRs of synchronous generators while the second level is actuated, in Italy, by the so-called automatic system for voltage regulation (SART) also called reactive power regulator (RPR), installed on most of power stations (for Italian Grid Code all those power stations with a nominal power above 100 MVA). Other devices are used to control the voltage in transmission grid like shunt reactors and capacitors, static VAR systems. These systems are not covered here, although for the implementation of most of them the presented platforms and technique can be profitably used.

These two levels are implemented using different control loops, each of them has a well-defined dominant time constant:

- Generator voltage control loops, implemented by AVRs, are characterized by a dominant step response time constant (closed loop) of about 0.5–1 s.
- Reactive power control loops of the power stations, implemented by SARTs, are characterized by a dominant step response time constant (closed loop) of about 5–10 s.
- Pilot nodes voltage control loops, implemented by regional voltage regulators (RVRs), are characterized by a dominant step response time constant (closed loop) of about 50 s or higher.

As a result, comparing the real-time requirements with the real-time performance of RTOSs presented, the most demanding task is the measurement and processing of the electrical quantities, i.e., 50 Hz three-phase voltages and currents, and not the dynamic performance of control loops, since the three control loops specified above are slow. In Section 9.2.1, it has been shown that a sampling time of 1.666 ms with RTAI has adequately low jitter and therefore the presented HW and SW platforms are suitable for the purpose.

In this section an implementation of an AVR and a SART is presented. Both systems have been developed with high-level tools as presented in Reference 29: from the mathematical model, a C source code is automatically generated ready to be compiled on the final hardware. For both devices, a functional scheme as in Figure 9.10 is implemented.

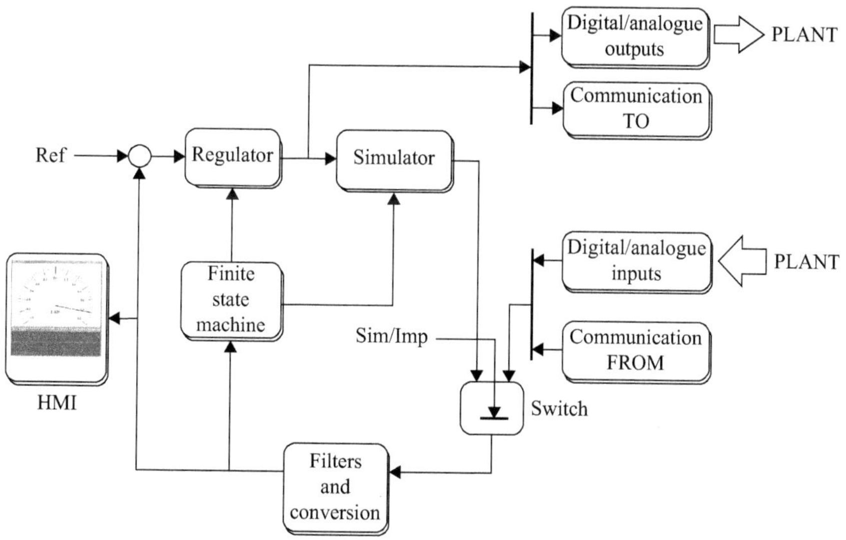

Figure 9.10 Functional scheme of a generic regulator

The device consists of several modules as follows:

- A real-time simulator of the physical plant (generator/grid) mathematically modelled using a high-level software tool, e.g., MATLAB/Simulink and Scilab/Scicos: once the regulator shows the desired behaviour, source files are generated and compiled to run in the regulator block of the final hardware device. The simulator code is left in the final code of the controller and runs in parallel with the control algorithm in real time. It can be used to facilitate factory commissioning and further maintenance or upgrade activities.
- In charge of the regulator, there is the main control task of the device under development (in both presented cases a P.I.).
- A finite state machine realizes all the logic functions to be handled. Traditional implementations make use of industrial programmable logic controller (PLC) for these functions. With the presented platform, it is possible to embed all the logic in the same model. That means a considerable saving in installation and maintenance costs.
- The HMI, which can be realized using the RTAI-Lab extension or other suitable libraries. The HMI could be displayed directly on the console of the industrial PC or remotely everywhere through the network capabilities offered by the RTOS. A very interesting advantage offered by the SW platform is the possibility to use all graphics libraries available on the OS, i.e., it is possible to use more sophisticated widget libraries to obtain an accurate interface.
- The I/O activities, which include drivers for both digital/analogue I/O cabled signals and communications from/to DCS, SCADA, SVR, and so on. These I/O channels are implemented using digital and analogue I/O boards (for cable signals) or using network boards (for remote digitalized communications): as the HW platform utilized is a standard PC, it is possible to use the widest range of I/O cards. In the presented implementations standard PCI cards were used: this choice gives an immediate economic benefit and also a future benefit for the maintenance in the long life cycle of the entire device.

9.3.1 Excitation control systems

ECSs of synchronous generators actuate the primary level of voltage regulation: they work to maintain constant the voltage of the output terminals of the generator (or another point electrical near to the generator) under various load conditions. This task is accomplished basically controlling the excitation field of the synchronous electrical machine.

Over time, these control systems have evolved to comprise other tasks related not only to the single generator but also to the overall electrical system as well. Besides the specific task of controlling the voltage of the synchronous generator at a certain node, an ECS performs other control functions:

- It maintains the working point of the generator inside the capability of the generator, namely, field excitation current, and under- and over-excitation limits.

- It compensates the voltage reactive drop controlling the voltage at a certain point of the system through the compound action.
- It adds damping to the electro-mechanical system oscillations. This function is called power system stabilizer (PSS).
- It protects the generator and the step-up transformers against high magnetic fluxes by controlling the ratio of per unit voltage to per unit frequency through the V/Hz limiter.

Technical specifications of the functionalities requested by an ECS can be found in Reference 30. The specifications provided in such a standard are independent from the technology adopted for the implementation.

An excitation control system (ECS) requires real-time signal processing and it has been historically implemented with analogue electronics and dedicated digital signal processors (DSPs) or microcontrollers. The embedding of digital systems with DSPs to implement algorithms for active and reactive power control of synchronous generators is described in Reference 31. There is also a consolidated experience in the use of microprocessor technology to implement ECSs that has been reviewed in Reference 32 (although the development cycle is very different from that here described).

The ECS has been developed according to the block representation depicted in Figure 9.11.

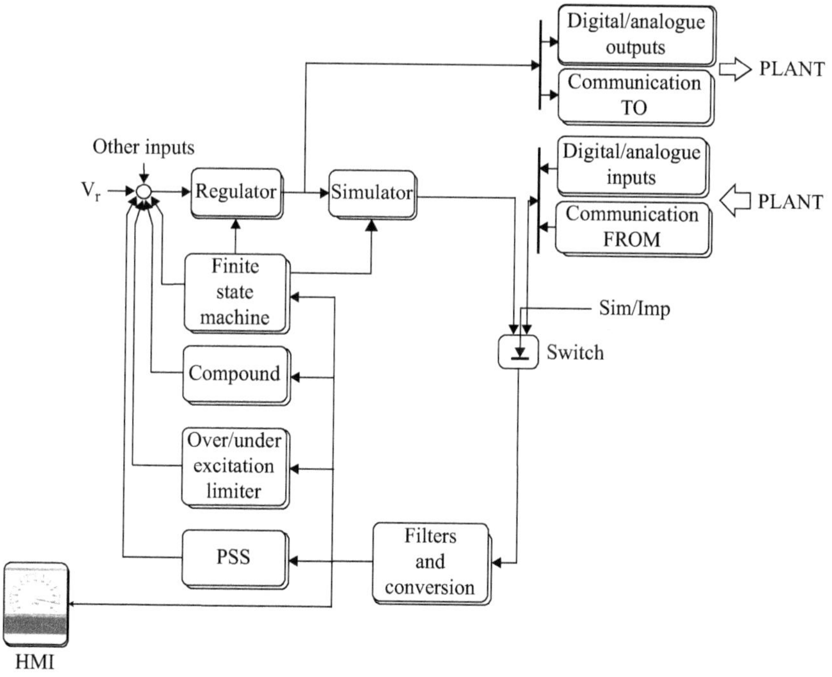

Figure 9.11 ECS implementation

The finite state machine realizes all the logic functions to be handled (start/stop sequences, the synchronization for parallel connection, protection functions, warnings and alarms). A voltage regulator is the proper control process and it is essentially implemented as a proportional integral type regulator. Line drop compensation, reactive droop compensation, PSSs are embedded software modules.

9.3.1.1 Redundancy

Control systems for generators require to have high integrity and high availability. Several architectures can be adopted to achieve a higher level of dependability. An example applied to an ECS can be found in Reference 33 where a triple-modular redundant architecture is shown. Three channelled subsystems, fully active and with equal authority, are coordinated by a median authority unit to derive a single control signal based on a majority principle. In our work, it has been decided to realize a redundant two-channel structure as shown in Figure 9.12.

Two digital control units are working in parallel, receiving the same signals from the plant. An arbiter logic function (hardware implemented) sets one of the two controllers as the master and the other as the slave. The two controllers communicate using TCP/IP (a simple protocol on UDP has been chosen) and exchange continuously information with each other to assure a bumpless transfer between the two controllers. The slave unit copies the state variables (both numerical and logic) memorized by the master. The communication channel is physically implemented as a full-duplex Ethernet communication over a cross Ethernet cable.

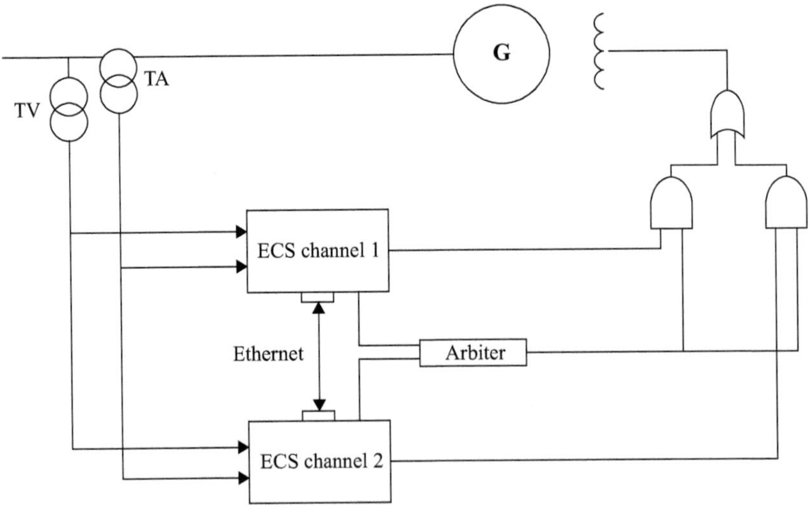

Figure 9.12 Illustration of the redundancy principle

9.3.2 Secondary voltage control

In order to improve voltage control, several countries have implemented hierarchical systems aimed at coordinating the management of reactive power resources. The main benefits recognized to a coordinated control are as follows:

- To improve the voltage profile of all high voltage nodes by reducing the variation around the desired values.
- To enhance system security by increasing the reserves of reactive power available to support the network during emergency conditions.
- To increase the active power transfer capability.
- To better utilize reactive resources and therefore reduce the total power system losses.

Studies on real-time automatic control of power system voltage and reactive power started in the 1970s and led to significant proposals and applications. Algorithms and control schemes have been initially proposed and discussed in References 34,35 and then exploited and developed in several countries under different approaches.

One of the implemented strategies of voltage control in EHV transmission networks consists in dividing a power system into several areas [36] and setting up a hierarchical voltage control system. Voltage hierarchical controls have been studied and developed in Italy [37] and France [38] starting in 1980s. Hence similar hierarchical controls for EHV transmission networks have also been adopted in Spain [39] and South Africa [40]. In general, however, each country implements its own voltage control. Some implementations are summarized in Reference 41.

In Italy, the first SVR has been implemented at the end of the 1980s by the Automation Research Centre (CRA) of the former Italian Electricity Board (ENEL) [37,42,43]. Since November 2005, the code for transmission, despatching, development and security of the Italian grid (Grid Code) has been enforced. This code regulates the relations between TERNA and grid users. Chapter 4 ("Dispatching Regulations") of the Grid Code states that each power plant participating to SVR must be equipped with a new generation apparatus named "SART", which must comply with the specifications given in the annex document A.16 [44].

At the time of writing this chapter, the coordinated automatic voltage control is operated by TERNA, namely, the Italian Transmission System Operator. The national EHV transmission system is divided into regions; each region is divided into areas; and each area is characterized by a pilot node [37, 45, 46]. The system controls pilot nodes voltage through a hierarchical structure made of three control levels. The first hierarchical level, namely, primary voltage regulation (PVR), is constituted by the conventional generator voltage control loops that are implemented by the AVRs. The voltage references of the AVRs are controlled by the outputs of higher level controllers. The second hierarchical level, namely, SVR, includes the reactive power control loops of each power station's generator, that are closed by local voltage and reactive power regulators named SART. Each power plant participating to the SVR receives from the RVRs a reactive level, expressed in percentage of the reactive

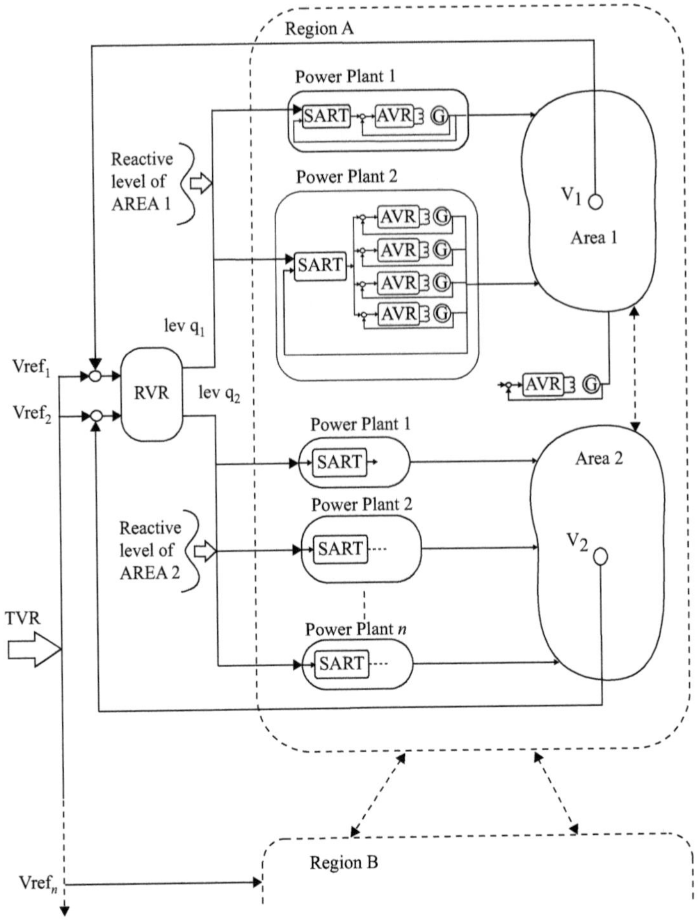

Figure 9.13 Schematic diagram of the voltage control hierarchical system

power limit. Such a limit is computed in real time as resulting from generator's capability curves. The SART apparatus actuates the reactive level by generating UP/DOWN commands (through width- or frequency-modulated pulses) to the voltage reference calibrators of the AVRs. The second level is completed by the pilot nodes voltage control loops, which are coordinated by the RVRs [37, 45, 46]. A scheme of principle representing the PVR and the SVR is shown in Figure 9.13.

9.3.2.1 SART: automatic system for voltage control

Considering the number of devices that would be needed for the Italian electricity market (i.e., an optimistic list of no more than 100 power stations), the high flexibility required for the implementation of some logics of the device and also of the overall architecture, economical constraints and maintenance goals, it has been decided to develop the SART for Italian power stations using RTAI on common industrial PCs.

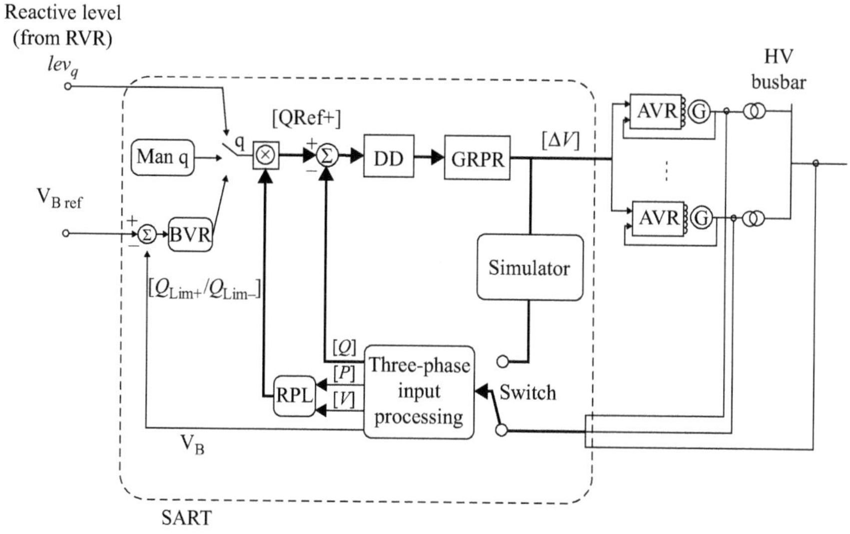

Figure 9.14 SART Central Unit Regulator block scheme

The SART is composed of the Central Unit (SART-CU) and the Peripheral Units (SART-PU). While the SART-CU has been in all cases implemented with an industrial PC endowed with RTAI, for the SART-PU different solutions have been proposed: an implementation using the same hardware and software adopted for the central unit or the use of commercial measuring devices integrated in the system.

A SART Central Unit

The SART-CU executes the real-time code that realizes all the following functionalities as embedded software modules. The regulator implements the control algorithm shown in Figure 9.14. It includes:

- A reactive power control loop for each generator.
- The power station high voltage busbar voltage control loop.

There are three different inputs of the regulator according to the functional behaviour chosen. The regulator receives a voltage control reference when operating as voltage busbar controller, or receives a reactive level signal when operating as reactive controller under the SVR. The third possible input is when the reactive power level (RPL) of the power plant is manually input by the operator through the HMI (normally this is done only during commissioning activities).

The reactive level lev_q received from the RVR is a scalar value varying in the range $[-1, +1]$. The value $+1$ represents a request to the power plant to produce the maximum value of reactive power, according to the actual voltage and active power. Similarly, the value -1 means the maximum reactive power absorption request according to the operating point of each generator in the power plant. From the RPL block comes out a value in pu for each generator, indicating the limit of the generator according to the actual limit curve. The working point sets a limit in the reactive

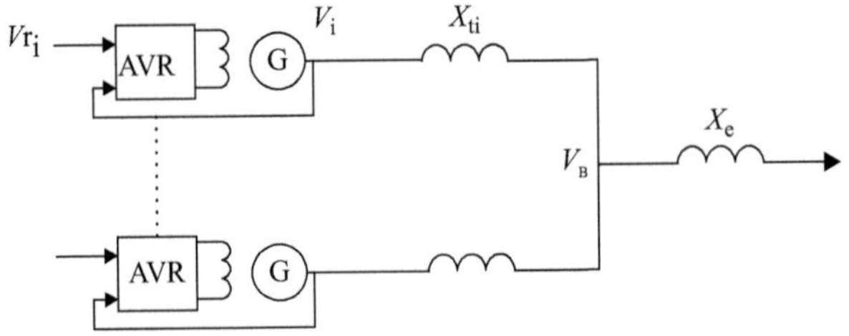

Figure 9.15 Dynamic decoupling matrix: relation between reactive power variations and voltage variations

power absorbed or injected into the grid by the generator. The maximum reactive power that can be delivered is called $Q_{\mathrm{Lim}+}$ and the maximum reactive power that can be absorbed is called $Q_{\mathrm{Lim}-}$. The SART system permits introducing several capability curves (according to the active cooling system) and to choose on the fly which one is active at a certain time.

The reference values of reactive power for each generator are calculated multiplying lev_q by $Q_{\mathrm{Lim}+}$ (or $Q_{\mathrm{Lim}-}$). The actual reactive power generated less this reference gives the error value q_{err} for the reactive power of each generator. Before applying this value to the chosen regulator (traditionally a P.I. is used), the value is elaborated through the dynamic decoupling (*DD*) matrix with the following equation:

$$[q_{\mathrm{ctrl}}] = [DD][q_{\mathrm{err}}] \tag{9.2}$$

The *DD* matrix is the inverse of the non-diagonal matrix which couples the vector of the generator terminal voltages to the vector of the generator reactive powers [47]. From observing the scheme shown in Figure 9.15, one can deduce the following relations:

$$\Delta Q_i = \frac{\Delta V_i - \Delta V_B}{x_{ti}} \tag{9.3}$$

$$\Delta V_B = \sum_{i=1}^{n} \Delta Q_i x_e \tag{9.4}$$

$$\Delta V_i = \Delta Q_i x_{ti} + \sum_{i=1}^{n} \Delta Q_i x_e \tag{9.5}$$

The relation between generator voltage variations and reactive power variations can be expressed with the following matrix expression:

$$\begin{bmatrix} \Delta V_1 \\ \vdots \\ \Delta V_n \end{bmatrix} = \begin{bmatrix} x_{t1} + x_e & x_e & x_e \\ x_e & x_{ti} + x_e & x_e \\ x_e & x_e & x_{tn} + x_e \end{bmatrix} \begin{bmatrix} \Delta Q_1 \\ \vdots \\ \Delta Q_n \end{bmatrix} \tag{9.6}$$

The q_{ctrl} values are finally used to generate AVRs voltage references by means of UP/DOWN commands. The choice to generate UP and DOWN commands is

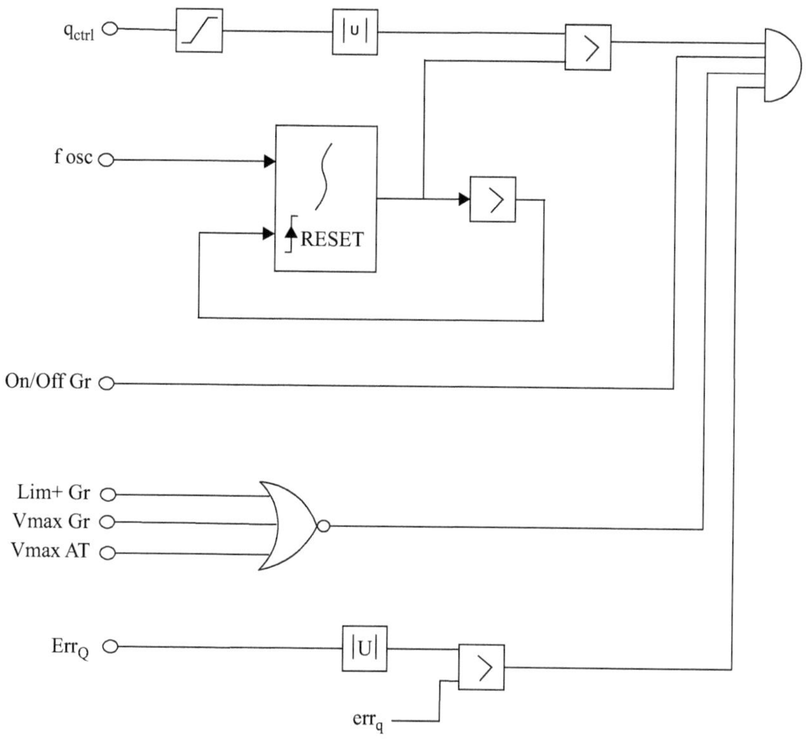

Figure 9.16 Generation of UP PWM pulses

mainly due to the constraint of integrating the SART system in existing power stations. Those commands are commonly available in all modern AVRs for synchronous generators. The SART device is designed to work with two different modulations for the UP/DOWN commands: pulse-frequency modulation (PFM) and PWM. With PWM, when q_{ctrl} is positive an UP pulse is generated and remains UP till the value q_{ctrl} remains positive. With PFM, a pulse of a specified length is generated.

An example of logical circuit generating the UP pulses in PWM is shown in Figure 9.16. A ramp signal at the frequency of the pulses to be generated with amplitude between 0 and 11 is compared with the absolute value of the q_{ctrl} signal and it generates the PWM pulse. The UP pulses are enabled only if all the following conditions are true:

- the generator is under control of SART: signal On/Off Gr is true;
- the generator has not reached its upper reactive limit: signal Lim + Gr is false;
- the generator has not reached its upper voltage limit: signal Vmax Gr is false;
- the AT busbar of the power plant has not reached its upper voltage limit: signal Vmax AT is false;
- the difference between the reactive power injected by the generator and the reference reactive power is lower than a set insensitivity value: Err_Q is lower than err_q.

A similar scheme can be drawn for the DOWN pulses considering symmetrical signals for lower limits in reactive power and voltage. The pulse would be an UP if q_{ctrl} is positive or a DOWN pulse if q_{ctrl} is negative.

It is noteworthy that SART can be utilized for power station stand-alone control, i.e., independently from the presence of the SVR. This can be done either by locally setting the reactive level signal (manually in Man q mode, as in Figure 9.14), or by overlapping the reactive power control loop by the power station HV busbar voltage control loop, through the busbar voltage regulator (BVR). Under BVR, daily memorized profiles of HV busbar voltage can be stored and actuated.

When inserted in a power plant, a SART needs to implement several different communication tasks:

- inputs of measured voltages, currents and active and reactive powers from peripheral units;
- I/O from/to all the AVRs;
- I/O from/to the SCADA system;
- state variables from master to slave central unit.

Figure 9.17 represents the principle scheme of the communication architecture. Such scheme is a real-world application case: a power station with 3 gas/steam combined cycles, for a total of six generators. Four SART-PUs are required to measure electrical quantities: three for the combined-cycle generators and one for the plant HV station busbar.

All communications described above are implemented over IP stack using different Ethernet-based physical layers. Using optical fibre converters and multi-mode optical fibre cables, distances of some hundred metres in the power plant are covered.

9.3.2.2 Experimental results

The SART system has been applied to different power stations connected to the Italian transmission system at 400 kV. Data were registered during commissioning operations with Italian TSO TERNA.†

Figure 9.18 shows the reactive power response of the three groups: it is possible to recognize the two 323 MVA gas turbines and the single 360 MVA steam generator. At $t = 20$ s the power station moves from the busbar regulation voltage mode (where the voltage of the high-voltage busbar is controlled) to a lev q mode (where the reactive powers follow the signal received from the TSO). At about $t = 34$ s, a $+10\%$ step is applied to the signal lev q. Every group changes its reactive power generated according to its capability curve and its active power generated (not shown). The three generators are driven by means of a series of digital pulses modulated in duration (PWM). Figure 9.19 shows the voltage profile of one of the groups with the UP and DOWN commands sent from the SART to the excitation system is. Pulses are width modulated.

†Data are sent from the SART apparatus through the Ethernet with a simple protocol using UDP/IP. An external laptop is connected to the SART LAN and collects all the data into file that can be elaborated on line or used for further elaborations. Data are collected at a frequency of 100 Hz, i.e., the sampling time is 10 ms.

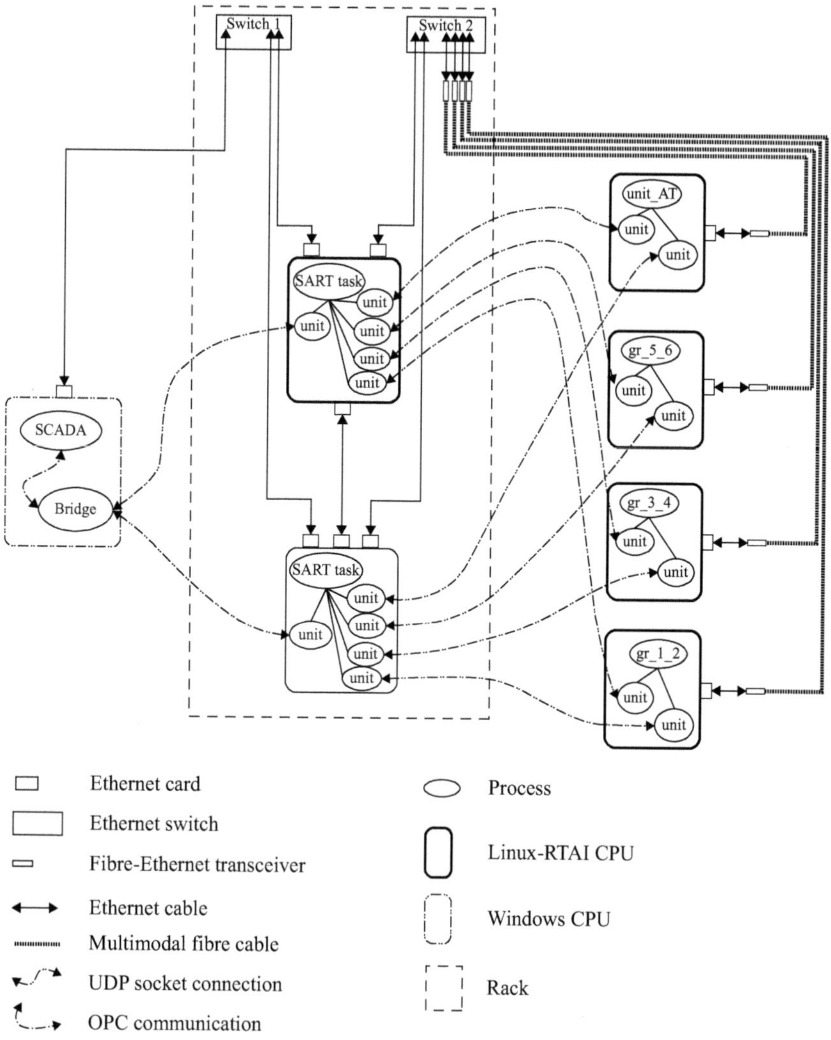

Figure 9.17 General communication diagram of SART

9.3.3 Voltage control with distributed generation

The penetration of power plants based on renewable energy sources (RES) is constantly and rapidly increasing due to several reasons. Among these, government policies support the decrease of greenhouse gas emissions. Moreover, the liberalization of electricity markets promotes technological innovations. RES are normally not modulated in active power production nor are modulated in reactive power. Once entered into production, RES tend to deliver their nominal power to the grid.

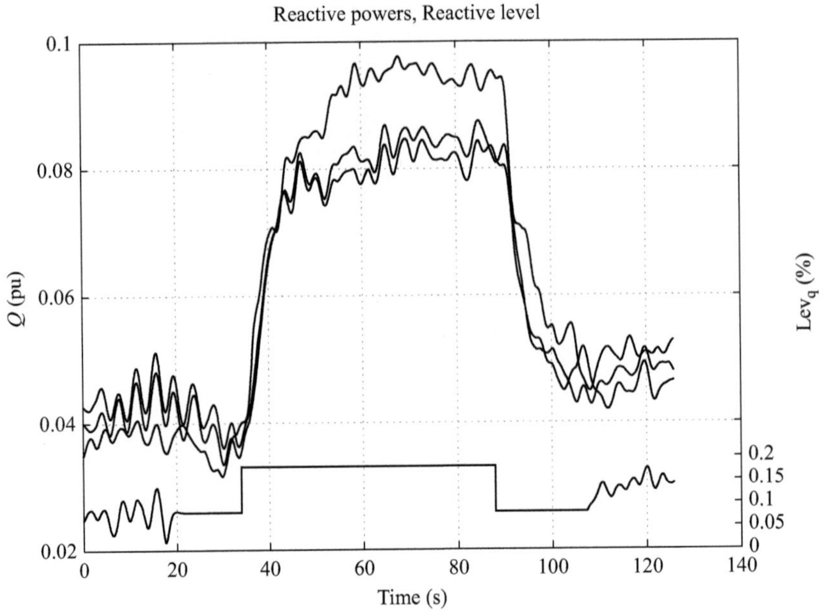

Figure 9.18 Reactive power response to a step in lev q *signal*

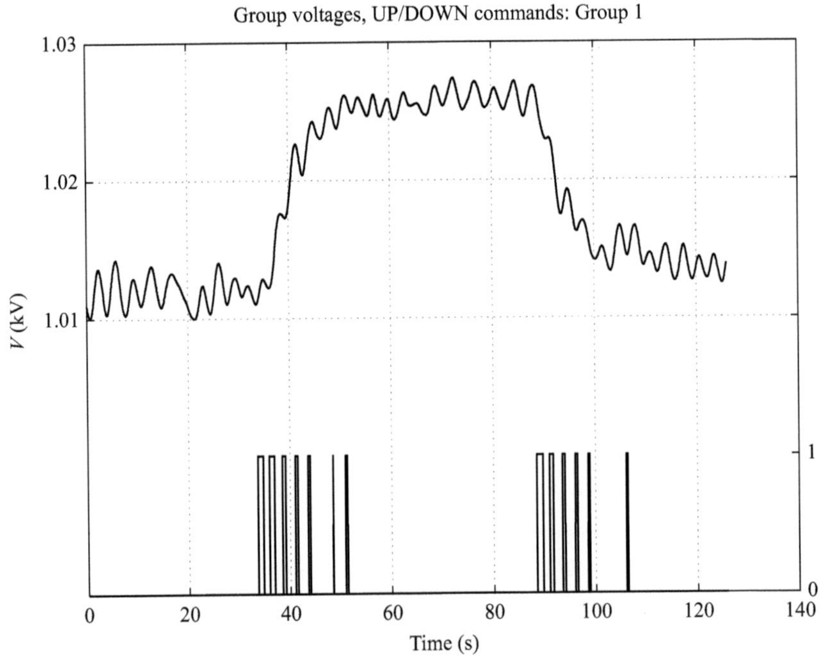

Figure 9.19 Voltage of a group and UP and DOWN commands

If RES are connected to distribution networks, a major impact comes from active power flows injected into the lines, which produces voltage variations of the buses. For particular situations, such voltage variations can become relevant and determine the so-called *voltage rise* [48]. If RES are connected to the transmission network, they can provide voltage regulation, but in practice, their effect is to decrease the total amount of reactive power resources of the overall system.

The integration of non-dispatchable renewable power plants into the voltage control architecture of the transmission network is currently a significant challenge. The development of innovative solutions for the hierarchical voltage control system (as the utilization of coordinated ECS into the Italian SVR algorithms) requires real-time platforms with embedded network capabilities. In this context, the Italian case seems to be of great interest because a strong penetration of renewable energy is developing in a transmission system where a pre-existing hierarchical voltage regulation is in use.

For the voltage rise problem, several studies propose techniques to make distributed generators exchange reactive power in a controlled way. For those plants connected directly – singularly or grouped in clusters – to the transmission grid, there are proposals for their participation to the secondary voltage control (where implemented). For both problems, the presented HW and SW platforms based on GPPs and RTOSs seem to be suitable.

9.3.3.1 Voltage rise mitigation in distribution networks

Actual distribution networks, designed to feed passive loads, are subject to different impacts [49] in case many active generators are connected. The voltage rise effect is depicted in Figure 9.20 and defined as follows. For short lines, the voltage drop (at steady state) between two buses, say 1 and 2, when power P_1 (active) and Q_1

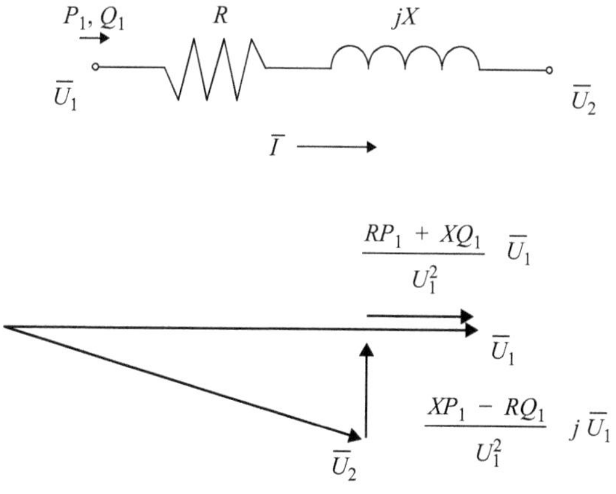

Figure 9.20 Voltage drop between two nodes

(reactive) flows from bus 1 to bus 2, is given by the approximation (powers are assumed positive when absorbed by loads):

$$U_1 - U_2 \simeq \frac{RP_1 + XQ_1}{U_1} \tag{9.7}$$

where

- U_1, U_2 are the voltages at buses 1 and 2 of the line;
- R and X are the resistance and reactance of the line;
- P_1 and Q_1 are active and reactive power flowing from bus 1 to bus 2.

In distribution grids, the R/X ratio is not negligible. Therefore, both RP_1 and XQ_1 terms of (9.7) stand, and voltage drop between buses 1 and 2 is a function of both active and reactive power flows. A relevant injection of active power, generated by DGs into the network, can therefore change the sign of $U_1 - U_2$. This causes voltage magnitudes to rise, from the substation bus to the end of the network, instead of dropping, as normally expected.

Different approaches to mitigate the voltage rise are described in the literature. Among those, we cite the operation of on-line tap changers (OLTCs) proposed in [50, 51]; the increase in conductor size; the limitation (or modulation) of active power delivered into the network by DGs; and the possibility of allowing DGs to exchange reactive power. The latter approach requires generators that can automatically control the reactive power of the DGs in a coordinated way (centralized) [52] or locally (or decentralized) [53]. For all regulations above, the real-time platform discussed in this chapter shows relevant advantages. The developed ECS is ready to be inserted in such regulations.

9.3.3.2 Reactive power control in distribution networks

In this subsection, we are interested in power plants where several medium-size generators are interconnected, forming a local medium-voltage network and that are connected to the high- or medium-voltage network through a single point of connection (POC). For these configurations, we propose a hierarchical control strategy similar to that described in Section 9.3.2.

To control the voltage of the POC, a central control unit coordinates the reactive power of each generator. This control unit implements the RPRs in a way similar to that described in the SART system. Such a hierarchy is shown to work properly for power plants based on different technologies, namely, solar photovoltaic [54], wind power [55] and hydro resources [56].

Figure 9.21 shows a synoptic scheme of the proposed control system, which is similar to the SART scheme given in Figure 9.14. The major difference is that generators are not connected in parallel to a busbar but, rather, are connected to a radial medium-voltage distribution network.

The regulator receives the reference voltage from the DSO or TSO (depending on the voltage of the POC). The pilot bus is the POC of the RES generation plant. An external voltage control loop, implemented by the BVR, as shown in Figure 9.21,

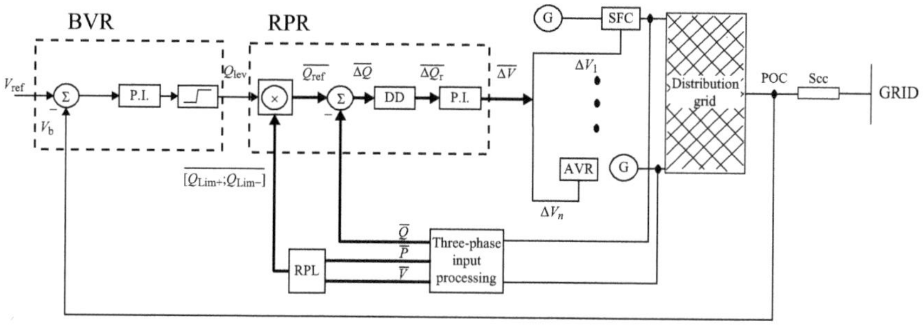

Figure 9.21 Synoptic scheme of the secondary voltage control. In the more general case, it is assumed the co-presence of generators under inverters (static frequency converter) with traditional synchronous generators

computes a RPL lev_q (between -1 and $+1$), which is then multiplied by each generator reactive power limit, as described in Section 9.3.2.1. This operation results in a vector of reference reactive powers, namely Q_{rif}. Every value of this vector compares with the actual reactive power of each generator. The vector of errors is multiplied by the *DD* matrix and sent to the generator reactive power regulators (GRPRs).

Given the generality of the internal medium-voltage network, it is not possible to use the simple expression (9.6) to compute *DD*, which has to be calculated instead of defining the sensitivities $\frac{\partial V}{\partial Q}$ that link the variation of the voltage of each node to the variation of the reactive power injected in each other node. The resulting control signals of the GRPRs are used as a reference for every generator. It is also possible to exclude the external voltage control loop and send a lev_q reference signal directly to the reactive power control loop. In this case, the generators can participate in a secondary voltage regulation as a single power plant.

A simple test to check the behaviour of the proposed control consists in simulating the response to a step variation of the reference voltage of the POC. At the time being, only simulations are available, as the implementation of this regulation in real-world RES is still a proposal.

For the sake of an example, we consider the response of a real-world 48 MW PV plant where an internal distribution network at 20 kV is used. The interested reader can find more details on this system in Reference 54. The system has been simulated using Dome [57]. It is assumed that the initial voltage reference of the POC is 1.005 pu, and that all generators initially produce same reactive power injections. At $t = 500$ s, the reference voltage jumps to 1.03 pu. As shown in Figure 9.22, the regulator imposes a new value of reactive power to be supplied by the generators. The dynamic behaviour of the proposed secondary voltage regulation scheme is the desired one. In fact, reactive power injections are equally distributed among all generators and the trajectories of the voltages do not oscillate and are able to perfectly track the reference signal.

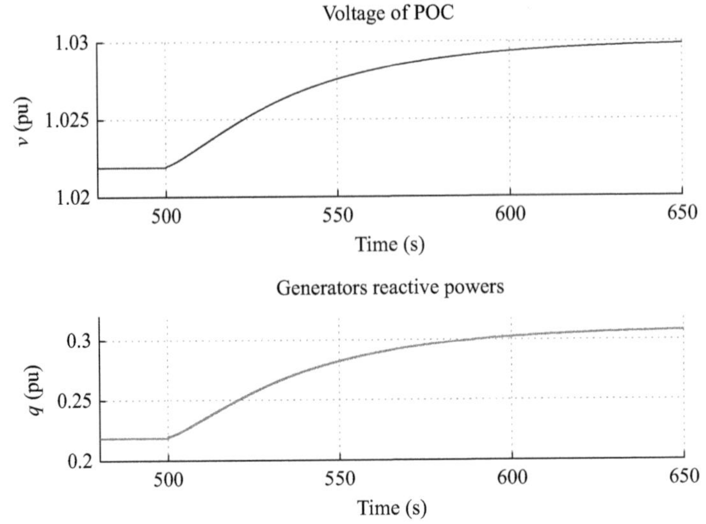

Figure 9.22 Voltage at the POC and reactive power of generators

9.4 Conclusions

This chapter discusses technical aspects of the implementation of digital controllers for power systems using GPPs and RTOSs. Two real-world examples are also described in detail. With this regard, we have presented our experience with an open-source RTOS, i.e., RTAI. However, the appraisal given in this chapter also applies to most commercial RTOSs. We consider that the main advantages of such platforms are the quick and easy prototyping, flexible software development, and low maintenance costs. Similar platforms and methodologies can be applied as well to regulate the voltage in distribution networks that include distributed generators. This chapter discusses the voltage rise phenomenon in distribution networks and illustrates promising simulation results obtained by implementing a secondary voltage control for power plants that are internally interconnected through a distribution network and have a single POC with the external grid.

Bibliography

[1] Kopetz H. *Real-Time Systems: Design Principles for Distributed Embedded Applications*, 2nd Edition. Berlin: Springer; 2011.

[2] Young SJ. *Real Time Languages: Design and Development*. New York, NY: Halstead Press; 1982.

[3] Wittenmark B, Nilsson J, Torngren M. "Timing problems in real-time control systems". In: *Proceedings of the American Control Conference*, WA: Seattle, vol. 3; 1995. p. 2000–2004.

[4] Lim S, Bae YH, Jang GT, *et al.* "An accurate worst case timing analysis for RISC processors". *IEEE Transactions on Software Engineering*. 1995 Jul;21: 97–108.

[5] Zhang N, Burns A, Nicholson M. "Pipelined processors and worst case execution times". *Real-Time Systems*. 1993 Oct;5:319–343.

[6] Basumallick S, Nilsen K. "Cache issues in real-time systems". In: *Proceedings of the ACM SIGPLAN Workshop on Language, Compiler and Tool Support for Real-Time Systems*; 1994.

[7] Labrosse JJ. *MicroC/OS-II: The Real-Time Kernel*, 2nd Edition. San Francisco, CA: CMP Books; 2002.

[8] Wells G. "A comparison of four microcomputer operating systems". *Real-Time Systems*. 1993;5:345–368.

[9] Barbalace A, Luchetta A, Manduchi G, Moro M, Soppelsa A, Taliercio C. "Performance comparison of VxWorks, Linux, RTAI, and Xenomai in a hard real-time application". *IEEE Transactions on Nuclear Science*. 2008;55: 435–439.

[10] Neto A, Sartori F, Piccolo F, Barbalace A, Vitelli R, Fernandes H. "Linux real-time framework for fusion devices". In: *Proceeding of the 25th Symposium on Fusion Technology Fusion Engineering and Design*, Germany: Rostock, vol. 84; 2009. p. 1408–1411.

[11] Chiandone M, Cleva S, Menis R, Sulligoi G. "Industrial motion control applications using Linux RTAI". In: *International Symposium on Power Electronics, Electrical Drives, Automation and Motion*, Italy: Ischia, 2008. p. 528–533.

[12] Chiandone M, Cleva S, Sulligoi G. "PC-based feedback acceleration control using Linux RTAI". In: *13th European Conference on Power Electronics and Applications*, Spain: Barcelona, 2009. p. 1–8.

[13] Chiandone M, Cleva S, Pavan AM, Sulligoi G. "Monitoring applications of electrical and climate data for PV systems using Linux RTAI". In: *International Conference on Clean Electrical Power*, Italy: Capri, 2009. p. 264–267.

[14] Proctor FM, Shackleford WP. "Real-time operating system timing jitter and its impact on motor control". *Proceedings of the SPIE Sensors and Controls for Intelligent Manufacturing II*. 2001;4563:10–16.

[15] Sulligoi G, Vicenzutti A, Chiandone M, Bosich D. "Generators electromechanical stability in shipboard grids with symmetrical layout: dynamic interactions between voltage and frequency controls". In: *AEIT*, Italy: Mondello, 2014. p. 1–5.

[16] Cerf VG, Kahn RE. "A protocol for network intercommunication". *IEEE Transactions on Communications*. 1974;COM-22:637–648.

[17] Chiueh VT. "Supporting real-time traffic on Ethernet". In: *Proceedings of Real-Time Systems Symposium*, San Juan, 1994. p. 282–286.

[18] Kiszka J, Wagner B. "RTnet – a flexible hard real-time networking framework". In: *Proceedings of 10th IEEE Conference on Emerging Technologies and Factory Automation*, Italy: Catania, 2005. p. 8–456.

[19] Ferrari P, Flammini A, Marioli D, Taroni A. "A distributed instrument for performance analysis of real-time Ethernet networks". *IEEE Transactions on Industrial Informatics*. 2008;4:16–25.

[20] Cruz RL. "A calculus for network delay. I. Network elements in isolation". *IEEE Transactions on Information Theory*. 1991;37:114–131.

[21] Cruz RL. "A calculus for network delay. II. Network analysis". *IEEE Transactions on Information Theory*. 1991;37:132–141.

[22] Georges JP, Rondeau E, Divoux T. "Evaluation of switched Ethernet in an industrial context by using the Network Calculus". In: *Fourth IEEE International Workshop on Factory Communication Systems*; 2002. p. 19–26.

[23] Bencivenni M, Bortolotti D, Carbone A, *et al.* "Performance of 10 Gigabit Ethernet using commodity hardware". *IEEE Transactions on Nuclear Science*. 2010;57:630–641.

[24] Wu JH, Zheng G, Liu GP, Li JG. "Real-time performance analysis of industrial embedded control systems using switched Ethernet". In: *IEEE International Conference on Networking, Sensing and Control*, London, 2007. p. 64–69.

[25] Lee KC, Lee S, Lee MH. "Worst case communication delay of real-time industrial switched Ethernet with multiple levels". *IEEE Transactions on Industrial Electronics*. 2006;53:1669–1676.

[26] de M Valentim RA, Morais AHF, Brandao GB, Guerreiro AMG. "A performance analysis of the Ethernet nets for applications in real-time: IEEE 802.3 and 802.3". In: *Proceedings of INDIN*, South Korea: Daejeon, 2008. p. 956–961.

[27] Lee KC, Lee S. "Performance evaluation of switched Ethernet for networked control systems". In: *Proceedings of INDIN*; 2002. p. 3170–3175.

[28] Angioni A, Sadu A, Ponci F, *et al.* "Coordinated voltage control in distribution grids with LTE based communication infrastructure". In: *IEEE 15th International Conference on Environment and Electrical Engineering*, Italy: Rome, 2015. p. 2090–2095.

[29] Bucher R, Balemi S. "Scilab/Scicos and Linux RTAI – a unified approach". In: *Proceedings of 2005 IEEE Conference on Control Applications*, Ontario: Toronto, 2005. p. 1121–1126.

[30] 421.4-2014 – IEEE Guide for the Preparation of Excitation System Specifications; April 21, 2014. DOI: 10.1109/IEEESTD.2014.6803835. E-ISBN: 978-0-7381-8984-0.

[31] Erceg G, Erceg R, Erceg I. "Concepts of synchronous generator's digital control". In: *Proceedings of IEEE International Symposium on Industrial Electronics*; 2004. p. 543–548.

[32] Arcidiacono V, Corsi S, Ottaviani G, Togno S, Baroffio G, C Raffaelli ER. "The ENEL's experience on the evolution of excitation control systems through microprocessor technology". *IEEE Transactions on Energy Conversion*. 1998 Sep;13:292–299.

[33] Hingston RS, Ham PAL, Green NJ. "Development of a digital excitation control system". In: *Proceedings of Fourth International Conference on Electrical Machines and Drives*, London, 1989. p. 125–129.

[34] Hano I, Tamura Y, Narita S, Matsumoto K. "Real time control of system voltage and reactive power". *IEEE Transactions on Power Apparatus and Systems*. 1969 Oct;PAS-88:1544–1559.
[35] Narita S, Hammam M. "A computational algorithm for real-time control of system voltage and reactive power. Part I – problem formulation". *IEEE Transactions on Power Apparatus and Systems*. 1971 Nov;PAS-90: 2495–2501.
[36] Nakamura Y, Okada T. "Voltage and reactive power control by dividing a system into several blocks". *IEE in Japan*. 1969;89:75–82.
[37] Arcidiacono V. "Automatic voltage and reactive power control in transmission systems". In: *Proceedings of 1983 CIGRE-IFAC Symposium*, Italy: Florence, 1983.
[38] Paul JP, Leost JY, Tesseron JM. "Survey of the secondary voltage control in France: present realization and investigations". *IEEE Transactions on Power Systems*. 1987 May;PWRS-2:75–82.
[39] Sancha JL, Fernandez JL, Cortes A, Abarca JT. "Secondary voltage control: analysis, solutions and simulation results for the Spanish transmission system". *IEEE Transactions on Power Systems*. 1996;11:630–638.
[40] Corsi S, Villiers FD, Vajeth R. "Power system stability increase by secondary voltage regulation applied to the South Africa grid". In: *Proceedings of 2010 IREP Symposium Bulk Power System Dynamics and Control*; Rio de Janeiro, p. 1–18.
[41] Mousavi OA, Cherkaoui R. "Literature survey on fundamental issues of voltage and reactive power control". EPFL, Lausanne, Switzerland: MARS Project; 2011.
[42] Corsi S, Chinnici R, Lena R, Vannelli G, Bazzi U, Cima E. "General application to the main ENEL's power plants of an advanced voltage and reactive power regulator for EHV network support". In: *Proceedings of 1998 CIGRE Conference*, Paris, 1998.
[43] Arcidiacono V, Corsi S, Natale A, Raffaelli C. "New developments in the application of ENEL transmission system voltage and reactive power automatic control". In: *Proceedings of 1990 CIGRE Conference*; 1990.
[44] Terna. Sistema Automatico per la Regolazione di Tensione (SART) per centrali elettriche di produzione GRTN document nr. DRRPX03019; 2003.
[45] Corsi S, Pozzi M, Sabelli C, Serrani A. "The coordinated automatic voltage control of the Italian transmission grid – Part I: reasons of the choice and overview of the consolidated hierarchical system". *IEEE Transactions on Power Systems*. 2004 Nov;3:1723–1732.
[46] Corsi S, Pozzi M, Sforna M, Dell'Olio G. "The coordinated automatic voltage control of the Italian transmission grid – Part II: control apparatus and field performance of the consolidated hierarchical system". *IEEE Transactions on Power Systems*. 2004 Nov;19:1733–1741.
[47] Arcidiacono V, Menis R, Sulligoi G. "Improving power quality in all electric ships using a voltage and VAR integrated regulator". In: *Proceedings of IEEE Electric Ship Technologies Symposium*, VA: Arlington, 2007. p. 322–327.

[48] Masters CL. "Voltage rise the big issue when connecting embedded generation to long 11 kV overhead lines". *Power Engineering Journal*. 2002;16: 5–12.

[49] Walling RA, Saint R, Dugan RC, Burke J, Kojovic LA. "Summary of distributed resources impact on power delivery systems". *IEEE Transactions on Power Delivery*. 2008;23:1636–1644.

[50] Viawan FA, Sannino A, Daalder J. "Voltage control with on-load tap changers in medium voltage feeders in presence of distributed generation". *Electric Power Systems Research 77*. 2007;10:1314–1322.

[51] Casavola A, Franze G, Menniti D, Sorrentino N. "A command governor approach to the voltage regulation problem in MV/LV networks with renewable generation units". In: *Proceedings of International Conference on Clean Electrical Power ICCEP*, Italy: Capri, 2009. p. 304–309.

[52] Madureira AG, Pecas-Lopes JA. "Coordinated voltage support in distribution networks with distributed generation and microgrids". *IET Renewable Power Generation*. 2008;3:439–454.

[53] Keane A, Ochoa LF, Vittal E, Dent CJ, Harrison GP. "Enhanced utilization of voltage control resources with distributed generation". *IEEE Transactions on Power Systems*. 2011;26:252–260.

[54] Chiandone M, Campaner R, Pavan AM, Arcidiacono V, Milano F, Sulligoi G. "Coordinated voltage control of multi-converter power plants operating in transmission systems. The case of photovoltaics". In: *Proceedings of International Conference on Clean Electrical Power ICCEP*, Italy: Taormina, 2015. p. 1–6.

[55] Chiandone M, Campaner R, Arcidiacono V, Sulligoi G, Milano F. "Automatic voltage and reactive power regulator for wind farms participating to TSO voltage regulation". In: *Proceedings of IEEE PowerTech*, NL: Eindhoven, 2015. p. 1–6.

[56] Campaner R, Chiandone M, Arcidiacono V, Sulligoi G, Milano F. "Automatic voltage control of a cluster of hydro power plants to operate as a virtual power plant". In: *Proceedings of IEEE 15th International Environment and Electrical Engineering (EEEIC)*, Italy: Rome, 2015. p. 1–6.

[57] Milano F. "A Python-based software tool for power system analysis". In: *Proceedings of IEEE Power and Energy Society General Meeting*, BC: Vancouver, 2013. p. 1–5.

Chapter 10

Optimal control processes in active distribution networks

Mario Paolone[1], Jean-Yves Le Boudec[2], Konstantina Christakou[1] and Dan-Cristian Tomozei[2]

Typical optimal controls of power systems, such as scheduling of generators, voltage control, losses reduction, have been so far commonly investigated in the domain of high-voltage transmission networks. However, during the past years, the increased connection of distributed energy resources (DERs) in power distribution systems results in frequent violations of operational constraints in these networks and has raised the importance of developing optimal control strategies specifically applied to these systems (e.g., References 1–6). In particular, two of the most important control functionalities that have not yet been deployed in active distribution networks (ADNs) are voltage control and lines congestion management [8]. Usually, this category of problems has been treated in the literature by means of linear approaches applied to the dependency between voltages and power flows as a function of the power injections, e.g., References 4,6,9,10.

On the one hand, recent progress in information and communication technologies, the introduction of new advanced metering devices (see Chapter 3) such as phasor measurement units and the development of real-time state estimation algorithms (see Chapter 6) present new opportunities and will, eventually, enable the deployment of processes for *optimal* voltage control and lines congestion management in distribution networks.

On the other hand, ADNs exhibit specific peculiarities that render the design of such controls compelling. In particular, it is worth noting that the solution of optimal problems becomes of interest only if it meets the stringent time constraints required by real-time controls and imposed by the stochasticity of DERs, in particular photovoltaic

[1]School of Engineering, École Polytechnique Fédérale de Lausanne (EPFL), Switzerland
[2]School of Computer and Communication Sciences, École Polytechnique Fédérale de Lausanne (EPFL), Switzerland

*As defined in Reference 7, the term ADNs has emerged to define power distribution grids that have systems in place to control a combination of DERs, defined as generators, loads and storage. Distribution network operators (DNOs) have the possibility of managing the electricity flows using a flexible network topology. DERs take some degree of responsibility for system support, which will depend on a suitable regulatory environment and connection agreement.

units (PVs), largely present in these networks. Moreover, control schemes are meaningful for implementation in real-time controllers only when convergence to an optimal solution is guaranteed. Finally, control processes for ADNs need to take into account the inherent multi-phase and unbalanced nature of these networks, as well as the non-negligible R/X ratio of longitudinal parameters of the medium and low-voltage lines, e.g., References 11,12, together with the influence of transverse capacitances.* Taking into consideration the aforementioned requirements, the distribution management systems (DMSs) need to be updated accordingly in order to incorporate optimization processes for the scheduling of the DERs [13].

This chapter starts with a general description of a centralized DMS architecture that includes voltage control and lines congestion management functionalities. Then, the formulation of the corresponding optimal control problems is described, based on a linearized approach linking control variables, e.g., power injections, transformers tap positions, and controlled quantities, e.g., voltages, current flows, by means of sensitivity coefficients. Computation processes for these sensitivity coefficients are presented in Sections 10.2 and 10.3. Finally, in Section 10.4, we provide case studies of optimal voltage control and lines congestion management targeting IEEE distribution reference networks suitably modified to integrate distributed generation.

10.1 Typical architecture of ADN grid controllers

10.1.1 Control architecture

Throughout this chapter, we consider an ADN equipped with a number of distributed controllable energy resources, a monitoring infrastructure that provides the DNO with field measurements and a centralized DMS adapted from Reference 13. The architecture of the considered DMS is shown in Figure 10.1.

Its main modules are the following:

- **State estimation**: The first step towards the development of optimal control schemes for ADNs is the knowledge of the system state. To this end, the state estimation (SE) module involves algorithms that process field measurements and provide the DNO with the state of the grid, i.e., the voltage phasors at the network buses. It is worth noting that control functionalities in distribution systems can be characterized by dynamics in the order of few seconds, since they might be associated to the dynamics of renewable energy resources (RERs), e.g., Reference 14. In this respect, we consider the presence of a real-time state estimator (RTSE) capable of assessing the ADNs' state within few tens/hundreds of milliseconds with relatively high levels of accuracy and refresh rate (e.g., Reference 15). Provided that the network admittance matrix is known, once the voltage phasors are obtained, the computation of the nodal power injections, as well as the flows of each line, is straightforward.

*Note that line shunt parameters are non-negligible in case of networks characterized by the presence of coaxial cables. These types of components are typical, for instance, in the context of urban distribution networks.

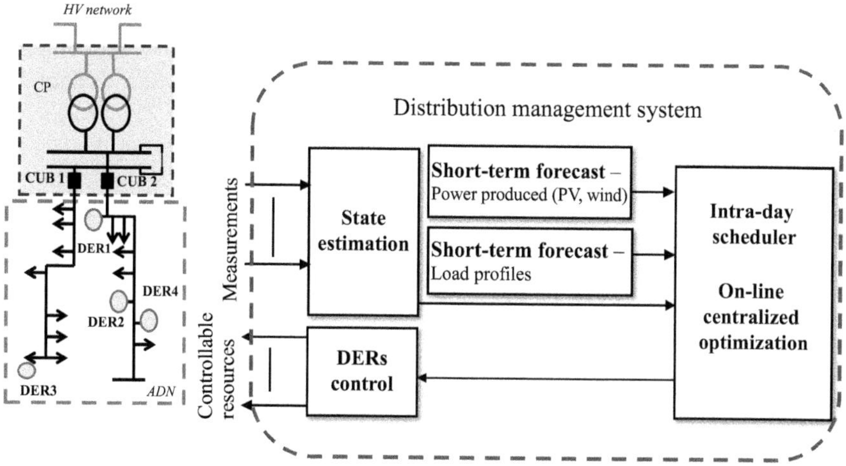

Figure 10.1 Distribution management system adopted for the proposed centralized controller

- **Short-term forecasts:**† This DMS module incorporates algorithms that are able to provide ultra-short-term forecasts for both the loads' consumption and the RERs' production, e.g., Reference 14. ADNs are characterized by increased penetration of highly volatile RERs. Therefore, the possibility to forecast as accurately as possible their power production can play a fundamental role, especially in cases where the RERs are requested to contribute to grid ancillary services. In the same direction, load forecasting is crucial especially in cases when demand-response actions are included in the control functionalities. This module is also useful in cases where multi-horizon optimization is used for the grid control, e.g., model predictive control [16], or uncertainty in loads and RERs' production is included in the control via, for instance, a robust optimization framework [17].
- **Intra-day scheduler and DERs control**: The intra-day scheduler module essentially comprises the real-time controller that acts in short-time intervals – in the order of few seconds to several minutes according to the control application. It uses the system state and the available short-term forecasts as inputs and formulates an optimization problem in order to obtain the optimal required power adjustments and the optimal variations in the under-load tap changers (ULTCs) positions which lead to the desired operation set point. Depending on the control application that the DNO wishes to implement, the objective function is modified accordingly. Typical examples of controls include resistive losses minimization, voltage deviations minimization, lines congestion management or energy supply cost minimization. Once the optimal set points are computed the DERs control module is responsible to communicate them to the controllable DERs.

†This functionality is of importance for energy management purposes. It is here mentioned for the sake of completeness but it is not used in the rest of the chapter.

In our case, the DNO is interested in minimizing the voltage deviations from the network-rated value and in maintaining the line current flows below their ampacity limits. In the following section, we focus on the actions that the online centralized controller performs to achieve these objectives.

10.1.2 Controller's actions

In what follows, the modules of state estimation, online centralized optimization and DERs control (see Figure 10.1) are adopted to formulate the control problem.

A network is considered composed of N_b buses, N_l lines and N_{DER} controllable resources. The rated value of the network voltage is denoted by V_o. The DNO wishes to compute the optimal DERs' active and reactive power variations (ΔP_i, ΔQ_i, $i = 1, \ldots, N_{DER}$), and the optimal ULTC positions Δn of the transformer interfacing the targeted ADN with the upper power grid layer[‡] to achieve primary voltage control and lines congestion management.

To guarantee convergence to an optimal solution and enable fast implementation of the control scheme in real-time controllers, we use a linearized approach to formulate the optimal control problem. The first step towards this direction is to linearize the dependencies of the voltage and line currents with respect to the nodal power injections and ULTC positions. With this aim, at each time step t, the DNO uses the network state computed by the SE module, i.e., the phase-to-ground voltage phasors $\bar{V}_i$ at each bus i (e.g., References 18,19) and, as a consequence, the branch current $\bar{I}_{ij}$ flowing from bus i to bus j.[§] To introduce the problem, we here make reference to direct sequence quantities. The extension to unbalanced three-phase systems is proposed in Section 10.3. Also, we assume that the system model, namely, the network admittance matrix, $[\bar{Y}]$ is known. Using this information, the DNO can compute, subsequently, the values of the voltage and current sensitivity coefficients with respect to absorbed/injected power of a bus l where a controllable resource is connected, as well as with respect to the transformer's ULTC positions:

$$K_{P_{i,l}} := \frac{\partial |\bar{V}_i|}{\partial P_l}, \quad K_{Q_{i,l}} := \frac{\partial |\bar{V}_i|}{\partial Q_l}$$

$$\bar{H}_{P_{ij,l}} := \frac{\partial \bar{I}_{ij}}{\partial P_l}, \quad \bar{H}_{Q_{ij,l}} := \frac{\partial \bar{I}_{ij}}{\partial Q_l}$$

$$K_{n_i} := \frac{\partial |\bar{V}_i|}{\partial V_o}, \quad \bar{H}_{n_k} := \frac{\partial \bar{I}_{ij}}{\partial V_o}$$

‡Note that we assume that transformer's ULTC is located in correspondence with the slack buses of the network because for distribution networks these represent the connections to external transmission or sub-transmission networks.

§In the rest of the chapter, complex numbers are denoted with a bar above (e.g., $\bar{V}$) and complex conjugates with a bar below (e.g., $\underline{V}$).

These sensitivities can be computed online by solving a linear system of equations (e.g., Reference 12). The details related to the computation of the sensitivities will be discussed in the following section.

Therefore, the following linear relations between variation in bus voltages, line currents and variations of active/reactive power $\Delta P_i, \Delta Q_i$ and ULTC $\boldsymbol{\Delta n}$ can be derived:

$$\Delta|\bar{V}|_i \approx \boldsymbol{K}_{P_i}\boldsymbol{\Delta P} + \boldsymbol{K}_{Q_i}\boldsymbol{\Delta Q} + \boldsymbol{K}_n\boldsymbol{\Delta n} \triangleq (\boldsymbol{K}_{P,Q,n}\boldsymbol{\Delta(P,Q,n)})_i$$
$$\Delta\bar{I}_{ij} \approx \bar{\boldsymbol{H}}_{P_{ij}}\boldsymbol{\Delta P} + \bar{\boldsymbol{H}}_{Q_{ij}}\boldsymbol{\Delta Q} + \bar{\boldsymbol{H}}_n\boldsymbol{\Delta n} \triangleq (\bar{\boldsymbol{H}}_{P,Q,n}\boldsymbol{\Delta(P,Q,n)})_{ij}$$

where

$$\boldsymbol{K}_{P_i} = \left[K_{P_{i,1}}, \ldots, K_{P_{i,N_{DER}}}\right]$$
$$\boldsymbol{K}_{Q_i} = \left[K_{Q_{i,1}}, \ldots, K_{Q_{i,N_{DER}}}\right]$$
$$\boldsymbol{K}_n = \left[K_{n_1}, \ldots, K_{n_{N_b}}\right]$$
$$\bar{\boldsymbol{H}}_{P_{ij}} = \left[\bar{H}_{P_{ij,1}}, \ldots, \bar{H}_{P_{ij,N_{DER}}}\right]$$
$$\bar{\boldsymbol{H}}_{Q_{ij}} = \left[\bar{H}_{Q_{ij,1}}, \ldots, \bar{H}_{Q_{ij,N_{DER}}}\right]$$
$$\bar{\boldsymbol{H}}_n = \left[\bar{H}_{n_1}, \ldots, \bar{H}_{n_{N_l}}\right]$$

Using the sensitivity coefficients $\boldsymbol{K}_{P,Q,n}$ and $\bar{\boldsymbol{H}}_{P,Q,n}$, the DNO can compute the optimal required power adjustments in the buses and the optimal ULTC positions $\{\boldsymbol{\Delta(P,Q,n)}^*\}$ which lead to the desired operation set point for optimal grid control. Depending on the grid's needs, the DNO can consider different objective functions. In this chapter, we assume that, at a given time step t, the DNO wishes to minimize the deviations of the voltage magnitudes in the network buses from the network-rated value, V_o, while keeping the line current flows below the ampacity limits, via the following constrained optimization problem (e.g., Reference 20):

$$\min_{\boldsymbol{\Delta(P,Q,n)}} \sum_i [(|\bar{V}_i| + (\boldsymbol{K}_{P,Q,n}\boldsymbol{\Delta(P,Q,n)})_i - |V_o|)^2 - \gamma^2]^+ \tag{10.1}$$

$$\text{subject to:} \quad |\bar{I}_{ij} + (\bar{\boldsymbol{H}}_{P,Q,n}\boldsymbol{\Delta(P,Q,n)})_{ij}| \leq I_{max}, \; i,j = 1,\ldots,N_b, \; i \neq j \tag{10.2}$$
$$(P_j, Q_j) \in \mathcal{H}_j, \; j = 1,\ldots,N_{DER} \tag{10.3}$$
$$n_{min} \leq n \leq n_{max} \tag{10.4}$$

where we have used the notation $[a]^+ \triangleq max(a,0)$. The constant γ in (10.1) represents the voltage threshold which defines the ranges outside of which the controller optimizes the voltage magnitudes. This avoids the minimization of the voltage deviations when they are within acceptable limits imposed by the DNO. Constraints (10.2) are the ampacity limits imposed on the line current flows. Constraints (10.3) represent the capability curves of the controllable resources. The last constraint (10.4) represents the minimum and maximum ULTC positions allowed.

Note that the formulation of the optimization problem in (10.1)–(10.4) is sufficiently generic. Indeed, according to the DNO's desire, the control problem can be modified to account for additional operational objectives. Also, in case the ULTCs are included in the control the problem becomes a mixed integer one, otherwise the corresponding sensitivity coefficients can be set to zero and control is achieved only through the scheduling of the DERs. In all cases, the key element for the formulation and solution of the linearized control problem is the computation of the sensitivity coefficients. With this aim, in the following sections, we recall the traditional way to compute sensitivity coefficients and we propose a method for the analytic derivation of these sensitivities, that is suitable for real-time network controllers.

10.2 Classic computation of sensitivity coefficients in power networks

Traditionally, there are two possible ways to calculate the sensitivity coefficients of our interest. The first method consists of estimating them by a series of load-flow calculations each performed for a small variation of a single control variable, i.e., nodal power injections, P_l, Q_l [6]:‖

$$\begin{aligned}
\frac{\partial|\bar{V}_i|}{\partial P_l} &= \left.\frac{\Delta|\bar{V}_i|}{\Delta P_l}\right|_{\substack{\Delta P_{i,i\neq l}=0\\ \Delta Q_{i,i\neq l}=0}} \qquad \frac{\partial|\bar{I}_{ij}|}{\partial P_l} = \left.\frac{\Delta|\bar{I}_{ij}|}{\Delta P_l}\right|_{\substack{\Delta P_{i,i\neq l}=0\\ \Delta Q_{i,i\neq l}=0}} \\
\frac{\partial|\bar{V}_i|}{\partial Q_l} &= \left.\frac{\Delta|\bar{V}_i|}{\Delta Q_l}\right|_{\substack{\Delta P_{i,i\neq l}=0\\ \Delta Q_{i,i\neq l}=0}} \qquad \frac{\partial|\bar{I}_{ij}|}{\partial Q_l} = \left.\frac{\Delta|\bar{I}_{ij}|}{\Delta Q_l}\right|_{\substack{\Delta P_{i,i\neq l}=0\\ \Delta Q_{i,i\neq l}=0}}
\end{aligned} \tag{10.5}$$

where $\bar{V}_i$ is the direct sequence phase-to-ground voltage of bus i and $\bar{I}_{ij}$ is the direct sequence current flow between buses i and j ($i,j \in \{1,\ldots,N_b\}$).

It is worth observing that such a method is computationally expensive as it entails several consecutive load-flow computations even for a small number of controllable power injections. Therefore, it cannot be adopted, in principle, for real-time implementation.

The second method uses the Newton–Raphson (NR) formulation of the load-flow calculation to directly infer the voltage sensitivity coefficients as sub-matrices of the inverted Jacobian matrix (e.g., References 21–25). Assuming that all the network buses are constant PQ-injection buses, no voltage-controlled bus is present in the

‖Note that, in what follows, we refer to the sensitivities of the current-flow magnitude with respect to active and reactive power injections as these quantities are real and they will be used for validation purposes and not to the complex quantities used in (10.1).

network and there is one slack bus in the system, the Jacobian matrix of the NR has the following form:

$$\boldsymbol{J} = \begin{bmatrix} \boldsymbol{J}_{PV} & \boldsymbol{J}_{P\delta} \\ \boldsymbol{J}_{QV} & \boldsymbol{J}_{Q\delta} \end{bmatrix} = \begin{bmatrix} \dfrac{\partial \boldsymbol{P}}{\partial |\bar{\boldsymbol{V}}|} & \dfrac{\partial \boldsymbol{P}}{\partial \boldsymbol{\delta}} \\ \dfrac{\partial \boldsymbol{Q}}{\partial |\bar{\boldsymbol{V}}|} & \dfrac{\partial \boldsymbol{Q}}{\partial \boldsymbol{\delta}} \end{bmatrix} \tag{10.6}$$

where $\boldsymbol{P}, \boldsymbol{Q}$ are the vectors of active and reactive nodal power injections/absorptions, and $|\bar{\boldsymbol{V}}|, \boldsymbol{\delta}$ are the magnitude and phase angle of the phase-to-ground voltage phasors. The elements of $\boldsymbol{J}$, i.e., the variation of the nodal active and reactive power injections as a function of the voltage magnitude and phase variations, are computed starting from the well-known non-linear power-flow equations. The detailed representation of these derivatives is given in (6.58)–(6.65) in Chapter 6.

The aforementioned Jacobian matrix, $\boldsymbol{J}$, is used in each iteration of the NR algorithm in order to linearly express the variations of active and reactive power as a function of the variations of the voltage magnitude and angles, in the following way:

$$\begin{bmatrix} \Delta \boldsymbol{P} \\ \Delta \boldsymbol{Q} \end{bmatrix} = \begin{bmatrix} \boldsymbol{J}_{PV} & \boldsymbol{J}_{P\delta} \\ \boldsymbol{J}_{QV} & \boldsymbol{J}_{Q\delta} \end{bmatrix} \times \begin{bmatrix} \Delta |\bar{\boldsymbol{V}}| \\ \Delta \boldsymbol{\delta} \end{bmatrix} \tag{10.7}$$

In our case, we are interested in the opposite link, namely expressing the variations of the voltage magnitude as a function of the active and reactive variations in the buses. This can be obtained in a straightforward manner, by inverting the Jacobian matrix:

$$\begin{bmatrix} \Delta |\bar{\boldsymbol{V}}| \\ \Delta \boldsymbol{\delta} \end{bmatrix} = \begin{bmatrix} \boldsymbol{J}_{PV} & \boldsymbol{J}_{P\delta} \\ \boldsymbol{J}_{QV} & \boldsymbol{J}_{Q\delta} \end{bmatrix}^{-1} \times \begin{bmatrix} \Delta \boldsymbol{P} \\ \Delta \boldsymbol{Q} \end{bmatrix} \tag{10.8}$$

At this point it is important to note that $\boldsymbol{J}$ is the Jacobian matrix of the whole network and that it also contains the elements corresponding to voltage angles. However, the optimal control problem as formulated in (10.1) requires only the sensitivities that correspond to specific controllable nodal injections and not the elements related to voltage angles. Therefore, the desired sensitivities correspond to sub-matrices of the inverted Jacobian matrix that need to be properly extracted.

It is worth observing that such a method does not allow to compute the sensitivities against the transformer's ULTC positions. Additionally, as known, the sub-matrix $\frac{\partial \boldsymbol{Q}}{\partial |\bar{\boldsymbol{V}}|}$ is usually adopted to express voltage variations as a function of reactive power injections when the ratio of longitudinal line resistance versus reactance is negligible. It is worth noting that such an assumption is no longer applicable to distribution systems that require, in addition, to take into account active power injections (e.g., Reference 26). Finally, such a method can be computationally expensive for very large networks. In these cases, this method requires the inversion of a large Jacobian matrix simply to extract a few columns corresponding to the controllable resources' power injections.

To overcome the aforementioned limitations, an efficient method for the computation of the desired sensitivities is given in the following section based on Reference 12.

10.3 Efficient computation of sensitivity coefficients of bus voltages and line currents in unbalanced radial electrical distribution networks

10.3.1 Voltage sensitivity coefficients

The analysis starts with the exact computation of the voltage sensitivity coefficients as a function of the network admittance matrix and its state. To this end, we derive mathematical expressions that link bus voltages to bus active and reactive power injections. For this purpose, a N_b-bus three-phase generic electrical network is considered. The following analysis treats each phase of the network separately and, thus, it can be applied to unbalanced networks (i.e., even for networks that cannot be decomposed with sequence components).

As known, the equations that link the voltage of each phase of the buses to the corresponding injected current are in total $M = 3N_b$ and they are given by:

$$[\bar{\boldsymbol{I}}_{abc}] = [\bar{\boldsymbol{Y}}] \cdot [\bar{\boldsymbol{V}}_{abc}] \tag{10.9}$$

where $[\bar{\boldsymbol{I}}_{abc}] = [\bar{I}_1^{a,b,c}, \ldots, \bar{I}_{N_b}^{a,b,c}]^T$ and $[\bar{\boldsymbol{V}}_{abc}] = [\bar{V}_1^{a,b,c}, \ldots, \bar{V}_{N_b}^{a,b,c}]^T$. We denote by a, b, c the three network phases. The $[\bar{\boldsymbol{Y}}]$ matrix is the so-called compound admittance matrix (e.g., Reference 27) and is formed as described in Section 6.1.1.1 of Chapter 6.

In order to simplify the notation, in what follows we will assume the following correspondences: $[\bar{\boldsymbol{I}}_{abc}] = [\bar{I}_1, \ldots, \bar{I}_M]^T$, $[\bar{\boldsymbol{V}}_{abc}] = [\bar{V}_1, \ldots, \bar{V}_M]^T$. For the rest of the analysis, we will consider the network as composed of S slack buses (the set of slack buses is $\mathcal{S}$) and N buses (the set of non-slack buses is $\mathcal{N}$) with PQ injections (i.e., $\{1, 2, \ldots, M\} = \mathcal{S} \cup \mathcal{N}$, with $\mathcal{S} \cap \mathcal{N} = \emptyset$). The PQ injections are considered constant and independent of the voltage.

The link between power injections and bus voltages reads

$$\underline{S}_i = \underline{V}_i \sum_{j \in \mathcal{S} \cup \mathcal{N}} \bar{Y}_{ij} \bar{V}_j, \quad i \in \mathcal{N} \tag{10.10}$$

The derived system of equations (10.10) holds for all the phases of each bus of the network. Since the objective is to calculate the partial derivatives of the voltage magnitude over the active and reactive power injected in the other buses, we have to consider separately the slack bus of the system. As known, the assumptions for the slack bus equations are to keep its voltage constant and equal to the network-rated value, by also fixing its phase equal to zero. Hence, for the three phases of the slack bus, it holds that:

$$\frac{\partial \bar{V}_i}{\partial P_l} = 0, \quad \forall i \in \mathcal{S} \tag{10.11}$$

At this point, by using (10.10) as a starting point, one can derive closed-form mathematical expressions to define and quantify voltage sensitivity coefficients with respect to active and reactive power variations in correspondence with the N_b buses of the network. To derive voltage sensitivity coefficients, the partial derivatives of the voltages with respect to the active and reactive power P_l and Q_l of a bus $l \in \mathcal{N}$ have to be computed. The partial derivatives with respect to active power satisfy the following system of equations:

$$\mathbb{1}_{\{i=l\}} = \frac{\partial \underline{V}_i}{\partial P_l} \sum_{j \in \mathcal{S} \cup \mathcal{N}} \bar{Y}_{ij} \bar{V}_j + \underline{V}_i \sum_{j \in \mathcal{N}} \bar{Y}_{ij} \frac{\partial \bar{V}_j}{\partial P_l} \tag{10.12}$$

where it has been taken into account that:

$$\frac{\partial \underline{S}_i}{\partial P_l} = \frac{\partial \{P_i - jQ_i\}}{\partial P_l} = \mathbb{1}_{\{i=l\}} \tag{10.13}$$

The system of equations (10.12) is not linear over complex numbers, but it is linear with respect to $\frac{\partial \bar{V}_i}{\partial P_l}, \frac{\partial \underline{V}_i}{\partial P_l}$; therefore, it is linear over real numbers with respect to rectangular coordinates. As we show next, it has a unique solution for radial networks and can therefore be used to compute the partial derivatives in rectangular coordinates to reduce the computational effort.

A similar system of equations holds for the sensitivity coefficients with respect to the injected reactive power Q_l. With the same reasoning, by taking into account that:

$$\frac{\partial \underline{S}_i}{\partial Q_l} = \frac{\partial \{P_i - jQ_i\}}{\partial Q_l} = -j\mathbb{1}_{\{i=l\}} \tag{10.14}$$

we obtain that:

$$-j\mathbb{1}_{\{i=l\}} = \frac{\partial \underline{V}_i}{\partial Q_l} \sum_{j \in \mathcal{S} \cup \mathcal{N}} \bar{Y}_{ij} \bar{V}_j + \underline{V}_i \sum_{j \in \mathcal{N}} \bar{Y}_{ij} \frac{\partial \bar{V}_j}{\partial Q_l} \tag{10.15}$$

By observing the above linear systems of equations (10.12) and (10.15), we can see that the matrix that needs to be inverted in order to solve the system is fixed independently of the power of the lth bus with respect to which we want to compute the partial derivatives. The only element that changes is the left-hand side of the equations.

Once $\frac{\partial \bar{V}_i}{\partial P_l}, \frac{\partial \underline{V}_i}{\partial P_l}$ are obtained, the partial derivatives of the voltage magnitude can be expressed as:

$$\frac{\partial |\bar{V}_i|}{\partial P_l} = \frac{1}{|\bar{V}_i|} Re\left(\underline{V}_i \frac{\partial \bar{V}_i}{\partial P_l} \right) \tag{10.16}$$

and similar equations hold for derivatives with respect to reactive power injections.

Theorem 10.1. *The system of equations (10.12), where l is fixed and the unknowns are $\frac{\partial \bar{V}_i}{\partial P_l}$, $i \in \mathcal{N}$, has a unique solution for every radial electrical network and for any operating point $(\bar{\boldsymbol{V}}, \bar{\boldsymbol{S}})$ where the load-flow Jacobian is invertible. The same holds for the system of equations (10.15), where the unknowns are $\frac{\partial \bar{V}_i}{\partial Q_l}$, $i \in \mathcal{N}$.*

Proof. Since the system is linear with respect to rectangular coordinates and there are as many unknowns as equations, the theorem is equivalent to showing that the corresponding homogeneous system of equations has only the trivial solution. The homogeneous system can be written as:

$$0 = \underline{\Delta}_i \sum_{j \in \mathcal{S} \cup \mathcal{N}} \bar{Y}_{ij} \bar{V}_j + \underline{V}_i \sum_{j \in \mathcal{N}} \bar{Y}_{ij} \bar{\Delta}_j, \quad \forall i \in \mathcal{N} \tag{10.17}$$

where $\bar{\Delta}_i$ are the unknown complex numbers, defined for $i \in \mathcal{N}$. We want to show that $\bar{\Delta}_i = 0$ for all $i \in \mathcal{N}$. Let us consider two electrical networks with the same topology, i.e., same $[\bar{\boldsymbol{Y}}_{abc}]$ matrix, where the voltages are given. In the first network, the voltages are

$$\begin{aligned} \bar{V}_i' &= \bar{V}_i, & \forall i \in \mathcal{S} \\ \bar{V}_i' &= \bar{V}_i + \epsilon \bar{\Delta}_i, & \forall i \in \mathcal{N} \end{aligned} \tag{10.18}$$

and in the second network they are

$$\begin{aligned} \bar{V}_i'' &= \bar{V}_i, & \forall i \in \mathcal{S} \\ \bar{V}_i'' &= \bar{V}_i - \epsilon \bar{\Delta}_i, & \forall i \in \mathcal{N} \end{aligned} \tag{10.19}$$

where ϵ is a positive real number.

Let $\underline{S}_i'$ be the conjugate of the absorbed/injected power at the ith bus in the first network, and $\underline{S}_i''$ in the second. Apply (10.10) to bus $i \in \mathcal{N}$ in the first network:

$$\begin{aligned} \underline{S}_i' &= \underline{V}_i' \sum_{j \in \mathcal{S} \cup \mathcal{N}} \bar{Y}_{ij} \bar{V}_j' \\ &= (\underline{V}_i + \epsilon \underline{\Delta}_i) \left(\sum_{j \in \mathcal{S}} \bar{Y}_{ij} \bar{V}_j + \sum_{j \in \mathcal{N}} \bar{Y}_{ij} (\bar{V}_j + \epsilon \bar{\Delta}_j) \right) \\ &= \underline{V}_i \sum_{j \in \mathcal{S} \cup \mathcal{N}} \bar{Y}_{ij} \bar{V}_j + \epsilon^2 \underline{\Delta}_i \sum_{j \in \mathcal{N}} \bar{Y}_{ij} \bar{\Delta}_j + \epsilon \underline{\Delta}_i \sum_{j \in \mathcal{S} \cup \mathcal{N}} \bar{Y}_{ij} \bar{V}_j + \underline{V}_i \sum_{j \in \mathcal{N}} \bar{Y}_{ij} \epsilon \bar{\Delta}_j \end{aligned}$$

Similarly, for the second network and for all buses $i \in \mathcal{N}$:

$$\underline{S}_i'' = \underline{V}_i \sum_{j \in \mathcal{S} \cup \mathcal{N}} \bar{Y}_{ij} \bar{V}_j + \epsilon^2 \underline{\Delta}_i \sum_{j \in \mathcal{N}} \bar{Y}_{ij} \bar{\Delta}_j - \epsilon \underline{\Delta}_i \sum_{j \in \mathcal{S} \cup \mathcal{N}} \bar{Y}_{ij} \bar{V}_j - \underline{V}_i \sum_{j \in \mathcal{N}} \bar{Y}_{ij} \epsilon \bar{\Delta}_j$$

Subtract the last two equations and obtain

$$\underline{S}_i' - \underline{S}_i'' = 2\epsilon\left(\underline{\Delta}_i \sum_{j\in\mathcal{S}\cup\mathcal{N}} \bar{Y}_{ij}\bar{V}_j + \underline{V}_i \sum_{j\in\mathcal{N}} \bar{Y}_{ij}\bar{\Delta}_j\right)$$

By (10.17), it follows that $\underline{S}_i' = \underline{S}_i''$ for all $i \in \mathcal{N}$. Thus, the two networks have the same active and reactive powers at all non-slack buses and the same voltages at all slack buses. As the load-flow Jacobian matrix is invertible according to Theorem 10.1 hypothesis, we can apply the inverse function theorem. As a consequence, the non-linear system of the power flow equations is locally invertible in a neighbourhood around the current operating point $(\bar{\boldsymbol{V}}, \bar{\boldsymbol{S}})$. Now, we take ϵ arbitrarily small, such that $\underline{V}_i'$ and $\underline{V}_i''$ belong to this neighbourhood where there is a one-to-one mapping between the powers and voltages. As the powers that correspond to $\underline{V}_i'$ and $\underline{V}_i''$ are exactly the same, then it follows that the voltage profile of these networks must be exactly the same, i.e., $\bar{V}_i - \epsilon\bar{\Delta}_i = \bar{V}_i + \epsilon\bar{\Delta}_i$ for all $i \in \mathcal{N}$ and thus $\bar{\Delta}_i = 0$ for all $i \in \mathcal{N}$. □

10.3.2 *Current sensitivity coefficients*

From the previous analysis, the sensitivity coefficients linking the power injections to the voltage variations are known. Thus, it is straightforward to express the branch current sensitivities with respect to the same power injections. Assuming to represent the lines that compose the network by means of π-model equivalents, the current flow $\bar{I}_{ij}$ between buses i and j can be expressed as a function of the phase-to-ground voltages of the relevant i,j buses as follows:

$$\bar{I}_{ij} = \bar{Y}_{ij}(\bar{V}_i - \bar{V}_j) + \bar{Y}_{i_0}\bar{V}_i \tag{10.20}$$

$$\bar{I}_{ji} = \bar{Y}_{ij}(\bar{V}_j - \bar{V}_i) + \bar{Y}_{j_0}\bar{V}_j \tag{10.21}$$

where $\bar{Y}_{ij}$ is the generic element of $[\bar{\boldsymbol{Y}}]$ matrix between bus i and bus j and $\bar{Y}_{i_0}$ is the shunt element on the receiving end of line $i-j$.

Since the voltages can be expressed as a function of the power injections into the network buses, the partial derivatives of the current with respect to the active and reactive power injections in the network can be expressed as:

$$\frac{\partial \bar{I}_{ij}}{\partial P_l} = \bar{Y}_{ij}\left(\frac{\partial \bar{V}_i}{\partial P_l} - \frac{\partial \bar{V}_j}{\partial P_l}\right) + \bar{Y}_{i_0}\left(\frac{\partial \bar{V}_i}{\partial P_l}\right), \quad \frac{\partial \bar{I}_{ij}}{\partial Q_l} = \bar{Y}_{ij}\left(\frac{\partial \bar{V}_i}{\partial Q_l} - \frac{\partial \bar{V}_j}{\partial Q_l}\right) + \bar{Y}_{i_0}\left(\frac{\partial \bar{V}_i}{\partial Q_l}\right) \tag{10.22}$$

$$\frac{\partial \bar{I}_{ji}}{\partial P_l} = \bar{Y}_{ij}\left(\frac{\partial \bar{V}_j}{\partial P_l} - \frac{\partial \bar{V}_i}{\partial P_l}\right) + \bar{Y}_{j0}\left(\frac{\partial \bar{V}_j}{\partial P_l}\right), \quad \frac{\partial \bar{I}_{ji}}{\partial Q_l} = \bar{Y}_{ij}\left(\frac{\partial \bar{V}_j}{\partial Q_l} - \frac{\partial \bar{V}_i}{\partial Q_l}\right) + \bar{Y}_{j0}\left(\frac{\partial \bar{V}_j}{\partial Q_l}\right) \tag{10.23}$$

Applying the same reasoning as earlier, the branch current sensitivity coefficients with respect to an active power P_l can be computed using the following expressions:

$$\frac{\partial|\bar{I}_{ij}|}{\partial P_l} = \frac{1}{|\bar{I}_{ij}|} Re\left(\underline{I}_{ij} \frac{\partial \bar{I}_{ij}}{\partial P_l} \right) \tag{10.24}$$

Similar expressions can be derived for the current coefficients with respect to the reactive power in the buses as:

$$\frac{\partial|\bar{I}_{ij}|}{\partial Q_l} = \frac{1}{|\bar{I}_{ij}|} Re\left(\underline{I}_{ij} \frac{\partial \bar{I}_{ij}}{\partial Q_l} \right) \tag{10.25}$$

10.3.3 Sensitivity coefficients with respect to transformer's ULTC

This subsection is devoted to the derivation of analytic expressions for the voltage sensitivity coefficients¶ with respect to tap positions of a transformer. We assume that transformers' tap changers are located in correspondence with the slack buses of the network as for distribution networks these represent the connections to external transmission or sub-transmission networks. As a consequence, the voltage sensitivities as a function of the tap positions are equivalent to the voltage sensitivities as a function of the slack reference voltage.** We assume that the transformers' voltage variations due to tap position changes are small enough so that the partial derivatives considered in the following analysis are meaningful. Furthermore, we assume that the power injections at the network buses are constant and independent of the voltage.

With the same reasoning as in Section 10.3.1, the analysis starts in (10.10). We write $\bar{V}_\ell = |\bar{V}_\ell| e^{j\delta_\ell}$ for all buses ℓ. For a bus $i \in \mathcal{N}$, the partial derivatives with respect to the voltage magnitude $|\bar{V}_k|$ of a slack bus $k \in \mathcal{S}$ are considered:

$$-\underline{V}_i \bar{Y}_{ik} e^{j\delta_k} = \underline{W}_{ik} \sum_{j \in \mathcal{S} \cup \mathcal{N}} \bar{Y}_{ij} \bar{V}_j + \underline{V}_i \sum_{j \in \mathcal{N}} \bar{Y}_{ij} \bar{W}_{jk} \tag{10.26}$$

where

$$\bar{W}_{ik} := \frac{\partial \bar{V}_i}{\partial |\bar{V}_k|} = \left(\frac{1}{|\bar{V}_i|} \frac{\partial |\bar{V}_i|}{\partial |\bar{V}_k|} + j \frac{\partial \delta_i}{\partial |\bar{V}_k|} \right) \bar{V}_i, \quad i \in \mathcal{N}$$

We have taken into account that:

$$\frac{\partial}{\partial |\bar{V}_k|} \sum_{j \in \mathcal{S}} \bar{Y}_{ij} \bar{V}_j = \bar{Y}_{ik} e^{j\delta_k} \tag{10.27}$$

¶Note that as shown earlier once the voltage sensitivities are obtained, the ones of currents can be computed directly.

**It is worth noting that even if the ULTCs have phase-shifting capabilities, we do not compute the corresponding sensitivities. The reason is that we consider that ULTCs are located in correspondence with the slack bus of the network. Therefore, any shift in the slack bus voltage phase directly implies that all the voltage angles of the network buses rotate by the same quantity.

and

$$\frac{\partial \underline{S}_i}{\partial |\bar{V}_k|} = 0 \tag{10.28}$$

The derived system of equations (10.26) is linear with respect to $\underline{W}_{ik}$ and $\bar{W}_{ik}$, and has the same associated matrix as the system in (10.12). Since the resulting homogeneous system of equations is identical to the one in (10.17), by Theorem 10.1, it has a unique solution.

After resolution of (10.26), we find that the sensitivity coefficients with respect to the tap position of the transformer at bus k are given by

$$\frac{\partial |\bar{V}_i|}{\partial |\bar{V}_k|} = |\bar{V}_i| Re\left(\frac{\bar{W}_{ik}}{\bar{V}_i}\right) \tag{10.29}$$

10.4 Application examples

10.4.1 Distribution network case studies

Two IEEE distribution test feeders have been used for the validation of the proposed method and for its application to the problem of voltage control and lines congestion management. The first adopted feeder is the IEEE 34-bus distribution test feeder. Its topology is shown in Figure 10.2 and it is based on the original feeder in Reference 28. It is a three-phase feeder and consists of 34 buses where bus 1 is the slack bus. The assumed line-to-line Root Mean Square (RMS)-rated voltage is equal to 24.9 kV and the network base power is 2.5 MVA. The used line configuration is the #300 of Reference 28 for each line of the feeder. The values of the resistance, reactance and susceptance, as well as the line lengths are given in Appendix B.

Table 10.1 shows the active and reactive power consumption of the loads in the three phases of each network bus.

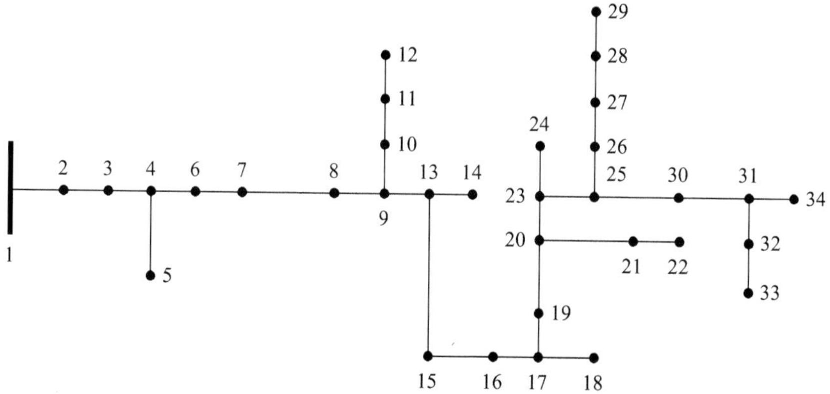

Figure 10.2 Adopted IEEE 34-bus distribution test feeder

Table 10.1 Load and generation data for the IEEE 34-bus distribution test feeder

Bus	P_α (kW)	Q_α (kvar)	P_β (kW)	Q_β (kvar)	P_c (kW)	Q_c (kvar)
1	0	0	0	0	0	0
2	0	0	−15	−7.5	−12.5	−7
3	0	0	0	0	0	0
4	0	0	0	0	0	0
5	0	0	−8	−4	0	0
6	0	0	0	0	0	0
7	0	0	0	0	0	0
8	0	0	0	0	0	0
9	0	0	0	0	0	0
10	0	0	0	0	0	0
11	−17	−8.5	0	0	0	0
12	−67.5	−35	0	0	0	0
13	0	0	−2.5	−1	0	0
14	0	0	−20	−10	0	0
15	0	0	0	0	−2	−1
16	−8.5	−4	−5	−2.5	−12.5	−5
17	0	0	0	0	0	0
18	25	0	25	0	25	0
19	0	0	0	0	0	0
20	0	0	0	0	0	0
21	0	0	0	0	0	0
22	−75	−37.5	−75	−37.5	−75	−37.5
23	25	0	25	0	25	0
24	25	0	25	0	25	0
25	2	−1	−7.5	−4	−6.5	−3.5
26	0	0	0	0	0	0
27	−72	−55	−67.5	−52.5	−67.5	−52.5
28	0	0	−12.5	−6	−10	−5.5
29	−10	−8	−21.5	−13.5	−10	−8
30	−18	−12	−20	−13	−65	−35.5
31	−15	−7.5	−5	−3	−21	−11
32	0	0	0	0	0	0
33	25	0	25	0	25	0
34	−13.5	−8	−15.5	−9	−4.5	−3.5

The same table gives the values of the active power injected by the DERs, installed in buses 18, 23, 24 and 33. The DERs do not inject any reactive power in the nominal case. We use the convention that negative values represent power consumption, whereas positive values power injection.

The second adopted feeder is the IEEE 13-bus distribution test feeder, shown in Figure 10.3. It is based as well on Reference 28. It is a three-phase feeder composed of 13 buses where bus 1 represents the connection to the sub-transmission network. The assumed network line-to-line RMS voltage is equal to 15 kV, the base power is 10 MVA and the lines are unbalanced. The used line configuration is the #602 of Reference 28 for each line of the feeder. The values of the resistance, reactance

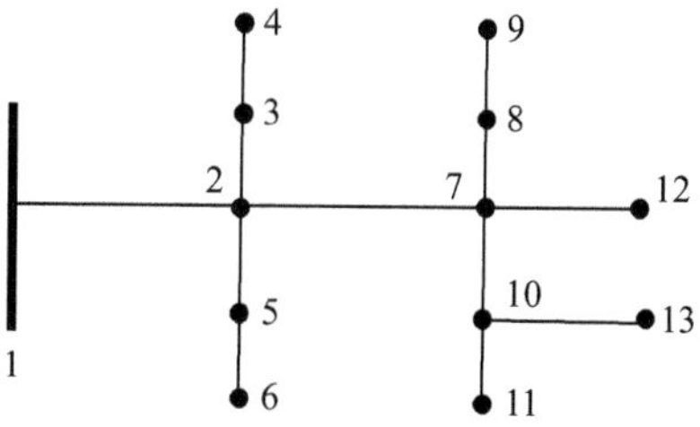

Figure 10.3 Adopted IEEE 13-bus distribution test feeder

Table 10.2 Load and generation data for the IEEE 13-bus distribution test feeder

Bus	P_α (kW)	Q_α (kvar)	P_β (kW)	Q_β (kvar)	P_c (kW)	Q_c (kvar)
1	0	0	0	0	0	0
2	−17	−10	−66	−38	−117	−68
3	0	0	0	0	0	0
4	−160	−110	−120	−90	−120	−90
5	0	0	−170	−125	0	0
6	0	0	−230	−132	0	0
7	−385	−220	−385	−220	−385	−220
8	0	0	0	0	−170	−151
9	−485	−190	−68	−60	−290	−212
10	0	0	0	0	0	0
11	0	0	0	0	−170	−80
12	0	0	0	0	0	0
13	−128	−86	0	0	0	0

and susceptance and the line lengths are given in Appendix B. The loads are also characterized by unbalanced power absorptions as can be observed in Table 10.2.

In the following sections, first the proposed method for the computation of the sensitivity coefficients is validated and then several application examples of voltage control and lines congestion management are shown.

10.4.2 Numerical validation

The numerical validation of the proposed method for the computation of voltage and current sensitivities is performed using the IEEE 13-bus test feeder and by using two different approaches. In particular, as the inverse of the load-flow Jacobian matrix provides the voltage sensitivities, the comparison reported below makes reference to such a method for the voltage sensitivities only. On the contrary, as the inverse of the load-flow Jacobian matrix does not provide current sensitivity coefficients, their accuracy is evaluated by using a numerical approach where the load-flow problem is solved by applying small injection perturbations into a given network (see Section 10.2). A similar approach is deployed to validate the sensitivities with respect to

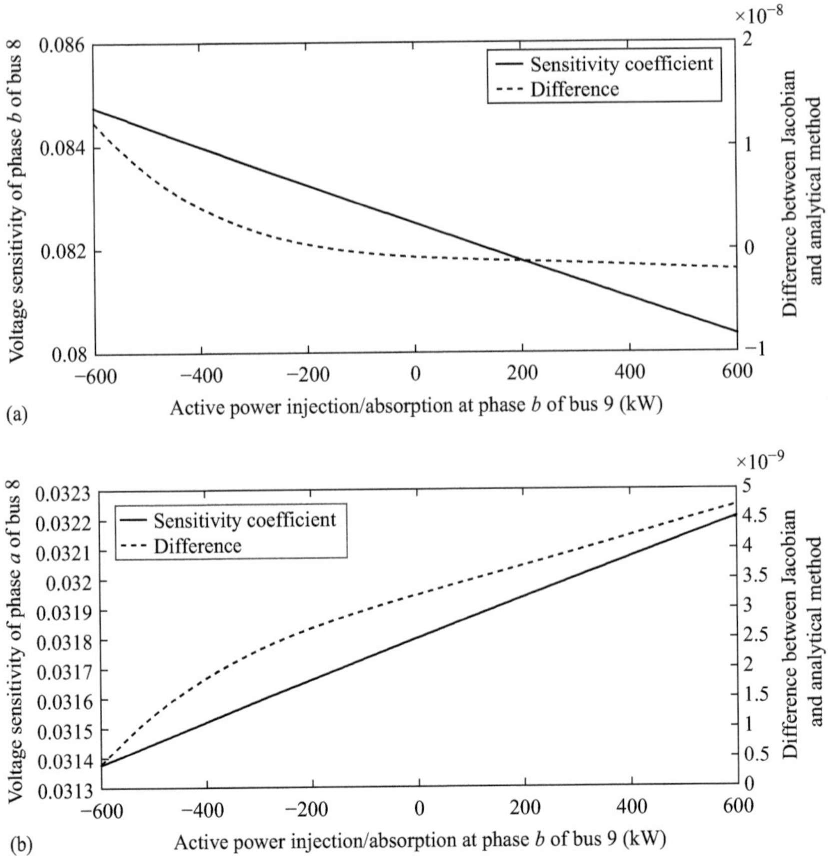

Figure 10.4 Voltage sensitivities of bus 8 as a function of active power injections at bus 9. (a) Voltage sensitivity of phase b of bus 8 with respect to active power injection at phase b of bus 9. (b) Voltage sensitivity of phase a of bus 8 with respect to active power injection at phase b of bus 9

ULTC positions of the transformers, i.e., small perturbations of the voltage magnitude of one phase of the slack bus and solution of the load-flow problem.

For brevity, we limit the validation of the proposed method to a reduced number of buses. In particular, we refer to the variation of voltages at bus 8 with respect to load/injection at bus 9:

$$\frac{\partial|\bar{V}_8^a|}{\partial P_9^b}, \quad \frac{\partial|\bar{V}_8^b|}{\partial P_9^b}, \quad \frac{\partial|\bar{V}_8^a|}{\partial Q_9^b}, \quad \frac{\partial|\bar{V}_8^b|}{\partial Q_9^b}$$

In Figure 10.4(a), the voltage sensitivity of phase b of bus 8 is shown with respect to active power absorption and generation at phase b of bus 9. Figure 10.4(b) shows

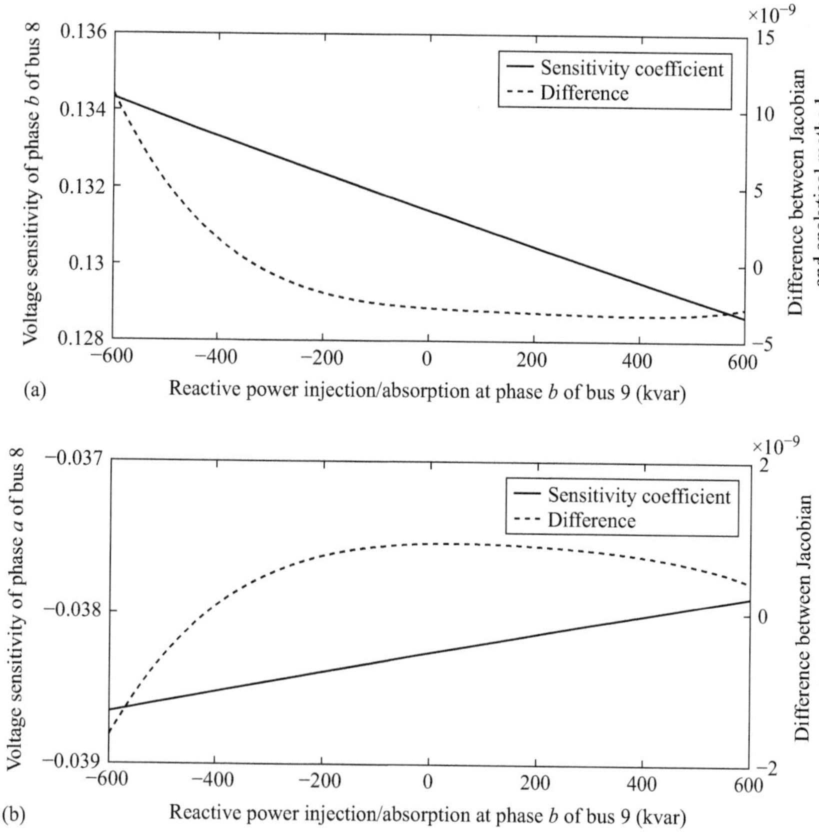

Figure 10.5 Voltage sensitivities of bus 8 as a function of reactive power injections at bus 9. (a) Voltage sensitivity of phase b of bus 8 with respect to reactive power injection at phase b of bus 9. (b) Voltage sensitivity of phase a of bus 8 with respect to reactive power injection at phase b of bus 9

for the same buses as Figure 10.4(b), same sensitivity but referring to voltage and power belonging to different phases. Additionally, Figure 10.5(a) and (b) shows the voltage sensitivity of bus 8 with respect to reactive power absorption and generation at bus 9. In all these four figures, the dashed line represents the difference between the traditional approach, i.e., based on the inverse of the Jacobian matrix, and the analytic method proposed here. As it can be observed, the overall differences are negligible, in the order of magnitude of 10^{-9}. In Figure 10.6(a) and (b), the current sensitivity coefficient of phase a of branch 10-13 is presented with respect to active and reactive power absorption/generation at phase a of bus 13. In the same figures, the dashed lines represent the difference between the analytic values and the numerical ones. Even for these coefficients extremely low errors are obtained, in the order of 10^{-5}.

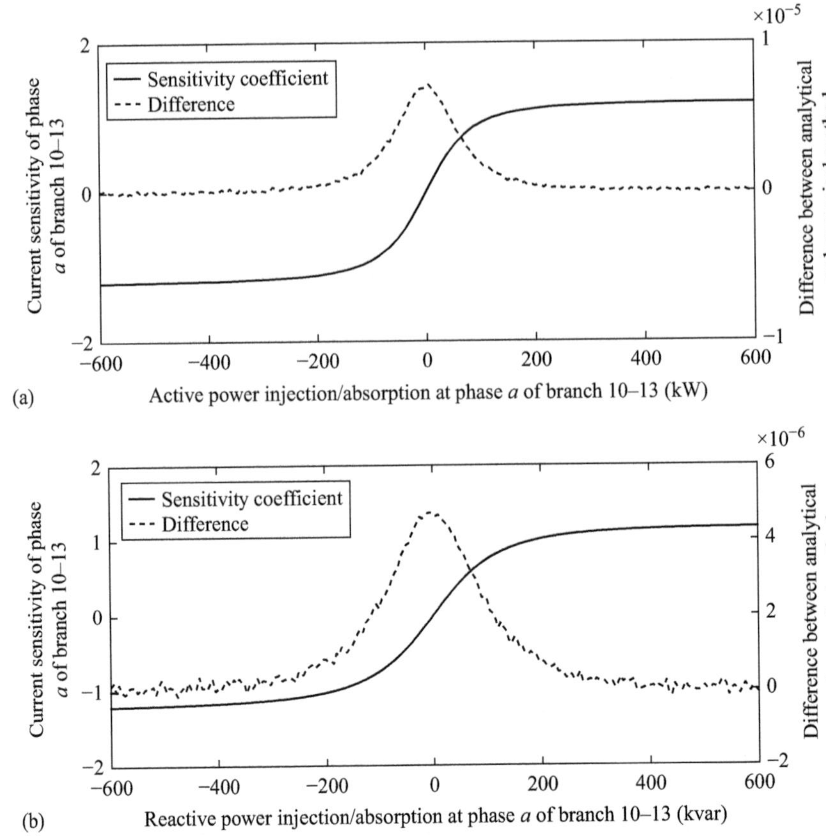

Figure 10.6 Current sensitivity of branch 10-13 as a function of power injections at bus 13. (a) Current sensitivity of phase a of branch 10-13 with respect to active power at phase a of bus 13. (b) Current sensitivity of phase a of branch 10-13 with respect to reactive power at phase a of bus 13

Concerning the validation of voltage sensitivities against tap-changer positions, we have made reference to the IEEE 13-bus test feeder where the slack bus and therefore the primary substation transformer is placed in correspondence with bus 1. We assume to vary the slack bus voltage of $\pm 6\%$ over 72 tap positions (where position "0" refers to the network-rated voltage). In Figure 10.7, the sensitivity of voltage in phase a of bus 7 is shown with respect to the tap positions in phases a, b and c of the slack. Also, in this case, the difference between the analytically inferred sensitivities and the numerical computed ones is negligible (i.e., in the order of magnitude of 10^{-6}).

It is worth observing that for the case of the voltage sensitivities, coefficients that refer to the voltage variation as a function of a perturbation (power injection or tap-changer position) of the same phase, show the largest coupling although a non-negligible cross dependency can be observed between different phases (see for instance Figures 10.4 and 10.5).

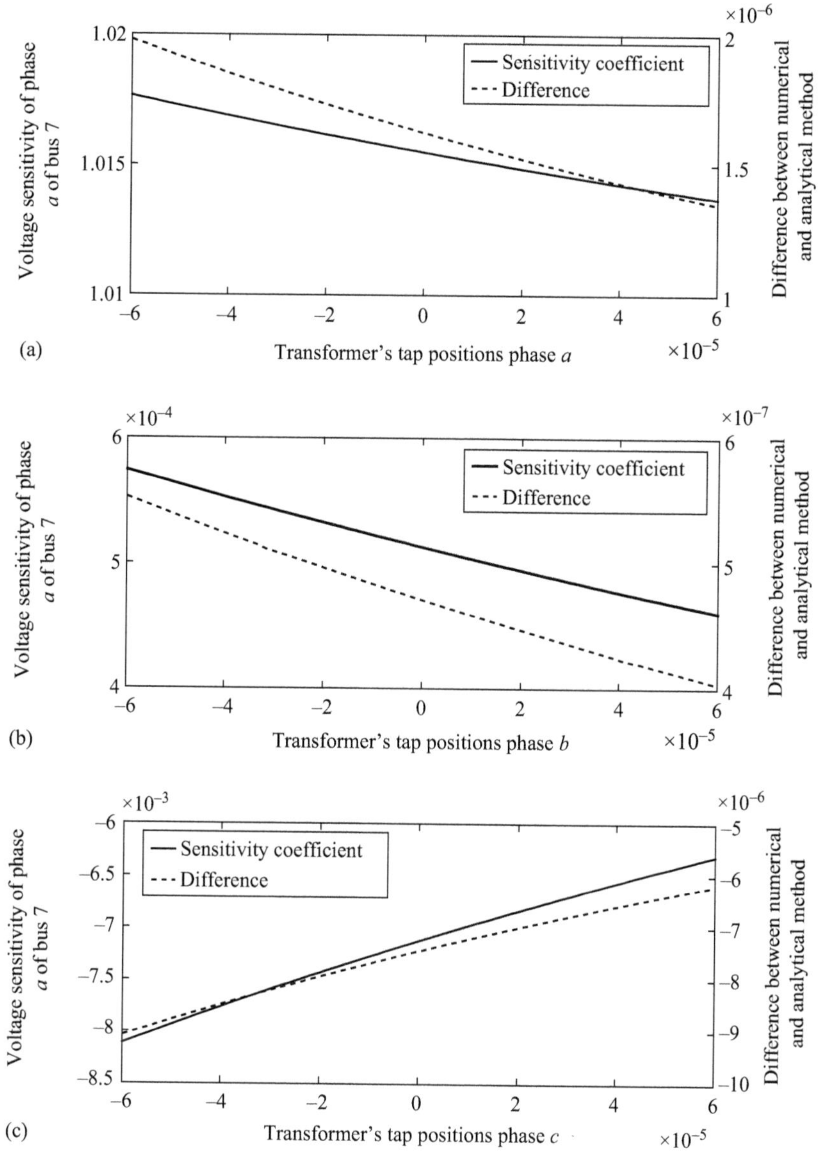

Figure 10.7 Voltage sensitivities of phase a of bus 7 as a function of transformer's ULTC positions. (a) Voltage sensitivity of phase a of bus 7 with respect to ULTC position at phase a of the slack bus. (b) Voltage sensitivity of phase a of bus 7 with respect to ULTC position at phase b of the slack bus. (c) Voltage sensitivity of phase a of bus 7 with respect to ULTC position at phase c of the slack bus

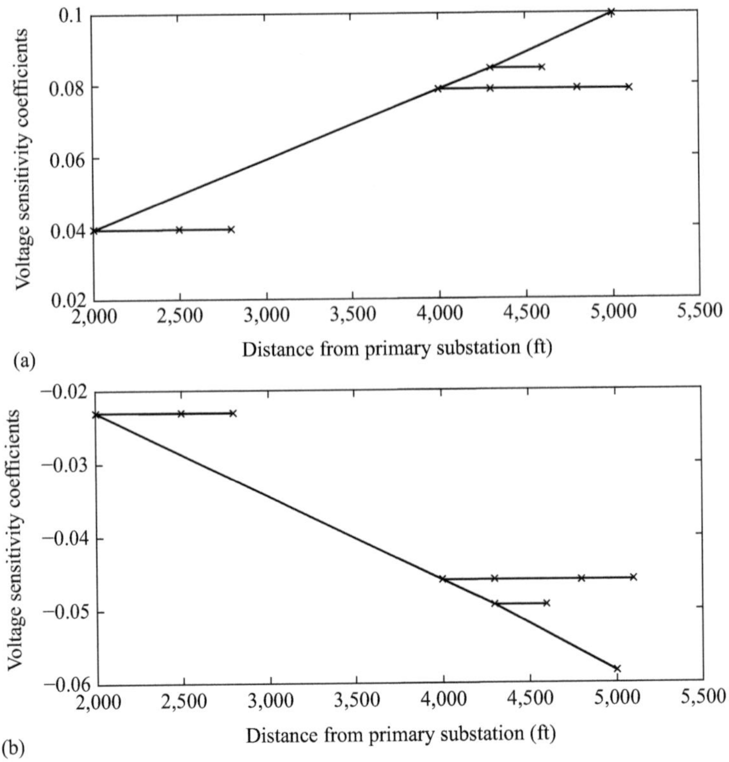

Figure 10.8 Voltage sensitivities with respect to active power absorption at bus 13 as a function of the distance from the slack bus. (a) Voltage sensitivities $\frac{\partial|\bar{V}_i^a|}{\partial P_{13}^a}$ *with respect to active power at phase a of bus 13 as a function of the distance from the slack bus. (b) Voltage sensitivities* $\frac{\partial|\bar{V}_i^b|}{\partial P_{13}^a}$ *with respect to active power at phase a of bus 13 as a function of the distance from the slack bus*

Finally, Figures 10.8 and 10.9 depict the variation of voltage sensitivity coefficients in all the network with respect to active and reactive power absorption at phase *a* of bus 13 as a function of the distance from the slack bus.

This type of representation allows to observe the overall network behaviour against specific PQ buses absorptions/injections. In particular, we can see that larger sensitivities are observed when the distance between the considered voltage and the slack bus increases. Furthermore, a lower, but quantified dependency between coefficients related to different phases can be observed. Also, as expected, reactive power has a larger influence on voltage variations although the active power exhibits a non-negligible influence.

From the operational point of view it is worth observing that, Figures 10.8 and 10.9 provide to network operators an immediate view of the response of the electrical

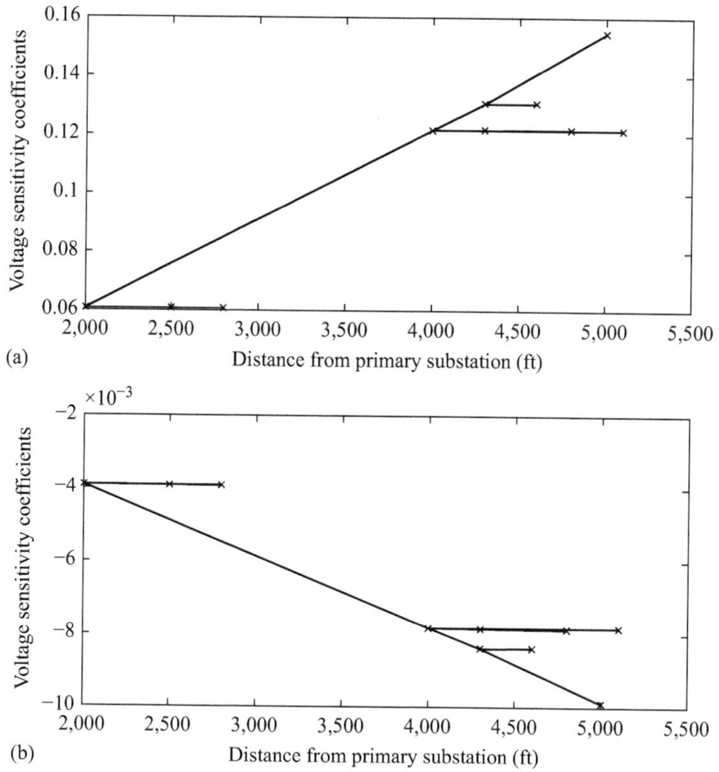

Figure 10.9 Voltage sensitivities with respect to reactive power absorption at bus 13 as a function of the distance from the slack bus. (a) Voltage sensitivities $\frac{\partial|\bar{V}_i^a|}{\partial Q_{13}^a}$ *with respect to reactive power absorption at phase a of bus 13 as a function of the distance from the slack bus. (b) Voltage sensitivities* $\frac{\partial|\bar{V}_i^b|}{\partial Q_{13}^a}$ *with respect to reactive power at phase a of bus 13 as a function of the distance from the slack bus*

network against specific loads/injections that could also be used for closed-loop control or contingency analysis.

10.4.3 Voltage control and lines congestion management examples

For the voltage control and lines congestion management application, the IEEE 34-bus test feeder is considered. Note that the regulators and shunt capacitors are excluded to make the network weaker. The network comprises a number of controllable distributed generation units. We consider three different application examples. In the first one, only voltage control is performed by coordinating the DERs' power production with the ULTC positions. In the second one, both voltage control and lines congestion management are included in the optimization problem and the control variables are

Table 10.3 Initial and maximum operational set points of the DERs and the ULTC in the IEEE 34-bus test feeder

	P_{init} (kW)	P_{max} (kW)	n_{init}	n_{min}	n_{max}
DER_{18}	210	600	0	−36	+36
DER_{23}	250	1 200			
DER_{24}	100	1 200			
DER_{33}	150	600			

solely the DERs' active and reactive power production. The third application example shows a 24 h case study of voltage control.

Example 10.1: In buses 18, 23, 24 and 33 of the IEEE 34-bus test feeder, we assume to have DERs that the DNO can control in terms of active and reactive power. Their initial operating values, as well as their rated power outputs, are shown in Table 10.3. For this case study, the loads shown in Table 10.1 are multiplied by a factor of 1.3 for phase a, 1.24 for phase b and 1.3 for phase c. Furthermore, the DNO has control on the transformer's ULTC positions.

In view of the above, the optimal control problem is formulated as a linearized one taking advantage of the voltage sensitivity coefficients. The controlled variables are the bus voltages and the control variables are the DER's active and reactive power injections and the ULTC positions under the control of the DNO, $\boldsymbol{\Delta x} = [\boldsymbol{\Delta P}_{DER}, \boldsymbol{\Delta Q}_{DER}, \boldsymbol{\Delta n}]$. It is important to state that, formally, this problem is a mixed integer optimization problem due to the inclusion of ULTC. However, for reasons of simplicity, in this example, the tap positions are considered pseudo-continuous variables which are rounded to the nearest integer once the optimal solution is reached. The objective of the linear optimization problem considered in this example is essentially the function in (10.1) with $\gamma = 0$ (in other words, the cost function of the optimization does not account for a deadband surrounding V_o):

$$\min_{\boldsymbol{\Delta(P,Q,n)}} \sum_i [(|\bar{V}_i| + (\boldsymbol{K}_{P,Q,n}\boldsymbol{\Delta(P,Q,n)})_i - |V_o|)^2] \tag{10.30}$$

The imposed constraints on the operational points of the DERs and the ULTC positions are the following:

$$\begin{aligned} 0 \leq P_{DER_i} &\leq P_{DER_{i_{max}}}, \quad i = 1, \ldots, N_{DER} \\ Q_{DER_{i_{min}}} \leq Q_{DER_i} &\leq Q_{DER_{i_{max}}}, \quad i = 1, \ldots, N_{DER} \\ n_{min} \leq n &\leq n_{max} \end{aligned} \tag{10.31}$$

In order to simplify the analysis, we have assumed that the DER capability curves are rectangular ones in the PQ plane. The minimum and maximum reactive power limits are −25% and 25% of the maximum active power values, respectively.

Table 10.4 Optimal operational set points of the DERs and the tap changers in the IEEE 34-bus test feeder when the system operator has control on their three-phase output

	P_{opt_1} (kW)	Q_{opt_1} (kvar)	n_{opt_1}
DER_{18}	410.6	−150	−1
DER_{23}	1 200	−300	
DER_{24}	92.1	−300	
DER_{33}	463.2	−150	

Table 10.5 Optimal operational set points of the DERs and the tap changers in the IEEE 34-bus test feeder when the system operator has control on each of the three phases independently

	P_{opt_2} (kW)	Q_{opt_2} (kvar)	n_{opt_2}
DER_{18}^{a}	127.29	50	0
DER_{18}^{b}	50	−50	
DER_{18}^{c}	37.30	−50	
DER_{23}^{a}	50.94	100	
DER_{23}^{b}	387	−100	
DER_{23}^{c}	400	93.74	
DER_{24}^{a}	0	100	
DER_{24}^{b}	0	−100	
DER_{24}^{c}	70.71	−100	
DER_{33}^{a}	8.25	50	
DER_{33}^{b}	106.65	−50	
DER_{33}^{c}	200	−8.5	

The formulated linearized problem is solved by using the classic linear least squares method. The method used to calculate analytically the sensitivity coefficients allows us to consider two different optimization scenarios. In the first (opt_1), the operator of the system is assumed to control the set points of the DERs considering that they are injecting equal power into the three phases, whereas in the second case (opt_2), it is assumed to have a more sophisticated control on each of the phases independently except for the tap-changer positions. It is worth noting that this second option, although still far from a realistic implementation, allows us to show the capability of the proposed method to deal with the inherent unbalanced nature of distribution networks. Tables 10.4 and 10.5 show the optimal operational set points corresponding to these cases.

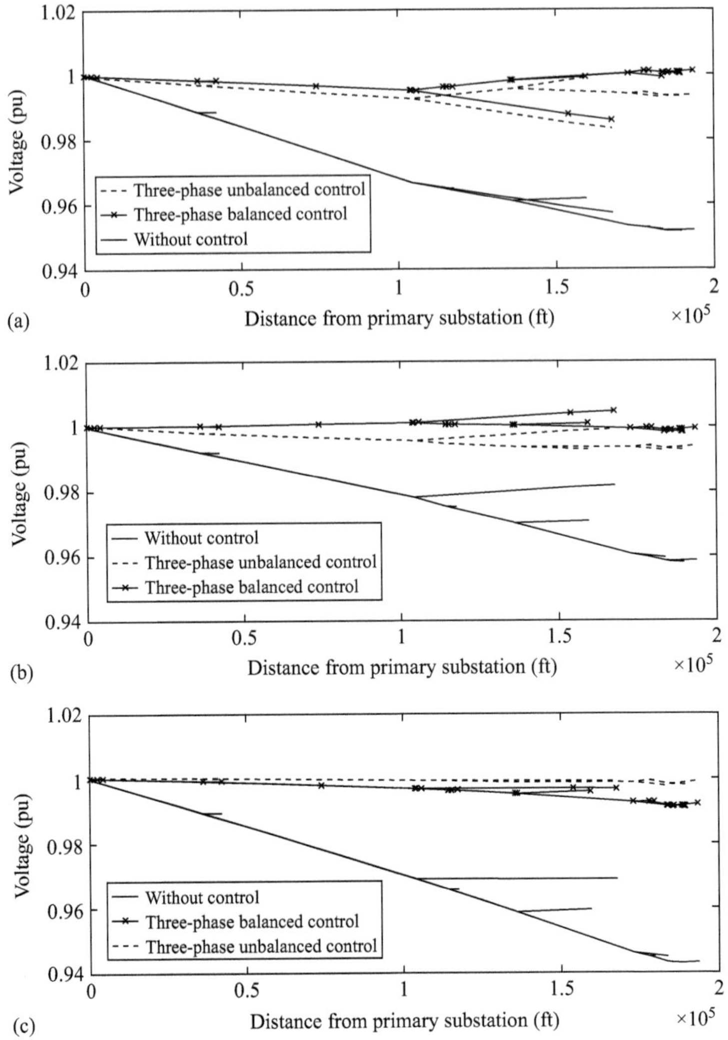

Figure 10.10 Initial and optimized voltage profile of the IEEE 34-bus test feeder. (a) Voltage profile of phase a of the buses, uncontrolled case (black solid line), balanced control (black line with markers) and three-phase unbalanced control (dashed line). (b) Voltage profile of phase b of the buses, uncontrolled case (black solid line), balanced control (black line with markers) and three-phase unbalanced control (dashed line). (c) Voltage profile of phase c of the buses, uncontrolled case (black solid line), balanced control (black line with markers) and three-phase unbalanced control (dashed line)

Table 10.6 Initial and maximum operational set points of the DERs in the IEEE 34-buss test feeder

	P_{init} (kW)	P_{max} (kW)
DER_{18}	630	1 200
DER_{23}	750	2 400
DER_{24}	300	2 400
DER_{33}	450	1 200

Additionally, in Figure 10.10, the voltage profile of the buses of the system is presented in the initial and the optimal cases. The solid line in the figures shows the initial voltage profile, the solid line with the markers shows the first case optimal scenario (opt_1) and the dashed line represents the second case where the DNO has full control in each of the phases of the DERS (opt_2). The offset in the graphs, observed in the slack bus, depicts the optimal ULTC position in each case. What can be observed is that, when there is a possibility to control each of the three phases of the DERs output, the optimal voltage profile is better than the one corresponding to control of the balanced three-phase output of the set points of the DERs.

Example 10.2: As in Example 10.1, in buses 18, 23, 24 and 33 of the IEEE 34-bus test feeder, we assume to have DERs that the DNO can control in terms of active and reactive power. Their initial operating values, as well as their rated power outputs, are shown in Table 10.6. The DERs do not inject any reactive power in the base case and their minimum and maximum reactive power limits are −25% and 25% of the maximum active power values, respectively. For this case study, the loads shown in Table 10.1 are multiplied by a factor of 1.6 for each phase *a*, *b* and *c*.

In this case, the DNO is interested to control the available distributed and centralized resources in order to improve the network voltage profile while guaranteeing that line current flows are below their ampacity limits. In this example, only three-phase balanced control of the DERs active and reactive power injections is considered and ULTC control is not taken into account. The optimal control problem is formulated as follows:

$$\min_{\Delta(P,Q)} \sum_i [(|\bar{V}_i| + (\boldsymbol{K}_{P,Q}\boldsymbol{\Delta(P,Q)})_i - |V_o|)^2 - \gamma^2]^+ \tag{10.32}$$

$$\text{subject to:} \quad |\bar{I}_{ij} + (\bar{\boldsymbol{H}}_{P,Q}\boldsymbol{\Delta(P,Q)})_{ij}| \leq I_{max}, \quad i,j = 1,\ldots,N_l, \ i \neq j \tag{10.33}$$

$$0 \leq P_{DER_i} \leq P_{DER_{imax}}, \quad i = 1,\ldots,N_{DER} \tag{10.34}$$

$$Q_{DER_{imin}} \leq Q_{DER_i} \leq Q_{DER_{imax}}, \quad i = 1,\ldots,N_{DER} \tag{10.35}$$

In this example, we tune the values γ and I_{max} in the optimal control problem formulation in order to evaluate the performances of the proposed method in the case of violations of the lines ampacity limits. In particular, we observe that in the base

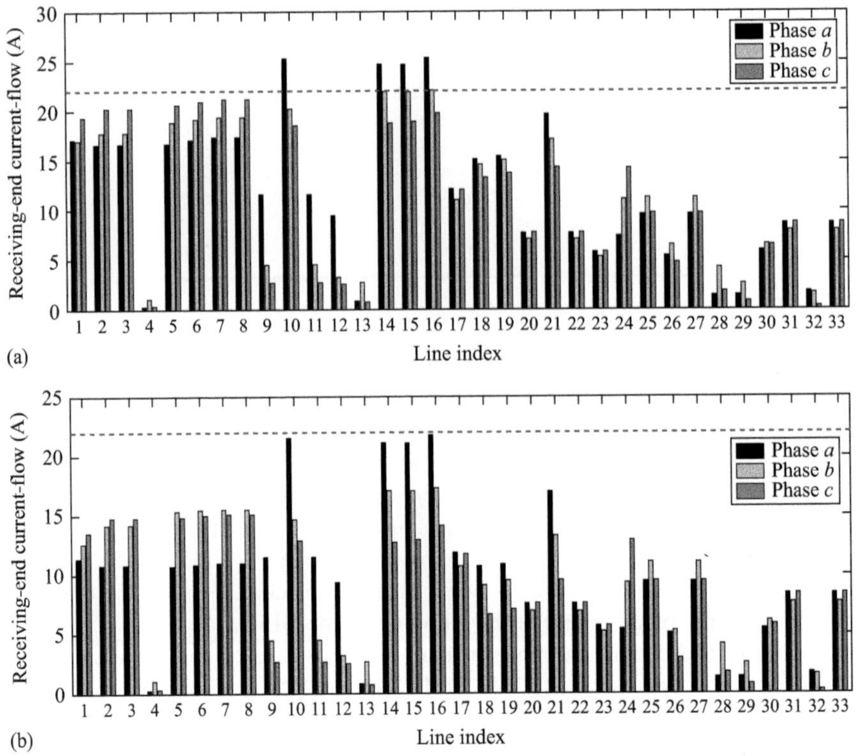

Figure 10.11 Uncontrolled and optimized lines current flows (receiving end). (a) Uncontrolled current flow at the receiving end of the network lines. (b) Optimized current flow at the receiving end of the network lines

case the maximum voltage deviation is in the order of 2% and the maximum line current flow is 25 A. Therefore, we set $\gamma = 10\%$ and $I_{max} = 20$ A to focus on the lines congestion management problem.

The results of this case study are shown in Figure 10.11 where the line current flows are shown for the receiving end of the network lines and in Figure 10.12 for the line current flows at the sending end of the lines. In both figures, on the top the uncontrolled network current profile is depicted where it can be observed that four lines violate the line ampacity limit. The same figures, on the bottom, show the results after the optimal control problem is solved. In this case, all lines satisfy, as expected, the maximum allowed current limit.

The network voltage profile for each phase is shown in Figure 10.13 before and after the control actions. In both cases, the voltage profiles are within $\pm 10\%$ of the network-rated value.

The optimal active and reactive power injections of the controllable resources for this case study are shown in Table 10.7.

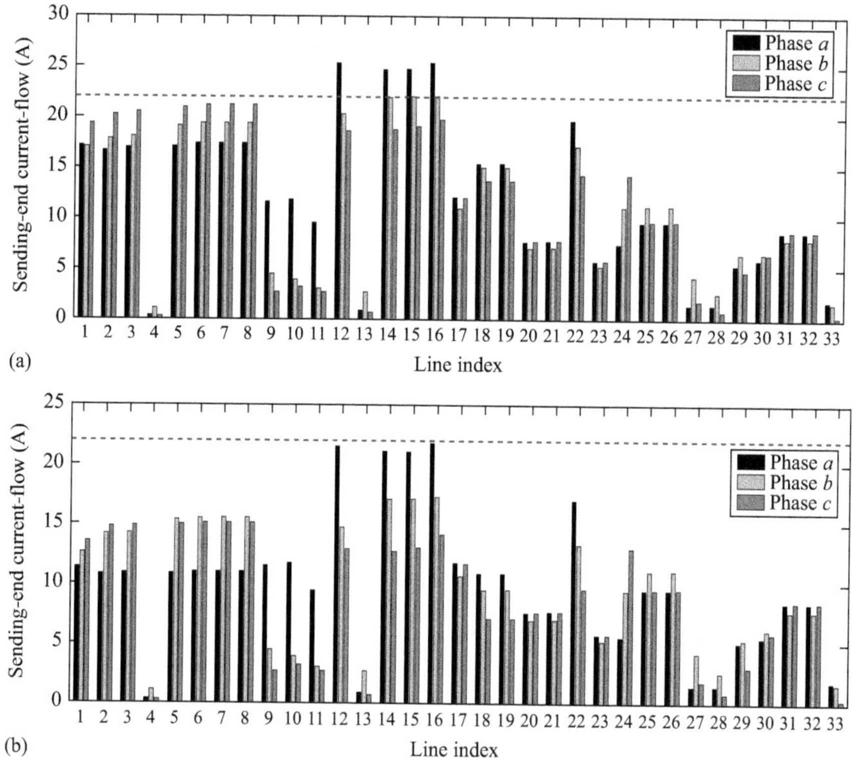

Figure 10.12 Uncontrolled and optimized lines current flows (sending end). (a) Uncontrolled current flow at the sending end of the network lines. (b) Optimized current flow at the sending end of the network lines

Example 10.3: We consider once again the IEEE 34-bus test feeder equipped with the four generators located in buses 18, 23, 24 and 33. In this example, we perform 24 h voltage control via scheduling of the active and reactive power of the DERs, as well as of the ULTCs. Only the case of balanced control is considered and lines congestion management is not taken into account.

The optimal control problem in this case is formulated as follows:

$$\min_{\boldsymbol{\Delta(P,Q,n)}} \sum_i (|\bar{V}_i| + (\boldsymbol{K_{P,Q,n}\Delta(P,Q,n)})_i - |V_o|)^2 + \psi(\boldsymbol{\Delta n})\boldsymbol{\Delta n}^2 \tag{10.36}$$

subject to:

$$0 \le P_{DER_i} \le P_{DER_{i_{max}}}, \quad i = 1,\ldots,N_{DER} \tag{10.37}$$

$$Q_{DER_{i_{min}}} \le Q_{DER_i} \le Q_{DER_{i_{max}}}, \quad i = 1,\ldots,N_{DER} \tag{10.38}$$

$$n_{min} \le n \le n_{max} \tag{10.39}$$

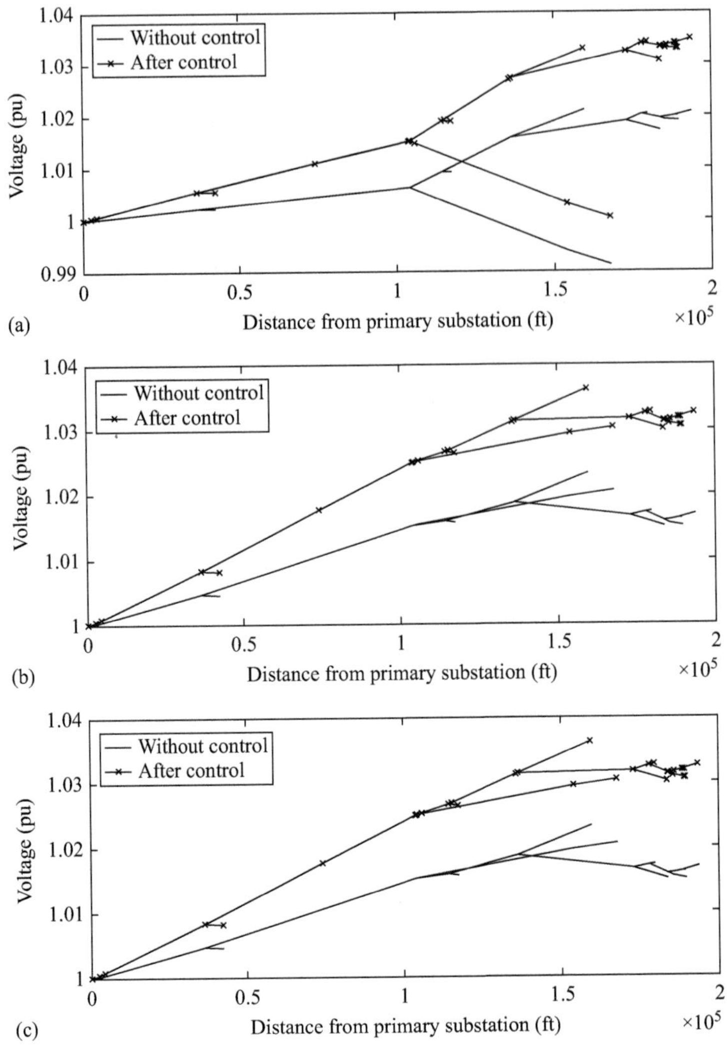

Figure 10.13 Initial and optimized voltage profile of the IEEE 34-bus test feeder. (a) Voltage profile of phase a of the buses, uncontrolled case (black solid line), balanced control (black line with markers). (b) Voltage profile of phase b of the buses, uncontrolled case (black solid line), balanced control (black line with markers). (c) Voltage profile of phase c of the buses, uncontrolled case (black solid line), balanced control (black line with markers)

Table 10.7 Optimal operational set points of the DERs

	P_{opt} (kW)	Q_{opt} (kvar)
DER_{18}	608.8	108.7
DER_{23}	729.4	109.8
DER_{24}	279.4	109.8
DER_{33}	429.4	110.0

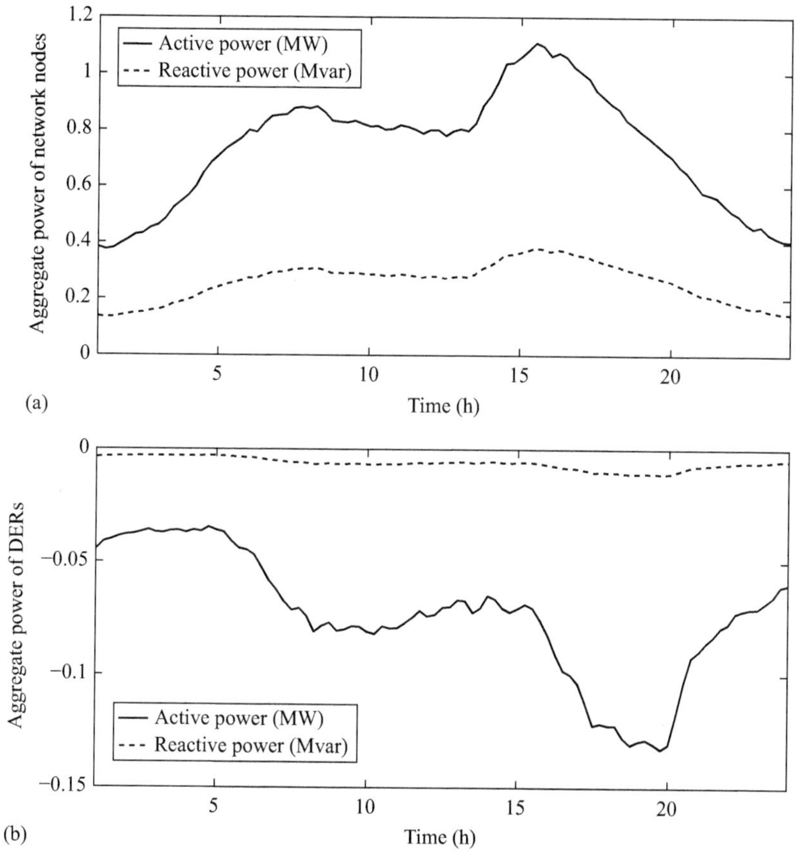

Figure 10.14 Twenty-four hour aggregate active and reactive power consumption/generation. (a) Twenty-four hour aggregate active and reactive power consumption. (b) Twenty-four hour aggregate active and reactive power generation of DERs

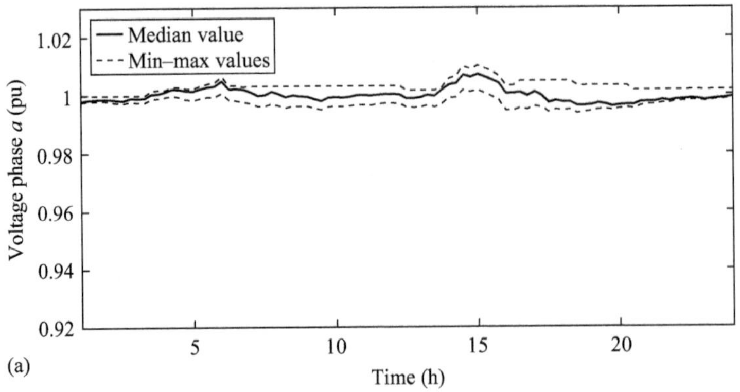

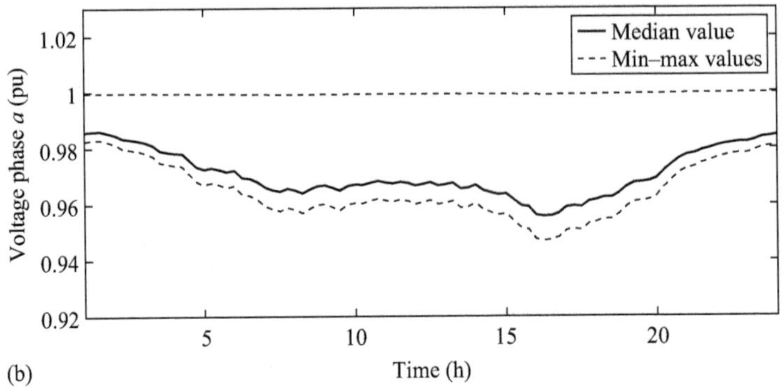

Figure 10.15 Network voltage profile of phase a before and after the control actions. (a) Uncontrolled voltage profile of phase a of the network, median value (solid line), minimum and maximum values (dashed lines). (b) Optimized voltage profile of phase a of the network, median value (solid line), minimum and maximum values (dashed lines)

where ψ is a penalty function for altering the tap-changer position.[††] The first term of (10.36) represents the voltage control cost function. The operator can perform this type of control by deploying solely the DERs or by coordinating control of the DERs and the ULTC positions. In the case where the tap changers are included, the DNO needs to account for the limited number of ULTC operations. This is represented by the term ψ of (10.36). This function multiplies the ULTC set points variation and

[††]As we deal with primary voltage control, (10.36) has to penalize the changes of ULTC as these devices are typically used by the DNO rarely.

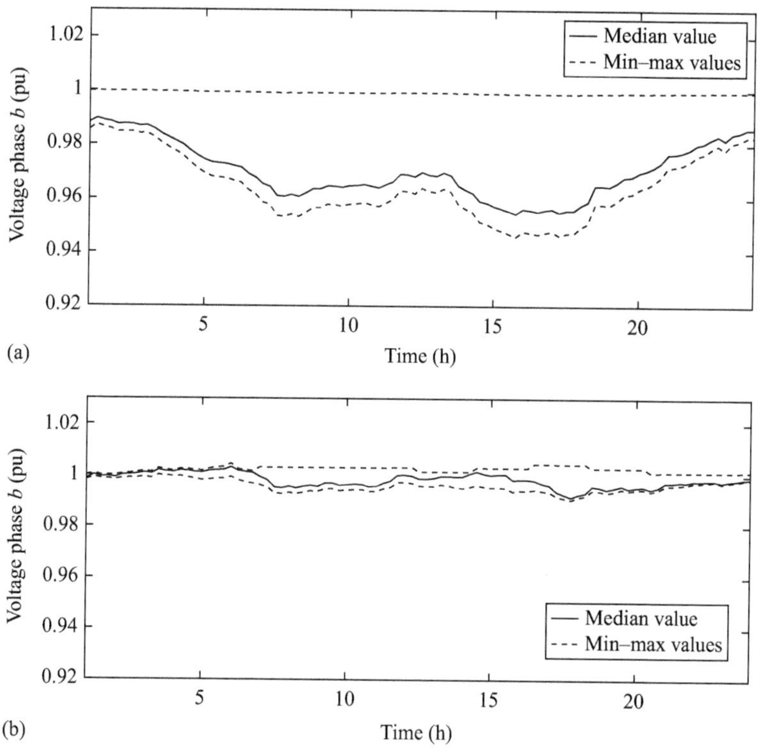

Figure 10.16 Network voltage profile of phase b before and after the control actions. (a) Uncontrolled voltage profile of phase b of the network, median value (solid line), minimum and maximum values (dashed lines). (b) Optimized voltage profile of phase b of the network, median value (solid line), minimum and maximum values (dashed lines)

increases with the number of ULTC operations in a given time window.[‡‡] Specifically, we have chosen:

$$\psi(\boldsymbol{\Delta n}) := \lambda \left(\sum_{s=0}^{W-1} |\boldsymbol{\Delta n}(t-s)| \right) \tag{10.40}$$

where λ is a constant. Such an expression of ψ allows to weight the accumulated number of ULTC changes within a given time window W.

[‡‡]By including this function the DNO has the option to upper-bound the total number of ULTC operations, thus respecting the nature and cost of these devices [29].

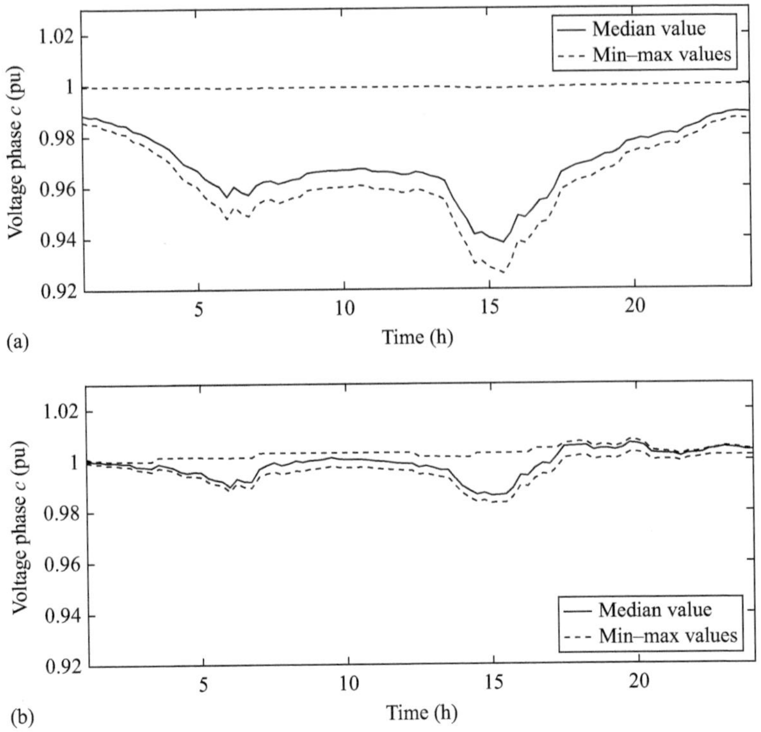

Figure 10.17 Network voltage profile of phase c before and after the control actions. (a) Uncontrolled voltage profile of phase c of the network, median value (solid line), minimum and maximum values (dashed lines). (b) Optimized voltage profile of phase c of the network, median value (solid line), minimum and maximum values (dashed lines)

For this application example, the aggregate load profile of the network is depicted in Figure 10.14 in terms of 24 h active and reactive power injections.

Figures 10.15–10.17 show, for each network phase, the 24 h voltage profile before any control action (top) and after (bottom) the scheduling of the DERs' active and reactive power and the ULTC positions. As it can be observed, the voltage profiles before the control actions are unbalanced and exhibit deviations from the network-rated value in the order of 5%–7%. After the scheduling of the controllable resources, the resulting voltage profiles for each of the three phases are flatter around 1 pu and exhibit maximum deviations from the network-rated value in the order of 1%–2%.

Figure 10.18 shows the 24 h active and reactive power output of the controllable DERs which is the result of the solution of the optimal control problem in (10.36)–(10.39), whereas Figure 10.19 shows the optimal ULTC positions along the day. It is worth noting that the obtained number of ULTC changes is compatible with a typical operation of such a device, i.e., less than 10 manoeuvres per day.

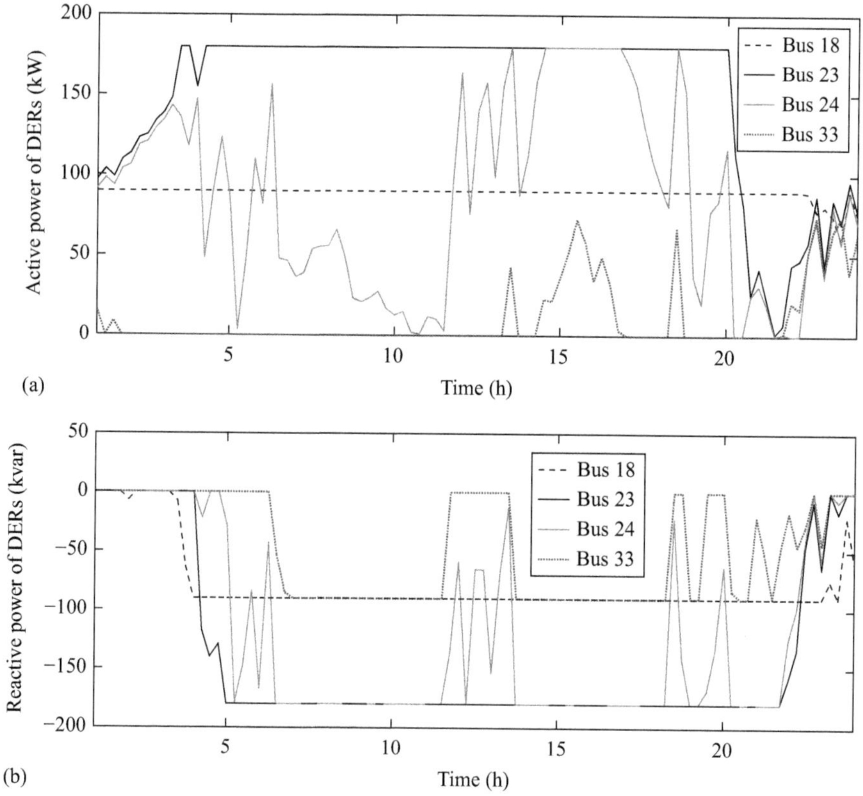

Figure 10.18 Twenty-four hour controlled active and reactive power output of the DERs. (a) Twenty-four hour three-phase active power profile of the DERs. (b) Twenty-four hour three-phase reactive power profile of the DERs

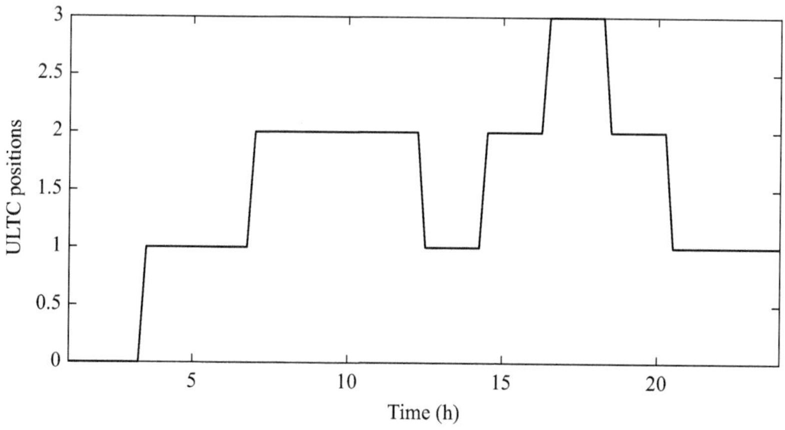

Figure 10.19 Twenty-four hour ULTC positions

10.5 Conclusions

In this chapter, we consider a centralized real-time control architecture for voltage regulation and lines congestion management in ADNs that is based on a linearized approach that links control variables (e.g., power injections and transformers tap positions) and controlled quantities (e.g., voltages and current flows) by means of sensitivity coefficients.

We validate the proposed analytic method by making reference to typical IEEE 13- and 34-bus distribution test feeders. The numerical validation of the computation of the coefficients is performed using the IEEE 13-bus test feeder and it shows that the errors between the traditional approaches, i.e., based on the inverse of the Jacobian matrix, and the analytic method are extremely low (in the order of magnitude of 10^{-6}–10^{-9}). The IEEE 34-bus test feeder is used to show application examples related to a possible integration of the proposed method for the problem of optimal voltage control and lines congestion management in unbalanced distribution systems. The simulation results show that the proposed algorithm is able to improve the voltage and current profiles in the network, and also that when each of the three phases of the DERs can be controlled independently of the others, the resulting optimal voltage and current profiles are better than the ones corresponding to the balanced control of the three-phase output of the set points of the DERs.

Bibliography

[1] Singh N, Kliokys E, Feldmann H, Kussel R, Chrustowski R, Joborowicz C. "Power system modelling and analysis in a mixed energy management and distribution management system". *IEEE Transactions on Power Systems*. 1998;13(3):1143–1149.

[2] Jenkins N, Allan R, Crossley P, Kirschen D, Strbac G. "Embedded generation". vol. 9. Johns AT, Warne DF, editors. Stevenage, UK: IET; 2000.

[3] James NG. *Control and Automation of Electrical Power Systems*. Hoboken, NJ: CRC Press; 2006.

[4] Zhou Q, Bialek J. "Generation curtailment to manage voltage constraints in distribution networks". *IET, Generation, Transmission & Distribution*. 2007;1(3):492–498.

[5] Senjyu T, Miyazato Y, Yona A, Urasaki N, Funabashi T. "Optimal distribution voltage control and coordination with distributed generation". *IEEE Transactions on Power Delivery*. 2008;23(2):1236–1242.

[6] Borghetti A, Bosetti M, Grillo S, *et al.* "Short-term scheduling and control of active distribution systems with high penetration of renewable resources". *IEEE Systems Journal*. 2010;4(3):313–322.

[7] D'adamo C, Abbey C, Baitch A, *et al.* "Development and operation of active distribution networks". *Cigré Task Force, Paris, France, Tech Brochure C*. 2011;6:1–6.

[8] Pilo F, Jupe S, Silvestro F, *et al.* "Planning and optimisation of active distribution systems: an overview of CIGRE Working Group C6. 19 activities". In: *Integration of Renewables into the Distribution Grid, CIRED 2012 Workshop*. Stevenage, UK: IET; 2012. p. 1–4.
[9] Khatod DK, Pant V, Sharma J. "A novel approach for sensitivity calculations in the radial distribution system". *IEEE Transactions on Power Delivery*. 2006;21(4):2048–2057.
[10] Conti S, Raiti S, Vagliasindi G. "Voltage sensitivity analysis in radial MV distribution networks using constant current models". In: *IEEE International Symposium on Industrial Electronics (ISIE)*. Italy: Bari, IEEE; 2010. p. 2548–2554.
[11] Czarnecki L, Staroszczyk Z. "On-line measurement of equivalent parameters for harmonic frequencies of a power distribution system and load". *IEEE Transactions on Instrumentation and Measurement*. 1996;45(2):467–472.
[12] Christakou K, LeBoudec J, Paolone M, Tomozei DC. "Efficient computation of sensitivity coefficients of node voltages and line currents in unbalanced radial electrical distribution networks". *IEEE Transactions on Smart Grid*. 2013;4(2):741–750.
[13] Bersani A, Borghetti A, Bossi C, *et al.* "Management of low voltage grids with high penetration of distributed generation: concepts, implementations and experiments". In: *Cigré 2006 Session*. EPFL-CONF-180111; 2006.
[14] Torregrossa D, Le Boudec JY, Paolone M. "Model-free computation of ultra-short-term prediction intervals of solar irradiance". *Solar Energy*. 2016;124:57–67.
[15] Sarri S, Zanni L, Popovic M, Boudec JYL, Paolone M. "Performance assessment of linear state estimators using synchrophasor measurements". *IEEE Transactions on Instrumentation and Measurement*. 2016 Mar;65(3):535–548.
[16] Camacho EF, Alba CB. *Model Predictive Control*. Berlin: Springer Science & Business Media; 2013.
[17] Ben-Tal A, El Ghaoui L, Nemirovski A. *Robust Optimization*. Princeton, NJ: Princeton University Press; 2009.
[18] Sarri S, Paolone M, Cherkaoui R, Borghetti A, Napolitano F, Nucci CA. "State estimation of active distribution networks: comparison between WLS and iterated Kalman-filter algorithm integrating PMUs". In: *Third IEEE PES Innovative Smart Grid Technologies (ISGT) Europe Conference*. Germany: Berlin, 2012.
[19] Pegoraro PA, Tang J, Liu J, Ponci F, Monti A, Muscas C. "PMU and smart metering deployment for state estimation in active distribution grids". In: *Energy Conference and Exhibition (ENERGYCON), 2012 IEEE International*. Italy: Florence, 2012. p. 873–878.
[20] De Carne G, Liserre M, Christakou K, Paolone M. "Integrated voltage control and line congestion management in active distribution networks by means of smart transformers". In: *Industrial Electronics (ISIE), 2014 IEEE 23rd International Symposium on Industrial Electronics (ISIE)*. Turkey: Istanbul, 2014. p. 2613–2619.

[21] Peschon J, Piercy DS, Tinney WF, Tveit OJ. “Sensitivity in power systems”. *IEEE Transactions on Power Apparatus and Systems*. 1968;(8):1687–1696.

[22] Shirmohammadi D, Hong H, Semlyen A, Luo G. “A compensation-based power flow method for weakly meshed distribution and transmission networks”. *IEEE Transactions on Power Systems*. 1988;3(2):753–762.

[23] Wood AJ, Wollenberg BF. *Power Generation, Operation, and Control*, vol. 2. New York, NY: Wiley; 1996.

[24] Marconato R. *Electric Power Systems*, vol. 2. Milano, Italy: CEI, Italian Electrotechnical Committee; 2002.

[25] Begovic MM, Phadke AG. “Control of voltage stability using sensitivity analysis”. *IEEE Transactions on Power Systems*. 1992;7(1):114–123.

[26] Christakou K, Tomozei DC, Bahramipanah M, Le Boudec JY, Paolone M. “Primary voltage control in active distribution networks via broadcast signals: the case of distributed storage”. *IEEE Transactions on Smart Grid*. 2014 Sep;5(5):2314–2325.

[27] Arrillaga J, Bradley D, Bodger P. *Power System Harmonics*. New York, NY: John Wiley; 1985.

[28] Kersting WH. “Radial distribution test feeders”. *IEEE Transactions on Power Systems*. 1991 Aug;6(3):975–985.

[29] Virayavanich CHKWS, Seiler A. “Reliability of on-load tap changers with special consideration of experience with delta connected transformer windings and tropical environmental conditions”. In: *Cigré, Paper*; 1996. p. 12–103.

Chapter 11

Control of converter interfaced generation

S. D'Arco[1], A. Monti[2], T. Heins[2] and J.A. Suul[1,3]

This chapter presents an overview of control solutions for converter interfaced generation. In the following, the focus is only on situations in which the energy resource is connected to the grid through a full-scale inverter. As a result, configurations such as doubly-fed induction generators for wind turbines are not considered. Under this assumption, the grid-connected inverter is the main controllable element that determines how power can flow from the energy source to the grid.

Actively controlled power converters have been commonly utilized as the grid interface of energy storage systems or regenerative loads requiring bidirectional power flow. However, the growth of generation from renewable energy sources (RESs) has widely expanded the application of grid connected power converters. Indeed, a power converter is included as the grid interface in most renewable generation units as wind turbines and photovoltaic panels, and applications to variable speed hydropower plants are also being evaluated. Thus, large-scale utilization of RES is a dominant contributor to the ongoing evolution of the modern power system towards a system with dominant presence of power electronics converters.

A power converter operated as the grid interface of a generation unit needs to be synchronized with the external grid to ensure a controllable flow of active and reactive power. However, the operational requirements and the corresponding synchronization mechanisms can differ, depending on the specific needs of the application and the local conditions of the power system. Thus, the applied synchronization strategy can also be used as a basis for defining and classifying control methods for power electronics converters. In the following, the operation of grid interface converters is classified within two main categories: "grid-following" and "grid-forming" control [1].

These two approaches are defined as different ways of integrating power converters in the classical frequency control architecture for power grids. The grid-following converters are integrated in the system by synchronizing to the voltage defined by the rest of the grid, while grid-forming converters are typically designed to operate according to similar synchronization mechanisms as synchronous generators.

[1]SINTEF Energy Research, Trondheim, Norway
[2]Institute for Automation of Complex Power Systems, RWTH Aachen University, Aachen, Germany
[3]Department of engineering cybernetics, Norwegian University of Science and Technology, Trondheim, Norway

A dominant presence of grid-following converters can weaken the frequency regulation and compromise the stability of the power system. However, introduction of grid-forming converters can compensate for the declining share of synchronous generators resulting from large-scale development of converter interfaced generation. The typical converter topologies utilized as the grid interface of RES generation and the basic principles for the most common implementations of grid-following control are presented in this chapter. Furthermore, an approach for designing grid-forming converter control by emulating the synchronization mechanism of synchronous machines is introduced. On this basis, different varieties of control strategies labeled as "virtual synchronous machines" (VSMs) are presented as examples of grid-forming control.

The grid-forming control strategies designed to emulate the synchronization mechanism of synchronous generators are explicitly intended to be compatible with the frequency control principles and operation strategies of traditional power systems. However, for future power system configurations based only on power electronic converters, the direct compatibility with traditional grids dominated by synchronous generators might not be required. Such conditions may already be appearing in microgrid scenarios when local islanding mode is activated. Thus, other strategies for synchronizing the operation of converters in a power system are also being widely research. As an example of a possible solution, this chapter presents an introduction to the virtual oscillator control (VOC). The VOC has been explored in recent years as a potential basis for designing power grids fully based on converter interfaced sources. In this case, the synchronization of operation is obtained by exploring characteristics of non-linear oscillators to self-synchronize their operation.

11.1 Hardware structure

Before entering in the control characteristics, we report here some simple considerations on the hardware structure of grid interface converters. Given the assumption reported in the introduction of the chapter, a typical hardware topology can be summarized as in Figure 11.1 [1,3].

The following elements are then considered:

1. DC capacitor creating the DC-bus interface with the generation source
2. Voltage source converter (VSC) module
3. Output LCL filter

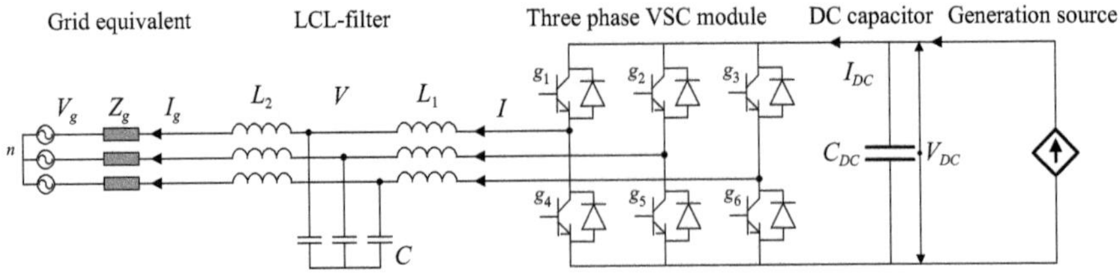

Figure 11.1 Hardware structure of a grid-connected inverter

The DC bus provides the natural interface between the generation source and the grid interface converter. In the case of photovoltaic applications, the power supplied to the DC bus is typically controlled by a DC/DC converter with a maximum power point tracking (MPPT) algorithm. For wind applications, the power flowing into the DC bus is the result of a conversion from variable frequency AC to DC. The generator frequency will then be controlled according to the current wind conditions. In both cases, the DC capacitor provides a decoupling between the switching operation generation side converter and the grid interface. As a result, the power injection from the generation side converter to the DC bus can be modeled as a controlled current source when considering the control of the grid interface converter.

The output inverter is usually realized in two main versions (see Figure 11.2):

- Standard two-level VSC
- Multilevel converters based on Neutral Point Clamped (NPC) topologies, typically in a three-level VSC configuration

Depending on the power level, these topologies can be utilized for both single-phase and three-phase configurations. More complex topologies, such as the modular multilevel converter (MMC) are adopted mostly for high voltage applications including high voltage DC (HVDC) transmission systems. Modulation strategies for the different topologies are out of scope of this chapter.

The last component is the output filter. Here we can consider three different options:

1. Simple inductance (L-filter)
2. LC filter
3. LCL filter

The simple L-filter mainly applied for low-power configurations, and it is applicable when the inverter is operated in current control mode, i.e., if a current reference is provided as the output of the higher level control. The LC filter can be applied for the same purposes as the first case, but can also be utilized when the converter is operated as a controlled voltage source. This case assumes that the grid can be seen

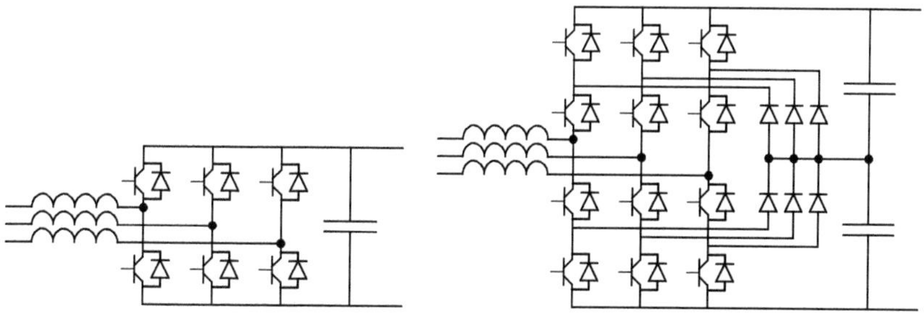

Figure 11.2 Two level and three level inverter structure

from the converter as a source with a significant impedance. The more complete case is the LCL filter. The LCL filter can be interpreted in two different ways:

1. A first LC-stage acting as a harmonic filter, followed by a decoupling inductance
2. A full third-order filter for frequency shaping

Moving from the case of a simple inductance to a full LCL configuration, the grid interface becomes a system of higher order moving from a first-order transfer function to a third-order case. While a simple PI control can offer full control of the interface for the case of a L-filter, different control strategies are necessary with the third-order LCL filter, as, for example, nested control loops or state space approach [4]. Alternatively, active damping methods can be utilized to suppress the oscillatory effects that can be triggered by the interaction among the inductors and the capacitance [5].

11.2 Grid-following and grid-forming control

As already mentioned, operation of grid connected converters can be generally classified according to the modes of operation enabled by the control system. Thus, converter control can in general terms be classified into "grid-following" and "grid-forming" strategies [6,7]. In this chapter, the classification of grid-forming and grid-following control is based on the most general characteristics associated with their operating principles. The following definitions are considered:

- *A grid-following converter relies on the presence of an external grid to be able to operate.*
  The converter requires synchronization to an external grid voltage and usually operates with a controlled current or power injection. A common example of a grid-following control strategy is the use of a phase-locked loop (PLL) for grid synchronization in combination with a fast current control loop.
- *A grid-forming converter can operate independently from the presence of an external grid voltage and is inherently able to establish a local grid.*
  The control of the converter is typically designed to obtain similar characteristics as a synchronous machine. This implies that the converter will be controlled so that the generation unit can be represented as an equivalent voltage source behind an impedance when seen from the external grid.

The grid interface converters for renewable power generation systems are conventionally controlled as grid-following units. Thus, their operation is fully dependent on the presence of a stable external grid voltage with a relatively tightly regulated frequency and would be compromised if this external grid is disconnected or experiences challenges in the frequency and voltage regulation. However, when a strong external grid is available, grid-following operation allows for accurate power flow control which is suitable for MPPT of RES-based generation. The grid-forming operation is instead implying requirements for a degree of dispatchability or a limited energy buffering capacity as part of the generation plant. This follows from the requirements for the converter to be able to participate in the definition of the grid frequency and voltage, which is gaining importance with the declining share of large synchronous generators in the power system. However, when operating in a strong grid without any challenges in the frequency regulation, grid-forming converters can provide controllable power injection in a similar way as synchronous machines, either with or without participation in the primary frequency control.

While the basic principles for grid-following control are well established [8], many different grid-forming control schemes have been proposed and implemented during the last two decades [6,7]. Early concepts for stand-alone operation of power electronic converters and control strategies for synchronization and load sharing among parallel connected converters in islanded power systems were first introduced for uninterruptible power supply (UPS) applications [9]. Later, similar control strategies have been widely studied for parallel operation of grid interface converters in microgrids [10–12]. In such converter-dominated systems, the main objectives of control have been to obtain stability, desirable dynamic response, and load sharing between the parallel units. However, the rapid development of converter interfaced generation in traditional large-scale power systems have introduced the need for also supporting the operation of the remaining synchronous generators and their frequency regulation. Thus, control of power converters to explicitly emulate the inertial characteristics of synchronous machines has become relevant for mitigating the effects of declining inertia in traditional power system [13–15]. Due to this development, the concept of grid-forming control is becoming increasingly important for transmission system operators (TSOs). Thus, definitions of functional requirements for grid-forming control are currently being developed, for instance by the European Network of Transmission System Operators for Electricity (ENTSO-E) [19].

To provide both inertia emulation and grid-forming capabilities by explicitly emulating the rotor dynamics of a synchronous machine, a wide range of control methods have been proposed [15–17]. Although several different names have been introduced for various implementations, these strategies are here generally referred to as VSMs. These control strategies can be considered as a specific subset of "grid-forming" control. However, several other implementations of grid-forming control with or without explicit inertia emulation have been proposed in the literature [7,18].

Beyond the general categories of grid-following and grid-forming control, it can also be relevant to consider converters defined with "grid-supporting" control [6]. This operating mode refers to a grid interfaced converter that provides supporting services to the grid as for example frequency support by participating in the primary frequency control or voltage support by reactive power control. However, in relatively strong grid conditions, such services can be provided by both grid-following and grid-forming converters independently of the synchronization mechanism and detailed implementation applied for the converter. Thus, specific details on grid supporting control will not be further discussed as a separate topic.

11.3 Implementations for grid-following control

Figure 11.3 presents the key building blocks of a conventional grid-following control. In this case, we show only an L-filter as part of the grid interface, but the same general

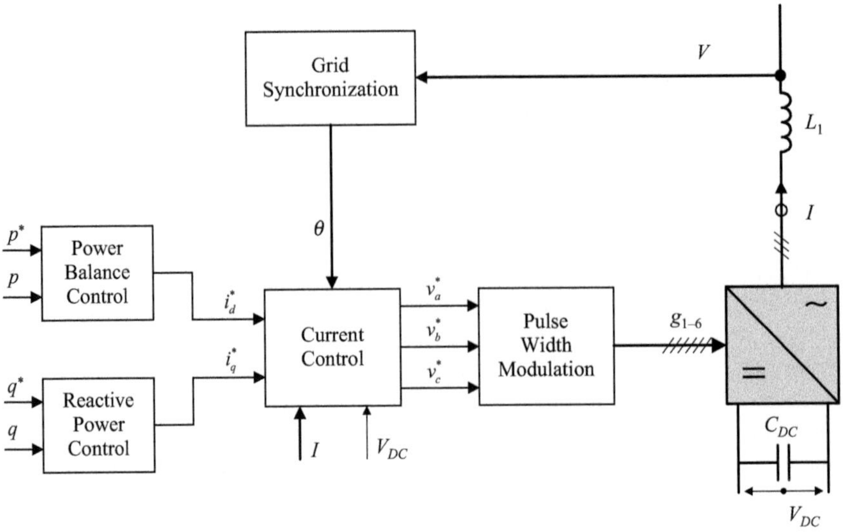

Figure 11.3 Block diagram of a grid-following implementation

structure of the control system can also be applied for converters with an LCL-filter. The control scheme is based on the implementation of two independent outer loops for active and reactive power control. The grid synchronization strategy is also a key component and determines the capability of the inverter to instantaneously follow the frequency and phase angle of the local grid voltage: indeed, this is the aspect determining the name of the control architecture as "grid-following" control.

The control is usually implemented in the *dq* domain by adopting a synchronous reference frame, aligned with the phase angle of the measured voltage. Furthermore, the system is assumed to be balanced. Details on how to approach the unbalanced situation are reported in [3].

The Park transformation is given by the following matrix:

$$T = \sqrt{\frac{2}{3}} \begin{bmatrix} \cos(\theta) & \cos\left(\theta - \frac{2}{3}\pi\right) & \cos\left(\theta + \frac{2}{3}\pi\right) \\ -\sin(\theta) & -\sin\left(\theta - \frac{2}{3}\pi\right) & -\sin\left(\theta + \frac{2}{3}\pi\right) \\ \frac{\sqrt{2}}{2} & \frac{\sqrt{2}}{2} & \frac{\sqrt{2}}{2} \end{bmatrix} \tag{11.1}$$

The matrix T describes an orthonormal transformation which means that the inverse is equal to the transposed of the matrix T and correspondingly that power is preserved across the two domains. Another significant option of transformation is given by the version preserving the amplitude of the transformed quantities: as a result, a scaling factor is necessary for power and energy calculation. The advantage is that the coefficients of the transformation do not involve root squares evaluations and then the matrix is simpler.

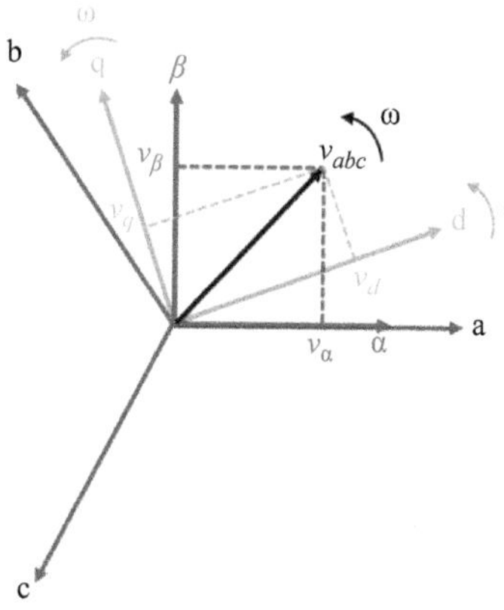

Figure 11.4 Vector diagram showing relation between phase quantities and Park components

As a result, considering the measurement of a three phase voltage, the following relation between the phase domain and Park domain can be written:

$$
\begin{aligned}
v_{abc} &= \begin{bmatrix} v_a \\ v_b \\ v_c \end{bmatrix} \quad v_{dq0} = \begin{bmatrix} v_d \\ v_q \\ v_0 \end{bmatrix} \\
v_{dq0} &= T_p v_{abc}
\end{aligned}
\tag{11.2}
$$

Assuming the zero sequence to be null, the d and q components of the voltage describe, in steady state, a vector rotating at the speed of the network frequency. The purpose of the PLL is to track this rotating vector. An example of a PLL implementation is shown in Figure 11.5. As can be seen from the figure, the PLL utilizes the estimated phase angle to transform the voltage measurements into the dq-frame. This phase angle is then obtained from the integration of a frequency estimate which is provided by a PI-controller using the phase angle of the voltage vector with respect to the d-axis as an input. Thus, the q-axis voltage component is controlled to zero when the PLL is locked to orientation of the voltage vector.

By using the same phase angle for dq transformation of the measured current as for the voltage, it is possible to define an equivalent current vector in the Park domain. Furthermore, by extending the definition of complex power to the Park domain, it is possible to define the following relation:

$$
\begin{aligned}
\bar{s} &= \bar{v}\bar{\iota}^* \\
\bar{s} &= \left(v_d + jv_q\right)\left(i_d - ji_q\right) \\
\bar{s} &= \left(v_d i_d + v_q i_q\right) + j\left(-v_d i_q + v_q i_d\right)
\end{aligned}
\tag{11.3}
$$

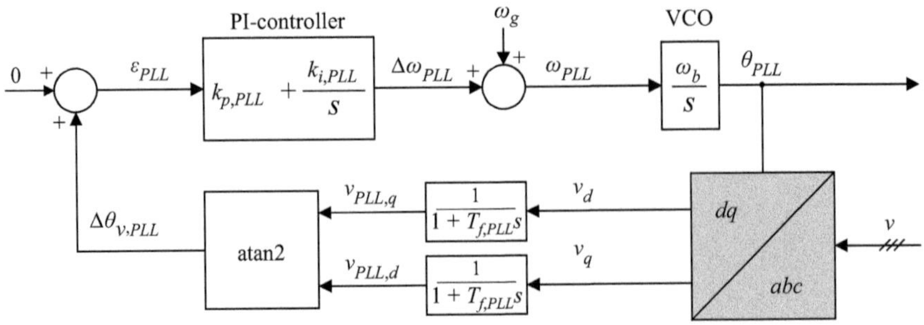

Figure 11.5 PLL diagram for grid-following inverter

where s is the complex representation of apparent power, $\bar{\imath}$ denotes a complex vector, and * denotes the complex conjugate of a variable. By taking real and imaginary parts, we have in general:

$$p(t) = v_d i_d + v_q i_q$$
$$q(t) = -v_d i_q + v_q i_d \tag{11.4}$$

The term $p(t)$ is the instantaneous power delivered by the inverter. In a steady-state balanced system, this quantity is constant and equal to the active power. The term $q(t)$ does not have an immediate physical interpretation, but in the case of steady-state conditions, the operation provides the reactive power delivered by the inverter.

As a result, the two instantaneous power components can be used in the control architecture as feedback measurement of active and reactive power. Given that the definition holds only for steady-state balanced conditions, the measurements are usually processed with a low-pass filter (in many cases first order). Since the PLL is designed to perfectly synchronize with the rotating voltage vector, the quadrature component of the voltage is null, and the expression of the power components is further simplified:

$$p(t) = v_d i_d$$
$$q(t) = -v_d i_q \tag{11.5}$$

This solution determines a decoupled control for the active and reactive power in which the direct component of the current is used only for the active power and the quadrature component for the reactive power. As a result, the current can be controlled by adopting the architecture reported in Figure 11.3, where the current controller is typically implemented as shown in Figure 11.6.

Given that the steady-state values of the two currents in a Park rotating reference frame are constant, a simple PI controller can be adopted to achieve zero steady-state error. The two controllers are also equipped with two cross-terms between the two current components and two feedforward terms compensating for the network voltage.

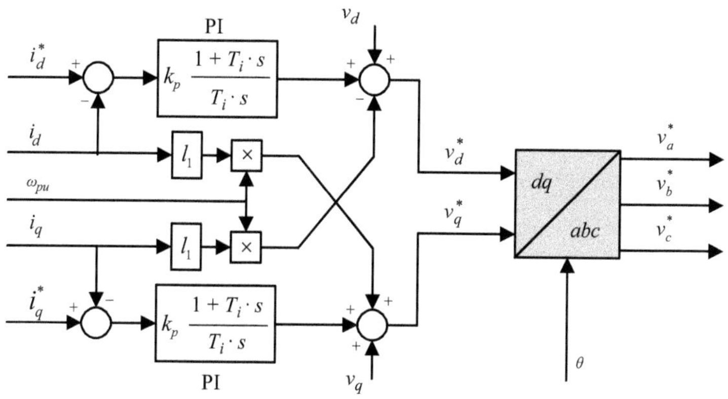

Figure 11.6 Current control in Park domain for a grid-following converter

These terms can be understood looking at the differential equation of a three-phase line in the Park domain:

$$
\begin{aligned}
v_d &= R_d i_d + L_d \frac{di_d}{dt} - \omega L_d i_q + v_{sd} \\
v_q &= R_q i_q + L_q \frac{di_q}{dt} + \omega L_q i_d + v_{sq}
\end{aligned}
\tag{11.6}
$$

In a balanced network, the resistances on the two axes are equal and equal to the line resistance, while the two inductances are also equal and their value is given by the different between the phase self-inductance and the mutual inductance between two phase.

Thanks to the decoupling terms and the feedforward component of the voltage, the overall transfer function for each of the current loops can be written as:

$$G(s) = 1/(R + sL) \tag{11.7}$$

As a result, the design of the PI controller can be rather simple. The time constant T_i is selected to cancel the pole of the transfer function. The resulting transfer function of the open loop including the controller is then given by a single integrator presenting a behavior in the Bode plan of a straight line at 20 db/decade as shown in Figure 11.7. The proportional coefficient is fixed to determine the zero-crossing of the open loop transfer function and correspondingly the speed of response of the current loop. Thanks to this choice, the loop presents a phase margin of 90° and it is then extremely robust.

As mentioned, the outer loops deal with the control of the active and reactive power flow. In the simplest implementation, the active and reactive power control can be achieved in open loop by rescaling the references with the amplitude of the measured voltage. However, in many RES applications, the grid interface converter is responsible for controlling the DC bus voltage. In this case, the power balance control in Figure 11.3 is replaced with a DC voltage controller. Similarly, in case of a high grid inductance, the reactive power controller can be replaced by AC voltage controller.

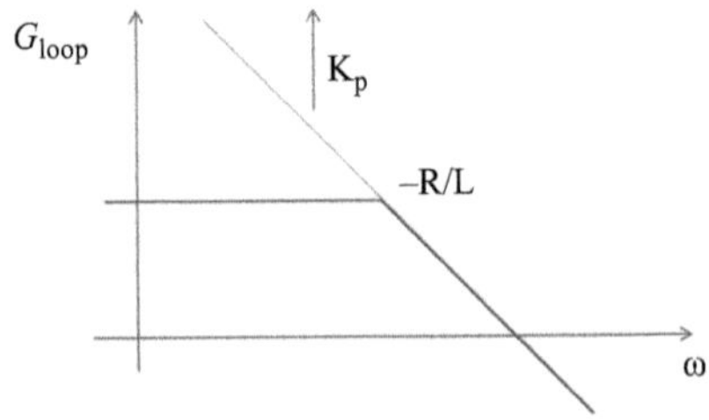

Figure 11.7 Open loop transfer function of the current controller after compensation

In all cases, PI controllers are typically adopted to provide accurate tracking of the reference signals. Recent studies also consider the use of model predictive control or other advanced control methods for optimizing the performance.

11.4 Implementations for grid-forming control

As already mentioned, the concept of grid-forming control is gaining importance in response to the growing needs for utilizing power electronic converters as the grid interface of RES-based generation units. Many implementations are feasible and a thorough classification or a comprehensive description of all relevant schemes is beyond the scope of this chapter. Instead, examples of the most common implementation schemes are described with emphasis on their main features and highlighting their conceptual similarities. An overview of a generic grid-forming scheme is presented in Figure 11.8. It should be noted that the phase angle used for generating the AC output voltages of the converters is not obtained by a PLL as in the grid-following schemes but can instead be integrated with the active power control. Thus, an overview of the main options for synchronization and phase angle generation for grid-forming control strategies is presented in the following.

11.4.1 Internal frequency generation

This control scheme arguably represents the simplest implementation for a grid-forming converter since there is no synchronization mechanism. Instead, the frequency is assumed to be given directly by an internal reference signal. The implementation could be based on a fixed voltage amplitude and frequency provided in open loop as indicated in Figure 11.9, or by introducing closed loop voltage control [20]. In case of a closed loop control, the controllers can operate in a synchronous rotating reference frame in a similar way as the grid-following scheme presented in Figure 11.3. However, the phase angle for the Park transformation is not synchronized with an external grid but rather generated internally from a fixed frequency reference as shown in Figure 11.9.

It should be noted that this scheme will not allow for connection to an external grid or for parallel operation of multiple converters. Thus, this implementation is suitable

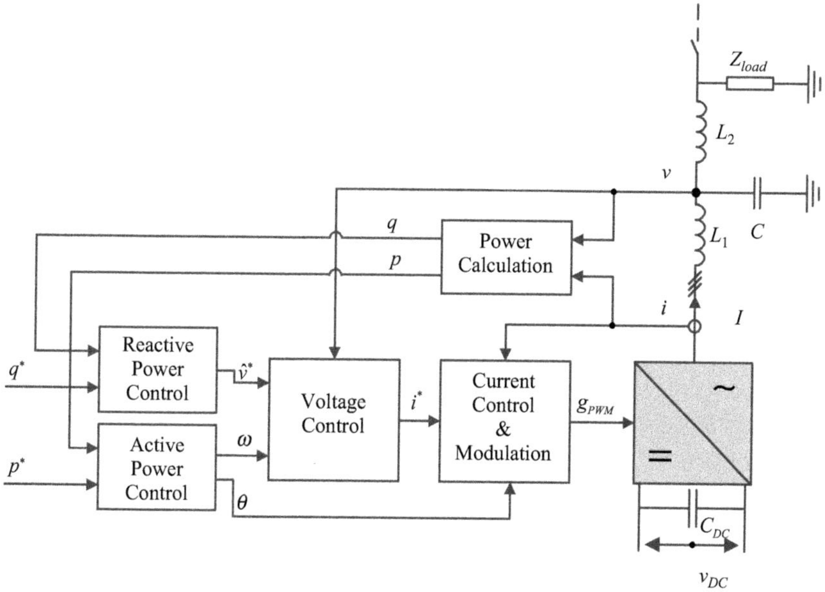

Figure 11.8 Overview of grid-forming control

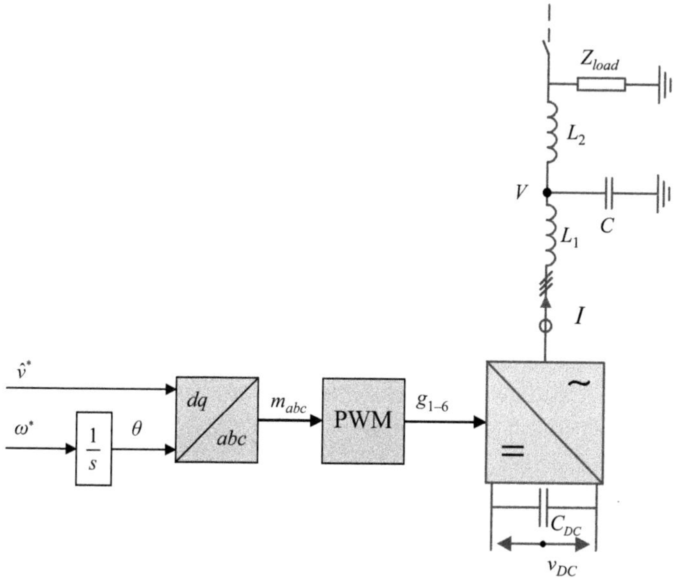

Figure 11.9 Scheme for grid-forming control with internal generation of the frequency

if the converter operates alone in a grid with mainly passive loads or if its size is dominant compared to other units. In this last condition, the converter determines the operating frequency acting as a master unit and other converter units or generators synchronize to it in a form of master–slave configuration.

11.4.2 Synchronization based on power flow

Synchronous machines are inherently grid-forming and can maintain their synchronization to an external grid without a dedicated synchronization control block as a PLL. These considerations have inspired the design of grid-forming control schemes that attempt to replicate the synchronization mechanism of synchronous machines. In practical terms, the synchronous machine can be viewed as a controlled voltage source connected to the grid via an impedance representing the armature windings as shown in Figure 11.10. The power transferred by a synchronous machine to the grid depends by the power angle, which is the angle between the grid voltage and the internally induced voltage of the machine. In stationary conditions, the relation is expressed by

$$p = \frac{v_c v_g}{x} \sin \delta \tag{11.8}$$

Assuming that the machine is connected and synchronized to the grid, the reaction to small perturbances of the power flow will result in a variation of the power angle and a transient response during the return to the stable steady-state operating condition. Indeed, if the perturbation causes the machine to rotate faster than the frequency defined by the grid, the power angle will increase and consequently the power transferred to the grid will increase. The increased power transfer to the grid will slow down the rotation as a result of the torque balance of the rotor according to:

$$J \frac{d}{dt} \omega = T_e - T_m \tag{11.9}$$

The opposite process will occur in the condition of the machine rotating slower than the frequency of the external grid. The effect will be to reestablish the synchronism with the grid and the balance between the mechanical input power and the electrical power delivered to the grid. It is clear that this synchronization mechanism is based on a link between the power angle and the power flow and that it can be reproduced by a power electronics converter controlled as a voltage source by specifying the voltage amplitude and voltage phase.

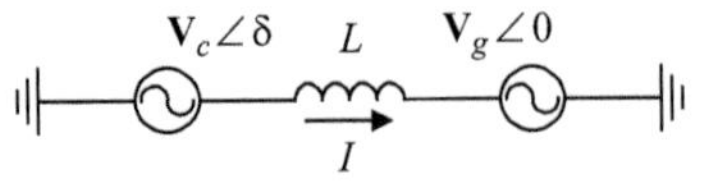

Figure 11.10 Scheme illustrating power flow between two voltage sources separated by an impedance

11.4.3 Power-frequency droop control

A power frequency droop characteristic can be easily included in the control of grid-following converters for contributing to the primary frequency control. This introduces a linear algebraic relation between the power and the frequency in the form:

$$p = p^* + \frac{1}{m_p}(\omega - \omega^*) \tag{11.10}$$

where p^* and f^* are the external references for the active power and the frequency and $1/m_p$ is the droop gain. In a grid-following unit where the frequency can be easily measured, the droop characteristics applied for influencing the power flow could also be equivalently used for defining the active current references.

The power-frequency droop characteristics applied to grid-following converters can be algebraically inverted in order to generate a frequency reference based on the power measured from the converter unit as:

$$\omega^* = \omega - m_p(p - p^*) \tag{11.11}$$

This frequency reference can be integrated to obtain an angle to be used in the Park transformation as shown in Figure 11.11. While the two droop formulations are equivalent in steady state, the expression in (11.11) establishes a link between the power flow and the power angle, as previously described, and can provide the same synchronization mechanism as described in section 11.4.2. Therefore, this scheme offers grid-forming features and the ability to establish a grid as well as to share load with other units operated in parallel [9–12]. A low-pass filter is normally added to the power measurement.

The control scheme in Figure 11.11 has been thoroughly investigated in literature and complemented with additional features for better control of voltages or harmonics. It should be noted that this scheme is particularly suited in the case multiple converters are controlled according to this approach. This would ensure that the load distribution between the units is always shared according to the droop characteristics. Furthermore, this provides resiliency against failure or disconnection of any of the units since the grid-forming property inherently allows the other units to remain in operation.

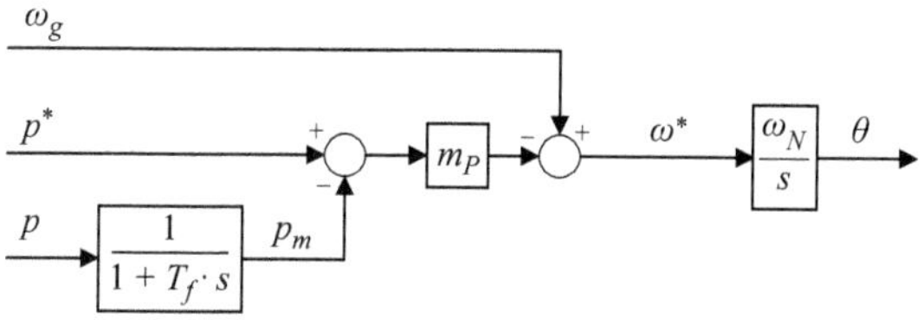

Figure 11.11 Generation of the angle in grid-forming power-frequency droop control scheme

11.4.4 Emulation of generator swing equation

In general, the objective of emulating the operation of a synchronous machine is not to have a high-fidelity representation of a specific machine with a specific set of parameters but rather to capture the features that are more relevant from a power system perspective. In this sense, the synchronous machine characteristics can be represented by a swing equation as:

$$\frac{d\omega}{dt} = \frac{p^*}{T_a} - \frac{p}{T_a} - \frac{k_d \cdot (\omega - \omega_g)}{T_a} \tag{11.12}$$

Including a swing equation model in the control of the converter aims also at replicating the synchronization mechanism inherently determining the operation of a synchronous machine. The swing equation model can be translated into a control scheme as shown in Figure 11.12. Implementing such a virtual swing equation provides grid-forming control properties, and will explicitly provide a virtual inertia as well as a damping effect equivalent to the function of the damper windings in a synchronous machine [13–15].

The grid-forming schemes based on power-frequency droop and the schemes based on a virtual swing equation can be proven to be equivalent [15, 21]. Indeed, the damping term and the inertia in the swing equation are linked to the droop gain and to the low-pass filter time constant in Figure 11.11 by the following relations

$$T_a = T_f \cdot \frac{1}{m_p}, \qquad k_d = \frac{1}{m_p} \tag{11.13}$$

11.4.5 Single integration of power difference

The schemes based on the emulation of the swing equation are characterized by the presence of two cascaded integrators. The first integrator translates the power difference into a virtual speed and the second integrator converts the speed into an angular position. The presence of a double integrator ensures the explicit emulation of inertia effects in the control but can also have negative impact on the power response and the overall stability. Moreover, the relation between the variation of power flow and the change of the phase angle is at the core of the synchronization mechanism, and this does not necessitate a double integration. These considerations are at the

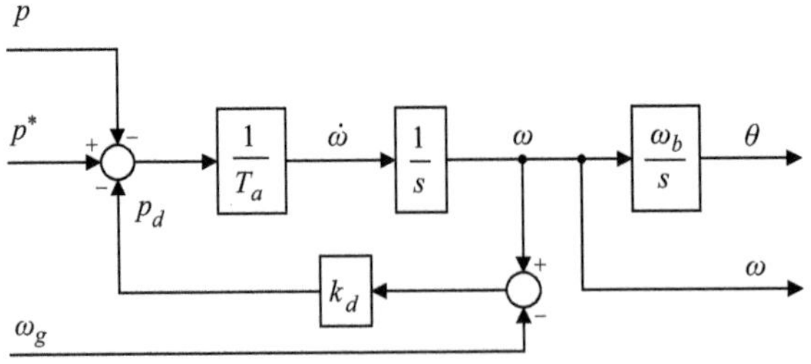

Figure 11.12 Generation of the angle in grid-forming control scheme based on virtual swing equation

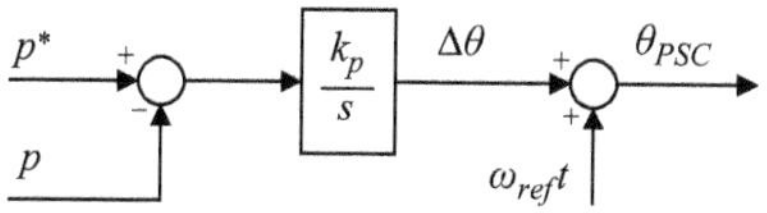

Figure 11.13 Power synchronization control-generation of the angle in power synchronization control scheme

base of other control schemes where the power error is integrated and directly added to the angle reference. A clear example is represented by the power synchronization controller presented in [22] and shown in Figure 11.13. This scheme was primarily introduced to ensure the operation of a VSC in a very weak grid where the presence of the PLL could introduce stability problems.

11.5 Virtual Synchronous Machines

As already mentioned, VSMs can be considered a subset of grid-forming control. Furthermore, this type of control includes a rather large number of different implementations whose main common principle is the explicit emulation of a synchronous machine swing equation. The general concept was first introduced in [13] under the acronym VISMA, but several other implementations have been later proposed in literature, included concepts labeled as virtual synchronous generators (VSGs) [14], synchronverters [23], synchronous power controllers [24], or generator emulation control [25,26].

The most common approach for VSM-based control is to implement the control strategy in a reference frame defined by the phase angle resulting from integration of the speed of the virtual swing equation. Thus, the virtual swing equation according to the diagram shown in Figure 11.12 is a part of the overall control scheme. The main parameters defining the synchronization mechanism and power control response of the VSM are then the inertia time constant T_a and the equivalent damping coefficient k_d, while the inputs to the swing equation are the power reference p^*, the measured output power p, and the grid frequency ω_g. An additional power-frequency droop can be introduced in the power reference to emulate the steady-state behavior of a governor in a synchronous generator.

While the basic structure of the virtual swing equation for VSMs in Figure 11.12 contains the main elements used in most implementations, several varieties can be identified on basis of how the damping term is implemented. Especially since the grid frequency ω_g in Figure 11.11 will not be directly available to the control system, the approach for calculating the damping power p_d varies between different implementations. One common option is to explicitly set ω_g to be equal to the nominal angular frequency or the frequency reference ω^* [23]. This approach, as shown in Figure 11.14 implies that the damping term will become equivalent to a frequency droop. Other alternatives for implementing the damping term includes estimation of the grid frequency by a PLL (or an FLL) as shown in Figure 11.15 [28] or by an internal estimation which could be based on a simple low pass filter applied to the virtual speed as shown in Figure 11.16 [27]. This choice can significantly influence

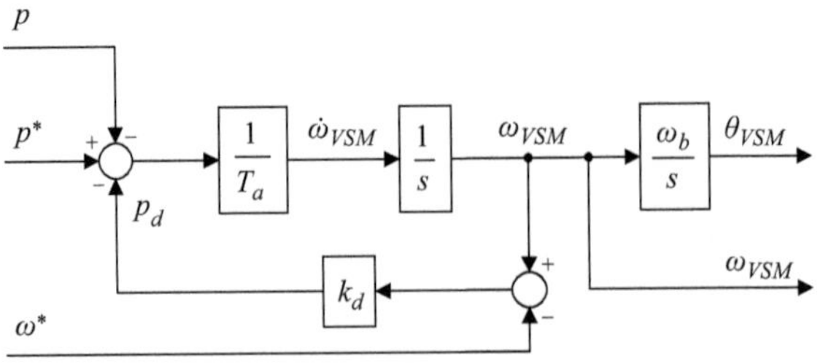

Figure 11.14 *Implementation of swing equation with damping term obtained from reference frequency*

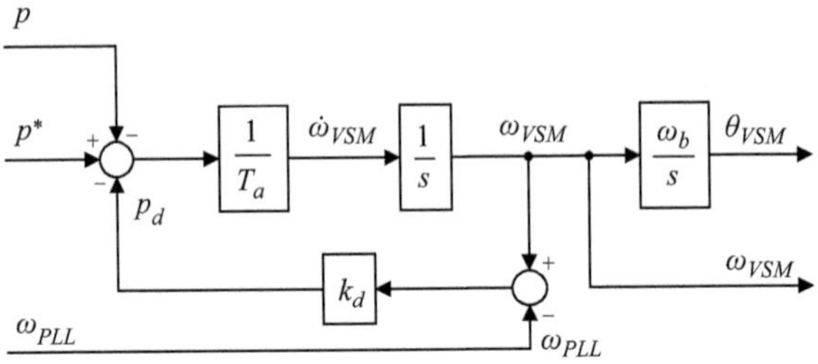

Figure 11.15 *Implementation of swing equation with damping term obtained from a PLL*

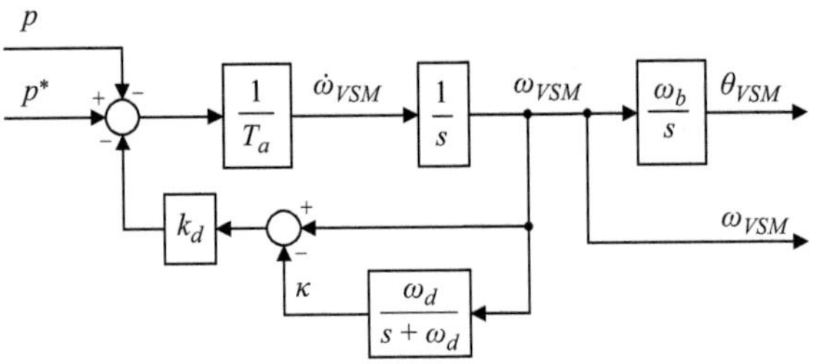

Figure 11.16 *Implementation of swing equation with damping term obtained by low-pass filtering the virtual inertia rotating speed*

the behavior of the control scheme under different operating conditions and should be carefully assessed.

The generation of the phase angle defining the reference frame orientation of the control system is a main feature of the VSM schemes, but the overall control system also requires a strategy for generating the amplitude of the output voltage from the converter. The generation of the voltage amplitude is often adapted to include an equivalent representation of the armature winding of a synchronous generator as a virtual impedance. However, it should be considered that the parameters are not restricted by physical design of a real synchronous machine. Thus, the virtual impedance characteristics can be defined more freely based on their effects on the overall behavior of the VSM. The effect of a virtual impedance on the output voltage of the converter when assuming a dynamic model can be expressed by complex space vectors in the *dq* frame as:

$$\mathbf{v}^* = \hat{v}^* - (r_v + j \cdot \omega l_v)\,\mathbf{i} - \frac{l_v}{\omega_b}\frac{d\mathbf{i}}{dt} \tag{11.14}$$

where r_v and l_v are the virtual resistance and the virtual inductance, respectively. However, for most cases when a VSM is controlled to actuate a voltage according to (11.14), it is more relevant to consider a quasi-stationary approximation to avoid dependency on the derivative of the current. This implies neglecting the last term in the equation and has been widely used for VSM implementations [21,28].

The virtual impedance can also be configured to provide a current reference for control of the converter. A dynamic implementation in the *dq* frame can then be obtained by integrating the expression for the current derivative that can be obtained from (11.14) [27]. If a quasi-stationary implementation is considered, the current reference signal after ignoring the current derivative can be directly obtained as an algebraic equation given by:

$$\mathbf{i}^* = \frac{\hat{v}^* - \mathbf{v}}{(r_v + j \cdot \omega\, l_v)} \tag{11.15}$$

However, for a quasi-stationary implementation of a virtual impedance providing a current reference, a small low-pass filter should be included in the voltage measurement **v** to avoid stability problems [27].

The implementation of the voltage reference generation for the switching operation of the converter another main aspect differing among various control schemes. The following three main implementation schemes have been commonly utilized [21]:

1. *Direct modulation schemes*: The voltage amplitude reference is combined with the phase angle from the virtual swing equation to create a voltage reference for the generation of the gate signals as shown in Figure 11.17. A virtual impedance and/or a reactive power droop may be added. In Figure 11.17, a quasi-stationary representation of the virtual impedance has been assumed.

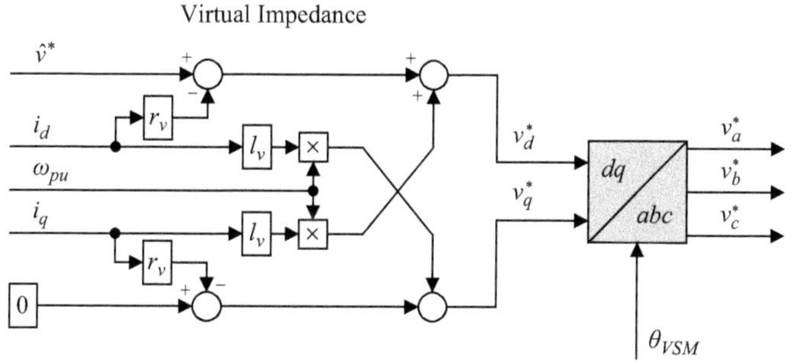

Figure 11.17 Direct connection of voltage reference to the modulator

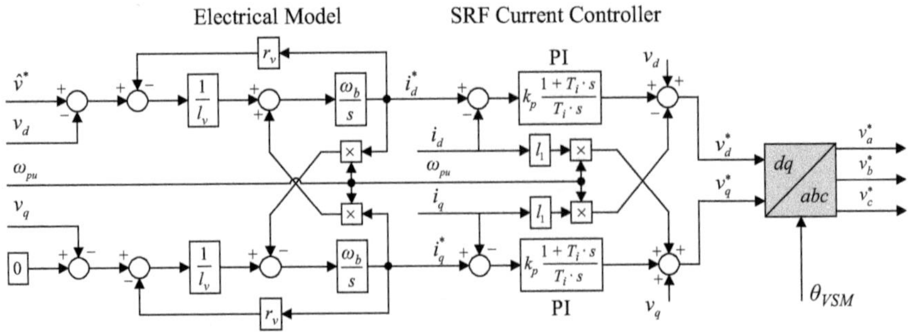

Figure 11.18 Generation of the voltage reference with current controller and virtual impedance

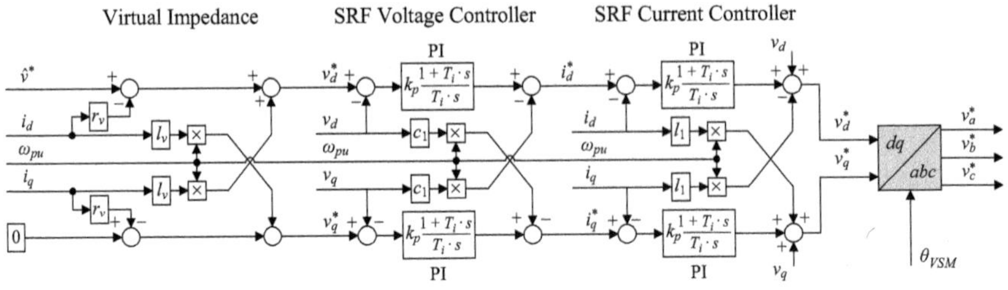

Figure 11.19 Generation of the voltage reference with cascaded voltage and current controller

2. *Current controlled VSM schemes*: The voltage amplitude is converted into a current reference by dividing with a virtual impedance [27]. The virtual impedance could be represented by a quasi-stationary equation or by a model including dynamics. The current reference is assumed as input for an inner loop current controller as shown in Figure 11.18. In the example, a dynamic representation of the virtual impedance has been assumed. A reactive power droop may be added.
3. *Voltage controlled VSM schemes*: The voltage amplitude is assumed as reference for a cascaded controller composed by an inner current control loop and an outer voltage control loop [28]. A virtual impedance and/or a reactive power droop may be added. The scheme in Figure 11.19 provides an example with a quasi-stationary representation of the virtual impedance.

The full implementation of a VSM control scheme should include both the elements for the generation of the phase angle and the voltage amplitude, described above. In addition, there can be additional features as active damping, saturation of reference signals for the controllers and other protective functions. The number of variations would become too large to be represented and for illustrative purposes a single example of a detailed scheme is presented in Figure 11.20 based on [28]. This is intended to offer an example of how the different parts that have been described in the chapter could be integrated together and to further highlight that many possibilities arise in terms of implementation of a VSM scheme.

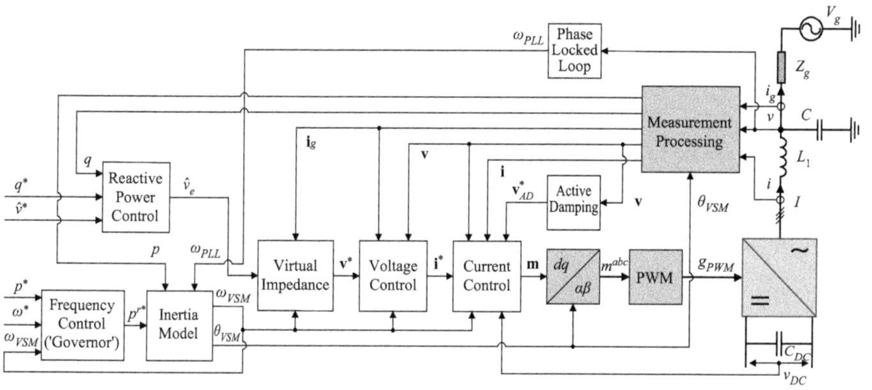

Figure 11.20 Overview of implementation scheme for VSM with cascaded voltage control

11.6 Virtual Oscillator Control

The control schemes presented in the previous sections are intended for integration in a classical power system. The programmability of power electronic converters also makes it possible to consider radical changes and development of completely new concepts. One example in this direction is given by the VOC realized by digitally programming the oscillator's differential equations onto inverters' microcontroller, for emulating the dynamics of nonlinear limit-cycle oscillators [33]. The output current is the main feedback signal fed into the oscillator equations, while the oscillator states are used to provide the voltage reference signal for the pulse-width modulation (PWM) [30]. The main idea is to establish with this process a limit cycle acting at a given frequency [33]. This approach has created most interest for micro-grid applications by creating a new concept of synchronization suited for low- or zero-inertia power systems.

The oscillator model is usually established with reference to the behavior of a parallel LC circuit: the desired dynamics are created by adding nonlinear current and voltage sources, and the measured inverter output current serves as a parallel current source. Several applications developed the concept starting from the Van der Pol oscillator [32, 34]. Recently the Andronov–Hopf oscillator (AHO) has been also proposed as an alternative [35].

The implementation presented in the following uses the dynamics of a Andronov–Hopf bifurcation in a two-dimensional system. An unstable limit cycle is created in a subcritical Andronov–Hopf bifurcation, whereas a supercritical bifurcation creates a unique and stable limit cycle with a specific radius [36]. Differently from the Van der Pol case, the limit cycle is in this case perfectly circular making it a preferable choice for power systems applications.

The AHO circuit model comprises a parallel LC-tank that sets the system frequency. The inductor current and capacitor voltage are used as the control states of the model. The desired oscillator dynamics are achieved through state-dependent non-linear voltage and current sources defined as:

$$v_m = \xi/\omega_n \left(2X_n^2 - \|x\|^2\right) \varepsilon i_l$$
$$i_m = \xi/\varepsilon\omega_n \left(2X_n^2 - \|x\|^2\right) v_c \tag{11.16}$$

where $\sqrt{2}X_n$ determines the amplitude of the oscillator limit cycle. $V_n = \kappa_v X_n$ and ω_n are the nominal system voltage and frequency, ξ governs the speed of convergence to steady-state, $||\cdot||$ is the Euclidean norm and x are the state variables. The filter inductor current is measured and compared to a reference current signal. The resulting current error is used as an input of the oscillator circuit after scaling with the current gain and applying an axis rotation with a rotation angle defined by the characteristic impedance of the network. The rotation matrix is given by:

$$R(\varphi) := \begin{bmatrix} \cos\varphi & -\sin\varphi \\ \sin\varphi & \cos\varphi \end{bmatrix} \tag{11.17}$$

The oscillator behavior can be translated into a circuit model, which is based on a parallel LC circuit with the resonant frequency ω_n. The two state variables (inductor current and capacitor voltage) are used to generate the reference voltage signal for the PWM of the inverter.

The complete topology of the controller is shown in Figure 11.21.

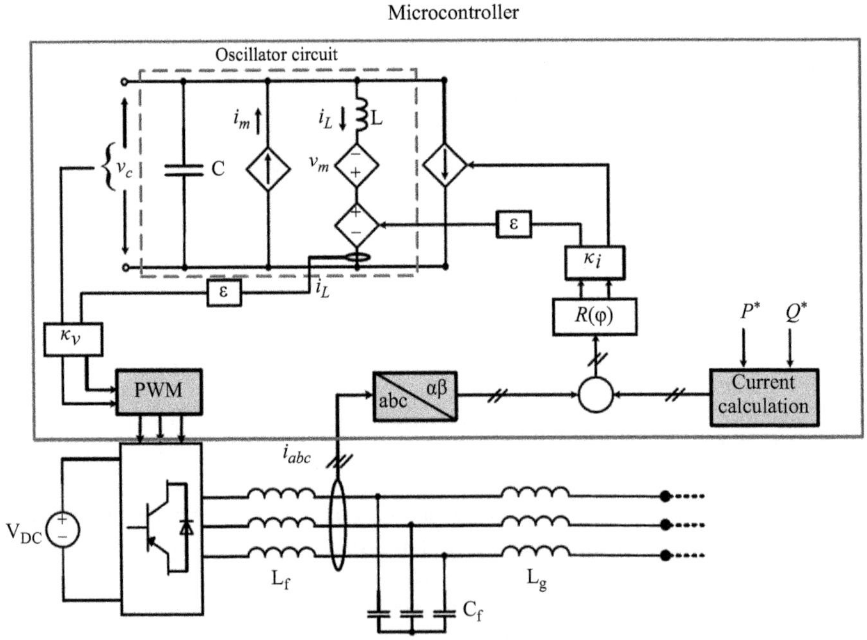

Figure 11.21 Block diagram of Andronov–Hopf oscillator control

As an input, the filter-side current is measured and compared against a reference current signal. The reference current is obtained from the active and reactive power set-points, P^* and Q^*, respectively, by using the output voltage signal according to:

$$\begin{bmatrix} i_a^* \\ i_j^* \end{bmatrix} = \frac{2}{3\left\|v_{\alpha\beta}\right\|^2}\begin{bmatrix} v_\alpha & v_\beta \\ v_\beta & -v_\alpha \end{bmatrix}\begin{bmatrix} P^* \\ Q^* \end{bmatrix} \tag{11.18}$$

Before entering the oscillator circuit, an axis rotation is applied with a rotation matrix and the signal is scaled by a current gain κ_i. The rotation matrix is the one already introduced before. The RMS voltage amplitude $V$and phase angle θ at the inverter terminal are defined as:

$$V = \frac{1}{\sqrt{2}}(v_\alpha^2 + v_\beta^2)^{\frac{1}{2}}, \ \theta = \arctan\left(\frac{v_\beta}{v_\alpha}\right) \tag{11.19}$$

At the same time, the overall dynamic behavior of the oscillator can be described by the following differential equations:

$$\dot{V} = \frac{\xi}{\kappa_r^2}V\left(2V_n^2 - 2V^2\right) - \frac{\kappa_v K_i}{3CV}(\sin\varphi\,(Q-Q^*) + \cos\varphi\,(P-P^*))$$

$$\dot{\theta} = \omega_n - \frac{\kappa_v \kappa_i}{3CV^2}(\sin\varphi\,(P-P^*) - \cos\varphi\,(Q-Q^*))\,. \tag{11.20}$$

The steady-state characteristics can be coherently derived as steady solutions of these equations setting the derivative to 0:

$$V = \frac{V_n}{\sqrt{2}}\left(1 + \sqrt{1 - \frac{2\kappa_i\kappa_v^3}{3C\xi V_n^4}(\sin\varphi\,(Q-Q^*) + \cos\varphi\,(P-P^*))}\right)^{\frac{1}{2}}$$

$$\omega = \omega_n^* - \frac{\kappa_v \kappa_i}{3CV^2}(\sin\varphi\,(P-P^*) - \cos\varphi^*\,(Q-Q^*)) \tag{11.21}$$

Also for this type of oscillator, the steady-state model expresses a droop-like voltage and frequency regulation. The Andronov–Hopf based VOC supports the grid, by adjusting its power output proportionally to the voltage and frequency deviation from their nominal values. The active and reactive power set-points define the operating point, at which nominal voltage and frequency are generated.

11.6.1 Power dispatch

In this section, the power dispatching capabilities of the virtual oscillator controller are analyzed. Being able to directly control an inverter's power output is important to ensure optimal operation of non-dispatchable DER, like PV systems. Assuming the adoption of such solution within a microgrid, a desirable characteristic of the AHO, therefore, would be to perform grid-feeding control in grid-connected mode and grid-forming control in islanded operation with the same controller.

Starting from the steady-state model reported above, it can be shown that the AHO decouples voltage and frequency control by using a rotation matrix. The deltas of active and reactive power are transformed via an axis rotation into two error terms that determine the voltage and frequency output in a droop like fashion. The following expressions state the general form of the errors that are controlled via voltage and frequency respectively:

$$V_\varepsilon = \sin\varphi\,(Q^* - Q) + \cos\varphi\,(P^* - P) \sim V - V_n,$$
$$\omega_\varepsilon = \sin\varphi^*\,(P^* - P) - \cos\varphi^*\,(Q^* - Q) \sim \omega - \omega_n, \tag{11.22}$$

Different methods have been proposed in literature for controlling active and reactive power output with the VOC.

In [37], a complex gain $K = |K|\angle\vartheta$ that scales the input current to control the inverter power output is introduced. This gain relates to the apparent power, where $|S|$ depends on $|K|$ and $arg(S)$ on ϑ. The value of K is calculated from the power set-points by using an analytical expression that relates apparent power output S and the complex gain K. To perform this calculation, the power flow equations have to be solved for $|V|$ and δ. The solution strongly depends on the estimation accuracy of the network impedance. Adaptive integrators are used to mitigate the steady-state error in the power output resulting from uncertainty of the system parameters. This method shows very good dynamic and decoupling performance compared to an implementation that uses PI-controllers to control K. The modification was employed on the Van der Pol VOC design presented in [34].

Simultaneous regulation of active and reactive power is achieved by controlling the voltage and current gains, κ_V and κ_I respectively, with two PI-controllers in [38]. The active power output is controlled with a PI-controller that tunes the voltage gain κ_V, while the reactive power output is regulated via the current gain κ_I. This control algorithm is carried out on a Van der Pol system. The main drawbacks of this method are that the control variables κ_V and κ_I have a coupled relation with P and Q and that abrupt changes in these variables may desynchronize the inverters and cause high circulating currents. To minimize these effects, the control response needs to be slow.

A dispatchable VOC (dVOC) that is designed from a top-down system-level was presented in [39]. The frequency and voltage regulation characteristics are very similar to the AHO, with a major difference being the presence of a voltage set-point, in addition to active and reactive power set-points. The dVOC guarantees global asymptotic stability and inverter power output according to the set-points, if all inverters in the grid are controlled by a dVOC and the set-points are consistent with the AC power flow equations. A precise control of active and reactive power is therefore only possible, if a system wide power flow calculation occurs and all inverters adopt the dVOC control strategy. Compatibility in systems with other control strategies or energy sources that are not inverter based is unclear. Set-point dispatch has only been shown for active power in inductive grids, simultaneous control of active and reactive power in non-inductive networks was not presented yet.

Grid-following VOCs that are presented in the literature inherit undesirable characteristics of the Van der Pol VOC have inferior dynamic performance or system wide requirements. In the following, a modified method for simultaneous control of active and reactive power, which uses the VOC and benefits from the AHO dynamics, is discussed.

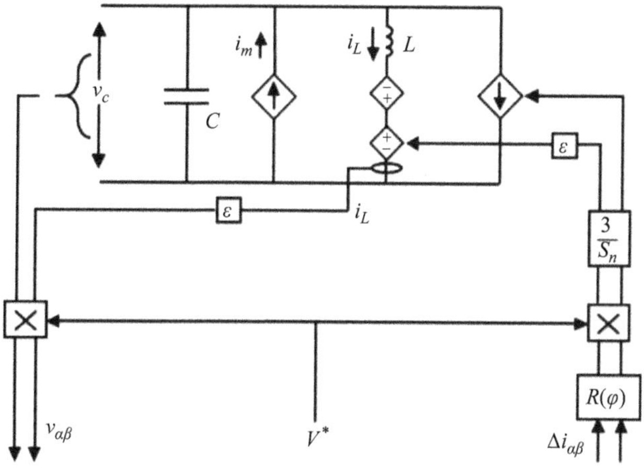

Figure 11.22 Modified version of the oscillator for voltage control

11.6.2 Modified Andronov–Hopf oscillator

This section introduces a modified version of the AHO that better suits the characteristics of power dispatch. The modified AHO is designed to maintain the inverter terminal voltage close to the nominal system voltage V_n. The power set-points adjust the voltage regulation by defining the nominal operating point, i.e., the operating point at which the inverter generates nominal voltage. Voltage deviations are proportional traded off with increased or decreased power output. As result a steady-state error between the power output and power set-point appears because the set-point is not reachable. To eliminate the tracking error, the inverter terminal voltage must be controlled to follow the power flow solution. To solve this problem, a dynamic voltage set-point is introduced to define the voltage amplitude that the oscillator generates at the power set-points.

The oscillator circuit is designed to provide a limit cycle with a normalized RMS amplitude of $X_n = 1V$ during an unforced oscillation, i.e. when $P = P^*$ and $Q = Q^*$. The circuit is interfaced to the system with the voltage and current gains, which are defined as [35]:

$$\begin{aligned} \kappa_v &= V_n \\ \kappa_i &= \frac{3V_w}{S_n} \end{aligned} \tag{11.23}$$

where S_n is the inverter rated apparent power. These gains are located at the input and the output of the oscillator circuit and scale the oscillator to the physical system, so that the output voltage is $V_n = \kappa_v X_n$. The dynamic voltage set-point replaces the static internal parameter V_n of the controller to enable control over the oscillator amplitude and the voltage associated with the power set-points.

The input and the output scaling of the oscillator circuit thus are dynamically adapted and κ_v and κ_i are replaced as explicit parameters.

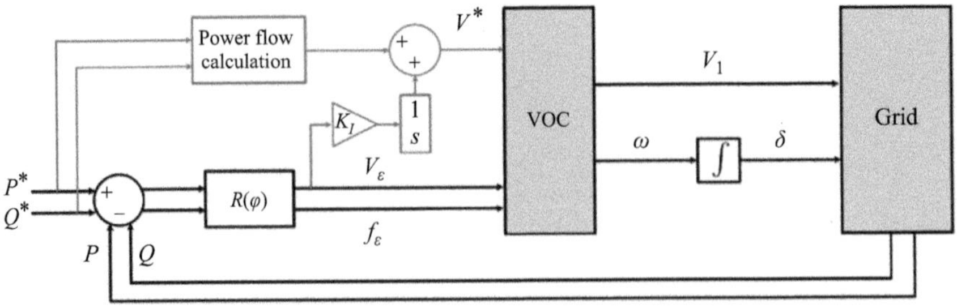

Figure 11.23 Compensation of uncertainty on power flow solution

The steady-state voltage and frequency regulation can be calculated as:

$$V = \frac{V^*}{\sqrt{2}}\left(1 + \sqrt{1 - \frac{2}{C\xi S_n}\left(\sin\varphi\,(Q - Q^*) + \cos\varphi\,(P - P^*)\right)}\right)^{\frac{1}{2}}$$

$$\omega = \omega_n - \frac{V^{*2}}{CV^2 S_n}\left(\sin\varphi\,(P - P^*) - \cos\varphi\,(Q - Q^*)\right) \tag{11.24}$$

V^* is controlled to follow the solution of the power flow equations to achieve zero steady state error. The value of V^* is determined in two steps:

1. Calculation of the power flow solution
2. Integration of the remaining error term

First, the solution of the power flow equations is calculated for the equivalent connected system and the initial value of the voltage set-point is provided. The result is improved by the addition of an integrator that eliminates the remaining error caused by system uncertainties.

Given that the full solution of the power flow can be difficult to have, an approximate solution can be calculated locally by solving the power flow equations for a very simple model given by the converter connected to the network described by a lumped impedance and equivalent voltage source. The following equations can be used for this purpose:

$$\frac{|V_1|^4}{R^2 + X^2} - \frac{|V_1|^2}{R^2 + X^2} * \left(|V_2|^2 + \frac{2}{3}(QX + PR)\right) + \frac{P^2 + Q^2}{3^2} = 0 \tag{11.25}$$

The receiving end voltage $|V_2|$ is assumed to be the nominal system voltage V_n determined by the main grid. The values of R and X are the real and imaginary parts of the equivalent impedance Z, which consists of the LCL-filter, distribution line impedances between the AHO and the PCC, as well as the impedance of the connected grid. The estimation of the grid impedance is based on the X/R-ratio and grid capacity. The grids inductance and resistance are calculated as

$$X = \frac{V_n^2}{S_{SCC}},$$
$$R = \frac{X}{(\text{X/R-ratio})}, \tag{11.26}$$

where S_{SCC} denotes the short-circuit capacity, which is also used for fault-studies or is determined by the rating of the transformer connection to a higher voltage level. The parameters of the LCL-filter are known and the distribution line impedance has to be estimated from system parameters of the utility. The calculation updates the voltage set-point simultaneously with the power set-points.

Since the exact system parameters are hard to obtain or unknown, the calculated voltage has a reduced accuracy and an error in the power output remains. Applying the following control law allows to integrate the remaining deviation and thereby eliminate the steady state error:

$$\frac{d}{dt}\Delta V^* = K_I * \underbrace{(\sin\varphi\,(Q - Q^*) + \cos\varphi\,(P - P^*))}_{=V_\varepsilon} \tag{11.27}$$

where K_I denotes the integrator gain. The controller input is the voltage power error term V_ε that needs to be controlled to zero; ΔV^* is the controller output. The voltage set-point is then calculated as

$$V^* = |V_1| + \Delta V^* \tag{11.28}$$

11.7 Concluding remarks

The chapter presented an overview of implementation options for grid connected converters. The evolution of the presented solutions also represent how the role of power electronic converters has been evolving over the years. The grid-following concept can be seen as the first stage of an ongoing development, in which the role of renewable power generation was initially considered marginal in the power system context. In the recent years, correspondingly with the growing penetration of wind and solar installations, the attention has been moved toward grid-forming approaches because of their capability to play a key role in grid dynamics. Solutions like VSM can be seen as a sort of continuity with the past but reinterpreted with the programmability of power electronics. The control freedom provided by power electronics is now offering the possibility to consider also completely new approaches that are intended for a fully electronic grid. The VOC is one example in this direction which, by exploiting non-linear dynamics, offers a new and more robust way to work in grids without significant physical inertia from rotating machines.

Bibliography

[1] A. Monti, F. Milano, E. Bompard, and X. Guillaud, "*Converter-Based Dynamics and Control of Modern Power Systems*," London: Academic Press, 2021.

[2] Y. Yazdani and R. Iravani, "*Voltage-Sourced Converters in Power Systems – Modeling, Control, and Applications*," Hoboken, NJ: Wiley/IEEE Press, 2010.

[3] R. Teodorescu, M. Liserre, and P. Rodríguez, *Grid Converters for Photovoltaic and Wind Power Systems*, Chichester, UK: Wiley, 2011.

[4] B. Li, M. Zhang, L. Huang, L. Hang, and L. M. Tolbert, "A new optimized pole placement strategy of grid-connected inverter with LCL-filter based on state variable feedback and state observer," in: *2013 Twenty-Eighth Annual IEEE Applied Power Electronics Conference and Exposition (APEC)*, 2013, pp. 2900–2906, doi: 10.1109/APEC.2013.6520710.

[5] M. Hanif, V. Khadkikar, W. Xiao, and J. L. Kirtley, "Two degrees of freedom active damping technique for LCL filter-based grid connected PV systems," *IEEE Transactions on Industrial Electronics*, vol. 61, no. 6, 2014, pp. 2795–2803, doi: 10.1109/TIE.2013.2274416.

[6] J. Rocabert, A. Luna, F. Blaabjerg, and P. Rodríguez, "Control of power converters in AC microgrids," *IEEE Transactions on Power Electronics*, vol. 27, no. 11, 2012, pp. 4734–4749.

[7] R. Rosso, X. Wang, M. Liserre, X. Lu, and S. Engelken, "Grid-forming converters: control approaches, grid-synchronization, and future trends—a review," *IEEE Open Journal of Industry Applications*, vol. 2, 2021, pp. 93–109.

[8] F. Blaabjerg, R. Teodorescu, M. Liserre, and A.V. Timbus, "Overview of control and grid synchronization for distributed power generation systems," *IEEE Transactions on Industrial Electronics*, vol. 53, no. 5, 2006, pp. 1398–1409.

[9] M. Chandorkar, D.M. Divan, and R. Adapa, "Control of parallel connected inverters in standalone ac supply systems," *IEEE Transactions on Industry Applications*, vol. 29, no. 1, 1993, pp. 136–143.

[10] J.M. Guerrero, J. Matas, L.G. de Vicuña, M. Castilla, and J. Miret, "Wireless control strategy for parallel operation of distributed-generation inverters," *IEEE Transactions on Industrial Electronics*, vol. 53, no. 5, 2006, pp. 1461–1470.

[11] K. De Brabandere, B. Bolsens, J. Van den Keybus, A. Woyte, J. Driesen, and R. Belmans, "A voltage and frequency droop control method for parallel inverters," *IEEE Transactions on Power Electronics*, vol. 22, no. 4, 2007, pp. 1107–1115.

[12] N. Pogaku, M. Prodanović, and T.C. Green, "Modeling, analysis and testing of autonomous operation of an inverter-based microgrid," *IEEE Transactions on Power Electronics*, vol. 22, no. 2, 2007, pp. 613–625.

[13] H.-P. Beck and R. Hesse, "Virtual synchronous machine," in: *Proceedings of the 9th International Conference on Electrical Power Quality and Utilisation*, Barcelona, Spain, 9–11 October 2007, 6 pp.

[14] K. Sakimoto, Y. Miura, and T. Ise, "Stabilization of a power system with a distributed generator by a virtual synchronous generator function," in: *Proceedings of the 8th International Conference on Power Electronics* – ECCE Asia, Jeju, Korea, 30 May–3 June 2011, 8 pp.

[15] S. D'Arco and J.A. Suul, "Virtual synchronous machines – classification of implementations and analysis of equivalence to droop controllers for microgrids," in: *Proceedings of the IEEE PowerTech Grenoble 2013*, Grenoble, France, 16–20 June 2013, 7 pp.

[16] H. Bevrani, T. Ise, and Y. Miura, "Virtual synchronous generators: a survey and new perspectives," *International Journal of Electric Power and Energy Systems*, vol. 54, 2014, pp. 244–254.

[17] H. Alrajhi Alsiraji and R. El-Shatshat, "Comprehensive assessment of virtual synchronous machine based voltage source converter controllers," *IET Generation, Transmission, Distribution*, vol. 11, no. 7, 2017, pp. 1762–1769.

[18] E. Rokrok, T. Qoria, A. Bruyere, B. Francois, and X. Guillaud, "Classification and dynamic assessment of droop-based grid-forming control schemes: application in HVDC systems," *Electric Power Systems Research*, vol. 189, 2020, p. 106765.

[19] ENTSO-E Technical Group on High Penetration of Power Electronic Interfaced Power Sources, Technical Report "High Penetration of Power Electronic Interfaced Power Sources and the Potential Contribution of Grid Forming Converters," 2020. Available from: https://euagenda.eu/upload/publications/untitled-292051-ea.pdf

[20] J. Liang, T. Jing, O. Gomis-Bellmunt, J. Ekanayake, and N. Jenkins, "Operation and control of multiterminal HVDC transmission for offshore wind farms," *IEEE Transactions on Power Delivery*, vol. 26, no. 4, 2011, pp. 2596–2604.

[21] S. D'Arco and J.A. Suul, "Equivalence of virtual synchronous machines and frequency-droops for converter-based microgrids," *IEEE Transactions on Smart Grid*, vol. 5, no. 1, 2014, pp. 394–395.

[22] L. Zhang, L. Harnefors, and H. Nee, "Power-synchronization control of grid connected voltage-source converters," *IEEE Transactions on Power Systems*, vol. 25, no. 2, 2010, pp. 809–820.

[23] Q.-C. Zhong and G. Weiss, "Synchronverters: inverters that mimic synchronous generators," *IEEE Transactions on Industrial Electronics*, vol. 58, no. 4, 2011, pp. 1259–1267.

[24] P. Rodriguez, I. Candela, and A. Luna, "Control of PV generation systems using the synchronous power controller," in: *Proceedings of the 2013 IEEE Energy Conversion Congress and Expo*, ECCE 2013, Denver, CO, 15–19 Sept. 2013, pp. 993–998.

[25] H. Alatrash, A. Mensah, E. Mark, G. Haddad, and J. Enslin, "Generator emulation controls for photovoltaic inverters," *IEEE Transactions on Smart Grid*, vol. 3, no. 2, pp. 996–1011, 2012.

[26] M. Guan, W. Pan, J. Zhang, Q. Hao, J. Cheng, and X. Zheng, "Synchronous generator emulation control strategy for voltage source converter (VSC) stations," *IEEE Transactions on Power Systems*, vol. 30, no. 6, 2015, pp. 3093–3101.

[27] O. Mo, S. D'Arco, and J.A. Suul, "Evaluation of virtual synchronous machines with dynamic or quasi-stationary machine models," *IEEE Transactions on Industrial Electronics*, vol. 64, no. 7, 2017, pp. 5952–5962.

[28] S. D'Arco, J.A. Suul, and O.B. Fosso, "A virtual synchronous machine implementation for distributed control of power converters in smartgrids," *Electric Power System Research*, vol. 122, 2015, pp. 180–197.

[29] B.B. Johnson, S.V. Dhople, A.O. Hamadeh, and P.T. Krein, "Synchronization of parallel single-phase inverters with virtual oscillator control," *IEEE Transactions on Power Electronics*, vol. 29, no. 11, 2014, pp. 6124–6138.

[30] B. Johnson, M. Rodriguez, M. Sinha, and S. Dhople, "Comparison of virtual oscillator and droop control," in: *2017 IEEE 18th Workshop on Control and Modeling for Power Electronics (COMPEL)*, 2017, pp. 1–6.

[31] M. Sinha, S. Dhople, B.B. Johnson, and N. Ainsworth, "Nonlinear supersets to droop control," in: *2015 IEEE 16th Workshop on Control and Modeling for Power Electronics (COMPEL)*, 2015, pp. 1–6.

[32] M. Sinha, F. Dörfler, B.B. Johnson, and S.V. Dhople, "Uncovering droop control laws embedded within the nonlinear dynamics of van der pol oscillators," *IEEE Transactions on Control of Network Systems*, vol. 4, no. 2, 2017, pp. 347–358.

[33] J. Moehlis, K. Josic, and E.T. Shea-Brown, "Periodic orbit," in: Scholarpedia 1.7 (2006). revision #153208, p. 1358. doi: 10.4249/scholarpedia.1358.

[34] B.B. Johnson, M. Sinha, N.G. Ainsworth, F. Dörfler, and S.V. Dhople, "Synthesizing virtual oscillators to control islanded inverters," *IEEE Transactions on Power Electronics*, vol. 31, no. 8, 2016, pp. 6002–6015.

[35] M. Lu, S. Dutta, V. Purba, S.V. Dhople, and B.B. Johnson, "A grid-compatible virtual oscillator controller: analysis and design," in: *2019 IEEE Energy Conversion Congress and Exposition (ECCE)*, 2019, pp. 2643–2649.

[36] Y. A. Kuznetsov, "Andronov–Hopf bifurcation," in: Scholarpedia 1.10 (2006). Revision #90964, p. 1858. doi: 10.4249/scholarpedia.1858.

[37] D. Raisz, T. T. Thai, and A. Monti, "Power control of virtual oscillator controlled inverters in grid-connected mode," *IEEE Transactions on Power Electronics*, vol. 34, no. 6, 2019, pp. 5916–5926.

[38] M. Ali, H. I. Nurdin, and J. Fletcher, "Dispatchable virtual oscillator control for single-phase islanded inverters: analysis and experiments," *IEEE Transactions on Industrial Electronics*, vol. 68, no. 6, June 2021, pp. 4812–4826.

[39] M. Colombino, D. Gro, and F. Drfler, "Global phase and voltage synchronization for power inverters: a decentralized consensus-inspired approach," in: *2017 IEEE 56th Annual Conference on Decision and Control (CDC)*, 2017, pp. 5690–5695.

Chapter 12

Combined voltage–frequency control with power electronics-based devices

Georgios Tzounas[1], Weilin Zhong[2], Mohammed A.A. Murad[3] and Federico Milano[2]

A relevant question related to the regulation provided by power electronics-based devices is how to utilize efficiently their active and/or reactive control loops and which control signals to dedicate to which control objectives. These devices are faster and, at least with respect to control capabilities, more versatile than conventional synchronous machines. Power-electronic converter represents thus both a great challenge and an unprecedented opportunity for the dynamic performance of power systems. This chapter focuses on the opportunities and describes a set of control schemes that improve the overall power system dynamic response through a combined voltage–frequency regulation strategy. The schemes described in this chapter are intended for any power electronics-based devices, including distributed energy resources (DERs) and energy storage systems (ESSs), as well as flexible AC transmission system (FACTS) devices. The performance of the control schemes in this chapter is illustrated through examples based on benchmark test systems as well as on a realistic model of the Irish transmission system.

12.1 Introduction

As the penetration of converter-based resources to the power grid increases, the total amount of available rotational inertia decreases. This leads to higher frequency and rate of change of frequency (RoCoF) variations which, in turn, increase the risk of a system-level collapse in case of a severe power imbalance [1]. On the other hand, the capability of converter-based generation, such as DERs, to regulate the frequency through the available power reserve is limited because (i) they are typically designed to achieve a (near) maximum power extraction; and (ii) the availability of a certain power reserve is hard to be ensured, since a large portion of converter-based generation is stochastic, e.g. wind and solar photo-voltaic [2]. Consequently, there has been intense

[1]ETH Zürich, Switzerland
[2]School of Electrical and Electronic Engineering, University College Dublin, Ireland
[3]DIgSILENT GmbH, Gomaringen, Germany

research in the last decade on novel solutions for providing frequency support to the grid.

Frequency regulation in power systems is traditionally provided by modifying the active power, while the reactive power is modified to regulate the voltage. This appears as an intuitive choice for conventional large-scale systems, where the active and reactive power flows are largely decoupled due to the highly inductive nature of transmission lines [3]. On the other hand, DERs are often integrated within distribution networks (DNs), where the resistance/reactance ratio of feeders is large, thus leading to a strong interaction of active and reactive power with voltage and frequency, respectively. In this vein, a solution that has been proposed is to artificially impose the active/reactive power decoupling through the control of power converters, see e.g. the virtual impedance control approach [4,5]. On the contrary, in this chapter, we focus on the potential of exploiting the coupling between the active and the reactive power for the design of efficient control loops that can improve the overall dynamic response of the power grid.

In this vein, some recent studies have explored the ability of power electronics based devices to regulate the frequency through voltage control. This concept of voltage-based frequency control (VFC) relies on effectively taking advantage of the sensitivity of loads to voltage variations. Relevant applications from the existing literature include, but are not limited to, resources in small isolated systems [6], DERs integrated within microgrids [7], smart transformers [8], and static var compensators (SVCs) installed at the transmission system level [9].

In this chapter, we describe three control schemes that can improve the dynamic response of power systems through a combined voltage-frequency regulation of power electronics-based resources. In particular, the chapter discusses:

- A VFC scheme provided by FACTS devices, in particular SVCs, connected (close) to load buses.
- A voltage feedback-based frequency control scheme for converter-based resources. This technique consists in modifying the reference of the device's voltage control loop.
- A fully-coupled control scheme for converter-based resources, in which both active and reactive power injections are modified to compensate both for frequency and voltage variations.

For the implementation of these schemes, we have chosen in this chapter to keep each control loop simple yet practical, by employing standard filters and controllers widely used in industrial applications. The effectiveness of the control schemes is discussed in Sections 12.5–12.7 through non-linear time-domain simulations. All simulation results in this chapter are obtained using the power system analysis software tool Dome [10].

12.2 Voltage-based frequency control through SVC devices

The main idea behind most VFC schemes proposed in the literature, including the controls in [6–9], is to exploit the voltage dependency of loads to vary the active power consumption through variation of the bus voltage magnitudes. Therefore, in this section, we first recall a common and simple way to model voltage-dependent loads and then, we discuss the implementation of a VFC scheme using a FACTS device, namely an SVC with local measurements and standard control filters [9].

12.2.1 Voltage dependency of loads

Industrial and residential loads are generally modeled as aggregated power consumption in the dynamic analysis of power systems. These aggregated models can be static or dynamic [11]. A common static model expresses the active and reactive powers as functions of the bus voltage magnitude, as follows*:

$$P_{\mathrm{L}} = P_{\mathrm{L},o}\left(\frac{v_{\mathrm{L}}}{v_{\mathrm{L},o}}\right)^{\alpha_p} \tag{12.1}$$

$$Q_{\mathrm{L}} = Q_{\mathrm{L},o}\left(\frac{v_{\mathrm{L}}}{v_{\mathrm{L},o}}\right)^{\alpha_q} \tag{12.2}$$

where P_{L}, Q_{L} are the active and reactive power demand and v_{L} is the load bus voltage magnitude; the superscript "o" denotes values at the initial operating condition. The exponents α_p and α_q vary depending on the load type [12]. Typical ranges for the exponents are $\alpha_p \in (0.9, 1.7)$ and $\alpha_q \in (1.9, 4)$ [13].

A change in the operation voltage, say Δv_{L}, results in the following change in power demand ΔP_{L}:

$$\Delta P_{\mathrm{L}} = \left[(v_{\mathrm{L}} + \Delta v_{\mathrm{L}})^{\alpha_p} - v_{\mathrm{L}}^{\alpha_p}\right]\frac{P_{\mathrm{L},o}}{(v_{\mathrm{L},o})^{\alpha_p}} \tag{12.3}$$

For example, assume $v_{\mathrm{L}} = v_{\mathrm{L},o} = 1$ pu and $\alpha_p = 1.5$. Then, a $\Delta v_{\mathrm{L}} = 5\%$ voltage increase will lead to $\Delta p_{\mathrm{L}} = 7.6\%$.

For the purpose of assessing a VFC scheme through simulations, it is acceptable to assume that the voltage dependency of loads is known. For the static model (12.1)–(12.2), this corresponds to assuming that the exponents α_p and α_q are known. However, we note also that, in practice, there is a need for a way to accurately identify and track the voltage dependency of loads in real-time, to ensure that the performance of the VFC is not compromised. With this regard, we refer the interested reader to the voltage-dependent load identification techniques discussed in [14–16].

12.2.2 Frequency control through SVC

In this section, we describe a VFC provided by means of a FACTS device, in particular through an SVC. An SVC is a combination of a capacitor and a variable shunt reactor controlled through thyristor-based power electronic switches. The SVC can generate and absorb reactive power and thus is conventionally utilized for voltage control at the bus to which it is connected. The control diagram of an SVC is shown in Figure 12.1, where the controlled variable is the susceptance b_{SVC}.

*The following chapter on smart transformers further discusses the interaction between power electronic controllers and voltage and frequency-dependent load models.

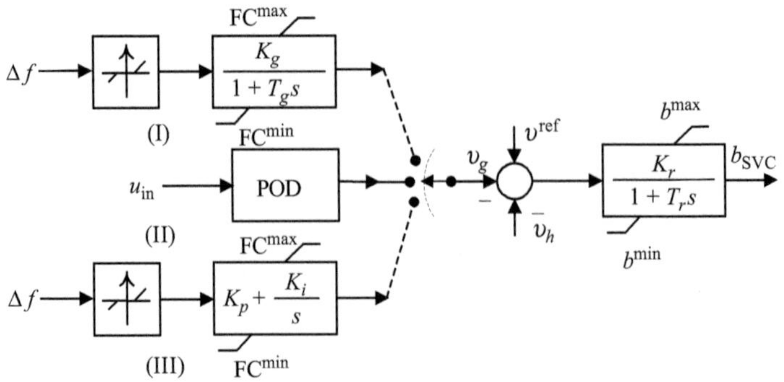

Figure 12.1 Block diagram of conventional SVC model with additional control loops: (I) droop frequency control, (II) POD control and (III) PI-based frequency control

The SVC model is defined by the following differential–algebraic equations [11]:

$$T_r \dot{b}_{\mathrm{SVC}} = -b_{\mathrm{SVC}} + K_r(v^{\mathrm{ref}} - v_h - v_g) \tag{12.4}$$

$$Q_{\mathrm{SVC}} = b_{\mathrm{SVC}} v_h^2 \tag{12.5}$$

where v^{ref}, v_h, K_r, T_r, and Q_{SVC} are, respectively, the reference voltage, the voltage at the bus to which the SVC is connected, the regulator gain, the regulator time constant, and the output reactive power generated by the SVC. In a conventional application, the frequency control loop of a SVC is utilized for damping low-frequency (electromechanical) oscillations (0.1–2 Hz). That is, the signal v_g is usually the output of a power oscillation damper (POD). Instead, in this section, we employ two different types of controllers to get v_g, which are designed for VFC. In particular, the following control methods are considered:

- a droop (lag) controller, and
- a proportional integral (PI) controller,

see (I) and (III) in Figure 12.1. For both types, the frequency error Δf is considered as a control input. The presence of the deadband (*db*) ensures that for a small variation of the frequency, the controller will not deteriorate the local voltage response.

In case that the droop controller is used, v_g is given by:

$$T_g \dot{v}_g = K_g \Delta f - v_g \tag{12.6}$$

where T_g and K_g are the time constant and gain of the lag controller, respectively; Δf is the frequency error, where $\Delta f = f^{\mathrm{ref}} - f_l$; f^{ref} is the reference frequency and f_l is the measured frequency signal at the SVC bus. The signal f_l can be obtained through a bus frequency estimation method, e.g., through a phase-locked loop (PLL).

If the PI controller is used, v_g is as follows:

$$v_g = K_p \Delta f + x_f \tag{12.7}$$

$$\dot{x}_f = K_i \Delta f \tag{12.8}$$

where K_p, K_i, and x_f are the proportional gain, the integral gain, and the state variable of the PI control, respectively.

To ensure that the bus voltage remains within its operational range, both types of frequency control constrain the output signal to its respective limits ($\mathrm{FC}^{\max}$ and $\mathrm{FC}^{\min}$). Moreover, anti-windup type limits are considered to get better overall transient response [17].

We note that the placement of the SVC in the power network is critical for the effectiveness of the VFC. In general, the VFC is expected to have better performance if located at a bus where voltage variations have a significant impact on the active power consumption, i.e., at or close to voltage-dependent load buses.

The effectiveness of the SVC-based frequency control is tested in Section 12.5 using the WSCC 9-bus system and a dynamic model of the All-island Irish Transmission System (AIITS). In particular, four scenarios are tested and compared by carrying out non-linear time domain simulations:

1. no SVC connected (NSVC);
2. conventional ($v_g = 0$) SVC (CSVC);
3. CSVC with lag frequency controller (LFC); and
4. CSVC with PI frequency controller (PIFC).

12.3 Frequency control of converter-based resources through modified voltage control reference

In this section, we examine the effectiveness of modifying the voltage reference of converter-based resources, by using a feedback signal that aims to mitigate the part of their injected active power that does not contribute to frequency regulation. The modified voltage control reference is derived in the next section. The focus is on the regulation provided by DERs and ESSs.

12.3.1 Modified voltage control reference

Our starting point are the well-known power flow equations. The power injection at the network buses of a power system can be described as:

$$\bar{\boldsymbol{S}}(t) = \boldsymbol{P}(t) + j\boldsymbol{Q}(t) = \bar{\boldsymbol{v}}(t) \circ \left(\bar{\mathbf{Y}}\, \bar{\boldsymbol{v}}(t)\right)^* \tag{12.9}$$

where $\boldsymbol{P}, \boldsymbol{Q} \in \mathbb{C}^{n\times 1}$ are the column vectors of bus active and reactive power injections, respectively; n is the number of network buses; $\bar{\mathbf{Y}} \in \mathbb{C}^{n\times n}$ is the network admittance matrix; $\bar{\boldsymbol{v}} \in \mathbb{C}^{n\times 1}$ is the vector of bus voltages; $\circ$ denotes the element-wise multiplication; and $*$ indicates the conjugate. The hth elements of $\boldsymbol{P}$ and $\boldsymbol{Q}$ can be written as:

$$\begin{aligned} P_h &= \sum_{k=1}^{n} P_{hk} = \sum_{k=1}^{n} v_h v_k (G_{hk} \cos\theta_{hk} + B_{hk} \sin\theta_{hk}) \\ Q_h &= \sum_{k=1}^{n} Q_{hk} = \sum_{k=1}^{n} v_h v_k (G_{hk} \sin\theta_{hk} - B_{hk} \cos\theta_{hk}) \end{aligned} \tag{12.10}$$

where the time dependency is omitted for simplicity; P_{hk}, Q_{hk} are the active and reactive power flows, respectively, from bus h to bus k; G_{hk} and B_{hk} are the real and imaginary parts of the (h, k) element of $\bar{\mathbf{Y}}$, i.e., $\bar{Y}_{hk} = G_{hk} + jB_{hk}$; v_k is the voltage magnitude at bus k, $k = 1, 2, \ldots, n$; and $\theta_{hk} = \theta_h - \theta_k$, where θ_h and θ_k are the voltage phase angles at buses h and k. Differentiation of (12.10) gives:

$$\begin{aligned} dP_h &= dP_{\theta,h} + dP_{v,h} \\ dQ_h &= dQ_{\theta,h} + dQ_{v,h} \end{aligned} \tag{12.11}$$

where

$$dP_{\theta,h} = \sum_{k=1}^{n} \frac{\partial P_h}{\partial \theta_{hk}} d\theta_{hk} \tag{12.12}$$

$$dP_{v,h} = \sum_{k=1}^{n} \frac{\partial P_h}{\partial v_k} dv_k \tag{12.13}$$

$$dQ_{\theta,h} = \sum_{k=1}^{n} \frac{\partial Q_h}{\partial \theta_{hk}} d\theta_{hk} \tag{12.14}$$

$$dQ_{v,h} = \sum_{k=1}^{n} \frac{\partial Q_h}{\partial v_k} dv_k \tag{12.15}$$

$dP_{\theta,h}$, $dQ_{\theta,h}$ are the quota of dP_h and dQ_h that depend on the voltage angles and consequently, the components of the active and reactive power that can be effectively used to regulate the frequency in the system. This observation is the key of the control presented in this section. To better illustrate this point, one can consider $dP_{\theta,h}$, $dQ_{\theta,h}$ with respect to time and substitute $\frac{d\theta_k}{dt} = \omega_o \omega_k$, where ω_k is the frequency at bus k and ω_o is the reference synchronous speed in rad/s. On the other hand, $dP_{v,h}$, $dQ_{v,h}$ are the quota of dP_h and dQ_h that depend on the voltage magnitudes and thus the components that can be used to modify the voltage response.

In standard DER voltage control schemes, the voltage reference is constant, at least for a given period, as follows:

$$v^{\text{ref}}(t) = v_o^{\text{ref}} = v_{h,o} \tag{12.16}$$

where v_o^{ref} is the desired reference voltage and $v_{h,o}$ denotes the value, in steady-state, of the voltage magnitude at bus h where the DER is connected.

The control scheme discussed in this section modifies (12.16) with the objective to reduce—ideally, nullify—the term $dP_{v,h}$. Since this term does not contribute to the frequency response of the system, the effect is to make $dP_h \approx dP_{\theta,h}$ and, hence, optimize the effectiveness of the frequency control. The control considered, in turn, is designed to impose the following constraint:

$$dP_{v,h} = 0 \tag{12.17}$$

Using (12.10), and assuming for simplicity a lossless transmission system, i.e., $G_{hk} = 0$, and defining $\tilde{B}_{hk} = B_{hk} \sin \theta_{hk}$, we can rewrite (12.13) as follows:

$$dP_{v,h} = \sum_{k=1}^{n} \tilde{B}_{hk}(t)\, d[v_h(t)\, v_k(t)] \tag{12.18}$$

or, equivalently,

$$dP_{v,h} = \sum_{k=1}^{n} \tilde{B}_{hk}(t)\, (v_k\, dv_h + v_h\, dv_k) \tag{12.19}$$

From (12.19), a sufficient condition so that (12.17) is satisfied reads as follows:

$$\sum_{k=1}^{n} (v_k(t)\, dv_h + v_h(t)\, dv_k) = 0 \tag{12.20}$$

The last equation is equivalent to:

$$v_h(t) \sum_{k=1}^{n} v_k(t) = c_o \tag{12.21}$$

where c_o is a constant, which, following from the system initialization, takes the value:

$$c_o = v_{h,o}(t) \sum_{k=1}^{n} v_{k,o}(t) \tag{12.22}$$

From the above analysis, we get that the modified voltage control reference that can be utilized to achieve the control objective (12.17) is:

$$v^{\mathrm{ref}}(t) = \frac{c_o}{\sum_{k=1}^{n} v_k(t)} \tag{12.23}$$

Note that (12.23) is valid also for lossy transmission systems, i.e., $G_{hk} \neq 0$, and is, thus, a general condition. Implementation of (12.23) requires measuring the voltage magnitudes at the buses to which the DER is connected which in turn depends on the topology of the system. Figure 12.2(a) shows a DER connected to the grid in antenna. In this case, only one remote measurement is needed, namely v_2. If the DER is connected to the grid through multiple buses, more measurements are required. For example, in the topology shown in Figure 12.2(b), one has to measure v_2 and v_3.

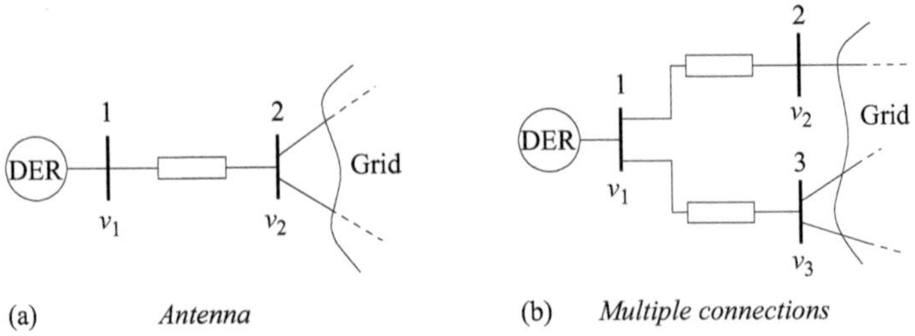

Figure 12.2 Examples of DER connectivity to the grid

12.3.2 DER and ESS models

We briefly describe here the basic control structure of the simplified DER and ESS model used in the case study of Section 12.6 to test the modified voltage control reference scheme described in Section 12.3.1.

The block diagram of the DER model is depicted in Figure 12.3. It consists of an inner control loop that regulates the components $\imath_{\mathrm{d}}$, $\imath_{\mathrm{q}}$ of the current in the dq reference frame, and two outer loops for frequency and voltage control, respectively. The frequency control loop filters the frequency error $\omega^{\mathrm{ref}} - \omega$ and implements a droop control with droop constant $\mathcal{R}$. The frequency control output is then added to the DER's active power reference. On the other hand, the voltage control loop receives the error $v^{\mathrm{ref}} - v_h$ and implements a PI control. The output of the voltage control loop is added to the DER's reactive power reference.

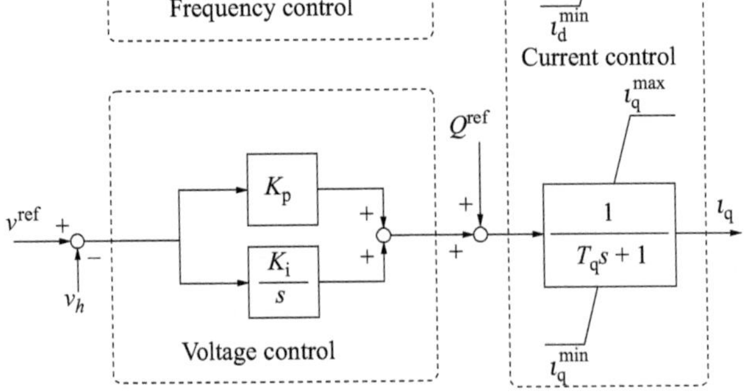

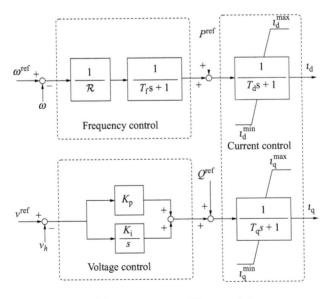

Figure 12.3 Simplified DER control diagram

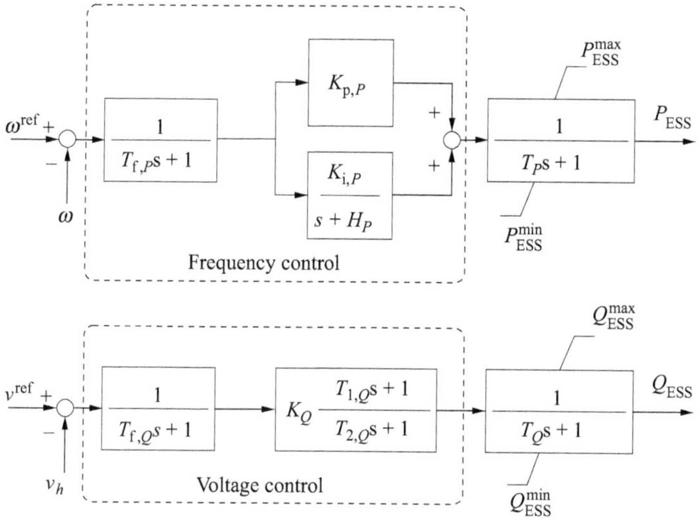

Figure 12.4 Simplified ESS control diagram

The control structure of the ESS model is shown in Figure 12.4. The frequency control loop filters the frequency error and implements a lead-lag regulator with droop constant $\mathcal{R} = H_P/(K_{i,P} + K_{p,P}H_P)$, the output of which is fed to the ESS active power dynamics. The voltage control loop filters the voltage error and implements a lead-lag control, the output of which is input to the ESS reactive power dynamics.

The modified voltage reference control scheme presented in the previous paragraph is tested in Section 12.6 using the DER and ESS models described above and based on a modified version of the WSCC 9-bus system. The following control modes are considered:

1. Constant power control (CPC), i.e., without the frequency and voltage control loops;
2. FC, i.e., with the frequency control connected and the voltage control disconnected;
3. FC+VC, i.e., with both frequency and voltage control connected and the voltage control reference given by (12.16);
4. FC+MRVC, i.e., with both frequency and voltage control connected and with the modified voltage control reference given by (12.23).

12.4 Coupled voltage–frequency control of DERs

In this section, we present a control scheme for DERs, in which both active and reactive power are varied to regulate both frequency and voltage. This is in contrast to current practice, where frequency and voltage controllers are decoupled.

12.4.1 Control structure

Considering a simplified DER model, the block diagram of the control scheme is depicted in Figure 12.5. The control scheme consists of an inner current control loop and two outer loops for frequency and voltage regulation, respectively. The current control loop regulates the d and q axis components of the current ($\imath_{\rm d}$, $\imath_{\rm q}$) in the dq reference frame. These components are limited between their minimum and maximum values through an anti-windup limiter. The frequency control loop receives the frequency error ϵ_ω and applies a droop control and a washout filter acting in parallel. On the other hand, the voltage control loop adjusts the bus voltage error ϵ_v by means of a PI controller and a washout filter also connected in parallel. The outputs of the frequency and voltage controllers are then added to the DER's active and reactive power references. It is worth observing that we have adopted simple conventional controllers on purpose, as these are most commonly implemented in practice. As a matter of fact, the main objective of this work is to show how combining the effect of different control channels impacts on the performance of the overall system.

The examined DER control scheme includes four channels that can be combined to formulate different active and reactive power control modes. A summary of the available control modes for the active power of the DER is as follows:

- FP: The active power is employed to regulate the frequency. The FP mode is the standard way to regulate the frequency in conventional power systems.
- VP: The active power is employed to regulate the voltage. In this mode, a voltage control channel acts by modifying the DER active power reference.
- FVP: The active power reference is modified to control both the frequency and the voltage. In this case, both FP and VP in Figure 12.5 are switched on.

Similarly, the available modes for the control of the DER reactive power can be summarized as follows:

- VQ: The reactive power is utilized to regulate the voltage. This is the classic approach, i.e., voltage regulation is conventionally realized by means of the VQ mode.
- FQ: The reactive power reference of the DER is modified to provide frequency regulation.
- FVQ: Both VQ and FQ are switched on in a combined control of the reactive power.

In this work, we study the effectiveness of frequency and voltage regulation provision through both the active and the reactive power of DERs, which leads to the combined scheme FVP+FVQ. In the case study of Section 12.7, the dynamic performance of this configuration is compared to other configurations, including the conventional approach to frequency–voltage control, i.e., FP+VQ.

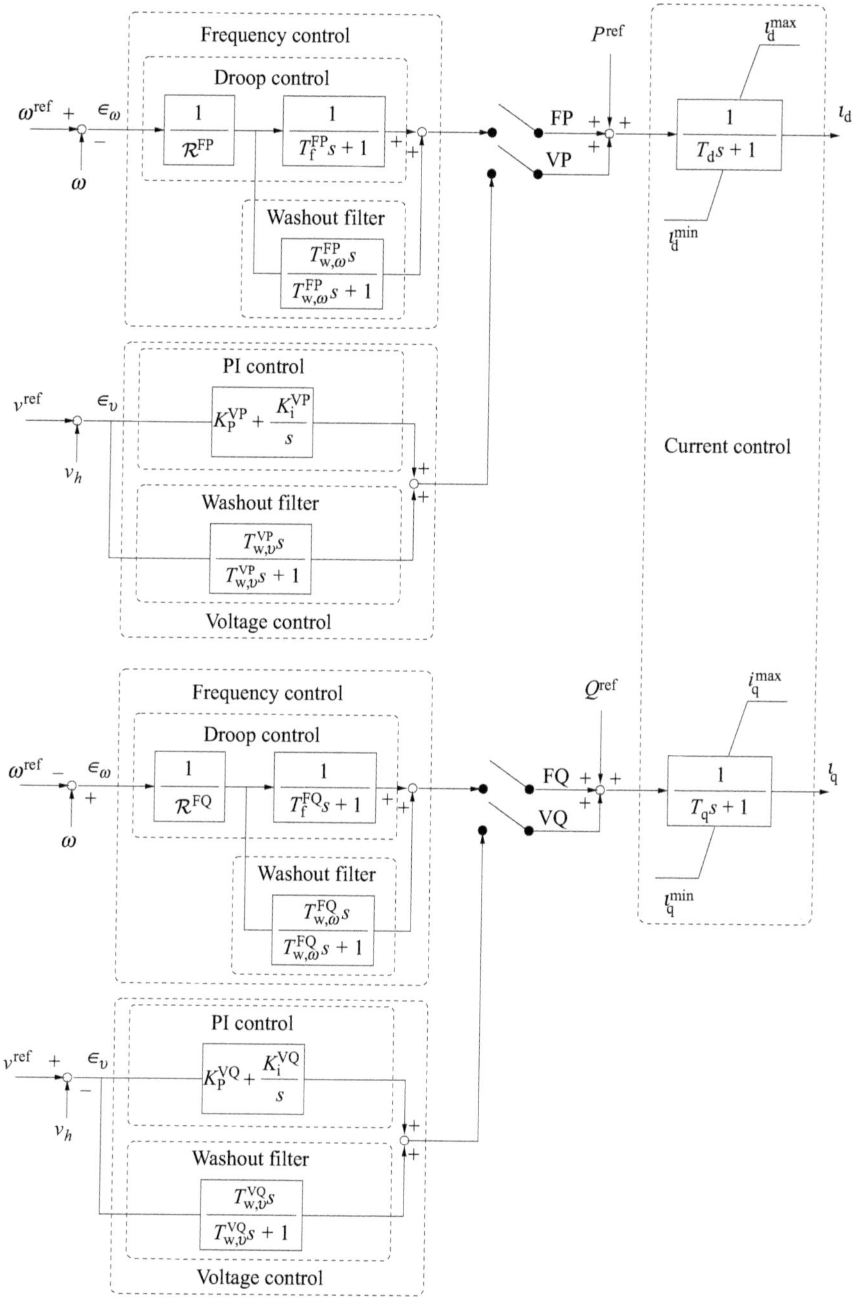

Figure 12.5 DER control scheme

12.4.2 Assessment metric

The examined control configurations are assessed using standard criteria, such as the deviation of frequency and voltages from their nominal values. As a complementary assessment tool, this section presents a scalar metric that is employed to capture the combined effect of frequency/voltage response provided at a bus of the power network. Consider (12.11). Then, the parts of $dP_{\theta,h}$, $dQ_{\theta,h}$, and $dP_{v,h}$, $dQ_{v,h}$ that are due to local variations of the frequency and the voltage at bus h, respectively, are given by the following expressions [18]:

$$\begin{aligned} dP^{\text{loc}}_{\theta,h} &= -Q_h d\theta_h\,, \quad dQ^{\text{loc}}_{\theta,h} = P_h d\theta_h\,, \\ dP^{\text{loc}}_{v,h} &= \frac{P_h}{v_h} dv_h\,, \quad dQ^{\text{loc}}_{v,h} = \frac{Q_h}{v_h} dv_h \end{aligned} \tag{12.24}$$

Then rewriting (12.24) using time derivatives, one has:

$$\begin{aligned} \frac{dP^{\text{loc}}_{\theta,h}}{dt} &= -Q_h \theta'_h\,, \quad \frac{dQ^{\text{loc}}_{\theta,h}}{dt} = P_h \theta'_h\,, \\ \frac{dP^{\text{loc}}_{v,h}}{dt} &= P_h u'_h\,, \quad \frac{dQ^{\text{loc}}_{v,h}}{dt} = Q_h u'_h \end{aligned} \tag{12.25}$$

where

$$\theta'_h = \frac{d\theta_h}{dt}\,, \qquad u'_h = \frac{1}{v_h}\frac{dv_h}{dt} \tag{12.26}$$

The first term, i.e., θ'_h, is the deviation of the bus frequency with respect to the synchronous frequency; whereas u'_h represents the transient rate of change of the voltage normalized with respect to the bus voltage magnitude. The latter quantity has the same unit as a frequency and is thus comparable with the frequency deviation θ'_h.

In the remainder of this section, we are interested in assessing the combined active/reactive injection effect on the voltage/frequency response provided at bus h. With this aim we utilize the quantity:

$$\mu'_h = \sqrt{(\theta'_h)^2 + (u'_h)^2} \tag{12.27}$$

In particular, we are interested in assessing the cumulative effect of μ'_h for a given time interval $[t_0, t]$. This interval is determined based on the time scale of the primary response of generators, which lasts from few seconds to few tens of seconds. Finally, the following quantity can be used as a metric to assess the joint frequency/voltage response at a given bus h of a power network:

$$\mu_h = \int_{t_0}^{t} \mu'_h \, dt \tag{12.28}$$

The metric in (12.28) possesses the property that the two components corresponding to the frequency and voltage are considered with the same weights, while having the same units, thus being summable and directly comparable. In this chapter, the metric is used in the case study of Section 12.7 to compare the effectiveness of different DER active/reactive control configurations. With this regard, note that smaller values of μ_h are in general obtained for smaller frequency and voltage variations, which in turn, indicates a better dynamic response at bus h.

12.5 Case study I: VFC through SVC

In this section, the WSCC 9-bus system and the AIITS are considered to study the performance of the SVC control scenarios discussed in Section 12.2.

12.5.1 WSCC 9-bus system

The WSCC 9-bus system is shown in Figure 12.6. It consists of three synchronous machines (SMs), three transformers, three loads, and six transmission lines. All SMs are equipped with automatic voltage regulators (AVRs) and turbine governors (TGs). The dynamic data of the network are provided in [19]. In this section, we assume that a SVC is connected to bus 8 of the system. The parameters used for the SVC controllers can be found in [9].

The system is simulated by applying a three phase fault at bus 6 at $t = 1$ s. The fault is cleared after 60 ms by tripping the line that connects buses 6 and 9. The trajectories of the frequency of the center of inertia (COI), as well as the voltage at bus 8 are depicted in Figures 12.7 and 12.8, respectively.

Figure 12.7 shows that the utilization of LFC and PIFC leads to a significant improvement of the initial frequency deviation. Concerning the voltage response (see Figure 12.8), after the disturbance the CSVC provides reactive power support and therefore, improves the bus voltage. This also leads to increase the load consumption, as imposed by (12.3). In this case, the CSVC leads to a relatively good frequency response even without frequency control loop. But this result is not always guaranteed. How effective is the CSVC for frequency control depends on the disturbance and cannot be determined *a priori*.

No secondary frequency control is considered in this study. Hence, the trajectories of frequency reach a post disturbance equilibrium with a non-zero steady-state error.

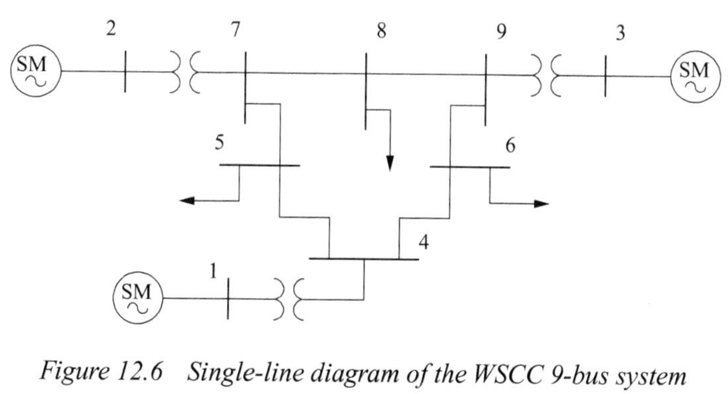

Figure 12.6 Single-line diagram of the WSCC 9-bus system

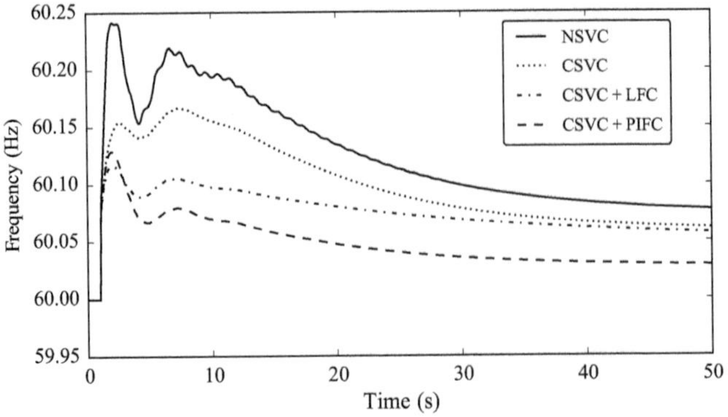

Figure 12.7 Frequency response in COI frame

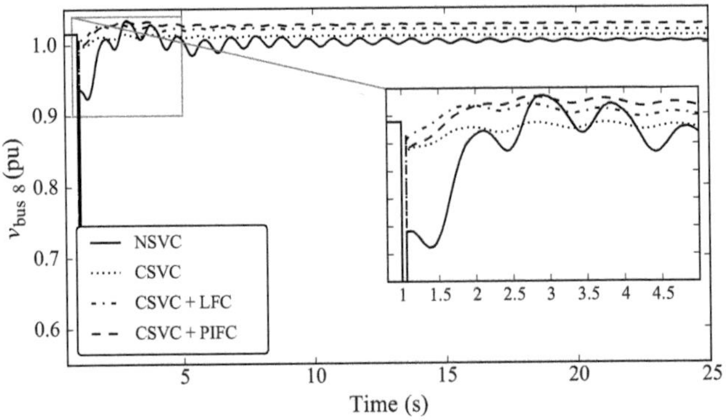

Figure 12.8 Response of the voltage at bus 8

Compared to all other cases, this error is the smallest when the PIFC is employed. The PIFC also leads to a higher steady-state voltage magnitude at the bus of the SVC. This is a consequence of the perfect tracking behavior of the PI control. Note that the PIFC integrator does not eliminate the steady-state frequency error, due to the deadband in the VFC input and the limits on the output.

Overall, the amount of frequency response improvement obtained in Figure 12.7 when either LFC or PIFC is included is significant (>0.1 Hz). It is clear that this improvement varies depending on several factors: size of the system, number of SVCs installed, location of the SVCs, etc. To better quantify the real impact of the VFC provided by SVCs, in the next section, we study the effect of inclusion of frequency control loops in SVCs installed at the AIITS.

12.5.2 All-island Irish system

The AIITS model is built based on static data provided by EirGrid Group, the Irish Transmission System Operator (TSO). Dynamic data are defined based on current knowledge about the capacities and technologies of power plants. The system consists of 1,479 buses, 1,851 transmission lines and transformers, 245 loads, 22 SMs with AVRs and TGs, 6 power system stabilizers (PSSs), 173 wind generators of which 139 are doubly-fed induction generators and 34 are constant speed wind turbines. The map of the AIITS is provided in Appendix F.

In order to carry out a realistic case study, we first validate the AIITS by applying a real severe over-frequency event. On February 28, 2018, the VSC-HVDC link East-West Inter-connector (EWIC) [20] that connects the AIITS with the Great Britain (GB) transmission system was tripped. At that moment, Ireland was exporting 470 MW to GB. Due to the loss of the EWIC, the frequency in the Irish grid rose to 50.42 Hz. Over-frequency protections were triggered and several wind farms were curtailed. In the AIITS model, we consider the 470 MW active power export as a constant load. A comparison of simulated and actual frequency response following the outage of the EWIC is shown in Figure 12.9. With a proper tuning of TG parameters, we obtain a satisfactory match between the simulated transient and the actual one. Note that for the purpose of the model validation, no VFC is considered.

We examine the frequency response of the AIITS model with inclusion of SVC based VFC. The same four scenarios discussed for the 9-bus system are considered. For our simulations, we consider that the AIITS includes five SVCs. More details about the setup and the parameters of the SVCs and PLLs can be found in [9].

The test system is simulated by applying the disturbance considered for the system's validation. The comparative trajectories of the frequency and the voltage at a SVC bus (Omagh Main) are shown in Figures 12.10 and 12.11, respectively. The NSVC case is the same as the simulated response shown in Figure 12.9. Compared to NSVC and CSVC, the utilization of CSVC with LFC and PIFC improves the

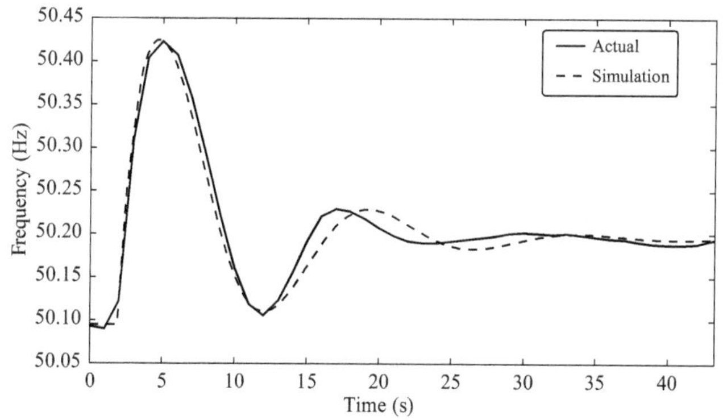

Figure 12.9 Simulated and actual frequency response due to loss of the EWIC

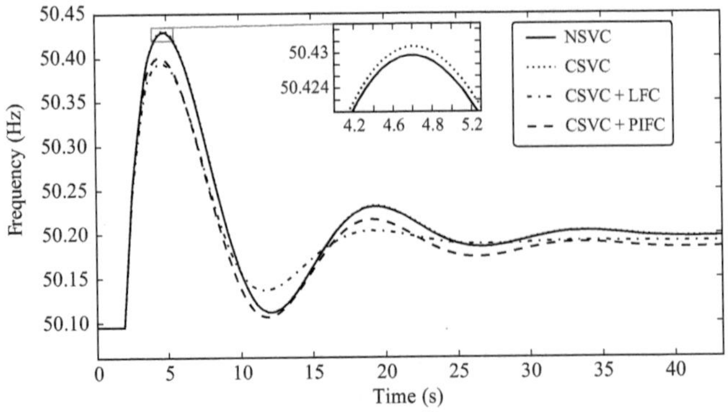

Figure 12.10 Frequency response following loss of the EWIC

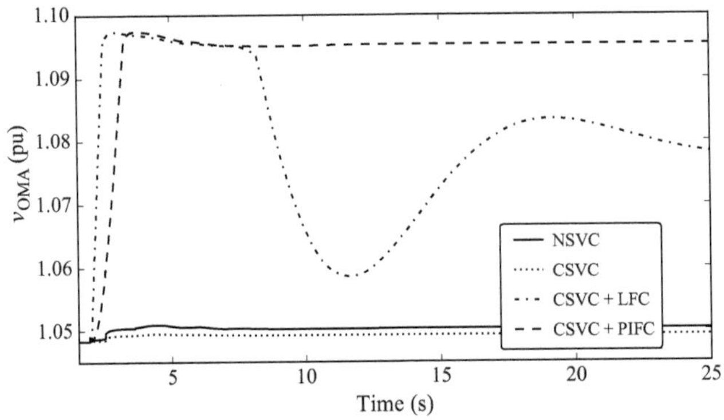

Figure 12.11 Voltage at bus Omagh Main following loss of the EWIC

frequency response. This improvement (for LFC $\approx$ 0.037 Hz and for PIFC $\approx$ 0.03 Hz) is relevant given that only 5 SVCs are utilized. At the time of the EWIC outage, the active power generation in the AIITS is greater than the demand. Hence, the bus voltage is increased by the SVC with LFC and PIFC to increase power consumption. On the other hand, the CSVC without frequency control ensures the best voltage control, which in turn slightly deteriorates the frequency response (see zoomed view in Figure 12.10). Due to the limits imposed in the VFC, voltage fluctuations remain within the maximum operating range (1.1 pu). Even though PIFC provides the minimum steady-state error, the overall transient response of the frequency is better when the LFC is used.

We now consider an under-frequency event. In particular, we assume that 155 MW of generation is disconnected. The results for the four examined scenarios are shown in Figures 12.12 and 12.13. An overall better response is achieved with LFC and PIFC

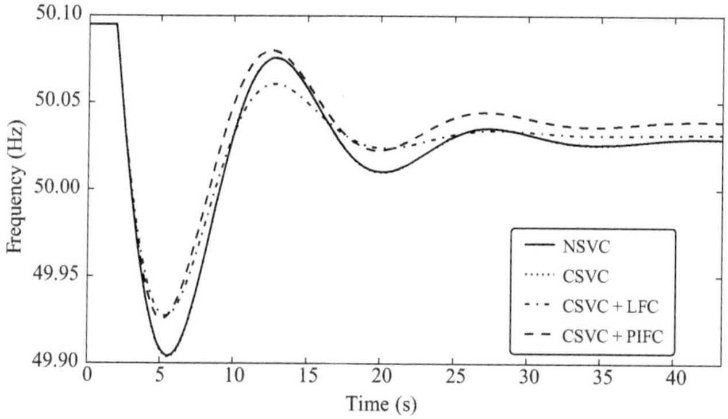

Figure 12.12 Frequency response following the loss of 155 MW generation

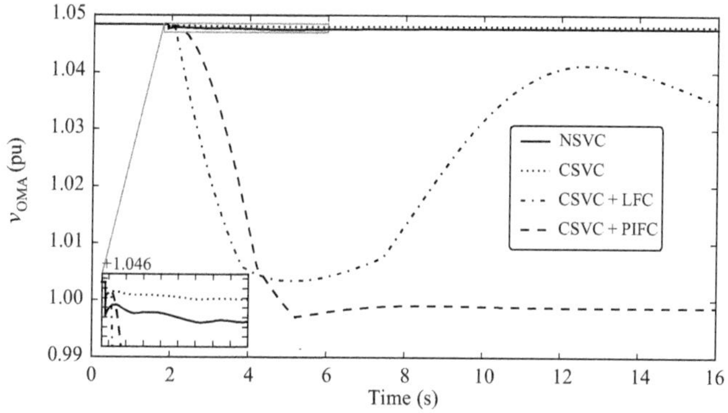

Figure 12.13 Voltage at bus Omagh Main following the loss of 155 MW generation

compared to NSVC and CSVC cases. The voltage also remains well within its lower bound (0.9 pu).

12.5.3 Discussion

In all scenarios considered in this case study, the SVCs are connected at the transmission level. This solution is the most flexible from the point of view of the transmission system operator (TSO) because it does not require that they engage with customers or distribution system operators. Moreover, no VFC strategy discussed here requires changing the existing infrastructure or to develop a communication framework. This is an added value of the VFC based on SVCs. Even if the TSO wanted to limit the utilization of SVCs exclusively for voltage control in normal operation, the VFC could be enabled following a large contingency. This can be achieved safely and automatically by simply setting a large *db* value in the input of the VFC.

12.6 Case study II: FC+MRVC scheme for DERs

This section presents simulation results based on a modified version of the WSCC 9-bus test system. The modifications with respect to the original system described in Section 12.5 are as follows. SMs at buses 2 and 3 have been replaced by DERs represented according to the model described in Section 12.3.1. Then, the mechanical starting time of the SM connected to bus 1 is decreased to 23 s. With the above changes, the inertia of the system has been reduced by 65% compared to the data of the original system. Finally, in Section 12.6.2, an ESS is connected to bus 5. The parameters used for the DER and ESS controllers are provided in [21].

12.6.1 DERs connected to buses 2 and 3

We test the impact of voltage reference (12.23) on the frequency regulation provided by the DERs connected to buses 2 and 3 of the modified WSCC 9-bus system. In this scenario, no ESS is included in the system.

The examined control modes that include frequency regulation consider as control input signal the frequency of the center of inertia, which, since there is only one SM left in the system, coincides with the rotor speed of the SM connected to bus 1 (ω_1). Moreover, for the FC+MRVC, the DERs at buses 2 and 3 are connected to the transmission system in antenna and thus require the voltage magnitudes at buses 7 (v_7) and 9 (v_9), respectively, to implement reference (12.23).

We consider the transient that follows the tripping, at $t = 1$ s, of the line that connects buses 4 and 6. Simulation results are presented in Figures 12.14–12.16. In particular, Figure 12.14 shows the trajectory of the rotor speed of the SM (in Hz) following the disturbance. The system dynamic behavior for the CPC and FC is not acceptable, while for the examined disturbance, inclusion of the voltage control loop significantly improves the frequency response. The FC+MRVC provides a further significant improvement in the frequency regulation (> 0.1 Hz) with respect to the FC. A comparison of the voltage references of the FC+MRVC and FC+VC for the

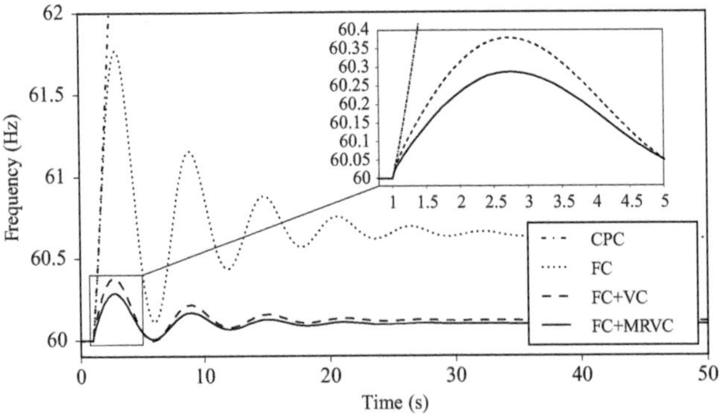

Figure 12.14 Frequency response after the outage of lines 4–6

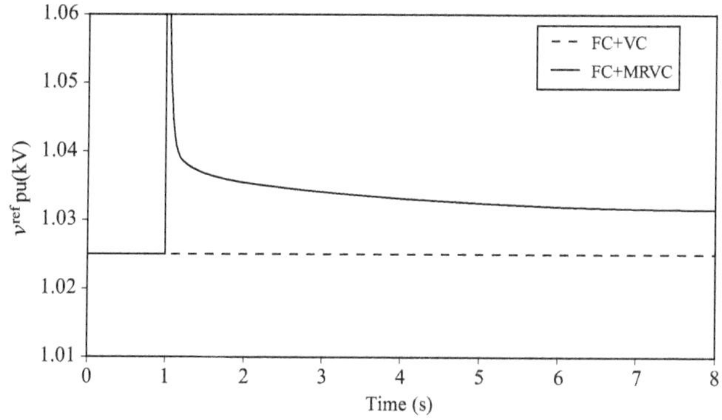

Figure 12.15 DER connected to bus 2: voltage reference after the outage of lines 4–6

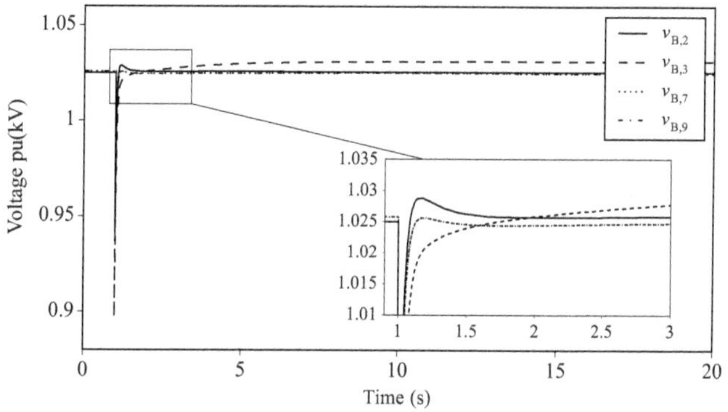

Figure 12.16 FC+MRVC: voltage response following the outage of lines 4–6

DER connected to bus 2 is shown in Figure 12.15. It is important to note that the improvement provided by the FC+MRVC comes without jeopardizing the voltage response of the system. This is shown in Figure 12.16. Overall, in the case of a line trip, the FC+MRVC provides an important improvement in the frequency regulation of the system. We have applied trips to other lines of the network and obtained similar results and same conclusions.

12.6.2 ESS connected to bus 5

In this scenario, an ESS is connected to bus 5 of the modified WSCC 9-bus system. The objective is to assess the impact of the FC+MRVC on the frequency response provided by the ESS. With this aim, in this scenario, the DERs connected to buses 2 and 3 do not provide frequency and voltage control (see CPC in Section 12.6.1).

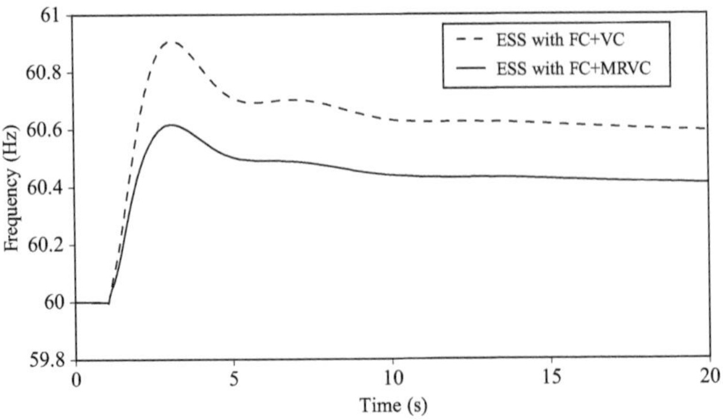

Figure 12.17 Frequency response following a three-phase fault

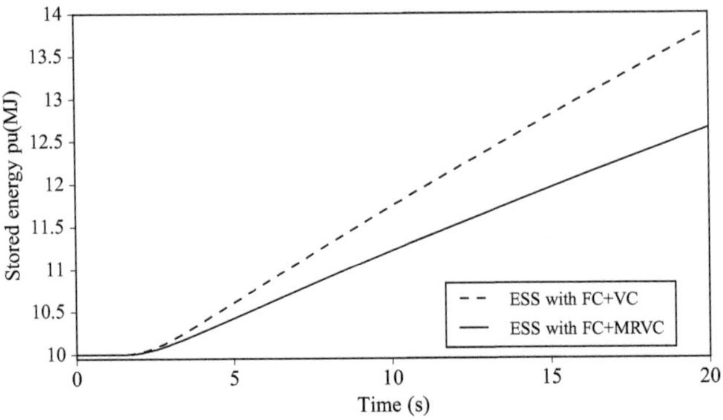

Figure 12.18 ESS stored energy following a three-phase fault

The test system is simulated considering a three-phase fault at bus 6. The fault occurs at $t = 1$ s and is cleared after 80 ms by tripping the line that connects buses 6 and 9. Two modes are compared for the ESS control: FC+VC and FC+MRVC. The voltage reference (12.23) of the FC+MRVC is implemented with the measurements of the voltage magnitudes at buses 4 and 7. The frequency response of the system is shown in Figure 12.17. The FC+MRVC provides a significant improvement in the frequency regulation of the system, while this is achieved in an economic way, since a lower variation of the stored energy is required (see Figure 12.18).

We have thoroughly tested the dynamic behavior of the modified 9-bus system with inclusion of the FC+MRVC for a variety of different contingencies. There exist scenarios for which the difference between the FC+MRVC and FC+VC is smaller than the cases shown in Figures 12.14 and 12.17. In Figure 12.19, we consider a 15% sudden increase of the susceptance and conductance of all system loads occurring at

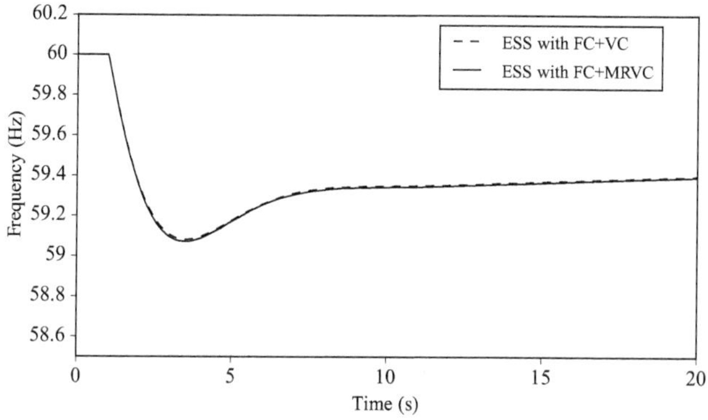

Figure 12.19 Frequency response following a load increase

$t = 1$ s. In this case, FC+VC and FC+MRVC show substantially the same dynamic behavior. In general, we can conclude that the FC+MRVC, in the cases where it is not beneficial, at least does not worsen the dynamic response of the system.

12.7 Case study III: FVP+FVQ control of DERs

This section presents simulation results based on the IEEE 39-bus benchmark system [22]. The system comprises 10 SMs (Gen 1-10), totaling 6,354.1 MW and 1,357.1 MVAr of active and reactive power generation. SMs are represented by fourth-order (two-axis) models and are equipped with automatic voltage regulators, turbine governors, and power system stabilizers. In this case study, SMs are also assumed to participate to secondary frequency regulation through an automatic generation control (AGC) scheme. The AGC is modeled as an integrator the output of which is used to update the active power set-points of the machines every 5 s. Loads are modeled using the ZIP model [11]. ZIP loads in this section consist of 20% constant power 10% constant current and 70% constant impedance consumption [23].

For the purpose of this case study, the system is modified to include a 30% penetration of non-synchronous generation. To this aim, the SMs Gen 5, 6, and 8 connected to buses 34, 35, and 37 are substituted by converter-based DERs. Considering the practical capacity of a single DER, the DERs connected to each bus here are not single generation sources but are modeled as a combination of several DERs. The single-line diagram of the modified IEEE 39-bus system is shown in Figure 12.20. DERs and their controls are modeled as described in Section 12.4.

The parameters of the frequency, voltage, and inner current controllers of the DERs are given in [24]. The first estimation of the control parameters of each filter has been obtained by setting the time constants of the corresponding differential equations based on the requirements for the time scale of their action, which leads most of the parameters to lie in a certain range. Then the final values of the parameters

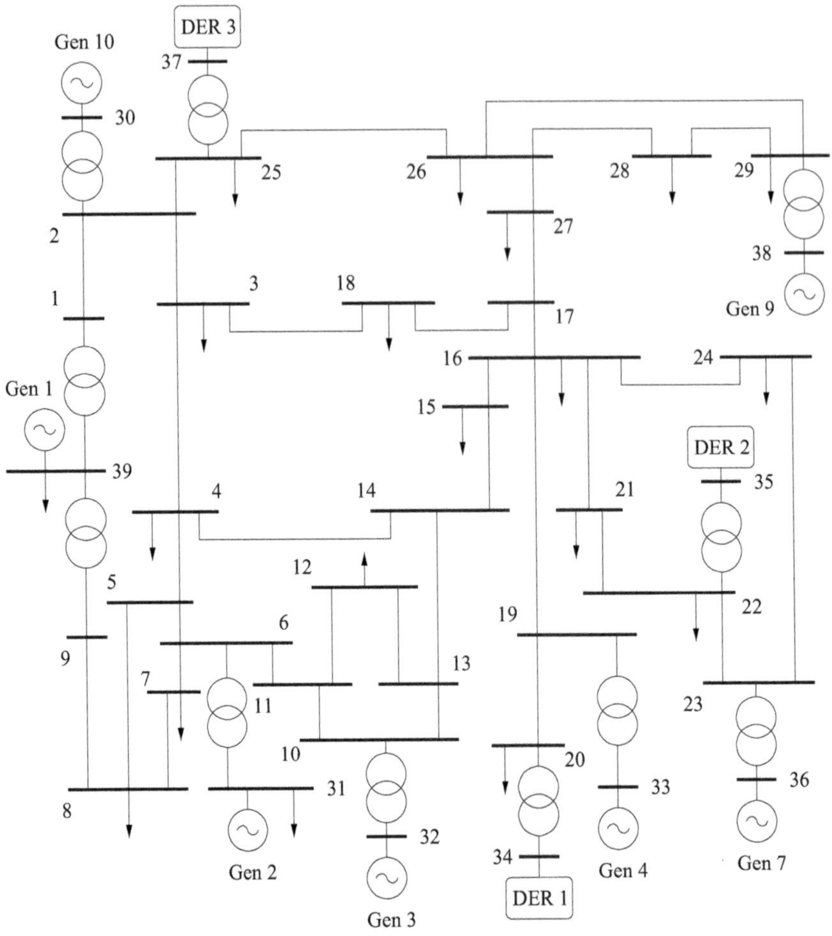

Figure 12.20 Single-line diagram of the modified IEEE 39-bus system

have been determined through a trial-and-error procedure. It is also relevant to note that the controllers employed here provide an acceptable response for a wide range of operating conditions. For example, for the droop constants of the primary frequency control, good results are obtained in the range $\mathcal{R} \in [10^{-2}, 10^{-1}]$.

To guarantee a fair comparison, different control modes are compared keeping constant control parameter settings. Note that, when preparing the case study, we have also tried different approaches, for example we tuned each control mode separately with an aim to achieve the best dynamic response. However, since, as discussed above, the controllers perform well for a relative large range of their parameters, their set up does not modify the main conclusions that are drawn in this section.

12.7.1 FQ and VP control modes

In this section, we consider the FQ and VP controls, which are the components that differentiate the FVP+FVQ control strategy from the classical approach, where frequency and voltage regulation are provided only by means of FP and VQ, respectively (see mode definitions in Section 12.4).

We first examine the FQ mode, i.e., the ability of DERs 1–3 to improve the dynamic response of the system by controlling the frequency through the reactive power. To this aim, the system is simulated for both positive/negative signs of the input control error assuming the tripping of Gen 10 at $t = 1$ s. Results are shown in Figure 12.21 where, for the sake of comparison, we have included the response of the system when DERs (i) do not provide any control and (ii) act based on the classic FP control. Figure 12.21 indicates that the FQ control improves the COI frequency response of the system if utilized with input error $\epsilon_\omega = \omega - \omega^{\text{ref}}$. The main reason for FQ's effectiveness in this case is that the DERs respond to the under-frequency by reducing their reactive power injection and thus the voltage levels at the network. Due to the voltage dependency of loads, the power demand level decreases, thus reducing the imbalance and helping the recovery of the frequency. Note, finally, that the improvement provided by the FQ mode is lower than the one of the classic FP. This result is as expected. Yet, as it will be seen in Section 12.7.5, the benefits of using FQ are more apparent when applied at the DN level.

We show the effect of regulating the voltage at the DER terminal bus through its active power injection, i.e. the VP mode. A simulation is carried out considering the outage of lines 15 and 16 and results are shown in Figure 12.22. The VP mode improves the transient behavior of the voltage when the control input error is $\epsilon_v = v^{\text{ref}} - v_h$. However, as shown in [24], this mode also introduces large deviations in the power sharing among the DERs connected to the system and thus, using it individually is not

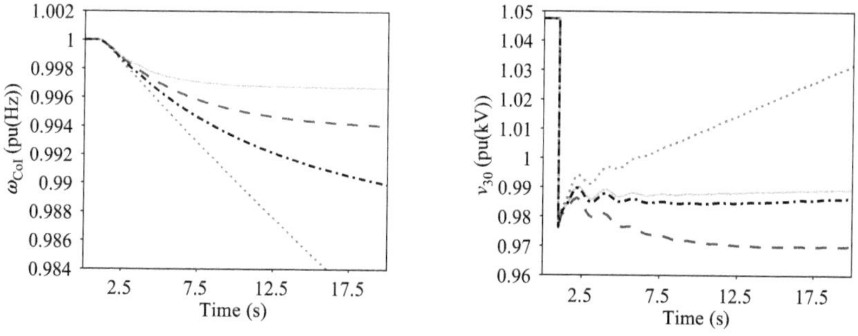

(a) COI frequency (b) Voltage at bus 30

Figure 12.21 Transient response following the loss of Gen 10

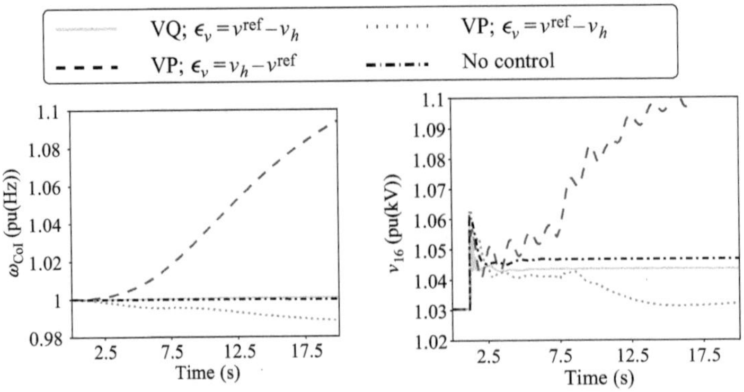

(a) COI frequency (b) Voltage at bus 16

Figure 12.22 Transient response following the outage of lines 15 and 16

suggested. The VP can still contribute to improve the overall system dynamic response if utilized with a relatively small gain and as an auxiliary control that coordinates with other modes.

The following remark on the signs of the control input errors ϵ_ω, ϵ_v is relevant. For the frequency response of the system to improve, ϵ_ω in FQ has to be the opposite from the one utilized in FP. The need for opposite actions when regulating the frequency through the active and reactive power, respectively, can be also observed in the structure of (12.25). On the other hand, to improve the voltage regulation, ϵ_v needs to be implemented with the same sign for both VP and VQ modes. This is again consistent with (12.25), which suggests that regulating the voltage variations requires actions in the same direction for both active and reactive power. Hence, $\epsilon_\omega = \omega - \omega^{ref}$ and $\epsilon_v = v^{ref} - v_h$ are chosen for FQ and VP in this case study.

12.7.2 Performance of FVP+FVQ control

In this subsection, we study the performance of the FVP+FVQ control scheme. This scheme is compared to the classic FP+VQ control, as well as to FP+FQ, VP+VQ, and FP+FVQ. A simulation is carried out considering the disconnection of the load at bus 3 ($P_3 = 3.22$ pu, $Q_3 = 0.024$ pu). The transient behavior of the system following the disturbance is presented in Figure 12.23 where the response of the system with all DER controls disconnected as well as that of the original IEEE 39-bus system serve as references for comparison.

The following remarks are relevant. (i) Compared to the original system, a 30% penetration of DERs worsens the overall dynamic behavior of the system, when these resources provide no restorative control actions. This result is as expected. (ii) The FP+FQ control shows a better frequency response than the classic FP+VQ, yet, it leads to a poor voltage behavior (see Figure 12.23(b)). (iii) Although the VP+VQ scheme shows a very good voltage response, it leads to a poor frequency response. (iv)

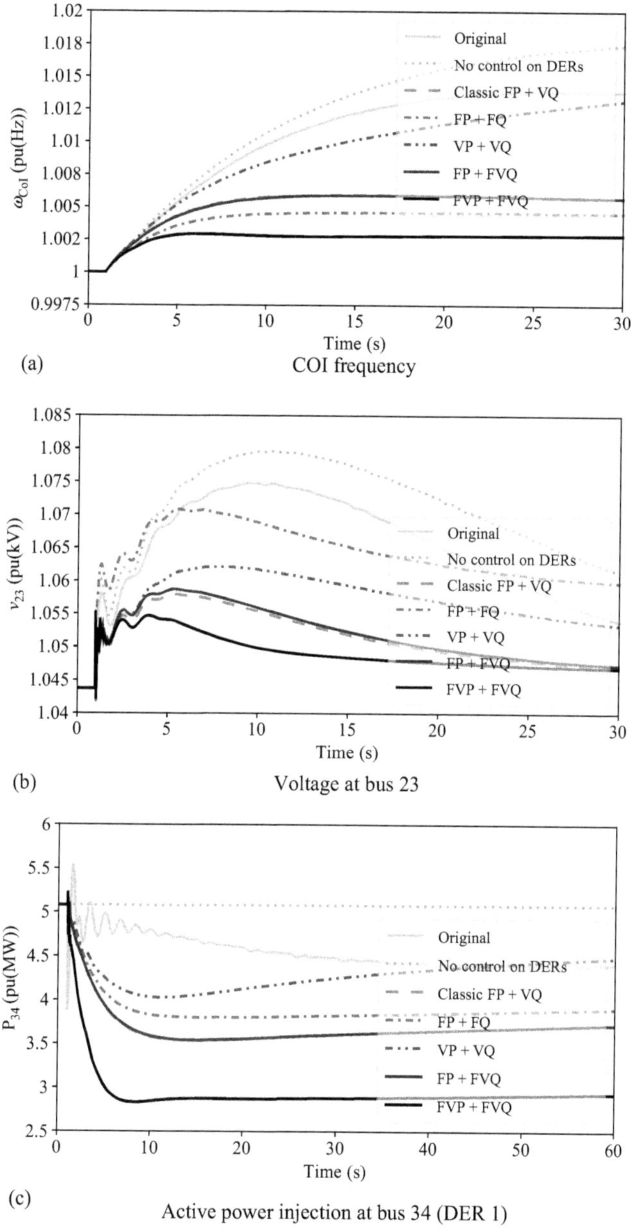

(a) COI frequency

(b) Voltage at bus 23

(c) Active power injection at bus 34 (DER 1)

Figure 12.23 Transient response after disconnection of load at bus 3

Combining the FP+FQ and VP+VQ modes in a single scheme leads to the FVP+FVQ which provides the best frequency and voltage dynamic response among the schemes compared.

To validate the tuning of the parameters of the controllers and build the trust of the adequateness of this tuning for the stability of the overall system, we have assessed the transient behavior of the system for a wide range of operating conditions and disturbance scenarios. With this aim, we have tested the FVP+FVQ control under a variety of disturbances, including generator tripping, line outages, short circuits, and load disconnections. Moreover, we have considered the impact of varying the voltage dependency of loads by considering a constant impedance load model. A summary of the results obtained is presented in Table 12.1, where $\Delta\omega$ refers to the relative variation of the frequency of the COI and Δv refers to the relative variation of a bus voltage magnitude that is local to the disturbance. For each scenario, the table provides the maximum relative variations, as well as the variations few seconds for primary frequency and voltage responses after the disturbance, i.e. at $t = 20$ s of the simulation for the frequency and at $t = 10$ s for the voltage. The smallest frequency/voltage variations obtained for each scenario are marked in bold.

Simulation results suggest that, overall, the FVP+FVQ control leads to an improvement of both primary frequency and voltage regulation of the system. This improvement is significant in case of an outage of synchronous generation, a load switching, or a line trip, while for short circuits, FVP+FVQ performs as the conventional FP+VQ. Finally, note that in contrast to commonly proposed solutions, the performance enhancement provided by FVP+FVQ comes in an inexpensive way, i.e. without the need to install any extra equipment, e.g., storage devices.

12.7.3 Performance of voltage/frequency response metric

We study the accuracy of metric μ_h, defined in Section 12.4.2, to assess the joint voltage/frequency response of DERs. In particular, Table 12.2 shows the value of the metric at bus 34, where DER 1 is connected, at $t = 15$ s and for the same disturbances considered in Table 12.1. The value of μ_{34} for each scenario is calculated using the local voltage and its time derivative (v_{34}, dv_{34}/dt) and the variation of the frequency of the COI ($\Delta\omega_{\mathrm{CoI}}$). Moreover, the results for all control modes are normalized so that the metric for FP+VQ at $t = 15$ s equals to 1. Comparison between Tables 12.2 and 12.1 indicates that the metric μ_h can capture the combined voltage/frequency response with good accuracy. With this regard, recall that smaller values of μ_h imply a better overall dynamic response. It is also worth noting that in the occurrence of a fault at bus 4 and for constant impedance loads, the FP+FQ control shows the worst dynamic response from the metric point of view, although its $\Delta\omega$ and Δv (Table 12.1) are not the worst. In fact, the voltage response for FP+PQ control in this scenario is worse than the other controllers at the first 4 s of the simulation, which is not shown in Table 12.1 and it is not observable unless we check the full time-domain response of both the frequency and the voltage. μ_h captures these effects and, hence, provides an accurate and convenient way to evaluate the joint frequency and voltage response of DERs. In this case study, μ_h is utilized as a tool to assess the performance of DER.

Table 12.1 Frequency/voltage deviations for different contingencies, control modes and load models

Load model	**ZIP**						**Constant impedance**					
Control	**Classic FP+VQ**		**FP+FQ**		**FVP+FVQ**		**Classic FP+VQ**		**FP+FQ**		**FVP+FVQ**	
$\Delta\omega$ (%)	**max**	**at 20 s**	**max**	**at 20 s**	**max**	**at 20 s**	**max**	**at 20 s**	**max**	**at 20 s**	**max**	**at 20 s**
Load 3 out.	0.5814	0.5946	0.4593	0.4593	**0.2934**	**0.2817**	0.5518	0.5516	0.4082	0.4075	**0.2717**	**0.2637**
Load 20 out.	1.1452	1.1318	0.9476	0.9456	**0.5533**	**0.5202**	1.1378	1.1355	0.8986	0.8985	**0.5501**	**0.5246**
Gen 4 out.	−1.1478	−1.1404	−0.7056	−0.7050	**−0.5141**	**−0.4727**	−1.1029	−1.1028	−0.6199	−0.6158	**−0.4946**	**−0.4644**
Gen 7 out.	−1.0663	−1.0495	−0.7422	−0.7348	**−0.5393**	**−0.4826**	−1.0408	−1.0325	−0.6711	−0.6685	**−0.5331**	**−0.4852**
Lines 8 and 9 out.	0.0231	0.0229	0.0180	0.0180	**0.0121**	**0.0118**	0.0804	0.0804	0.0657	0.0655	**0.0537**	**0.0526**
Lines 21 and 22 out.	0.1610	0.1598	0.1544	0.1544	**0.1390**	**0.1354**	0.1994	0.1993	0.1840	0.1837	**0.1612**	**0.1564**
Fault at bus 4	**0.3066**	**−0.0201**	0.3970	−0.0217	0.3505	−0.0238	**0.3543**	0.0294	0.4585	**0.0133**	0.4245	−0.0294
Fault at bus 8	**0.2527**	**0.0042**	0.3182	0.0049	0.2989	0.0059	**0.2865**	0.0276	0.3590	0.0211	0.3257	**0.0150**
Δv (%)	**max**	**at 10 s**	**max**	**at 10 s**	**max**	**at 10 s**	**max**	**at 10 s**	**max**	**at 10 s**	**max**	**at 10 s**
Load 3 out.	1.2867	0.7845	1.7309	1.5417	**1.2533**	**0.6042**	1.2126	0.8275	1.5831	1.4612	**1.1632**	**0.6719**
Load 20 out.	2.6464	1.5891	6.7898	6.7588	**2.6464**	**1.4299**	2.5097	1.5937	6.4399	6.4291	**2.5097**	**1.4550**
Gen 4 out.	−7.4634	−3.5033	−10.319	−10.264	**−7.4634**	**−3.2346**	−6.0130	−3.3522	−8.7384	−8.7016	**−6.0130**	**−3.1234**
Gen 7 out.	−4.8425	−2.9453	−8.0907	−8.0907	**−4.8425**	**−2.8840**	−3.6201	−2.7031	−6.9113	−6.9113	**−3.6201**	**−2.6622**
Lines 8 and 9 out.	−2.8946	−1.9047	−2.8946	−1.9217	**−2.8946**	**−1.9044**	**2.3011**	1.7715	2.3376	1.7594	2.3220	**1.7590**
Lines 21 and 22 out.	−6.2882	−4.2634	−6.6219	**−4.1781**	**−6.2882**	−4.2203	−5.5680	−4.0392	−6.1302	**−3.9151**	**−5.5680**	−4.0117
Fault at bus 4	−100	−2.1525	−100	**−2.1329**	−100	−2.1749	−100	−2.0802	−100	**−2.0603**	−100	−2.1052
Fault at bus 8	−100	−1.3188	−100	**−1.3138**	−100	−1.3244	−100	−1.2711	−100	**−1.2645**	−100	−1.2762

Table 12.2 Metric μ_{34} (DER 1)

Load model	ZIP			Constant impedance		
Control	**FP+VQ**	**FP+FQ**	**FVP+FVQ**	**FP+VQ**	**FP+FQ**	**FVP+FVQ**
μ_{34}	**at 15 s**	**at 15 s**	**at 15 s**	**at 15 s**	**at 15 s**	**at 15 s**
Load 3 out.	1	0.7891	**0.5467**	1	0.7543	**0.5528**
Load 20 out.	1	0.8283	**0.5327**	1	0.8029	**0.5384**
Gen 4 out.	1	0.6278	**0.4953**	1	0.5816	**0.5187**
Gen 7 out.	1	0.7171	**0.5511**	1	0.6857	**0.5665**
Lines 8 and 9 out.	1	0.8671	**0.7708**	1	0.8192	**0.7315**
Lines 21 and 22 out.	1	0.9544	**0.9101**	1	0.9155	**0.8602**
Fault at bus 4	**1**	1.5189	1.1792	**1**	3.4706	1.5001
Fault at bus 8	**1**	1.5611	1.2027	**1**	1.9155	1.0742

12.7.4 Application to aggregated power generation

This scenario assumes that the converter-based resources connected to buses 34, 35, and 37 consist of several smaller DERs, the power generation and control modes of which are coordinated through a power aggregation mechanism. This mechanism can be implemented in practice as a virtual power plant (VPP) [25,26]. For the purpose of this study, we consider that a varying percentage of the DERs that compose the VPP utilize the FVP+FVQ scheme, and the rest act based on the classic FP+VQ control.

Figure 12.24 shows the results for two disturbances, (a) outage of the lines 21 and 22 and (b) disconnection of the load connected to bus 20. Results are compared by means of the joint frequency/voltage response metric at bus 34 (μ_{34}), where DER 1 is connected. The metric is calculated as discussed in Section 12.4.2. As expected, the FVP+FVQ mode has the best performance for both disturbances, which confirms the results shown in Table 12.1. Interestingly, the classic FP+VQ mode combined with 20–40% FVP+FVQ control worsens the transient response for disturbance (a). We conclude that, depending on practical requirements, the VPP operator can design its assets to apply and/or switch between different control modes.

12.7.5 Impact of resistance/reactance line ratio

We study the performance of the examined control when applied to DERs integrated within DNs. To study the DN effect, the R/X ratios of the feeders that connect the DERs to the system are altered so that $R/X \approx 1$. The evolution of μ_{34} for different control modes is presented in Figure 12.25, where we have considered the loss of Gen 10. Moreover, the results are normalized so that for FP+VQ μ_{34} equals to 1 at $t = 30$ s when $R/X \ll 1$. When $R \approx X$, all DER control modes have a better performance. It is interesting to observe that for $R \approx X$, VP+VQ has the largest improvement among the examined modes. The same effect can also be observed under

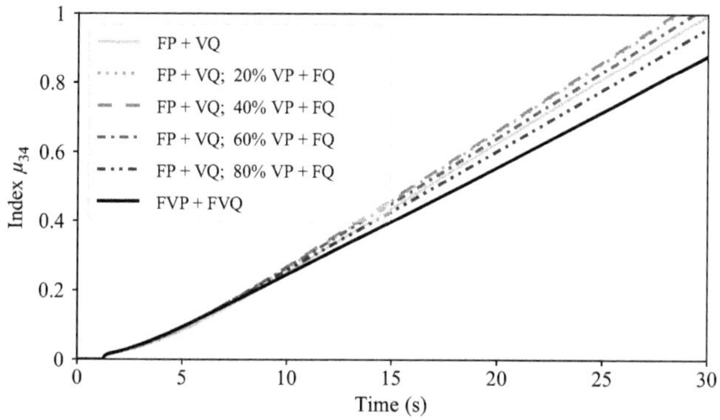

(a) Transient response following the outage of lines 21 and 22

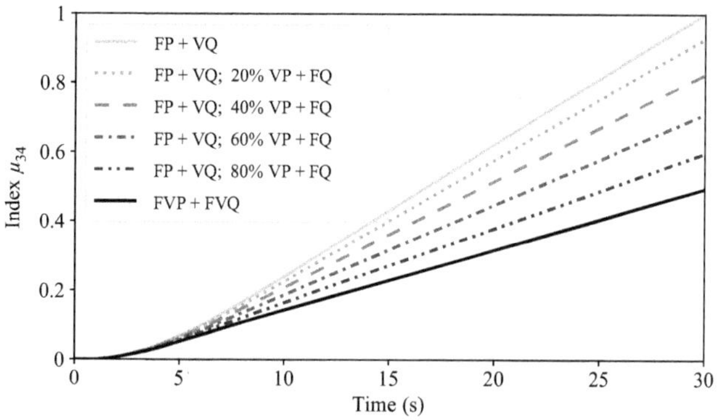

(b) Transient response following the load disconnection at bus 20

Figure 12.24 μ_{34} *(DER 1) for FVP+FVQ applied to a portion of VPP assets*

different disturbance scenarios in this test system. Finally, the FVP+FVQ control shows the best overall dynamic response among the modes compared.

12.7.6 Impact of DER penetration level

In this scenario, we study the impact of the share of DERs to the total generation mix of the system on the performance of the examined control scheme. To this aim, and in addition to the DERs at buses 34, 35, and 37, DERs are also connected to buses 36 and 38, by replacing the local SMs. As a consequence, the penetration of DERs to the modified IEEE 39 bus system increases to 50%. A time-domain simulation of the system is carried out by applying the loss of Gen 10 at $t = 1$ s and results are presented in Figure 12.26. As it can be seen, increasing the DER penetration from 30% to 50%, although it leads to a worse voltage response, it does not deteriorate the frequency regulation of the system, which interestingly, slightly improves. Compared to the classic FP+VQ, the FP+FQ outperforms in terms of frequency, but leads to a poor voltage response. As expected, the dual effect holds when the VP+VQ scheme is

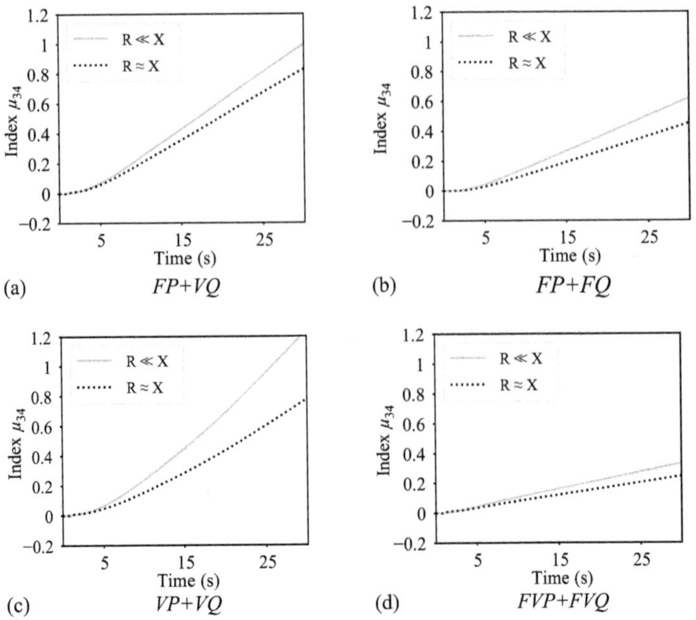

Figure 12.25 Transient response following the loss of Gen 10

applied. Most importantly, the FVP+FVQ control leads to the best dynamic behavior among the examined control modes.

12.7.7 Impact of system granularity

In this section, we focus on the effect of the system's granularity and further evaluate the examined control when employed for resources connected to the DN level. To this aim, a more detailed modeling of the DN and loads is considered. In particular, each of the SMs at buses 32–38 and loads at neighbor buses is substituted with the 8-bus, 38 kV DN shown in Figure 12.27 [27] (note that, for illustration, in Figure 12.27, only one DN is shown). As a byproduct, the instantaneous power generation by DERs is increased to 70%. The behavior of loads in this example is represented using the dynamic load model proposed in [28]. Moreover, to account for the proximity of loads, for potential imbalances, as well as for possible harmonics of the power converters, noise has been added on the voltage angle at every DN bus. Noise is modeled as an Ornstein-Uhlenbeck's process with Gaussian distribution [29].

We carry out a time-domain simulation considering the disconnection of the load at bus 3. Comparison of the FP+VQ and FVP+FVQ modes is presented in Figure 12.28 and indicates that FVP+FVQ leads to a better dynamic behavior. Note that the control considered is in general expected to be more effective and thus lead to larger improvement of the system's response, the higher is the coupling between

(a)   COI frequency, 30% DERs
(b)   COI frequency, 50% DERs
(c)   Bus 30 voltage, 30% DERs
(d)   Bus 30 voltage, 50% DERs

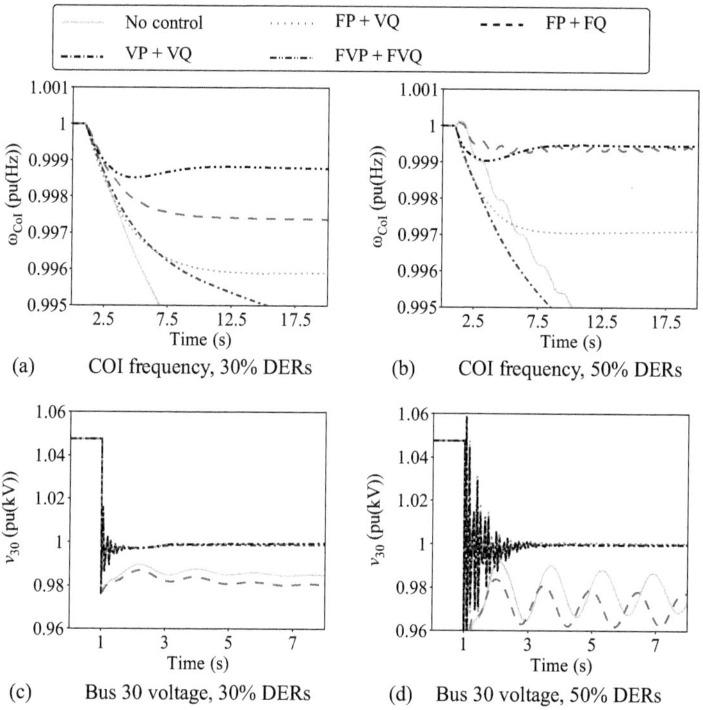

Figure 12.26   Transient response following the loss of Gen 10

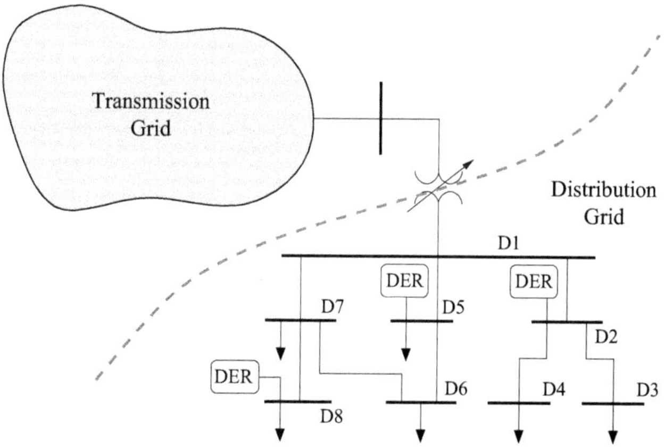

Figure 12.27   Topology of DN model used in Section 12.7.7

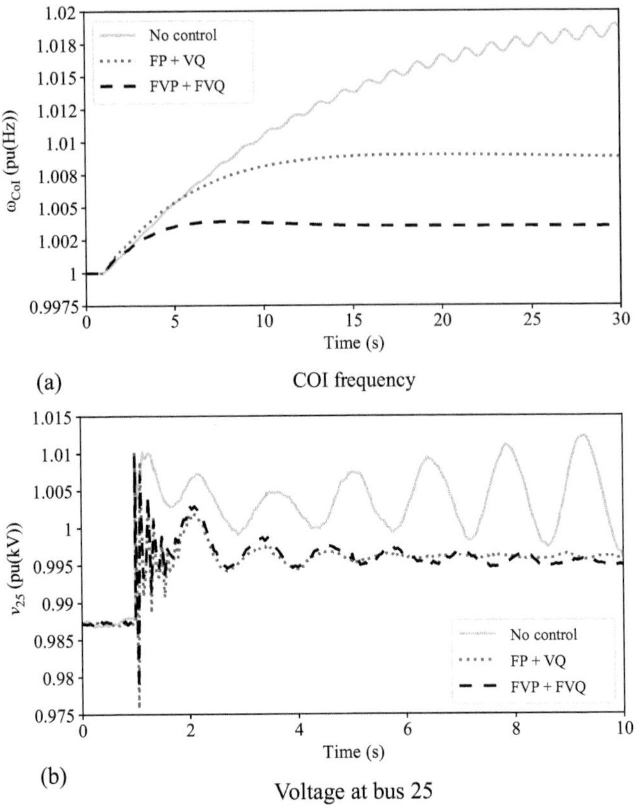

Figure 12.28 Transient response after disconnection of load at bus 3

the active and reactive power flows, i.e., at lower voltage levels and DN applications. This is confirmed by Figure 12.28, when compared to results discussed in previous sections of this case study, for example with Figure 12.23.

Bibliography

[1] Milano F, Dörfler F, Hug G, Hill DJ, Verbič G. Foundations and challenges of low-inertia systems. In: *Power Systems Computation Conference (PSCC)*, Dublin, Ireland, 2018, pp. 1–25.

[2] Mauricio JM, Marano A, Gómez-Expósito A, Martínez Ramos JL. Frequency regulation contribution through variable-speed wind energy conversion systems. *IEEE Transactions on Power Systems*, 2009;24(1):173–180.

[3] Kundur P. *Power System Stability and Control*. New York, NY: McGraw-Hill, 1994.

[4] Guerrero JM, de Vicuna LG, Matas J, Castilla M, Miret J. Output impedance design of parallel-connected UPS inverters with wireless load-sharing control. *IEEE Transactions on Industrial Electronics*, 2005;52(4): 1126–1135.

[5] Li YW, Kao C. An accurate power control strategy for power-electronics interfaced distributed generation units operating in a low-voltage multibus microgrid. *IEEE Transactions on Power Electronics*, 2009;24(12):2977–2988.

[6] Delille G, Yuan J, Capely L. Taking advantage of load voltage sensitivity to stabilize power system frequency. In: *2013 IEEE Grenoble Conference, IEEE*, 2013, pp. 1–6.

[7] Farrokhabadi M, Cañizares CA, Bhattacharya K. Frequency control in isolated/islanded microgrids through voltage regulation. *IEEE Transactions on Smart Grid*, 2017;8(3):1185–1194.

[8] De Carne G, Buticchi G, Liserre M, Vournas C. Load control using sensitivity identification by means of smart transformer. *IEEE Transactions on Smart Grid*, 2018 July;9(4):2606–2615.

[9] Murad MAA, Tzounas G, Liu M, Milano F. Frequency control through voltage regulation of power system using SVC devices. In: *Proceedings of the IEEE PES General Meeting*, 2019, pp. 1–5.

[10] Milano F. A Python-based software tool for power system analysis. In: *Proceedings of the IEEE PES General Meeting*, 2013, pp. 1–5.

[11] Milano F. *Power System Modelling and Scripting*. London: Springer, 2010.

[12] Ballanti A, Ochoa LN, Bailey K, Cox S. Unlocking new sources of flexibility: CLASS: the world's largest voltage-led load-management project. *IEEE Power and Energy Magazine*, 2017;15(3):52–63.

[13] Collin AJ, Tsagarakis G, Kiprakis AE, McLaughlin S. Development of low-voltage load models for the residential load sector. *IEEE Transactions on Power Systems*, 2014;29(5):2180–2188.

[14] Ortega Á, Milano F. Estimation of voltage dependent load models through power and frequency measurements. *IEEE Transactions on Power Systems*, 2020;35(4):3308–3311.

[15] De Carne G, Liserre M, Vournas C. On-line load sensitivity identification in LV distribution grids. *IEEE Transactions on Power Systems*, 2016;32(2): 1570–1571.

[16] Milano F. Complex frequency. *IEEE Transactions on Power Systems*, 2021;36:1–1.

[17] Murad MAA, Ortega Á, Milano F. Impact on power system dynamics of PI control limiters of VSC-based devices. In: *2018 Power Systems Computation Conference (PSCC)*, 2018, pp. 1–7.

[18] Arrillaga J, Watson NR. 4. In: *Computer Modelling of Electrical Power Systems*, New York, NY: John Wiley & Sons, 2001, pp. 81–128.

[19] Anderson PM, Fouad AA. *Power System Control and Stability*, New York, NY: Wiley-IEEE Press, 2003.

[20] Egan J, O'Rourke P, Sellick R, Tomlinson P, Johnson B, Svensson S. Overview of the 500MW EirGrid East-West Interconnector, considering System Design

and execution-phase issues. In: *Universities Power Engineering Conference (UPEC)*, 2013, pp. 1–6.
[21] Tzounas G, Milano F. Improving the frequency response of DERs through voltage feedback. In: *Proceedings of the IEEE PES General Meeting*, 2021, pp. 1–5.
[22] Illinois Center for a Smarter Electric Grid (ICSEG). *IEEE 39-Bus System*; URL: http://publish.illinois.edu/smartergrid/ieee-39-bus-system/.
[23] Milanovic JV, Yamashita K, Villanueva SM, Djokic SŽ, Korunović LM. International industry practice on power system load modeling. *IEEE Transactions on Power Systems*, 2012;28(3):3038–3046.
[24] Zhong W, Tzounas G, Milano F. Improving the power system dynamic response through a combined voltage-frequency control of distributed energy resources. *IEEE Transactions on Power Systems*, published in Feb. 2022 (Early Access), available at https://doi.org/10.1109/TPWRS.2022.3148243.
[25] Zhong W, Murad MAA, Liu M, Milano F. Impact of virtual power plants on power system short-term transient response. *Electric Power Systems Research*, 2020;189:106609.
[26] Zhong W, Chen J, Liu M, Murad MAA, Milano F. Coordinated control of virtual power plants to improve power system short-term dynamics. *Energies*, 2021;14(4):1182.
[27] Murphy C, Keane A. Local and remote estimations using fitted polynomials in distribution systems. *IEEE Transactions on Power Systems*, 2016;32(4):3185–3194.
[28] Jimma K, Tomac A, Vu K, Liu C. A study of dynamic load models for voltage collapse analysis. In: *Proceedings of the Bulk Power System Voltage Phenomena, Voltage Stability and Security NFS Workshop*, Deep Creek Lake, MD, August 1991.
[29] Milano F, Zárate-Miñano R. A systematic method to model power systems as stochastic differential algebraic equations. *IEEE Transactions on Power Systems*, 2013;28(4):4537–4544.

Chapter 13

Smart transformer control of the electrical grid

Giovanni De Carne[1], Marco Liserre[2] and Felix Wald[1]

This chapter introduces the smart transformer (ST) concept, a power electronics-based transformer that, in addition to the voltage transformation, can offer enhancing services to the grid. The chapter begins with a brief introduction on the ST concept and gives an overview of the offered services. In the second section, the ST architecture and control for each transformation stage are described, analyzing different topology alternatives. Basic and more advanced services provided to the grid are described in the third section that includes the innovative concept to regulate the grids power consumption by means of controlled voltage or frequency variations. The chapter closes with an overview of innovative grid concepts, such as DC grids.

13.1 The smart transformer concept

The first idea of a power electronics transformer (PET) can be found in 1968, when W. Mc Murray patented an idea of a DC transformer having a high frequency link [1]. The idea referred to the possibility to control the DC voltage under different amplitude outputs by means of solid-state switches and electronic controls. In the 1980*s*, Brook proposed an innovative transformer [2] where a "voltage shaping" capability of the DC transformer has been considered, moving the focus from the hardware configuration to the possible services that the PET could provide.

Despite an increasing attention to the topic, larger PET prototypes have only been realized since the beginning of the 2000′. The large-scale application of insulated-gate bipolar transistor (IGBT) first, and silicon carbide-/galium nitride-based Mosfets then allowed to reduce the PET's volume and weight, while increasing its efficiency. First examples of PET transformers were realized for traction applications, with the goal to increase the efficiency of the transformation stage, while reducing volume and weight [3,4]. Despite technically successful (the desired goals were achieved), the system was never put on the market, due to the restricted range of possible business cases.

[1]Institute for Technical Physics, Karlsruhe Institute of Technology, Germany
[2]Chair of Power Electronics, Kiel University, Germany

Only in recent years, the PET has been considered as an alternative for conventional transformers in medium-voltage (MV)/low-voltage (LV) substations. The need for power scalability, higher grid controllability, and availability of DC ports made the PET an interesting solution for grid operators. Several academic partners began to develop first prototypes for grid applications, such as the FREEDM center in US [5], focusing on demonstrating the PET performance in terms of efficiency and power routing capability for AC and DC microgrid applications [6].

Starting from the PET's advanced features, the HEART project, carried out at the Kiel University in Germany, has developed the ST concept [7,8]. The HEART project aims at demonstrating the advanced functionalities and the services that a ST can offer to the distribution grid.

To identify which possible services a ST is able to provide to the distribution grid, three different grid levels can be considered, all of them are shown in Figure 13.1: MV AC grid, LV AC grid, and DC grid (assumed as future grid scenario).

The MV AC/DC converter works as front-end converter for the MV grid, controlling the active power flow in order to keep the MV DC link voltage stable. As additional degree of freedom, the ST is able to regulate the injection of reactive

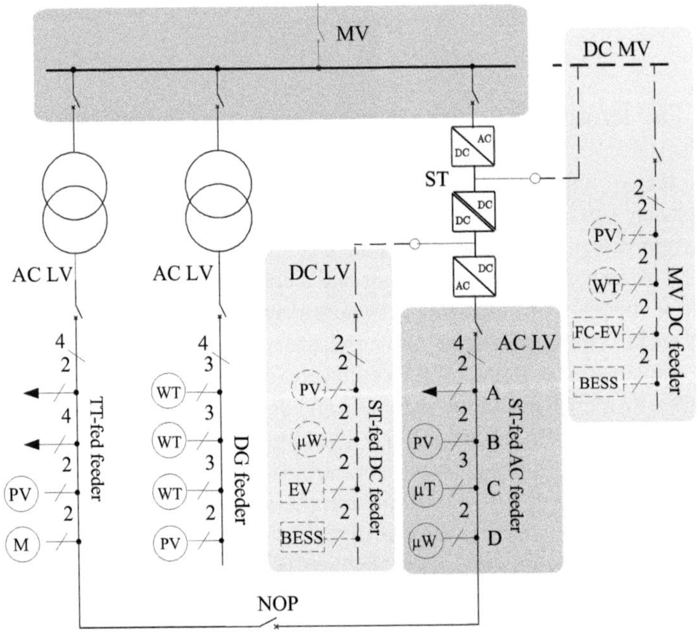

Figure 13.1 ST concept to provide services in distribution grids: load identification and control in LV grids (green area), active and reactive power support in MV grids (red area), enabled DC connectivity in MV and LV side (blue area)

power in the MV grid, and, at the same time, offer harmonic compensation capability. It can provide reactive power compensation services at the MV busbar level, offer voltage regulation support in the MV grid, and compensate the current and voltage harmonic content coming from large industrial loads (Figure 13.1 red box).

As mentioned above, the ST provides a first interface for AC grids with future DC grids (Figure 13.1, blue box). Depending on the architectures, this connection can be realized at both LV and MV levels. If realized at MV DC level, this connection can create a new concept of regional distribution systems, enabling the connection of MW-scale DC loads (e.g., electrolyzers, fast charging electric vehicle stations) or generators (e.g., photovoltaic, wind parks, and large size energy storage systems). In LV DC, small generators and loads, like electric vehicles and batteries, can be hosted. Due to a reduction in the power conversion AC/DC stages, the system losses can be reduced.

At the LV AC side, the ST works as stiff grid-forming converter, controlling the voltage waveform in the fed grid (Figure 13.1, green box). Due to this feature, the ST can offer both basic and advanced services: as an example of basic service, the ST shall provide always a three-phase balanced voltage independently from the current waveforms, that can be unbalanced or affected by harmonic content. An advanced service, that we are going to discuss in this chapter, is the capability to modify the power consumption of voltage- and frequency-dependent loads, acting on the voltage and frequency magnitude. This feature makes a variety of services possible, that are not available with conventional solutions (e.g., iron transformer): the on-line identification of the load power sensitivity to voltage and frequency variations; the voltage- and frequency-based load control to shape the power consumption, upwards and downwards, acting on a controlled variation of the voltage and frequency and on the load sensitivity to these variables; the provision of primary frequency regulation support, by adapting the load consumption depending on the measured MV frequency.

13.2 ST architectures and control

The ST interfaces three different grid topologies that vary on voltage levels as well as on supply system, either AC or DC (Figure 13.2): a MV AC grid, a LV AC grid, and two optional DC grids, both at MV and LV level (assumed as futuristic scenario). As a consequence, the ST three stage operations are interlaced: the MV converter controls the MV AC active current to regulate the MV DC link voltage at the nominal value. It can regulate the reactive power injection independently from the active power, remaining in the converter ampacity limits. The DC/DC converter transfers the power from the MV to the LV DC link in order to control the voltage in the LV DC link. The LV side converter controls the AC voltage waveform to be sinusoidal and balanced, and, upon request, it can shape the load consumption acting on voltage parameters (e.g., magnitude and frequency).

In the following sections, the topology and control of each stage has been described in detail, in order to give the reader a clear overview of the STs' features.

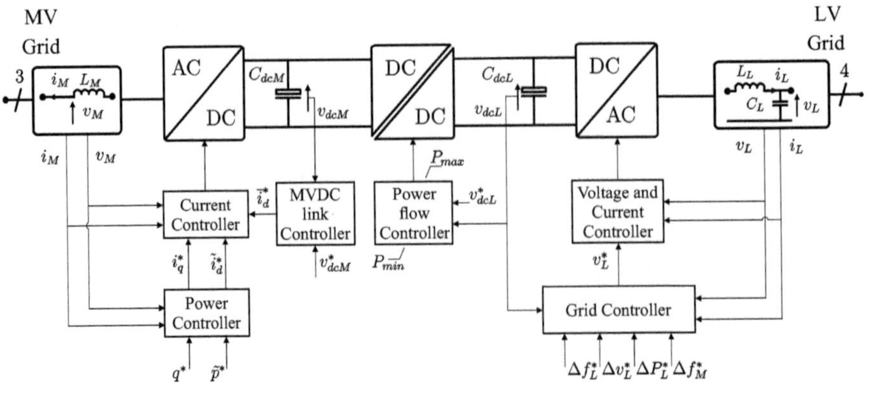

Figure 13.2 ST control overview

13.2.1 MV AC/DC converter

The ST MV stage interfaces the MV DC link to the MV AC grid. Depending on the countries, this stage works at relatively high voltage (> 10 kV), and due to the limited power for MV applications, at low current (< 100 A). This requirement led to investigate different topologies that can respect this high-voltage/low-current need. For these reasons, multi-level solutions have been considered for ST applications, and among the existing ones, good candidates are the one shown in Figure 13.3:

(a) *Neutral point clamped (NPC):* simple multi-level solution, largely adopted in industry, that provides DC-link access in the middle-point. As main drawback, the reduced number of levels implies large output filters and relatively high switching frequencies, increasing the losses.
(b) *Cascaded half-bridge (CHB):* modular design, low switching frequency, and low control complexity make the CHB a valid solution for ST. However, this topology does not provide direct access to the MV DC link (like the MMC) and it requires isolated DC supplies for each cell.
(c) *Modular multi-level converter (MMC):* the performance of the MMC is similar to the one of a CHB, with the addition of the availability of a MV DC link connection. On the down side, the MMC requires a more complex control structure and bulky DC capacitors.

At the MV stage, the ST control consists of two layers (Figure 13.2): in the first one, the MV DC link voltage and the AC current are regulated, while in the second one, grid services can be provided by means of a power controller (e.g., voltage support, power factor compensation, and active filtering). In the first layer, the MV DC link controller compiles the active current reference i^*_{dM} for the current controller (Figure 13.4), which is needed to maintain the DC voltage v_{dcM} at the nominal value. The ST current controller, absorbing the desired AC active current, is able to keep the DC link voltage constant, complying with the LV-side load energy request plus

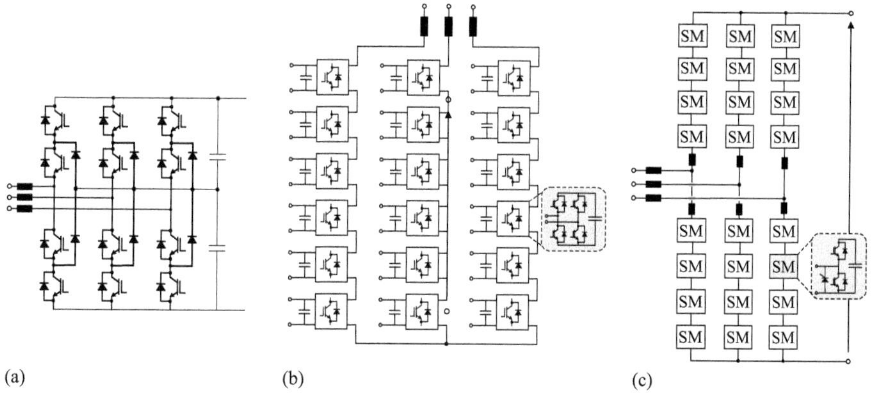

Figure 13.3 Common MV AC/DC converter topologies for a ST: (a) NPC converter, (b) CHB, and (c) MMC

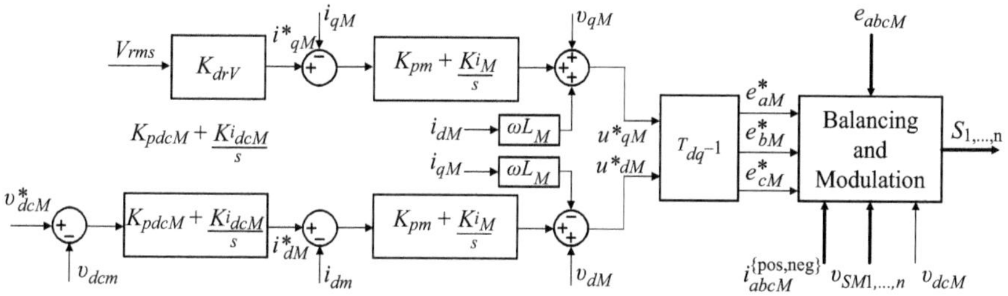

Figure 13.4 Generic ST MV AC/DC converter control structure in dq rotating framework

the ST losses. As shown in Figure 13.4, the power controller can provide both active and reactive power. As an example of this capability, the outer power loop controller can deliver reactive current (i_{qM}) depending on an external signal (e.g., the MV grid *rms* voltage V_{rms}) and the implemented control law (in this example a simple droop characteristic K_{drV}).

The internal control loop regulates the output current, minimizing the error between the current references coming from the power controllers i^*_{dM} and i^*_{qM}, and the measured grid currents i_{dM} and i_{qM}, respectively. The control output is then sent to a modulation block, which will create the new switching signals for the power semiconductors $S_{1,\dots,N}$. Depending on the chosen topology (e.g., for CHB or MMC), additional balancing algorithms may be inserted in series or parallel to this control structure.

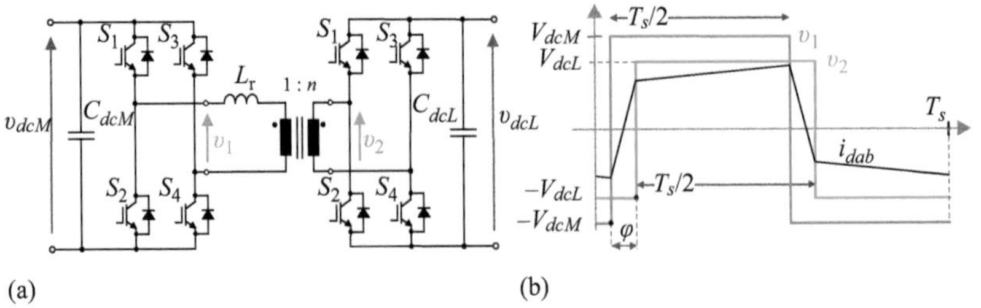

Figure 13.5 (a) Schematic circuit of a DAB and (b) voltages and currents at the primary and secondary side of DAB windings

13.2.2 DC/DC conversion stage

The ST can play a role in the future grid as enabler of DC distribution grids (Figure 13.1, blue box). The presence of a double DC link, at LV and MV level, offers a natural connection point for DC technologies. MV DC grids can in the future enable the direct connection of MW-scale DC loads or generators, like PV and wind parks, large size BESS and fast charging electric vehicle (FCEV) stations, without the need of additional AC/DC conversion stages. At LV level, DC grids provide a valid alternative to AC grids to integrate smaller generators and loads, like electric vehicles (EV) and battery energy storage systems (BESS), reducing the conversion losses caused by the additional AC/DC stages at the user side. These two DC stages are interfaced and controlled by means of a DC/DC converter that shall offer galvanic insulation, voltage transformation, and power flow controllability. The DC/DC converter technologies currently investigated for ST applications are the dual (or Multiple) active bridge (DAB) and series resonant converter (SRC).

(a) *DAB*: The DAB technology (Figure 13.5) regulates the power flow P_{dabL} between primary and secondary windings, acting on the primary v_{dcM} and secondary v_{dcL} voltage waveform phase shift ratio d:

$$P_{dabL} = \frac{v_{dcL} v_{dcM}}{2N f_s L_r} d\,(1-d) \tag{13.1}$$

where N is the transformer turning ration, f_s is the DAB switching frequency, and L_r is the leakage inductance of the transformer. It follows that the power flow controller regulates the LV DC link voltage v_{dcL} at the nominal value, by controlling the power flow between the MV and LV DC link. The main advantage of this topology is the possibility to directly control the power flow and perform control strategies on the output voltage. As main drawback, the control complexity and losses are increased with respected to the SRC solution.

(b) *Series resonant converter (SRC)*: The SRC (Figure 13.6(a)) is an open-loop controlled DC/DC converter, that is based on a resonant tank for transferring energy

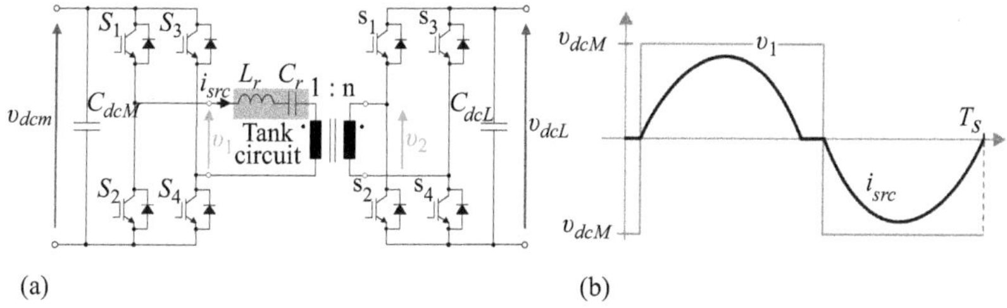

Figure 13.6 (a) Schematic circuit of an SRC and (b) voltages and currents at the primary side of an SRC

between the primary and the secondary winding. Effectively, the SRC works as a high-efficiency ideal DC transformer, transforming the DC voltage in open loop (i.e., without voltage feedback), as shown in Figure 13.6(b). The main advantage is the simplicity of the control solution and a lower need for measurements on the secondary side. On the other hand, it does not allow a power flow control, and for some topologies, bi-directional power flow.

13.2.3 LV DC/AC converter

The ST LV DC/AC stage synthesizes a balanced voltage waveform, working as grid forming converter for the LV grid. With respect to classical grid-forming converters for electric drive applications, the ST shall offer a neutral conductor return path, due to the presence of single-phase loads in the LV grid.

In order to offer the neutral conductor return path, several topologies, shown in Figure 13.7, have been considered for the ST LV side [9]:

(a) *NPC*: As described in the MV converter solutions, the NPC offers a multi-level solution, with the availability of the neutral path connection to the DC link mid-point.
(b) *Two-level Four-leg converter*: The classical two-level converter is the most common solution for LV applications, due to its simple control, low conduction losses, and low number of required switches. The main drawback is the need for a higher switching frequency in order to achieve a clean current output waveform.
(c) *Three-level T-type converter*: Offers the same control simplicity of the two-level converter, while it achieves an improved quality of voltage and low switching losses, typical of an NPC converter.

The voltage control in the ST grid is achieved by means of a cascaded voltage and current control loop. The control can be performed in static *abc*, $\alpha\beta$, or rotating *dq*0 frames. The advantage of using the *dq*0 rotating frame is the reduction of the control equations from 3 to 2 sets. However, this advantage is minimized in this case due to the presence of the neutral conductor connection that increases the number of independent phases. A possible control scheme is shown in Figure 13.8, where

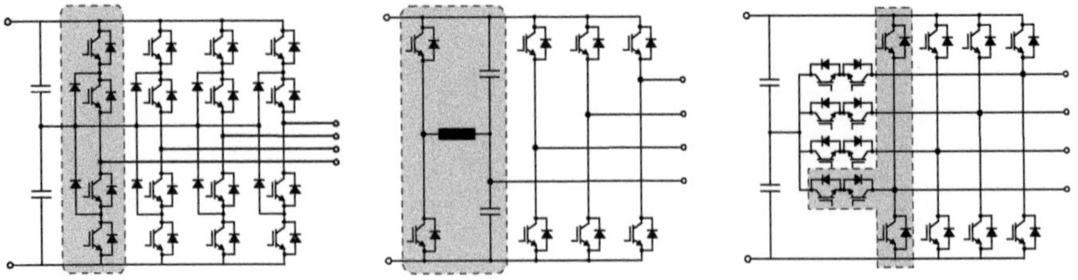

Figure 13.7 ST LV converter topologies: (a) NPC converter, (b) two-level four-leg converter, and (c) three-level T-type converter

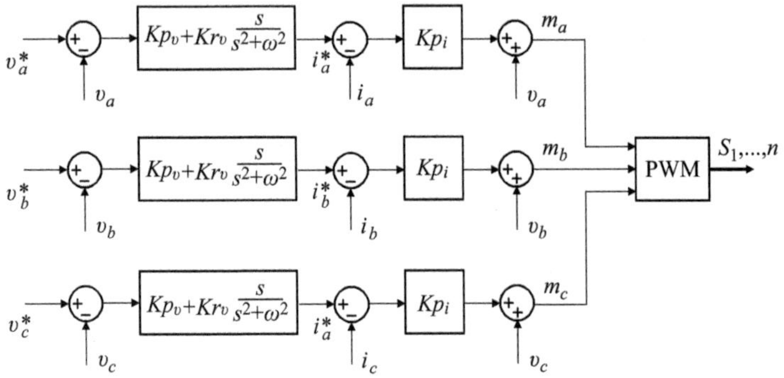

Figure 13.8 ST LV DC/AC converter controller scheme, example in an abc static reference frame

the voltage and current control loops are developed in *abc* static frame by means of proportional(-resonant) controllers. The voltage reference of each phase (e.g., v_a^*) is compared to the measured voltage, and their mismatch minimized, setting the new current reference i_a^*. An internal current loop is added normally, either with a simple gain (e.g., Kp_i in Figure 13.8) or with a resonant controller, to offer additional current damping.

13.3 Services provision to AC grids

A main property of the ST is the AC power flow decoupling between the MV and LV grids, thanks to the presence of one or more DC stages. This decoupling allows to control the two grids independently, with the only connection link represented by their energy exchange.

This section provides an overview of the services that the ST can offer to the AC grids thanks to the AC power flow decoupling property.

13.3.1 Disturbance rejection

Conventional transformers are characterized by a passive transformation of the voltage, independently from the power quality status. If a temporary LV condition or a voltage swell occurs in the MV grid, it is transferred to the LV grid. At the same time, if the LV grid load shows unbalanced or harmonic current absorption, its impact is seen also at the MV side. Thanks to the AC power flow decoupling capability, the ST is able to reject voltage or current disturbances from both primary and secondary sides, avoiding to transfer it to the other grid.

In Figure 13.9, several common disturbances at both low and medium voltage side have been considered, and the capacity of the ST of rejecting them has been depicted:

(a) *Voltage sag in MV grid:* during faults it may occur that the voltage drops below the normal operative conditions. In this case, the ST avoids to transfer the disturbance to the secondary side. However, as can be seen in the first column of Figure 13.9, the MV DC-link voltage drops sharply after the voltage drop. This is due to the need to deliver the required energy in the LV grid, while the available power in MV grid is reduced. After few cycles, the DC-link controller is able to restore the voltage at the nominal value.

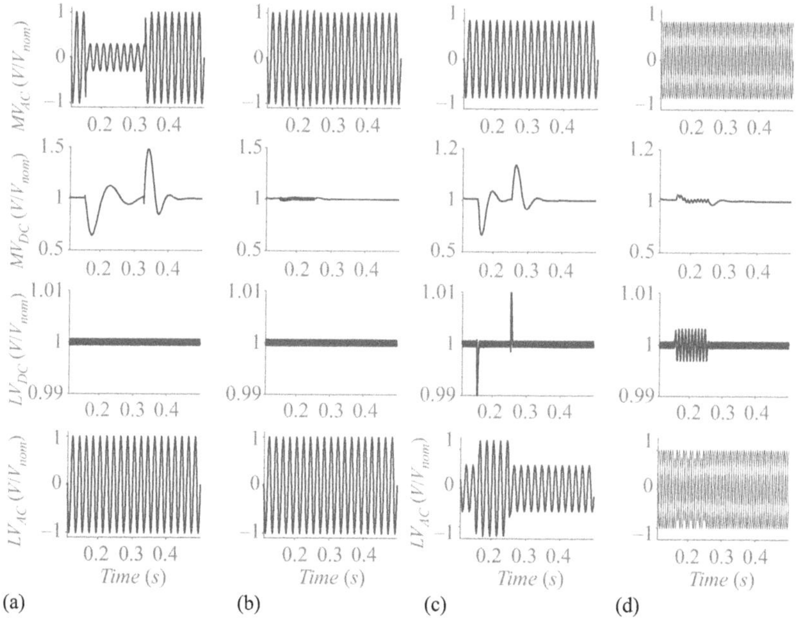

Figure 13.9 Voltages and/or currents of different stages for (a) a voltage sag on MV side, (b) harmonics on MV side, (c) load step on LV side, and (d) unbalanced loads on LV side

(b) *Harmonic voltage in MV grid:* large industrial loads can inject high harmonic currents, that can affect the grid voltage. In the second column of Figure 13.9, a 5th harmonic voltage of 5% has been inserted in the voltage waveform. As can be noticed, the LV grid voltage is perfectly sinusoidal, and only a fourth-order voltage ripple can be seen on the MV DC-link.
(c) *Load increase in LV grid:* fast power variations are common in LV grids, due to the sudden power decrease (e.g., after local faults, cloud passing on photovoltaic plants), or increase (e.g., during fast charging of electric vehicles). The ST is able to quickly damp the fast power change, leaving the MV grid voltage unaffected. As can be seen in the third column of Figure 13.9, the energy to cope with the sudden power increase is provided by the MV DC-link, while its controller restores the DC-link voltage after several tens of milliseconds.
(d) *Unbalanced load in LV grid:* due to the presence of single-phase loads, the LV grid power demand can be unbalanced. A conventional transformer tends to transfer such a unbalanced condition to the primary side, while the ST is able to appear as a perfectly balanced load at the primary side, as shown in the fourth column of Figure 13.9. As only effect, a second-order voltage oscillation can be noted in both MV and LV DC link, due to the 100 Hz power component in unbalanced load conditions.

13.3.2 Load sensitivity identification

Any control action involving the voltage or frequency [10,11], without a proper estimation of the loads active and reactive power response to voltage and frequency variations, can lead to incorrect control actions. A typical example of such a controller is the conservation voltage reduction (CVR) method [12,13]. Operating on the transformer tap-changer, the load power consumption can be decreased, achieving a relatively low-cost measure for energy-saving purposes.

In industrial practice, the load sensitivities to voltage and frequency variations are assumed *a priori*, using historical, statistical, or technical analysis, but they are not estimated in real time. If the real sensitivity changes during operation, any voltage- or frequency-based corrective action could under-perform.

To overcome this limitation, the on-line load identification approach has been developed in [14]. The approach applies an intentional small perturbation in voltage or frequency magnitude, while measuring the resulting load power variation. This enables the estimation of the load sensitivity in real time. Since only a small variation is necessary (i.e., about 1–2 %) the experiment can be repeated as often as deemed necessary, e.g., every hour or every 10 min.

The load sensitivity to voltage and frequency expresses the variation in power consumption during a voltage amplitude and frequency change in the grid. The load power consumption depends on four parameters: voltage amplitude, frequency, time, and the initial operating point. The load power consumption can be mathematically generalized as follows:

$$\begin{aligned} P &= P(V, f, t, P_0) \\ Q &= Q(V, f, t, Q_0) \end{aligned} \tag{13.2}$$

where V and f are the voltage amplitude and the frequency, respectively; t represents the power dependency on the time, due to the capability of certain loads to restore their power in time (neglected in this analysis for matter of simplicity); P_0 and Q_0 are the initial operating conditions.

To represent the relation between voltage, frequency and measured power mathematically, the exponential model, described in (13.3), has been chosen:

$$P = P_0\left(\frac{V}{V_0}\right)^{K_p}\left(1 + K_{fp}\frac{f - f_0}{f_0}\right)$$
$$Q = Q_0\left(\frac{V}{V_0}\right)^{K_q}\left(1 + K_{fq}\frac{f - f_0}{f_0}\right) \tag{13.3}$$

where K_p, K_{fp}, K_q, and K_{fq} are the active and reactive power sensitivity coefficients on voltage and frequency, respectively. Although any mathematical model (e.g., ZIP) can fit the sensitivity identification purpose, this model has been chosen due to the following characteristics [15]:

- It is independent of the initial voltage and it does not require initialization.
- Only one parameter identifies the relation between active and reactive power, voltage and frequency.
- The exponent is equal to the load sensitivity to the voltage.

Initially only the voltage sensitivities will be considered, assuming the grid frequency constant, the derivative of (13.3) with respect to V results in:

$$\frac{dP}{dV} = K_p P_0\left(\frac{V}{V_0}\right)^{K_p-1}\frac{1}{V_0}$$
$$\frac{dQ}{dV} = K_q Q_0\left(\frac{V}{V_0}\right)^{K_q-1}\frac{1}{V_0} \tag{13.4}$$

and with the assumption of $V{=}V_0$, we obtain:

$$\frac{dP/P_0}{dV/V_0} = K_p$$
$$\frac{dQ/Q_0}{dV/V_0} = K_q \tag{13.5}$$

Although the voltage reference V_0 is known (i.e., the nominal voltage), the variable P_0 in (13.5) depends on the nominal conditions, that are usually not met in normal grid operations. However, in the case of exponential loads (13.3), the voltage base V_0 and P_0 can be chosen arbitrarily as any voltage V_1 and P_1:

$$\frac{P_1}{V_1^{K_p}} = \frac{P_0}{V_0^{K_p}}$$
$$\frac{Q_1}{V_1^{K_q}} = \frac{Q_0}{V_0^{K_q}} \tag{13.6}$$

It follows that the active and reactive power sensitivities to voltage variations can be calculated independently from the ith operating point:

$$\frac{dP/P_i}{dV/V_i} = K_p$$
$$\frac{dQ/Q_i}{dV/V_i} = K_q \tag{13.7}$$

Finally, (13.7) can be discretized for any time step t_k, where the chosen base is represented as previous time step t_{k-1}:

$$\frac{\frac{P(t_k)-P(t_{k-1})}{P(t_{k-1})}}{\frac{V(t_k)-V(t_{k-1})}{V(t_{k-1})}} = K_p$$
$$\frac{\frac{Q(t_k)-Q(t_{k-1})}{Q(t_{k-1})}}{\frac{V(t_k)-V(t_{k-1})}{V(t_{k-1})}} = K_q \tag{13.8}$$

Similar to the voltage sensitivity coefficients, also the frequency sensitivity coefficients can be estimated considering the voltage at the nominal value $V = V_0$, so that (13.3) becomes:

$$P = P_0\left(1 + K_{fp}(f - f_0)\right)$$
$$Q = Q_0\left(1 + K_{fq}(f - f_0)\right) \tag{13.9}$$

With respect to the voltage sensitivities case, now the relation between power and frequency is assumed linear. The frequency sensitivity of load power is computed as:

$$\frac{dP/P_0}{df} = K_{fp}$$
$$\frac{dQ/Q_0}{df} = K_{fq} \tag{13.10}$$

Repeating the same steps as for the voltage sensitivity coefficients, and thus discretizing (13.10) at the time t_k, the frequency sensitivity coefficients can be estimated:

$$\begin{aligned}
\frac{\frac{P(t_k)-P(t_{k-1})}{P(t_{k-1})}}{\frac{f(t_k)-f(t_{k-1})}{f_0}} &= K_{fp} \\
\frac{\frac{Q(t_k)-Q(t_{k-1})}{Q(t_{k-1})}}{\frac{f(t_k)-f(t_{k-1})}{f_0}} &= K_{fq}
\end{aligned} \tag{13.11}$$

To estimate the load sensitivity to voltage and frequency in real conditions, the ST applies a controlled voltage (or frequency) disturbance in the grid. As an example of this disturbance, a trapezoidal profile has been adopted as controlled voltage variation, as shown in Figure 13.10 in a real measurement campaign performed at the Energy Smart Home Lab at the Karlsruhe Institute of Technology. The considered load is a common household, where different house appliances are connected at the same time, and the voltage is supplied by the LV DC/AC converter of the ST. As can be seen, the ST imposes a voltage variation and measures the load power consumption, in order to perform the calculation of the load dependence on the voltage mathematically. The trapezoidal shape used in this study varies the voltage amplitude of 0.02 pu for 0.4 s, with a voltage decreasing ramp equal to 0.05 pu/s. The choice of the disturbance characteristics, such as voltage amplitude and ramp-rate, comes from the compromise to limit the impact on the power quality in the grid (deep or long-lasting voltage variations) and to create a voltage variation that cannot be confused with the one caused by the stochastic load switching noise.

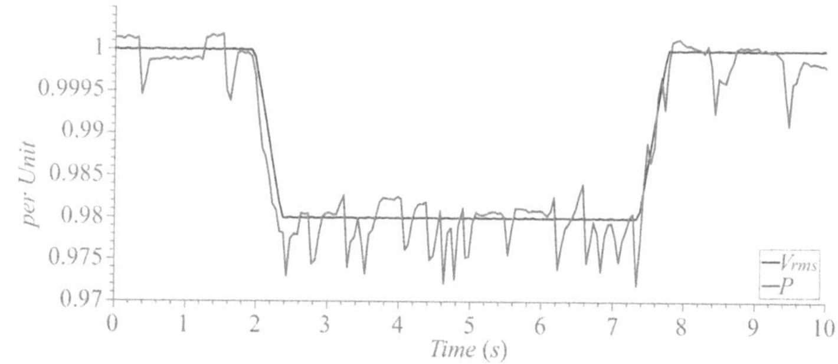

Figure 13.10 Example of an on-line load identification performed in a real experiment at the Energy Smart Home Lab at the Karlsruhe Institute of Technology. Measured load rms voltage and active power are plotted with a blue and red line, respectively.

13.3.3 Voltage-based load control

As demonstrated in the previous section, the ST is able to influence active power consumption of voltage-sensitive loads in a LV grid, by regulating the grid voltage amplitude. This section introduces a voltage-based load control [16] that, exploiting the voltage sensitivity coefficient estimation of the previous section, is able to achieve highly accurate load consumption variations.

Assuming that the LV grid is an unbalanced three-phase system with loads of different nature, each phase can have different load sensitivity coefficients. Let us consider that the load identification algorithm evaluates the voltage sensitivity coefficients K_p for each phase separately. To achieve this estimation, the ST introduces a controlled balanced three-phase voltage disturbance, that allows to measure the load power response in each phase independently. The total power variation ΔP for the three phases can be calculated as:

$$\Delta P = \Delta P_A + \Delta P_B + \Delta P_C \tag{13.12}$$

where P_A, P_B, P_C are the pre-disturbance consumed active powers for each phase, and ΔP_A, ΔP_B, ΔP_C are the corresponding phase load variations.

Let us consider the normalized load power sensitivity, as in (13.8) and let us apply the voltage-based load control to all three phases:

$$\begin{aligned}
\Delta P_A &= \frac{P_A}{V_A} K_{pA}(V - V_A) \\
\Delta P_B &= \frac{P_B}{V_B} K_{pB}(V - V_B) \\
\Delta P_C &= \frac{P_C}{V_C} K_{pC}(V - V_C)
\end{aligned} \tag{13.13}$$

Combining (13.13) and (13.12) and isolating the voltage which is applied for achieving a specified power variation ΔP, results in:

$$V = \frac{\Delta P + \left(P_A K_{pA} + P_B K_{pB} + P_C K_{pC}\right)}{\frac{P_A}{V_A} K_{pA} + \frac{P_B}{V_B} K_{pB} + \frac{P_C}{V_C} K_{pC}} \tag{13.14}$$

The ST main control feature allows to synthesize a balanced and sinusoidal three-phase set of voltages in the grid that are independent from the load current waveform demand (e.g., distorted or unbalanced). Under this assumption, let us simplify (13.14), imposing a voltage magnitude in each phase equal to the nominal one $V_A = V_B = V_C = V_0$.

It follows that the per-unit voltage amplitude V_{LV}^* that we have to impose to achieve the desired power change is defined with the formula:

$$\frac{V_{LV}^*}{V_0} = 1 + \frac{\Delta P}{P_A K_{pA} + P_B K_{pB} + P_C K_{pC}} \tag{13.15}$$

As can be noted from Figure 13.12, following a request to modify the connected load consumed active power ΔP, the ST applies a new voltage magnitude V^*_{LV} (Figure 13.11) that depends on the pre-calculated active power voltage sensitivity K_p. After an initial transient, the ST will restore the nominal voltage in the LV grid, in order to not impact on the load power quality for a long period.

An interesting application of the voltage-based load control is the support of primary frequency regulation to synchronous generators [17]. As shown in the example of Figure 13.13, the ST can measure the MV frequency f_{MV}, and it can adjust grid power ΔP following a certain characteristic:

$$\Delta P = \frac{1}{R_f} \Delta f_{MV} \tag{13.16}$$

controlling the LV amplitude V_{LV}, the load consumption of voltage-dependent loads P_{LV} varies depending on the direction and magnitude of the MV frequency variation. This service can provide power control instantaneously, and it is meant to provide

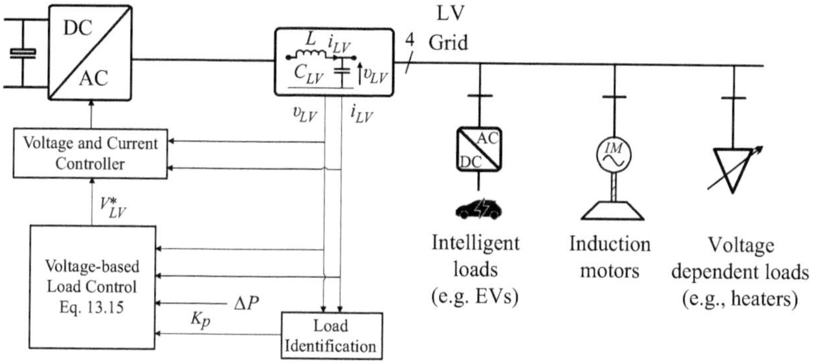

Figure 13.11 Voltage-based load control concept: interaction with intelligent loads, induction motors, and voltage-dependent loads (e.g., electric heaters)

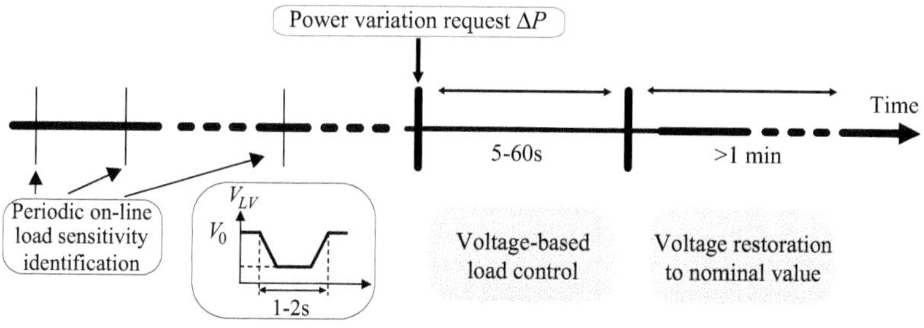

Figure 13.12 Timeline of the voltage-based load control actions

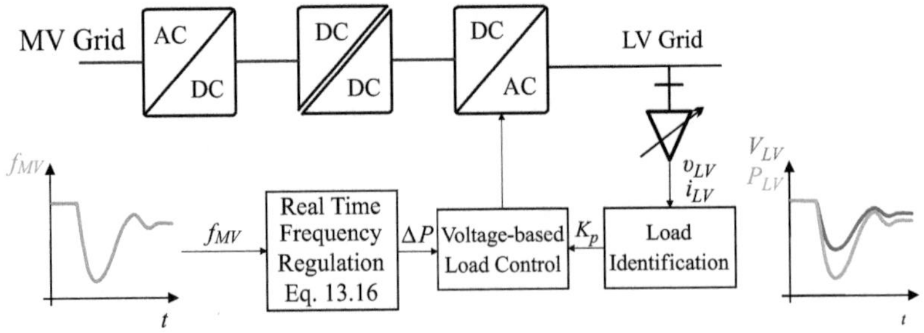

Figure 13.13 Real-time frequency regulation concept: blue line is the MV frequency f_{MV}; red line is the LV voltage amplitude V_{LV}; green line is the LV grid active power consumption P_{LV}

initial assistance to synchronous generators and larger energy storage systems (e.g., batteries) during their power ramp-up/-down. In a large-scale study in Ireland [18], it has been demonstrated that this feature can improve the system stability, while propelling the integration of wind turbines in the Irish system of an additional 10 %.

13.3.4 Frequency-based load control

In classical power system studies, the grid frequency is a variable that has to be kept constant in order to guarantee the stability of the system. This concept derives from the use of synchronous machines, which need a constant grid frequency.

However, the ST decouples the AC power flow between the primary and the secondary side, allowing an independent control of the frequency of the fed grid with respect to the main synchronous system. This possibility allows the use of controlled frequency variations as signals for local loads and generators to vary their consumption/production set-points.

Different appliances in AC systems are indeed equipped with power/frequency characteristics, in order to support the grid during power imbalances. Three common examples of these loads and generators can be found in Figure 13.14: an energy storage system used for example for photovoltaic applications, that can provide frequency regulation support to the grid; a synchronous machine, connected as example with a micro-gas turbine, that regulates the power output depending on the governor set-point; a domestic PV plant, equipped with derating power/frequency droop characteristic, if the grid frequency increases more than, e.g., in Germany, 50 Hz.

The ST can vary the secondary grid frequency in a controlled way, in order to vary the power set-point of the aforementioned energy resources, and thus regulate the fed grid net load [19]. Similar to (13.15), the frequency set-point f_{LV}^* can be decided with the frequency/active power sensitivity coefficients determined by (13.11):

$$\frac{f_{LV}^*}{f_0} = 1 + \frac{\Delta P}{P_A K_{pfA} + P_B K_{pfB} + P_C K_{pfC}} \tag{13.17}$$

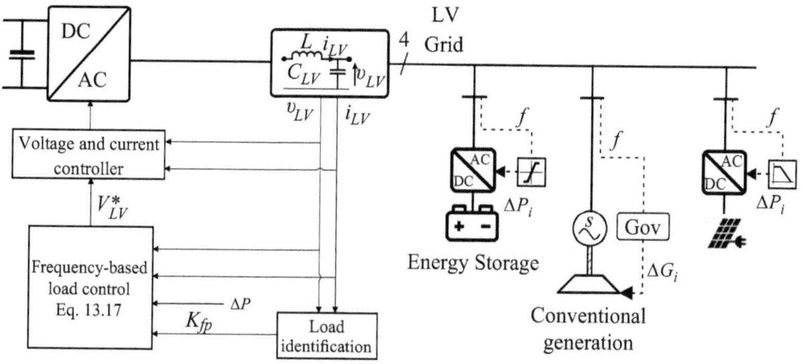

Figure 13.14 Frequency-based load control concept: interaction with energy storage systems, synchronous generators, distributed generator (e.g., PV plants)

This feature can extend the power controllability of the ST-connected load. As demonstrated in recent works, the frequency-based load control can be used, as an example, to avoid the reverse power flow in MV grid [19], or to avoid overloading a ST during high peak demand [20].

13.4 Enabling services for future DC grids

The ST offers a natural connection point for the future DC grids, where existing (e.g., photovoltaic) and new resources (e.g., energy storage systems and electric vehicle charging stations) can be integrated without additional AC/DC conversion stages [21,22].

A possible future grid architecture is shown in Figure 13.15, where the AC and the DC system co-exist. The resources, that by their nature work in DC, can decide to switch over to the DC grid, instead of remaining connected with the AC one, skipping a power transformation stage. A typical example is the charging of an electric vehicle: the AC connection can offer slow charging services (e.g., at home), reducing the size of the AC/DC converter, while the DC connection allows higher power capability, and thus more suitable for fast-charging (e.g., in public spots, such as commercial centers or offices).

In a future scenario, the loads can share both an AC and a DC connection (as shown in Figure 13.15) and re-route their power flow depending on efficiency reasons or grid needs. In [23], efficiency improvements up to 6 % and line losses reduction up to 22 % have been achieved using this grid topology.

As mentioned before, a DC connection can be realized directly from the ST conversion stages. At this regard, Figure 13.16 introduces two different architecture approaches to create a DC connection for LV grids [24]:

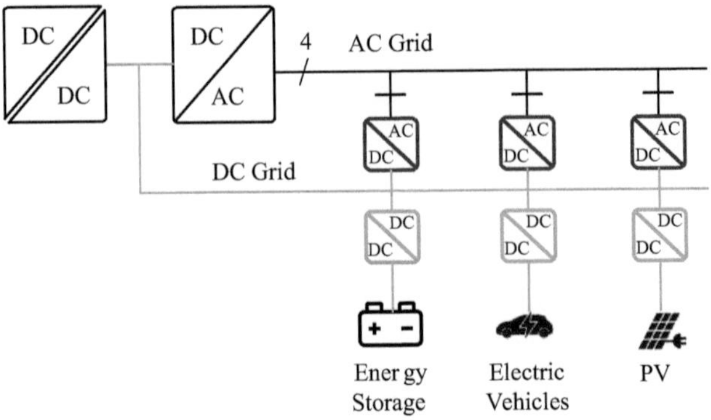

Figure 13.15 Hybrid DC/AC grid concept: the existing AC grid (black line) runs in parallel with the DC grid (blue line), allowing different connection possibilities to DC loads and generators

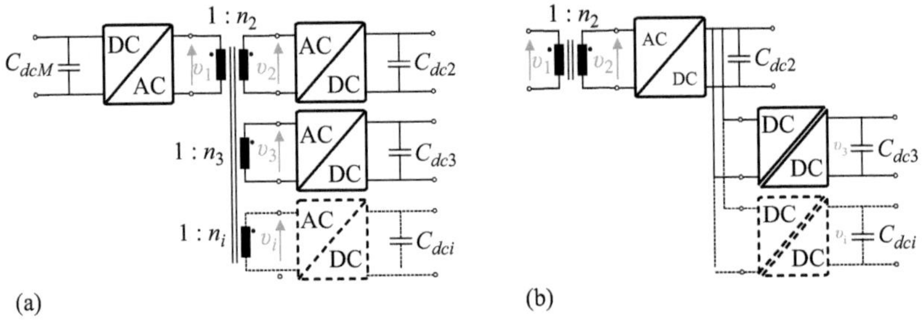

Figure 13.16 DC ports architecture: (a) magnetic coupling and (b) galvanic coupling

(a) *Magnetic connection*, where the medium frequency isolation transformer is realized with additional windings directly connected to the main transformer core. The power transfer does not occur at DC link level, but directly in the magnetic coupling. Despite a more complex transformer design, this architecture allows to reduce the number of semiconductor switches, decreasing the overall costs. With this topology, the number of DC ports is pre-determined at design level, and they cannot be scaled up.

(b) *Galvanic connection*, where the DC grid is directly connected to the ST LV DC link by means of a DC/DC converter. It allows great flexibility in adapting the voltage at different levels, and to scale up the number of DC connection ports. The downside is that for each new connection port, an isolated DC/DC converter is needed for voltage safety reasons.

13.5 Conclusions and future outlook

The ST, as a power electronics-based transformer, can increase the system controllability and offer services to the distribution grid. This chapter has provided an overview of the ST hardware and control features. It has introduced a new concept of estimating the load sensitivity to voltage and frequency, and how this information can be used to control the load consumption by means of controlled voltage and frequency variations. This chapter also described future DC grid scenarios based on ST, highlighting how a hybrid AC/DC distribution system can lower grid losses and improve system efficiency.

The ST topic is still at the beginning. More work is needed to integrate this technology into the distribution grid:

- Standards for asynchronously connected grids (i.e., the ST secondary side) shall be developed in order to integrate the ST safely into distribution grids.
- Business cases shall be proposed, in order to make the ST an attractive solution for distribution system operators and network utilities.
- Hardware efficiency shall be increased, that, despite the recent improvements in the switching technology, is still not at the same level of the conventional iron transformer.
- New control strategies shall be introduced, in order to fully exploit the increased power controllability of the ST.

Acknowledgment

The work of Giovanni De Carne and Felix Wald was supported by the Helmholtz Association and the Helmholtz Young Investigator Group under the program "Energy System Design" and "Hybrid Networks" (VH-NG-1613), respectively. The work of Marco Liserre has been supported by the German Federal Ministry of Education and Research (BMBF) within the Kopernikus Project ENSURE "New ENergy grid StructURes for the German Energiewende" under Grant 03SFK1I0-2.

Bibliography

[1] W. Mcmurray. Power converter circuits having a high frequency link, 1968, Patent, US3517300A.

[2] J.L. Brook, R.I. Staab, J.C. Bowers, and H.A. Niehaus. Solid state regulated power transformer with waveform conditioning capability, 1980, Patent, US4347474A.

[3] D. Dujic, A. Mester, T. Chaudhuri, A. Coccia, F. Canales, and J.K. Steinke. Laboratory scale prototype of a power electronic transformer for traction applications. In *Proceedings of the 2011 14th European Conference on Power Electronics and Applications*, Aug 2011, pp. 1–10.

[4] J.E. Huber and J.W. Kolar. Volume/weight/cost comparison of a 1mva 10 kv/400 v solid-state against a conventional low-frequency distribution transformer. In *2014 IEEE Energy Conversion Congress and Exposition (ECCE)*, Sept 2014, pp. 4545–4552.

[5] A.Q. Huang, M.L. Crow, G.T. Heydt, J.P. Zheng, and S.J. Dale. The future renewable electric energy delivery and management (freedm) system: the energy internet. *Proceedings of the IEEE*, 99(1):133–148, 2011.

[6] X. She, X. Yu, F. Wang, and A.Q. Huang. Design and demonstration of a 3.6-kv: 120-v/10-kva solid-state transformer for smart grid application. *IEEE Transactions on Power Electronics*, 29(8):3982–3996, 2014.

[7] M. Liserre, G. Buticchi, M. Andresen, G. De Carne, L.F. Costa, and Z.-X. Zou. The smart transformer: impact on the electric grid and technology challenges. *IEEE Industrial Electronics Magazine*, 10(2):46–58, 2016.

[8] L.F. Costa, G. De Carne, G. Buticchi, and M. Liserre. The smart transformer: a solid-state transformer tailored to provide ancillary services to the distribution grid. *IEEE Power Electronics Magazine*, 4(2):56–67, 2017.

[9] M. Schweizer and J.W. Kolar. Design and implementation of a highly efficient three-level t-type converter for low-voltage applications. *IEEE Transactions on Power Electronics*, 28(2):899–907, 2013.

[10] P. Aristidou, G. Valverde, and T. Van Cutsem. Contribution of distribution network control to voltage stability: a case study. *IEEE Transactions on Smart Grid*, PP(99):1–1, 2015.

[11] G. Delille, B. Francois, and G. Malarange. Dynamic frequency control support by energy storage to reduce the impact of wind and solar generation on isolated power system's inertia. *IEEE Transactions on Sustainable Energy*, 3(4): 931–939, 2012.

[12] K.P. Schneider, J. Fuller, F. Tuffner, and R. Singh. Evaluation of conservation voltage reduction (cvr) on a national level. *Pacific Northwest National Laboratory Report*, 2010.

[13] Z. Wang and J. Wang. Review on implementation and assessment of conservation voltage reduction. *IEEE Transactions on Power Systems*, 29(3): 1306–1315, 2014.

[14] G. De Carne, M. Liserre, and C. Vournas. On-line load sensitivity identification in lv distribution grids. *IEEE Transactions on Power Systems*, 32(2): 1570–1571, 2017.

[15] T. Van Cutsem and C. Vournas. *Voltage Stability of Electric Power Systems*. Kluwer Academic Publishers, Amsterdam, 1998.

[16] G. De Carne, G. Buticchi, M. Liserre, and C. Vournas. Load control using sensitivity identification by means of smart transformer. *IEEE Transactions on Smart Grid*, 9(4):2606–2615, 2018.

[17] G. De Carne, G. Buticchi, M. Liserre, and C. Vournas. Real-time primary frequency regulation using load power control by smart transformers. *IEEE Transactions on Smart Grid*, 10(5):5630–5639, 2019.

[18] J. Chen, M. Liu, G. De Carne, *et al.* Impact of smart transformer voltage and frequency support in a high renewable penetration system. *Electric Power Systems Research*, 190:106836, 2021.

[19] G. De Carne, G. Buticchi, Z. Zou, and M. Liserre. Reverse power flow control in a st-fed distribution grid. *IEEE Transactions on Smart Grid*, 9(4): 3811–3819, 2018.

[20] G. De Carne, G. Buticchi, M. Liserre, and C. Vournas. Frequency-based overload control of smart transformers. In *2015 IEEE Eindhoven PowerTech*, 2015, pp. 1–5.

[21] X. She, A.Q. Huang, S. Lukic, and M.E. Baran. On integration of solid-state transformer with zonal dc microgrid. *IEEE Transactions on Smart Grid*, 3(2):975–985, 2012.

[22] X. Yu, X. She, X. Zhou, and A.Q. Huang. Power management for dc microgrid enabled by solid-state transformer. *IEEE Transactions on Smart Grid*, 5(2):954–965, 2014.

[23] D. Das, V.M. Hrishikesan, C. Kumar, and M. Liserre. Smart transformer-enabled meshed hybrid distribution grid. *IEEE Transactions on Industrial Electronics*, 68(1):282–292, 2021.

[24] L.F. Costa, G. Buticchi, and M. Liserre. Optimum design of a multiple-active-bridge dc–dc converter for smart transformer. *IEEE Transactions on Power Electronics*, 33(12):10112–10121, 2018.

Chapter 14

On the interactions between plug-in electric vehicles and the power grid

Ekaterina Dudkina[1], Luca Papini[1], Emanuele Crisostomi[1] and Robert Shorten[2]

Growing concerns over climate changes have driven regulatory pressures to reduce urban pollution, emissions from CO_2 and other particulate matters, and city noise, which have motivated intense activity in the search for alternative road transportation propulsion systems. In this context, plug-in electric vehicles (PEVs), either as purely electric vehicles or as plug-in hybrid vehicles, have become a significant fraction of the overall transportation fleet in many countries worldwide.

This ever-increasing penetration level of PEVs is posing significant challenges to the existing power grids' infrastructures, as a significant load in terms of the required energy (e.g., during night charging of the vehicles connected for charging) or in terms of the required power (e.g., during fast charging events). The uncertainty of the charging events (in terms of time, space and required energy) further challenge the ability of the power grids to seamlessly handle such new electrical loads in addition to the already existing base load. At the same time, however, the ability of PEVs to operate in a vehicle-to-grid (V2G) mode may be also exploited to provide ancillary regulatory services to facilitate the general operation of the power grid. The optimal trade-off between these two contrasting aspects of the charging problem is still a subject of intense study in the power research community.

Accordingly, this chapter reviews the most recent research and technological advances in the charging process of PEVs; it presents through simple simulations the importance of controlled charging, and the advantages of utilizing decentralized schemes for practical implementation; also, it identifies some particularly interesting new trends that can be observed at the intersection between the transportation and the power networks, and it outlines interesting future directions in the context of electric vehicles.

[1]Department of Energy, Systems, Territory and Constructions Engineering, University of Pisa, Italy
[2]Dyson School of Design Engineering, Imperial College London, UK

14.1 Introduction

The amount of electrical vehicles (EVs) on the roads is stably growing despite the pandemic-related crises and economic downturn. In 2020, the number of electric cars registration increased by 41% while the overall car sales dropped by 16%. This increase is registered despite the gradual reduction of government subsidies over the past 5 years [1].

In numbers, around 3 million new electric cars were registered in 2020, and, for the first time, Europe (with 1.4 million) is leading in the ranking, followed by China with 1.2 million registrations, and United States with 295,000 new electric cars. The European leaders in newly registered electrical cars are Germany (395,000), France (186,000), United Kingdom (175,000), and Norway (106,000) [2]. While taking the fourth position in EV sales, in terms of share of EV among all cars sold, Norway is unconditional leader—seven out of ten cars sold are plug-in electric vehicles (PEVs) [3]. In Iceland, the share of PEVs exceeded 50% and 30% in Sweden [2].

Such increasing numbers are seen with concern by grid operators. The significant electrification of the mobility sector may endanger the stability of the grid, especially if vehicles are charged in an uncoordinated fashion. For instance, it was estimated in [4] that if $8-9\%$ of the vehicles of New England are connected to the grid for charging within an hour, then the electric grid could collapse. This scenario may actually occur assuming that a significant proportion of vehicles may get connected for charging around the same time, for instance in the evening when the owners go back home after work.

At the same time, however, it is also known that PEVs may actually represent a convenient resource for the power network. In fact, if vehicles are left connected to the grid when idle, then the power grid may use them as a large virtual storage device, which can be charged and discharged as it is deemed more convenient for the grid [5]. The potential of such V2G and ancillary services has not been fully explored yet, but may in principle change the way power grids are currently managed, and push the transition towards peer-to-peer (P2P) energy exchanges among conventional energy users (i.e., the PEV owners).

This chapter is organized as follows: Section 14.2 is dedicated to reviewing the state-of-the-art. This includes a review of the most recent technologies that are adopted for charging PEVs, and also of the algorithms that have been developed to optimize their integration in the power grid. Section 14.4 briefly overviews classic charging strategies, as uncontrolled charging, centralized controlled charging, and some promising decentralized charging strategies that have been recently proposed for PEV charging. Section 14.5 shows simple simulation results that highlight the differences and the advantages of charging strategies. Section 14.6 is then dedicated to briefly outline and discuss some new and open control problems related to PEVs which are expected to gather more interest in the next few years. Finally, we conclude and summarize the contents of the chapter in the last Section 14.7.

14.2 Review of the state-of-the-art

14.2.1 Technological aspects

14.2.1.1 Charging stations

In order to deal with the growing numbers of PEVs, the number of electric vehicle service equipment (EVSE) and charging stations (CSs) has been growing consequently. Even if the majority of the charging stations are located at home or in the proximity of working places [1], public charges are an essential enabler for long trips, allowing to provide a quick charging point and, in general, to reduce range anxiety of PEV owners. The charging station are nowadays split into two main categories:

- slow chargers, that takes 5–8 h to fully charge the vehicle battery;
- fast chargers, which permits to reach 80% of the battery capacity within 30 min.

By 2020, the number of public charging stations reached 1.3 million (of which 30% are fast chargers). The leader of number of publicly accessible charging stations is China – 500,000 slow chargers and 310,000 fast chargers. Europe is following with around 250,000 and 38,000 units of slow and fast chargers, respectively, while for the United Stated the corresponding numbers are 82,000 of slow and 17,000 of fast chargers, where nearly 60% are superchargers by Tesla Inc. (Palo Alto, CA, USA) [1]. Despite the continuous growth, the majority of countries have not reached the targets set by Alternative Fuel Infrastructure Directive (AFID, [6]) which is 1 public charger every 10 PEVs by 2020. In Europe, the average ratio is 0.09, where Netherlands (0.22) and Italy (0.13) overtook the target, while Norway and Iceland have the lowest EVSE per PEV ratio – 0.03. There are two main reasons for such low values: first, these countries are sparsely populated and have a high share of detached houses, thus the majority of PEV owners use private charging stations; second, the share of fast chargers is relatively high – 40% in Iceland and 31% in Norway (vs. 3% in Netherlands and 9% in Italy) [1].

All of the aforementioned EVSE is based on the nowadays most common charging method—which is, conductive charging via cable. This method is widely applied due to its relatively low cost, high efficiency and possibility of charging with high rates (therefore, allows for high-speed charging). Considering its efficiency, it is expected that cable charging will remain the main charging solution in the nearest future. Nevertheless, novel options are being developed.

The second most known charging technique is wireless charging [7], which exploits principles of electromagnetic induction. This technique has a number of advantages, mainly related to the possibility of *ubiquitous charging*, for instance during short idle times (e.g., at traffic lights). Many pilot projects are currently investigating this solution at different levels of technological readiness [8], though fewer examples can be found of functioning implementations. In particular, the main concerns relate to the low efficiency due to higher energy losses during a charge, which can be even more significant when the coils of the car and of the CS are not properly aligned. In addition, the cost of the charging station, the low charging rates, and concerns on possible harmful effects of high-intensity electromagnetic emissions close to the vehicle, still jeopardize a more widespread utilization of wireless charging. More details and a broader perspective regarding this technology are given in Section 14.6.1.

Alternative solutions have been proposed with the aim of reducing the charging time or eliminate the need for human interactions, as it would be essential for self-driving vehicles, but they are still on the proof-of-concept phase. Among these solutions there are battery swap [9,10], i.e., when, on the station, the empty battery can be replaced with the fully charged one. This could have a positive impact in creating a circular economy system around the battery market. The scarcity and the unsustainable extraction of raw materials arise the need for an improved battery recycling technology and economical system which allows to properly manage the used one. In fact, the so-called "second-life" of batteries is a practice which sees the used PEV batteries applied to build stationary storage system, aiming to alleviate the burden of energy demand from the grid, and the Amsterdam arena is an example of such solution [11]. Furthermore, the management of the exhaust batteries recycling process can be simplified if managed by a provider which is in charge of extraction and re-use of the high-valued minerals (e.g., cobalt, lithium) of which the PEV batteries are made of. Another solution can be found in automatic charging systems [12] where the cable is leaded by an automatic system, thus not requiring human interactions.

14.2.1.2 Charging modes

The duration of a charging process depends on the battery capacity (or energy requirement) as well as on the charging power provided by the CS. Currently, a wide range of charging rates can be found, associated with different charging modes. The standard for electric vehicle conductive charging systems (in particular, IEC 61851-1 [13]) defines four types of charging modes or levels [14,15]:

- Mode 1 assumes direct connection of the vehicle to the plug without special safety systems and is mainly applied to charge light vehicles, like electrical scooters or bikes. This mode has certain restrictions as it is nowadays forbidden for public chargers in Italy, Switzerland, Denmark, Norway, France, and Germany and it is forbidden at all in United States, Israel, and England. The rated values for current and voltage are limited by 16 A and 250 V in single-phase alternating current (AC) and 16 A and 480 V in three-phase AC.
- Mode 2 can be used with standard domestic and industrial sockets; it also includes in-cable control and protective device, named Control box. Some restrictions are applied in United States, Canada, Switzerland, Denmark, France, and Norway. The rated values for current and voltage should not be higher than 32 A and 250 V in single-phase AC while 32 A and 480 V in three-phase AC. The chargers using the first two modes are categorized as slow charging modes.
- Mode 3 is permanently incorporated to the electric system and has specifically dedicated circuit and socket outlets. This mode is mainly used in public charges (in Italy, e.g., it is the only mode allowed for public charging stations based on AC). There is no limit set on the rated current and voltage, but the most common values do not overcome the charging values of 32 A and 250 V in single-phase

AC and 32 A and 480 V in three-phase AC. This mode can provide both slow and quick charging speeds.

- Mode 4 is the only mode based on direct current (DC). As in mode 3, the charger is built into the charging point and equipped with a control box; the station is usually a bigger installation due to the presence of AC/DC converters. There are two connector standards—the Japanese CHAdeMO and the European Combo 2 or CCS. The mode 4 allows reaching very fast charging speeds and referred as fast DC—even if there are no legislative restrictions, it usually allows charging up to 200 A and 400 V.

Nowadays, the AC-based charging is the most used type of charger, as in this case, battery can be recharged anywhere, via a standard electrical plug. On the other hand, DC-based fast chargers have been spreading faster and faster. Publicly available fast chargers are the crucial factor to permit longer trips and encourage hesitating drivers, even those without a private charging point, to move to PEV.

14.2.1.3 Driving ranges

The biggest historical concern over PEVs, in addition to their higher price, was the concern related to their short range when compared to conventional internal combustion engine (ICE) vehicles and combined with the scarce availability of charging points. This situation has given rise to the so-called range anxiety (i.e., the fear that the battery of the PEV may be depleted before reaching the desired destination or a charging point). However, newest models of vehicles allow for considerably long drives, in a range between 500 and 850 km (see [16]). Accordingly, the increased driving ranges are, on one side, broadening the cases when PEVs may be conveniently adopted. On the other hand, this trend is affecting the required number of charging operations and their frequency.

14.3 Optimized charging of the vehicles

In parallel to researches aiming at improving the enabling technologies for electric mobility, there is also a great interest in developing novel algorithms and procedures to optimally use the existing infrastructures. Most notably, these researches aim at finding optimized solution to share the available power among the PEVs that need charging.

14.3.1 Charging strategies

The impact of the increasing number of electrical vehicles on the distribution power grid is expected to become even more significant in the near future as the electrification of the mobility sector is steadily advancing (the PEV electricity consumption has reached 80 TWh by 2020 and is expected to reach 525–860 TWh according to different expected scenarios already by 2030 [1]). The impact may be seen from two different points of view.

Although the rapid growth of the number of electrical vehicles causes extra burdens to the grid, the implementation of bidirectional charging alongside with the implementation of smart charging strategies could not only mitigate the burden on the grid, caused by the charging vehicles, but also support its operation by improving the electric grid efficiency and reliability. This could help to avoid additional investments in grid expansion/ reinforcement and help in accommodation of intermittent renewable energy sources [5].

Two types of charging strategies can be distinguished [17]:

1. *Uncoordinated charging:* In this case, all car owners decide individually when and where to charge their vehicle, and the charging process starts as soon as the vehicle is plugged-in. Even if this may be regarded as the most comfortable mode for PEV holders, it has a range of serious drawbacks. The first one is overloading – simulations and case studies on uncoordinated charging showed that penetration of PEV may significantly increase the load peak, and, depending on the grid, this value can even overcome 60% with 30% of PEV penetration [18]. The raise of peak demand is caused by similar driving patterns and daily routines of vehicle drivers. As an example, consider that the majority of users are charging PEVs at home at the end of a working day [4]. The charging requests may overlap and this would in turn lead to a load peak increase, lines and transformers overloading events, voltage deviations, an increase of electricity cost and related CO_2 emission by the electricity system [5]. The last point is particularly interesting, given that CO_2 reduction is named as the main advantage of PEVs (despite such CO_2 increase could still occur where energy is generated, which is usually outside of city centers and far from where most people live). However, an uncontrolled large number of vehicles charging simultaneously, in addition to requiring a more expensive and less environmental friendly energy mix, does not also fully benefit of the generated intermittent renewable energy, which is not generally synchronized with the peak demand. Thus, additional load balancing must be assured—and single-phase connection may cause further line current imbalances [19,20]. Accordingly, uncontrolled charging is often considered as the "worst-case" scenario [21], and it is not guaranteed to be realizable with an increasing number of PEVs.
2. *Coordinated or smart charging*: In the coordinated or smart charging, the choice of charging a single vehicle is taken considering the state of the power grid (e.g., how many other vehicles are connected for charging). From a mathematical point of view, the vehicle is connected by evaluating the *most convenient* moment in time, where the convenience is evaluated in terms of objective functions that can be either minimized or maximized depending on the context. Examples of these are the minimization of power losses, voltage deviations, charging costs, CO_2 (or other pollutant) emissions. Other examples are maximization of social welfare or comfort. The coordination may be either implemented in a centralized fashion [22], where a (possibly third-party) grid operator serves as an aggregator for a number of PEVs, and is accountable for PEV charging and choosing the best rate and/or charging time to achieve economic and technical advantages; or in a decentralized way [4] where single PEV owners take the most convenient choice, usually exploiting available information. In this case, a signal is usually shared,

which could be the charging price, or the CO_2 emissions level, and individuals adapt their charging behaviour on the basis of the signal. The simplest example is the day/night tariff: during the night, when the electricity demand is low, the prices are also lowered.

Sections 14.4 and 14.5 further discuss the differences between such two charging strategies and extensively illustrate their differences through explicative examples.

The previous two types of charging pertain unidirectional active power flows from the grid to the vehicles. More recently, however, there is a growing interest in the ability of vehicles to also provide active power to the grid, or to exchange reactive power. Such possibilities are better illustrated in the following section.

14.3.2 V2G and vehicle-to-everything (V2X)

A more sophisticated example of smart charging is V2G exchange, when the PEVs do not just behave as conventional loads, but similarly to storage devices, they also have the ability to inject power in the grid, if needed by the grid, and if energy is not required by their owners in the short term. In this case, additional technical equipment for the vehicle and the charges is required to enable bi-directional flows*: V2G enabled vehicle, bi-directional charger, communication system or "protocol" and control system. In this case, the PEV can flatten the load profile not only by reducing peak but also by providing back-up power to the load. Among the studies aiming at maximizing the utility of PEVs for the grid, a great part is dedicated to ancillary services: since PEVs may provide power with fast dispatch, the most attractive applications include frequency control [23], load balance [24], and spinning reserves. V2G mode is seen as the most attractive one, thanks to its flexibility, and it can be tackled with different control algorithms and problem formulations. Examples of these are, for example, quadratic and dynamic programming [25,26], and, distributed ones, such as a distributed multi-agent system (MAS), stochastic decentralized control [4], game-theory, or particle swarm optimization [27].

By 2021, there are now over 90 V2G projects in 22 countries [28], which are providing a range of services to the grid. Most of them are proof-of-concept trials, however small commercial projects are also being rolled out. Following V2G, a range of PEV operational modes has emerged, in particular, vehicle-to-building (V2B), vehicle-to-home (V2H), vehicle-to-community (V2C), and vehicle-to-load (V2L). All these applications are often collectively gathered under the umbrella term V2X – "vehicle-to-everything" [29].

Even if V2G is a promising mode for PEV operation, there are some social and technical barriers. The first concern is battery degradation. V2G application increases the number of charge and discharge cycles, which is known to potentially reduce the battery life [30]. The effects of batteries state-of-the-charge limitations on V2G implementation are discussed in [21]. Even if in the end, the assessed profit of V2G services is overcoming the cost of batteries [31], the increase of the final cost of services and the possible ways to minimize degradation of storage should

*https://www.v2g-hub.com/

be taken into account in V2G operation modeling. Among the possible solutions aiming at mitigating this degradation, there are algorithms to prolong the life of the batteries, for example, by intelligent control and optimization of charging time [18]. An alternative solution is to reuse batteries from EVs as a storage for microgrids for providing frequency regulation or balancing services [32].

While the majority of V2G studies assess the technical side of V2G integration, often there is an assumption that V2G enrollment reaches 100% and all the owners of the vehicles are system "optimizers" [33]. However, there are some concerns about consumers' acceptance of V2G concept, as their lifestyle, willingness to have control over their vehicles and homes are not always considered.

14.3.3 Automatic adaptation of charge rates

Assuming that the power grid has the ability to change charge rates, or to decide when a charging process may start and when it may finish, it is important to determine what the observed variable should be or, in other words, how the grid should determine when charge rates should be changed. Following the discussion and the findings of [4], there are three main alternatives, i.e., the power grid may monitor the frequency, the voltages or the power consumption. In particular, only control procedures based on power consumption have been found to be efficient, as it is described below:

1. Frequency-based strategies, which include the center-of-inertia frequency (also called system frequency) and the local bus frequency. The former indicates how balanced the system is, while the local bus frequency locally assesses the balance conditions on a specific bus, which is a more feasible and easily measurable value. It is worth noticing that, considering the ramp rate of electrical vehicles, these two values are similar to each other, which was also confirmed in a simulation study [4]. However, the application of frequency-based control strategies is not always efficient since instabilities may occur even without pronounced frequency variations. Accordingly, if the power grid monitors frequency values to decide when unstable situations are about to take place, it may likely occur that charge rates would be reduced when it is too late to prevent unstable situations from occurring [4].
2. Voltage magnitudes may be used to indicate possible grid malfunctions and are related to the transmission capacities of the lines. Unlike frequency variations, which could be noticeable only when it is too late, voltage magnitude fluctuations may already occur without any stability issues in the grid. This happens due to the presence of different voltage regulation devices, such as under-load tap changers or static VAR compensators. Accordingly, if the power grid monitors voltage values, it may reduce charge rates even when there is no actual reason to do so.
3. Power-based control strategies are probably the most suitable to prevent power grid collapses from occurring. Similar to what is already happening for other conventional electric loads, the power grid may utilize known load thresholds and historical load values to determine the actual risk of a power network, and to reduce charging rates if required.

14.4 Charging strategies

In this section, we provide some simple examples to briefly illustrate and discuss the impact of different charging strategies for PEVs, as earlier introduced in Section 14.3.1. All scenarios are based on unidirectional charging and do not consider the V2G concept, although algorithms could be easily modified to take it account, if possible.

14.4.1 Uncontrolled charging

In the absence of coordination and control strategies, it is assumed that each PEV will start charging as soon as it is plugged in. In this case, the power consumed by a fleet of PEVs is proportional to the number of connected PEVs:

$$P(t) = \sum_{i=1}^{N(t)} c_i(t), \tag{14.1}$$

where $N(t)$ is the number of PEVs connected at time t, and $c_i(t)$ is the charging rate of the i'th vehicle. This strategy may give rise to issues to the power grid if many vehicles are connected for charging at the same time, for instance, as already mentioned, during the evening when owners come back home from work.

14.4.2 Controlled charging—centralized solutions

The uncontrolled scenario depicted in Section 14.1 does not scale well with the number of PEVs, and it may require significant investments in the power infrastructure to allow for a significantly larger electric load. Accordingly, to minimize the impact on the distribution system and, at the same time, to provide a fair charging to all the connected vehicles, it is envisaged that charging stations should implement a dynamic modification of the charging rates. This could have several advantages, as in addition to shaving the peak of consumption, similar solutions may be also adopted to reduce CO_2 emissions by trying to match power demand with power generation from renewable sources [34]. A simple example of a coordinated charging strategy is to limit the total charging power dedicated to PEVs, by reducing the charging rates of all PEVs connected to the grid. Such a limitation may be caused by a large number of connected vehicles, by a reduced availability of generated power, or by an increased power demand by other non-deferrable loads. If we let $\bar{P}(t)$ denote the maximum power that can be devoted to PEVs (i.e., as the difference between the generated power and the baseline load), and $\bar{c}_i$ denotes the maximum charging rate of the ith vehicle (it may either depend on the technology of the CS, or on the technology of the vehicle), then a centralized controlled scenario would allocate a time-varying charging rate $c_i(t)$ to each vehicle according to the following equation (e.g., see [35]):

$$c_i(t) = \min\left\{\frac{\bar{P}(t)}{N(t)}, \bar{c}_i(t)\right\}. \tag{14.2}$$

This solution optimally shares the available power among the connected PEVs, and allocates the same charging rates to each vehicle (apart from possible physical constraints due to different adopted technologies).

However, this solution can be hardly implemented in practice, as it would require to exactly know at each time step the number of connected vehicles ($N(t)$ in (14.2)) to optimally share the available power. Thus, a huge amount of information should be transmitted in both directions almost in real time (i.e., vehicles communicating their connection/disconnection, and the power grid communicating the time-varying charging rates $c_i(t)$ to all vehicles).

14.4.3 Controlled charging—decentralized solutions

As explained in Section 14.4.2, centralized control of PEV charging requires a significant amount of bidirectional communications that need to be exchanged among the vehicles and the infrastructure. This raises serious concerns regarding the ability of such algorithms to scale well with the number of vehicles. In addition, centralized solutions pose all the burden on a central node that has to compute all the charging rates in real time, which is usually not a robust solution against possible failures in the central node. A possible solution to mitigate such concerns is to resort on distributed or decentralized solutions. For instance, AIMD algorithm [4,35] solves a classic problem of resources distribution, where several agents are requesting a share of limited resources. This algorithm was originally used to manage congestion on the Internet by sharing the bandwidth among users [36,37]. As the PEV charging problem may be seen as case of sharing energy among PEVs, the AIMD algorithms has been also applied to develop PEV charging strategies [4,35,38–40]. The similarity also refers to the random and unpredictable connection and disconnection of users in the Internet case, which resembles one of the main concerns of the charging strategies when PEVs are connected and disconnected from the power system. Indeed, it is desirable not to know the precise amount of plugged vehicles, and let the owners have complete freedom to manage their vehicles.

The AIMD algorithm consists of two stages: a first called additive increase (AI) that corresponds to a linear increase of PEVs charging rates by a constant value ($\alpha > 0$ in kW). All PEVs keep increasing their charging rates until the overall power consumed by all PEVs reaches the maximum available power or, in other words, until a capacity event occurs. In this case, a second stage, called multiplicative decrease (MD), takes place. In fact, as soon as the maximum capacity has been reached, a signal is sent to all plugged PEVs (or equivalently to all charging stations), and all charging rates are reduced by a multiplicative factor ($0 < \beta < 1$).

The pseudocode of the AIMD algorithm is presented for convenience in Algorithm 1.

The advantage of such an AIMD procedure is that a broadcast signal is sent out to all CSs (or to all PEVs) only when the power system (e.g., a distribution system operator (DSO)) senses that the total power used to charge PEVs is equal to the maximum one (i.e., in a power-based control strategy, as described in Section 14.3.3). When CSs receive such signal, all charging rates are reduced so that the total power

Algorithm 1: AIMD algorithm

Initialisation $t = 1$;
  while $t < t_{simulation}$ **do**
    if $\sum_{i}^{N(t)} c_i(t) < \bar{P}(t)$ **then**
      $c_i(t+1) = min\{c_i(t) + \alpha, \bar{c}_i(t)\}$; ▷ AI step
    else
      $c_i(t+1) = \beta c_i(t)$ ▷ MD step
    end if
    $t = t + 1$
  end while
End

remains below the desired threshold. Conversely, if only few PEVs are connected for charging, they are always charged at the maximum possible charging rate.

14.4.4 Controlled charging—prioritized decentralized solutions

The assumption of the AIMD algorithm is that all PEVs reduce the charging rate in a multiplicative fashion as soon as a capacity event occurs. In some cases, however, it is desired to prioritize some vehicles with respect to others, for instance assuming that some PEVs need to charge more in a shorter time, or simply that some owners are willing to subscribe more expensive contracts to be charged in an uncontrolled fashion (i.e., with maximum allowed charging rates). This can be implemented again in a completely decentralized fashion, by using the so-called unsynchronized AIMD algorithm (e.g., see [41]), where only the charging rates of a subset of PEVs are reduced, thus ensuring a fast charging to prioritized vehicles. The unsynchronized AIMD case implements such kind of optimization by assigning a different probability of charging rate decrease to different groups of vehicles, depending on their charging priorities. This is usually performed by assigning a probability π_i of performing the MD step, that depends on the history of previous charging rates c_i, and on some utility function assigned to that vehicle (e.g., based on its willingness to pay for charging, or the energy requirements). Then, during a capacity event, only a share of connected PEVs switches to the multiplicative decrease step, and the implementation of unsynchronized AIMD algorithm is given in Algorithm 2.

The advantage of this last algorithm is that it gives flexibility to PEV aggregators, power systems, PEV owners, to automatically adapt their charging rates taking into account the constraints of the power grid, as well as the preferences of the PEV owners.

14.5 Simulations

The charging strategies described in Section 14.4 are now compared and discussed through simple and illustrative simulations. We consider here a similar scenario to the

Algorithm 2: Unsynchronized AIMD algorithm

Initialisation $t = 1$;
 while $t < t_{simulation}$ **do**
  if $\sum_i^{N(t)} c_i(t) < \bar{P}(t)$ **then**
   $c_i(t+1) = min\{c_i(t) + \alpha, \bar{c}_i(t)\}$;
  else
   $$c_i(t+1) = \begin{cases} \beta c_i(t), & \text{with probability } \pi_i(t) \\ min\{c_i(t) + \alpha, \bar{c}_i(t)\}, & \text{with probability} 1 - \pi_i(t); \end{cases}$$
  end if
  $t = t + 1$
 end while
End

one analyzed in [4], i.e., an evening scenario where potentially a large number of PEVs are plugged for charging for the next day. For simplicity, we assume that all charging rates may provide up to 3.3 kW (i.e., $\bar{c}_i = 3.3$ kW). We consider a time step of 10 s, and we assume that a random number of vehicles is connected for charging at each time step, so that after 1 h approximately 18,000 vehicles are connected overall for charging. In addition, we assume that each PEV requires an average amount of energy equal to 6.6 kWh, which implies that, on average, each PEV would be fully charged in about 2 h if it is charged at the maximum charging rate (i.e., this corresponds to assuming that the average daily trips cause a discharge of about 6.6 kWh that can be compensated in a two-hour charge at the maximum charging rate). In our simulations, we do not consider the baseline loads (i.e., other loads different from PEVs), which corresponds to assuming that $\bar{P}$, the maximum power that can be allocated to PEVs, corresponds to the difference between the maximum power that can be used for all loads minus the baseline loads. All simulations are stochastic, since both the number of vehicles that arrive at each time step and their required energy take stochastic values.

Figure 14.1(a) and b shows what happens in the uncontrolled scenario when all vehicles are charged at the maximum charging rate, in terms of charged vehicles (Figure 14.1(a)) and power demand (Figure 14.1(b)), respectively. In the first hour, the power demand by PEVs increases as PEVs are being plugged to the power network for charging. After 1 h, all vehicles have been plugged, and, after about another hour, some vehicles have already finished their charging process. Overall, it takes about 3 h to fully charge all vehicles (the last vehicles arrive after 1 h, and require 2 h of charging on an average). Accordingly, the down slope starting at the end of the second hour corresponds to the gradual disconnection of the fully charged vehicles that were plugged first. Obviously, such a scenario is feasible as long as the power system can accommodate any number of PEVs and can charge all of them at the maximum charging rate. If we assume, however, that $\bar{P}(t)$ is actually lower than the peak of required power (e.g., a constant $\bar{P}(t) = 40$ MW in Figure 14.1(b)), then the uncontrolled scenario is not feasible anymore. As the number of PEVs is increasing

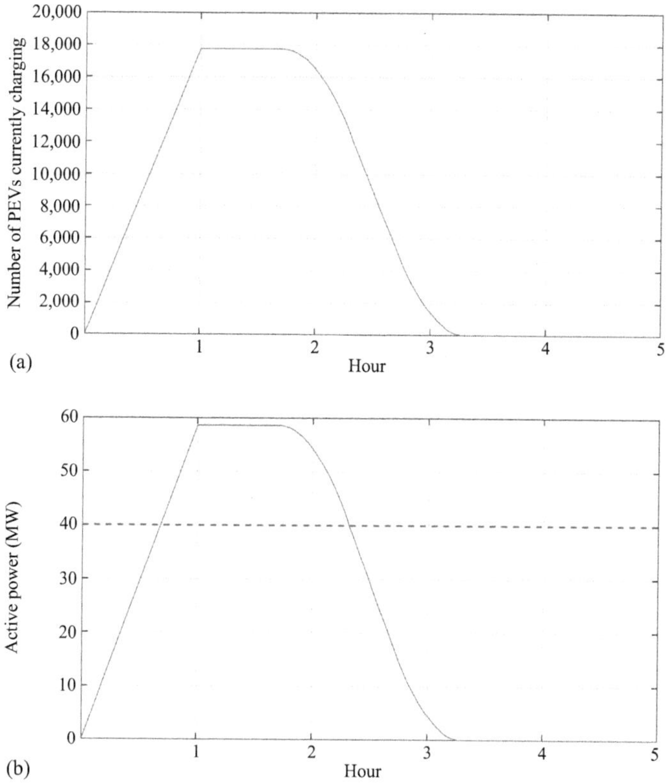

Figure 14.1 Number of electrical vehicles charging (above) and active power consumption (below) in the uncontrolled scenario

worldwide, this is the main concern of power operators, and alternatives to building new power infrastructure and increase $\bar{P}$ is highly desirable.

In the centralized scenario, it is assumed that the power grid knows at each time step how many vehicles are connected everywhere, and can instantaneously adjust all charging rates, so that the required power is never greater than the maximum allowed one. The outcome of this new simulation is now shown in Figure 14.2. In particular, in Figure 14.2(b), it is possible to appreciate that the overall power now never exceeds the available one. Of course, this implies that the fleet of vehicles will be charged in a longer horizon of time. In particular, all the vehicles get charged after 3 h and 50 min, i.e., 35 min later than in the uncontrolled scenario.

In the decentralized scenario, it is assumed that the PEVs either increase or decrease their charging rates according to a broadcast signal that alerts all vehicles when a congestion event occurs. This variation is implemented as the AIMD algorithm with parameters $\alpha = 18$ W, and $\beta = 0.9$. With this value of α, assuming a newly

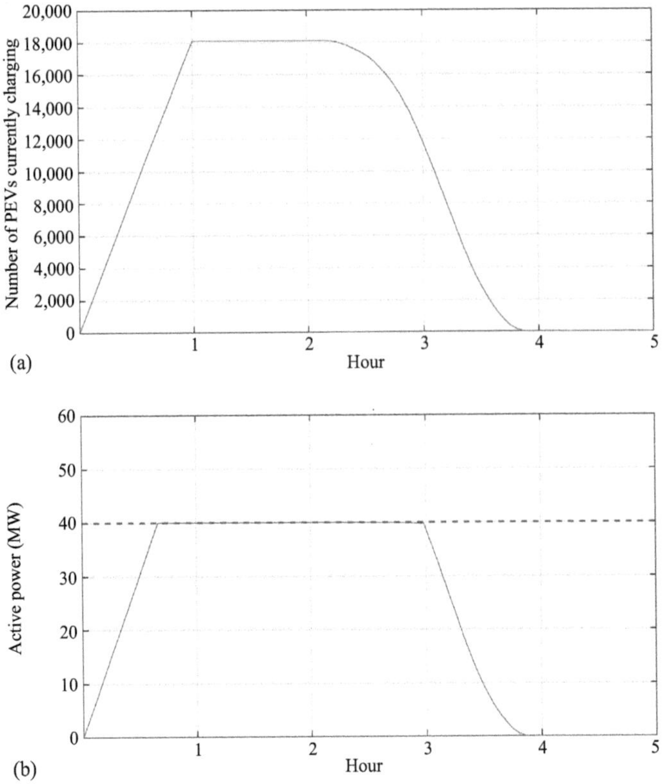

Figure 14.2 Number of electrical vehicles charging (above) and active power consumption (below) in the centralized controlled scenario

connected vehicle is initially charged with half the maximum value, then it would start being charged with full charge (if possible) in about 15 min. Differently from the ideal centralized case, all vehicles are fully charged after about 4 h, i.e., about 10 min later than in the ideal controlled case. This delay is due to the fact that the power allocated to PEVs oscillates below the maximum allowed value, while in the ideal case it is always exactly equal to such a maximum value, as can be seen from Figure 14.3(b).

A single vehicle charging rate is presented in Figure 14.4. Here the vehicle is connected to the charging station at the beginning of the simulation and gradually reaches the peak of 3.3 kW. When too many vehicles are connected for charging, it has to reduce its charging rate according to the AIMD algorithm, until it is fully charged.

As an example, for the unsynchronized AIMD algorithm, here for simplicity we assume that each PEV randomly belongs to one of the four categories with different

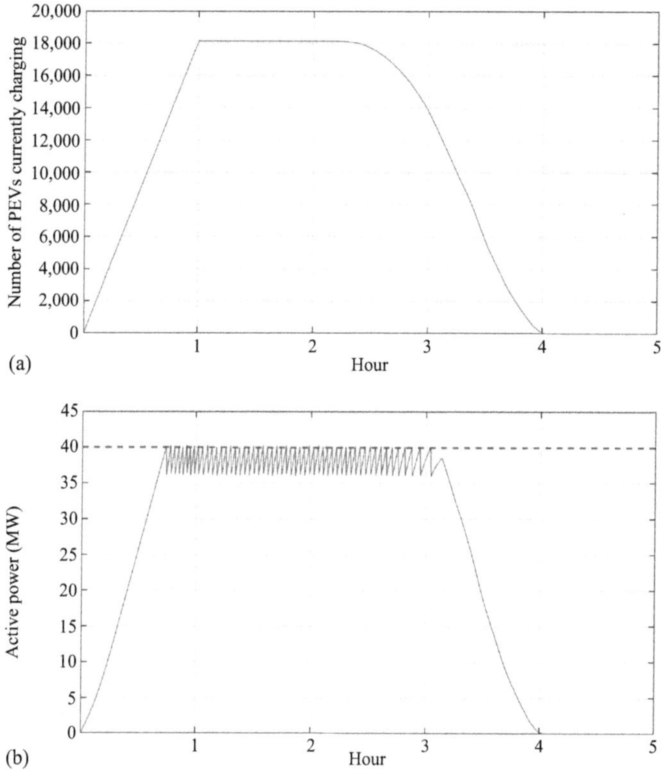

Figure 14.3 Number of electrical vehicles charging (above) and active power consumption (above) in the decentralized scenario with AIMD

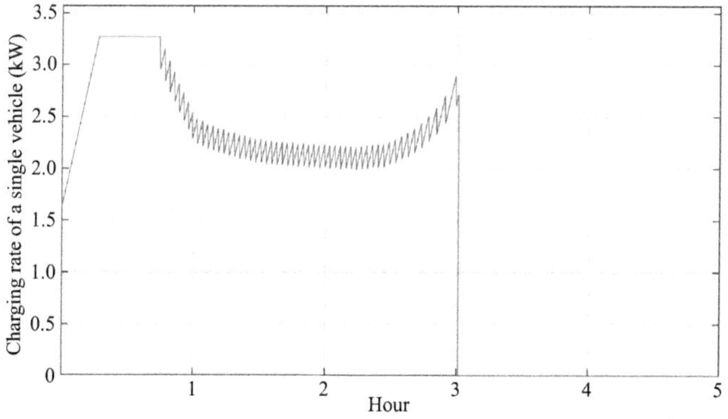

Figure 14.4 Power consumption of a single vehicle in the decentralized scenario

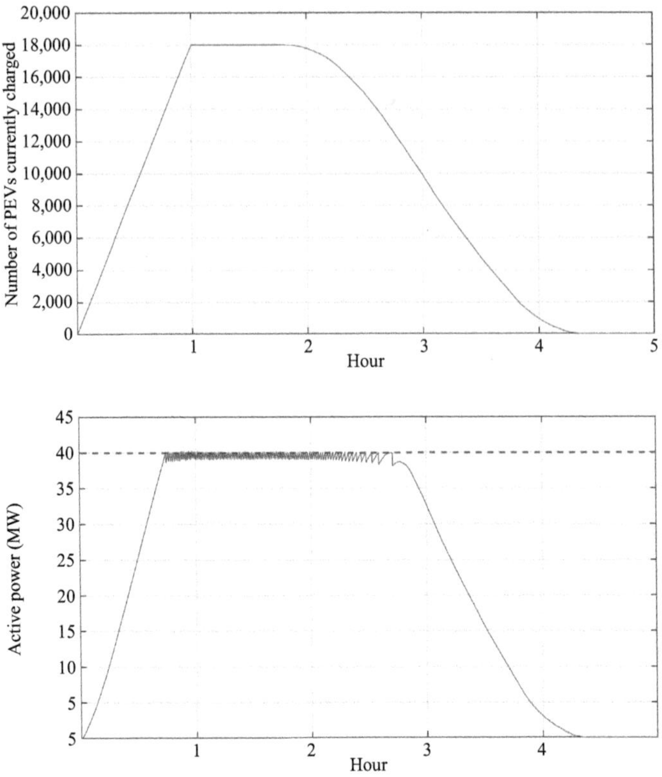

Figure 14.5 Number of electrical vehicles charging (above) and active power consumption (below) in the unsynchronized AIMD scenario

priorities. In particular, PEVs belonging to the highest class of vehicles will be charged without every performing the MD step (i.e., similar to the uncontrolled case), while the other classes perform the MD step with a different probability depending on the category of the vehicle. Figure 14.5 shows the outcome of the simulation, and one can appreciate that similar to the synchronized case, PEVs are charged without having to compute and communicate the number of plugged vehicles, and yet the threshold of power is not exceeded. Details of the charging rates of vehicles belonging to different groups are shown in Figure 14.6 (for the sake of clarity, the figure reports the charging rates of only the first 100 vehicles). The number of the group corresponds to the priority, and vehicles with the highest priority are always charged with the highest possible charging rate. Conversely, PEVs belonging to the first two groups are charged with a lower charging rate on average and finish their charging process later than the others.

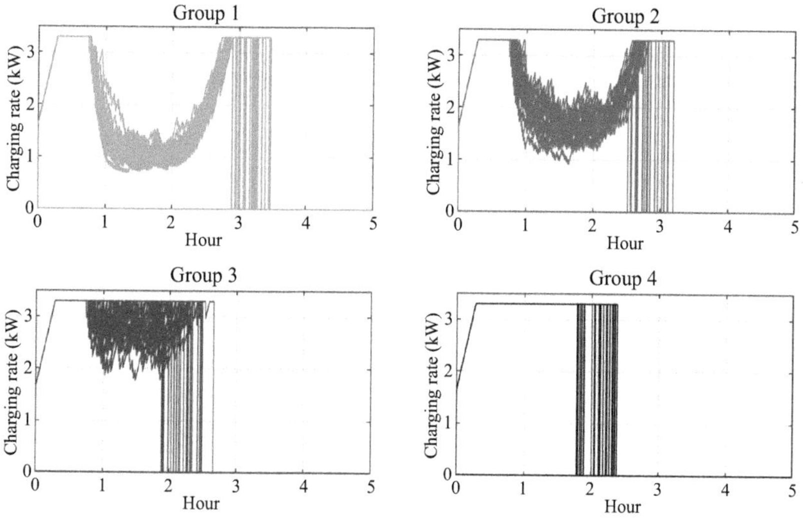

Figure 14.6 *Charging rates of vehicles belonging to different groups. PEVs with highest priority (bottom right) never reduce their charging rate during congestion events, and finish charging before the other PEVs. The PEVs with lower priority (upper left) significantly reduce their charging rates to accommodate the requirements of the power grid*

14.6 Research interests and future trends

This section is devoted to identify some active areas of research around PEVs and to illustrate their related scientific interest and commercial potential. Most of the current research interests lie in improving the technology around PEVs, most notably to increase their driving range, decreasing their weight, improving the efficiency of batteries, and increasing charging rates. However, there is also a number of new topics that may be of utmost importance in the next few years, and here we briefly overview some of them.

14.6.1 Wireless power charging

Wireless power transfer (WPT) is a general term reflecting transmission of energy between emitter and receiver without any physical link but through electromagnetic field. This technology is already spread in our everyday life—radio frequency identification (RFID) tags, toothbrushes, mobile phone chargers, are among the possible applications. This technology enforces mobility and safety, therefore, convenience of devices.

WPT technology has been known for years already. The first implementation of WPT in automotive industry was performed by the soviet electrical engineer Babat, who in 1943 built an electric car, supplied via inductive power transmission. The efficiency of charging at that time with 20 cm air gap was 4% [42]. Partners for advanced transit and highways (PATH) is usually named as the first project, which developed the operating prototype of the dynamic wireless charging of PEVs. Conducting experiments from 1980s to 1990s, they managed to reach 60% efficiency with 8 cm air gap. However, due to the low efficiency and economic feasibility, this project was not commercialized. Nowadays, the rapid spread of PEVs reinforces the interest to wireless charging of PEVs.

The WPT technology can be classified into near- and far-field transmissions: the former are based on magnetic fields and inductive techniques and work on distances shorter than 1 m, while the second ones and far-field electromagnetic transmission (e.g., lasers, microwave and radio-wave transmissions) [43]. Near-field WPT systems are also used for EV charging to wirelessly transfer high-powers, mitigate the main EV operation drawbacks (such as range anxiety) and, in the future, they may replace the currently existing conductive systems [42,44]. The high frequency required for efficient power transmission leads to a drop in the efficiency due to the relevant power losses that occur in the power electronic modules, soft ferromagnetic cores, and wires in which the device consists in, and thus requiring the implementation of thermal management solutions for the system to operate. Litz wires are nowadays state-of-the-art for WPT devices, allowing to reduce the losses that arise from skin effects and proximity effects in the conductors. The widespread of wide band-gap devices (SiC power modules [45]) has enabled a significant efficiency increase of the power electronics moving towards higher switching frequencies.[†] Moreover, the research on novel materials and advanced manufacturing techniques, such as additive manufacturing and powder pressed technology, is envisioned to reduce the power losses of the iron cores operating with high-frequency time-varying field, thus leading to improved efficiency, reduced need for thermal management and a potential increase in the power and range for the transmitted power. High-power WPT for EV charging can be grouped in static and dynamic wireless charging.

Static wireless charging is simpler and more user-friendly than traditional conductive charging. As the static is referring to the position of the car—it can be charged when it is parked or being in stationary modes, for example, in garage or in charging station. For this, the receiver of the vehicle should be perfectly aligned with the transmitter of the charging station, thus requiring an accurate position control system in order to achieve high performance, which increases the complexity of the system. Dynamic wireless charging exploits a distributed charger that is installed underneath the road surface and allows charging of PEVs during the trip. Dynamic charging may mitigate the main barriers to transition to PEVs, as it permits the extension of driving ranges [46], the reduction of battery size [47], and, as a consequence, a higher acceptance of PEVs by drivers [42]. In this case, the position of the car is not of utmost importance, and the efficiency of charging is determined mainly by the traffic flow and the speed of the vehicle. Artificial intelligence and advanced learning algorithms can be developed to track and optimize the power transfer mode in accordance

[†]https://www.autocarpro.in/news-international/bosch-develops-faster-emergency-assistance-for-motorcyclists-56501

with the motion of the receiver through GPS tracking systems. A particular case of dynamic charging is the so-called quasi-dynamic type, which allows charging during the stops, or when the PEV moves slowly, for example, during deceleration or acceleration events. This type of charging is particularly practical, for example, on a bus route, as in the city center, several bus routes are overlapped [48], or in the busy traffic lights. Therefore, frequency of traffic light and bus stops are critical parameters to assess feasibility of quasi-dynamic system performance.

There are numerous examples of successful implementation of such technologies. Among all is worth mentioning the online electric vehicle (OLEV) system [44] developed by the Korea Advanced Institute of Science and Technology (KAIST) which demonstrated the technology in Gumi city on a 24 km route for public transportation, namely electric buses, being capable to supply 100 kW power with 85% of transfer efficiency. Moreover, there are numerous start-ups [8] and companies (i.e., [49]) working on the implementations of wireless charging. Static WPT technology already accepts charging at up to 20 kW and reached 92% efficiency, which is equivalent to mode 2 of cable charging. Some other companies are already providing wireless charging infrastructure for domestic [50], industry, and fleet of vehicles [49].

14.6.2 Peer-to-peer energy exchange

The deployment of privately owned distributed generation based on renewable energy sources pushed the development of peer-to-peer (P2P) energy markets, i.e., an emerging and promising trading alternative to the current centralized energy system. The main traders are the prosumers, which are both consumers of energy, buying the electricity, and sellers of the excess energy generated. The underlying idea is that instead of selling this excess to the grid, they can trade it locally and directly, for example, with their neighbors. The elimination of third party actors and long distance energy transfer may bring several benefits. For example, since the demand is expected to be covered locally, less investments are needed to build and maintain the infrastructure for transmission and upstream generation leading to a reduction on the overall grid losses, which positively reflects on the cost of the energy at the consumers end. Moreover, P2P helps to promote the integration of renewable energy sources and to raise general awareness of energy field among buyers who can choose, for example, between green source of energy and the electricity produced by their neighbors.

Participation of electrical vehicles in P2P markets can be seen from different points of view. On the one side, PEV could be seen as a storage system behaving as a buffer to balance the intermittent renewable energy sources and reduce transactions with the main grid. Indeed, although there are several solutions for storage system operation in P2P trading [51], the need of additional batteries in the local markets is almost unavoidable. Therefore, accommodation of PEVs for these purposes may help to avoid investments on expensive equipment. From the other perspective, PEV can be seen as an independent market participant. P2P energy trading involving PEV could be beneficial during the day time, when vehicles are usually parked while solar radiation is the strongest. Therefore, the possibility to store and trade energy generated by using PV panels may provide benefits to the owners of PEVs as well as for other parties [52] and avoid curtailment of PV generation.

Another example of P2P interaction is on dynamic wireless charging lanes. Assuming that the PEV accepts extra discharging during the trip [53], the PEV can sell the excess energy to the grid, without any discomfort for the driver. In this case, the trading is not limited to certain hours, when the car is parked, but also allows some transactions when the vehicle is moving.

14.6.3 Optimal utilization of electric charge points

The problem of associating PEVs with CSs may be seen as a special instance of the more general problem of associating users with (fewer) shared resources. While the problem may be addressed by increasing the number of CSs (with a large economic effort for building expensive infrastructure), it is interesting to optimally exploit the available infrastructure. In particular, by improving the possibility of conveniently charging a vehicle during the day may alleviate the need to charge all vehicles in a synchronous fashion during the night time. A better utilization of the available charging infrastructure may be obtained by developing suitable software and hardware solutions along two main lines of research:

1. Optimally associating PEVs with CSs. This result can be obtained by enforcing the utilization of reservation schemes, which implies that when a PEV goes to a CS is sure that the charging process will start immediately. Alternatively, some optimized assignment schemes that do not require reservations have been proposed in [54].
2. Sharing private CSs. Private CSs, such as home charging stations, are usually only used by owners for night charging of their own vehicles. Accordingly, such CSs are not used during the day, or in general when the owner is not at home. This represents a waste of available infrastructure, which may be actually shared with other drivers. PEV owners would be interested in this opportunity, as they would get revenues by allowing other people to utilize their own charging infrastructure. In this case, the main problem relies in building devices for actually allowing the utilization of personal charging points, equipped with appropriate paying devices (e.g., see [55]). Also, another concern is the design of appropriate mechanisms to guarantee that the utilization of private charging points does not occur during protected time slots (i.e., when the owner is supposed to be at home).

From this perspective, the problem of optimally utilizing electric charge points shares many points in common to the problem of optimally assigning conventional vehicles to parking spaces, and the rich literature on this latter problem may be used to develop similar algorithms also for the PEVs case. Also, the development of Internet of Things (IoT), and in general the nascent concepts of smart homes, smart cities, and smart everything [56], appear as the key enabling technologies to facilitate the development of such reservation schemes, or mechanisms for easily sharing private equipment. In this context, the connectivity between PEVs fleets, the related interactive infrastructures and the effective management of their interactions is an emerging application of the IoT technology [54].

14.6.4 Distributed ledger technologies

As we have seen, an uncontrolled electrification of the mobility sector may require a significant investment in power generation to handle the increased electric demand. Conversely, a smart management of the power grid may defer such investments but would require a general rethinking of how power grids currently work (e.g., to implement time-varying charge rates; V2G power flows; active/reactive management of the electric fleet; reservation schemes for charging processes; and up to P2P energy exchanges among PEV owners).

Overall, a new management of the power grid requires users to be involved in the energy sector, from a more basic role of energy aware users, i.e., who are willing to shift their energy requirements to more convenient moments in time, for instance in terms of non-dispatchable renewable energy generation [34], to a more advanced role of energy traders and sellers in a fully established P2P market.

Accordingly, the new paradigm requires new policies (e.g., in some countries private citizens are not allowed to sell energy directly to other citizens) and new tools and algorithms to enforce social compliance to such rules. From this perspective, an emerging role is being played by the so-called distributed ledger technologies (DLTs), which is the general term for blockchain and related technologies [57].

In general applications in the domain of smart cities, DLTs are the candidate technology to leverage a number of distributed business models, including peer to peer trading, access to share resources, protect the rights of individuals, without requiring central arbiters. Supported by the recent advances in IoT, applications of interest in the context of PEVs are found in the realization of dockChain Adapters [58] and in the optimized assignment of electric vehicles to charging stations [54], to promote virtuous behaviors of PEV owners and to discourage opportunist actions (e.g., respecting reservation assignments).

14.7 Conclusions

PEVs both pose challenges and raise opportunities to the existing power grids. While challenges are expected to be faced if uncontrolled charging schemes are adopted, PEVs may also represent a large virtual battery at disposal of the power grid, if PEVs are left plugged-in when not in operation, which can be used to deliver services to the grid, and defer new investments to strengthen the existing facilities.

Most of the research on PEVs is currently focused in improving the enabling technologies, most notably batteries and charging stations, but in order to fully unleash their potential, it is important also to change the behavior of PEV owners and to take them into account in the operation of the grid.

This chapter has reviewed the state-of-the-art regarding the charging process of PEVs and their interaction with the power grid. The importance and the impact of different charging strategies have been extensively showcased through illustrative examples. Finally, the last part of the chapter has been devoted to identify and discuss some new and relevant research problems which are supposed to play an important role in the context of PEVs in the next few years.

Bibliography

[1] Global EV Outlook 2021 – Analysis. Available from: https://www.iea.org/reports/global-ev-outlook-2021.

[2] CO_2 Targets Propel Europe to 1st Place in eMobility Race, Transport and Environment, 2021. Available from: https://www.transportenvironment.org/sites/te/files/publications/2020%20EV%20sales%20briefing.pdf.

[3] Worldwide Number of Electric Cars. Available from: https://www.statista.com/statistics/270603/worldwide-number-of-hybrid-and-electric-vehicles-since-2009/.

[4] Moschella M, Murad MAA, Crisostomi E, *et al*. Decentralized charging of plug-in electric vehicles and impact on transmission system dynamics. *IEEE Transactions on Smart Grid*. 2021;12(2):1772–1781.

[5] García-Villalobos J, Zamora I, San Martín JI, *et al.* Plug-in electric vehicles in electric distribution networks: a review of smart charging approaches. *Renewable and Sustainable Energy Reviews*. 2014;38:717–731. Available from: https://www.sciencedirect.com/science/article/pii/S1364032114004924.

[6] Directive2014/94/EU of the European Parliament and of the Council of 22 October 2014 on the Deployment of Alternative Fuels Infrastructure Text with EEA Relevance; 2014. Code Number: 307. Available from: http://data.europa.eu/eli/dir/2014/94/oj/eng.

[7] Lu X, Wang P, Niyato D, *et al.* Wireless charging technologies: fundamentals, standards, and network applications. *IEEE Communications Surveys & Tutorials*. 2015;18(2):1413–1452.

[8] Top Wireless Charging Startups. Available from: https://tracxn.com/d/trending-themes/Startups-in-Wireless-Charging/.

[9] Infante W, Ma J, Liebman A. Operational strategy analysis of electric vehicle battery swapping stations. *IET Electrical Systems in Transportation*. 2018; 8(2):130–135.

[10] Battery SwapTechnologies. Section: E-mobility. Available from: https://www.nio.com/news/nio-es8-launches-nio-house-oslo.

[11] Amsterdam Arena: Used EV Batteries Storage System. Section: E-mobility. Available from: https://www.climateaction.org/news/amsterdam-arena-installs-major-new-battery-storage.

[12] New Technologies For EV Charging | DazeTechnology. Section: E-mobility. Available from: https://www.dazetechnology.com/technologies-charging-for-electric-vehicles/.

[13] IEC 61851-1:2017 | IEC Webstore. Available from: https://webstore.iec.ch/publication/33644.

[14] The 4 Electric Vehicle Charging Modes | DazeTechnology. Section: E-mobility. Available from: https://www.dazetechnology.com/charging-modes-for-ev/.

[15] Chen T, Zhang XP, Wang J, *et al.* A review on electric vehicle charging infrastructure development in the UK. *Journal of Modern Power Systems and Clean Energy*. 2020;8(2):193–205.

[16] Doll S. Longest-Range Electric Vehicles (EVs) You Can Buy in 2021, 2021. Available from: https://electrek.co/2021/07/03/longest-range-evs-2021/.

[17] Crisostomi E, Shorten R, Stüdli S, *et al. Electric and Plug-in Hybrid Vehicle Networks: Optimization and Control.* Boca Raton, FL: CRC Press, 2017.

[18] Yilmaz M, Krein PT. Review of the impact of vehicle-to-grid technologies on distribution systems and utility interfaces. *IEEE Transactions on Power Electronics*. 2013;28(12):5673–5689.

[19] Kikhanadi MR, Hajizadeh A, Shahirinia A. Charging coordination and load balancing of plug-in electric vehicles in unbalanced low-voltage distribution systems. *IET Generation, Transmission & Distribution*. 2020 Feb;14(3):389–399. Available from: http://www.scopus.com/inward/record.url?scp=85078203134&partnerID=8YFLogxK.

[20] Gou F, Yang J, Zang T. Ordered charging strategy for electric vehicles based on load balancing. In: 2017 IEEE Conference on Energy Internet and Energy System Integration (EI2); 2017. p. 1–5.

[21] Quinn C, Zimmerle D, Bradley TH. An evaluation of state-of-charge limitations and actuation signal energy content on plug-in hybrid electric vehicle, vehicle-to-grid reliability, and economics. *IEEE Transactions on Smart Grid*. 2012;3(1):483–491.

[22] Hussain MT, Sulaiman DNB, Hussain MS, *et al.* Optimal Management strategies to solve issues of grid having electric vehicles (EV): a review. *Journal of Energy Storage*. 2021;33:102114. Available from: https://www.sciencedirect.com/science/article/pii/S2352152X20319435.

[23] Pillai JR, Bak-Jensen B. Vehicle-to-grid systems for frequency regulation in an Islanded Danish distribution network. In: 2010 IEEE Vehicle Power and Propulsion Conference, 2010, pp. 1–6. ISSN: 1938-8756.

[24] Druitt J, Früh WG. Simulation of demand management and grid balancing with electric vehicles. *Journal of Power Sources*. 2012;216:104–116. Available from: https://www.sciencedirect.com/science/article/pii/S0378775312008907.

[25] Clement-Nyns K, Haesen E, Driesen J. The impact of charging plug-in hybrid electric vehicles on a residential distribution grid. *IEEE Transactions on power systems*. 2009;25(1):371–380.

[26] Xu J, Wong VW. An approximate dynamic programming approach for coordinated charging control at vehicle-to-grid aggregator. In: 2011 IEEE International Conference on Smart Grid Communications (SmartGridComm). IEEE, 2011. pp. 279–284.

[27] Leemput N, Van Roy J, Geth F, *et al.* Comparative analysis of coordination strategies for electric vehicles. In: 2011 2nd IEEE PES International Conference and Exhibition on Innovative Smart Grid Technologies. IEEE, 2011, pp. 1–8.

[28] V2G Hub | V2G Around the World: V2G Hub | V2G Around the World. Available from: https://www.v2g-hub.com/insights/.

[29] Thompson AW, Perez Y. Vehicle-to-everything (V2X) energy services, value streams, and regulatory policy implications. *Energy Policy*. 2020; 137:111136. Available from: https://www.sciencedirect.com/science/article/pii/S0301421519307244.

[30] Han X, Lu L, Zheng Y, *et al.* A review on the key issues of the lithium ion battery degradation among the whole life cycle. *eTransportation*. 2019;1:100005. Available from: https://www.sciencedirect.com/science/article/pii/S2590116819300050.

[31] Han S, Han S, Sezaki K. Economic assessment on V2G frequency regulation regarding the battery degradation. In: 2012 IEEE PES Innovative Smart Grid Technologies (ISGT). IEEE, 2012, pp. 1–6.

[32] Beer S, Gomez T, Dallinger D, *et al.* An economic analysis of used electric vehicle batteries integrated into commercial building microgrids. *IEEE Transactions on Smart Grid*. 2012;3(1):517–525.

[33] Sovacool BK, Noel L, Axsen J, *et al.* The neglected social dimensions to a vehicle-to-grid (V2G) transition: a critical and systematic review. *Environmental Research Letters*. 2018;13(1):013001.

[34] Gu Y, Häusler F, Griggs W, *et al.* Smart procurement of naturally generated energy (SPONGE) for PHEVs. *International Journal of Control*. 2016;89(7):1467–1480.

[35] Stüdli S, Crisostomi E, Middleton R, *et al.* A flexible distributed framework for realising electric and plug-in hybrid vehicle charging policies. *International Journal of Control*. 2012;85(8):1130–1145.

[36] Chiu DM, Jain R. Analysis of the increase and decrease algorithms for congestion avoidance in computer networks. *Computer Networks and ISDN Systems*. 1989;17(1):1–14.

[37] Srikant R. *The Mathematics of Internet Congestion Control, in Systems & Control: Foundations & Applications*. New York, NY: Birkäuser, 2003.

[38] Beil L, Hiskens I. A distributed wireless testbed for plug-in hybrid electric vehicle control algorithms. In: Proceedings of the North American Power Symposium. Champaign, IL, USA, 2012, pp. 1–5.

[39] Ravasio M, Incremona GP, Colaneri P, *et al.* Distributed nonlinear AIMD algorithms for electric bus charging plants. *Energies*. 2021;14(4389):1–17.

[40] Ucer E, Kisacikoglu MC, Yuksel M. Analysis of decentralized AIMD-based EV charging control. In: IEEE Power & Energy Society General Meeting. Atlanta, GA, 2019, pp. 1–5.

[41] Wirth FR, Stüdli S, Yu JY, *et al.* Nonhomogeneous place-dependent Markov Chains, unsynchronised AIMD, and optimisation. *Journal of the ACM*. 2019;66(4):1–37.

[42] Cirimele V, Diana M, Freschi F, *et al.* Inductive power transfer for automotive applications: state-of-the-art and future trends. *IEEE Transactions on Industry Applications*. 2018;54(5):4069–4079.

[43] Jang YJ. Survey of the operation and system study on wireless charging electric vehicle systems. *Transportation Research Part C: Emerging Technologies*. 2018;95:844–866.

[44] Song K, Koh KE, Zhu C, *et al.* A review of dynamic wireless power transfer for in-motion electric vehicles. In: *Wireless Power Transfer—Fundamentals and Technologies*. Rijeka, Croatia: InTech, 2016, pp. 109–128.

[45] Silicon Carbide Power Modules. Section: WPT. Available from: https://www.infineon.com/cms/en/product/technology/silicon-carbide-sic/.

[46] Chopra S, Bauer P. Driving range extension of EV with on-road contactless power transfer: a case study. *IEEE Transactions on Industrial Electronics*. 2011;60(1):329–338.

[47] Bi Z, Song L, De Kleine R, *et al.* Plug-in vs. wireless charging: life cycle energy and greenhouse gas emissions for an electric bus system. *Applied Energy*. 2015;146:11–19.

[48] Hwang I, Jang YJ, Ko YD, *et al.* System optimization for dynamic wireless charging electric vehicles operating in a multiple-route environment. *IEEE Transactions on Intelligent Transportation Systems*. 2017;19(6):1709–1726.

[49] Wireless Power EV Charging for Domestic and Industrial Applications. Section: E-mobility. Available from: https://witricity.com/wireless-charging-solutions/.

[50] Evatran. Meet Plugless | The Wireless EV Charging Station. Available from: https://www.pluglesspower.com/.

[51] Lüth A, Zepter JM, del Granado PC, *et al.* Local electricity market designs for peer-to-peer trading: the role of battery flexibility. *Applied Energy*. 2018;229:1233–1243.

[52] Aznavi S, Fajri P, Shadmand MB, *et al.* Peer-to-peer operation strategy of pv equipped office buildings and charging stations considering electric vehicle energy pricing. *IEEE Transactions on Industry Applications*. 2020;56(5):5848–5857.

[53] Nguyen DH. Electric vehicle—wireless charging–discharging lane decentralized peer-to-peer energy trading. *IEEE Access*. 2020;8:179616–179625.

[54] Moschella M, Ferraro P, Crisostomi E, *et al.* Decentralized assignment of electric vehicles at charging stations based on personalized cost functions and distributed ledger technologies. *IEEE Transactions on Internet of Things*. 2021;8(14):11112–11122.

[55] Thompson E, Ordóñez Hurtado RH, Griggs W, *et al.* On charge point anxiety and the sharing economy. In: IEEE International Conference on Intelligent Transportation Systems (ITSC). Yokohama, Japan, 2017.

[56] Khan MA, Salah K. IoT security: review, blockchain solutions, and open challenges. *Future Generation Computer Systems*. 2018;82:395–411.

[57] Ferraro P, King C, Shorten R. Distributed ledger technology for smart cities, the sharing economy, and social compliance. *IEEE Access*. 2018;6:62728–62746.

[58] O' Connell J, Cardiff B, Shorten R. dockChain: A solution for electric vehicles charge point anxiety. In: IEEE International Conference on Intelligent Transportation Systems (ITSC), 2018.

Part III

Stability Analysis

Chapter 15

Time-domain simulation for transient stability analysis

Massimo La Scala[1]

It has been widely recognized that time-domain simulation is the most accurate method to describe power system transient behaviour since it can represent 'as they are' controls, non-linearities, saturation, strong dissipative effects and the 'silent sentinels', i.e., the protection system. To counterbalance this interesting feature of the approach, there is the formidable computational burden associated to the simulation of real systems when real-time framework is required for dynamic security assessment, control, etc. However, the structure of the problem presents some interesting characteristics which allowed the use of parallel/distributed computing. This chapter synthesizes the results obtained by the authors in this field.

15.1 Introduction

Time-domain simulation techniques are widely used in the power industry because of their versatility and accuracy. The basic principle consists in formulating circuit equations according to the topology and the physical nature of elementary components and to solve them by numerical computation.

It has been widely recognized that time-domain simulation is the most accurate method to describe power system transient behaviour since it represents controls, non-linearities, saturations, strong dissipative effects and the so-called silent sentinels (i.e., the protection system) 'as they are'. This interesting feature is counterbalanced by the formidable computational burden associated to the simulation of real systems especially when real-time calculations are required. Consequently, sparsity preservation is an important feature for simulation of large-scale systems. Power community gave a fundamental contribution to the development of the sparsity technology initially solving large steady-state problems [1] and successively applying these techniques to dynamical problems such as transient stability assessment (TSA) by time-domain simulation. In the latter case, it is widely recognized that formulating the set of equations through the so-called sparse formulation based on a set of differential-algebraic equations (DAEs) [2] is a very efficient way to perform calculations.

[1]Department of Electrical Engineering and Information, Politecnico di Bari, Italy

Furthermore, it was observed, in the last decades, that most blackouts have been due to poor dynamic performances under large disturbances and instabilities occurred in nowadays stressed power systems. The main causes of instability on the transient time scale can be associated to two phenomena, namely: rotor angle instability and voltage instability. Power engineers face the problem that, while stability is increasingly a limiting factor in secure system operation, a detailed simulation of system dynamic response is very time consuming.

Another feature of time-domain simulation is linked to the 'stiffness' associated to very different time constants usually considered in power system representation. A typical example is the simulation of the dynamic behaviour associated with voltage instability problems which involve a time horizon from seconds to minutes. In this case, it is desirable to combine transient stability simulations and long-term analysis into a single computer program differently from the usual formulations which tend to separate the dynamic behaviour of power systems into different time horizons. Large effort has been spent in 1990s in this direction [3–5]. In all cases, this formidable stiff problem can be solved by the use of variable step size and variable order integration algorithms. Singular perturbation theory and time-scale decoupling were proposed to solve stiff problems, and this approach is implicitly adopted in most transient stability codes where very fast dynamics, such as the one associated to electrical transients of lines, are represented by algebraic equations assumed instantaneous, stable and in steady-state conditions whereas very slow time response (boiler, turbine, etc.) are assumed still in steady-state conditions [6]. Relying on time-scale decomposition, more recently, a quasi-steady-state approximation of long-term dynamics, which, basically, replaces faster phenomena by their equilibrium conditions, was effectively introduced for reducing the complexity of the system representation while increasing the computation efficiency of time simulations. This approach resulted fruitful in time-domain simulation of voltage stability problems in long-term studies [7].

Each power system stability analysis involves the step-by-step solution in the time domain of perhaps several thousands of non-linear DAEs and, even worse, this analysis needs to be performed for tens or hundreds of cases, possibly in real time. In fact, TSA concerns the simulation of faults and anomalous conditions (contingencies), which can produce instability in power networks. In particular, the *online TSA* is an extremely challenging computing problem [8], requiring in real time (up to 10–15 min) the preventive simulation of the system under a set of fault conditions and outages. A number of probable contingencies must be simulated in a short-time horizon in order to evaluate possible instability conditions and plan appropriate corrective actions. These preventive simulation and planning actions are constantly repeated until a system operator, by an online evaluation of the power network state, detects unsafe operating conditions. Thus, when a contingency really occurs, appropriate corrective actions are known and can be triggered. The TSA is a sub-problem of the more general dynamic security assessment (DSA). The DSA function consists basically of a contingency screening, a simulation of a set of faults and outages, a post-contingency analysis and a security assessment. This chapter refers to the TSA simulation engine and the transient time scale in the range of seconds.

The need for online stability analysis in control centres has motivated researchers to develop algorithms and computer architectures which could allow hundreds of

TSA simulations to be performed in real time. In late 1970s, the possibility to explore potentials of traditional algorithms in exploiting innovative computer architecture was pursued. It was shown, at that time, that only a limited amount of parallelism could be exploited [9–12].

New classes of algorithms and discretization rules were introduced in late 1980s to take advantage of high-performance computing due to parallel/vector processing and, later, to distributed computing.

Experiments running different TSA algorithms both on parallel, vector and distributed machines were conducted at that time. Since computer power and technologies are continuously evolving, it seems, here, more interesting to focus on the structure of the time-domain simulation than architecture and computing details. The structure of the problem presents some interesting features which can take advantage of ever-changing computing and communication technologies for real-time applications.

15.2 Time-domain simulations and transient stability

Conventional power system stability studies compute the system response to a sequence of large disturbances, usually a generator outage or a network short circuit, followed by protective branch switching operations. The process is a direct simulation in the time domain of duration varying between 1 s and 20 min or more. System modelling reflects the fact that different components of the power system have their greatest influences on stability at different stages of the response. In fact, short-term models emphasize the rapidly responding system electrical components whereas long-term models deal with the representation of slowly oscillatory system power balance, assuming that the fast electromechanical transients have damped out. A classification which is usually adopted in this field consists in defining 'transient stability' the short-term problem covering the post-disturbance times of up to 5–10 s whereas anything in the long term is associated to the concepts of frequency stability and long-term voltage stability.

Long-term stability analysis can be implemented in real time on ordinary computing resources in control centres whereas the online implementation of the TSA requires unconventional computers and algorithms depending on the number of simulation and time requirements. Here, the focus is on power system transient stability representation and related algorithms although methods are general enough to be applied to long-term stability analysis as well.

Power system TSA leads to the solution of non-linear systems of DAEs as well as in many electrical problems involving transient network analysis and continuous system simulation. This formulation, known as sparse formulation in the power system realm, derives from the strong need to preserve the sparsity of large-scale networks [2].

In general, initial value problems (IVPs) for DAEs can be formulated as follows:

$$\boldsymbol{F}(t,\boldsymbol{y},\dot{\boldsymbol{y}}) = 0 \tag{15.1}$$

where $\boldsymbol{F}, \boldsymbol{y}, \dot{\boldsymbol{y}}$ are s-dimensional vectors and $\boldsymbol{F}$ is assumed suitably differentiable.

It is possible to associate the matrix pencil $\left(\frac{\partial F}{\partial \dot{y}}, \frac{\partial F}{\partial y}\right)$ to problem (15.1) [13]. Its nilpotency index depends on $t, y, \dot{y}$; in many cases of practical importance, however, the structure of the pencil is fixed.

The power system equations have well-known structures, even for their detailed representation in transient stability analysis [2], and it can be assumed valid over a wide range of possible specific details to be represented.

For typical power system representations, the system (15.1) can be arranged as follows:

$$\dot{\mathbf{x}}(t) = \mathbf{f}(\mathbf{x}(t), \mathbf{V}(t)) \tag{15.2a}$$

$$\mathbf{g}(\mathbf{x}(t), \mathbf{V}(t)) = 0 \tag{15.2b}$$

$$\mathbf{x}(t_0) = \mathbf{x}_0 \tag{15.2c}$$

where the s-dimensional vector $\mathbf{y}$ is partitioned in two subvectors $\mathbf{x}$ and $\mathbf{V}$, namely the vector of state variables and nodal voltages such that $\mathbf{y} = [\mathbf{x}^{\mathrm{T}}, \mathbf{V}^{\mathrm{T}}]^{\mathrm{T}}$ and $\mathbf{V}$ is a vector of the same dimension as $\mathbf{g}$.

In particular, $\mathbf{x}$ represents the state vector with dimension $p = \sum_{i=1}^{m} p_i$, being p_i the state vector dimension of each synchronous generator, and $\mathbf{V}$ is the $2n$-dimensional vector of the nodal voltages being n the number of busbars. In (15.2c), $\mathbf{x}_0$ denotes the state vector at the initial time t_0 in steady-state conditions. Equation (15.1) represents the dynamic model of synchronous machines, whereas (15.2) denotes the coupling effect of the power system network.

It has been already specified how grid dynamics is typically an electrical phenomenon and characterized by a faster response than the electromechanical time response since it depends on inductance and capacitance of different components (lines, transformers, etc.). Thus, the differential equations associated to the grid degenerate into algebraic equations [2, 6].

The problem (15.2) is characterized by the nilpotency index equal to 1 for all t, $\mathbf{x}$ and $\mathbf{V}$ such that the inverse of the Jacobian $\frac{\partial g}{\partial x}$ exists and is bounded [13]. In power system TSA the index 1 condition is assured for most of practical cases with the exception of operating conditions and trajectories close to voltage collapse.

Typical properties of DAEs utilized in transient stability analysis are the stiffness and the mildly non-linearity. Equations (15.2a) and (15.2b) can be written in a more compact form

$$\boldsymbol{E}\dot{\boldsymbol{y}} = \boldsymbol{h}(t, \boldsymbol{y}) \tag{15.3}$$

where $\boldsymbol{E}$ = block diag[$\mathbf{I}$, $\mathbf{0}$] and $\boldsymbol{h} = [\mathbf{f}^{\mathrm{T}}, \mathbf{g}^{\mathrm{T}}]^{\mathrm{T}}$ and $\boldsymbol{y}$ and $\boldsymbol{h}$ are s-dimensional vectors.

In order to solve DAEs as in (15.3), it is necessary to adopt an integration rule. Different integration rules were adopted to reach this scope as it can observed by the abundant literature in this field reviewed in Reference 2. Codes developed for transient stability simulation usually adopt the trapezoidal rule: it is numerically stable but, differently from many other methods [2], does not yield falsely stable system responses in the presence of fast, non-quiescent components. This is a fundamental issue during the rapid electrical response of the power system after a large discontinuity, when the excitation control system, one of the fastest components in TSA, is heavily involved. It is also very important to prevent falsely stable trajectories while simulating unstable cases, because the main goal of the TSA is the design of control actions to enhance power system stability.

An interesting generalization of implicit integration rules has been provided in Reference 14 where a large class of discretization rules, the so-called boundary value methods (BVMs), has been applied to DAEs. These methods were applied for the first time to non-linear DAEs problems in Reference 14 although the transformation of an IVP into a boundary value one can be referred to the work of Miller [15] and Olver [16] and the solution of ordinary differential equations (ODEs) by applying a BVM was treated in References 17–19. These discretization rules can describe a large class of existing implicit methods and may introduce new methods which share interesting numerical properties such as 0-stability, A-stability as shown in Reference 14. For these reasons, they have been adopted here for the following developments.

The concept of A-stability is particularly useful for solving stiff differential equations. Methods characterized by this property have no stability-imposed restrictions on the step size h. Consequently, the stability of the real phenomena is kept also for the simulated trajectory. In the case of A-stability, a step size longer than the shortest time constants of the system will produce no divergent oscillation induced by numerical problems.

It should be recalled that A-stability is judged on the basis of a very simple first-order test equation:

$$\begin{aligned} &\dot{\mathbf{y}}(t) = \boldsymbol{\lambda}\mathbf{y}(t), \quad t \in [t_0, t_f], \quad \mathrm{Re}(\lambda) < 0 \\ &\mathbf{y}(t_0) = \mathbf{y}_0 \end{aligned}$$

For classical linear-step formulas applied to the IVP, the solution space is the linear combination of $\mathbf{z}_j^i$ where superscript 'i' denotes the time step of the integration rule and $\mathbf{z}_j, j = 1, \ldots, k$ are the roots of the characteristic polynomial. Only the root $z_1 \sim \mathrm{e}^{\lambda h}$ with the smallest modulus is interesting, while the others are considered parasitic. This is related to the classical theory of Dahlquist [20] which essentially requires that all the other roots have modulus smaller than one. Moreover, this is a hard drawback since the regions of complex plane where this requirement is fulfilled are usually relatively small for explicit methods. They become large for implicit A-stable methods but, in this case, the accuracy order of the methods cannot be higher than two (Dahlquist barrier). Alternative selection criterion could be chosen, as solving an IVP by means of BVMs with k_1 initial and k_2 final conditions. In this case, stability is preserved if k_1 roots are inside and k_2 roots are outside the unit circle as stated by a theorem reported in Reference 14. The main advantage of this approach is that the Dahlquist barrier can be overcome as in some BVM methods such as the extended trapezoidal rule (ETR) and reverse Adams methods (RAMs) which will be discussed in the followings. Consequently, if a fixed accuracy is needed, larger time steps can be used and the property of A-stability is preserved.

Another important feature, while solving ODEs and DAEs, is to avoid the so-called hyperstability, consisting in obtaining a stable numerical solution for a physically unstable phenomenon. This numerical condition must be prevented especially for TSA because the objective of the simulation is to single out and isolate the large disturbances (contingencies) which cause the physical instability and this would be impossible in case of ‘hyperstability’. For conventional algorithms, the solution is provided by the reduction of the step size during simulations according to the end user’s experience or the evaluation of eigenvalues; this second possibility implies the availability of an efficient eigenvalue computation tool for large-scale systems.

The only classical multistep formula which is perfectly A-stable is the trapezoidal rule; another class of linear multistep methods for DAEs are the backward differentiation formulas (BDFs). It is interesting to remind that the first implementation of the code EUROSTAG [21], relied on the general Gear algorithm, made intensive use of BDFs method. It was observed that BDFs of order 1 and 2 suffered hyperstability whereas the BDFs of order greater than 2 were not A-stable. Subsequently, the code was modified with successfully adopting mixed Adams-BDF variable step-size algorithm to avoid hyperstability [21].

In Reference 14, it was demonstrated that an A-stable method prevents hyperstability if and only if it is perfectly A-stable.

Let us define for simplicity a constant mesh over the time interval $[t_0, t_f]$; let $t_i = t_0 + ih$, for $i = 1, \ldots, n$, be the mesh points, with $t_n \geq t_f$. The continuous problem (15.3) is then discretized by means of a k-step linear formula with k_1 initial and $k_2 = k - k_1$ final conditions. It assumes the form:

$$\boldsymbol{y}_0, \boldsymbol{y}_1, \ldots, \boldsymbol{y}_{k_1-1}, \quad \text{given}$$

$$\boldsymbol{y}_{n-k_2+1}, \ldots, \boldsymbol{y}_n \quad \text{given}$$

$$\sum_{j=0}^{k} \alpha_j \boldsymbol{E}\boldsymbol{y}_{i-k_1+j} = h \sum_{j=0}^{k} \beta_j \boldsymbol{h}_{i-k_1+j}, \quad i = k_1, \ldots, n - k_2 \tag{15.4}$$

where $\mathbf{y}_i$ is an approximation of $\mathbf{y}(t_i)$ and $\mathbf{h}_i = \mathbf{h}(t_i, \mathbf{y}_i)$.

For each index i, this formula uses approximations of the solution at k_1 left points and k_2 right points for providing an approximation of the solution at t_i with a given order p. The initial additional conditions $[\boldsymbol{y}_0, \boldsymbol{y}_1, \ldots, \boldsymbol{y}_{k_1-1}]$ and the final additional conditions $[\boldsymbol{y}_{n-k_2+1}, \ldots, \boldsymbol{y}_n]$ are implicitly approximated by means of auxiliary multistep methods (of order p).

A k-step BVM with k_1 initial and $k_2 = k - k_1$ final conditions applied to (15.3) can be formulated as:

$$\boldsymbol{y}_0 \quad \text{given}$$

$$\sum_{j=0}^{k^r} \alpha_j^r \boldsymbol{E}\boldsymbol{y}_j = h \sum_{j=0}^{k^r} \beta_j^r \boldsymbol{h}_j, \quad r = 1, \ldots, k_1 - 1 \tag{15.5a}$$

$$\sum_{j=0}^{k} \alpha_j \boldsymbol{E}\boldsymbol{y}_{i-k_1+j} = h \sum_{j=0}^{k} \beta_j \boldsymbol{h}_{i-k_1+j}, \quad i = k_1, \ldots, n - k_2 \tag{15.5b}$$

$$\sum_{j=0}^{k^r} \alpha_j^r \boldsymbol{E}\boldsymbol{y}_{n-k^r+j} = h \sum_{j=0}^{k^r} \beta_j^r \boldsymbol{h}_{n-k^r+j}, \quad r = k_1 + 1, \ldots, k \tag{15.5c}$$

The $k-1$ formulas (15.5a) and (15.5c) are named, respectively, initial and final methods and their number does not depend on the size of the mesh. The number of steps adopted for the auxiliary methods (k^r) and the coefficients α_j^r and β_j^r must be selected to ensure an order p accuracy. Equation (15.5b) defines the basic method which has been proved to determine the behaviour of the solution when n is large. From (15.5), it can be observed that a main peculiarity of BVMs consists in the possibility of concurrently compute the numerical solution of a given order at all points of a prescribed integration interval. In next sections, it is shown how this feature is very attractive especially for parallel-in-time implementations.

The BVM approach can yield different classes of methods since many integration rules can derive from (15.5). One of them consists in the so-called ETR methods that can preserve, at any order, the stability properties of the trapezoidal rule overcoming the Dahlquist barrier [14]. ETR methods are perfectly A-stable, i.e., are A-stable methods which prevent hyperstability.

ETR methods can be defined as follows:

$$k = 2\nu + 1, \quad k_1 = \nu + 1, \quad k_2 = \nu, \quad \nu = 1, 2, \ldots$$

y_0 given

$$\boldsymbol{E}(\boldsymbol{y}_r - \boldsymbol{y}_{r-1}) = h\sum_{j=0}^{k} \beta_j^r \boldsymbol{h}_j, \quad r = 1, \ldots, \nu \tag{15.6a}$$

$$\boldsymbol{E}(\boldsymbol{y}_i - \boldsymbol{y}_{i-1}) = h\sum_{j=0}^{\nu} \beta_j(\boldsymbol{h}_{i-\nu-1+j} + \boldsymbol{h}_{i+\nu-j}), \quad i = \nu + 1, \ldots, n - \nu \tag{15.6b}$$

$$\boldsymbol{E}(\boldsymbol{y}_{n-\nu+r} - \boldsymbol{y}_{n-\nu+r-1}) = h\sum_{j=0}^{k} \beta_{k-j}^r \boldsymbol{h}_{n-k+j}, \quad r = 1, \ldots, \nu \tag{15.6c}$$

In order to ensure that these methods have order $k+1$, right coefficients can be chosen as in Table 15.1.

It can be observed that if $\nu = 0$ is assumed, the trapezoidal rule (ETR2), largely adopted for time-domain simulation in transient stability, can be obtained.

The formulation (15.5) can also lead to the k-step RAMs of order $k+1$. These methods can be formulated as:

$$k_1 = 1, \quad k_2 = k - 1$$

Table 15.1 Coefficients of ETR of order 4 (ETR4)

j	β_j^1	β_j
0	9/24	−1/12
1	19/24	13/24
2	−5/24	13/24
3	1/24	−1/24

Table 15.2 Coefficients of RAM of order 3

j	β_j	β_j^2
0	5/12	−1/12
1	8/12	8/12
2	−1/12	5/12

y_0 given

$$\boldsymbol{E}(\boldsymbol{y}_i - \boldsymbol{y}_{i-1}) = h\sum_{j=0}^{k}\beta_j\boldsymbol{h}_{i-1+j}, \quad i = 1,\dots,n-k+1 \tag{15.7a}$$

$$\boldsymbol{E}(\boldsymbol{y}_{n-k+r} - \boldsymbol{y}_{n-k+r-1}) = h\sum_{j=0}^{k}\beta_j^r\boldsymbol{h}_{n-k+j}, \quad r = 2,\dots,k \tag{15.7b}$$

Coefficients to ensure an order 3 are provided in Table 15.2.

The convergence properties of BVMs applied to ODEs were extended to DAEs of the form (15.2) in Reference 14.

On the basis of (15.2), the problem (15.5) may be recast in matrix notation as:

$$(\boldsymbol{A}\otimes\boldsymbol{I})\hat{\boldsymbol{x}} - h(\boldsymbol{B}\otimes\boldsymbol{I})\hat{\boldsymbol{f}} - \hat{\boldsymbol{b}} = 0 \tag{15.8}$$

$$(\boldsymbol{B}\otimes\boldsymbol{I})\hat{\boldsymbol{g}} = 0$$

where $\boldsymbol{A}$ and $\boldsymbol{B}$ have entries α_j^r, α_j and β_j^r, β_j, respectively, $\hat{\boldsymbol{x}} = [\mathbf{x}_i^{\mathrm{T}},\dots,\mathbf{x}_n^{\mathrm{T}}]^{\mathrm{T}}$, $\hat{\boldsymbol{f}} = [\mathbf{f}_i^{\mathrm{T}},\dots,\mathbf{f}_n^{\mathrm{T}}]^{\mathrm{T}}$, $\hat{\boldsymbol{g}} = [\mathbf{g}_i^{\mathrm{T}},\dots,\mathbf{g}_n^{\mathrm{T}}]^{\mathrm{T}}$, the vector $\hat{\boldsymbol{b}}$ contains the known terms in $\boldsymbol{y}_0$ and $\otimes$ denotes Kronecker tensor product.

Formulation in (15.8) is very general and applies to a large class of BVMs but, for the specific methods introduced here, it should be observed that the matrix **B** appearing in (15.8) is invertible. Thus, it is possible to simplify the approach into the following problem:

$$(\boldsymbol{A}\otimes\boldsymbol{I})\hat{\boldsymbol{x}} - h(\boldsymbol{B}\otimes\boldsymbol{I})\hat{\boldsymbol{f}} - \hat{\boldsymbol{b}} = 0 \tag{15.9}$$

$$\hat{\boldsymbol{g}} = 0$$

This formulation is particularly useful since it enhance the sparsity of the problem.

In order to evaluate the error associated to these discretization formulas the following theorem applies.

Theorem 15.1. *Let us suppose that the system (15.2) satisfies:*

I. *$(\delta\mathbf{g}/\delta\boldsymbol{V})(t,\boldsymbol{x},\boldsymbol{V})$ is invertible in a neighbourhood of the solution of (15.2);*
II. *the initial values $\boldsymbol{x}_0$ and $\boldsymbol{V}_0$ are consistent, that is $\mathbf{g}(t_0,\boldsymbol{x}_0,\boldsymbol{V}_0) = 0$;*
III. *the BVM is of order p and characterized by an invertible matrix B.*

Then the solution of (15.12) $\hat{y} = [\boldsymbol{y}_1^T,\boldsymbol{y}_2^T,\dots,\boldsymbol{y}_i^T,\dots,\boldsymbol{y}_n^T]^T$ has global error:

$$\|\boldsymbol{y}_n - \boldsymbol{y}(\boldsymbol{t}_n)\| = O(h^P), \quad t_n - t_0 = nh \le constant.$$

Proof is reported in Reference 14.

15.3 Transient stability and high-performance computing

Power industry from the very beginning of its history has always been worried about the computational barrier, i.e., the consciousness that many things could not be easily implemented since large-scale power systems are the most complex machine ever built by mankind; it is large, complex, stiff and sometimes ... not perfectly known.

In the 1960s, the power community was one of the first scientific and industrial realities to see the importance of the computer era and, consequently, it developed its own codes, computer architectures, software environment and sometimes it invented its own mathematics which was later on used in other fields. Overcoming the fear for the computational complexity, the industry created a new industry of the computer application in power. Eventually, software, computing tools, bunches of relays, actuators and sensors improved the quality of supervision and control and the reliability of the power systems.

In the new millennium, it is conceivable that high-performance computing can play for the power industry the same role of the first computers in the 1960s. Setting power people free of the 'mind barrier' linked to the complexity and the computational effort can give rise to new applications and, perhaps, a new vision of the power business.

A major driver which pushed research towards high-performance computer applications in the power industry was the potential to solve the challenging problem of the DSA since, it was observed that, transmission system security is strictly linked to dynamic phenomena which are particularly severe because of the fast time response and the undesirable characteristic to widespread effects on a large geographical scale. The aim of accurately characterizing transient stability limits is generally part of a broader issue related to the DSA which also includes the assessment of limits for voltage instability and collapse, medium- and long-term instabilities, etc.

Here, the analysis is limited to transient stability problems and, consequently, TSA is treated as a part of DSA. It should be observed, however, that time-domain simulation methods on the transient time scale can also be used to detect and assess limits of phenomena such as transient voltage instability and uncontrolled rotor angle instability which, finally, results in voltage instability.

The TSA problem results particularly challenging in terms of computational requirements since it is generally assumed that in an interval of at most 15 min between two successive DSA/TSA cycles, hundreds of contingencies and transient stability simulations should be performed for at least 8–10 s, on real networks characterized by thousands of nodes. Figure 15.1 shows how domain decomposition of contingencies is inherently parallel and, naturally, can be implemented on multiprocessors.

The literature reports a large number of methods for transient stability time using variants of transient energy function techniques and equal area criterion extensions. These schemes use simplified models of the system and controller dynamics and are applied primarily to contingency screening and filtering. Since these methods are not prone to adopt a detailed representation of the system and cannot take into account actual controllers and protection systems, their potentials, in evaluating the impact of real-time control strategies on security, experience some limitations.

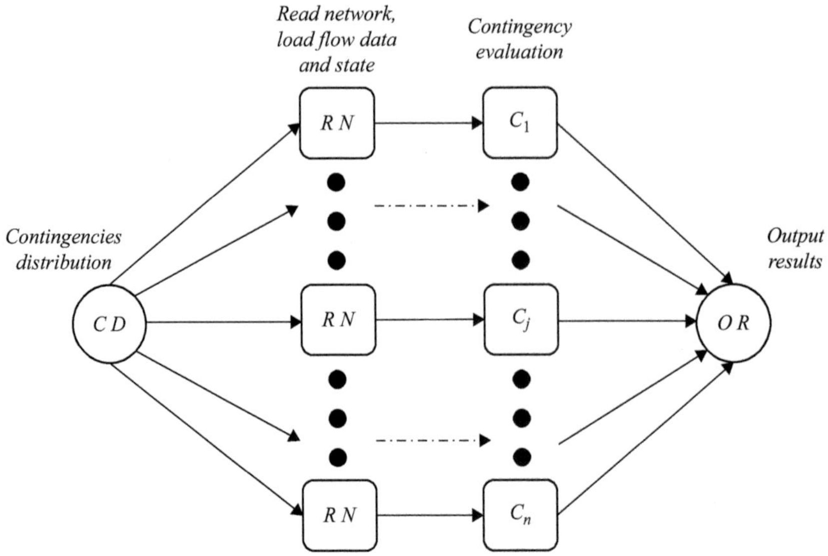

Figure 15.1 Domain decomposition of contingency analysis in DSA/TSA

The need for contingency screening and analysis and appropriate preventive/corrective control actions implemented for detecting and correct poor dynamic system performances was made more evident after 2003–2004 the worldwide 'annus horribilis' for blackouts such as, to remember the major ones:

- August 14, 2003, blackout of North East USA and Ontario 63 000 MW load loss affecting 50 million people;
- September 23, 2003, blackout of South Sweden and East Denmark 6 500 MW load loss affecting 4 million people;
- September 28, 2003, blackout of Italy 50 000 MW load unsupplied affecting 60 million people;
- August 12, 2004, blackout of three Australian States: Queensland, NSW and Victoria load loss 1 000 MW.

At that time, the power community received a strong wake up about the relevance of dynamics in stressed operating conditions and power industry discussed once more the importance of security, poor dynamic performances, protection schemes and control settings, emergency operation, etc. During the season of blackouts, it was discussed the opportunity of introducing more secure operating protocols (such as '$N-2$' security) in order to take into account, prudently, the 'randomness' or the 'volatility' of energy markets and the need of more detailed simulations.

The need for a more detailed representation of protection and emergency systems was risen. Without such description of the system, every consideration about dynamics can become useless since it does not take into account a large amount of strongly non-linear devices which play a significant role under large disturbances. Among various issues, there was a debate about the 'third-zone' of distance relays as a cause of improper distance-relay switching and the need of modelling protections in detail in security assessment codes.

Despite recent progresses, TSA remains a challenging topic, and the volatility of operating conditions in a competitive environment makes this challenge considerably harder. High-performance computing was proposed to win the barrier of the complexity integrating in nowadays software, probabilistic methods which can be considered unavoidable in presence of hidden complexities such as the market unforeseeable behaviour. This idea has been adopted to implement new codes for Total Transfer Capacity evaluation based on both the explicit treatment of dynamic constraints through the Dynamic Optimization approach [22] and a probabilistic approach to take into account the unavoidable uncertainties linked to the market behaviour and dynamic parameters [23]. That experience showed the computational complexity of the approach and could be overcome with efficient methods for TSA and high-performance computing.

Competitive market of electricity pushes the power industry to review and improve controls in order to ensure system security while pushing towards a higher exploitation of the infrastructures. Furthermore, unbundling of the basic parts of the power system infrastructure (generation, transmission, distribution) and the fragmentation of each one of them obliged to a profound revisiting of the control paradigms adopted in vertically integrated utilities. As an example, since generation or load is not directly controllable by the same company a more flexible and 'active' transmission system is required.

New controls will make a great use of the new advances in power systems technology such as: substation automation, new communication protocols to make each component visible at substation as well as control centre level such as IEC 61850, flexible AC transmission systems (FACTS) devices and dispersed resources. This new environment requires again new detailed models which cannot easily be utilized with direct methods for stability assessment and, consequently, pushes the use of time-domain simulations on the transient time scale. As an example, response-based control can be adopted for controlling dangerous cascade events driven by electromechanical transients. Usually, the transient time scale is considered too fast for today's computers and preventive control is considered the only way to avoid severe effects. In a competitive market, preventive control for very unlikely phenomena cannot be considered practical due to its heavy effect on economical transactions. Differently, corrective control for TSA, ideal for the market, is a computationally intensive problem.

In addition, it is well known that some phenomena cannot be properly controlled using local information only, but they need a Wide-Area Vision of the network. The term 'vision' here is deliberately used since it implies something more than measurement, something connected to the overall understanding of the system behaviour trough diagnostic tools, evaluation of synthetic indicators, Artificial Intelligence (AI), etc. Today's phasor measurement units (PMUs) technology allows us to have very fast and synchronized measurements with thousands of dollars per substation. But vision without control is useless. New protection schemes, computer architectures, algorithms, software can flourish allowing more and more the exploitation of the power system infrastructure (highly desired by the competitive market) while ensuring the required level of reliability. Rapid TSA is an unavoidable task to allow calculation on-the fly while measuring the dynamic of the system.

Dynamic optimization approach introduced in Reference 24 showed to be promising also for Wide-Area Control under the following assumptions: corrective control is operated as a response-based method and real-time trajectories are obtained through wide area measurement systems (WAMS) and some mathematical manipulations [25, 26].

The deployment of distributed computing for TSA/DSA has been partially explored considering the issue of the computational performances [27]. In order to achieve the computational power required by the online DSA, it was assumed that existing resources can be combined in a single heterogeneous meta-computer [28]. However, new issues need to be addressed. As an example, the application of distributed computing of the DSA task at geographically dispersed centres and decentralized control may help in making the computation of this control function tractable. Furthermore, distributed computing and advances in communications technology can play an important role on the deployment of faster and more advanced controls. The reformulation of new control laws coupled with the enhancement of analytical tools and software are key issues for enabling the migration of today's manual operator-initiated controls to automatic controls and take full advantages from supercomputing and faster communications.

In order to reach all these goals and exploit the potentials of these technologies, it is unavoidable to face the problem of computational complexity of time-domain simulation and introduce new real-time or faster-than-real methods exploiting the potentials of high-performance computing and communication.

15.4 A new class of algorithms: from step-by-step solutions to parallel-in-time computations

Time-domain solution of large sets of coupled algebraic and ODE is probably the most relevant task in most applications for power system analysis and control. Fast and reliable solvers are essential for the development of online DSA functions whenever transient and long-term analysis have to be integrated in a unique code. The challenge of making available to control centre operators online tools dedicated to stability analysis pushed researchers into developing new algorithms which exploit some sort of parallelism on parallel/vector machines so that 'two-order of magnitude' speed-up could be reached in computation. However, in the late 1970s and early 1980s, the power system community arrived to the conclusion that traditional algorithms for simulating power system dynamics were not appropriate for parallel/vector computing because only a limited amount of parallelism could be achieved [29–32].

Research in the dynamic simulation of very large scale integrated (VLSI) circuits showed that the use of relaxation-type methods could result in speed improvements over direct methods such as the Newton algorithm when several thousands of equations were treated [33]. Waveform relaxation method was applied to power systems

[34] after being tested for dynamic simulations of VLSI circuits. This method was introduced to solve ODEs but its use was broadened for treating DAEs [35]. An overview of these methods shows how they had been based on exploiting that kind of parallelism which was called 'parallelism in space' or 'across the system'. Such approach is based on the concept of partitioning 'geographically' the original physical problem so that more sub-problems can be solved separately thanks to the intrinsic parallelism of the adopted algorithm itself.

Despite the intrinsically sequential character of the IVP, which derives from the discretization of ODEs, a sort of parallelism can also be detected 'across the time'. Parallel-in-time algorithms, introduced in Reference 36, proved to be good candidates for parallel processing implementation in power system dynamic simulations. In parallel-in-time methods, the simulation time window is divided into series of blocks with each block containing a number of steps at which solutions to system equations are to be found. A critical presentation of various concrete parallel approaches is exhaustively surveyed in power system literature in Reference 8.

In References 14, 37–42, some relaxation-based algorithms have been proposed showing that the solution of this large problem is feasible, convergence is ensured and, most importantly, a large degree of parallelism can be achieved. These algorithms, although very efficient from the viewpoint of parallel computation due to the simple structure of the parallelism-in-time implementation, had the drawback of being characterized by heavier computational costs with respect to the most efficient sequential algorithms such as the so-called step-by-step Very Dishonest Newton (VDHN) Method [43].

The algorithms presented in References 14, 38–41, belonging to the class of relaxation/Newton parallel-in-time algorithms, resulted to be the most promising since they had been conceived in such way that the degree of parallelism offered by IVP could be expanded through block-implicit methods and, at the same time, better performances could be achieved through a mix of relaxation (non-linear iterations) and Newton iterations (linear iterations) with respect to pure relaxation algorithms (characterized by non-linear iterations on both space and time). In fact, with this kind of approach, the overall algebraic-differential set of equations describing the system is transformed by a stable implicit integration rule into a unique algebraic set of equations at each time step. All these equations are then solved concurrently by a composite linear/non-linear method.

In Figure 15.2, a pictorial representation of the multiplying effect on the speed up of parallelism-in-time is presented. It should be considered that the simple unidirectional structure of time gives rise to a simpler communication pattern among processors than in parallelism-in-space. This is an efficient way for concurrent computing as it was shown by many actual implementations which compared on different parallel/vector computing architectures various parallel-in-time and parallel-in-space algorithms as reported in References 41, 42.

Being able to treat all time steps concurrently, parallel-in-time algorithms provided the opportunity to develop new integration rules. For instance, parallel-in-time BVMs were proposed for power system TSA. These methods are characterized by some interesting properties such as the possibility to achieve higher accuracy, the

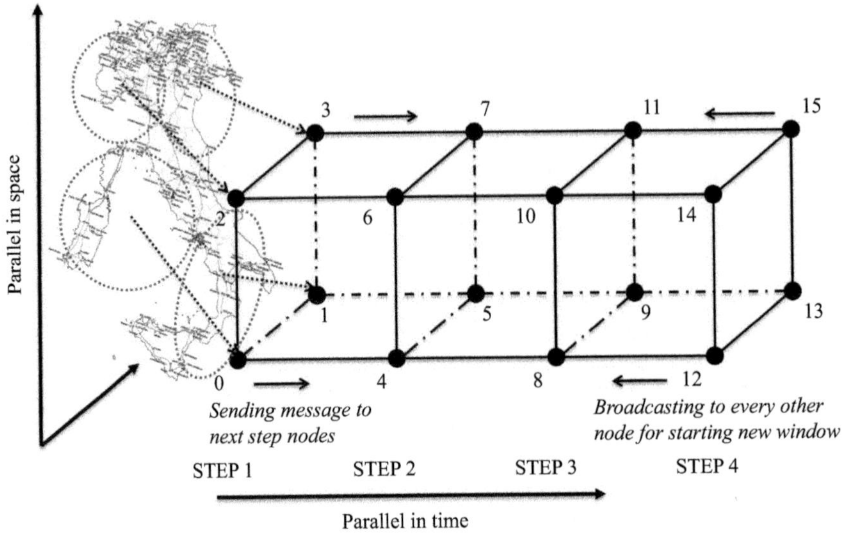

Figure 15.2 A pictorial example of parallelism-in-space and parallelism-in-time

capacity to solve efficiently, with the same method, stable and unstable problems and the ability to treat stiff problems. Moreover, it was showed how they can be implemented efficiently on vector/parallel computers [14].

In this chapter, in order to show the potentials and summarize the whole class of parallel-in-time algorithms, a BVM-based parallel-in time algorithm is illustrated. Its mathematical formulation is very general and can include a vast class of algorithms and, hopefully, provide hints for developing and boosting new ones.

By applying the integration rules, represented synthetically by (15.5a)–(15.5c), it is possible to formulate the IVP as a concurrent time-step solution of the following set of algebraic equations:

$$\hat{\boldsymbol{H}}(\hat{\boldsymbol{y}}) = 0 \tag{15.10}$$

where $\hat{\boldsymbol{H}} = [\boldsymbol{H}_1^T, \boldsymbol{H}_2^T, \ldots, \boldsymbol{H}_i^T, \ldots, \boldsymbol{H}_n^T]^T$ and $\boldsymbol{H}_i$ is the set of discretized DAEs that describes the transient stability problem at the ith time step.

Equation (15.10), consisting of $n \times s$ algebraic equations, also describes the transient behaviour of the system. Taking into account (15.5), (15.10) at a generic ith time step can be written as

$$\boldsymbol{H}_i(\hat{\boldsymbol{y}}) = \frac{1}{h}\sum_{j=0}^{k^i} \alpha_j^i \boldsymbol{A}\boldsymbol{y}_{i-k_1^i+j} - \sum_{j=0}^{k^i} \beta_j^i \boldsymbol{h}_{i-k_1^i+j} = 0, \quad i = 1, \ldots, n \tag{15.11}$$

where the k^i steps coincide with k^r or k according to the auxiliary and basic integration formulas appearing in (15.5) and the indices k_1^i correspond to i, k_1 and $k^r - k + r$, respectively, for initial, basic and final integration formulas.

Let us adopt a Newton's iteration for solving (15.10):

$$\boldsymbol{J}_{\hat{H}}[\hat{\boldsymbol{y}}^{(s)}][\hat{\boldsymbol{y}}^{(s+1)} - \hat{\boldsymbol{y}}^{(s)}] = -\hat{\boldsymbol{H}}[\hat{\boldsymbol{y}}^{(s)}], \quad \hat{\boldsymbol{y}}^{(0)} \in \boldsymbol{S}_{\rho 0}(\bar{\boldsymbol{y}}) \tag{15.12}$$

where $\bar{y} = [y^T(t_1), \ldots, y^T(t_n)]^T$ is the exact solution on the mesh, $\hat{y}^{(0)}$ is an initial approximation of the solution, $S_{\rho 0}(\bar{y})$ is the ball cantered in $\bar{y}$ with radius ρ defined as

$$S_{\rho 0}(\bar{y}) = \left\{ \hat{w} = [w_1^T, \ldots, w_n^T]^T : \|w_i - y(t_i)\| \leq \rho, 1 \leq i \leq n \right\} \qquad (15.13)$$

and $J_{\hat{H}}$ denotes the Jacobian matrix of $\hat{H}$. It should be observed that the Jacobian $J_{\hat{H}}$ has a block banded structure characterized by blocks of dimensions $s \times s$.

The existence of a unique solution of (15.10) is proved in Reference 14, where also the convergence properties of the Newton iteration (15.12) are reported. It should be observed that the inversion (or better the LU factorization) of a very large matrix of dimensions $sn \times sn$, such as the Jacobian matrix in (15.12), is not practical because of its very high dimensions. Consequently, a secondary iterative procedure based on block-Newton–Gauss–Jacobi or block-Newton–Gauss–Seidel has been proposed to solve the linear problem which is originated by the primary iteration of the Newton type. Each block processed in the Gauss–Seidel or Gauss–Jacobi scheme is characterized by the dimensions of the problem associated to each time step, that is $s \times s$: this is the dimension of the problem usually solved by conventional step-by-step codes with state-of-the-art sparsity techniques.

The number of secondary iterations depends on the block diagonal dominance of the Jacobian matrix. It is expected that the overall iterative process may experience a higher number of iterations with regard to the pure Newton method. Even though the convergence properties of the method degrade with respect to the pure Newton method, asymptotic linear convergence is granted under very light assumptions as demonstrated in Reference 14 and reported here for purpose of lucidity:

Theorem 15.2. *Let us suppose that $\hat{y}$ is the solution of (15.10). Assume that $D(\hat{y})$, $-L(\hat{y})$ and $-U(\hat{y})$ be the block diagonal, the block-strictly lower and strictly upper triangular parts of $J_{\hat{H}}(\hat{y})$ where $D(\hat{y})$ is non-singular and defined as follows:*

$$D(\hat{y}) = \textbf{blockdiag}\left[\frac{\delta H_i}{\delta y_i}\right]$$

Let us define the following matrices:

$$\begin{aligned} M_{GJ}(\hat{y}) &= [D(\hat{y})]^{-1}[U(\hat{y}) + L(\hat{y})] \\ M_{GS}(\hat{y}) &= [D(\hat{y}) - L(\hat{y})]^{-1}U(\hat{y}), \end{aligned}$$

If $\rho[M_{GJ}(\hat{y})] < 1$, $\rho[M_{GS}(\hat{y})] < 1$ then there exists an open ball of $\hat{y}$ where the multistep-Newton–Gauss–Jacobi (multistep-Newton–Gauss–Seidel) iteration is well defined and $\hat{y}$ is a point of attraction of the iterative scheme. The asymptotic convergence rate is at least linear.

In References 14, 41, 44, it was shown through actual implementation how the multistep-Newton–Gauss–Jacobi and the multistep-Newton–Gauss–Seidel algorithms perform efficiently for the solution of these problems. The number of secondary iterations is fixed to a constant value giving rise to the multistep attribute in the name of the methods.

High-order integration rules can be employed if Newton/relaxation algorithms are adopted. Of course, the computational complexity of the overall problem increases. As an example, when adopting an ETR of order 4, the Jacobian matrix is characterized by a block-quadri-diagonal structure where each block has dimension $s \times s$. This implies that the computational cost of each secondary iteration can become considerable since matrix by vector multiplication (even exploiting sparsity) involves four blocks for each row of the blocked-Jacobian.

Automatic selection of the step size is an important feature of codes for the solution of DAEs which need to be addressed. Parallel-in-time algorithms share the important property to generate the whole trajectory on the simulation interval.

An approximation of the maximum value of the local truncation error (LTE) can be assessed at the first Newton iteration on the basis of an initial guess grid. This value can be used to choose the constant time-step size which ensures an LTE compatible with the overall error due to the iterative process. At the second iteration, the solution obtained on the first grid is interpolated (or eventually extrapolated) on the new grid allowing the iterative process to be continued. The procedure can then be applied to successive iterations and a uniform grid can be adaptively updated during the iterative process. It can be observed by numerical tests that the values of LTE obtained at the first iteration were good enough to choose the right step size. Thus, a uniform grid can be chosen on the basis of LTEs assessed at the first iteration avoiding adjustments of the time step at successive iterations. The same procedure can be used for generating a non-uniform grid. Often it is useful to solve the overall time interval by the solution of successive subintervals (time windowing) [11]. In this case, a preliminary iteration is performed on the overall time interval to assess the most effective grid for a fixed LTEs.

Another opportunity provided by parallel-in-time algorithms is the possibility to obtain an approximation of the trajectory during the iterative process allowing the implementation of Multigrid methods. These methods were introduced for the solution of Discretized Partial Equations in order to accelerate the convergence of iterative solvers as described in References 45, 46. In References 36, 37, these methods were applied for the first time in transients stability problems for the solution of IVPs applied to DAEs.

Multigrid techniques recursively solve a system of discrete equations with a fixed time step by using a hierarchy of fine grids and the relation existing between different discretizations of the same continuous problem. There are different approaches to implement this idea [46]. A first approach is based on the use of coarser grids as correction grids which accelerate convergence of a relaxation scheme on finer grids by eliminating the so-called smooth error components, a typical example of this approach is the coarse Grid Correction Iteration [46]. Another method looks at the finer grids as correction grids improving accuracy on coarser grids and refining the solution only infrequently on the finer levels (an example is the so-called Nested Iteration Method). Many different schemes have been proposed to combine these two different approaches by implementing algorithms such as the V-cycle, W-cycle, the Full Multigrid, the Full Approximated Storage for non-linear systems [46]. The differences between these algorithms are mainly in the number of cycles between coarser grids and finer grids and also the method adopted to interpolate/extrapolate the solution from a processed grid to the next one.

The simplest method is the so-called Nested Iteration method where a better initial guess on a grid with time step h can be obtained by an approximate solution on a coarser grid with time step $2h$ and so on. Numerical tests have shown that in many cases this recursive initialization technique can eliminate many relaxation sweeps required by naive guesses as, for example, a flat start with all the state variables and voltages set at the steady-state value [36, 37].

As an example of interpolation rules adopted for nested iteration, let y_i^{2h}, $i = 1, 2, \ldots, n$ be the solution on a n-time-steps coarse grid. Then an interpolation operator on the finer grid with $2n$ time steps is given by:

$$\begin{aligned} y_{2i}^h &= y_i^{2h} \\ y_{2i+1}^h &= \tfrac{1}{2}\left(y_i^{2h} + y_{i+1}^{2h}\right) \end{aligned} \tag{15.14}$$

where 'i' denotes the time step and the superscript $2h$ or h indicate, respectively, the coarse and the fine grid by the time step length equal to $2h$ and h, respectively.

Provided the basic features of the class of parallel-in-time algorithms, the outline of a general algorithm which summarizes the implemented procedures is reported in the followings:

Algorithm Outlines

Step 1. Define the window size.

The overall integration interval is divided into a number of adjacent subintervals where integration rule (15.9) is applied. The introduction of such a decomposition permits the solution of systems characterized by a relatively small size compared with those associated to the entire number of mesh points. The choice of the window size is basically linked to the assessment of the computational burden associated to the BVM solver and the gains which can be obtained from high-performance computing. Of course, the number of time steps in each window has to exceed the number of steps of the chosen BVM multistep integration formula. In parallel-in-time implementations of BVMs, the proposed task partitioning consists in associating a single processor to the solution of a single time step.

Step 2. Fix the step size.

Different approaches can be pursued:

(1) it may be an input datum,
(2) or, as described above, it can be preliminary chosen on the basis of an assessment of the LTE and a prescribed tolerance of the error,
(3) or it can be fixed on the basis of any Multigrid strategy.

Step 3. In each window solve the non-linear system (15.9).

This goal can be obtained through the application of the Newton–Gauss–Jacobi or Newton–Gauss–Seidel methods [14, 38, 39, 41] or more in general any other relaxation scheme.

In order to capture the implementation features of step 3, a simple example is shown corresponding to the choice of an ETR of order 2 (i.e., the classical trapezoidal rule) when applying both a Newton–Gauss–Jacobi or Newton–Gauss–Seidel strategy. For purpose of lucidity, (15.9) is solved concurrently for four time steps per window assuming four processors.

With reference to (15.10), the following equations need to be solved:

$$\begin{aligned} &\boldsymbol{H}_1(\boldsymbol{y}_1) = 0 \\ &\boldsymbol{H}_2(\boldsymbol{y}_1, \boldsymbol{y}_2) = 0 \\ &\boldsymbol{H}_3(\boldsymbol{y}_2, \boldsymbol{y}_3) = 0 \\ &\boldsymbol{H}_4(\boldsymbol{y}_3, \boldsymbol{y}_4) = 0 \end{aligned} \tag{15.15}$$

The Jacobian of $\hat{\boldsymbol{H}}$, say $\boldsymbol{J}_{\hat{H}} = (\boldsymbol{J}_{ij}),\quad i,\ j = 1, \ldots, 4$ is clearly block-lower-bidiagonal. Applying to previous equations the modified Newton method, it is possible to obtain the following iterative process:

$$\boldsymbol{J}_{\hat{H}}[\hat{\boldsymbol{y}}^{(s+1)} - \hat{\boldsymbol{y}}^{(s)}] = -\hat{\boldsymbol{H}}[\hat{\boldsymbol{y}}^{(s)}] \tag{15.16}$$

where the Jacobian $\boldsymbol{J}_{\hat{H}}$ is assumed constant during the iterative process as long as the number of iterations does not exceed a fixed number. The previous linear system (15.16) can be finally solved using a fixed number of block-Gauss–Jacobi (GJ) or block-Gauss–Seidel (GS) iterations. More precisely, the coefficient matrix $\boldsymbol{J}_{\hat{H}}$ being block-lower bi-diagonal, one iteration of GS will solve (15.16) exactly, while the number of GJ iterations depends on the iterative process and the desired accuracy.

The Gauss–Jacobi iterative scheme is based on two levels of iterations: the primary and the secondary iterations. The inner iterative process (i.e., the one relative to the secondary iterations) can be written as:

$$\begin{aligned} \Delta\boldsymbol{y}_1^{(r+1)} &= -\boldsymbol{J}_{11}^{-1}\boldsymbol{H}_1\left[\boldsymbol{y}_1^{(s)}\right] \\ \Delta\boldsymbol{y}_2^{(r+1)} &= -\boldsymbol{J}_{22}^{-1}\left\{\boldsymbol{H}_2\left[\boldsymbol{y}_1^{(s)}, \boldsymbol{y}_2^{(s)}\right] + \boldsymbol{J}_{21}\Delta\boldsymbol{y}_1^{(r)}\right\} \\ \Delta\boldsymbol{y}_3^{(r+1)} &= -\boldsymbol{J}_{33}^{-1}\left\{\boldsymbol{H}_3\left[\boldsymbol{y}_2^{(s)}, \boldsymbol{y}_3^{(s)}\right] + \boldsymbol{J}_{32}\Delta\boldsymbol{y}_2^{(r)}\right\} \\ \Delta\boldsymbol{y}_4^{(r+1)} &= -\boldsymbol{J}_{44}^{-1}\left\{\boldsymbol{H}_4\left[\boldsymbol{y}_3^{(s)}, \boldsymbol{y}_4^{(s)}\right] + \boldsymbol{J}_{43}\Delta\boldsymbol{y}_3^{(r)}\right\} \end{aligned} \tag{15.17}$$

When convergence is reached for this scheme or the number of GJ iterations exceeds a fixed maximum threshold $\bar{\boldsymbol{r}}$, the primary iteration can be updated as follows:

$$\hat{\boldsymbol{y}}^{(s+1)} = \hat{\boldsymbol{y}}^{(s)} + \Delta\hat{\boldsymbol{y}}^{(\bar{r})} \tag{15.18}$$

The computation is carried out as shown in Figure 15.3 where the arrows denote data exchange between processors and a Work Unit (WU) is equal to the central processing unit (CPU) time required to perform a single time-step computation. Each processor is devoted to the computation of the variables of a single time step for all the iterations required to achieve the convergence. More time steps may be committed to the same processor if the number of window steps is greater than the number of available processors. Processors are synchronized after completing the parallel computations relative to the current iteration to check the convergence.

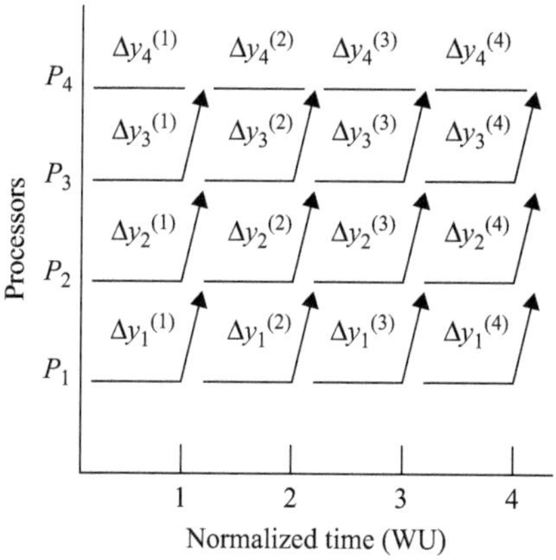

Figure 15.3 Sequence of parallel operations for the Newton–Gauss–Jacobi approach

In the GS method, the solution of (15.16) is obtained as a limit point in the following linear iteration scheme:

$$
\begin{aligned}
\boldsymbol{y}_1^{(s+1)} &= \boldsymbol{y}_1^{(s)} - \boldsymbol{J}_{11}^{-1}\boldsymbol{H}_1\left[\boldsymbol{y}_1^{(s)}\right]\\
\boldsymbol{y}_2^{(s+1)} &= \boldsymbol{y}_2^{(s)} - \boldsymbol{J}_{22}^{-1}\left\{\boldsymbol{H}_2\left[\boldsymbol{y}_1^{(s)},\boldsymbol{y}_2^{(s)}\right] + \boldsymbol{J}_{21}\left[\boldsymbol{y}_1^{(s+1)} - \boldsymbol{y}_1^{(s)}\right]\right\}\\
\boldsymbol{y}_3^{(s+1)} &= \boldsymbol{y}_3^{(s)} - \boldsymbol{J}_{33}^{-1}\left\{\boldsymbol{H}_3\left[\boldsymbol{y}_2^{(s)},\boldsymbol{y}_3^{(s)}\right] + \boldsymbol{J}_{32}\left[\boldsymbol{y}_2^{(s+1)} - \boldsymbol{y}_2^{(s)}\right]\right\}\\
\boldsymbol{y}_4^{(s+1)} &= \boldsymbol{y}_4^{(s)} - \boldsymbol{J}_{44}^{-1}\left\{\boldsymbol{H}_4\left[\boldsymbol{y}_3^{(s)},\boldsymbol{y}_4^{(s)}\right] + \boldsymbol{J}_{43}\left[\boldsymbol{y}_3^{(s+1)} - \boldsymbol{y}_3^{(s)}\right]\right\}
\end{aligned}
\tag{15.19}
$$

At first glance, the Gauss–Seidel scheme appears to be strictly sequential. It is possible, however, to implement the Gauss–Seidel method in a parallel way as referred to in Reference 39. There, the approach was named pipelining-in-time. The computational arrangement is shown in Figure 15.4. After computing $\boldsymbol{y}_1^1$ on processor P_1 (i.e., the variables relative at the first time step during the first iteration), the calculation of the second step value $\boldsymbol{y}_2^1$ can be carried out by processor P_2 in parallel with the determination of $\boldsymbol{y}_1^2$ by P_1. As computation proceeds, the number of time-step calculations to be carried out concurrently increases up to a maximum equal to the number of processors. Processors are synchronized after completing a set of concurrent time-step evaluations (each time step referring to a different iteration). This synchronization scheme ensures the repeatability of results.

The advantage of the Gauss–Seidel consists, in general, in a reduction of the number of iterations necessary to reach the convergence with regard to the Gauss–Jacobi version. The drawback consists in a speed up lower than theoretical one obtained by the use of the Gauss–Jacobi scheme since the solution at different time steps can be updated only sequentially.

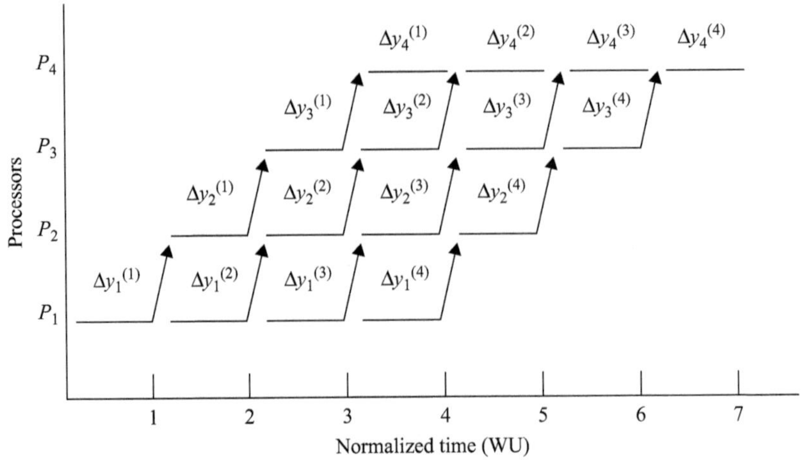

Figure 15.4 Sequence of parallel operations for the Newton–Gauss–Seidel approach

When solving (15.9), either by the Gauss–Jacobi or by the Gauss–Seidel method, it can be observed that the convergence pattern exhibits a remarkable directionality due to the uni-directionality of the time [36–39]. Namely, the first time steps of a window converge before the subsequent ones. A computational expedient consists in replacing the first converged step of a window with the first unassigned time step, as proposed in Reference 41. The visual effect is that of having something like a window moving across the integration time step towards the end of the simulation interval. This approach, named Travelling Window, can allow a significant reduction of the computational effort since it eliminates unnecessary iterations on already converged steps.

15.5 Performances in parallel-in-time computations

Parallel-in-time algorithms can be implemented on parallel machines yielding a speed up due to concurrent processing of time steps. These methods are characterized by a high efficiency in implementing codes on parallel computers with regard to the performances of parallelized versions of step-by-step codes. This is due to the 'coarse grain approach' involved in the parallelism-in-time where there is a large amount of calculations associated to each processor compared with the overhead due to communications. The main drawback of the parallel-in-time approach is the software overhead due to the fact these methods solve a larger set of equations than step-by-step codes.

It should be noted that the Gauss–Seidel versions outperform their respective Jacobi counterparts in a serial environment. This is not necessarily true when applying parallel processing on message passing machine because of the simpler structure of the information flow in Jacobi-type algorithms as proved in References 38, 39, 41, 42.

In general, the slow down due to Gauss–Jacobi versions is so high that it would be very difficult to get high performance if a limited number of processors are adopted. Thus, in the following, the implementation of ETR4GS and ETR2GS is compared with the fastest sequential algorithm, i.e., the VDHN Method.

The results of one of the possible implementations of these algorithms [14, 41] is presented here and briefly discussed to summarize the potentials of this approach. A four-processor vector/parallel computer was adopted for the implementation of this class of algorithms. These experiences are exemplary since high performance was achieved by the combined use of parallel and vector processing. Algorithmic parallelism was exploited in the two ways:

- by the concurrent execution of different time steps calculations on the available processors;
- by vector processing which is used to speed up the computation of a single time step through the operation pipelining of the vectorizable DO loops.

Load balancing problems were minimal because of the coarse grain approach due to time-parallelism. As a matter of fact, each CPU has to solve a single time step whose computational burden was heavy due to the dimensionality of the problem associated to a single step (thousands of DAEs) and practically constant.

For speed up evaluation, the reference program VDHN was compiled both enabling the automatic vectorization of the code (version VDHN-V) and disabling vectorization (version VDHN-S).

Consequently, two speed up values can be considered for performance evaluation:

- $S_1 = \dfrac{\text{Run time of the VDHN–S method on 1 CPU}}{\text{Run time of the BVM}}$
- $S_2 = \dfrac{\text{Run time of the VDHN–V method on 1 CPU}}{\text{Run time of the BVM}}$

The second speed up S_2 is introduced to take into account the possibility that the best competitor, i.e., the VDHN method, can also be vectorized on the same machine and is a more fair evaluation of the advantages due to the BVMs.

It is interesting to separately assess the effects that vector and parallel processing have in determining the final speed up value. The speed up S_1 is, consequently, expressed as the product of three distinct figures S_s, S_v and S_p which measures, respectively, the algorithmic, vectorization and parallelization speed up, where:

$$S_s = \frac{\text{Run time of the VDHN method}}{\text{Run time of the BVM}}$$

$$S_v = \frac{\text{BVM run time on CPU 1}}{\text{BVM to run enabling vectorization on 1 CPU}}$$

$$S_p = \frac{\text{BVM to run enabling vectorization on 1 CPU}}{\text{BVM to run enabling vectorization on } n \text{ CPU}}$$

Table 15.3 Performances of some parallel-in-time methods

Integration rule	LTE max (pu) x 10^{-3}	CPUs	S_s	S_v	S_p	S_1	S_2	R
ETR2GS ($h = 0.02$ s)	2.021	4	0.39	8.73	3.54	12	8.2	5.4
		2	0.46	8.62	1.82	7.2	4.9	3.2
		1	0.51	8.44	1	4.3	2.9	1.9
ETR4GS ($h = 0.02$ s)	0.015	4	0.35	8.82	3.61	11.1	7.57	5
ETR4GS ($h = 0.06$ s)	0.103	4	0.79	8.85	3.65	25.5	17.4	11.4

Tests were performed for a 1 s simulation of a 662-bus network with 91 generators characterized by the so-called detailed modelling [2]. Table 15.3 summarizes some results obtained in a test campaign reported in References 14, 41 using parallel/vector computing.

In order to show the potentials of the approach, in Table 15.3 the previous defined performance indicators, the maximum value of LTE and the ratio R between the simulated time interval and computational time are reported. It should be observed that R gives an idea of how much faster-than-real a single transient stability simulation can be run.

Testing ETR2GS, it can be realized that as long as the number of time steps processed concurrently increases, the sequential speed up S_s decreases due to the increase of the size of the problem but the cumulative effect of S_v and S_p is such that the overall speed up always increases. ETR4GS, run with a step size h of 0.02 s, exhibited a sequential speed up S_s lower than ETR2GS since a higher number of calculations were needed to implement a more complex integration rule. This overhead was partially counterbalanced by slightly higher vector and parallel speed ups due to a higher number of calculations associated to ETR4GS. As computations increases, the grain of the problem increases reducing communication problems, consequently improving parallelization performances.

As it can be observed from Table 15.3, best results can be obtained with ETR4GS run on a grid characterized by a step size of 0.06 s. It should be remarked that the choice of this step size when adopting ETR4 guarantees better LTE than the one obtained by ETR2 applied to a grid with step size 0.02 s. Consequently, the comparison is fair since it is based on a higher accuracy on the largest time step.

ETR4 outperforms ETR2 because of a higher sequential speed up due to the utilization of a coarser grid. A slight improvement of vector and parallel speed up is also due to the fact that more computations are needed to evaluate a single step of ETR4GS than the ones necessary when using ETR2GS. Based on these results, it seems that in the considered class of parallel-in-time algorithms the best choice is the extended trapezoidal rule of order four (ETR4) with the implementation of a block-Newton–Gauss–Seidel algorithm.

In this experience, the synergy between vectorization and parallelism-in-time yielded a speed up in excess of 25 which was one of the best result reached up with these algorithms. Also, the speed up evaluated taking into account the vectorization of the VDHN code is still significant and equal to 17. Wallclock timings obtained with ETR4GS are also interesting for online applications since they show the possibility to simulate transients an order of magnitude faster than real and run several hundreds of contingencies every 10–15 min, which is what usually required in DSA. The joint use of a moderate degree of parallelism-in-time and vectorization increases several times the speed up obtainable by traditional techniques of vectorization. The synergism obtainable by the two approaches permits to reach a sufficiently high speed up for online implementations of power system TSA.

15.6 Conclusions

Time-domain simulation for transient stability is an essential tool for TSA analysis and the implementation of reliable algorithms for controlling dynamic performances on the transient scale.

The experience accumulated in developing new pieces of software for fast simulation of transients exploiting high-performance computing has been summarized in this chapter. Pushed by the industry need for real-time transient simulation and the challenge of large-scale preventive/corrective control based on extended real-time or perhaps faster-than-real simulations new methods were developed in 1990s. Among those methods, a new way to exploit parallelism across time was introduced and showed to be effective in multiplying the speed up obtainable from methods which, before its introduction, could only rely on spatial decomposition.

The experience on these methods showed that speed ups of the order of some tens could be achieved through high-performance computing with regard to the best sequential algorithms utilized by the power industry. The results conducted to effective methods which, incidentally, how it often happens in Science, paved the way to a new class of algorithms whose mathematical formulation is at the basis of dynamic optimization algorithms which yield a practical solution to new problems as shown in Chapter 5.

Bibliography

[1] N. Sato and W. F. Tinney, 'Techniques for exploiting the sparsity of the network admittance matrix', *IEEE Transactions on Power Apparatus and Systems*, vol. 82, pp. 944–950, Dec. 1963.

[2] B. Stott, 'Power system dynamic response calculations', *Proceedings of IEEE*, vol. 67, pp. 219–240, Feb. 1979.

[3] M. Stubbe, A. Bihain, J. Deuse, and J. C. Baader, 'STAG: a new unified software program for the study of the dynamic behaviour of electrical power systems', *IEEE Transactions on Power Systems*, vol. 4, pp. 129–138, Feb. 1989.

[4] H. Fankhauser, K. Aneros, A. Edris, and S. Torseng, 'Advanced simulation techniques for the analysis of power system dynamics', *IEEE Computer Applications in Power*, vol. 3, pp. 31–36, Oct. 1990.

[5] J. Y. Astic, A. Bihain, and M. Jerosolimski, 'The mixed Adams-BDF variable step size algorithm to simulate transient and long-term phenomena in power systems,' *IEEE Transactions on Power Systems*, vol. 9, pp. 929–935, May 1994.

[6] P. V. Kokotovic and H. K. Khalil, 'Singular perturbation methods in system and control', *IEEE Press Selected Reprint Series*, IEEE, Piscataway, NJ, 1986.

[7] M.-E. Grenier, D. Lefebvre, and T. Van Cutsem, 'Quasi steady-state models for long-term voltage and frequency dynamics simulation', *Proceedings of IEEE Power Tech Conference*, St. Petersburg Russia, June 2005, pp. 1–8.

[8] D. J. Tylavsky, A. Bose, F. Alvarado, *et al.* 'IEEE Committee Report by a Task Force of the Computer and Analytical Methods Subcommittee of the Power Systems Engineering Committee "Parallel processing in power systems computation"', *IEEE Transactions on Power Systems*, vol. 7, no. 2, pp. 629–638, 1992.

[9] D. E. Barry, C. Pottle, and K. Wirgau, 'A technology assessment study of near term computer capabilities and their impact on power flow and stability simulation program', *EPRI-TPS-77-749, Final Report*, 1978.

[10] J. Fong and C. Pottle, 'Parallel processing of power system analysis problems via simple parallel microcomputer structures', *EPRI-EL-566SR*, 1977.

[11] R. Podmore, M. Liveright, S. Virmani, N. M. Peterson, and J. Britton, 'Application of an array processor for power system network computations', *IEEE Power Industry Computer Applications Conference*, Cleveland, OH, May 1979, pp. 325–331.

[12] F. M. Brasch, J. E. Van Ness, and S. C. Kang, 'Design of multiprocessor structures for simulation of power system dynamics', *EPRI-EL-1756,* 1981.

[13] C. W. Gear and L. R. Petzold, 'ODE methods for the solution of differential/algebraic systems', *SIAM Journal of Numerical Analysis*, vol. 21, no. 4, pp. 716–728, Aug. 1984.

[14] F. Iavernaro, M. La Scala, and F. Mazzia, 'Boundary values methods for time domain simulations of power system dynamic behaviour', *IEEE Transactions on Circuits and Systems I. Fundamental Theory and Applications*, vol. 45, pp. 50–63, ISSN: 1057-7122, 1998.

[15] J. C. P. Miller, 'Bessel functions – Part II', *Mathematical Tables, X, British Association for the Advancement of Sciences*, Cambridge University Press, Cambridge, 1952.

[16] F. W. Olver, 'Numerical solution of second order linear difference equations', *Journal of Research,* vol. 71B, pp. 111–129, 1967.

[17] A. O. H. Axelsson and J. G. Verwer, 'Boundary value techniques for initial value problems in ordinary differential equations', *Mathematics of Computation*, vol. 45, no. 171, pp. 153–171, Jul. 1985.

[18] L. Lopez and D. Trigiante, 'Boundary value methods and BV-stability in the solution of initial value problems', *Applied Numerical Mathematics*, vol. 11, pp. 225–239, 1993.

[19] P. Amodio and F. Mazzia, 'Boundary value methods for the solution of differential-algebraic equations', *Numerische Mathematik*, vol. 66, pp. 411–421, 1994.

[20] J. D. Lambert, *Computational Methods in Ordinary Differential Equations*, Wiley, New York, NY, 1985.

[21] J. Y. Astic, A. Bihain, and M. Jerosolimski, 'The mixed Adams-BDF variable step size algorithm to simulate transient and long-term phenomena in power systems', *IEEE Transactions on Power Systems*, vol. 9, pp. 929–935, May 1994.

[22] S. Bruno, M. La Scala, P. Scarpellini, and G. Vimercati, 'Probabilistic evaluation of ATC in a market characterized by intense bilateral contracts', *Proceedings of Second Probabilistic Methods Applied to Power Systems (PMAPS)*, 22–26 September 2002, Naples, Italy, pp. 729–735.

[23] E. De Tuglie, M. Dicorato, M. La Scala, and F. Torelli, 'Dynamic parameter estimation for dynamic security assessment', *Proceedings of Second Probabilistic Methods Applied to Power Systems (PMAPS)*, 22–26 September 2002, Naples, Italy, pp. 603–608.

[24] M. La Scala, M. Trovato, and C. Antonelli, 'On line dynamic preventive control: an algorithm for transient security dispatch', *IEEE Transactions on Power Systems*, vol. 13, no. 2, pp. 601–608, May 1998.

[25] A. Bose, S. Bruno, M. De Benedictis, and M. La Scala, 'A dynamic optimization approach for wide-area control of transient phenomena', *CIGRÈ General Meeting*, Paris, 29 August–3 September 2004.

[26] A. Bose, M. Bronzini, S. Bruno, M. De Benedictis, and M. La Scala, 'Load shedding scheme for response-based control of transient stability', *IREP Symposium 2004 – Bulk Power System Dynamics and Control*, Cortina D'Ampezzo, 22–27 August 2004.

[27] G. Alosio, M. Bochicchio, M. La Scala, and R. Sbrizzai, 'A distributed computing approach for real-time transient stability analysis', *IEEE Transactions on Power Systems*, vol. 12, no. 2, pp. 981–987, May 1997.

[28] G. Alosio, M. Bochicchio, and M. La Scala, 'Metacomputing for on-line stability analysis in power systems: a proposal', *International Conference and Exhibition on High-Performance Computing & Networking, HPCN Europe 1996*, Bruxelles, 15–19 April, 1996, Lecture Notes on Computer Science, Springler-Verlag, Berlin.

[29] D. E. Barry, C. Pottle, and K. Wirgau, 'A technology assessment study of near term computer capabilities and their impact on power flow and stability simulation program', *Final Report EPRI-TPS-77-749*, 1978.

[30] J. Fong and C. Pottle, 'Parallel processing of power system analysis problems via simple parallel microcomputer structures. Exploring applications of parallel processing to power system analysis problems', *Report EPRI-EL-566-SR,* 1977.

[31] R. Podmore, M. Liveright, S. Virmani, N. M. Peterson, and J. Britton, 'Application of an array processor for power system network computations', *Proceedings of PICA Conference*, Cleveland, OH, May 1979, pp. 325–331.

[32] F. M. Brasch, J. E. Van Ness, and S. C. Kang, 'Design of multiprocessor structures for simulation of power system dynamics', *Report EPRI-EL-1756*, 1981.
[33] J. K. White and A. L. Sangiovanni-Vincentelli, *Relaxation Techniques for the Simulation of VLSI Circuits*, Kluwer, Boston, MA, 1987.
[34] M. Ilic-Spong, M. L. Crow, and M. A. Pai, 'Transient stability simulation by waveform relaxation methods', *IEEE Transactions on Power Systems*, vol. 2, no. 4, pp. 943–952, Nov. 1987.
[35] M. L. Crow and M. Ilic, 'The parallel implementation of the waveform relaxation method for transient stability simulations', *IEEE Transactions on Power Systems*, vol. 5, no. 3, pp. 922–932, Aug. 1990.
[36] M. La Scala, A. Bose, D. J. Tylavsky, and J. S. Chai, 'A highly parallel method for transient stability analysis', *IEEE Transactions on Power Systems.*, vol. 5, no. 4, pp. 1439–1445, 1990.
[37] M. La Scala, A. Bose, and D. J. Tylavsky, 'A relaxation type multigrid parallel algorithm for power system transient stability analysis', *Proceedings International Symposium on Circuits & Systems 1989 Conference,* Portland, Oregon, 8–11 May 1989, vol. 3, pp. 1954–1957.
[38] M. La Scala, M. Brucoli, F. Torelli, and M. Trovato, 'A Gauss–Jacobi–Block Newton method for parallel transient stability analysis', *IEEE Transactions on Power Systems*, vol. 5, no. 4, pp. 1168–1175, 1990.
[39] M. La Scala, R. Sbrizzai, and F. Torelli, 'A pipelined-in-time parallel algorithm for transient stability analysis', *IEEE Transactions on Power Systems*, vol. 6, no. 2, pp. 715–722, 1991.
[40] M. La Scala, G. Sblendorio, and R. Sbrizzai, 'Parallel-in-time implementation of transient stability simulations on a transputer network', *IEEE Transactions on Power Systems*, vol. 9, no. 2, pp. 1117–1125, May 1994.
[41] G. P. Granelli, M. Montagna, M. La Scala, and F. Torelli, 'Relaxation-Newton methods for transient stability analysis on a vector/parallel computer', *IEEE Transactions on Power Systems*, vol. 9, no. 2, pp. 637–643, May 1994.
[42] M. La Scala, G. Sblendorio, A. Bose, and J. Q. Wu, 'Comparison of algorithms for transient stability simulations on shared and distributed memory multiprocessors', *IEEE Transactions on Power Systems*, vol. 11, no. 4, pp. 2045–2050, Nov. 1996.
[43] 'Extended transient-midterm stability package: technical guide for the stability program', *EPRI EL-2000-Computer Code Manual-Project* 1208, Jan. 1987, prepared for EPRI, Palo Alto, CA.
[44] M. La Scala and A. Bose, 'Relaxation/Newton methods for concurrent time step solution of differential-algebraic equations in power system dynamic simulations', *IEEE Transactions on Circuits Systems*, vol. 40, no. 5, pp. 317–330, May 1993.
[45] A. Brandt, 'Multi-level adaptive solutions to boundary value problems', *Mathematics of Computation*, vol. 31, n. 138, Apr. 1977, Providence (RI), American Mathematical Society, pp. 333–390.
[46] S. F. McCormick (Ed.), *Multigrid Methods*, SIAM, Philadelphia, PA, 1987.

Chapter 16
Voltage security in modern power systems

Roberto S. Salgado[1], *Cristian Bovo*[2] *and Alberto Berizzi*[2]

16.1 Introduction

The variations of bus voltage magnitude, resulting from load and intermittent generation changes and/or possible contingencies, emphasise the need of controlling properly the reactive power supply as well as the voltage magnitude (QV) during the daily operation of the power network. The main objectives of this strategy are usually the management of the production and consumption of reactive power, to keep reasonable levels of bus voltage magnitude, satisfying pre-specified limits and reducing the active power transmission losses [1]. Analytically, the QV-control is considered a non-linear and multi-objective problem, of high dimension, with distinct levels of hierarchy. The complex nature of this issue shows the necessity of monitoring procedures more advanced than those used in the case of active power-frequency control [2]. In order to improve the electrical system security and/or to reduce the active power generation cost (by adjusting the magnitude of the bus voltage for the reduction of transmission losses), strategies to control QV were proposed in Spain, France and Italy in the late 1970s [3]. The coordination of such control strategy is organised in three hierarchical levels [4–6]:

1. Primary voltage control (PVC), which compensates fast and random variations of the voltage magnitude, maintaining these variables as close as possible to reference values. Only local information is used to define the control actions, whose time constant is in the range of a few seconds.
2. Secondary voltage control (SVC), which regulates slow (but considerable) variations in the voltage magnitude (resulting from the load change, for instance). The actions determined by this type of control are based on the reference values of the voltage magnitude in a specific load bus, named *pilot bus*. This level consists of regional regulators of the voltage magnitude as well as local reactive power regulators. The time constant of this type of control is of the order of a one or more minutes.
3. Tertiary voltage control (TVC), which determines the control actions to optimise pilot bus voltage (set points), aiming usually at economic and security

[1]Department of Electrical and Electronic Engineering, Federal University of Santa Catarina, Brazil
[2]Energy Department, Politecnico di Milano, Italy

benefits. The information of the complete system, including all hierarchical levels, is used in computational applications such as the optimal power flow (OPF). This type of control is based on centralised monitoring, which results in a time constant ranging from some minutes to several hours. The tertiary level of voltage regulation receives periodically (generally every 5–15 min) information about the state of the power network. This information is used in a computational program (in general the OPF) to calculate the reference voltage magnitude (*set points*) of the pilot buses, based on the short-term load forecast. The values of the reference voltage are transmitted to the secondary control level.

Automatic voltage regulators are directly affected in individual generators in the PVC. All regulators of the same region are coordinated by the SVC, and both (the PVC and the SVC) are coordinated by the TVC [2]. According to the hierarchy of the QV-control, the SVC level receives set points from the TVC level, so as to provide suitable actions to the primary level. The following sections are focused on the SVC based on their traditional formulation and its effect on the power system operation. The SVC uses regional information to modify properly the controllers, which are aimed at adjusting the reference voltage magnitude of the pilot buses. The power system is divided into areas, for which a set of generating units (*regulating units* or *regulators*) is selected to control the voltage magnitude level of a pre-selected bus of the area. The SVC coordinates automatically the reactive power resources, so as to establish a satisfactory voltage level, with an adequate reactive power margin of the QV-control devices. This allows to keep a suitable voltage level in the transmission system when the load changes (in hourly time scale) and/or after modifications in the topology of the power system.

Figure 16.1 represents the structure of the hierarchical voltage.

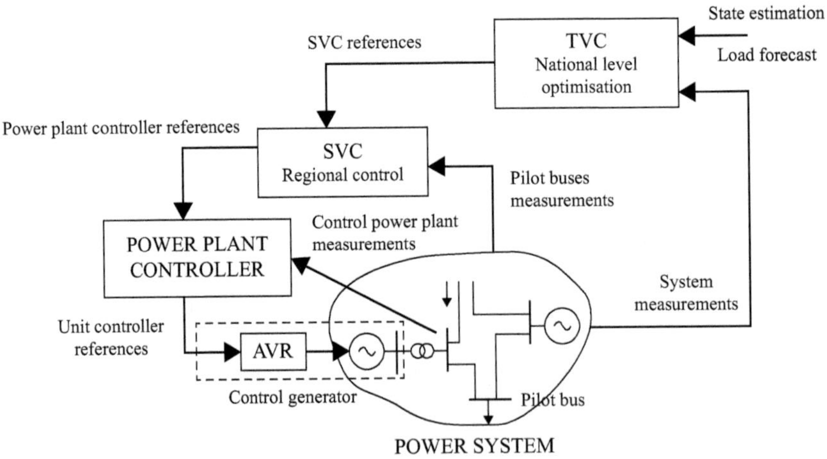

Figure 16.1 Structure of the hierarchical voltage reactive power control system

The strategy of QV hierarchical control has been employed in several countries (Italy, France, Belgium, Spain, Romania, etc.) [1]. It is recognised that this scheme provides a better distribution of the reactive power generation and enables a faster recovery of the voltage magnitude to the satisfactory level after a contingency. In addition, keeping a suitable margin of reactive power on each QV-control device prevents the problems of voltage instability and increases the security level of the power system. However, there are some difficulties to implement the hierarchical QV-control. First, the coupling between the buses, which is a drawback when areas of the network have to be defined, and so the pilot buses and their set points. Other difficulties are related to the monitoring in real time of the regulating generators, the adjustment of the voltage set point of the pilot buses to follow the load variations as well as eventual contingencies, and the effect of other QV-control devices, such as shunt capacitors and reactors.

Two aspects are extremely important for the efficiency of the SVC. The first is the selection of the pilot buses. These buses are associated with electrically coherent zones and their voltage magnitude is supposed to be the *best* representation of the voltage magnitude level in the area [4]. The procedures usually described in the literature are based on computer simulations. In addition, sensitivity relationships based on the linearisation of the power flow (PF) equations are often used, despite the inaccuracies of this model if the electrical network operates too close to its limits. Once selected, these buses are kept in this function for every operation condition. In spite of any modifications, the imposition of the reference voltage magnitude in this selected set of buses can add difficulties the process of solution of the PF equations. The second aspect concerns the selection of the regulating generators, whose main functions are to control the voltage magnitude of the pilot buses and maintaining an adequate margin of reactive power. This is accomplished through a control scheme, such that all generators of an area operate at the same per unit reactive power injected, i.e., proportionally to their maximum reactive power capability (alignment constraints). The difficulties of this choice are related to availability, capacity and location of these generators.

France was the first country to physically implement the hierarchical control, in 1979. The strategy used overlapped the regulators of the PVC and the QV-control of the SVC. However, the voltage magnitude regulators were not specifically installed in regulating buses, such that the centralised control was necessary. This methodology provided satisfactory results for some years [7]. As the French electrical system increased, the coupling between the QV-control devices became too strong, which brought extra difficulties to impose a uniform reactive power level in the regulating buses, and to control the voltage magnitude of pilot buses. In order to overcome this difficulty, the French operator considered the possibility to control more than one pilot bus in each area. In addition, for the purpose of static analysis of the SVC, an optimisation problem was formulated in which the objective function takes into account both the voltage magnitude at pilot buses and the reactive power level of the regulating generators.

Differently from the French strategy, the Italian scheme makes use of a device installed on each regulating generator. Additionally, there are specific control units referred to as regional voltage regulator (RVR) for every area, where the QV-control is implemented. These units also control shunt capacitors and reactors, tap transformers and other devices used to avoid saturation of reactive resources of the generators of each area. From the point of view of static analysis, performance indices for minimising losses, minimising the voltage magnitude deviation and maximising the loadability are used to determine the optimal set points [6]. This strategy has a degree of complexity smaller than that of the French power system, being also a reference to the electrical network of other European countries [7].

Since the 1990s, with hierarchical control effectively implemented, several studies have been proposed to analyse the performance of the SVC, as well as to overcome the difficulties found until then. Reference 4 presents a security analysis of operation under decentralised control and under coordinated control (studied separately), compared to systems with only the PVC. The methodology presented is based in three steps: selection, analysis and corrective control of contingencies. The systems under SVC were found to show better performance at steady state, from the point of view of re-establishing the voltage stability with respect to power systems with the only PVC. In Reference 8, the behaviour of the system under voltage instability is analysed, demonstrating the benefits of SVC in comparison with power systems that have only PVC. In addition, Reference 8 proposes an emergency control strategy, more efficient than that based on the use of pilot buses in conditions of imminent voltage instability. The improvement brought by SVC with respect to the performance of the power system operation under voltage instability is also described in Reference 9.

The problem of the adequate choice of the pilot buses and the pre-specification of their voltage magnitude is addressed in Reference 5. Depending on the power system loadability, different sets of pilot buses are demonstrated to be more suitable. In Reference 10, a comparison between the power system security versus economy is presented, focusing on the use of the SVC and the use of shunt compensation. The results presented in that paper show that SVC is a very effective solution from the point of view of voltage stability and, for the cases shown in the paper, economically more advantageous. Reference 11 focuses on the use of first- and second-order information to determine sensitivity factors which are able to address preventive and/or corrective actions in the presence of SVC. In Reference 3, a study based on the Singular Value Decomposition of the Jacobian matrix of the PF equations considering the SVC is proposed.

The optimisation of the power flow solution including the SVC constraints is presented in Reference 12. The analysis from the point of view of the Italian power system operation is shown, which takes into account both economical and security issues. The multi-objective programming including SVC is also shown in Reference 13, with emphasis on both the contingency analysis and the performance with respect to the voltage stability. The analysis of the stability margin is also presented in Reference 14. Reference 15 shows preliminary studies of the implementation of SVC in the Brazilian power system, more specifically the south-eastern region. The dynamic behaviour of the power system is analysed under several conditions, including load changes, contingencies in the transmission system and increase of the loadability margin. In order to observe the influence on the other power system variables, pilot buses were selected. The strategy used in the hierarchical control is based on the squared deviation of the voltage magnitude of the pilot buses from a pre-specified value. This error is sent to the automatic voltage regulators (AVRs), which operate aiming at correcting this voltage magnitude deviation. The reactive power control has not been considered with respect to the alignment equations, which makes this scheme simpler from the practical point of view, but less efficient [16].

Recently, power system voltage stability and control are receiving revived attention due to important factors: (a) the massive increasing of intermittent Renewable Energy Sources (RES), that make it more difficult to anticipate power system operating conditions and security margins and (b) the possibility to suitably face this increased uncertainty by means of monitoring and control structures. In particular, both the spreading of devices able to provide online voltage stability indices, like phasor measurement units (PMUs) [17,18], and the implementation of control structures like Hierarchical Voltage Control (including SVC) [19–21] help the transmission system operators (TSOs) to identify and quantify timely the risk of voltage collapse and to find out the most effective countermeasures to avoid deterioration of voltage profiles.

This chapter focuses on the analysis of power systems in the presence of SVC, that has proved to be an efficient countermeasure against the deterioration of voltage profiles [22]. It is well known that many problems related to the QV-control arise as load increases, such that the parameters mostly affecting loadability margins are to be identified. In fact, as the system gets stressed as a result of load increase, the reactive output of generators becomes larger, which makes very important to take into account capability limits of generators that sustain voltage profiles. Some approaches, particularly related to countries like France, Italy and Belgium, tried to investigate the effect of SVC for different patterns of the system loadability based on optimisation methods [23,24]. Here, analytical models of the PF and OPF considering the inclusion of the *SVC* constraints are developed. The main aspects presented in this Chapter are:

1. The formulation of the PF and OPF problems in rectangular coordinates, which is proved: (a) to provide benefits from the numerical point of view (robustness, no trigonometric evaluations, computational speed); (b) to make readily available important information as by-product of the optimisation.
2. The analysis of the numerical method to find the power network solutions as well as the quality of these solutions from the point of view of power system operation.
3. The assessment of the effect of the constraints introduced by the SVC in the determination of power flow solutions corresponding to optimised performance indices, such as the deviation of a reference value and the loadability margin.

Results and comparisons are provided based on: (a) the New England 39 buses network with three control areas and (b) a model of the extra high voltage (EHV) Italian transmission system (about 959 buses), which includes SVC organised in 13 voltage control areas.

16.2 The power flow problem in rectangular coordinated

Denoting by n_b the total number of buses of the power system and assuming that rectangular coordinates are used to model the complex bus voltages, the set of $n_{eq} = 2n_b - 2$ of non-linear PF equations is expressed as,

$$\begin{aligned}
\Delta P_i &= P_{g_i} - P_{d_i} - e_i \sum_{j=1}^{n} (G_{ij} e_j - B_{ij} f_j) + f_i \sum_{j=1}^{n} (G_{ij} f_j + B_{ij} e_j) \\
\Delta Q_i &= Q_{g_i} - Q_{d_i} - f_i \sum_{j=1}^{n} (G_{ij} e_j - B_{ij} f_j) - e_i \sum_{j=1}^{n} (G_{ij} f_j + B_{ij} e_j) \\
\Delta V_i &= V_i^{r^2} - (e_i^2 + f_i^2)
\end{aligned} \tag{16.1}$$

where P_{g_i} and Q_{g_i}, P_{d_i} and Q_{d_i} are the real and reactive power generation and demand at bus i, respectively; the third and fourth terms on the right-hand side are the injections of real and reactive power expressed in terms of the real and imaginary components of the complex voltages (e_i, f_i); V_i^r is the magnitude of the specified reference voltage, that is, $V_i^r = \sqrt{e_i^2 + f_i^2}$; and G_{ij} and B_{ij} are the elements of the admittance bus matrix, which depend only on the parameters of the transmission system. For a pre-specified demand level, represented by the variables $P_{d_i}^0$ and $Q_{d_i}^0$, the conventional modelling of the power flow problem requires the pre-specification of some variables, such as P_{g_i}, Q_{g_i} and V_i^r. For any convergent solution, the magnitude of ΔP_i, ΔQ_i and ΔV_i (named *mismatches*) must satisfy a pre-specified tolerance (usually 10^{-3} per unit) at the final solution. There are $(n_b - 1)$ equations related to the active power mismatches. The equations related to ΔV_i correspond to a set of buses with QV-control devices. The number of equations related to the reactive power and voltage magnitude mismatches ΔQ_i and ΔV_i is $(n_b - 1)$. In compact form,

$$\begin{aligned}
\Delta P_i &= P_{g_i} - P_{d_i}^0 - P_i(e, f) \\
\Delta Q_i &= Q_{g_i} - Q_{d_i}^0 - Q_i(e, f) \\
\Delta V_i &= V_i^{r^2} - (e_i^2 + f_i^2)
\end{aligned} \tag{16.2}$$

where the superscripts 0 and r indicate pre-specified and references variables, respectively; and $P_i(e, f)$ and $Q_i(e, f)$ are the injections of real and reactive power expressed in terms of the real and imaginary components of the complex voltages. Traditionally, these components compose a set of $n_{vr} = 2n_b - 2$ variables of the power flow problem.

Equations (16.1) and (16.2) can be re-written in compact form as,

$$\boldsymbol{g}(\boldsymbol{x}) = \boldsymbol{y}_s - \boldsymbol{g}_0(\boldsymbol{x}) = \boldsymbol{0} \tag{16.3}$$

where $\boldsymbol{g}(\cdot)$ is a column vector of dimension n_{eq}, whose components are the algebraic functions representing the power and quadratic voltage magnitude mismatches, and $\boldsymbol{x}$ represents the vector of power flow variables; $\boldsymbol{y}_s$ and $\boldsymbol{g}_0(\boldsymbol{x})$ are vectors of dimension $((n_b - 2) \times 1)$, with components $(P_{g_i} - P_{d_i}^0)$, $(Q_{g_i} - Q_{d_i}^0)$ and $V_i^{r^2}$, and $P_i(e, f)$, $Q_i(e, f)$ and $(e_i^2 + f_i^2)$, respectively.

The linear system to be solved at each iteration of Newton's method is given by:

$$\begin{bmatrix} \boldsymbol{\Delta P} \\ \boldsymbol{\Delta Q} \\ \boldsymbol{\Delta V} \end{bmatrix} = \begin{bmatrix} \boldsymbol{J}_1 & \boldsymbol{J}_2 \\ \boldsymbol{J}_3 & \boldsymbol{J}_4 \\ \boldsymbol{J}_5 & \boldsymbol{J}_6 \end{bmatrix} \begin{bmatrix} \boldsymbol{\Delta e} \\ \boldsymbol{\Delta f} \end{bmatrix} \tag{16.4}$$

where $\boldsymbol{J}_1$, $\boldsymbol{J}_2$, $\boldsymbol{J}_3$, $\boldsymbol{J}_4$, $\boldsymbol{J}_5$ and $\boldsymbol{J}_6$ are the matrices of first derivatives of (16.3) with respect to the variables ($\boldsymbol{e}$, $\boldsymbol{f}$); $\boldsymbol{\Delta P}$, $\boldsymbol{\Delta Q}$ and $\boldsymbol{\Delta V}$ are vectors with components ΔP_i, ΔQ_i and ΔV_i, respectively; and $\boldsymbol{\Delta e}$ and $\boldsymbol{\Delta f}$ are vectors with components Δe_i and Δf_i, respectively.

The Jacobian matrices are given by:

$$\begin{aligned}
\boldsymbol{J}_1 &= \mathrm{diag}(\boldsymbol{e})\boldsymbol{G} + \mathrm{diag}(\boldsymbol{f})\boldsymbol{B} + \mathrm{diag}(\boldsymbol{Ge} - \boldsymbol{Bf}) \\
\boldsymbol{J}_2 &= \mathrm{diag}(\boldsymbol{f})\boldsymbol{G} - \mathrm{diag}(\boldsymbol{e})\boldsymbol{B} + \mathrm{diag}(\boldsymbol{Gf} + \boldsymbol{Be}) \\
\boldsymbol{J}_3 &= \mathrm{diag}(\boldsymbol{f})\boldsymbol{G} - \mathrm{diag}(\boldsymbol{e})\boldsymbol{B} - \mathrm{diag}(\boldsymbol{Gf} + \boldsymbol{Be}) \\
\boldsymbol{J}_4 &= -\mathrm{diag}(\boldsymbol{e})\boldsymbol{G} - \mathrm{diag}(\boldsymbol{f})\boldsymbol{B} + \mathrm{diag}(\boldsymbol{Ge} - \boldsymbol{Bf}) \\
\boldsymbol{J}_5 &= \mathrm{diag}(2\boldsymbol{e}) \\
\boldsymbol{J}_6 &= \mathrm{diag}(2\boldsymbol{f})
\end{aligned} \tag{16.5}$$

where the operator diag(·) applied to a vector $\boldsymbol{x}$ with dimension $n_x \times 1$, represents a diagonal matrix of order $n_x \times n_x$ whose terms are x_i. If this operator is applied to a matrix $\boldsymbol{A}$ of order $n_x \times n_x$, it provides the vector $\boldsymbol{a}$ whose components a_i are the main diagonal of matrix $\boldsymbol{A}$. The components of the vectors $\boldsymbol{e}$ and $\boldsymbol{f}$ are the real and imaginary parts of the complex bus voltages, generically e_i and f_i for the ith bus. $\boldsymbol{G}$ and $\boldsymbol{B}$ are the real and imaginary components of the bus admittance matrix, respectively.

16.2.1 The power flow with SVC constraints

In order to model the constraints related to the *SVC* in the PF problem, it is necessary to increase the number of equations, such as to represent the behaviour of the reference voltage magnitude at some pre-selected buses (one at each area) and the individual reactive power level of all QV-control devices of the area. For this purpose, further than the conventional types of buses ($V\delta$, PV and PQ), it is necessary to define the following two types:

- *PQV bus*: it represents analytically the so-called *pilot buses*, that is, the net real and reactive power injections as well as the voltage magnitude are specified a priori. Each area must have a pilot bus, and then the number of pilot buses is equal to the number of areas.
- *P bus*: it models the regulating buses, where QV-control devices (generator or synchronous compensator, SVC, etc.) are supposed to be installed. Only the active power injection is pre-specified in this type of bus. The number of *P buses* is equal to the number of buses previously selected to control the voltage magnitude of the pilot bus.

The constraints related to the reactive power margin of the QV-control devices are represented by the so-called *alignment constraints*. These constraints impose that the reactive power margin of each regulating unit with respect to its capacity limit must be equal to the total reactive power margin of the area with respect to the total capacity of the area, that is also named *reactive level* of the area $k(q_k)$. The level can be considered the pu expression of the reactive power injected in the area, in pu with respect to the area reactive power capability; $(1 - q_k)$ can be considered the pu area reactive margin. The generating units belonging to the same control area are characterised by the same value of the reactive level. Analytically, this is expressed by the following equation:

$$\frac{Q_{gi}}{Q_{gi}^M} = \frac{\sum\limits_{j\in\Omega_k} Q_{gj}}{\sum\limits_{j\in\Omega_k} Q_{gj}^M} \qquad \text{for any } i \in \Omega_k \tag{16.6}$$

where Ω_k represents the set of regulation generators of the k area, $\sum_{j\in\Omega_k} Q_{gj}$ is the total reactive power generated in the k area and the superscript M represents the maximum limit. The number of alignment equations is equal to the number of regulating buses.

In summary, the SVC constraints are handled according to References 2,5,6, such that the resulting set of PF equations is given by,

$$\begin{aligned}
&P_{gi} - P_{d_i}^0 - P_i(e,f) = 0 && \text{P, PV, PQ and PVQ buses}\\
&Q_{gi} - Q_{d_i}^0 - Q_i(e,f) = 0 && \text{PQ and PQV buses}\\
&V_i^{r2} - (e_i^2 + f_i^2) = 0 && \text{PV and PQV buses}\\
&\frac{Q_{gi}}{Q_{gi}^M} - \frac{\sum\limits_{j\in\Omega_k} Q_{gj}}{\sum\limits_{j\in\Omega_k} Q_{gj}^M} = 0 && \text{P buses}
\end{aligned} \tag{16.7}$$

Assuming that n_{pq} is the number of PQ buses (excluding the pilot buses), n_{pv} is the number of PV buses, n_{pi} is the number of pilot buses and n_{gr} is the number of regulating buses. The number of equations n_{eqs} and the number of variables n_{var} to be determined in the power flow problem with SVC constraints are, respectively,

$$\begin{aligned}
n_{eqs} &= n_b - 1 + n_{pq} + n_{pi} + n_{pv} + n_{pi} + n_{gr} = 2n_b - 2 + n_{pi}\\
n_{var} &= 2n_b - 2
\end{aligned} \tag{16.8}$$

where the number of pilot buses is equal to the number of areas. Note that the difference between the number of equations resulting from the inclusion of the SVC constraints with respect to that of the conventional power flow model is equal to the number of pilot buses (or areas), which is generally small if compared to the total number of buses. Observe also that the linear system corresponding to the solution of (16.7) through Newton's method has more equations than variables, such that the coefficient matrix is rectangular, with rank equal to or smaller than $(2n_b - 2)$, and whose solution (using least squares based methods) is useless for practical power system purposes.

In order to overcome this difficulty, consider that the fourth row of (16.7) can be re-written as,

$$Q_{g_i} = \frac{\sum\limits_{(j \neq i) \in \Omega_k} Q_{g_j}}{\left(\frac{\sum\limits_{j \in \Omega_k} Q_{g_j}^M}{Q_{g_i}^M} - 1 \right)} \tag{16.9}$$

and the reactive power of the jth regulating bus is expressed as,

$$Q_{g_j}(e,f) = Q_{d_j} + Q_j(e,f) \tag{16.10}$$

such that (16.9) can be re-written as,

$$Q_{g_i} = \frac{\sum\limits_{(j \neq i) \in \Omega_k} \left(Q_{d_j} + Q_j(e,f) \right)}{\left(\frac{\sum\limits_{j \in \Omega_k} Q_{g_j}^M}{Q_{g_i}^M} - 1 \right)} \tag{16.11}$$

and the reactive power mismatch equation of the ith regulating bus is expressed as,

$$\frac{\sum\limits_{(j \neq i) \in \Omega_k} \left(Q_{d_j} + Q_j(e,f) \right)}{\left(\frac{\sum\limits_{j \in \Omega_k} Q_{g_j}^M}{Q_{g_i}^M} - 1 \right)} - Q_{d_i} - Q_i(e,f) = 0 \tag{16.12}$$

or, in compact form,

$$Q_{g_{ri}}(e,f) - Q_{d_i} - Q_i(e,f) = 0 \tag{16.13}$$

Equation (16.13) ensures that the reactive power mismatch equations as well as the alignment equations are simultaneously satisfied in the solution of (16.7). It must be pointed out that (16.13) expresses the reactive power mismatch equation of the ith bus as a function of the reactive power generation of the j-area. Now, the analytical model of the power flow problem with SVC constraints is given by,

$$\begin{array}{ll} P_{g_i} - P_{d_i}^0 - P_i(e,f) = 0 & \text{P, PV, PQ and PVQ buses} \\ Q_{g_i} - Q_{d_i}^0 - Q_i(e,f) = 0 & \text{PQ, PQV buses} \\ Q_{g_{ri}} - Q_{d_i}^0 - Q_i(e,f) = 0 & \text{P buses} \\ V_i^{r2} - (e_i^2 + f_i^2) = 0 & \text{PV and PQV buses} \end{array} \tag{16.14}$$

where $Q_{g_{ri}}$ is given by (16.13).

With respect to the number of equations of the conventional power flow model, the inclusion of the pilot buses brings an increase of n_{pi} equations, as indicated by (16.14). This implies in a rectangular Jacobian matrix (if Newton's method is used), with the number of rows greater than the number of columns. Therefore, an auxiliary variable (for each area j), here denoted by z_j, is added in the third row of (16.14), that is,

$$Q_{gri} - Q^0_{d_i} - Q_i(e,f) + z_j = 0$$

Now, the number of equations and the number of variables are

$$\begin{aligned} n_{eqs} &= n_b - 1 + n_{pq} + n_{pi} + n_{gr} + n_{pv} + n_{pi} + n_{gr} = 2n_b - 2 + n_{pi} \\ n_{var} &= 2n_b - 2 + n_{pi} \end{aligned} \tag{16.15}$$

such that the linear system to be solved at each iteration becomes:

$$\begin{bmatrix} \Delta P \\ \Delta Q \\ \Delta V \end{bmatrix} = \begin{bmatrix} J_1 & J_2 & 0 \\ J_3^* & J_4^* & F_n \\ J_5 & J_6 & 0 \end{bmatrix} \begin{bmatrix} \Delta e \\ \Delta f \\ \Delta z \end{bmatrix} \tag{16.16}$$

where Δz is a column vector, of dimension n_{pi}, whose components are the increments corresponding to the auxiliary variables previously mentioned.

The computation of matrices J_3^* and J_4^* is similar to that of matrices J_3 and J_4 previously presented in Section 16.2, except for the terms corresponding to (16.12), that is,

- If bus i is not a regulating bus:

$$\begin{aligned} J_3^*(i,j) &= J_3(i,j) \\ J_4^*(i,j) &= J_4(i,j) \end{aligned}$$

- If bus i is a regulating bus:

$$\begin{aligned} J_3^*(i,i) &= J_3(i,j) \\ J_3^*(i,j) &= \frac{1}{\left(\dfrac{\sum\limits_{(j\neq i)\in\Omega_k} Q_{gj}^M}{Q_{gi}^M}\right)} J_3(i,j) \end{aligned}$$

$$\begin{aligned} J_4^*(i,i) &= J_4(i,j) \\ J_4^*(i,j) &= \frac{1}{\left(\dfrac{\sum\limits_{(j\neq i)\in\Omega_k} Q_{gj}^M}{Q_{gi}^M}\right)} J_4(i,j) \end{aligned}$$

Since the number of auxiliary variables is equal to the number of areas and the number of regulating buses is usually greater than the number of areas, F_n is a sparse matrix, with dimension $(n_{pq} + n_{pi} + n_{gr}) \times n_{pi}$ and rank equal to n_{pi}, whose components are defined by,

$$\begin{aligned} F_n(i,j) &= 1, \quad \text{if the regulating device } i \text{ belongs to area } j \\ F_n(i,j) &= 0, \quad \text{if the regulating device } i \text{ does not belong to area } j \end{aligned} \tag{16.17}$$

Note that in the solution of the power flow problem with SVC constraints, the alignment equations of the regulating buses are satisfied and the reactive power of these buses is computed through (16.13). This imposes a zero value to the components of vector z in the end of the iterative process, indicating that these variables do not have any influence if the quality of the power flow solution.

The reactive power of the regulating devices must also satisfy the capability condition expressed as,

$$Q_{gi}^{m} \leq Q_{g_i} \leq Q_{gi}^{M} \tag{16.18}$$

such that during the iterations the traditional procedure of bus conversion is used to handle these limits. In the presence of SVC, P buses and PV buses can be eventually converted into PQ buses. The PV–PQ bus conversion is applied to one bus at each iteration, which requires the interchange of the quadratic voltage magnitude mismatch equation and the reactive power mismatch equation of the PV bus which has reached the reactive power limit. In case of the P buses, the reactive power limits of the regulating buses of an area are simultaneously reached, and then the conversion P–PQ is applied to the whole set of P buses. From this point on, these devices cannot control the voltage magnitude of the pilot bus anymore, which requires that the PQV bus of the area is converted into PQ bus. This situation is characterised by a reactive level equal to 1.

16.3 The OPF with SVC constraints

The general OPF problem with SVC constraints can be expressed as

Minimise $f_o(e,f)$
subject to

$$\begin{aligned}
&\Delta P_i = P_{g_i} - P_{d_i}^{0} - P_i(e,f) = 0\\
&\Delta Q_i = Q_{g_i} - Q_{d_i}^{0} - Q_i(e,f) = 0\\
&\Delta Q_{ri} = Q_{g_{ri}} - Q_{d_i}^{0} - Q_i(e,f) = 0\\
&\Delta V_i = V_i^{r2} - (e_i^2 + f_i^2) = 0\\
&P_{gi}^{m} \leq P_{g_i} \leq P_{gi}^{M}\\
&Q_{gi}^{m} \leq Q_{g_i} \leq Q_{gi}^{M}\\
&Q_{g_{ri}}^{m} \leq Q_{g_{ri}} \leq Q_{g_{ri}}^{M}\\
&V_i^{m} \leq V_i \leq V_i^{M}
\end{aligned} \tag{16.19}$$

where $f_o(e,f)$ represents the objective function and all variables have been previously defined. Note that the equality constraints are the power flow equations with SVC constraints stated by (16.14). The voltage magnitude quadratic constraint includes the pilot buses, and the active and reactive power constraints represent the capacity limits.

Two of the most common performance indices used in the problem stated by (16.19) are [25]:

- Minimisation of the squared deviation of bus voltage magnitude from a reference value, which is analytically represented by the following equation:

  $$f_o(e,f) = \sum (V_i - V_i^r)^2$$

  where $V_i = \sqrt{e_i^2 + f_i^2}$ is the voltage magnitude at bus i and V_i^r is the reference voltage magnitude at the ith bus.
- Maximisation of the loadability in a pre-specified direction, in which the load is parameterised in terms of the scalar ρ, that is,

  $$\begin{aligned} P_{d_i} &= P_{d_i}^0 + \rho^2 \Delta P_{d_i} \\ Q_{d_i} &= Q_{d_i}^0 + \rho^2 \Delta Q_{d_i} \end{aligned}$$

  where $P_{d_i}^0$ and $Q_{d_i}^0$ are the real and reactive power load in the base case, and ΔP_{d_i} and ΔQ_{d_i} represent the pattern of variation of the load; and ρ is the load parameter, whose squared form ensures that the power load is not reduced, even for negative values of the load parameter.

16.3.1 *The maximum loadability with SVC constraints*

The determination of the maximum loadability for which there is a real power flow solution with SVC constraints can be modelled by the following optimisation problem:

$$\begin{aligned} &\text{Maximise } \rho \\ &\text{subject to} \\ &\qquad P_{g_i} - (P_{d_i}^0 + \rho^2 \Delta P_{d_i}) - P_i(e,f) = 0 \\ &\qquad Q_{g_i} - (Q_{d_i}^0 + \rho^2 \Delta Q_{d_i}) - Q_i(e,f) = 0 \\ &\qquad V_i^{r2} - (e_i^2 + f_i^2) = 0 \\ &\qquad \frac{Q_{gj}}{Q_{gj}^M} - \frac{\sum_{k\in\Omega_j} Q_{gk}}{\sum_{k\in\Omega_j} Q_{gk}^M} = 0 \end{aligned} \tag{16.20}$$

where all variables have been previously defined. Note that the equality constraints are those stated by (16.7), which can be replaced alternatively by the simplified form of (16.14). Additionally, the reactive power generation must satisfy a set of inequality constraints given by,

$$Q_{gi}^m \leq Q_{gi} \leq Q_{gi}^M \tag{16.21}$$

where i is any bus with a reactive power device, including the buses where the regulating generators are connected.

The inclusion of both the alignment equations and the voltage magnitude constraints of the pilot buses brings additional difficulties to obtain power flow solutions at the critical loadability level. These extra restrictions change the path of the power flow solution from the base case to the critical loadability and increase of the number of stationary points corresponding to local minima.

The traditional procedure based on bus conversion is adopted to deal with the inequality stated by (16.21): if necessary, PV and P buses are converted into PQ buses. The conversion PV–PQ bus is applied to each bus individually, which requires the change of the equations of the quadratic voltage magnitude and reactive power balance of the PV bus whose limit has been reached. In case of P-type buses, the controlling generators of the same area reach simultaneously the reactive power limit according to the alignment equations, and hence the P–PQ conversion is applied to the complete set of the area controlling generators. Consequently, the controlling generators of this area can no longer be used to control the voltage magnitude of the pilot bus. Accordingly, the voltage equality constraint of that pilot bus is excluded from the set of PF equations, and the PQV bus is converted into PQ bus, i.e., a normal load bus. In general, there is a trend of the P-type and PV-type to reach the reactive power generation limit, particularly when the loading of the system increases. Due to this way of dealing with such limits, the bus conversions P–PQ and PQV–PQ result in a large number of buses without (direct or indirect) control of the voltage magnitude.

The loadability levels at which the reactive power generation of the regulating generators of an area reach the limit can be seen as *points of constraint interchange* of the pilot buses, in an analogous way to that presented in Reference 26. After one of these points is reached, the voltage magnitude of the pilot bus is free, such that only the power balance equations remain in the set of equality constraints. This condition is similar to that of the PV buses in Limit-Induced Bifurcation studies, when the main objective is to obtain and checking the nature of the constraint-switching points. In the SVC case, it is possible to apply a similar test, based on sensitivity relationships, to check the features of the PQV–PQ constraint interchange point. For details, the interested reader is referred to Reference 26.

16.3.2 Minimisation of the squared deviation of the bus voltage magnitude from a reference value

In this case, the optimisation variables are the real and imaginary components of the complex bus voltages and the auxiliary variables z_j, similarly to the problem stated by (16.14). Thus, the reactive power generation is computed as a function of the optimisation variables and only the real power generation of the reference bus is supposed to be modified. Besides, the QV-control devices not selected as regulators are treated as voltage magnitude control buses, that is, they are modelled similarly to the PV buses of the conventional power flow.

The optimisation problem that represents the minimisation of the squared deviation of the bus voltage magnitude from a reference value with SVC constraints is given by,

$$
\begin{array}{lll}
\text{Minimise} & \frac{1}{2}\sum_{i=1}^{n}\left(V_i - V_i^r\right)\left(V_i - V_i^r\right) & \\
\text{subject to} & & \\
& \Delta P_i = P_{g_i} - P_{d_i}^0 - P_i(e,f) = 0 & \text{P, PV, PQ and PQV buses} \\
& \Delta Q_i = Q_{g_i} - Q_{d_i}^0 - Q_i(e,f) = 0 & \text{PQ, PQV and P buses} \\
& \Delta V_i = V_i^{r^2} - (e_i^2 + f_i^2) = 0 & \text{PV and PQV buses} \\
& Q_{gi}^m \leq Q_{g_i} \leq Q_{gi}^M & \text{PV and P buses} \\
& V_i^m \leq V_i \leq V_i^M & \text{P and PQ buses}
\end{array}
\qquad (16.22)
$$

where all variables have been previously defined and, for the sake of simplicity the inequality constraint corresponding to the real power of the reference bus is not included.

Supposing that the solution of this optimisation problem is obtained through the interior point method described in Section 16.4.1, the Lagrangian function of the problem stated by (16.22) is given by,

$$
\begin{aligned}
\mathcal{L}(\boldsymbol{y}) &= f(\boldsymbol{x}) + \boldsymbol{\lambda}_p^t \Delta \boldsymbol{P} + \boldsymbol{\lambda}_q^t \Delta \boldsymbol{Q} + \boldsymbol{\lambda}_v^t \Delta \boldsymbol{V} \\
&\quad + \boldsymbol{\pi}_{Qg}^t \left(\boldsymbol{F}_{Qg}\boldsymbol{Q}_g + \boldsymbol{s}_{Qg} - \boldsymbol{Q}_g^{lim}\right) + \boldsymbol{\pi}_v^t \left(\boldsymbol{F}_v \boldsymbol{V}^2 + \boldsymbol{s}_v - \boldsymbol{V}^{lim}\right) \\
&\quad + \mu \sum \ln s_{Qg} + \mu \sum \ln s_v
\end{aligned}
$$

where $\boldsymbol{y}$ includes all primal and dual variables; $\boldsymbol{\lambda}_p$, $\boldsymbol{\lambda}_q$ and $\boldsymbol{\lambda}_v$ are the Lagrange multipliers related to the equality constraints; $\boldsymbol{\pi}_{Q_g}$ and $\boldsymbol{\pi}_v$ are the Lagrange multipliers related to the inequality constraints; and $\boldsymbol{s}_{Q_g}$ and $\boldsymbol{s}_v$ are the slack variables corresponding to the inequalities. Vectors $\boldsymbol{Q}_g$ and $\boldsymbol{V}^2$ are expressed as,

$$
\boldsymbol{Q}_g = \boldsymbol{Q}_d^0 + \boldsymbol{Q}(\boldsymbol{e},\boldsymbol{f}) \quad \text{and} \quad \boldsymbol{V}^2 = \boldsymbol{e}^2 + \boldsymbol{f}^2
$$

and the other variables have been previously defined.

The Karush–Kuhn–Tucker (KKT) first-order optimality conditions are:

$$
\begin{aligned}
\nabla_e \mathcal{L}(\boldsymbol{y}) &= \nabla_e f(\boldsymbol{x}) + \boldsymbol{J}_1^t \boldsymbol{\lambda}_p + (\boldsymbol{J}_3^*)^t \boldsymbol{\lambda}_q + \boldsymbol{J}_5{}^t \boldsymbol{F}_v^t \boldsymbol{\pi}_v \\
\nabla_f \mathcal{L}(\boldsymbol{y}) &= \nabla_f f(\boldsymbol{x}) + \boldsymbol{J}_2^t \boldsymbol{\lambda}_p + (\boldsymbol{J}_4^*)^t \boldsymbol{\lambda}_q + \boldsymbol{J}_6{}^t \boldsymbol{F}_v^t \boldsymbol{\pi}_v \\
\nabla_z \mathcal{L}(\boldsymbol{y}) &= \boldsymbol{F}_n \\
\nabla_{\lambda_p} \mathcal{L}(\boldsymbol{y}) &= \Delta \boldsymbol{P} \\
\nabla_{\lambda_q} \mathcal{L}(\boldsymbol{y}) &= \Delta \boldsymbol{Q} \\
\nabla_{\lambda_v} \mathcal{L}(\boldsymbol{y}) &= \Delta \boldsymbol{V}
\end{aligned}
$$

$$\begin{aligned}
\nabla_{\pi_{Q_g}} \mathcal{L}(\boldsymbol{y}) &= \boldsymbol{F}_{Q_g}\left(\boldsymbol{Q}_d^0 + \boldsymbol{Q}(\boldsymbol{e},\boldsymbol{f})\right) + \boldsymbol{s}_{Q_g} - \boldsymbol{Q}_g^{lim} \\
\nabla_{\pi_v} \mathcal{L}(\boldsymbol{y}) &= \boldsymbol{F}_v(\boldsymbol{e}^2 + \boldsymbol{f}^2) + \boldsymbol{s}_v - \boldsymbol{V}^{lim} \\
\nabla_{s_{Q_g}} \mathcal{L}(\boldsymbol{y}) &= \boldsymbol{S}_{Q_g}\boldsymbol{\pi}_{Q_g} - \mu\boldsymbol{u} \\
\nabla_{s_v} \mathcal{L}(\boldsymbol{y}) &= \boldsymbol{S}_v\boldsymbol{\pi}_v - \mu\boldsymbol{u}
\end{aligned} \tag{16.23}$$

such that

$$\nabla_e f(\boldsymbol{x}) = 2\boldsymbol{e} \circ \left(\boldsymbol{u} - \left(\boldsymbol{V}^r./\sqrt{(\boldsymbol{e}.^2 + \boldsymbol{f}.^2)}\right)\right)$$

$$\nabla_f f(\boldsymbol{x}) = 2\boldsymbol{f} \circ \left(\boldsymbol{u} - \left(\boldsymbol{V}^r./\sqrt{(\boldsymbol{e}.^2 + \boldsymbol{f}.^2)}\right)\right)$$

$$\boldsymbol{Q}_g^{lim} \triangleq \begin{bmatrix} \boldsymbol{Q}_g^M \\ -\boldsymbol{Q}_g^m \end{bmatrix}_{(2n_g \times 1)}$$

$$\boldsymbol{V}^{lim} \triangleq \begin{bmatrix} \boldsymbol{V}^{M^2} \\ -\boldsymbol{V}^{m^2} \end{bmatrix}_{(2n_s \times 1)}$$

$$\boldsymbol{F}_{Q_g} \triangleq \begin{bmatrix} \boldsymbol{I} \\ -\boldsymbol{I} \end{bmatrix}_{(2n_g \times n_g)}$$

$$\boldsymbol{F}_v \triangleq \begin{bmatrix} \boldsymbol{I} \\ -\boldsymbol{I} \end{bmatrix}_{(2n_s \times n_s)}$$

$$\boldsymbol{\pi}_{Q_g} \triangleq \begin{bmatrix} \boldsymbol{\pi}_{Q_g}^M \\ \boldsymbol{\pi}_{Q_g}^m \end{bmatrix}_{(2n_g \times 1)}$$

$$\boldsymbol{\pi}_v \triangleq \begin{bmatrix} \boldsymbol{\pi}_v^M \\ \boldsymbol{\pi}_v^m \end{bmatrix}_{(2n_s \times 1)}$$

$$\boldsymbol{S}_{Q_g} \triangleq \begin{bmatrix} \boldsymbol{s}_{Q_g}^M \\ \boldsymbol{s}_{Q_g}^m \end{bmatrix}_{(2n_g \times 1)}$$

$$\boldsymbol{S}_v \triangleq \begin{bmatrix} \boldsymbol{s}_v^M \\ \boldsymbol{s}_v^m \end{bmatrix}_{(2n_s \times 1)}$$

where $\circ$ is the Hadamard or element-wise product; ./ indicates element-wise division; $.^2$ represents the element-wise square; n_g is the number of generating units; n_s is the total number of PQ and P buses; $\boldsymbol{u}$ is an unitary vector of dimension $n_b \times 1$; $\boldsymbol{S}_{Q_g}$ and $\boldsymbol{S}_v$ are diagonal matrices, whose components are the elements of vectors $\boldsymbol{s}_{Q_g}$ and $\boldsymbol{s}_v$;

$\boldsymbol{I}$ is a diagonal matrix with zero components, except for those corresponding to the regulating buses, which are the unity; $\boldsymbol{J}_1, \boldsymbol{J}_2, \boldsymbol{J}_3, \boldsymbol{J}_4, \boldsymbol{J}_5$ and $\boldsymbol{J}_6$ are computed as shown in Section 16.2.1.

The solution of (16.24) through Newton's method requires the solution of the following linear system at each iteration:

$$
\begin{bmatrix}
\boldsymbol{G}_1 & \boldsymbol{G}_2 & \boldsymbol{0} & \boldsymbol{J}_1^t & (\boldsymbol{J}_3^*)^t & \boldsymbol{J}_5^t & \boldsymbol{H}_{q1}^t & \boldsymbol{H}_{v1}^t & \boldsymbol{0} & \boldsymbol{0} \\
\boldsymbol{G}_2^t & \boldsymbol{G}_3 & \boldsymbol{0} & \boldsymbol{J}_2^t & (\boldsymbol{J}_4^*)^t & \boldsymbol{J}_6^t & \boldsymbol{H}_{q2}^t & \boldsymbol{H}_{v2}^t & \boldsymbol{0} & \boldsymbol{0} \\
\boldsymbol{0} & \boldsymbol{0} & \boldsymbol{0} & \boldsymbol{J}_{tp}^t & \boldsymbol{J}_{qq}^t & \boldsymbol{J}_{tv}^t & \boldsymbol{H}_{q3}^t & \boldsymbol{H}_{v3}^t & \boldsymbol{0} & \boldsymbol{0} \\
\boldsymbol{J}_1 & \boldsymbol{J}_2 & \boldsymbol{J}_{tp} & \boldsymbol{0} & \boldsymbol{0} & \boldsymbol{0} & \boldsymbol{0} & \boldsymbol{0} & \boldsymbol{0} & \boldsymbol{0} \\
\boldsymbol{J}_3^* & \boldsymbol{J}_4^* & \boldsymbol{J}_{qq} & \boldsymbol{0} & \boldsymbol{0} & \boldsymbol{0} & \boldsymbol{0} & \boldsymbol{0} & \boldsymbol{0} & \boldsymbol{0} \\
\boldsymbol{J}_5 & \boldsymbol{J}_6 & \boldsymbol{J}_{tv} & \boldsymbol{0} & \boldsymbol{0} & \boldsymbol{0} & \boldsymbol{0} & \boldsymbol{0} & \boldsymbol{0} & \boldsymbol{0} \\
\boldsymbol{H}_{q1} & \boldsymbol{H}_{q1} & \boldsymbol{H}_{q1} & \boldsymbol{0} & \boldsymbol{0} & \boldsymbol{0} & \boldsymbol{0} & \boldsymbol{0} & \boldsymbol{I}_q & \boldsymbol{0} \\
\boldsymbol{H}_{v1} & \boldsymbol{H}_{v1} & \boldsymbol{H}_{v1} & \boldsymbol{0} & \boldsymbol{0} & \boldsymbol{0} & \boldsymbol{0} & \boldsymbol{0} & \boldsymbol{0} & \boldsymbol{I}_v \\
\boldsymbol{0} & \boldsymbol{0} & \boldsymbol{0} & \boldsymbol{0} & \boldsymbol{0} & \boldsymbol{0} & \boldsymbol{S}_{Qg} & \boldsymbol{0} & \boldsymbol{\Pi}_{Qg} & \boldsymbol{0} \\
\boldsymbol{0} & \boldsymbol{0} & \boldsymbol{0} & \boldsymbol{0} & \boldsymbol{0} & \boldsymbol{0} & \boldsymbol{0} & \boldsymbol{S}_v & \boldsymbol{0} & \boldsymbol{\Pi}_v
\end{bmatrix}
\cdot
\begin{bmatrix}
\Delta \boldsymbol{e} \\ \Delta \boldsymbol{f} \\ \Delta \boldsymbol{z} \\ \Delta \boldsymbol{\lambda}_p \\ \Delta \boldsymbol{\lambda}_q \\ \Delta \boldsymbol{\lambda}_v \\ \Delta \boldsymbol{\pi}_{Qg} \\ \Delta \boldsymbol{\pi}_v \\ \Delta \boldsymbol{s}_{Qg} \\ \Delta \boldsymbol{s}_v
\end{bmatrix}
= -
\begin{bmatrix}
\nabla_e \mathcal{L}(\boldsymbol{y}) \\ \nabla_f \mathcal{L}(\boldsymbol{y}) \\ \nabla_z \mathcal{L}(\boldsymbol{y}) \\ \nabla_{\lambda_p} \mathcal{L}(\boldsymbol{y}) \\ \nabla_{\lambda_q} \mathcal{L}(\boldsymbol{y}) \\ \nabla_{\lambda_v} \mathcal{L}(\boldsymbol{y}) \\ \nabla_{\pi_{Qg}} \mathcal{L}(\boldsymbol{y}) \\ \nabla_{\pi_v} \mathcal{L}(\boldsymbol{y}) \\ \nabla_{s_{Qg}} \mathcal{L}(\boldsymbol{y}) \\ \nabla_{s_v} \mathcal{L}(\boldsymbol{y})
\end{bmatrix}
\tag{16.24}
$$

where $\boldsymbol{I}_q$ and $\boldsymbol{I}_v$ are identity matrices of suitable dimension, and $\boldsymbol{u}$ is an unitary vector. The other matrices are given by,

$$
\begin{aligned}
\boldsymbol{G}_1 &= \nabla_{ee}^2 f(\boldsymbol{x}) + \text{diag}(\boldsymbol{\lambda}_p)\boldsymbol{G} + \boldsymbol{G}\text{diag}(\boldsymbol{\lambda}_p) - (\text{diag}(\boldsymbol{\lambda}_q)\boldsymbol{B} \\
&\quad + \boldsymbol{B}\text{diag}(\boldsymbol{\lambda}_q)) + 2\text{diag}(\boldsymbol{\lambda}_v) \\
\boldsymbol{G}_2 &= \nabla_{ef}^2 f(\boldsymbol{x}) + \text{diag}(\boldsymbol{\lambda}_p)\boldsymbol{B} - \boldsymbol{B}\text{diag}(\boldsymbol{\lambda}_p) + (\text{diag}(\boldsymbol{\lambda}_q)\boldsymbol{G} - \boldsymbol{G}\text{diag}(\boldsymbol{\lambda}_q)) \\
\boldsymbol{G}_3 &= \nabla_{ff}^2 f(\boldsymbol{x}) + \boldsymbol{G}_1 \\
\boldsymbol{J}_{tp} &= \boldsymbol{0}(nb, nb) \\
\boldsymbol{J}_{qq} &= \boldsymbol{I}(nb, nb) \\
\boldsymbol{J}_{tv} &= \boldsymbol{0}(nb, nb) \\
\boldsymbol{H}_q &= \boldsymbol{F}_q \left[\boldsymbol{J}_3^*(bge, bsf) \boldsymbol{J}_4^*(bge, bsf) \boldsymbol{J}_{qq}(bge, :) \right] \\
\boldsymbol{H}_v &= \boldsymbol{F}_v \left[\boldsymbol{J}_5(bsp, bsf) \boldsymbol{J}_6(bsp, bsf) \boldsymbol{J}_{tv}(bsp, :) \right] \\
\nabla_{ef}^2 f(\boldsymbol{x}) &= \text{diag}\left((\boldsymbol{V}_r \circ \boldsymbol{e} \circ \boldsymbol{f}) ./ \boldsymbol{V}.^3 \right) \\
\nabla_{ee}^2 f(\boldsymbol{x}) &= \text{diag}\left(2 \left(\boldsymbol{u} - \boldsymbol{V}_r ./ \boldsymbol{V} + (\boldsymbol{V}_r \circ \boldsymbol{e}.^2) ./ \boldsymbol{V}.^3 \right) \right) \\
\nabla_{ff}^2 f(\boldsymbol{x}) &= \text{diag}\left(2 \left(\boldsymbol{u} - \boldsymbol{V}_r ./ \boldsymbol{V} + (\boldsymbol{V}_r \circ \boldsymbol{f}.^2) ./ \boldsymbol{V}.^3 \right) \right)
\end{aligned}
\tag{16.25}
$$

where $\mathbf{0}$ is a null matrix of dimension $n_b \times n_b$; $\boldsymbol{H}^t_{q1}$, $\boldsymbol{H}^t_{q2}$ and $\boldsymbol{H}^t_{q3}$ are submatrices of $\boldsymbol{H}_q$, computed as the product of vector $\boldsymbol{F}_q$ times the matrices $\boldsymbol{J}_3(bge, bsf)$, $\boldsymbol{J}_4(bge, bsf)$ and $\boldsymbol{J}_{qq}(bge, :)$, respectively. Similarly, $\boldsymbol{H}^t_{v1}$, $\boldsymbol{H}^t_{v2}$ and $\boldsymbol{H}^t_{v3}$ are submatrices of $\boldsymbol{H}_v$, obtained as the product of vector F_v and the matrices $\boldsymbol{J}_5(bge, bsf)$, $\boldsymbol{J}_6(bge, bsf)$ and $\boldsymbol{J}_{tv}(bge, :)$, respectively; bge denotes the set of all generation buses, and bsf represents the set of all buses except the reference bus; symbol (:) denotes the set of all buses.

16.3.3 Constrained maximisation of the loadability with SVC

In this case, the optimisation variables are the real and imaginary components of the complex bus voltages, the auxiliary variables z_j, and the loadability factor ρ. The constraints of this optimisation problem are the same as in the previous case. The optimisation problem that represents the constrained load maximisation is given by,

$$
\begin{aligned}
&\text{Maximise} \quad \rho \\
&\text{subject to} \\
&\qquad \Delta P_i = P_{g_i} - (P^0_{d_i} + \rho^2 \Delta P_{d_i}) - P_i(e,f) = 0 \\
&\qquad \Delta Q_i = Q_{g_i} - (Q^0_{d_i} + \rho^2 \Delta Q_{d_i}) - Q_i(e,f) = 0 \\
&\qquad \Delta Q_{ri} = Q_{g_{ri}} - Q^0_{d_i} - Q_i(e,f) = 0 \\
&\qquad \Delta V_i = V_i^{r2} - (e_i^2 + f_i^2) = 0 \\
&\qquad P^m_{gi} \le P_{g_i} \le P^M_{gi} \\
&\qquad Q^m_{gi} \le Q_{g_i} \le Q^M_{gi} \\
&\qquad V^m_i \le V_i \le V^M_i
\end{aligned}
\tag{16.26}
$$

where all variables have been previously defined.

The KKT first-order optimality conditions are

$$
\begin{aligned}
\nabla_e \mathcal{L}(\boldsymbol{y}) &= \boldsymbol{J}_1^t \boldsymbol{\lambda}_p + (\boldsymbol{J}_3^*)^t \boldsymbol{\lambda}_q + \boldsymbol{J}_{5_p}^t \boldsymbol{\lambda}_v + \boldsymbol{J}_{5_b}^t \boldsymbol{F}_v^t \boldsymbol{\pi}_v \\
\nabla_f \mathcal{L}(\boldsymbol{y}) &= \boldsymbol{J}_2^t \boldsymbol{\lambda}_p + (\boldsymbol{J}_4^*)^t \boldsymbol{\lambda}_q + \boldsymbol{J}_{6_p}^t \boldsymbol{\lambda}_v + \boldsymbol{J}_{6_b}^t \boldsymbol{F}_v^t \boldsymbol{\pi}_v \\
\nabla_z \mathcal{L}(\boldsymbol{y}) &= \boldsymbol{F}_n \\
\nabla_\rho \mathcal{L}(\boldsymbol{y}) &= 1 + 2\rho \boldsymbol{\Delta P}_d^t \boldsymbol{\lambda}_p + 2\rho \boldsymbol{\Delta Q}_d^t \boldsymbol{\lambda}_q \\
\nabla_{\lambda_p} \mathcal{L}(\boldsymbol{y}) &= \Delta \boldsymbol{P} \\
\nabla_{\lambda_q} \mathcal{L}(\boldsymbol{y}) &= \Delta \boldsymbol{Q} \\
\nabla_{\lambda_v} \mathcal{L}(\boldsymbol{y}) &= \Delta \boldsymbol{V}_p \\
\nabla_{\pi P_g} \mathcal{L}(\boldsymbol{y}) &= \boldsymbol{F}_{P_g} \tilde{\boldsymbol{P}}_g + \boldsymbol{s}_{P_g} - \boldsymbol{P}_g^{lim} \\
\nabla_{\pi Q_g} \mathcal{L}(\boldsymbol{y}) &= \boldsymbol{F}_{Q_g} \tilde{\boldsymbol{Q}}_g + \boldsymbol{s}_{Q_g} - \boldsymbol{Q}_g^{lim} \\
\nabla_{\pi_v} \mathcal{L}(\boldsymbol{y}) &= \boldsymbol{F}_v(\boldsymbol{e}^2 + \boldsymbol{f}^2) + \boldsymbol{s}_v - \boldsymbol{V}^{lim} \\
\nabla_{s_{P_g}} \mathcal{L}(\boldsymbol{y}) &= \boldsymbol{S}_{P_g} \boldsymbol{\pi}_{P_g} - \mu \boldsymbol{u} \\
\nabla_{s_{Q_g}} \mathcal{L}(\boldsymbol{y}) &= \boldsymbol{S}_{Q_g} \boldsymbol{\pi}_{Q_g} - \mu \boldsymbol{u} \\
\nabla_{s_v} \mathcal{L}(\boldsymbol{y}) &= \boldsymbol{S}_v \boldsymbol{\pi}_v - \mu \boldsymbol{u}
\end{aligned}
\tag{16.27}
$$

where all matrices have been defined in Section 16.3.2.

At each iteration of the solution of (16.24) through Newton's method, the following linear system is solved:

$$
\begin{bmatrix}
\boldsymbol{G}_1 & \boldsymbol{G}_2^t & \boldsymbol{0} & \boldsymbol{0} & \boldsymbol{J}_1^t & (\boldsymbol{J}_3^*)^t & \boldsymbol{J}_5^t & \boldsymbol{H}_{p_1}^t & \boldsymbol{H}_{q_1}^t & \boldsymbol{H}_{v_1}^t & \boldsymbol{0} & \boldsymbol{0} & \boldsymbol{0} \\
\boldsymbol{G}_2 & \boldsymbol{G}_1 & \boldsymbol{0} & \boldsymbol{0} & \boldsymbol{J}_2^t & (\boldsymbol{J}_4^*)^t & \boldsymbol{J}_6^t & \boldsymbol{H}_{p_2}^t & \boldsymbol{H}_{q_2}^t & \boldsymbol{H}_{v_2}^t & \boldsymbol{0} & \boldsymbol{0} & \boldsymbol{0} \\
\boldsymbol{0} & \boldsymbol{0} & \boldsymbol{0} & \boldsymbol{0} & \boldsymbol{0} & \boldsymbol{F}_n^t & \boldsymbol{0} & \boldsymbol{0} & \boldsymbol{F}_n^t & \boldsymbol{0} & \boldsymbol{0} & \boldsymbol{0} & \boldsymbol{0} \\
\boldsymbol{0} & \boldsymbol{0} & \boldsymbol{0} & 2\boldsymbol{r}^t \boldsymbol{\lambda}_m & 2\rho \boldsymbol{\Delta P}_d^t & 2\rho \boldsymbol{\Delta Q}_d^t & \boldsymbol{0} & 2\rho \boldsymbol{\Delta P}_d^t & 2\rho \boldsymbol{\Delta Q}_d^t & \boldsymbol{0} & \boldsymbol{0} & \boldsymbol{0} & \boldsymbol{0} \\
\boldsymbol{J}_1 & \boldsymbol{J}_2 & \boldsymbol{0} & 2\rho \boldsymbol{\Delta P}_d & \boldsymbol{0} & \boldsymbol{0} & \boldsymbol{0} & \boldsymbol{0} & \boldsymbol{0} & \boldsymbol{0} & \boldsymbol{0} & \boldsymbol{0} & \boldsymbol{0} \\
\boldsymbol{J}_3^* & \boldsymbol{J}_4^* & \boldsymbol{F}_n & 2\rho \boldsymbol{\Delta Q}_d & \boldsymbol{0} & \boldsymbol{0} & \boldsymbol{0} & \boldsymbol{0} & \boldsymbol{0} & \boldsymbol{0} & \boldsymbol{0} & \boldsymbol{0} & \boldsymbol{0} \\
\boldsymbol{J}_5 & \boldsymbol{J}_6 & \boldsymbol{0} & \boldsymbol{0} & \boldsymbol{0} & \boldsymbol{0} & \boldsymbol{0} & \boldsymbol{0} & \boldsymbol{0} & \boldsymbol{0} & \boldsymbol{0} & \boldsymbol{0} & \boldsymbol{0} \\
\boldsymbol{H}_{p_1} & \boldsymbol{H}_{p_2} & \boldsymbol{0} & 2\rho \boldsymbol{\Delta P}_d & \boldsymbol{0} & \boldsymbol{0} & \boldsymbol{0} & \boldsymbol{0} & \boldsymbol{0} & \boldsymbol{0} & \boldsymbol{I}_p & \boldsymbol{0} & \boldsymbol{0} \\
\boldsymbol{H}_{q_1} & \boldsymbol{H}_{q_2} & \boldsymbol{F}_n & 2\rho \boldsymbol{\Delta Q}_d & \boldsymbol{0} & \boldsymbol{0} & \boldsymbol{0} & \boldsymbol{0} & \boldsymbol{0} & \boldsymbol{0} & \boldsymbol{0} & \boldsymbol{I}_q & \boldsymbol{0} \\
\boldsymbol{H}_{v_1} & \boldsymbol{H}_{v_2} & \boldsymbol{0} & \boldsymbol{0} & \boldsymbol{0} & \boldsymbol{0} & \boldsymbol{0} & \boldsymbol{0} & \boldsymbol{0} & \boldsymbol{0} & \boldsymbol{0} & \boldsymbol{0} & \boldsymbol{I}_v \\
\boldsymbol{0} & \boldsymbol{0} & \boldsymbol{0} & \boldsymbol{0} & \boldsymbol{0} & \boldsymbol{0} & \boldsymbol{0} & \boldsymbol{S}_{P_g} & \boldsymbol{0} & \boldsymbol{0} & \boldsymbol{\Pi}_{P_g} & \boldsymbol{0} & \boldsymbol{0} \\
\boldsymbol{0} & \boldsymbol{0} & \boldsymbol{0} & \boldsymbol{0} & \boldsymbol{0} & \boldsymbol{0} & \boldsymbol{0} & \boldsymbol{0} & \boldsymbol{S}_{Q_g} & \boldsymbol{0} & \boldsymbol{0} & \boldsymbol{\Pi}_{Q_g} & \boldsymbol{0} \\
\boldsymbol{0} & \boldsymbol{0} & \boldsymbol{0} & \boldsymbol{0} & \boldsymbol{0} & \boldsymbol{0} & \boldsymbol{0} & \boldsymbol{0} & \boldsymbol{0} & \boldsymbol{S}_v & \boldsymbol{0} & \boldsymbol{0} & \boldsymbol{\Pi}_v
\end{bmatrix}
\cdot
\begin{bmatrix}
\Delta \boldsymbol{e} \\ \Delta \boldsymbol{f} \\ \Delta z \\ \Delta \rho \\ \Delta \boldsymbol{\lambda}_p \\ \Delta \boldsymbol{\lambda}_q \\ \Delta \boldsymbol{\lambda}_v \\ \Delta \boldsymbol{\pi}_{P_g} \\ \Delta \boldsymbol{\pi}_{Q_g} \\ \Delta \boldsymbol{\pi}_v \\ \Delta \boldsymbol{s}_{P_g} \\ \Delta \boldsymbol{s}_{Q_g} \\ \Delta \boldsymbol{s}_v
\end{bmatrix}
= -
\begin{bmatrix}
\nabla_{\tilde{P}_g} \mathcal{L}(\boldsymbol{y}) \\ \nabla_{\tilde{Q}_g} \mathcal{L}(\boldsymbol{y}) \\ \nabla_e \mathcal{L}(\boldsymbol{y}) \\ \nabla_f \mathcal{L}(\boldsymbol{y}) \\ \nabla_{\lambda_p} \mathcal{L}(\boldsymbol{y}) \\ \nabla_{\lambda_q} \mathcal{L}(\boldsymbol{y}) \\ \nabla_{\lambda_v} \mathcal{L}(\boldsymbol{y}) \\ \nabla_{\pi P_g} \mathcal{L}(\boldsymbol{y}) \\ \nabla_{\pi Q_g} \mathcal{L}(\boldsymbol{y}) \\ \nabla_{\pi_v} \mathcal{L}(\boldsymbol{y}) \\ \nabla_{s P_g} \mathcal{L}(\boldsymbol{y}) \\ \nabla_{s Q_g} \mathcal{L}(\boldsymbol{y}) \\ \nabla_{s_v} \mathcal{L}(\boldsymbol{y})
\end{bmatrix}
\tag{16.28}
$$

where

$$G_1 = \mathrm{diag}(\boldsymbol{\lambda}_p)\boldsymbol{G} + \boldsymbol{G}\mathrm{diag}(\boldsymbol{\lambda}_p) - (\mathrm{diag}(\boldsymbol{\lambda}_q)\boldsymbol{B} + \boldsymbol{B}\mathrm{diag}(\boldsymbol{\lambda}_q)) + 2\mathrm{diag}(\boldsymbol{\lambda}_v) + 2\mathrm{diag}(\boldsymbol{\lambda}_v) \quad (16.29)$$

$$G_2 = \mathrm{diag}(\boldsymbol{\lambda}_p)\boldsymbol{B} - \boldsymbol{B}\mathrm{diag}(\boldsymbol{\lambda}_p) + \left(\mathrm{diag}(\boldsymbol{\lambda}_q)\boldsymbol{G} - \boldsymbol{G}\mathrm{diag}(\boldsymbol{\lambda}_q)\right) \quad (16.30)$$

$$\boldsymbol{H}_p = \boldsymbol{F}_p\left[\boldsymbol{J}_1(bge,:)\boldsymbol{J}_2(bge,:)\boldsymbol{J}_{tp}(bge,:)\right] \quad (16.31)$$

$$\boldsymbol{H}_q = \boldsymbol{F}_q\left[\boldsymbol{J}_3^*(bge,:)\boldsymbol{J}_4^*(bge,:)\boldsymbol{J}_{qq}(bge,:)\right] \quad (16.32)$$

$$\boldsymbol{H}_v = \boldsymbol{F}_v\left[\boldsymbol{J}_5(bsp,:)\boldsymbol{J}_6(bsp,:)\boldsymbol{J}_{tv}(bsp,:)\right] \quad (16.33)$$

where bsp denotes the whole set of buses PV and PQV.

16.4 Solution of the optimisation problem

16.4.1 Primal-dual interior point method

Consider the following constrained optimisation problem:

$$\begin{array}{ll} \text{Minimise} & f(\boldsymbol{x}) \\ \text{subject to} & \\ & \boldsymbol{g}(\boldsymbol{x}) = 0 \\ & \boldsymbol{h}(\boldsymbol{x}) \leq \boldsymbol{h}^{lim} \end{array} \quad (16.34)$$

where $\boldsymbol{x}$ is the vector of the primal optimisation variables, and the vectorial equations represent the equality and inequality constraints.

The solution of the problem stated by (16.34) through the non-linear primal-dual version of the interior point method can be summarised in the following steps:

- conversion of the inequality constraints into equality constraints through slack variables; this provides,

$$\begin{aligned} \boldsymbol{h}(\boldsymbol{x}) + \boldsymbol{s} - \boldsymbol{h}^{lim} &= \boldsymbol{0} \\ \boldsymbol{s} &> \boldsymbol{0} \end{aligned} \quad (16.35)$$

  where $\boldsymbol{s}$ is a vector whose components s_i are the slack variables corresponding to the inequality constraints;
- inclusion of the logarithmic function barrier in the objective function, that is,

$$f(\boldsymbol{x}) - \mu \sum_i \ln s_i$$

- application of the optimality conditions in the modified optimisation problem, which is expressed as,

$$\begin{aligned} &\text{Minimise} \quad f(\boldsymbol{x}) - \mu \textstyle\sum_i \ln s_i \\ &\text{subject to} \\ &\qquad \boldsymbol{g}(\boldsymbol{x}) = \boldsymbol{0} \\ &\qquad \boldsymbol{h}(\boldsymbol{x}) + \boldsymbol{s} - \boldsymbol{h}^{lim} = \boldsymbol{0} \\ &\qquad \boldsymbol{s} > \boldsymbol{0} \end{aligned} \tag{16.36}$$

The Lagrangian function of the problem stated by (16.36) is

$$\mathcal{L}(\boldsymbol{x}, \boldsymbol{s}, \boldsymbol{\lambda}, \boldsymbol{\pi}) = f(\boldsymbol{x}) - \mu \sum_i \ln s_i + \boldsymbol{\lambda}^t \boldsymbol{g}(\boldsymbol{x}) + \boldsymbol{\pi}^t \left(\boldsymbol{h}(\boldsymbol{x}) + \boldsymbol{s} - \boldsymbol{h}^{lim}\right) \tag{16.37}$$

where all terms have been previously defined.

The application of KKT first-order optimality conditions to (16.37) provides,

$$\begin{aligned} \nabla_x \mathcal{L}(\boldsymbol{x}, \boldsymbol{s}, \boldsymbol{\lambda}, \boldsymbol{\pi}) = \boldsymbol{0} &= \nabla_x f(\boldsymbol{x}) + \nabla_x \boldsymbol{g}(\boldsymbol{x})^t \boldsymbol{\lambda} + \nabla_x \boldsymbol{h}(\boldsymbol{x})^t \boldsymbol{\pi} \\ \nabla_s \mathcal{L}(\boldsymbol{x}, \boldsymbol{s}, \boldsymbol{\lambda}, \boldsymbol{\pi}) = \boldsymbol{0} &= -\mu \boldsymbol{e} + \boldsymbol{S}\boldsymbol{\pi} \\ \nabla_\lambda \mathcal{L}(\boldsymbol{x}, \boldsymbol{s}, \boldsymbol{\lambda}, \boldsymbol{\pi}) = \boldsymbol{0} &= \boldsymbol{g}(\boldsymbol{x}) \\ \nabla_\pi \mathcal{L}(\boldsymbol{x}, \boldsymbol{s}, \boldsymbol{\lambda}, \boldsymbol{\pi}) = \boldsymbol{0} &= \boldsymbol{h}(\boldsymbol{x}) + \boldsymbol{s} - \boldsymbol{h}^{lim} \end{aligned} \tag{16.38}$$

where $\nabla_x f(\boldsymbol{x})$ and $\nabla_x \boldsymbol{h}(\boldsymbol{x})$ represent the first derivative of $f(\boldsymbol{x})$ and $\boldsymbol{h}(\boldsymbol{x})$, respectively; $\nabla_x \boldsymbol{g}(\boldsymbol{x}) = \boldsymbol{J}(\boldsymbol{x})$ is the Jacobian matrix of $\boldsymbol{g}(\boldsymbol{x})$; $\boldsymbol{e}$ is an unity vector of suitable dimension; and $\boldsymbol{S}$ is a diagonal matrix composed by the elements of vector $\boldsymbol{s}$.

The conditions stated by (16.38) are increased by the non-negativity constraints corresponding to both the slack variables and the sign, relative to the dual multipliers, that is,

$$\boldsymbol{s} \geq \boldsymbol{0}, \qquad \boldsymbol{\pi} \geq \boldsymbol{0} \tag{16.39}$$

The stationary point of the problem represented by (16.36) is obtained by solving (16.38). By applying Newton's method, the following linear system must be solved at each iteration:

$$\begin{aligned} \boldsymbol{H}(\boldsymbol{x}, \boldsymbol{\lambda}, \boldsymbol{\pi}_l, \boldsymbol{\pi}_u)\boldsymbol{\Delta x} + \boldsymbol{J}(\boldsymbol{x})^t \boldsymbol{\Delta\lambda} + \nabla_x \boldsymbol{h}(\boldsymbol{x})^t \boldsymbol{\Delta\pi} &= -\boldsymbol{t} \\ \boldsymbol{\Pi}\boldsymbol{\Delta s} + \boldsymbol{S}\boldsymbol{\Delta\pi} &= -(-\mu \boldsymbol{e} + \boldsymbol{S}\boldsymbol{\pi}) \\ \boldsymbol{J}(\boldsymbol{x})\boldsymbol{\Delta x} &= -\boldsymbol{g}(\boldsymbol{x}) \\ \nabla_x \boldsymbol{h}(\boldsymbol{x})\boldsymbol{\Delta x} - \boldsymbol{\Delta s} &= -(\boldsymbol{h}(\boldsymbol{x}) + \boldsymbol{s} - \boldsymbol{h}^{lim}) \end{aligned} \tag{16.40}$$

where

$$\boldsymbol{H}(\boldsymbol{x}, \boldsymbol{\lambda}, \boldsymbol{\pi}_l, \boldsymbol{\pi}_u) = \nabla_x^2 f(\boldsymbol{x}) - \sum_i \lambda_i \nabla_x^2 g_i(\boldsymbol{x}) - \sum_j (\boldsymbol{\pi}_l + \boldsymbol{\pi}_u) \nabla_x^2 h_i(\boldsymbol{x})$$

is the matrix of second derivatives of the Lagrangian function (Hessian) with respect to the optimisation variables; $\nabla_x^2 f(\boldsymbol{x})$, $\nabla_x^2 g_i(\boldsymbol{x})$ and $\nabla_x^2 h_i(\boldsymbol{x})$ are the matrices of second derivatives of $f(\boldsymbol{x})$, $g_i(\boldsymbol{x})$ and $h_i(\boldsymbol{x})$, respectively;

$$\boldsymbol{t} = \nabla_x \mathcal{L}(\boldsymbol{x}, \boldsymbol{s}, \boldsymbol{\lambda}, \boldsymbol{\pi}) = \nabla_x f(\boldsymbol{x}) + \boldsymbol{J}(\boldsymbol{x})^t \boldsymbol{\lambda} + \nabla_x \boldsymbol{h}(\boldsymbol{x})^t \boldsymbol{\pi}$$

and $\boldsymbol{\Pi}$ is a diagonal matrix composed by the elements of vector $\boldsymbol{\pi}$.

Equation (16.40) can be re-written in the matrix form, that is,

$$\boldsymbol{W}(\boldsymbol{x},\boldsymbol{s},\boldsymbol{\lambda},\boldsymbol{\pi})\begin{bmatrix}\Delta\boldsymbol{x}\\ \Delta\boldsymbol{s}\\ \Delta\boldsymbol{\lambda}\\ \Delta\boldsymbol{\pi}\end{bmatrix} = -\begin{bmatrix}\boldsymbol{t}\\ (-\mu\boldsymbol{e}+\boldsymbol{S\pi})\\ \boldsymbol{g}(\boldsymbol{x})\\ \boldsymbol{h}(\boldsymbol{x})+\boldsymbol{s}-\boldsymbol{h}^{lim}\end{bmatrix} \tag{16.41}$$

where the matrix $\boldsymbol{W}(\boldsymbol{x},\boldsymbol{s},\boldsymbol{\lambda\pi})$ is given by

$$\begin{bmatrix}\boldsymbol{H}(\boldsymbol{x},\boldsymbol{\lambda},\boldsymbol{\pi}_l,\boldsymbol{\pi}_u) & \boldsymbol{0} & \boldsymbol{J}(\boldsymbol{x})^t & \nabla_x\boldsymbol{h}(\boldsymbol{x})^t\\ \boldsymbol{0} & \boldsymbol{\pi} & \boldsymbol{0} & \boldsymbol{S}\\ \boldsymbol{J}(\boldsymbol{x}) & \boldsymbol{0} & \boldsymbol{0} & \boldsymbol{0}\\ \nabla_x\boldsymbol{h}(\boldsymbol{x}) & \boldsymbol{I} & \boldsymbol{0} & \boldsymbol{0}\end{bmatrix} \tag{16.42}$$

where $\boldsymbol{I}$ is the identity matrix. Note that this matrix, originally non-symmetrical, can be made symmetrical by multiplying the second row by $\boldsymbol{S}$.

The solution of (16.41) provides the increments of the primal and dual variables. In order to ensure the non-negativity of both the slack variables and the sign of the dual multipliers, a step size is used, which is computed in the primal and dual spaces as,

$$\begin{aligned}\gamma_p &= \min\left[\min_{\Delta s_i<0}\frac{s_i}{|\Delta s_i|}\quad 1,0\right]\\ \gamma_d &= \min\left[\min_{\Delta\pi_j>0}\frac{\pi_j}{|\Delta\pi_j|}\quad 1,0\right]\end{aligned} \tag{16.43}$$

The primal and dual variables are updated according to the following expressions:

$$\begin{aligned}\boldsymbol{x}^{k+1} &= \boldsymbol{x}^k+\sigma\gamma_p\boldsymbol{\Delta x}^k\\ \boldsymbol{\lambda}^{k+1} &= \boldsymbol{\lambda}^k+\sigma\gamma_d\boldsymbol{\Delta\lambda}^k\\ \boldsymbol{s}^{k+1} &= \boldsymbol{s}^k+\sigma\gamma_p\boldsymbol{\Delta s}^k\\ \boldsymbol{\pi}^{k+1} &= \boldsymbol{\pi}^k+\sigma\gamma_d\boldsymbol{\Delta\pi}^k\end{aligned} \tag{16.44}$$

where σ is a constant which aims at ensuring non-zero values to the variables $\boldsymbol{s}$ and $\boldsymbol{\pi}$. According to Reference 27, $\sigma = 0.9995$ is recommended.

Therefore, the main target of the step factors $\sigma\gamma_p$ and $\sigma\gamma_d$ is to ensure the non-negativity of the slack variables as well as to provide a suitable reduction in the performance index represented by the Lagrangian function.

The barrier parameter μ is updated at each iteration by the following expression:

$$\mu = \frac{\boldsymbol{s}^t\boldsymbol{\pi}}{2l\beta} \tag{16.45}$$

where l is the number of inequality constraints.

The algorithm to solve an optimisation problem through the non-linear version of the primal-dual interior point method is summarised in the following steps:

1. Initialise the primal and dual variables.
2. Compute the gradient vector of the augmented Lagrangian function, i.e., (16.38).
3. Perform the convergence test: check the Euclidean norm of the gradient vector and the value of the barrier parameter μ with the respective tolerances. If the convergence criterion is satisfied, the iterative process is finished.
4. Compute the matrix $\boldsymbol{W}$ and solve the linear system, i.e., (16.41).
5. Determine the step factors in the primal and dual spaces, i.e., (16.43).
6. Update the optimisation variables, i.e., (16.45).
7. Compute the new value of the barrier parameter μ, i.e., (16.45). Return to Step 2.

16.4.2 Reduction of the linear system

Equation (16.41) can be significantly reduced, such that its dimension becomes independent of the number of inequality constraints [27]. For this purpose, the following vectors are defined:

$$\begin{aligned} \boldsymbol{v}_l &= -\mu \boldsymbol{e} + \boldsymbol{S}\boldsymbol{\pi} \\ \boldsymbol{y}_l &= \boldsymbol{h}(\boldsymbol{x}) + \boldsymbol{s}_u - \boldsymbol{h}^{lim} \end{aligned} \tag{16.46}$$

such that (16.40) can be re-written as,

$$\begin{aligned} \boldsymbol{\Pi}\boldsymbol{\Delta s} + \boldsymbol{S}\boldsymbol{\Delta}\pi &= -\boldsymbol{v}_l \\ \nabla_x \boldsymbol{h}(\boldsymbol{x})\boldsymbol{\Delta x} + \boldsymbol{\Delta s} &= -\boldsymbol{y}_l \end{aligned} \tag{16.47}$$

The second row of (16.47) can be expressed as,

$$\boldsymbol{\Delta s} = -\nabla_x \boldsymbol{h}(\boldsymbol{x})\boldsymbol{\Delta x} - \boldsymbol{y}_u \tag{16.48}$$

The replacement of (16.48) in the first row of (16.47) provides

$$\boldsymbol{\Delta\pi} = \boldsymbol{S}^{-1}(-\boldsymbol{v}_l + \boldsymbol{\Pi}\boldsymbol{y}_l) + \boldsymbol{S}^{-1}\boldsymbol{\Pi}\nabla_x \boldsymbol{h}(\boldsymbol{x})\boldsymbol{\Delta x} \tag{16.49}$$

The replacement of (16.49) in the first row of (16.40) results,

$$\begin{aligned} &\left(\boldsymbol{H}(\boldsymbol{x},\boldsymbol{\lambda},\boldsymbol{\pi}) + \nabla_x \boldsymbol{h}(\boldsymbol{x})^t \boldsymbol{S}_l^{-1}\boldsymbol{\Pi}\nabla_x \boldsymbol{h}(\boldsymbol{x})\right)\boldsymbol{\Delta x} + \boldsymbol{J}(\boldsymbol{x})^t\boldsymbol{\Delta\lambda} \\ &\quad = -\boldsymbol{t} - \nabla_x \boldsymbol{h}(\boldsymbol{x})^t \boldsymbol{S}^{-1}(-\boldsymbol{v}_l + \boldsymbol{\Pi}\boldsymbol{y}_l) \end{aligned} \tag{16.50}$$

By defining,

$$\begin{aligned} \widetilde{\boldsymbol{H}}(\boldsymbol{x},\boldsymbol{s},\boldsymbol{\lambda},\boldsymbol{\pi}) &= \boldsymbol{H}(\boldsymbol{x},\boldsymbol{\lambda},\boldsymbol{\pi}) + \nabla_x \boldsymbol{h}(\boldsymbol{x})^t \boldsymbol{S}_l^{-1}\boldsymbol{\Pi}\nabla_x \boldsymbol{h}(\boldsymbol{x}) \\ \widetilde{\boldsymbol{t}} &= -\boldsymbol{t} - \nabla_x \boldsymbol{h}(\boldsymbol{x})^t \boldsymbol{S}^{-1}(-\boldsymbol{v}_l + \boldsymbol{\Pi}\boldsymbol{y}_l) \end{aligned} \tag{16.51}$$

Equation (16.50) becomes

$$\widetilde{\boldsymbol{H}}(\boldsymbol{x},\boldsymbol{s},\boldsymbol{\lambda},\boldsymbol{\pi})\boldsymbol{\Delta x} + \boldsymbol{J}(\boldsymbol{x})^t\boldsymbol{\Delta\lambda} = \widetilde{\boldsymbol{t}} \tag{16.52}$$

which, together with the third row of (16.40), composes the reduced linear system expressed as,

$$\begin{bmatrix} \widetilde{\boldsymbol{H}}(\boldsymbol{x},\boldsymbol{s},\boldsymbol{\lambda},\boldsymbol{\pi}) & \boldsymbol{J}(\boldsymbol{x})^t \\ \boldsymbol{J}(\boldsymbol{x}) & \boldsymbol{0} \end{bmatrix} \begin{bmatrix} \boldsymbol{\Delta x} \\ \boldsymbol{\Delta\lambda} \end{bmatrix} = \begin{bmatrix} \widetilde{\boldsymbol{t}} \\ -\boldsymbol{g}(\boldsymbol{x}) \end{bmatrix} \tag{16.53}$$

Table 16.1 Bus type for the New England test case

Bus	Area	Type	Bus	Area	Type	Bus	Area	Type
1	2	0	14	1	0	27	2	0
2	2	0	15	3	0	28	3	0
3	2	3	16	3	0	29	3	0
4	1	0	17	2	0	30	2	4
5	1	0	18	2	0	31	1	2
6	1	0	19	3	0	32	1	4
7	1	3	20	3	0	33	3	4
8	1	0	21	3	0	34	3	1
9	1	0	22	3	3	35	3	4
10	1	0	23	3	0	36	3	1
11	1	0	24	3	0	37	2	4
12	1	0	25	2	0	38	3	1
13	1	0	26	2	0	39	1	4

The dimension of the reduced linear system of (16.53) is independent of the number of inequality constraints. It is equal to the summation of the number of optimisation variables plus the number of equality constraints.

16.5 Numerical results

In order to highlight the quality of the solutions provided by the analytical models described in the previous sections, and also to show some aspects of the convergence and numerical characteristics of the solution method, this section presents some numerical results obtained from the computational programs implemented in MatLab. Two test systems were used for this purpose. The first is the New England, which is composed by 39 buses, and the second is a model of the Italian transmission system which has 959 buses.

The numerical results have been obtained by solving the following OPF problems:

- Minimisation of the squared deviation of the bus voltage magnitude from a reference value with SVC constraints (referred to as *MSD problem*).
- Maximisation of the loadability with SVC constraints (referred to as *ML-SVC problem*).
- Maximisation of the loadability without SVC constraints (referred to as *ML-noSVC problem*).

16.5.1 The New England 39 buses network case

Aiming at taking into account the SVC scheme, the New England test system has been divided into three secondary control areas and three pilot buses (one for each control area) are defined. Table 16.1 shows the area and the type of each bus for the SVC purposes. No strategy other than the connections shown in the topology was

Table 16.2 Numerical results (MSD problem)

Bus	Area	Type	V (pu)	V^r (pu)	P_g (MW)	Q_g (Mvar)	q
30	2	4	0.991	0.993	250	35.838	0.089595
32	1	4	0.959	0.998	650	168.44	0.561467
33	3	4	0.986	0.995	632	168.24	0.67296
35	3	4	1.003	1.002	650	201.89	0.672967
37	2	4	0.995	0.990	540	22.399	0.089596
39	1	4	1.031	0.997	1 000	168.44	0.561467

used to divide the test system. The pilot buses of the areas 1, 2 and 3 are 3, 7 and 22, respectively, with pre-specified reference voltage magnitude of 0.998 pu, 0.988 pu and 1.004 pu. Bus types 0, 1, 2, 3 and 4 denote, respectively, the PQ, PV, reference, pilot and regulating buses. There are 6 generating buses, 3 PV buses, 1 reference bus and 29 load buses (including the pilot buses).

Table 16.2 presents some variables of the controlling buses at the optimal solution of the MSD problem, whose solution was found in 11 iterations and no particular numerical problems in the iterative process. Columns 4 and 5 show the voltage magnitude at the optimal solution (denoted V) and the corresponding reference value (denoted V^r). P_g and Q_g are the active and reactive power generation and the reactive level (q) represents the ratio between the reactive power produced by each controlling unit and its maximum reactive power. Note that the controlling generators of the same control area have the same values of the reactive level q. The numerical results show that the reactive level for each control area is smaller than 1.0, which indicates that no reactive power generation limit has been reached at the optimal solution. The voltage magnitude of the pilot buses is equal to the pre-specified reference value previously mentioned.

Table 16.3 shows the main variables of generation buses at the optimal solution in case of the maximisation of the loadability without including the SVC constraints, that is, the ML-noSVC problem. In this case, since the SVC scheme is not adopted, each controlling generators is modelled as a voltage magnitude control bus. The iterative process has converged in 13 iterations. On the other hand, since there are no equality constraints on the voltage magnitude, the pilot buses (3, 7 and 22) are modelled as PQ buses, and thus in the optimal solution their voltage magnitude (1.016 pu, 1.0005 pu and 1.026 pu, respectively) is different from the pre-specified value previously mentioned (0.998 pu, 0.988 pu and 1.004 pu).

In this case, the reactive power generation of bus 31 (349.99 Mvar), which is the reference bus, reaches its limit (350.00 Mvar). The maximum loadability factor determined as the solution of the OPF problem is 1.269, which indicates that an increase of 26.9% in the load can be supplied according to the steady-state operational conditions. The limitation factor of the loadability is the voltage constraint in the bus 1 (1.05 pu), which once reached does not allow additional increase in the loadability.

When the SVC constraints are introduced in the model, the maximum loadability factor is 1.171; that is, only an increase of 17.1% is allowed but all constraints

Table 16.3 Numerical results (ML-noSVC problem)

Bus	Area	Type	V (pu)	P_g (MW)	Q_g (Mvar)	q
30	2	1	1.0296	450	58.971	0.147428
31	1	2	1.0186	700	349.99	0.999971
32	1	1	1.05	850	241.95	0.8065
33	3	1	1.0272	800	43.262	0.173048
34	3	1	0.998	650	150.37	0.75185
35	3	1	1.05	800	106.74	0.3558
36	3	1	1.002	700	25.282	0.105342
37	2	1	1.05	680	91.719	0.366876
38	3	1	0.985	980	55.578	0.18526
39	1	1	1.0435	1 200	−40.803	−0.13601

Table 16.4 Numerical results (ML-SVC problem)

Bus	Area	Type	V (pu)	P_g (MW)	Q_g (Mvar)	q
30	2	4	1.0138	450	128.68	0.3217
32	1	4	0.98849	547.11	260.52	0.8684
33	3	4	0.99808	743.05	250	1
35	3	4	1.0164	734.44	300	1
37	2	4	1.0152	680	80.423	0.3217
39	1	4	1.05	1 200	260.93	0.869767

(including the alignment constraints) are satisfied. The last column of Table 16.4 indicates that the control area 3 reaches its maximum reactive capacity (regulating buses 33 and 35), and therefore there is no additional reactive power to control the magnitude of the pilot bus in this area. Moreover, the loadability factor is also limited by the maximum value of the voltage magnitude in the controlling generator 39, which reaches its boundary, as shown in the fourth column of Table 16.4.

Although the controlling units of area 3 have reached their maximum reactive power generation limit, at the optimal solution point the voltage magnitude of the pilot buses is equal to the pre-assigned value. This means that in case of an additional increase in the loadability it would be necessary to change the reference voltage value for the pilot bus of area 3. In other words, when the reactive level q of a generating unit is equal to 1, this indicates that all reactive resources of the control area are used to adjust the voltage magnitude of the pilot bus. Any other further increase of the loadability requires a change in the voltage magnitude of the pilot bus of the control area with reactive level q equal to 1. In this last case, the iterations necessary to find the solution are 11.

Figure 16.2 shows the voltage magnitude profile of the network for these three cases. It is possible to observe that the voltage profile obtained as a solution of

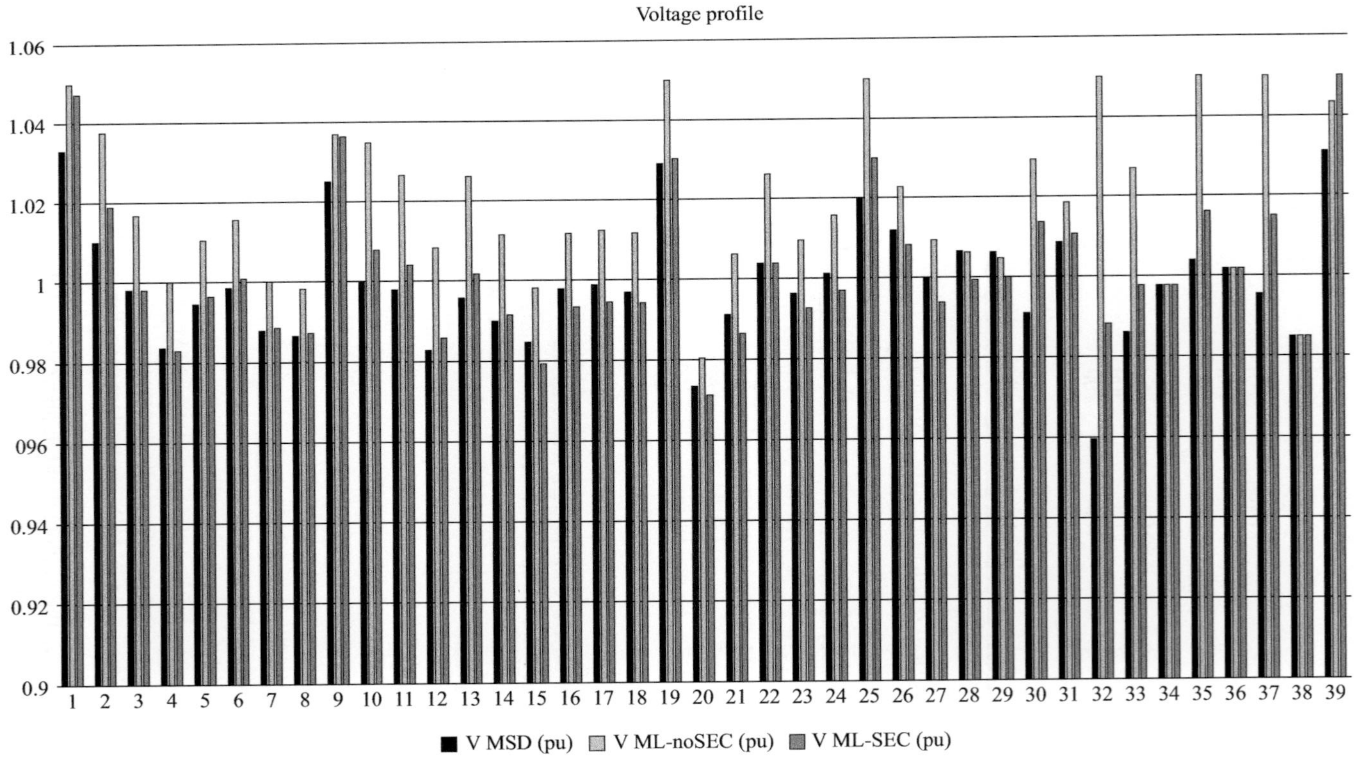

Figure 16.2 Voltage profile given by different OPF problems

Figure 16.3 Italian transmission system control areas

the MSD problem presents voltage magnitudes close to the reference value of the PQ buses. In this case, the average voltage magnitude of the system is 0.9992 pu, whilst in case of the ML-noSVC solution, there is an increment of the voltage magnitude level (the average voltage magnitude is 1.0192). If SVC constraints are introduced, the average voltage magnitude is 1.0027.

16.5.2 The Italian case

In order to assess the characteristics of the convergence properties as well as the robustness of the interior point method used to solve the OPF problems, tests with a reduced model of the Italian transmission system (230 and 400 kV) have been performed. In the network model, there are 959 buses, 1 099 lines and transformers and 163 generating units. Moreover, according to the SVC scheme implemented in Italy, there are 13 secondary control areas and 99 controlling generators (Figure 16.3

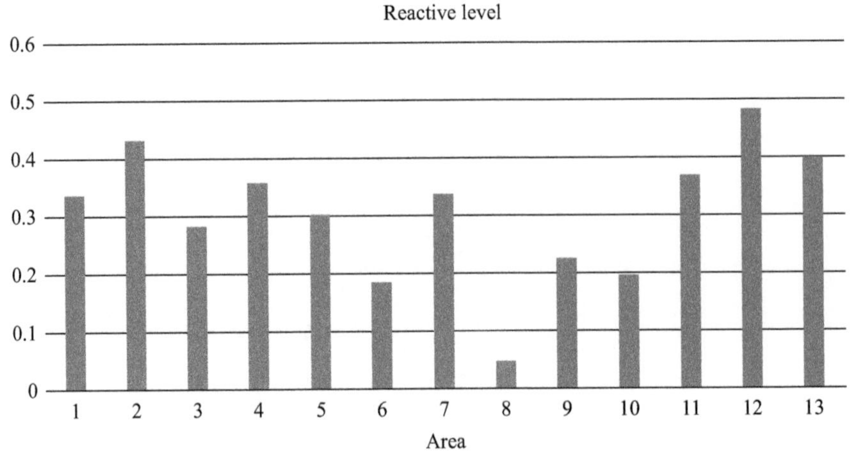

Figure 16.4 Reactive level: Italian case – ML-noSVC problem

shows the transmission system control areas adopted in Italy). For this network, optimal solutions of the MSD, ML-noSVC and MS-SVC problems have been obtained.

In the first test, the solution point of the MSD problem was determined. No control area has reached the saturation (the reactive levels have been always smaller than 1 and positive, as indicated in Figure 16.4). The solution is found in 13 s and eight iterations.

When the loadability of the system is maximised without the secondary voltage constraints, the maximum load increment is equal to 1.2609. Fourteen iterations are necessary to find the solution in about 32 s. At the optimal solution point, the Lagrangian multipliers for each power flow constraint are available. It is possible to find out which buses limit a further increment of the loadability. In particular, the highest (in magnitude) Lagrangian multiplier associated to the voltage constraint (−22.35 negative) is the for PV bus number 112. In this case, the voltage magnitude is fixed in the value defined in the network data file. Subsequently, in order to keep the voltage magnitude at the specified value, the reactive power produced by the generating units of the area reaches their maximum value. To avoid the action of the PVC in this bus, it is not possible to further increase the network load. Finally, from the analysis of the Lagrangian multiplier associated to the constraints of the PF equation, it is possible to observe the highest value (23.39) associated to the reactive power balance equation of bus 780.

The last test performed here was to determine the solution of the ML-SVC problem in case of the Italian system (see Figure 16.5). Now, the secondary control scheme is taken into account, and therefore the alignment constraints are introduced in the OPF model. Due to these constraints, the loadability factor is reduced to 1.14, and 25 interactions are necessary to find the solution. In this case, the reactive level of

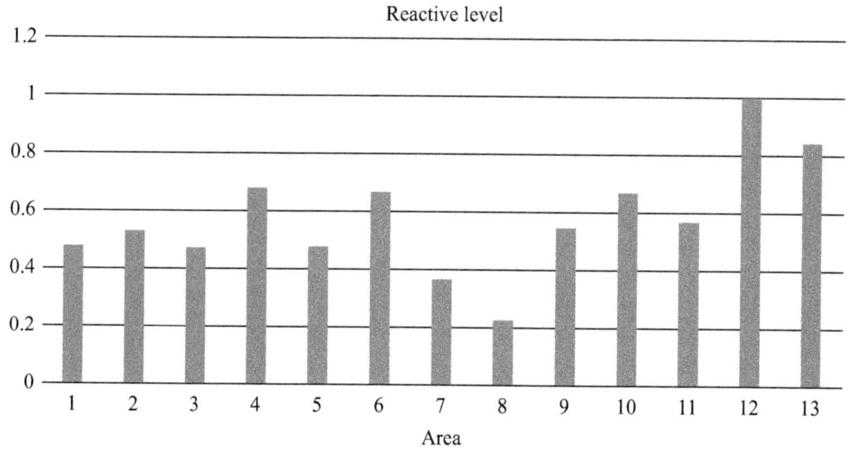

Figure 16.5 Reactive level: Italian case – ML-SVC problem

control area 12 reaches its maximum value ($q = 1$). The analysis of the Lagrangian multipliers at the optimal solution point reveals that the highest value for the voltage magnitude constraint corresponds to the PV bus 111, where the reactive power produced by the generating unit of this bus reaches its maximum value. On the other hand, bus 567 is characterised by the highest value for the Lagrangian multiplier associated to the reactive power flow balance equation. In the solution point, no bus reaches its maximum or minimum values.

16.6 Conclusions

The control of voltages and reactive power has become more and more critical in the power system operation, in recent years, due to the presence of the electricity market and the strong penetration of the RES that push system operators and electrical utilities to operate the transmission networks as close as possible to their maximum capacity. To improve voltage control in transmission grids, many projects have been developed around the world. The Hierarchical Voltage Control System, which is based on network area and resources subdivision, although developed by vertically integrated utilities in the past, is widely recognised as a viable solution and was adopted in several countries around the world.

This chapter deals with the computation of an optimal voltage profile using different optimisation strategies. For this purpose, the mathematical model of the optimisation problem is defined and described considering two issues: (i) defining the constraints of the optimisation problem in order to fulfil the actual operating condition of the SVC system and (ii) testing different objective functions. A primal dual interior point method is proposed to solve the OPF problem and the structure of the matrices used by the method is described in detail. In particular, in the OPF models, a quadratic formulation of the PF equations is adopted. In this way, no trigonometrical equations are adopted. The main advantage of this formulation is the robustness of the algorithm.

Different objective functions are presented and discussed with and without the presence of SVC scheme and comparative tests were made on two different network models. The first one is a small test system (the New England test system modified to include the SVC), and the second one is a model of the real Italian transmission system; the main characteristics of these different OPF models are emphasised for these particular grids. In both of these cases, and for each objective function, the solution of the OPF problem is found within few iterations and no particular numerical problems are identified. In general, the described approaches show good convergence properties also for systems of high dimension.

Bibliography

[1] Lin M, Rayudu RK, Samarasinghe S. "A review of voltage-var control". In: *Proceedings of the Australian Universities Power Engineering Conference (AUPEC)*. Christchurch, New Zealand; 2003. p. 1–6.

[2] Popovic DS, Levi VA. "Extension of the load flow model with remote voltage control". *Electric Power System Research*. 1992;25:207–212.

[3] Berizzi A, Bovo C, Delfanti M, Merlo M, Tortello F. "Singular value decomposition for an ORPF formulation in presence of SVR". In: *Proceedings of the IEEE Electrotechnical Conference*. Mediterranean, Malaga; 2006. p. 1–6.

[4] Popovic DS, Calovic MS, Levi VA. "Voltage reactive security analysis in power systems with automatic secondary voltage control". *Proceedings of the IET Generation, Transmission and Distribution*. 1994;141(3):177–183.

[5] Conejo A, Aguilar MJ. "Secondary voltage control: nonlinear selection of pilot buses, design of an optimal control law, and simulation results". *Proceedings of the IET Generation, Transmission and Distribution*. 1998 January; 145(1):77–81.

[6] Berizzi A, Bovo C, Delfanti M, Merlo M. *Optimisation Advances in Electric Power Systems*. New York, NY: Nova Science Publishers; 2008.

[7] Mousavi OA, Cherkaoui R. *Literature Survey on Fundamental Issues of Voltage and Reactive Power Control*, Deliverable of the MARS Project financially supported by "swisselectric research", École Polytechnique Federale de Lausanne, Switzerland, 10 June 2011.

[8] Popovic DS. "Impact of secondary voltage control on voltage stability". *Electric Power System Research*. 1997;40:51–62.

[9] Kwatny HG. "Stability enhancement via secondary voltage regulation". In: *Bulk Power Systems Voltage Phenomena II – Voltage Stability and Security*. Deep Creek, USA; 1991. p. 147–156.

[10] Cañizares C, Cavallo C, Pozzi M, Corsi S. "Comparing secondary voltage regulation and shunt compensation for improving voltage stability and transfer capability in the Italian power system". *Electric Power System Research*. 2005;40:67–76.

[11] Berizzi A, Finazzi P, Dosi D, Marannino P, Corsi S. "First and second order methods for voltage collapse assessment and security enhancement". *IEEE Transactions on Power Systems*. 1998 May;13(2):543–551.

[12] Ilea V, Bovo C, Merlo M, Berizzi A, Eremia M. "Reactive power flow optimisation in the presence of secondary voltage control". In: *Proceedings of the IEEE PowerTech Conference*. Bucharest, Romania; 2009. p. 1–6.

[13] Sung Y. *Design of Secondary Voltage and Stability Controls with Multiple Control Objectives*. Georgia Institute of Technology, Atlanta, GA; 2008.

[14] Rehtanz C. *Autonomous Systems and Intelligent Agents in Power System Control and Operation*. Berlin: Springer; 2003.

[15] Taranto GN, Martins N, Martins ACB, Falcão DM, Santos MG. "Benefits of applying secondary voltage control schemes to the Brazilian system". In: *Power Engineering Society Summer Meeting*. Seattle, WA; 2000. p. 937–942.

[16] Corsi S. *Voltage Control and Protection in Electrical Power Systems*. Berlin: Springer; 2015.

[17] Bai C, Begovic M, Nuqui R, Sobajic D, Song Y. "On voltage collapse stability monitoring with voltage instability predictors". In: *Proceedings of the 2013 IREP Symposium*. Crete, Greece; 2013. p. 1–8.

[18] Ning J, Liu X, Venkatasubramanian M. "Distributed real-time stability monitoring algorithms using synchrophasors". In: *Proceedings of the 2013 IREP Symposium*. Crete, Greece; 2013. p. 1–5.

[19] Berizzi A, Bignotti A, Finazzi P, Dosi D, Marannino P. "An automatic procedure for evaluating and improving operating margin against voltage collapse". In: *Proceedings of the IEEE Power System Computation Conference*. Dresden, Germany; 1996. p. 644–650.

[20] Hecke JV, Janssens N, Deuse J, Promel F. "Coordinated voltage control experience in Belgium". In: *Proceedings of the CIGRE Symposium*. Paris, France; 2000. p. 1–7.

[21] Marannino P, Zanellini F, Berizzi A, Medina D, Merlo M, Pozzi M. "Steady state and dynamic approaches for the evaluation of the loadability margins in the presence of the secondary voltage regulation". *IEEE Transactions on Power Systems*. 2004 May;19(2):1048–1057.

[22] Berizzi A, Merlo M, Marannino P, Zanellini F, Corsi S, Pozzi M. "Dynamic performances of the hierarchical voltage regulation: the Italian EHV system case". In: *Proceedings of the Power System Computation Conference*. Liege, Belgium; 2005. p. 1–7.

[23] Cañizares C. "Voltage stability assessment: concepts, practices and tools". *IEEE Power System Stability*, Sub-committee Special Publication SP101 PSS, ISBN 0780378695, Published in June 2003.

[24] Cutsem TV. "A method to compute reactive power margins with respect to voltage collapse". *IEEE Transactions on Power Systems*. 1991 February; 6(1):145–156.

[25] Ilea V, Bovo C, Merlo M, Berizzi A, Marannino P. "Reactive power flow optimisation in power systems with hierarchical voltage control". In: *Proceedings of the IEEE Power System Computation Conference*. Stockholm, Sweden; 2011. p. 1–6.

[26] Kataoka Y, Shinoda Y. "Voltage stability limit of electric power systems with generation reactive power constraints considered". *IEEE Transactions on Power Systems*. 2005 May;20(2):951–962.

[27] Grainger JJ, Stevenson WD. *Power System Analysis*. New York, NY: McGraw-Hill; 1994.

Chapter 17

Small-signal stability and time-domain analysis of delayed power systems

Georgios Tzounas[1], Muyang Liu[2], Ioannis Dassios[3], Rifat Sipahi[4] and Federico Milano[3]

This chapter describes the impacts that time delays in feedback control loops have on the small-signal as well as on the transient stability of power systems. We present a power system model comprising of delay differential algebraic equations (DDAEs) and describe general techniques to compute the spectrum and numerically integrate such model. The focus is on delays arising in measured signals, e.g., remote frequency measurements for power system stabilizers (PSSs) of synchronous machines (SMs). Several examples are discussed based on the IEEE 14-bus benchmark system as well as on a realistic model of the Irish transmission system.

17.1 Introduction

Time delays are intrinsic of many physical processes and control systems mainly because it takes time, first, to acquire and transmit information and, second, to formulate and implement decisions based on given information. Nevertheless, in standard power system models for voltage and transient stability analysis, time delays are conventionally neglected or approximated with simple lag blocks. With the exception of very long transmission lines [1], the behavior of most power system devices, including transformers and SMs, is in fact not affected by delays. However, regulators are affected by delays, and, in recent years, the ubiquitous presence of communication systems and remote measurements, e.g., phasor measurement units (PMUs), has attracted the attention of researchers in academia and industry to the impact of delays on control signals and on the stability of the overall power grid.

[1]ETH Zürich, Switzerland
[2]School of Electrical Engineering, Xinjiang University, China
[3]School of Electrical and Electronic Engineering, University College Dublin, Ireland
[4]Department of Mechanical and Industrial Engineering, Northeastern University, USA

Communication delays result from a series of processes along the data communication, from the measurement device to the control center, including long-distance data delivery, data packet dropouts, noise, communication network congestion, etc. [2,3]. Due to these phenomena, such delays are in principle time-variant. Yet, for the sake of simplicity, many power system studies have considered delays as constant [4,5]. The constant delay model typically provides a conservative stability assessment of the system, which in many cases is acceptable. On the other hand, precise estimation of the system's stability margin requires more detailed delay models, such as the ones described, for example, in [2,3,6].

The focus of most of the research on power systems with inclusion of delays is devoted to the design of robust controllers that are able to reduce the impact of communication latency. For example, the work in [6–8] deals with the development of robust wide-area control schemes that improve the damping of inter-area oscillations. However, design for improved robustness against time delays often sacrifices the overall performance of the control, e.g., see [9]. Accordingly, instead of a delay-robust design, some studies develop methods for delay compensation, the main objective of which is to generate a control signal that mimics well the original, delay-free signal. For example, proposed methods for delay compensation include derivative and predictive-based controllers, see [10–13].

Existing studies on small-signal stability of power systems with delays develop model equations and analyze them using proper numerical methods. The latter can be divided into two main categories: (i) time-domain methods and (ii) frequency-domain methods. Relevant contributions to these two approaches are briefly reviewed below.

17.1.1 Time-domain methods

These methods are based on Lyapunov–Krasovskii's stability theorem and Razumikhin's theorem, see for example, [14–19]. The application of time-domain methods allows defining robust controllers (e.g., H_∞ control) and dealing with uncertainties and time-varying delays. While the conditions of Lyapunov–Krasovskii's stability theorem and Razumikhin's theorem provide strong tools for the stability analysis and control of many problems, including the study of time-varying effects and nonlinearities, these conditions are only sufficient and hence, in the context of linear stability, many studies prefer frequency-domain tools in order to capture necessary and sufficient conditions of stability. In this context, computation of the delay margin—the largest delay less than which the system is stable—has been one of the main research topics. Another challenge with time-domain methods could be that it is necessary to find a Lyapunov functional or, according to Razumikhin's theorem, a Lyapunov function that bounds the Lyapunov functional. For time delay systems, developing these functionals indeed requires deep expertise, and this may pose challenges in analyzing non-linear DDAE systems; see [20] for applications to small-scale power systems.

If the DDAE system is linear or is linearized around an equilibrium point, finding the Lyapunov function, in turn, implies solving a linear matrix inequality (LMI) problem [21]. A drawback of this approach is that the size and the computational burden of LMIs highly increase with the size of the DDAE system and it is only in the last two decades that such calculations have become tractable [21].

17.1.2 Frequency-domain methods

These methods are mainly based on the evaluation of the roots of the characteristic equation of the corresponding linear time-invariant (LTI) system [16,22–28]. This approach in principle follows necessary and sufficient conditions of linear stability; however, due to the difficulty in determining the roots of the characteristic equation (see Section 17.2.1), the analysis is challenging. Therefore, while there are attempts to define an exact analytic solution for oversimplified power system models [29], an explicit solution cannot be found in general. Outside the field of power systems, many developments have been published, yet the majority of these results are limited to low-dimensional problems or assume explicit knowledge of the system's characteristic equation see for example [28,30]. But there also exist methods that by providing an approximation of the system enable us to analyze the stability of higher-dimensional systems, without a need for knowledge of the characteristic equation.

This chapter considers four different approaches that approximate the solution of the small-signal stability of DDAEs. These are: (i) a Chebyshev discretization of a set of partial differential equations (PDEs) that are equivalent to the original DDAEs [31]; (ii) a discretization of the time integration operator (TIO) as proposed in [32]; a linear multi-step (LMS) approximation which has been proposed in [33] and is implemented in the open-source software tool DDE-BIFTOOL [34]; and the well-known Padé approximants [35].

A common characteristic of the above techniques is the high computational burden, which, unfortunately, increases more than linearly with the size of the problem. Hence, proper numerical schemes and implementations have to be used. One option is to use graphics processing unit (GPU)-based numerical libraries. For example, all simulation results discussed in this chapter are obtained based on MAGMA, which provides an efficient GPU-based parallel implementation of LAPACK functions and QR factorization for solving the linear eigenvalue problem (LEP) [36]. Regarding spectral discretization techniques, another option is to consider in the discretization only the part of the system that includes the retarded variables, thus largely reducing the overall computational burden of the problem [37].

We note that the above list of techniques is not meant to be exhaustive, but rather to present a set of state-of-the-art options for solving benchmark problems. An exhaustive study is thus left to future work. For example, the following techniques, although not covered here, have been widely used in the literature, namely TRACE-DDE [38] and Lambert W function approach [39]. The backbone of TRACE-DDE also makes use of Chebyshev discretization and has been successfully implemented on problems with various multiple-delay models. As for the Lambert W function approach, it has been utilized to approximate the rightmost roots of DDEs, by mapping the infinite-dimensional eigenvalue problem to a Lambert W function representation, for which efficient solvers exist.

17.2 A general model for power systems with time delays

The conventional power system model used for voltage and transient stability analyses consists of a set of differential algebraic equations (DAEs), as follows [40]:

$$\begin{aligned} \dot{\boldsymbol{x}} &= \boldsymbol{f}(\boldsymbol{x}, \boldsymbol{y}, \boldsymbol{u}) \\ \boldsymbol{0} &= \boldsymbol{g}(\boldsymbol{x}, \boldsymbol{y}, \boldsymbol{u}) \end{aligned} \tag{17.1}$$

where $\boldsymbol{f}$ ($\boldsymbol{f} : \mathbb{R}^{n+m+p} \mapsto \mathbb{R}^n$) and $\boldsymbol{g}$ ($\boldsymbol{g} : \mathbb{R}^{n+m+p} \mapsto \mathbb{R}^m$) are non-linear functions that define the differential and algebraic equations, respectively, $\boldsymbol{x}$ ($\boldsymbol{x} \in \mathbb{R}^n$) are the state variables, $\boldsymbol{y}$ ($\boldsymbol{y} \in \mathbb{R}^m$) the algebraic variables, and $\boldsymbol{u}$ ($\boldsymbol{u} \in \mathbb{R}^p$) are discrete variables modelling events, e.g., line outages and faults. The time dependency of the system has been omitted in (17.1) for the sake of simplicity.

The DDAE formulation is obtained by introducing time delays in (17.1). Assume for now that the system includes a single delay $\tau > 0$, and let

$$\boldsymbol{x}_d = \boldsymbol{x}(t - \tau), \qquad \boldsymbol{y}_d = \boldsymbol{y}(t - \tau) \tag{17.2}$$

be the delayed state and algebraic variables, respectively, where t is the current simulation time. In the remainder of this chapter, the main focus is on small-signal stability analysis. In order to capture the fundamental roles of time delays on stability, they are assumed to be constant here.

If some state or algebraic variables in (17.1) are affected by a time delay as represented in (17.2), one obtains:

$$\begin{aligned} \dot{\boldsymbol{x}} &= \boldsymbol{f}(\boldsymbol{x}, \boldsymbol{y}, \boldsymbol{x}_d, \boldsymbol{y}_d, \boldsymbol{u}) \\ \boldsymbol{0} &= \boldsymbol{g}(\boldsymbol{x}, \boldsymbol{y}, \boldsymbol{x}_d, \boldsymbol{u}) \end{aligned} \tag{17.3}$$

which is the index-1 Hessenberg form of DDAEs given in [41]. Note that $\boldsymbol{g}$ does not depend on $\boldsymbol{y}_d$. As discussed in [4], this model is adequate for most power system applications. Finally, it is straightforward to extend (17.3) to the multiple-delay case: it suffices to define as many vectors of state and algebraic variables (17.2) as the number of delays present in the system. For simplicity and without lack of generality, in Sections 17.2.1 and 17.3.2, we only consider the single-delay case.

17.2.1 Steady-state DDAEs

Small-signal stability analysis deals with power system stability when it is subject to small disturbances around its equilibrium points. For the model (17.3), assume that a stationary solution is known and has the form:

$$\begin{aligned} \boldsymbol{0} &= \boldsymbol{f}(\boldsymbol{x}_0, \boldsymbol{y}_0, \boldsymbol{x}_{d0}, \boldsymbol{y}_{d0}, \boldsymbol{u}_0) \\ \boldsymbol{0} &= \boldsymbol{g}(\boldsymbol{x}_0, \boldsymbol{y}_0, \boldsymbol{x}_{d0}, \boldsymbol{u}_0) \end{aligned} \tag{17.4}$$

Note that in steady state, $\boldsymbol{x}_{d0} = \boldsymbol{x}_0$ and $\boldsymbol{y}_{d0} = \boldsymbol{y}_0$. Moreover, discrete variables $\boldsymbol{u}_0$ are assumed to be constant in the remainder of this chapter. Then, differentiating (17.3) at the stationary solution yields:

$$\Delta\dot{\boldsymbol{x}} = \boldsymbol{f}_x \Delta\boldsymbol{x} + \boldsymbol{f}_{x_d} \Delta\boldsymbol{x}_d + \boldsymbol{f}_y \Delta\boldsymbol{y} + \boldsymbol{f}_{y_d} \Delta\boldsymbol{y}_d \tag{17.5}$$

$$\boldsymbol{0} = \boldsymbol{g}_x \Delta\boldsymbol{x} + \boldsymbol{g}_{x_d} \Delta\boldsymbol{x}_d + \boldsymbol{g}_y \Delta\boldsymbol{y} \tag{17.6}$$

Without loss of generality, we ignore singularity-induced bifurcation points and it can be assumed that $\boldsymbol{g}_y$ is non-singular. Substituting (17.6) into (17.5), one obtains*:

$$\Delta\dot{\boldsymbol{x}} = \boldsymbol{A}_0 \Delta\boldsymbol{x} + \boldsymbol{A}_1 \Delta\boldsymbol{x}(t-\tau) + \boldsymbol{A}_2 \Delta\boldsymbol{x}(t-2\tau) \tag{17.7}$$

which describes a system of DDEs of *retarded* type, since the delay affects the states but not the state derivatives. In the above equation, we have:

$$\boldsymbol{A}_0 = \boldsymbol{f}_x - \boldsymbol{f}_y \boldsymbol{g}_y^{-1} \boldsymbol{g}_x \tag{17.8}$$

$$\boldsymbol{A}_1 = \boldsymbol{f}_{x_d} - \boldsymbol{f}_y \boldsymbol{g}_y^{-1} \boldsymbol{g}_{x_d} - \boldsymbol{f}_{y_d} \boldsymbol{g}_y^{-1} \boldsymbol{g}_x \tag{17.9}$$

$$\boldsymbol{A}_2 = -\boldsymbol{f}_{y_d} \boldsymbol{g}_y^{-1} \boldsymbol{g}_{x_d} \tag{17.10}$$

The first matrix $\boldsymbol{A}_0$ is the well-known state matrix that is computed for standard DAEs of the form (17.1). The interested reader can refer to Appendix D.1 for a proof of (17.9) and (17.10). The other two matrices are not null matrices since the system is of retarded type. The matrix $\boldsymbol{A}_1$ is found in any DDEs, while $\boldsymbol{A}_2$ appears specifically in DDAEs, although it can be null if either $\boldsymbol{f}$ does not depend on $\boldsymbol{y}_d$ or $\boldsymbol{g}$ does not depend on $\boldsymbol{x}_d$. If any of these two conditions are satisfied, then (17.13) becomes [4]:

$$\Delta(\lambda) = \lambda \boldsymbol{I}_n - \boldsymbol{A}_0 - \boldsymbol{A}_1 e^{-\lambda\tau} \tag{17.11}$$

Equation (17.7) is a particular case of the standard variational form of the linear DDEs:

$$\Delta\dot{\boldsymbol{x}} = \boldsymbol{A}_0 \Delta\boldsymbol{x}(t) + \sum_{i=1}^{\nu} \boldsymbol{A}_i \Delta\boldsymbol{x}(t-\tau_i) \tag{17.12}$$

which are studied in various forms; see the references in Section 17.1.2. As we mentioned above, (17.12) describes a retarded-type system with multiple delays if $\nu > 1$. In the special case that $\nu = 1$, the system is known to be of single-delay type since all the states are affected by the same delay τ_1. Substituting a sample solution of the form $e^{\lambda t}\boldsymbol{v}$, with $\boldsymbol{v}$ being a non-trivial possibly complex vector of order n, the *characteristic equation* of (17.12) can be stated as follows:

$$\det \Delta(\lambda) = 0 \tag{17.13}$$

where

$$\Delta(\lambda) = \lambda \boldsymbol{I}_n - \boldsymbol{A}_0 - \sum_{i=1}^{\nu} \boldsymbol{A}_i e^{-\lambda\tau_i} \tag{17.14}$$

*The interested reader can find in [4] the details on how to obtain (17.8)–(17.10) from (17.5) and (17.6).

is called the *characteristic matrix* or the *matrix pencil* of (17.12), see [22,42,43]. In (17.14), $\boldsymbol{I}_n$ is the identity matrix of order n. Similar to the analysis of ordinary differential equations, i.e., the case for which $\boldsymbol{A}_i = \boldsymbol{0}, \ \forall i = 1, \ldots, \nu$, the solutions of (17.13) are called the *characteristic roots*. The stability of (17.12) can be defined based on the location of these roots on the complex plane; i.e., the equilibrium is stable if and only if all roots have negative real parts, and unstable otherwise [22,23].

Due to the presence of the exponential function in (17.14), (17.13) is not in polynomial form in λ, but transcendental, and thus it has infinitely many roots. In the literature, (17.13) is often referred to as a *quasi-polynomial* [44]. In general, the explicit solution of (17.13) is not known and only a subset of this solution can be approximated numerically, as will be discussed in Section 17.3.3. What is critical in this approximation is to make sure that one approximates the solutions that are relevant from stability and system performance points of view; that is, one must approximate the dominant/rightmost characteristic roots. This problem does not have a trivial solution; see also Section 17.1.2.

17.3 Numerical techniques for DDAEs

This section introduces a number of techniques for the numerical analysis of DDAEs. The section begins with the well-known Padé approximants, which allow approximating a DDAE system with a finite set of DAEs. Then, the implementation of the time-domain integration schemes of DDAEs are briefly discussed. Finally, the section describes three discretization techniques to approximate the spectrum of DDAEs with inclusion of multiple delays.

17.3.1 Padé approximants

Padé approximants are the most common and simplest implementation of time delays for the numerical analysis of dynamical systems. Roughly speaking, Padé approximants allow representing time delay systems through a set of linear ordinary differential equations. The higher the order of such equations, the more precise the representation. Hence, DDAEs can be rewritten as a set of higher-dimensional DAEs. The rationale behind Padé approximants is briefly discussed below.

First, let us recall the well-known *time shifting* property of the Laplace transform:

$$f(t-\tau)\, u(t-\tau) \quad \xrightarrow{\mathcal{L}} \quad e^{-\tau s}F(s) \tag{17.15}$$

where s is the Laplace variable obtained via the Laplace transform $\mathcal{L}$, often referred to as the *complex frequency*; $u(t)$ is the unit step function; and $F(s)$ is the Laplace transform of the function $f(t)$. The approach based on Padé approximants consists in defining a rational polynomial transfer function, say $P(s)$, that approximates $e^{-\tau s}$. Then, the inverse Laplace transform $\mathcal{L}^{-1}$ allows obtaining the approximated time-domain function $\phi(t)$ that leads to an approximated solution of the DAEs:

$$e^{-\tau s}F(s) \approx \quad P(s)F(s) \xrightarrow{\mathcal{L}^{-1}} \phi(t) \tag{17.16}$$

Such an approximation can be obtained using the Taylor's expansion of $e^{-\tau s}$ around $\tau = 0$:

$$e^{-\tau s} = 1 - \tau s + \frac{(\tau s)^2}{2!} - \frac{(\tau s)^3}{3!} + \cdots \approx \frac{b_0 + b_1 \tau s + \cdots + b_q(\tau s)^q}{a_0 + a_1 \tau s + \cdots + a_p(\tau s)^p} \tag{17.17}$$

where the coefficients $a_1, \ldots, a_p$ and $b_1, \ldots, b_q$ are obtained by imposing that the first $p + q$ coefficients of the Taylor expansion of $e^{-\tau s}$ fit the expansion of the polynomial ratio in (17.17). Note that s has a different meaning than λ in (17.13). In fact, λ takes an infinite number of discrete values that solve (17.13), whereas s is the continuous independent variable of the Laplace transform.

Generally, $p \geq q$ is imposed in (17.17). If $p = q$, the coefficients a_i and b_i are obtained by the following iterative formula:

$$a_0 = 1, \quad a_i = a_{i-1} \frac{p - i + 1}{i\,(2p - i + 1)}, \quad b_i = (-1)^i\, a_i \tag{17.18}$$

The case $p = q$ is noteworthy as the amplitude of the frequency response of the Padé approximant is exact, only the phase is affected by an error. $p = q = 6$ is a common choice in numerical simulations.

The higher the order of the Padé approximant, the lower the phase error (see, e.g., the discussion on Padé approximants in [45]. However, for *small* delays, i.e., delays of the order of milliseconds (which are common in power systems), there is no point in considering high order Padé approximants. For example, let $p = 9$ and $\tau = 10^{-3}$ s. Then, one obtains $a_9 = -b_9 = 5.6679 \cdot 10^{-11}$ and $\tau^9 = 10^{-27}$, which leads to $a_9 \cdot \tau^9 = 5.6679 \cdot 10^{-38}$. This number is critically close to the minimum positive value that can be represented by the single-precision binary floating-point defined by the IEEE 754 standard, i.e., $2^{-126} \approx 1.18 \cdot 10^{-38}$. High-order Padé approximants may also show unstable poles or *defects* (i.e., a pair of a pole and a zero that are very close but not equal [35]). Hence, the floating point representation binds the maximum value of p as $p^{\max} = q^{\max} = 10$, which is the most commonly used upper limit.

As an example on how to use Padé in practice, let us obtain the Padé approximant for a unit step function $u(t)$. For the sake of simplicity, we consider the case with $p = q$. The approximant u_d of order p, in time-domain, of $u(t - \tau)$ given by (17.18) is as follows:

$$u_d = \tilde{x}_1 + b_1 \tau \tilde{x}_2 + \cdots + b_{p-1} \tau^{p-1} \tilde{x}_p + b_p \tau^p \dot{\tilde{x}}_p \tag{17.19}$$

where:

$$\dot{\tilde{x}}_i = \tilde{x}_{i+1}, \quad i = 1, 2, \ldots, p - 1 \tag{17.20}$$

and

$$a_p \tau^p \dot{\tilde{x}}_p = u - (a_0 \tilde{x}_1 + a_1 \tau \tilde{x}_2 + \cdots + a_{p-1} \tau^{p-1} \tilde{x}_p) \tag{17.21}$$

Knowing these coefficients, we can easily obtain the time-domain function of (17.17).

Note that (17.19)–(17.21) are linear and introduce p state variables per delay. Clearly, there is no limitation to the number of delays that can be included in the system, and there is no structural difference between the single-delay and the multiple-delay cases. Moreover, since the system is approximated through a set of DAEs, conventional time-domain integration and small-signal stability analysis can be used.

17.3.2 Numerical integration of DDAEs

While Padé approximants avoid the need to implement numerical methods that deal with time delays, they also introduce numerical issues. A well-known issue is that they may cause the birth of spurious high-frequency oscillations. Specific methods to deal with delays can thus be desirable. However, time-domain integration of DDAEs is not an easy task. For example, despite being A-stable for standard DAEs, the implicit trapezoidal method may show numerical issues when applied to DDAEs. Thus, high-order time-domain integration methods for DDAEs have been developed. Interested readers can find an excellent discussion on this topic in [46].

Although the general case can show interesting numerical issues, in this chapter, we focus only on the index-1 Hessenberg form (17.3). Furthermore, we only consider implicit integration schemes, for two reasons: (i) they are the most adequate to deal with stiff DAEs in general and power system models in particular; and (ii) most state-of-the-art power system software tools implement implicit integration methods. We thus provide the modifications that are required to adapt a general implicit integration scheme up to second order to integrate (17.3).

We start with DAEs and then we will show how the method can be extended to DDAEs. While using implicit methods, each step of the numerical integration is obtained as the solution of a set of non-linear equations. At a generic time t, and assuming a step length h, one has to solve:

$$\begin{aligned} \mathbf{0} &= \boldsymbol{p}(\boldsymbol{x}(t+h), \boldsymbol{y}(t+h), \boldsymbol{u}(t+h), h) \\ \mathbf{0} &= \boldsymbol{q}(\boldsymbol{x}(t+h), \boldsymbol{y}(t+h), \boldsymbol{u}(t+h), h) \end{aligned} \tag{17.22}$$

where $\boldsymbol{p}$ ($\boldsymbol{p} : \mathbb{R}^{n+m+p} \mapsto \mathbb{R}^n$) and $\boldsymbol{q}$ ($\boldsymbol{q} : \mathbb{R}^{n+m+p} \mapsto \mathbb{R}^m$) are non-linear functions that depend on the DAEs and on the implicit numerical method. In particular, $\boldsymbol{p}$ accounts for differential equations, and $\boldsymbol{q}$ for algebraic ones.

Since (17.22) are non-linear, their solution is generally obtained using a direct solver, e.g., Newton's method, which in turn, consists in iteratively computing the increments $\Delta\boldsymbol{x}^{(i)}$ and $\Delta\boldsymbol{y}^{(i)}$ and updating state and algebraic variables:

$$\begin{bmatrix} \Delta\boldsymbol{x}^{(i)} \\ \Delta\boldsymbol{y}^{(i)} \end{bmatrix} = \begin{bmatrix} \boldsymbol{p}_x^{(i)} & \boldsymbol{p}_y^{(i)} \\ \boldsymbol{q}_x^{(i)} & \boldsymbol{q}_y^{(i)} \end{bmatrix} \begin{bmatrix} \boldsymbol{p}^{(i)} \\ \boldsymbol{q}^{(i)} \end{bmatrix} = -[\boldsymbol{A}^{(i)}]^{-1} \begin{bmatrix} \boldsymbol{p}^{(i)} \\ \boldsymbol{q}^{(i)} \end{bmatrix} \tag{17.23}$$

$$\begin{bmatrix} \boldsymbol{x}^{(i+1)}(t+h) \\ \boldsymbol{y}^{(i+1)}(t+h) \end{bmatrix} = \begin{bmatrix} \boldsymbol{x}^{(i)}(t+h) \\ \boldsymbol{y}^{(i)}(t+h) \end{bmatrix} + \begin{bmatrix} \Delta\boldsymbol{x}^{(i)} \\ \Delta\boldsymbol{y}^{(i)} \end{bmatrix}$$

For simplicity, we ignore the functional dependence on variables and iteration indexes. Then, a general expression for $\boldsymbol{p}$ and $\boldsymbol{q}$ that is able to represent the backward Euler method (BEM), the implicit trapezoidal method (ITM), and backward differentiation formula (BDF) is as follows:

$$\boldsymbol{p} = \boldsymbol{\xi} - \beta h(\boldsymbol{f} + \kappa \boldsymbol{f}_t) \tag{17.24}$$
$$\boldsymbol{q} = -\boldsymbol{g}$$

where $\boldsymbol{f}_t$ is known vector of differential equations at time t and

$$\boldsymbol{\xi} = \boldsymbol{x} - \sum_{\ell=1}^{\nu} \gamma_\ell \boldsymbol{x}(t - (\ell - 1)h) \tag{17.25}$$

and without a substantial loss of generality, we assume a constant step length h for $\boldsymbol{x}(t - (\ell - 1)h)$ values. The Jacobian matrix of (17.24) is given by:

$$\boldsymbol{A} = \begin{bmatrix} \boldsymbol{I}_n - \beta h \boldsymbol{f}_x & -\beta h \boldsymbol{f}_y \\ -\boldsymbol{g}_x & -\boldsymbol{g}_y \end{bmatrix} \tag{17.26}$$

The coefficients γ_ℓ and β are computed according to a straightforward procedure given in [47]. Table 17.1 summarizes the coefficients for the BEM, ITM and order-2 BDF. The BEM and order-2 BDF are L-stable, where the BEM can be, in occasions, hyperstable, while the ITM is A-stable. Implicit methods of order higher than two are not considered in this chapter. A comprehensive discussion on implicit integration schemes and their properties can be found in [48].

To account for delays, we need to expand the set of $\boldsymbol{p}$ and $\boldsymbol{q}$. Let us define two general functional expressions:

$$\boldsymbol{0} = \boldsymbol{\phi}(\boldsymbol{x}, \boldsymbol{x}_d, t) = \hat{\boldsymbol{x}}(\boldsymbol{\alpha}(\boldsymbol{x}, t)) - \boldsymbol{x}_d \tag{17.27}$$
$$\boldsymbol{0} = \boldsymbol{\psi}(\boldsymbol{y}, \boldsymbol{y}_d, t) = \hat{\boldsymbol{y}}(\boldsymbol{\beta}(\boldsymbol{y}, t)) - \boldsymbol{y}_d \tag{17.28}$$

where $\boldsymbol{\alpha}(\boldsymbol{x}, t)$ and $\beta(\boldsymbol{y}, t)$ represent the functional dependence of state and algebraic variables on the delays. For the constant time delay (17.2), we have:

$$\boldsymbol{\alpha}(\boldsymbol{x}, t) = t - \tau, \qquad \boldsymbol{\beta}(\boldsymbol{y}, t) = t - \tau \tag{17.29}$$

but, of course, more complex expressions can be considered [46].

Table 17.1 Coefficients of BEM, BDF, and ITM

Scheme	Order	γ_1	γ_2	β	κ
BEM	1	1	–	1	0
BDF	2	4/3	−1/3	2/3	0
ITM	2	1	–	0.5	1

Applying the same rule to (17.3) and using the functional equations (17.24), (17.27) and (17.28), one obtains:

$$A = \begin{bmatrix} I_n - \beta h f_x & -\beta h f_y & -\beta h f_{x_d} & -\beta h f_{y_d} \\ g_x & g_y & g_{x_d} & 0 \\ \phi_x & 0 & \phi_{x_d} & 0 \\ 0 & \psi_y & 0 & \psi_{y_d} \end{bmatrix} \tag{17.30}$$

where the superscript i has been omitted to simplify the notation. From (17.27) and (17.28), $\phi_{x_d} = -I_{n_d}$ and $\psi_{y_d} = -I_{m_d}$ are negative identity matrices, while ϕ_x and ψ_y can be obtained using the chain rule:

$$\phi_x = \text{diag}\left\{\dot{\hat{x}}(x,t)\right\}\alpha_x \tag{17.31}$$

$$\psi_y = \text{diag}\left\{\dot{\hat{y}}(y,t)\right\}\beta_y \tag{17.32}$$

where $\dot{\hat{x}}(x,t)$ and $\dot{\hat{y}}(y,t)$ are the rate of change of x and y at time $\alpha(x,t) = t - \tau$ and $\beta(y,t) = t - \tau$, respectively. While $\dot{\hat{x}}(x,t)$ is easy to obtain by simply storing $\dot{x}$ during the time-domain integration, $\dot{\hat{y}}(y,t)$ requires an extra computation, i.e., solving the following equation at each time t:

$$0 = g_x f + g_y \dot{y} + g_{x_d} \dot{\hat{x}} \alpha_t \tag{17.33}$$

from which $\dot{y}$ can be obtained, if g_y is not singular, and stored. Observe that $\dot{y}$ can be discontinuous.

The simple structure of the Jacobians of ϕ and ψ allows rewriting (17.30) as:

$$A = \begin{bmatrix} I_n - \beta h(f_x + f_{x_d}\phi_x) & -\beta h(f_y + f_{y_d}\psi_y) \\ g_x + g_{x_d}\phi_x & g_y \end{bmatrix} \tag{17.34}$$

Equation (17.34) is general and can be used for various types of time-varying delays. In the case of constant time delays, i.e., (17.2), it is straightforward to observe that $\alpha_x = 0$ and $\beta_y = 0$ and, hence, $\phi_x = 0$ and $\psi_y = 0$. Therefore, for constant delays, (17.26) and (17.34) match. This result was expected since, at a given time t, both x_d and y_d, i.e., state and algebraic variables delayed by τ, are constants.

17.3.3 Methods to approximate the characteristic roots of DDAEs

As discussed above, a common approach to define the small-signal stability of the DDAEs is to use Padé approximants, which in turn lead to study of the stability of a set of standard DAEs. In this section, we discuss three other commonly-used numerical methods to approximate the eigenvalues of a DDAE system. These are (i) a Chebyshev discretization scheme of equivalent PDEs that resemble the original DDAEs; (ii) an approximation of the TIO; and (iii) a LMS discretization of the DDAEs based on a high-order implicit time-domain integration scheme.

17.3.3.1 Chebyshev discretization scheme

This approach consists in transforming the original problem of computing the roots of a system of retarded functional differential equations into a matrix eigenvalue problem of a PDE system of infinite dimension. No loss of information is involved in this step. Then the dimension of the PDEs is made tractable using a discretization based on a finite element method. The Chebyshev discretization scheme has been successfully applied to systems with both single and multiple, constant and time-varying, deterministic and stochastic delays, see, for example [43] and the references therein.

Consider the single-delay case first. Let $\boldsymbol{D}_N$ be the Chebyshev differentiation matrix of order N (see the Appendix D.2 for details) and define

$$\boldsymbol{M} = \left[\begin{array}{c} \hat{\boldsymbol{C}} \otimes \boldsymbol{I}_n \\ \hline \boldsymbol{A}_1 \ \boldsymbol{0} \ \ldots \ \boldsymbol{0} \ \boldsymbol{A}_0 \end{array}\right] \tag{17.35}$$

where $\otimes$ indicates the *tensor product* or Kronecker product (see Appendix D.3 for details), $\boldsymbol{I}_n$ is the identity matrix of order n; and $\hat{\boldsymbol{C}}$ is a matrix composed of the first $N-1$ rows of $\boldsymbol{C}$ defined as follows:

$$\boldsymbol{C} = -2\boldsymbol{D}_N/\tau \tag{17.36}$$

Then, the eigenvalues of $\boldsymbol{M}$ are an approximated spectrum of (17.11). As it can be expected, the number of points N of the grid affects the precision and the computational burden of the method, as it is further discussed in the case study.

The matrix $\boldsymbol{M}$ is the discretization of a set of PDEs where the continuum is represented by the interval $\xi \in [-\tau, 0]$. The continuum is discretized along a grid of N points and the position of such points is defined by the Chebyshev polynomial interpolation. The last n rows of $\boldsymbol{M}$ impose the boundary conditions $\xi = -\tau$ (i.e., $\boldsymbol{A}_1$) and $\xi = 0$ (i.e., $\boldsymbol{A}_0$), respectively.

Figure 17.1 illustrates matrix (17.35) through a graphical representation. Each element of the grid is an $n \times n$ matrix and there are N^2 elements. Light gray blocks are defined by the Chebyshev discretization and are very sparse. Dark gray blocks represent the state matrix $\boldsymbol{A}_0$ and delayed matrix $\boldsymbol{A}_1$ that appear in (17.11). Finally, white blocks indicate zero matrices.

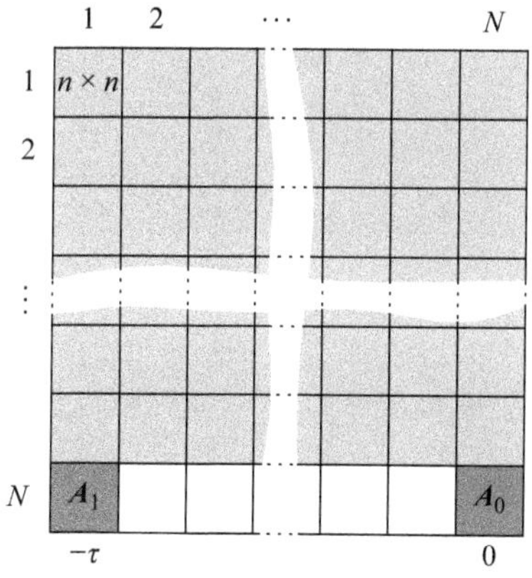

Figure 17.1 Representation of matrix ***M*** *for a system with a single delay* τ *and characteristic equation (17.11)*

Let us now consider the general multiple-delay case of the characteristic equation (17.13) and, thus, let us assume that there are ν delays, with $\tau_1 < \tau_2 < \cdots < \tau_{\nu-1} < \tau_\nu$. Each point of the Chebyshev grid corresponds to a delay $\theta_k = (N-k)\Delta\tau$, with $k = 1, 2, \ldots, N$ and $\Delta\tau = \tau_\nu/(N-1)$. Hence, $k = 1$ corresponds to the matrix $\boldsymbol{A}_\nu$, which corresponds to the maximum delay τ_ν; and $k = N$ is taken by the non-delayed state matrix $\boldsymbol{A}_0$. If a delay $\tau_i = \theta_k$ for some $k = 2, \ldots, N-1$, then the corresponding matrix $\boldsymbol{A}_i$ takes the position k in the grid. Of course, in general, the delays of the system will not match the points of the grid. For such cases, a linear interpolation is considered in this chapter, as follows. Let the time delay τ_i, $i \neq k$, satisfy the condition:

$$\theta_k < \tau_i < \theta_{k+1} \tag{17.37}$$

Then, the matrices that will be added to the positions k and $k+1$ are, respectively:

$$\boldsymbol{A}_{k,i} = \frac{\tau_i - \theta_k}{\Delta\tau}\boldsymbol{A}_i\,, \quad \boldsymbol{A}_{k+1,i} = \frac{\theta_{k+1} - \tau_i}{\Delta\tau}\boldsymbol{A}_i \tag{17.38}$$

Next, the resulting matrix of each point k of the grid is computed as the sum of the contributions of each delay that overlaps that point:

$$\boldsymbol{A}_k = \sum_{i \in \Omega_i} \boldsymbol{A}_{k,i} \tag{17.39}$$

where Ω_i is the set of delays τ_i that satisfies (17.37). Other more sophisticated interpolation schemes can be used. For example, a Lagrange polynomial interpolation is implemented in [49]. Figure 17.2 illustrates the Chebyshev discretization approach for the multiple-delay case.

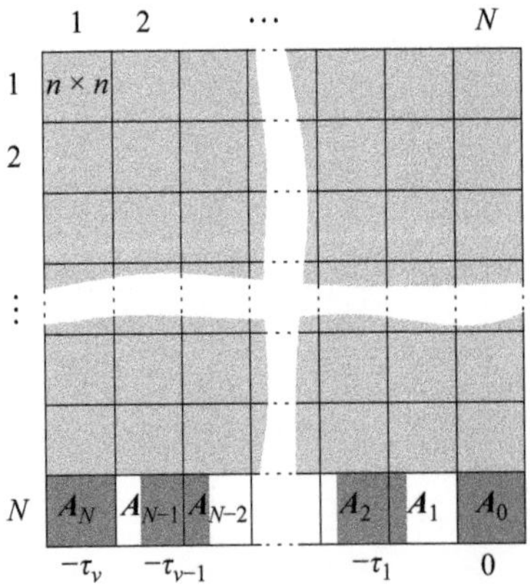

Figure 17.2 Representation of the Chebyshev discretization for a system with ν delays $\tau_1 < \tau_2 < \cdots < \tau_{\nu-1} < \tau_\nu$. In the general case, the delays do not exactly match the grid. Hence, an interpolation between consecutive points of the grid is required

17.3.3.2 Discretization of the TIO

The discretization of the TIO that is proposed in [49] is similar to the approach above, but instead of defining the discretization of an equivalent PDE system, it directly discretizes the set of original DDEs. In the interest of clarity, first consider the single-delay case and the following system:

$$\Delta\dot{\boldsymbol{x}}(t) = \boldsymbol{A}_0\Delta\boldsymbol{x}(t) + \boldsymbol{A}_1\Delta\boldsymbol{x}(t-\tau) \tag{17.40}$$

which is obtained from (17.7) by assuming that $\boldsymbol{A}_2 = \boldsymbol{0}$. The algorithm includes the following steps: (i) dividing the interval $[-\tau, 0]$ into a mesh of N intervals with constant step size $h = \tau/N$; and (ii) applying an integration scheme, e.g., a Runge–Kutta (RK) method, to the mesh that approximates the continuous solution of (17.40). Then the discrete counterpart of (17.40) is given by:

$$\boldsymbol{z}^{i+1} = \boldsymbol{S}_N\boldsymbol{z}^i \tag{17.41}$$

where $\boldsymbol{z} \in \mathbb{R}^{n \cdot r \cdot N}$, and $\boldsymbol{S}_N$ is the following $(n \cdot r \cdot N) \times (n \cdot r \cdot N)$ matrix:

$$\boldsymbol{S}_N = \begin{bmatrix} \boldsymbol{B}_0 & \boldsymbol{0} & \dots & \boldsymbol{0} & \boldsymbol{B}_1 \\ \boldsymbol{I}_{nr} & \boldsymbol{0} & \dots & \boldsymbol{0} & \boldsymbol{0} \\ \boldsymbol{0} & \boldsymbol{I}_{nr} & \dots & \boldsymbol{0} & \boldsymbol{0} \\ \vdots & \vdots & \ddots & \vdots & \vdots \\ \boldsymbol{0} & \boldsymbol{0} & \dots & \boldsymbol{I}_{nr} & \boldsymbol{0} \end{bmatrix} \tag{17.42}$$

where

$$\begin{aligned} \boldsymbol{B}_0 &= \boldsymbol{R} \cdot (\boldsymbol{1}_r \boldsymbol{e}_r^T \otimes \boldsymbol{I}_n) \\ \boldsymbol{B}_1 &= h\boldsymbol{R} \cdot (\mathcal{A} \otimes \boldsymbol{A}_1) \end{aligned} \tag{17.43}$$

and

$$\begin{aligned} \boldsymbol{R} &= (\boldsymbol{I}_{nr} - h\mathcal{A} \otimes \boldsymbol{A}_0)^{-1} \\ \boldsymbol{1}_r &= (1, \dots, 1)^T \\ \boldsymbol{e}_r &= (0, \dots, 0, 1)^T \end{aligned} \tag{17.44}$$

and $\mathcal{A}$ is the matrix of the Butcher's tableau that defines the integration scheme, as follows:

$$\begin{array}{c|c} \mathcal{C} & \mathcal{A} \\ \hline & \mathcal{B} \end{array} = \begin{array}{c|cccc} c_1 & a_{11} & a_{12} & \dots & a_{1r} \\ c_2 & a_{21} & a_{22} & \dots & a_{2r} \\ \vdots & \vdots & \vdots & \ddots & \vdots \\ c_r & a_{r1} & a_{r2} & \dots & a_{rr} \\ \hline & b_1 & b_2 & \dots & b_r \end{array} \tag{17.45}$$

and $\boldsymbol{I}_{nr}$ is the identity matrix of order $n \cdot r$. Note that $\mathcal{A}$ must be invertible, which means that an implicit scheme has to be used (e.g., BDF and Radau methods).

The single-delay case can be extended to the multiple-delay one by modifying the first row of the matrix $\boldsymbol{S}_N$ in (17.42). Assume that there are ν delays, with

$\tau_1 < \tau_2 < \cdots < \tau_{\nu-1} < \tau_\nu$. Then, the first and the last elements of the first row of (17.42) are occupied by $\boldsymbol{B}_0$ and $\boldsymbol{B}_\nu$, where $\boldsymbol{B}_0$ is defined as in (17.43) and $\boldsymbol{B}_\nu$ is:

$$\boldsymbol{B}_\nu = h\boldsymbol{R}(h\boldsymbol{A}_0)(\mathcal{A} \otimes \boldsymbol{A}_\nu) \tag{17.46}$$

The matrices associated with the remaining $\nu - 1$ delays are fitted to the grid through a linear interpolation similar to that described in Section 17.3.3.1.

17.3.3.3 LMS approximation

Another possible discretization is the one proposed in [33] and implemented in the software tool DDE-BIFTOOL. The TIO is discretized using a LMS method with polynomial interpolation to evaluate the delayed terms. Applying a k-step LMS method to (17.12), one obtains:

$$\sum_{j=0}^{k} \alpha_j \boldsymbol{x}_{L+j} = h \sum_{j=0}^{k} \left[\beta_j \boldsymbol{A}_0 \boldsymbol{x}_{L+j} + \sum_{i=0}^{\nu} \left(\boldsymbol{A}_i \tilde{\boldsymbol{x}}(t_{L+j} - \tau_i) \right) \right] \tag{17.47}$$

where α_j and β_j are the coefficients of the LMS method and $\tilde{\boldsymbol{x}}(t_{L+j} - \tau_i)$ are approximations of the values of the delayed state variables. These are computed using the Nordsieck interpolation, as follows:

$$\tilde{\boldsymbol{x}}(t_p - \epsilon h) = \sum_{\ell=-\rho}^{\sigma} P_\ell(\epsilon) \boldsymbol{x}_{p+\ell}, \quad \epsilon \in [0, 1) \tag{17.48}$$

where

$$P_\ell = \prod_{k=-\rho, k \neq \ell}^{\sigma} \frac{\epsilon - k}{\ell - k} \tag{17.49}$$

The resulting method is explicit whenever $\beta_0 = 0$ and $\min\{\tau_i\} > \sigma h$. Further details on this technique can be found in the DDE-BIFTOOL documentation and source code [34].

The LMS method forms an approximation of the TIO over the time step h, hence the eigenvalues $\boldsymbol{\mu}$ of the Jacobian matrix of (17.47) are an approximation of the exponential transforms of the roots $\boldsymbol{\lambda}$ of (17.13):

$$\boldsymbol{\mu} = \exp(h\boldsymbol{\lambda}) \tag{17.50}$$

The size of the resulting eigenvalue problem is inversely proportional to the step length h used in the discretization. The choice of h is heuristic and is a critical aspect of this technique. If the step length is too small, the size K of the problem can be huge, i.e., $K \gg n$. If h is too large, the approximation of the roots of (17.13) might not be accurate. The heuristic method for estimating h described in [33] leads to precise results although it might be conservative. Larger values of h can be obtained using the approach given in more recent works, e.g., [50]. A root is discarded if the following condition is satisfied:

$$\mathrm{abs}(\mu_j) > \exp(h \cdot \max\{\tau_i\}), \quad j = 1, 2, \ldots, K \tag{17.51}$$

17.4 Impact of delays on power system control

Time delays are commonly considered to have destabilizing effects on dynamical systems. This is actually the most common effect in most power system applications. However, many studies also demonstrate that delays do not necessarily destabilize a dynamical system, see for example, [51, Chapter 11] [22,28,52]. This "duality" characteristic of delays [16] has inspired numerous studies, where conditions under which stabilization can be achieved were investigated.

Inspired by this duality characteristic, here the focus is on the selecting particular controllers, namely power system stabilizers (PSSs), affected by time delays, and on demonstrating how proper design of a PSS despite the delay can stabilize the control loop, even if the delay is relatively large. This is in line with the above cited studies, yet it requires one to take several steps to reveal the controller parameters in the presence of delays. If carefully engineered, the stable closed loop can even produce desirable performance in the presence of delays. While this section deals with a particular case on the equilibrium dynamics through linear stability analysis, the discussion below allows drawing general conclusions, the most important of which is, in our opinion, that non-linear systems can always show unexpected behaviors.

In the remainder of this section, we consider a relevant example that allows an analytical assessment of its stability when delays in feedback are considered, i.e., the one-machine infinite-bus (OMIB) system with inclusion of a simplified PSS. We show the conditions for which the stability of the linearized OMIB equations is guaranteed independently from the magnitude of the controller's delay and present how system response time as measured by the concept of *σ-stability* can be understood in view of recent results [53].

17.4.1 OMIB system with simplified PSS

Consider the simplest example of a dynamical power system model, i.e., one SM connected through a transmission line to a bus of constant voltage and frequency (infinite bus). We assume the well-known simplified electromechanical model to represent the SM [54]:

$$\begin{aligned} \dot{\delta} &= \omega_b(\omega - 1) \\ M\dot{\omega} &= P_{\mathrm{m}} - P_{\mathrm{e}}(\delta) - D(\omega - 1) \end{aligned} \tag{17.52}$$

where δ and ω are the rotor angle and the speed of the SM; P_{m}, P_{e} are the mechanical, electrical power output of the SM; M is the SM's mechanical starting time and D its damping coefficient; and ω_b is the nominal synchronous angular frequency in rad/s. The electrical power P_{e} is defined as:

$$P_{\mathrm{e}}(\delta) = \frac{e'_{\mathrm{q}} v}{X_{\mathrm{tot}}} \sin(\delta - \theta) \tag{17.53}$$

where v, θ are the voltage magnitude and angle at the infinite bus; e'_{q} is the SM internal electromotive force, which is assumed constant. X_{tot} is the sum of the SM transient reactance (X_d') and the line reactance (X), where the latter is referred to the SM power base. Linearizing (17.52) around an equilibrium $[\delta_o \ \ \omega_o]^T$ leads to:

$$\Delta\dot{\delta} = \omega_b \Delta\omega \tag{17.54}$$

$$M\Delta\dot{\omega} = -\frac{e_q' v\cos(\delta_o - \theta)}{X_{tot}}\Delta\delta - D\Delta\omega \tag{17.55}$$

Equations (17.54) and (17.55) can be rewritten as a second-order LTI system, as follows:

$$\Delta\ddot{\delta} + d\Delta\dot{\delta} + b\Delta\delta = 0 \tag{17.56}$$

where $b = \frac{\omega_b e_q' v\cos(\delta_o)}{MX_{tot}}$, $d = \frac{D}{M}$.

In its simplest form, a PSS measures the SM's speed and introduces artificial damping into (17.54). The linearized closed-loop system can thus be written as:

$$\Delta\ddot{\delta} + d\Delta\dot{\delta} + b\Delta\delta = -u(\Delta\dot{\delta}) \tag{17.57}$$

To study the effect of delay on the stability of the OMIB system, we model the PSS as a proportional controller with two control channels, one with and one without delay. The output of such proportional retarded (PR) PSS can be described as:

$$u = K_p\Delta\dot{\delta} - K_r\Delta\dot{\delta}(t - \tau_r) \tag{17.58}$$

Merging (17.54), (17.57), and (17.58) leads to the following closed-loop system representation:

$$\Delta\ddot{\delta} + \left(d + \frac{K_p}{\omega_b}\right)\Delta\dot{\delta} + b\Delta\delta - \frac{K_r}{\omega_b}\Delta\dot{\delta}(t - \tau_r) = 0 \tag{17.59}$$

17.4.1.1 Stability analysis

We are interested in the spectral properties of (17.59). Applying the Laplace transform and substituting the initial conditions $\Delta\delta(0) = \Delta\dot{\delta}(0) = 0$ yield the characteristic equation:

$$q(s, \tau_r, K_r) = 0 \tag{17.60}$$

where

$$q(s, \tau_r, K_r) = s^2 + \left(d + \frac{K_p}{\omega_b}\right)s + b - \frac{K_r}{\omega_b}se^{-s\tau_r} \tag{17.61}$$

is the characteristic quasi-polynomial.

The system's spectral characteristics can be revealed by tracing the domains of stability that correspond to specified exponential decay rates σ (known as σ-stability). Applying to (17.61) the change of variable $s \rightarrow (s - \sigma)$ yields the following quasi-polynomial:

$$\tilde{q}(\sigma, s, \tau_r, K_r) = \tilde{q}_0(\sigma, s) + \tilde{q}_1(\sigma, s)\frac{K_r}{\omega_b}e^{\sigma\tau_r}e^{-s\tau_r} \tag{17.62}$$

where

$$\tilde{q}_0(\sigma, s) = (s-\sigma)^2 + \left(d + \frac{K_p}{\omega_b}\right)(s-\sigma) + b, \qquad \tilde{q}_1(\sigma, s) = -(s-\sigma) \quad (17.63)$$

Because the roots of the characteristic equation change continuously with respect to variations of system parameters and time delays[42], the system can change from stable to unstable, and vice versa, only if a root (or a pair of roots) crosses the imaginary axis of the complex plane. Hence, the σ-stability of the system can be assessed by finding the set of crossing points $(\tau_r^{\text{cr}}, K_r^{\text{cr}})$ that satisfy:

$$\tilde{q}(\sigma, \jmath, \tau_r^{\text{cr}}, K_r^{\text{cr}}) = 0 \quad (17.64)$$

where $s = \jmath\omega$. The set $(\tau_r^{\text{cr}}, K_r^{\text{cr}})$ can be determined by considering the magnitude and the argument of (17.64), as follows [55]:

$$\tau_r^{\text{cr}} = \frac{1}{\jmath w}\left(\text{Arg}(\tilde{q}_1(\sigma, \jmath w)) - \text{Arg}(\tilde{q}_0(\sigma, \jmath w)) + \frac{\pi}{2}(4\mu + \nu + 1)\right) \quad (17.65)$$

$$K_r^{\text{cr}} = \nu\, e^{-\sigma \tau_r^{\text{cr}}} \left| \frac{\tilde{q}_0(\sigma, \jmath w)}{\tilde{q}_1(\sigma, \jmath w)} \right| \omega_b \quad (17.66)$$

where $\nu = \pm 1$, $\mu = 0, \pm 1, \pm 2, \ldots$. Then, (17.65) and (17.66) allow drawing the system's σ-*stability map* in the (τ_r, K_r) space.

Note that if the delayed control channel is not utilized, i.e., $K_r = 0$, then the closed-loop OMIB system behavior is determined by the polynomial $\tilde{q}_0(\sigma, s)$. In this case, dissipative terms are defined by the coefficient of s:

$$c = d + \frac{K_p}{\omega_b} \quad (17.67)$$

Here, the coefficient d defines the damping of the open-loop system oscillatory mode, while K_p defines the amount of non-delayed artificial damping introduced by the simplified PSS to the system.

17.4.1.2 Delay-independent stability

The system is stable regardless the magnitude of the time delay τ_r, i.e., delay-independent stable, provided that certain conditions on the gain $K_r \in \mathcal{K}$, $\mathcal{K} \subset \mathbb{R}$, are satisfied. For a given set $\mathcal{K}$, a necessary condition for delay-independent stability is that the roots of the characteristic equation never cross the imaginary axis, or equivalently:

$$q(\jmath w, \tau_r, K_r) \neq 0\,,\ \forall \tau_r \geq 0,\ \forall K_r \in \mathcal{K} \quad (17.68)$$

Using (17.67) in (17.68) yields:

$$-w^2 + c\jmath w + d - \frac{K_r}{\omega_b}\jmath w e^{-\jmath w \tau_r} \neq 0 \Rightarrow \frac{\omega_b c}{K_r} + \jmath\frac{\omega_b}{K_r}(w - \frac{d}{w}) \neq e^{-\jmath w \tau_r} \quad (17.69)$$

Note that the real part of (17.69) does not depend on w, and thus, in the complex plane, the left-hand side defines the vertical line with abscissa $\omega_b c / K_r$. In addition, $e^{-\jmath w \tau_r}$ defines in the complex plane a unit circle centred at $(0, 0)$. Then, the critical condition for delay-independent stability is that the line $\omega_b c/K_r$ is tangent to the unit circle. Equivalently:

$$\frac{\omega_b c}{K_r} = \pm 1 \Rightarrow c = \pm \frac{K_r}{\omega_b} \tag{17.70}$$

From (17.70), the following cases are of interest [56]:

- If $c = -\frac{K_r^0}{\omega_b} < 0$, $K_r = K_r^0 > 0$, the system is delay-independent unstable in $\mathcal{K} = (-K_r^0, K_r^0)$. Moreover, since $c < 0$, the system is unstable around the origin of the τ_r-K_r plane. Hence, even if stable regions exist, these regions are guaranteed to be disconnected.
- If $c = 0$, there are no delay-independent stable or unstable regions.
- If $c = \frac{K_r^0}{\omega_b} > 0$, $K_r = K_r^0 > 0$, the system is delay-independent stable in $\mathcal{K} = (-K_r^0, K_r^0)$. The existence of a delay-independent stable region around the zero gain guarantees that there is a large connected stable domain in the τ_r-K_r plane. This feature is very important for two reasons: (i) there is the possibility that the dynamics can be characterized by high exponential decay rates for large delay values; (ii) the presence of a delay-independent stable region indicates that there exists at least one large, "connected" stable region from zero to infinite delay.

17.4.2 Numerical example

We provide a numerical example. Let $e'_{\rm q} = 1.22$ pu, $v = 1$ pu, $\theta = 0$ rad, $P_{\rm m} = 1$ pu, $X_{\rm tot} = 0.7$ pu. Then, the initial value δ_o of the rotor angle is given by:

$$\delta_o = \arcsin\left(\frac{P_{\rm m} X_{\rm tot}}{v e'_{\rm q}}\right) \tag{17.71}$$

The examined equilibrium is hence $[0.61\,,\,1]^{\rm T}$. Let also $M = 5$ MWs/MVA, and $\omega_b = 100\pi$ rad/s (50 Hz system). Then, $b = 89.756$ pu in (17.56). In the following, we discuss the σ-stability map of the system for the three cases of negative, zero and positive values of c.

Case 1

For $c = -0.4 < 0$, the stability map is shown in Figure 17.3. The map has a symmetric delay-independent unstable region obtained for $K_r \in (-125.6, 125.6)$. In addition, the controller can stabilize the system, provided that the delay is $\tau_r < 0.131$ s and a proper $K_r > 0$ is selected (see e.g., point $\Sigma_1(0.05, 729)$).

There also exist stable regions of the map in Figure 17.3 for delays higher than 0.131 s. For example, the system is stable around the point $\Sigma_3(0.30, -763.4)$. Note, however, that obtaining the equilibrium of a delayed system implies that a time equal to the maximum delay included in the system has elapsed but, meanwhile, the system may have been already rendered unstable. Indeed, Figure 17.3 indicates that there is no path to Σ_3 without crossing the system stability boundary, which implies that the system necessarily becomes unstable before actually reaching Σ_3.

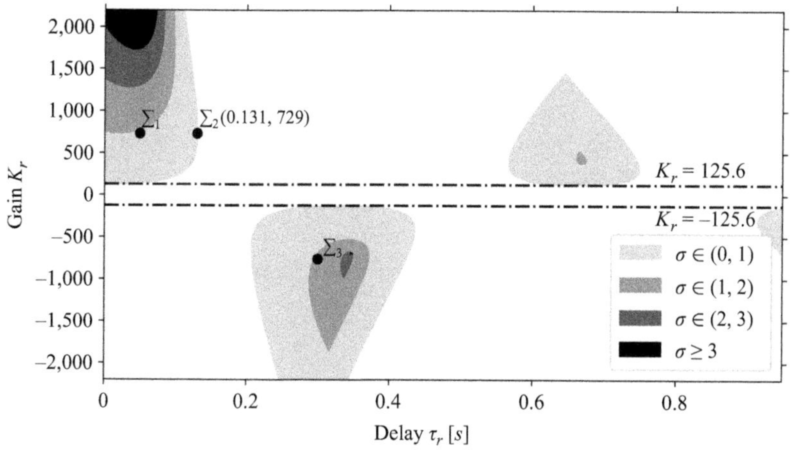

Figure 17.3 Closed-loop OMIB system: σ-stability map, $c = -0.4$

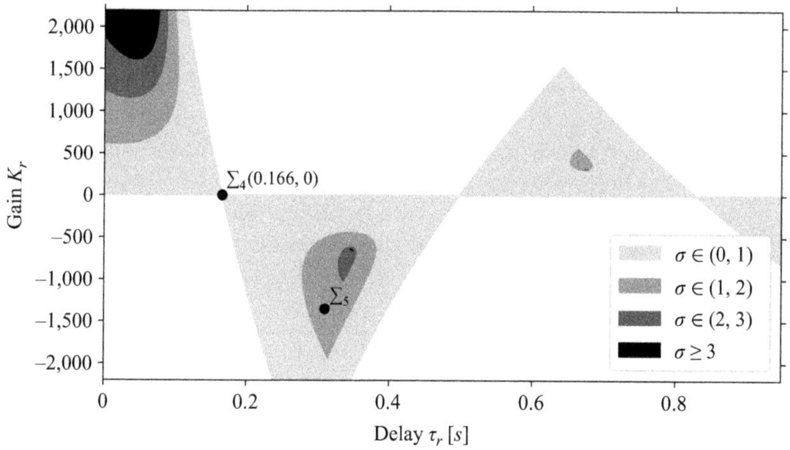

Figure 17.4 Closed-loop OMIB system: σ-stability map, $c = 0$

The effect of crossing the stability boundary of the closed-loop OMIB system is illustrated with a time-domain simulation. Suppose that system (17.52) with the inclusion of (17.58) operates around the stable equilibrium defined by the point Σ_1 of Figure 17.3.

Case 2

The σ-stability map for $c = 0$ is presented in Figure 17.4. In this case, the stability of the system depends on the magnitude of the delay, regardless of the value of the gain K_r. In fact, the horizontal line $K_r = 0$ comprises bifurcation points. The delay-free closed-loop system is stable for $K_r > 0$ and unstable for $K_r < 0$. Provided that a proper positive K_r value is selected and that $\tau_r < 0.166$ s (see point Σ_4), the delayed system is stable.

There also exist stable regions for $\tau_r > 0.166$ s. For example, the system is stable around Σ_5. However, similar to the discussion of Case 1, the system will likely lose stability before actually reaching e.g., Σ_5. An exception occurs if the system crosses Σ_4, which is a bifurcation point that connects two stable regions. In this scenario, the first-order information provided by the linearized system in Figure 17.4 is inconclusive on the feasibility of operating at Σ_5.

Case 3

The stability map for $c = 0.4 > 0$ is shown in Figure 17.5. In this case, the stable region is compact. For $K_r \in (-125.6, 125.6)$, the system is stable regardless of the magnitude of the delay τ_r. Moreover, all points of Figure 17.5 with $\sigma > 0$ represent stable and feasible stationary points of the linearized OMIB system. For example, such points are $\Sigma_6(0.13, 400)$ and $\Sigma_7(0.35, -410)$.

Overall, proper design of the PSS given by the PR control law (17.58) allows unifying the σ-stable regions, and thus allows one to operate the OMIB system under the presence of large delays. In particular, this is achieved by properly adjusting the control parameter K_p which introduces delay-free artificial damping to the system.

Finally, the delay τ_r in this example was assumed to be a fully controlled parameter. However, the above discussion is relevant also for systems with inherent delays. For the sake of example, consider again point Σ_6 of Figure 17.5. Suppose that the corresponding delay, i.e., 0.13 s, represents an uncontrolled physical phenomenon, e.g., the latency of a measurement transmitted through a communication system. In

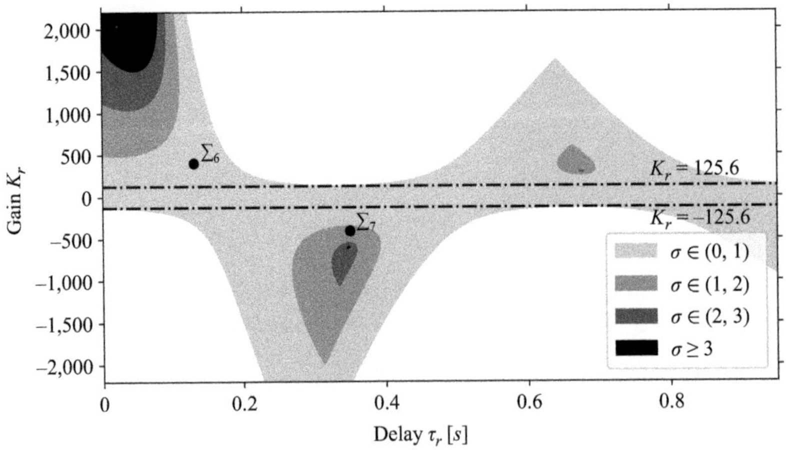

Figure 17.5 Closed-loop OMIB system: σ-stability map, c = 0.4

power systems, this situation describes, for example, the behavior of a wide area measurement system [3]. In such a scenario, the parameter τ_r can be adaptively adjusted to add an artificial delay, which ensures that the system under the total delay $0.13 + \tau_r$ always operates at a region of high exponential decay rate. Along these lines, we refer to the idea of delay scheduling, e.g., see [57].

17.5 Case studies

In this section, we consider two examples. The well-known IEEE 14-bus system is presented first to provide proof of concepts of the numerical techniques illustrated above. The chapter is then completed by a comparative study on a large real-world power system of the techniques to approximate the spectrum of DDAEs.

17.5.1 IEEE 14-bus system

The IEEE 14-bus system consists of 2 generators, 3 synchronous compensators, 2 two-winding and 1 three-winding transformers, 15 transmission lines, 11 loads, and 1 shunt capacitor (see Figure 17.6). The system also includes generator controllers, such

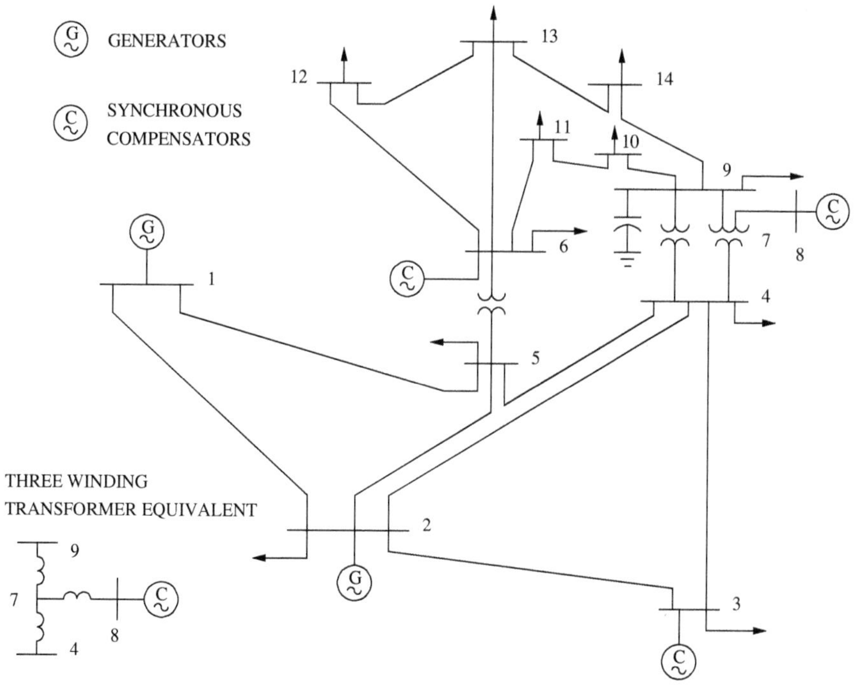

Figure 17.6 IEEE 14-bus test system. The system includes a PSS connected to generator 1

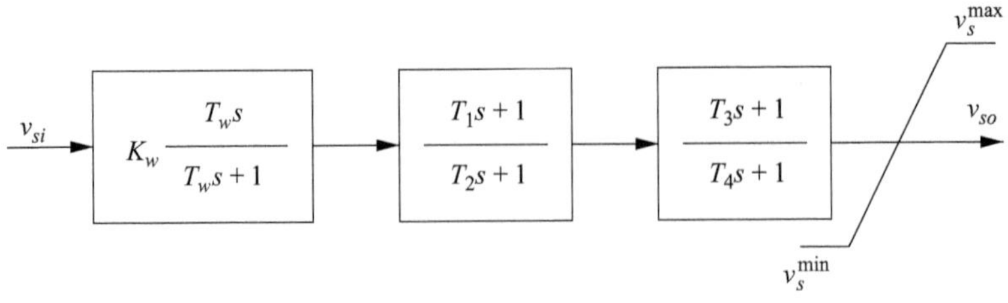

Figure 17.7 Power system stabilizer control diagram [54]

as primary voltage regulators. All dynamic data of this system as well as a detailed discussion of its transient behavior can be found in [54].

In the numerical analyses presented below, we assume that the IEEE 14-bus system includes a PSS connected to the automatic voltage regulator (AVR) of generator 1. A standard PSS control scheme includes a washout filter and a series of lead-lag blocks, as shown in Figure 17.7. The input signal of the PSS is typically the SM rotor speed which, in our formulation, is a state variable. In most cases, the rotor speed is measured locally, i.e., it is the rotor speed of the machine where the PSS is installed. However, there exist wide-area damping controllers (WADCs) that utilize remote signals, e.g., the frequency of a pilot bus, usually with the aim of damping an interarea oscillation [7,58]. Local measurements have at most a few milliseconds of delay while remote measurements can be affected by a delay of 100 ms or more [7]. For the sake of example, we consider that the input signal of the PSS of generator 1 is a remote measurement obtained through a wide area measurement systems (WAMS). Observe that the DDAEs that describe the PSS satisfy the index-1 Hessenberg form (17.3) where $x_d = v_{si} = \omega(t - \tau_\omega)$.

The input signal v_{si} "propagates" into the PSS equations, as follows:

$$
\begin{aligned}
\dot{v}_1 &= -(K_w v_{si} + v_1)/T_w \\
\dot{v}_2 &= \left(\left(1 - \frac{T_1}{T_2}\right)(K_w v_{si} + v_1) - v_2\right)/T_2 \\
\dot{v}_3 &= \left(\left(1 - \frac{T_3}{T_4}\right)\left(v_2 + \left(\frac{T_1}{T_2}(K_w v_{si} + v_1)\right)\right) - v_3\right)/T_4 \\
0 &= v_3 + \frac{T_3}{T_4}\left(v_2 + \frac{T_1}{T_2}(K_w v_{si} + v_1)\right) - v_{so}
\end{aligned}
\tag{17.72}
$$

where v_1, v_2, and v_3 are state variables introduced by the PSS washout filter and by lead-lag blocks, and other parameters are illustrated in Figure 17.7. Observe that (17.72) are in the form of (17.3) with $\boldsymbol{x} = [v_1 \ \ v_2 \ \ v_3]^{\mathrm{T}}$, and $y = v_{so}$.

Table 17.2 Critical eigenvalue of the IEEE 14-bus system for different values of τ_ω

τ_ω (s)	λ_c	τ_ω (s)	λ_c
0	$-1.76765 \pm 12.35628\jmath$	0.060	$-0.25208 \pm 12.07820\jmath$
0.010	$-1.49003 \pm 12.37513\jmath$	0.070	$-0.05721 \pm 11.96773\jmath$
0.020	$-1.21555 \pm 12.36373\jmath$	0.073	$-0.00251 \pm 11.93290\jmath$
0.030	$-0.95056 \pm 12.32437\jmath$	0.074	$0.01534 \pm 11.92114\jmath$
0.040	$-0.69982 \pm 12.26074\jmath$	0.075	$0.03300 \pm 11.90932\jmath$
0.050	$-0.46645 \pm 12.17721\jmath$	0.080	$0.11851 \pm 11.84932\jmath$

17.5.1.1 Steady-state analysis and delay margin

From [59], it is known that the IEEE 14-bus system shows undamped oscillations if the loading level is increased by 20% with respect to the base case and lines 2–4 outage occurs.† It is also well known that such oscillations can be properly damped through the PSS shown in Figure 17.7 in the excitation control scheme of the machine connected to bus 1.

In this section, we consider the bifurcation analysis for the IEEE 14-bus system using as bifurcation parameter the time delay τ_ω of the frequency signal $v_{si} = \omega(t - \tau_\omega)$ that enters into the PSS; and (ii) the loading level of the system. The parametrization based on the system delay was proposed for the first time in [60].

Table 17.2 shows the critical eigenvalue λ_c of the IEEE 14-bus system as a function of the PSS frequency measure time delay τ_ω, which is varied in the interval $[0, 80]$ ms. Results shown in Table 17.2 are obtained using the Chebyshev discretization discussed in Section 17.3.3.1. However, due to the small size of the IEEE 14-bus system, the same results can be obtained with many different techniques, including those discussed in Sections 17.3.1 and 17.3.3. Table 17.2 indicates that, as the delay increases, the difference between the non-delayed DAEs and the DDAEs becomes quite evident. A Hopf Bifurcation (HB) occurs for $\tau_\omega \approx 73$ ms, which is thus the *delay margin* of the PSS. In other words, if the PSS is fed by a remote frequency measurement signal, the communication system has a margin of 73 ms before the system becomes unstable.

17.5.1.2 Time-domain analysis

As discussed above, for a 20% increase of the loading level with respect to the base case, a HB occurs for $\tau_\omega > 73$ ms. Repeating the same analysis for lines 2–4 outage, we observe that the HB occurs for $\tau_\omega \approx 68.6$ ms. Thus, setting $73 > \tau_\omega > 69$ ms, it has to be expected that the transient following lines 2–4 outage is unstable, while the initial equilibrium point without contingency is stable, though poorly damped.

† See also Chapter 12 of this book for a comprehensive discussion on the limit cycles originated by perturbing the IEEE 14-bus system.

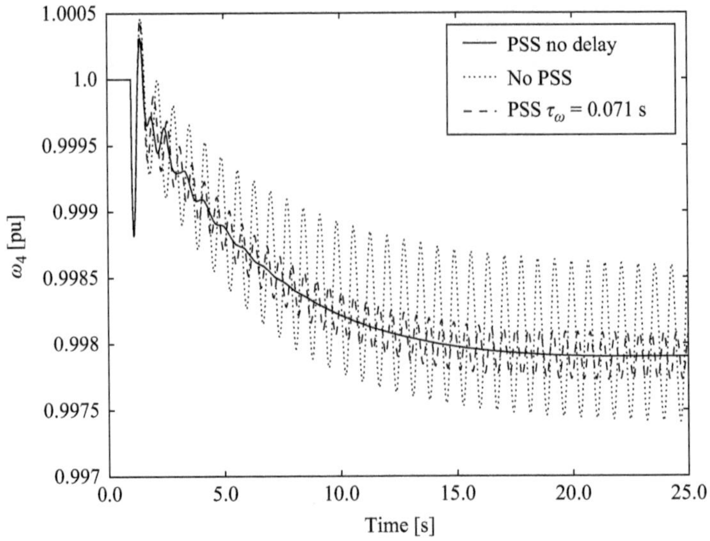

Figure 17.8 Rotor speed ω of machine 5 for the IEEE 14-bus system with a 20% load increase and for different control models following line 2–4 outage at t = 1 s

Figure 17.8 shows the time response of the IEEE 14-bus system without PSS, with PSS and with retarded PSS with $\tau_\omega = 71$ ms. As already known from [59], the trajectory of the system without PSS enters into a stable limit cycle after the line outage while the system with PSS is asymptotically stable. The behavior of the system with delayed PSS is similar to the case without PSS, i.e., presents a limit cycle trajectory. This result was expected as can be justified as follows. For $\tau_\omega \to \infty$, the PSS control loop behaves like an open-loop system, which is unstable. Therefore, there exists at least one critical delay value between zero and infinity which leads the zero-delay stable PSS to transition to instability. This critical delay value can be computed via the small-signal stability analysis corresponding to HB. Moreover, the added value of the time-domain simulations is to show that the system trajectory enters into a limit cycle rather than diverging.

The time-domain simulations for the retarded system are obtained using the ITM, adapted to include delays as discussed in Section 17.3.2. The same results can be obtained using the Padé approximants with $p = q = 6$, which, as outlined in Section 17.3.1, is the standard order used in commercial software.

It has to be noted that, for standard delay-free DAE power system models, HBs, which are co-dimension one local bifurcations, are *generic*. In other words, HBs are expected to occur given certain loading conditions and SM controllers. However, the case of infinite-dimensional dynamics such as delay systems requires further analysis to conclude on the genericity of the bifurcation points. This is currently an open field of research.

17.5.1.3 Small-signal stability maps

In this section we show the impact of PSS on the small-signal stability characteristics of the IEEE 14-bus system through stability maps in the delay-control gain parameter space. For a second-order LTI system with PR control, such as the one discussed in Section 17.4, one can analytically identify the parameter regions with specified exponential decay rates, as well as the conditions for delay-independent stability. However, real-world dynamical systems are larger in size and much more complex. Capturing the impact of delays on the behavior of more detailed power system models, such as the IEEE 14-bus system, can be achieved only by carrying out a numerical analysis and using techniques such as the ones discussed in Section 17.3.

In this example, an eigenvalue analysis is carried out for each point of the parameter space by applying the Chebyshev discretization technique. The approximate spectrum is calculated using the QR algorithm. Then, comparison allows obtaining the most critical eigenvalues by looking at their damping ratios. In analogy to the concept of σ-stability, we will refer to a system with critical oscillation damping ratio ζ, as ζ-*stable*. Using this concept allows building a map in a delay versus control gain space and identify regions with specified values of damping ζ. In the remainder of the chapter, such map is referred to as the ζ-*stability map* [56].

In the following we consider the IEEE 14-bus system and examine two damping control configurations, (i) a conventional PSS with delayed input signal; and (ii) a PSS that consists of two channels, one delayed and one non-delayed. In both cases, the damping controller is installed at the AVR of the SM connected at bus 1.

One-channel PSS with delayed input signal

The employed PSS model is as shown in Figure 17.7, where as control input signal is considered the delayed local rotor speed measurement $v_{si} = \omega(t - \tau)$. The dynamic order of the closed-loop system is 54. Setting the number of points of the Chebyshev differentiation matrix to $N = 10$, 540 eigenvalues are found in total.

The system's ζ-stability map in the τ-K_w plane is shown in Figure 17.9. The map consists of distinct and not compact stable regions, which stems from the fact that, without the PSS, the system is unstable. For $K_w \in (-0.55, 0.65)$, the system is unstable regardless of the magnitude of the delay. The delay margin of the system is 0.104 s and is obtained for $K_w = 1.5$. Thus, operation under the presence of a large delay, e.g. 0.35 s, is infeasible.

A technique commonly employed to mitigate destabilizing delay effects and increase the stability margin of a dynamical system is delay compensation. As already mentioned in Section 17.1, the main idea of delay compensation is to apply a control block which generates a signal that is similar to the original, non-delayed signal. The compensated signal is then fed to the controller. Let consider that delay compensation is applied to the PSS input signal v_{si}. To this aim, we employ the Proportional-Derivative (PD) control-based delay compensation method described in [10,12]. Then, the signal compensated by the PD method is:

$$v_{si}^{\text{com}} = v_{si} + K_\tau \dot{v}_{si} \qquad (17.73)$$

where K_τ is the compensation gain.

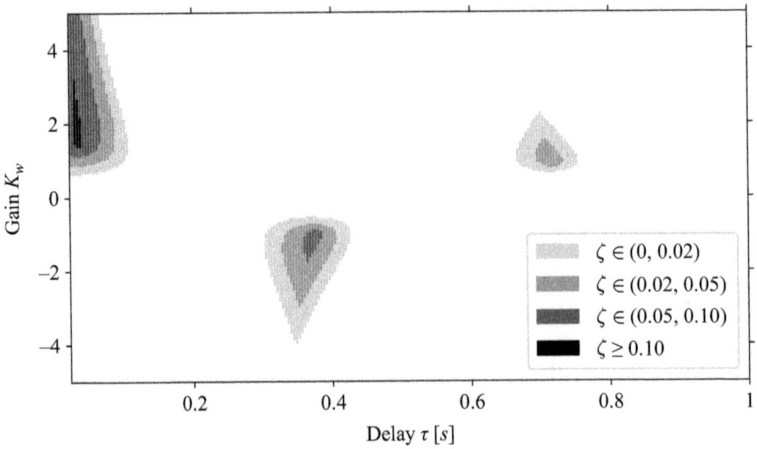

Figure 17.9 IEEE 14-bus system: ζ-stability map in the τ-K_w plane

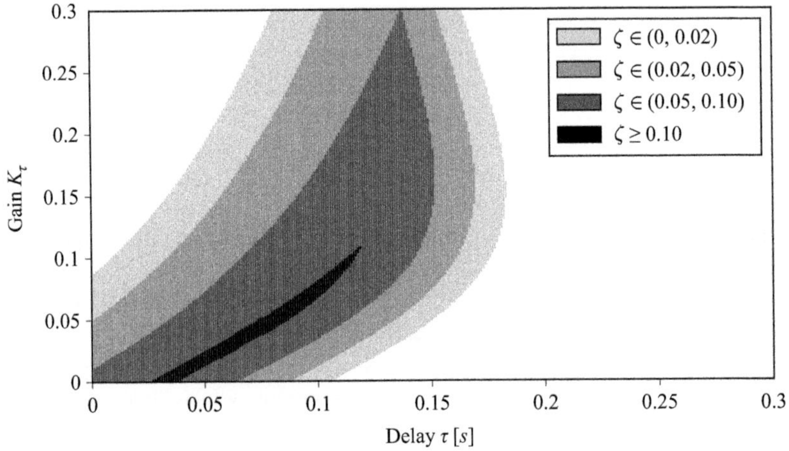

Figure 17.10 IEEE 14-bus system with delay compensation: ζ-stability map

With PSS gain $K_w = 1.5$, the ζ-stability map in the delay vs compensation gain (τ-K_τ) is shown in Figure 17.10. The inclusion of the PD compensation to the conventional PSS with delayed input signal increases the delay margin of the system to 0.184 s, while oscillations have damping ratio $\zeta \geq 0.10$ but only for delays smaller than 0.121 s. That is, delay compensation cannot handle well very large communication delays.

Dual-channel PSS

In the OMIB system example discussed in Section 17.4 a compact stable region in the delay-control gain plane can be achieved by employing a PR-based PSS scheme, tuned to operate the system at a point with good damping characteristics.

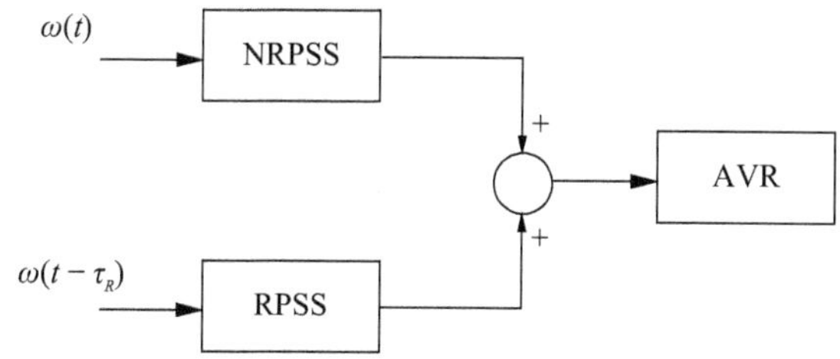

Figure 17.11 Dual-channel PSS configuration

Table 17.3 Analogy between the dual-channel PSS and the PR controller of Section 17.4

Channel	OMIB system	IEEE 14-bus system
Non-retarded	Proportional K_p	NRPSS
Retarded	K_r, τ_r	RPSS

We apply the same principle in the IEEE 14-bus system. To this aim, we test a PSS with two control channels: the first channel is not delayed; the second channel is delayed. The dual-channel PSS configuration is shown in Figure 17.11.

The first channel, namely Not Retarded PSS (NRPSS), is tuned to render the non-delayed system stable. The control input of NRPSS is the local rotor speed $\omega(t)$. The second channel, namely Retarded PSS (RPSS), tunes the delay dynamics so that the system operates at a point with good damping characteristics. The input signal of the RPSS is the delayed rotor speed $\omega(t - \tau_R)$, where $\tau_R \geq 0$ is the magnitude of the delay. In addition, $K_{w,P}$ and $K_{w,R}$ denote the gains of NRPSS and RPSS, respectively. An analogy between the dual-channel PSS configuration and the PR-based PSS of the OMIB system example of Section 17.4 is given in Table 17.3.

The NRPSS gain is tuned so that the system without delayed control is stable. For $K_{w,P} = 5$, $K_{w,R} = 0$, small-signal stability analysis shows that the rightmost pair of eigenvalues is $-0.1376 \pm 0.0203\jmath$. The most poorly damped pair is $-0.5171 \pm 7.2516\jmath$, which yields a damping ratio 0.071.

Considering $K_{w,P} = 5$, the ζ-stability map of the system is constructed in the $K_{w,R}$-τ_R plane. In this case, the dynamic order of the system is 57 and, using $N = 10$, 570 eigenvalues are in total calculated to obtain each point of the map. The resulting map, presented in Figure 17.12, shows that the stable region is compact, while the area with $K_{w,R} \in (-2.4, 2.5)$ is delay-independent stable. In Figure 17.12, the maximum damping is 0.178 and is achieved for $\tau_R = 0.34$ s, i.e. a relatively large delay value.

We further discuss the relevance of the dual-channel PSS for wide area measurement systems and control applications. A WADC typically employs a signal that is remote and that, thus, has to be transmitted to the control actuator through a communication network, which introduces an inherent and unavoidable delay. Considering

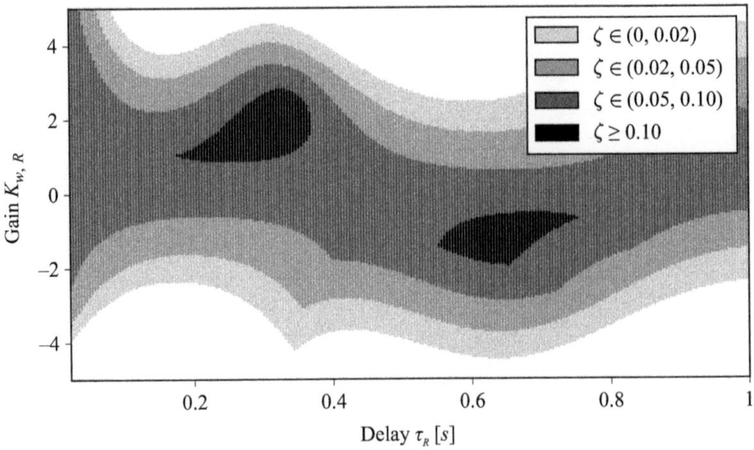

Figure 17.12 IEEE 14-bus system: ζ-stability map in the τ_R-$K_{w,R}$ plane

that the delay fed to the dual-channel PSS is inherent, the structure of Figure 17.12 implies that if a proper artificial delay is injected on top of the inherent delay, the system can be led to a region of better damping characteristics[‡]. This extra delay can be introduced, for example, by a properly designed controller that adjusts both the delay and gain values following a stable path, through consecutive quasi-steady state shifts of the system equilibrium.

The inherent delay can create a severe stability issue to the conventional PSS (see, for example, Figure 17.9), since the delay margin in this case is relatively small (0.104 s). For an inherent delay that has a small magnitude, a delay-dependent design of the standard PSS allows increasing the delay margin and may be adequate to avoid instability. In fact, in Figure 17.9, the region of the highest damping $\zeta \geq 0.10$ is obtained for a non-zero delay value, with the maximum damping ratio value being $\zeta_{\max} = 0.11$. The closed-loop loci related to the critical system mode have an angle of departure closer to 180° when $\tau = 0.03$ s. In other words, the phase shift introduced by the PSS is optimal when a small delay is present.

17.5.2 All-island 1479-bus Irish system

The numerical techniques presented in Section 17.3 work satisfactorily for *small* size systems, e.g., few tens of state variables and few tens of delays. For example, basically identical results are obtained for the IEEE 14-bus system discussed in the previous subsection regardless of the adopted technique for both time-domain integration and small-signal stability analysis. Based on our experience, Chebyshev discretization and Padé approximants always provide good results for small systems. TIO is also accurate

[‡]Outside the power systems literature there is theoretical and experimental evidence of this ‘delay scheduling’ idea [57].

provided that N is increased with respect to the Chebyshev method. For example, $N = 7$ is acceptable for small power systems. On the other hand, LMS provides good results if h is relatively small. For example, $h = 0.1$ s appears acceptable for small power systems. However, standard benchmarks are too small, not allowing us to draw sensible conclusions on the robustness and the accuracy of the techniques discussed in Section 17.2 for large-scale eigenvalue problems. Stiffness and numerical rounding errors play a crucial role as the size of the problem scales up, as shown in this section.

In this case study, the techniques in Section 17.3 are compared through a dynamic model of the All-Island Irish Transmission System (AIITS) set up at the UCD Electricity Research Centre. The model includes 1479 buses, 1851 transmission lines and transformers, 245 loads, 22 conventional synchronous power plants with AVRs and turbine governors, 6 PSSs and 176 wind power plants. The topology and static data of the transmission system are provided by the Irish TSO, EirGrid, but dynamic data are guessed based on the current knowledge of the technology of power plants.

17.5.2.1 Small-signal sensitivity analysis

The objective of this subsection is to compare the robustness of different methods for the small-signal stability analysis of a large DDAE system. With this aim, constant time delays are artificially included in most regulators of the AIITS, as follows. All bus terminal voltage measurements of the AVRs of the SMs include delays in the range $\tau_{\mathrm{AVR}} \in (5, 15)$ ms [4]. The input frequency signal of PSS devices is delayed in the range $\tau_{\mathrm{PSS}} \in (50, 250)$ ms [7]. The reheater of the turbine governors of thermal power plants is modelled as a pure delay in the range $\tau_{\mathrm{RH}} \in (3, 11)$ s.

The model of some variable-speed wind turbines includes a frequency regulation that receives as input the frequency of the Center of Inertia (COI) of the system. The model of the frequency regulator is based on the transient frequency control described in [61]. The frequency signal is assumed to be similar to those of PSS devices, hence $\tau_{\mathrm{TFC}} \in (50, 250)$ ms.

Finally, 20% of the loads are assumed to provide frequency regulation. In other words, 20% of loads are assumed to be equivalent thermostatically controlled heating systems. The dynamic model of these loads and their control is based on [62] and [63], respectively. Again, the input frequency signal is delayed and, in analogy with PSS devices, delays are chosen in the interval $\tau_{\mathrm{TCL}} \in (50, 250)$ ms.

The delay ranges considered in this case study are summarized in Table 17.4. In total, the system model contains 296 delays ranging in the interval (0.005, 11) s. This wide range is chosen with the purpose of determining the accuracy and performance of the methods presented in Section 17.2. The resulting DDAEs are *stiff*, as device and regulator time constants span several orders of magnitude, i.e., from tens of milliseconds to tens of seconds.

The order of the system, i.e., the number of state and algebraic variables, depends on the model. Table 17.5 shows system statistics for four different models, namely, no delays; constant delays; Padé approximant with $p = q = 6$; and Padé approximant with $p = q = 10$. The only DDAE model is the model where delays are implemented as in (17.3), as Padé approximants transform the delays into a set of linear differential equations.

Table 17.4 Ranges of time delays included in the AIITS

Device	Delayed signal	Delay	Range [s]
Primary voltage regulator	Bus voltage	τ_{AVR}	$(0.005, 0.015)$
Power system stabilizer	Frequency	τ_{PSS}	$(0.05, 0.25)$
Reheater of steam turbines	Steam flow	τ_{TG}	$(3, 11)$
Wind turbine frequency regulation	Frequency	τ_{TFR}	$(0.05, 0.25)$
Thermostatically controlled load	Frequency	τ_{TCL}	$(0.05, 0.25)$

Table 17.5 Statistics for the AIITS

Model	Type	State variables	Algebraic variables
No delays	DAE	2,239	7,478
Constant delays	DDAE	1,935	7,338
Padé approx. ($p = q = 6$)	DAE	3,415	7,929
Padé approx. ($p = q = 10$)	DAE	4,399	7,929

It is noteworthy that the DDAE model is also the model with the lowest number of variables. This is due to the fact that, in the standard model with no delays, delays are actually modelled as a simple lag transfer function, each of which introduces a state variable. Note also that, the lag transfer function is, in turn, the Padé approximant with $p = 1$ and $q = 0$. Higher order Padé approximants lead to a substantial increase of the order of the system, and hence of the computational burden of the initialization of system variables and time-domain simulations.[§] Transient analysis is out of the scope of this chapter but the latter remark has to be kept in mind when choosing the power system models.

All simulations are obtained using Dome [64]. The Dome version used in this case study is based on Python 3.4.1, NVidia Cuda 7.0, Numpy 1.8.2, CVXOPT 1.1.7, MAGMA 1.6.1, and has been executed on a 64-bit Linux Fedora 21 operating system running on a two Intel Xeon 10 Core 2.2 GHz CPUs, 64 GB of RAM, and a 64-bit NVidia Tesla K20X GPU.

Table 17.6 shows the 20 rightmost eigenvalues for the AIITS using different system models and techniques. For reference, the first column also shows the 20 rightmost eigenvalues of the non-delayed model. This system does not show any poorly damped mode, i.e., a mode whose damping is below 5%. Columns 2–5 of Table 17.6 show the results obtained using the Chebyshev discretization, the discretization of the TIO, the LMS approximation, and the Padé approximants. Two cases are shown for the

[§] Padé approximants also lead to increase the number of algebraic variables because the output u_d of the approximated transfer function (17.17) is algebraic, as shown by (17.19).

Table 17.6 20 rightmost eigenvalues for the AIITS

No delay	**Chebyshev discretization** $N = 7$	**Discretization of TIO** $N \cdot r = 21$	**LMS approximation** $h = 0.2$ s	**Padé approximant** $p = q = 6$	**Padé approximant** $p = q = 10$
-0.00010	-0.00010	-3.16992	0.91568	-0.00010	14,370.508
-0.02500	-0.02500	-3.46994	0.82393	-0.02500	2,166.5568
-0.02646	-0.02650	-3.54846	0.58361	-0.02848	1,545.1549
$-0.03780 \pm 0.32935\jmath$	$-0.03780 \pm 0.32935\jmath$	-3.79015	0.36998	$-0.03780 \pm 0.32935\jmath$	1,540.3456
-0.05475	-0.05475	-3.79481	0.29701	-0.05475	1,445.2436
-0.06615	$-0.06100 \pm 0.32755\jmath$	-3.85081	0.10980	-0.06615	1,434.9052
$-0.08759 \pm 0.10409\jmath$	-0.06615	-3.86392	-0.00327	$-0.08759 \pm 0.10409\jmath$	1,019.4456
-0.11681	$-0.08759 \pm 0.10409\jmath$	-4.25558	-0.05199	$-0.10759 \pm 0.33539\jmath$	891.50938
$-0.12665 \pm 0.34150\jmath$	$-0.11445 \pm 0.78025\jmath$	-4.33068	-0.09677	-0.11681	795.91920
$-0.13055 \pm 0.17132\jmath$	-0.11681	-4.52052	-0.13551	$-0.12906 \pm 0.34552\jmath$	724.39851
-0.13922	$-0.12818 \pm 0.34639\jmath$	-4.68635	-0.15511	$-0.13380 \pm 0.17103\jmath$	648.25856
-0.13950	$-0.13455 \pm 0.17176\jmath$	-4.80909	-0.23989	-0.13417	625.18431
-0.13978	-0.17139	-4.84030	-0.32102	$-0.17474 \pm 0.27121\jmath$	593.37327
-0.14008	$-0.17358 \pm 0.27051\jmath$	$-5.24457 \pm 0.35652\jmath$	-0.34557	-0.17504	587.83144
-0.14027	-0.17504	-5.26514	-0.45854	$-0.18411 \pm 0.78161\jmath$	533.95381
-0.14048	$-0.18208 \pm 0.81259\jmath$	$-5.67946 \pm 0.81568\jmath$	-0.55539	-0.18562	528.11686
-0.14062	$-0.18316 \pm 0.81807\jmath$	-5.74580	-0.67482	-0.18892	519.93536
-0.14081	-0.18562	-5.80760	-0.73128	-0.20000	497.91600
-0.14104	$-0.18877 \pm 0.81637\jmath$	-5.98648	-0.95327	$-0.20483 \pm 0.87988\jmath$	456.93850
-0.14119	-0.18892	-6.10122	-0.97517	$-0.20944 \pm 0.36519\jmath$	420.89130

latter, namely, $p = q = 6$ and $p = q = 10$. Both Chebyshev and TIO discretizations use a grid of order $N = 7$, which provides a good trade-off between accuracy and computational burden. For the discretization of the TIO, a fifth-order Radau IIA method is used, with $r = 3$ along with the following Butcher's tableau:

$$
\begin{array}{c|ccc}
\frac{2}{5} - \frac{\sqrt{6}}{10} & \frac{11}{45} - \frac{7\sqrt{6}}{360} & \frac{37}{225} - \frac{169\sqrt{6}}{1800} & -\frac{2}{225} + \frac{\sqrt{6}}{75} \\
\frac{2}{5} + \frac{\sqrt{6}}{10} & \frac{37}{225} + \frac{169\sqrt{6}}{1800} & \frac{11}{45} + \frac{7\sqrt{6}}{360} & -\frac{2}{225} - \frac{\sqrt{6}}{75} \\
1 & \frac{4}{9} - \frac{\sqrt{6}}{36} & \frac{4}{9} + \frac{\sqrt{6}}{36} & \frac{1}{9} \\
\hline
 & \frac{4}{9} - \frac{\sqrt{6}}{36} & \frac{4}{9} + \frac{\sqrt{6}}{36} & \frac{1}{9}
\end{array}
$$

Finally, an Adams-Bashforth sixth-order method is used for the LMS approximation, with the following coefficients:

$$
\begin{aligned}
\boldsymbol{\alpha} &= [1,\ -1,\ 0,\ 0,\ 0,\ 0] \\
\boldsymbol{\beta} &= [0,\ 1901/720,\ -1387/360,\ 109/30,\ -637/360,\ 251/720]
\end{aligned}
$$

A time step $h = 0.2$ s is used for the LMS approximation.

To complete the comparison of the four techniques whose results are provided in Table 17.6, Table 17.7 shows the computational burden of these techniques using the GPU-based MAGMA library. The information given in Table 17.7 is the time required to setup the full matrix for the eigenvalue analysis, the order of this matrix, and the time required to solve the full LEP.

While all methods are necessarily approximated, the most accurate one to estimate the spectrum of the DDAEs can be expected to be the Chebyshev discretization scheme. As indicated in [49], this approach shows a fast convergence. Moreover, simulation results on large-scale systems indicate that the Chebyshev discretization does not require N to be high [4]. The accuracy of other methods can be thus defined based on a comparison with the results obtained through the Chebyshev discretization method. As shown in Table 17.7, the lightest computational burden is provided by

Table 17.7 Computational burden of different methods to compute eigenvalues using the GPU-based MAGMA library

Model	Settings	Matrix setup	Matrix order	LEP sol.
No delays		1.18 s	2,239	11.91 s
Cheb. discretization	$N = 7$	29.4 s	13,545	12.69 m
Discretization of TIO	$N \cdot r = 21$	7.07 h	40,635	50.73 s
LMS approximation	$h = 0.2$ s	7.48 m	32,895	20.83 s
Padé approximant	$p = q = 6$	2.01 s	3,415	35.21 s
Padé approximant	$p = q = 10$	2.78 s	4,399	76.75 s

Padé approximants. However, the solution obtained with $p = q = 6$ shows some differences with respect to the Chebyshev discretization, e.g., two poorly damped modes, namely $-0.26201 \pm 6.3415i$ and $-0.2746 \pm 5.9609i$ do not appear in the solution based on the Chebyshev discretization. Both modes show a damping lower than 5% and, through the analysis of participation factors, both are strongly associated with fictitious state variables introduced by the Padé approximant, e.g., (17.20) and (17.21). This effect has to be expected as extraneous oscillations are a well-known drawback of Padé approximants. Finally, observe that for the Padé approximant with $p = q = 10$, results are fully unsatisfactory due to numerical issues. These are due to the extremely small values taken by coefficients in (17.19), (17.21). For the considered case study, numerical problems show up for $p = q \geq 8$.

Tables 17.6 and 17.7 also show that the techniques based on the TIO discretization and LMS approximation are both highly inaccurate and time consuming. In particular, the TIO discretization requires a huge time to setup up the matrix $\boldsymbol{S}_N$ of (17.42). It is likely that the implementation of the algorithm that builds $\boldsymbol{S}_N$ can be improved using parallelization, which was not exploited in this case. However, the inaccuracy of the results makes improving the implementation of this technique unnecessary. Note also that the size of the computational burden of the LMS approximation strongly depends on the time step h used in (17.47). The smaller the time step h, the more precise the approximation, but the higher the computational burden. However, for $h < 0.2$, the size of the eigenvalue problem becomes too big and the MAGMA solver fails returning a memory error. Unfortunately, in this case, $h = 0.2$ s is too large to obtain precise results.

A rationale behind the poor results shown by the TIO and LMS methods compared to the Chebyshev discretization and Padé method, is as follows. The Chebyshev discretization method works directly with (17.13) and is concerned solely with approximating the $e^{-\lambda\tau_i}$ terms. As such, the quality of the results depends only on the quality of this approximation. In the same vein, Padé approximants are concerned with the approximation of $e^{-s\tau_i}$ terms in (17.15). On the other hand, the TIO and LMS methods work with (17.12) and require approximations of both the $\dot{x}$ term and the delay terms. Hence, the TIO and LMS approximations require a step size small enough that the resulting difference equation is stable.

17.5.2.2 Time-domain simulation

This section compares the numerical robustness of the time integration schemes for DDAEs discussed in Section 17.3.2 as well as the DAE model based on Padé approximants. With this aim, Figure 17.13 shows the speed of the COI for the AIITS with stochastic wind speeds and a short-circuit occurring at bus 1238 at $t = 1$ s and cleared after 50 ms by removing line 1633. ITM and BDF schemes provide the same results using a time step of 0.01 s. BEM results are consistent with those of ITM and BDF but slightly more damped, as was expected due to the latent hyperstability of the BEM. On the other hand, the results obtained with a DAE based on Padé approximants with $p = q = 6$ and ITM are not satisfactory. The trajectory appears unstable and starts diverging even before the occurrence of the fault. Note that this divergence is caused

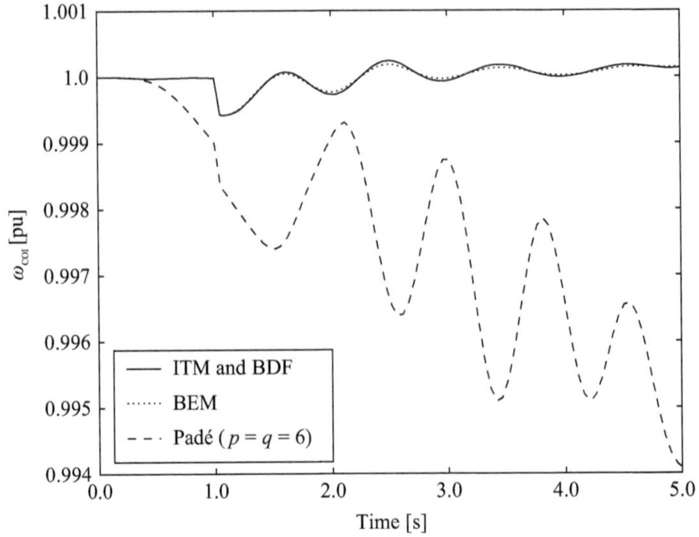

Figure 17.13 COI speed for the AIITS with stochastic wind speeds and a short-circuit occurring at bus 1238 at t = 1 s and cleared after 50 ms by removing line 1633

by numerical instabilities. In fact, the coefficients of the Padé approximants given by (17.18) make the Jacobian matrix of the resulting DAE ill-conditioned. Similar to the case of small-signal stability analysis discussed in the previous section, such ill-conditioning does not occur for relatively small systems with a reduced number of state variables, e.g., the IEEE 14-bus system and most IEEE dynamic benchmarks.

Bibliography

[1] Nutaro J, Protopopescu V. A new model of frequency delay in power systems. *IEEE Transactions on Circuits and Systems – II: Express Briefs*. 2012 Nov;59(11):840–844.

[2] Wang S, Meng X, Chen T. Wide-area control of power systems through delayed network communication. *IEEE Transactions on Control Systems Technology*. 2011;20(2):495–503.

[3] Liu M, Dassios I, Tzounas G, Milano F. Stability analysis of power systems with inclusion of realistic-modeling WAMS delays. *IEEE Transactions on Power Systems*. 2019 Jan;34(1):627–636.

[4] Milano F, Anghel M. Impact of time delays on power system stability. *IEEE Transactions on Circuits and Systems – I: Regular Papers*. 2012 Apr;59(4): 889–900.

[5] Ye H, Liu Y, Zhang P. Efficient eigen-analysis for large delayed cyber-physical power system using explicit infinitesimal generator discretization. *IEEE Transactions on Power Systems*. 2015;31(3):2361–2370.

[6] Padhy BP, Srivastava SC, Verma NK. A wide-area damping controller considering network input and output delays and packet drop. *IEEE Transactions on Power Systems*. 2016;32(1):166–176.

[7] Wu H, Tsakalis KS, Heydt GT. Evaluation of time delay effects to wide-area power system stabilizer design. *IEEE Transactions on Power Systems*. 2004 Nov;19(4):1935–1941.

[8] Yao W, Jiang L, Wen J, Wu QH, Cheng S. Wide-area damping controller of FACTS devices for inter-area oscillations considering communication time delays. *IEEE Transactions on Power Systems*. 2014 Jan;29(1):318–329.

[9] Ghosh S, Folly KA, Patel A. Synchronized versus non-synchronized feedback for speed-based wide-area PSS: effect of time-delay. *IEEE Transactions on Smart Grid*. 2016;9(5):3976–3985.

[10] Tian YC, Levy D. Compensation for control packet dropout in networked control systems. *Information Sciences*. 2008;178(5):1263–1278.

[11] Zheng Y, Brudnak MJ, Jayakumar P, Stein JL, Ersal T. A predictor-based framework for delay compensation in networked closed-loop systems. *IEEE/ASME Transactions on Mechatronics*. 2018;23(5):2482–2493.

[12] Liu M, Dassios I, Tzounas G, Milano F. Model-independent derivative control delay compensation methods for power systems. Energies. 2020;13(2):342.

[13] Yao W, Jiang L, Wen J, Wu Q, Cheng S. Wide-area damping controller for power system interarea oscillations: A networked predictive control approach. *IEEE Transactions on Control Systems Technology*. 2014;23(1):27–36.

[14] Gu K, Kharitonov VL, Chen J. *Stability of Time-Delay Systems*. Boston: Birkhauser; 2003.

[15] Loiseau JJ, Niculescu SI, Michiels W, Sipahi R. *Topics in Time Delay Systems, Analysis, Algorithms and Control*. Heidelberg: Springer-Verlag; 2009.

[16] Niculescu SI. *Delay Effects on Stability: A Robust Control Approach. Vol. 269 of Lecture notes in Control and Information Science*. Springer Verlag; 2001.

[17] Halanay A. *Differential Equations; Stability, Oscillations, Time Lags*. New York: Academic Press; 1966.

[18] Hale JK, Lunel SMV. *Introduction to Functional Differential Equations*. New York: Springer-Verlag; 1993.

[19] Krasovskii NN. *Stability of Motion*. Stanford University Press; 1963.

[20] Ting L, Min W, Yong H, Weihua C. New delay-dependent steady state stability analysis for WAMS assisted power system. In: Proceedings of the 29th Chinese Control Conference. Beijing, China; 2010. p. 29–31.

[21] Wu M, He Y, She J. *Stability Analysis and Robust Control of Time-Delay Systems*. New York: Springer; 2010.

[22] Stépán G. *Retarded Dynamical Systems: Stability and Characteristic Functions*. Vol. 210 of Pitman Research Notes in Mathematics Series. New York: Longman Scientific & Technical, Co-publisher John Wiley & Sons, Inc.; 1989.

[23] Datko R. A procedure for determination of the exponential stability of certain differential-difference equations. *Quarterly Applied Math.* 1978; 36(3).

[24] Olgac N, Sipahi R. An exact method for the stability analysis of time-delayed linear time-invariant (LTI) systems. *IEEE Transactions on Automatic Control.* 2002;47(5):793–797.

[25] Michiels W, Niculescu SI. *Stability and Stabilization of Time-Delay Systems. An Eigenvalue Approach.* Philadelphia: SIAM; 2007.

[26] Gu K, Niculescu SI, Chen J. On stability of crossing curves for general systems with two delays. *Journal of Mathematical Analysis and Applications.* 2005;311:231–253.

[27] Sipahi R, Olgac N. Complete stability map of third order LTI multiple time delay systems. *Automatica.* 2005;41:1413–1422.

[28] Sipahi R, Niculescu SI, Abdallah CT, Michiels W, Gu K. Stability and stabilization of systems with time delay, limitations and opportunities. *IEEE Control Systems Magazine.* 2011;31(1):38–65.

[29] Ayasun S, Nwankpa CO. Probability of small-signal stability of power systems in the presence of communication delays. In: International Conference on Electrical and Electronics Engineering (ELECO). vol. 1. Bursa, Turkey; 2009. p. 70–74.

[30] Vyhlídal T, Zítek P. Mapping based algorithm for large-scale computation of quasi-polynomial zeros. *IEEE Transactions on Automatic Control.* 2009;54(1):171–177.

[31] Jarlebring E, Meerbergen K, Michiels W. A Krylov method for the delay eigenvalue problem. *SIAM Journal on Scientific Computing.* 2010;32(6): 3278–3300.

[32] Bueler E. Error bounds for approximate eigenvalues of periodic-coefficient linear delay differential equations. *SIAM Journal of Numerical Analysis.* 2007 Nov;45:2510–2536.

[33] Engelborghs K, Roose D. On stability of LMS methods and characteristic roots of delay differential equations. *SIAM Journal of Numerical Analysis.* 2002;40(2):629–650.

[34] Engelborghs K, Luzyanina T, Samaey G. DDE-BIFTOOL v. 2.00: A MATLAB Package for Bifurcation Analysis of Delay Differential Equations. Department of Computer Science, K.U.Leuven, Leuven, Belgium; 2001. Technical Report TW-330.

[35] Baker Jr GA, Graves-Morris P. *Padé Approximants – Part I: Basic Theory.* Reading, MA: Addison-Wesley; 1981.

[36] Horton M, Tomov S, Dongarra J. A Class of Hybrid LAPACK Algorithms for Multicore and GPU Architectures. In: Symposium for Application Accelerators in High Performance Computing. Knoxville, TN; 2011.

[37] Li C, Chen Y, Ding T, Du Z, Li F. A sparse and low-order implementation for discretization-based eigen-analysis of power systems with time-delays. *IEEE Transactions on Power Systems.* 2019;34(6):5091–5094.

[38] Breda D, Maset S, Vermiglio R. Pseudospectral differencing methods for characteristic roots of delay differential equations. *SIAM Journal of Scientific Computing*. 2006;27:482–495.

[39] Yi S, Nelson PW, Ulsoy AG. *Time-Delay Systems: Analysis and Control Using the Lambert W Function*. World Scientific Publishing Company; 2010.

[40] Hiskens IA. *Power System Modeling for Inverse Problems. IEEE Transactions on Circuits and Systems – I: Regular Papers*. 2004 Mar;51(3):539–551.

[41] Zhu W, Petzold LR. Asymptotic stability of Hessenberg delay differential-algebraic equations of retarded or neutral type. *Applied Numerical Mathematics*. 1998;27:309–325.

[42] Michiels W, Niculescu S. *Stability and Stabilization of Time-Delay Systems*. Philadelphia: SIAM; 2007.

[43] Milano F, Dassios I, Liu M, Tzounas G. *Eigenvalue Problems in Power Systems*. CRC Press, Taylor & Francis Group; 2020.

[44] Niculescu SI. *Delay Effects on Stability: A Robust Control Approach*. Heidelberg: Springer-Verlag; 2001.

[45] Ali Pourmousavi A, Hashem Nehrir M. Introducing dynamic demand response in the LFC model. *IEEE Transactions on Power Systems*. 2014 Jul;29(4):1562–1572.

[46] Bellen A, Zennaro M. *Numerical Methods for Delay Differential Equations*. Oxford: Oxford Science Publications; 2003.

[47] Brayton RK, Gustavson FG, Hachtel GD. A new efficient algorithm for solving differential-algebraic systems using implicit backward differentiation formulas. *Proceedings of the IEEE*. 1972 Jan;60(1):98–108.

[48] Cellier FE, Kofman E. *Continuous System Simulation*. London, UK: Springer; 2006.

[49] Breda D. Solution operator approximations for characteristic roots of delay differential equations. *Applied Numerical Mathematics*. 2006;56: 305–317.

[50] Verheyden K, Luzyanina T, Roose D. Efficient computation of characteristic roots of delay differential equations using LMS methods. *Journal of Computational and Applied Mathematics*. 2008;214:209–226.

[51] Richard J, Goubet-Bartholomeüs A, Tchangani P, Dambrine M. Nonlinear delay systems: tools for a quantitative approach to stabilization. *Stability and Control of Time-Delay Systems*. 1998;p. 218–240.

[52] Abdallah C, Dorato P, Benites-Read J, Byrne R. Delayed positive feedback can stabilize oscillatory systems. In: Proceedings of American Control Conference. IEEE; 1993. p. 3106–3107.

[53] Ramírez A, Garrido R, Mondié S. Velocity control of servo systems using an integral retarded algorithm. *ISA Transactions*. 2015;58:357–366.

[54] Milano F. Power system modelling and scripting. London: Springer; 2010.

[55] Ramírez A, Mondié S, Garrido R, Sipahi R. Design of proportional-integral-retarded (PIR) controllers for second-order LTI systems. *IEEE Transactions on Automatic Control*. 2016;61(6):1688–1693.

[56] Tzounas G, Sipahi R, Milano F. Damping power system electromechanical oscillations using time delays. *IEEE Transactions on Circuits and Systems I: Regular Papers*. 2021;68(6):2725–2735.

[57] Olgac N, Ergenc AF, Sipahi R. "Delay scheduling": a new concept for stabilization in multiple delay systems. *Journal of Vibration and Control*. 2005;11(9):1159–1172.

[58] Tzounas G, Liu M, Murad MAA, Milano F. Stability analysis of wide area damping controllers with multiple time delays. IFAC-PapersOnLine. 2018;51(28):504 – 509. 10th IFAC Symposium on Control of Power and Energy Systems.

[59] Milano F. Power System Modelling and Scripting. London: Springer; 2010.

[60] Jia H, Cao X, Yu X, Zhang P. A Simple Approach to Determine Power System Delay Margin. In: Proceedings of the IEEE PES General Meeting. Montreal, Quebec; 2007. p. 1–7.

[61] Mauricio JM, Marano A, Gómez-Expósito A, Ramos JLM. Frequency regulation contribution through variable-speed wind energy conversion systems. *IEEE Transactions on Power Systems*. 2009 Feb;24(1):173–180.

[62] Hirsch P. Extended Transient-Midterm Stability Program (ETMSP) Ver. 3.1: User's Manual. EPRI, TR-102004-V2R1; 1994.

[63] Mathieu JL, Koch S, Callaway DS. State estimation and control of electric loads to manage real-time energy imbalance. *IEEE Transactions on Power Systems*. 2013;28(1):430–440.

[64] Milano F. A Python-based Software Tool for Power System Analysis. In: Proceedings of the IEEE PES General Meeting. Vancouver, BC; 2013.

Chapter 18

Shooting-based stability analysis of power system oscillations

Federico Bizzarri[1], Angelo Brambilla[1] and Federico Milano[2]

In the first chapter of [1], which is an excellent monograph on bifurcation and stability analysis of non-linear dynamical systems, the author lists "some fundamental numerical methods" that are used through his book. These are the Newton method, the integration of differential equations, the calculation of eigenvalues, and differential equation boundary value problems (BVPs). The former three methods are largely used in power system analysis. The Newton method, in fact, is the backbone of power flow analysis, most OPF solvers, and all implicit time-integration schemes. Then the integration of differential equations and the calculation of eigenvalue problems are clearly fundamental methods for transient and small-signal stability analyses. However, BVP techniques are almost completely absent from the literature on power systems analysis.

This chapter is an attempt to fill this gap as it illustrates some applications of the time-domain shooting method (TDSM), which is a particular case of the family of BVP methods, to accurately and efficiently locate periodic stationary solutions of high-voltage transmission systems. This technique has been successfully applied to low-power electronics circuits but is unconventional for the study of the electromechanical steady-state periodic behavior of power systems.

The chapter also discusses the inherent idiosyncrasies of the conventional formulation of the power system model and provides an alternative one to accommodate the hypotheses and mathematical requirements of the time-domain shooting method. Several phenomena are taken into account, including Hopf bifurcations arising from high load conditions and poorly damped electromechanical oscillations as well as resonance phenomena driven by the discrete voltage control of under-load tap changers. The chapter shows that the shooting method is able to properly detect both stable and unstable limit cycles and to quantitatively define their stability based on Floquet multipliers. For non-smooth limit cycles, the saltation matrix is also introduced to properly compute the system fundamental matrix. The effectiveness and numerical

[1]Department of Electronics, Information Technology and Bioengineering, Polytechnic of Milan, Italy
[2]School of Electrical and Electronic Engineering, University College Dublin, Ireland

efficiencies of the proposed technique are discussed through several examples based on well-known benchmark systems.

18.1 Introduction

The mechanism and the physical causes that lead to the birth of limit cycles in power systems have been object of intense research in the last three decades [2–6]. However, existing literature mainly focuses on the detection of Hopf bifurcations, which can be tackled through the analysis of stationary points and parametric small-signal stability analysis [7–10]. Existing literature does not focus on the systematic determination of generic limit-cycles. On the other hand, in other branches of electrical engineering research, the study of oscillators and limit cycles was defined and formalized in a systematic way (see, for instance [11–14] and references therein).

Relevant techniques proposed in the last three decades to determine periodic solutions of power system models (PSMs) can be found in [15–21]. Most of the references above describe small systems, and modeling issues are solved using *ad hoc* formulations. In [16, 18], the authors define the properties of limit-cycles using a dynamic system reduction based on the centre manifold. This approximation captures the behaviour of the system in a *neighborhood* of the equilibria but does not define the precise trajectory of the limit cycle. A more general approach to identify the trajectories of unstable limit cycles is given in [19, 21]. In these references, the authors compute trajectories iteratively departing from points *inside* and *outside* the stable manifold of the limit-cycle. The trajectory of the unstable limit-cycle is then determined through the interpolation among those trajectories. The major drawback of this technique is the poor computational efficiency (since it is basically a trial-and-error technique) and, more sensibly, the difficulty to define *a priori* which points belong or not the stable manifold of the limit cycle.

In this chapter, the formalism of two well-assessed techniques, namely the time-domain shooting method (TDSM) [22, 23] and probe-insertion technique (PIT) [24, 25], is applied to the systematic determination of limit cycles in power systems. Limit cycles (stable and unstable) can be originated by both super- and sub-critical Hopf bifurcations or by different mechanisms such as fold or flip bifurcations [26]. Furthermore, limit cycles corresponding to non-self-sustained oscillations can be detected too. They can be found in periodically driven systems; as an example one may think of power system oscillation caused by wind power fluctuation [6].

The TDSM and PIT require that the power system differential algebraic equations (DAEs) are written in a form suitable for the identification of limit-cycles by resorting to the sensitivity of the evolution of the system state variables with respect to their initial conditions. In particular, the system fundamental matrix and the monodromy matrix are two key ingredients [27]. With this aim, as a first step, the chapter discusses why the conventional formulation of power system synchronous machine models and the use of a synchronous reference for the machine rotor speeds and angles has to be revised to allow applying the TDSM and PIT. In fact, the conventional formulation presents a “non-physical” singularity in the Jacobian matrix of the Newton iteration scheme used to numerically solve the BVP which is the core of the TDSM and hence of the PIT.

The TDSM to locate limit cycles consists in the solution of a BVP with periodic constraints through an iterative scheme suitable for the solution of systems of algebraic non-linear equations, e.g., the Newton method. If the limit cycle is the periodic steady state solution of an autonomous system, the Jacobian matrix of the non-linear algebraic system to be iteratively solved exhibits a null eigenvalue and hence it is singular as soon as it converges to the solution. This jeopardize the convergence of the iterative scheme itself. In this case, the BVP must be augmented with a proper phase condition [26] to remove such a singularity. On the contrary, if the system is non-autonomous, the Jacobian has full rank. If the classical formulation of the PSM is considered, i.e., generator speeds are relative to a synchronous reference, the aforementioned Jacobian matrix is characterized by one additional null eigenvalue; rank-2 and rank-1 deficiencies are thus observed in the autonomous and non-autonomous case, respectively.

Examples can be found in the literature where the aforementioned limitations are not encountered. For instance, in [15, 20], the authors determine the steady-state solution of a very simple PSM composed of a single machine connected to an infinite bus. The latter provides a phase reference frame that avoids the singularity of the Jacobian matrix. However, the infinite bus model is not appropriate for real-world PSMs.

The chapter shows how power system DAEs can be properly reformulated without modifying their inherent meaning in such a way that the TDSM and the PIT can be successfully applied to a system of arbitrary size to locate limit cycles in the system state space. In particular, the proposed enhanced model yields a suitable full-rank bordered Jacobian matrix and hence the iterative Newton scheme can safely converge [28]. The proposed approach relies on a *generalization* of the well-known concept of center of inertia (COI) [29] and, thus, it can be easily implemented in existing power system simulators. With this aim, the TDSM is reviewed and it is highlighted why the standard formulation of PSM does not allow locating periodic steady-state solutions through the TDSM.

The main contribution of the chapter is the description of a systematic approach, namely the PIT, to determine both stable and unstable limit cycles exhibited by the PSM. The determination of the latter is important as they provide insights on the boundary of the region of attraction of the stable equilibrium point [19, 21]. With respect to other techniques proposed in the literature, the PIT provides the exact trajectory of an unstable limit cycle that, as shown in Section 18.4.2, is not necessarily a “simple” curve.

The chapter is organized as follows. Section 18.2 introduces in details the TDSM, which is the basis of the PIT discussed in Section 18.2.4. Section 18.3 is devoted to the power systems transient stability model; some inherent drawbacks of the conventional model are presented and overcome to accommodate the assumptions and mathematical requirements of the TDSM. Section 18.4 illustrates the proposed model and technique through three well-known test systems, namely the IEEE 14-bus; the WSCC 9-bus; and the 2-area networks. Furthermore, the TDSM is adopted to describe the onset of periodic oscillations due to the interaction among three different actors, namely two under-load tap changers (ULTCs) and a switching capacitor bank, in a radial network.

18.2 Mathematical background

18.2.1 The time-domain shooting method

The TDSM allows solving a BVP as a small number of simpler initial value problems (IVPs) [22, 23]. In general, an *n*th order ordinary differential equation (ODE) allows setting n independent boundary conditions – these can be initial conditions (as in IVPs), final conditions, or a mix of the two. To clarify things out, let us consider the following ODE

$$\dot{\boldsymbol{x}} = \boldsymbol{f}(\boldsymbol{x}(t), t) \tag{18.1}$$

where $\boldsymbol{x} \in \mathbb{R}^n$ and the time-varying Jacobian matrix of $\boldsymbol{f}$ is denoted by $\boldsymbol{f_x}(t)$.

To grasp the idea inherent in the TDSM, it may be useful to recall a classic (simplified) ballistic problem where a gunman has a fixed position of the cannon and of the target (i.e., the boundary conditions) but has freedom in the tilt of the cannon and does not care about the angle of arrival of the cannon ball. If a first shot misses the target, the gunner evaluates how much closer or farther the cannon ball gets from his objective and finally adjusts the tilt in order to (hopefully) hit the target with the next shot. The key of the gunman's method is the perturbation of the initial guess of the tilt and the evaluation of the *sensitivity* of the solution (i.e., the arrival position of the cannon ball) to this perturbation. The location of limit cycles can be visualized as a variant of the ballistic problem where the gunman is using a boomerang instead of a cannon. In this case, the boundary conditions represent a periodicity constraint.

For the (18.1) ODE, a boundary condition $\boldsymbol{x}(T_\gamma + t_0) = \boldsymbol{x}(t_0)$ for some time value T_γ defines, if it exists, a T_γ-periodic solution, say $\boldsymbol{\gamma}$, such that $\boldsymbol{\gamma}(t) = \boldsymbol{\gamma}(t + T_\gamma)$. If system (18.1) is autonomous, i.e., $\boldsymbol{f}$ does not explicitly depend on t, period T_γ must also be determined. This is normally done using a proper *phase condition* used to augment the TDSM equations as explained in Section 18.2.3 (for more specific details see [27, 26]). On the other side, if the system is non-autonomous and driven by a periodic term, the T_γ period of the steady state periodic solution (generically) coincides with the forcing signal one.

Any standard numerical integration method can be used for finding the solution of the IVP from time t_0, with some initial conditions $\boldsymbol{x}(t_0)$, to time $t_0 + T_\gamma$. Moreover, computing also the evolution of the $\boldsymbol{\Phi}(t, t_0)$ state transition matrix associated with (18.1), i.e., the solution of the *variational equation* [27]

$$\begin{cases} \dot{\boldsymbol{\Phi}}(t, t_0) = \boldsymbol{f_x}(t)\boldsymbol{\Phi}(t, t_0) \\ \boldsymbol{\Phi}(t_0, t_0) = \mathbf{1}_n \end{cases} \tag{18.2}$$

where $\mathbf{1}_n$ is the $n \times n$ identity matrix, provides the sensitivity of the solution of the IVP problem to its initial conditions. The state transition matrix (or fundamental solution matrix) $\mathbf{\Phi}(t, t_0)$ is defined as the unique matrix that satisfies the relation

$$\boldsymbol{x}'(t) - \boldsymbol{x}(t) = \mathbf{\Phi}(t, t_0)(\boldsymbol{x}'(t_0) - \boldsymbol{x}(t_0)) + O(\|\boldsymbol{x}'(t_0) - \boldsymbol{x}(t_0)\|^2) \tag{18.3}$$

for any possible infinitesimal perturbation $\boldsymbol{x}'(t_0)$ of $\boldsymbol{x}(t_0)$. At first order, it can be thus used to compute the time evolution of $\boldsymbol{x}'(t)$ with respect to the reference trajectory $\boldsymbol{x}(t)$. One of the important properties of the state transition matrix is the *composition* property

$$\mathbf{\Phi}(t_2, t_0) = \mathbf{\Phi}(t_2, t_1)\mathbf{\Phi}(t_1, t_0) \tag{18.4}$$

where $t_0 < t_1 < t_2$. By introducing the *state transition function* $\boldsymbol{\varphi}(\boldsymbol{x}(t_0), t_0, t) \equiv \boldsymbol{x}(t)$ to make evident as the final value reached by the solution of the IVP depends on $\boldsymbol{x}(t_0)$ and t_0, the periodicity condition $\boldsymbol{x}(T_\gamma + t_0) = \boldsymbol{x}(t_0)$ can be reformulated in terms of a non-linear algebraic function $\boldsymbol{R}$ to be zeroed as

$$\boldsymbol{R}(\boldsymbol{x}(t_0)) = \boldsymbol{\varphi}(\boldsymbol{x}(t_0), t_0, t_0 + T_\gamma) - \boldsymbol{x}(t_0) = \mathbf{0} \tag{18.5}$$

and numerically solved, for example, by using the Newton method

$$\boldsymbol{R}_{\boldsymbol{x}(t_0)}(\boldsymbol{x}^i(t_0))\left[\boldsymbol{x}^{i+1}(t_0) - \boldsymbol{x}^i(t_0)\right] = -\boldsymbol{R}(\boldsymbol{x}^i(t_0))$$

where i is the iteration index and the $\boldsymbol{R}_{\boldsymbol{x}(t_0)}(\boldsymbol{x}^i(t_0))$ Jacobian matrix of $\boldsymbol{R}(\boldsymbol{x}(t_0))$ is given by

$$\left.\frac{\partial \boldsymbol{\varphi}(\boldsymbol{x}_0, t_0, t)}{\partial \boldsymbol{x}_0}\right|_{\begin{cases} t = t_0 + T_\gamma \\ x_0 = x_0^i \end{cases}} - \mathbf{1}_n \equiv \mathbf{\Phi}^i(t_0 + T_\gamma, t_0) - \mathbf{1}_n \tag{18.6}$$

Using the information embedded in $\mathbf{\Phi}(t_0 + T_\gamma, t_0)$, the initial conditions of the IVP are corrected at next step $i + 1$ and a new IVP is solved until the condition $\boldsymbol{x}(T + t_0) = \boldsymbol{x}(t_0)$ is (approximately) met. When this happens, the solution of the IVP with the *right* initial conditions on the interval $[t_0, t_0 + T_\gamma]$ is $\boldsymbol{\gamma}$. Once the Newton procedure converges toward a periodic solution $\boldsymbol{\gamma}$, the $\mathbf{\Phi}(t_0 + T, t_0)$ matrix converges towards a matrix $\mathbf{\Psi}$ (called *monodromy matrix* or *principal matrix*) whose (complex) eigenvalues are the μ_k $(k = 1, ..., n)$ *characteristic multipliers*. If the condition $|\mu_k| \leq 1\ \forall k$ holds, then $\boldsymbol{\gamma}$, which is a limit cycle, is stable. If at least one eigenvalue has modulus greater than 1 then the limit cycle is unstable. To study how the multipliers vary as a function of the system parameters (entering and exiting the unit circle in the complex plane) allows to identify possible bifurcations of the limit cycle [26].

The numerical procedure described above is a direct mathematical transposition of the gunman problem: the Jacobian matrix of the Newton scheme represents, through the system state transition matrix, a measure of the sensitivity of the final conditions with respect to initial ones, and the sequence of IVPs, solved for proper different values of the initial conditions, is equivalent to a sequence of tilt adjustments. As the number of iterations grows, each IVP is expected to converge to a final value of the trajectory closer and closer to the adjusted initial one. This suggests that the TDSM is not suited to determine unstable limit cycles if the initial conditions of the IVP are not sufficiently close to the repeller. The PIT introduced in the following subsection overcomes this limitation.

It is straightforward to extend the conclusions provided in this section from the ODE case to the more general (semi-explicit index-1) DAE [30], which is typically adopted to describe the PSM. As a matter of fact, if (18.1) is generalized and re-written as

$$\begin{aligned} \dot{\boldsymbol{x}} &= \boldsymbol{f}(\boldsymbol{x}(t), \boldsymbol{y}(t), t) \\ \boldsymbol{0} &= \boldsymbol{g}(\boldsymbol{x}(t), \boldsymbol{y}(t), t) \end{aligned}$$

where $\boldsymbol{y} \in \mathbb{R}^m$ and $\boldsymbol{g}$ represents a set of algebraic constraints, the application of TDSM requires that it is possible to derive $\boldsymbol{y}(t) = \iota(\boldsymbol{x}(t), t)$, for all the involved IVPs, and use them in the evaluation of $\boldsymbol{f}$. This means that $\boldsymbol{g}(\boldsymbol{x}(t), \boldsymbol{y}(t), t)$ must be differentiable and $\boldsymbol{g}_y$ invertible.

It is not necessary to adjust the choice of the initial conditions of the DAE that guarantee to locate its periodic steady-state solution since $\boldsymbol{y}(t_0) = \iota(\boldsymbol{x}(t_0), t_0)$. As far as the state transition matrix $\boldsymbol{\Phi}(t, t_0)$ is concerned, provided that $\boldsymbol{f}(\boldsymbol{x}(t), \boldsymbol{y}(t), t)$ is differentiable too, the variational problem (18.2) becomes

$$\begin{cases} \dot{\boldsymbol{\Phi}}(t, t_0) = \left[\boldsymbol{f}_x(t) - \boldsymbol{f}_y(t)\boldsymbol{g}_y^{-1}(t)\boldsymbol{g}_x(t)\right] \boldsymbol{\Phi}(t, t_0) \\ \boldsymbol{\Phi}(t_0, t_0) = \boldsymbol{1}_n \end{cases}$$

In Sections 18.2.2 and 18.2.3, only the ODE case is discussed with no lack of generality.

18.2.2 The state transition matrix for hybrid dynamical systems

The state transition matrix $\boldsymbol{\Phi}(t, t_0)$ is defined for system (18.1) if the $\boldsymbol{f}$ vector field is *smooth*, i.e., the $\boldsymbol{f}_x$ Jacobian matrix is defined at any point of the trajectory $\varphi(\boldsymbol{x}(t_0), t_0, t)$. In other words, $\boldsymbol{f}$ must be a Lipschitz continuous function.

This is not true for hybrid dynamical systems that can be viewed as consisting of piece-wise defined continuous time evolution processes interfaced with some logical or decision making process [31]. A basic example of such systems is

$$\begin{cases} \dot{\boldsymbol{x}}(t) = \boldsymbol{f}(\boldsymbol{x}(t)) = \begin{cases} \boldsymbol{f}_1(\boldsymbol{x}(t)), & \boldsymbol{x}(t) \in \mathcal{R}_1(t) \\ \boldsymbol{f}_2(\boldsymbol{x}(t)), & \boldsymbol{x}(t) \in \mathcal{R}_2(t) \end{cases} \\ \boldsymbol{x}(t_0) = \boldsymbol{x}_0 \end{cases}$$

where $\boldsymbol{x} \in \mathbb{R}^n$, $\boldsymbol{f}_1$ and $\boldsymbol{f}_2$ are assumed to be smooth in $\mathbb{R}^n$, the subspaces $\mathcal{R}_1$ and $\mathcal{R}_2$ are not fixed but can vary at different time instants since they are separated by the $\mathcal{H}(t)$ time-varying surface so that, for every $t \geq t_0$, $\mathbb{R}^n = \mathcal{R}_1(t) \cup \mathcal{H}(t) \cup \mathcal{R}_2(t)$. An event manifold $h(\boldsymbol{x}, t)$ is adopted to verify whether or not a generic point $\boldsymbol{x}(t)$ belongs to either $\mathcal{H}$, or R_1 or R_2. More specifically

$$\begin{aligned} \mathcal{R}_1(t) &= \{\boldsymbol{x} \in \mathbb{R}^n | h(\boldsymbol{x}, t) < 0\} \\ \mathcal{H}(t) &= \{\boldsymbol{x} \in \mathbb{R}^n | h(\boldsymbol{x}, t) = 0\} \\ \mathcal{R}_2(t) &= \{\boldsymbol{x} \in \mathbb{R}^n | h(\boldsymbol{x}, t) > 0\} \end{aligned} \tag{18.7}$$

The manifold $h(\boldsymbol{x}, t)$ is assumed to be differentiable with respect to time, i.e., $\frac{\partial}{\partial t}h(\boldsymbol{x}, t)$ exists, and admit $\boldsymbol{\nu} = \nabla_x h(\boldsymbol{x}, t)$ as normal vector for all $\boldsymbol{x}(t) \in \mathcal{H}(t)$, moreover, for all $t \geq t_0$,

$$\left[\boldsymbol{\nu}^{\mathrm{T}}(\boldsymbol{x}, t)\boldsymbol{f}_1(\boldsymbol{x}(t))\right]\left[\boldsymbol{\nu}^{\mathrm{T}}(\boldsymbol{x}, t)\boldsymbol{f}_2(\boldsymbol{x}(t))\right] > 0,$$

which means that every trajectory reaches the surface $\mathcal{H}$ transversally so that sliding motion [32] is not allowed. The system described by (18.7) is a switching dynamical system since the vector field governing its dynamics switches between $\boldsymbol{f}_1$ and $\boldsymbol{f}_2$.

The system becomes an impact switching system if a mapping function $\boldsymbol{\Gamma}(\boldsymbol{x}) : \mathbb{R}^n \to \mathbb{R}^n$ is considered such that, whenever a trajectory crosses the surface $\mathcal{H}$, for instance from $\mathcal{R}_1$ to $\mathcal{R}_2$ at $\boldsymbol{x}(t_1) = \boldsymbol{x}^-(t_1)$, the variable $\boldsymbol{x}$ is *instantaneously* mapped into $\boldsymbol{x}^+(t_1) = \boldsymbol{\Gamma}(\boldsymbol{x}^-(t_1))$ in $\mathcal{R}_2$.

The dynamics of the system undergoes an *impact* at t_1 (i.e., $\boldsymbol{x}^+(t_1) \neq \boldsymbol{x}^-(t_1)$) and a *switch* (since $\boldsymbol{x}^+(t_1) \in \mathcal{R}_2$ implies $\boldsymbol{f} = \boldsymbol{f}_2$ for $t > t_1$). The Jacobian matrix of $\boldsymbol{\Gamma}(\boldsymbol{x})$ will be referred to as $\boldsymbol{\Gamma}_x(\boldsymbol{x})$.

With reference to Figure 18.1, assuming that $\boldsymbol{x}_0 \in \mathcal{R}_1$, the composition property of the state transition matrix allows to derive

$$\boldsymbol{\Phi}(t_2, t_0) = \boldsymbol{\Phi}(t_2, t_1)\boldsymbol{S}\boldsymbol{\Phi}(t_1, t_0)$$

being $\boldsymbol{S}$ the saltation matrix operator, i.e., a correction factor that must be taken into account to match the solutions of system (18.2) at the crossing time, to properly consider the effect of both the switch and the impact occurring for $t = t_1$. The analytic expression of such a matrix is

$$\begin{aligned}\boldsymbol{S} = \ & \boldsymbol{\Gamma}_x(\boldsymbol{x}^-(t_1)) \\ & + \left[\boldsymbol{f}_2(\boldsymbol{\Gamma}(\boldsymbol{x}^-(t_1)), t_1) - \boldsymbol{\Gamma}_x(\boldsymbol{x}^-(t_1))\boldsymbol{f}_1(\boldsymbol{x}^-(t_1), t_1)\right]\Xi(\boldsymbol{x}^-(t_1), t_1)\end{aligned} \tag{18.8}$$

where

$$\Xi(\boldsymbol{x}^-(t_1), t_1) = \frac{\boldsymbol{\nu}^{\mathrm{T}}(\boldsymbol{x}^-(t_1), t_1)}{\boldsymbol{\nu}^{\mathrm{T}}(\boldsymbol{x}^-(t_1), t_1)\boldsymbol{f}_1(\boldsymbol{x}^-(t_1), t_1) + \frac{\partial h}{\partial t}\Big|_{(\boldsymbol{x}^-(t_1), t_1)}} \tag{18.9}$$

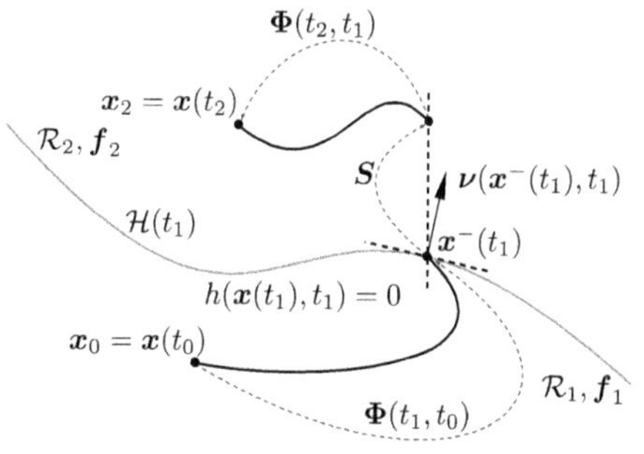

Figure 18.1 An example of impact/switch phenomenon in $\mathbb{R}^2 = \mathcal{R}_1 \cup \mathcal{H} \cup \mathcal{R}_2$

If $\boldsymbol{\Gamma}(\boldsymbol{x}) = \boldsymbol{x}$ (i.e., no impact occurs), (18.8) reduces to

$$\boldsymbol{S} = \mathbf{1}_n + \left[\boldsymbol{f}_2(\boldsymbol{x}(t_1), t_1) - \boldsymbol{f}_1(\boldsymbol{x}^-(t_1), t_1)\right] \boldsymbol{\Xi}(\boldsymbol{x}^-(t_1), t_1)$$

It is worth noticing that, in both cases, if the manifold h is not time-varying,

$$\boldsymbol{\Xi}(\boldsymbol{x}^-(t_1), t_1) = \frac{\boldsymbol{\nu}^{\mathrm{T}}(\boldsymbol{x}^-(t_1), t_1)}{\boldsymbol{\nu}^{\mathrm{T}}(\boldsymbol{x}^-(t_1), t_1)\boldsymbol{f}_1(\boldsymbol{x}^-(t_1), t_1)} \tag{18.10}$$

All the details concerning the analytic derivation of (18.8) can be found in [32].

The saltation matrix operator, viz. the first-order approximation of the so-called discontinuity mapping operator [33, 32], can be easily used to derive the state transition matrix of hybrid systems when the TDSM is applied to locate a periodic steady-state solution. During the numerical solution of the IVP involved in each step of the TDSM, the variational problem (18.2) is solved simultaneously by inserting the proper correction factor whenever the computed trajectory hits one of the manifolds that delimit the state space.

A special class of events that can be found when dealing with hybrid dynamical systems is made up of *delayed events*. They occur whenever the trajectory in the state space hits a given manifold but the effect of the intersection (switch, impact or both) is observed after a fixed amount of time. For these events, the expression of the saltation matrix is not as straightforward as in the aforementioned examples. A correction of the state transition matrix is given in the following and it is valid in case of a delayed switching event triggered by a manifold that is not time-varying. This is a typical situation that occurs when dealing with the PSM if ULTCs transformers are considered. The extension to more general cases is possible but is beyond the scope of this chapter. The interested reader can find more details on advanced applications of the saltation matrix in [34].

Let us consider the delayed switching event depicted in Figure 18.2. The trajectory originating from $\boldsymbol{x}_0$ at $t = t_0$ hits the manifold $\mathcal{H}$ at the point $\boldsymbol{x}_1$ at $t = t_1$. Nevertheless the vector field $\boldsymbol{f}_1$ continues to govern the dynamics of the system until a fixed delay, say Δt, is passed. At $t = t_1 + \Delta t$, it can be assumed that the trajectory hits at $\boldsymbol{x}_2$ the $\mathcal{H}'$

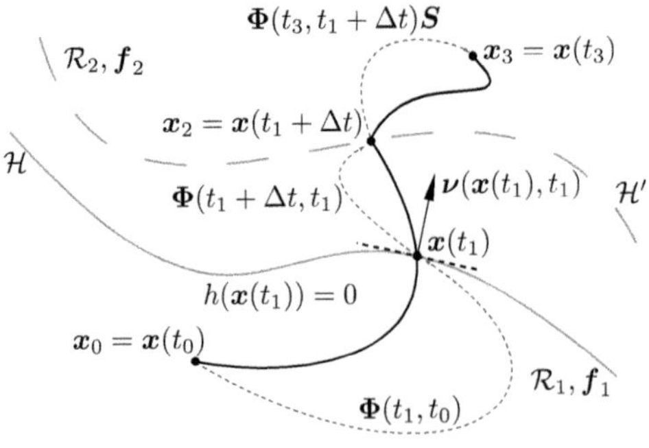

Figure 18.2 An example of delayed switching event in $\mathbb{R}^2 = \mathcal{R}_1 \cup \mathcal{H}' \cup \mathcal{R}_2$

virtual manifold identified by locus of points such that $\boldsymbol{x} - \boldsymbol{\varphi}(\boldsymbol{x}(t_0), t_0, t_1 + \Delta t) = 0$. We are interested in computing

$$\boldsymbol{\Phi}(t_3, t_0) = \boldsymbol{\Phi}(t_3, t_1 + \Delta t)\boldsymbol{S}\boldsymbol{\Phi}(t_1, t_0)$$

by resorting to a proper expression of the $\boldsymbol{S}$ saltation matrix that matches the solutions of system (18.2) between t_1 and $t_1 + \Delta t$. It can be written as [34]

$$\boldsymbol{S} = \boldsymbol{\Phi}(t_1 + \Delta t, t_1) + [\boldsymbol{f}_2(\boldsymbol{x}_2, t_1 + \Delta t) - \boldsymbol{f}_1(\boldsymbol{x}_2, t_1 + \Delta t)]\, \Xi(\boldsymbol{x}_1, t_1). \tag{18.11}$$

It deserves some attention the fact that Ξ must be computed when the trajectory hits $\mathcal{H}$, whereas the variation in the vector field is evaluated at $\boldsymbol{x}_2$ on the virtual manifold $\mathcal{H}'$. It is worth noticing that (18.11) reduces to (18.10) as soon as Δt goes to 0.

18.2.3 Bordering the Jacobian

An extremely important remark must be done if (18.1) represents an *autonomous* system, i.e., it does not explicitly depend on time. In this case, a multiplier of $\boldsymbol{\gamma}$, say μ_1, is equal to 1 [26]. Hence, the Jacobian (18.6) of the Newton scheme introduced in the previous section tends to be singular as the iterative procedure goes toward $\boldsymbol{\gamma}$ and one cannot safely compute $\boldsymbol{x}_0^{i+1}$ by inverting $\boldsymbol{R}_{\boldsymbol{x}(t_0)}(\boldsymbol{x}^i(t_0))$.*

A well-known technique to obtain a full rank matrix from a singular one is through addition of as many extra independent rows and columns as the rank deficiency [28]. This concept is briefly outlined below. Let us assume that $\boldsymbol{B}$ is a rank-1 deficient matrix, as it happens for $\boldsymbol{R}_{\boldsymbol{x}(t_0)}(\boldsymbol{x}^i(t_0))$ when the Newton iteration scheme converges towards $\boldsymbol{\gamma}$. Let $\boldsymbol{u}_1$ be the right eigenvector of $\boldsymbol{B}$ corresponding to the $\lambda_1 = 0$ null eigenvalue, i.e., $\boldsymbol{B}\boldsymbol{u}_1 = \boldsymbol{0}$ and $\boldsymbol{v}_1$ the left eigenvector, i.e., $\boldsymbol{B}^{\mathrm{T}}\boldsymbol{v}_1 = 0$. $\boldsymbol{u}_1$ spans the null space of $\boldsymbol{B}$ and is orthogonal to the image of $\boldsymbol{B}^{\mathrm{T}}$.

If $\boldsymbol{B}$ is bordered with an extra row and column as follows:

$$\boldsymbol{A} = \begin{bmatrix} \boldsymbol{B} & \boldsymbol{u}_1 \\ \boldsymbol{u}_1^{\mathrm{T}} & \alpha \end{bmatrix} \tag{18.12}$$

where $\alpha \in \mathbb{R}$, one obtains that $\det(\boldsymbol{A}) \neq 0$. In fact, consider $\boldsymbol{A}^{\mathrm{T}}$ which has the same eigenvalues of $\boldsymbol{A}$; if $\boldsymbol{A}^{\mathrm{T}}$ were singular it would admit the $\boldsymbol{\xi}$ null space that would satisfy

$$\begin{bmatrix} \boldsymbol{B}^{\mathrm{T}} & \boldsymbol{u}_1 \\ \boldsymbol{u}_1^{\mathrm{T}} & \alpha \end{bmatrix} \overbrace{\begin{bmatrix} \boldsymbol{\xi}_1 \\ \xi_2 \end{bmatrix}}^{\boldsymbol{\xi}} = \boldsymbol{0} \tag{18.13}$$

Consider the condition $\boldsymbol{B}^{\mathrm{T}}\boldsymbol{\xi}_1 + \boldsymbol{u}_2 = \boldsymbol{0}$. Since $\boldsymbol{u}_1$ is orthogonal to the image of $\boldsymbol{B}^{\mathrm{T}}$, it can be satisfied only when $\boldsymbol{\xi}_1 = \boldsymbol{v}_1$ and $\xi_2 = 0$ but this implies $\boldsymbol{u}_1^{\mathrm{T}}\boldsymbol{\xi}_1 = \boldsymbol{u}_1^{\mathrm{T}}\boldsymbol{v}_1 \neq 0$ and

*Actually, only when $\boldsymbol{R}_{\boldsymbol{x}(t_0)}(\boldsymbol{x}^{\bar{i}}(t_0)) = \Psi - \boldsymbol{1}_n$ the Jacobian matrix of the Newton scheme is singular. This happens for the (ideal) final iteration step $\bar{i}$. Nevertheless, from a numerical standpoint, the condition number of $\boldsymbol{R}_{\boldsymbol{x}(t_0)}(\boldsymbol{x}^i(t_0))$ increases as i tends toward $\bar{i}$. Thus inverting $\boldsymbol{R}_{\boldsymbol{x}(t_0)}(\boldsymbol{x}^i(t_0))$ becomes an issue as i increases.

hence (18.13) is not satisfied.† This concludes the simple proof that bordering the $\boldsymbol{B}$ matrix with its null space $\boldsymbol{u}_1$ removes its singularity.

From a practical point of view, to solve the problem (18.5) if the system is autonomous, one must include a new equation (involving of course a new unknown). The new equation must not modify the dynamics of the system. This is typically achieved by introducing a *phase condition* $q(\boldsymbol{x}(t_0), T_\gamma) = 0$ that removes the singularity and introduces the unknown T_γ period among the unknowns [26]. The non-linear algebraic function (18.5) becomes $\boldsymbol{R}(\boldsymbol{x}(t_0), T_\gamma)$.

In this way, a Newton iterative scheme is obtained whose Jacobian matrix becomes:

$$\begin{bmatrix} \boldsymbol{R}_{\boldsymbol{x}(t_0)} & \dfrac{\partial \boldsymbol{R}}{\partial T_\gamma} \\ \nabla^{\mathrm{T}}_{\boldsymbol{x}(t_0)} q & \dfrac{\partial q}{\partial T_\gamma} \end{bmatrix}$$

where $\nabla^{\mathrm{T}}_{\boldsymbol{x}(t_0)} q$ is the gradient of the scalar function $q(\boldsymbol{x}(t_0), T_\gamma)$ computed with respect to $\boldsymbol{x}(t_0)$. The resulting bordering is quite different from (18.12) and, to remove the singularity, the following conditions have to be satisfied:

C1. $\dfrac{\partial \boldsymbol{R}}{\partial T_\gamma}$ belongs to a subspace spanned *also* by the null space of $\boldsymbol{R}_{\boldsymbol{x}(t_0)}$

C2. $\nabla^{\mathrm{T}}_{\boldsymbol{x}(t_0)} q \cdot \boldsymbol{v}_1 \neq 0$

where $\boldsymbol{v}_1$ is the left nullspace (eigenvector) of $\boldsymbol{R}_{\boldsymbol{x}(t_0)}$.‡

Since it can be shown that the right eigenvector of $\boldsymbol{\Psi}$ associated to μ_1 is $\boldsymbol{f}(\boldsymbol{x}(t_0)) \equiv \boldsymbol{u}_1$ [27], a typical choice for q is the constraint imposing that the correction vector $\boldsymbol{x}^{i+1}(t_0) - \boldsymbol{x}^{i}(t_0)$ is orthogonal to $\boldsymbol{f}(\boldsymbol{x}^{i}(t_0))$ [35, 36]. In this way, if the obtained pseudo-Newton scheme converges to the *right* set of initial conditions and period of the limit cycle, it can be shown that

$$\frac{\partial \boldsymbol{R}}{\partial T_\gamma} = \nabla_{\boldsymbol{x}(t_0)} q = \boldsymbol{u}_1 \tag{18.14}$$

where $\boldsymbol{u}_1$ is the right nullspace (eigenvector) of $\boldsymbol{R}_{\boldsymbol{x}(t_0)}$.

†The $\boldsymbol{B}$ matrix can be decomposed as $\boldsymbol{B} = \boldsymbol{UDV}$, where $\boldsymbol{D}$ is the diagonal matrix of its eigenvalues, $\boldsymbol{U}$ is the matrix whose columns are the right eigenvectors and $\boldsymbol{V} = \boldsymbol{U}^{-1}$ is the matrix of the left eigenvectors. As a consequence, being δ_{jk} the Kronecker delta, $\boldsymbol{v}_j^{\mathrm{T}} \boldsymbol{u}_k \mathrm{T} = \delta_{jk}$ $(j, k = 1, \ldots, n)$.

‡The C1 and C2 conditions are justified as follows. We consider once more (18.12) and now we write

$$\widehat{\boldsymbol{A}} = \begin{bmatrix} \boldsymbol{B} & (\boldsymbol{u}_1 + \boldsymbol{\kappa}_1) \\ (\boldsymbol{u}_1 + \boldsymbol{\kappa}_2)^{\mathrm{T}} & \alpha \end{bmatrix}$$

where $\frac{\partial \boldsymbol{R}}{\partial T_\gamma} = \boldsymbol{u}_1 + \boldsymbol{\kappa}_1$ and $\nabla^{\mathrm{T}}_{\boldsymbol{x}(t_0)} q = \boldsymbol{u}_1^{\mathrm{T}} + \boldsymbol{\kappa}_2$, with $\boldsymbol{u}_1^{\mathrm{T}} \boldsymbol{\kappa}_1 = 0$ and $\boldsymbol{\kappa}_2^{\mathrm{T}} \boldsymbol{u}_1 = 0$. If $\widehat{\boldsymbol{A}}$ were singular we would have $\boldsymbol{B}^{\mathrm{T}} \boldsymbol{\xi}_1 + (\boldsymbol{u}_1 + \boldsymbol{\kappa}_1)\, \xi_2 = \boldsymbol{0}$ but as before, since $\boldsymbol{u}_1$ is orthogonal to the image of $\boldsymbol{B}^{\mathrm{T}}$, the only way to satisfy this condition is to choose $\boldsymbol{\xi}_1 = \boldsymbol{v}_1$ and $\xi_2 = 0$. This makes irrelevant the addition of the $\boldsymbol{\kappa}_1$ vector. Now we consider the $(\boldsymbol{u}_1 + \boldsymbol{\kappa}_2)^{\mathrm{T}} \boldsymbol{v}_1$ product; the C2 condition ensures that it is not null and forces $\boldsymbol{\xi}_1 = \boldsymbol{0}$, as before.

Concerning $\frac{\partial \boldsymbol{R}}{\partial T_\gamma}$, it can be easily shown that at the *i*th step of the Newton iterative scheme it coincides with $\boldsymbol{f}(\boldsymbol{x}(T_\gamma^i + t_0))$. As soon as the iterative scheme converges, $\boldsymbol{f}(\boldsymbol{x}(T_\gamma^i + t_0))$ tends to $\boldsymbol{f}(\boldsymbol{x}(T_\gamma + t_0)) = \boldsymbol{f}(\boldsymbol{x}(t_0)) \equiv \boldsymbol{u}_1$ [27].

If the periodic solution $\boldsymbol{\gamma}$ is characterized by more than one unitary multiplier, i.e., the Jacobian of the Newton scheme has more than one null eigenvalue, a single phase condition is no longer sufficient. In this case, further bordering equations are required, as discussed in the following sections.

As a final remark to this section, we observe that a relevant and largely used application of the bordering technique discussed in this section is the *continuation equation* used in homotopy methods such as the well-known continuation power flow analysis. The interested reader can find more details and further references on this topic in [29].

18.2.4 The probe-insertion technique

The probe-insertion technique (PIT), originally used in harmonic balance, i.e., in the frequency domain [37–39], is a valid tool to extend the application of the TDSM to power systems. A time domain version of the PIT was recently proposed in [24, 25]. The remainder of this section shows how to use the PIT to reliably locate limit cycles in the PSM.

To illustrate the basic idea underpinning the PIT, it is assumed that the generic dynamic system described by (18.1) is autonomous and admits a stable or unstable $\boldsymbol{\gamma}$ periodic solution whose period is T_γ. Then the coefficients of the real Fourier expansion of a component of $\boldsymbol{\gamma}$ can be determined as

$$
\begin{aligned}
v_R &= \frac{2}{T_\gamma}\int_{t_0}^{t_0+T_\gamma} \cos(\omega_\gamma \tau)\boldsymbol{r}^{\mathrm{T}}\boldsymbol{\gamma}(\tau)d\tau \\
v_I &= \frac{2}{T_\gamma}\int_{t_0}^{t_0+T_\gamma} \sin(\omega_\gamma \tau)\boldsymbol{r}^{\mathrm{T}}\boldsymbol{\gamma}(\tau)d\tau
\end{aligned}
$$

where $\omega_\gamma = {}^{2\pi}/_{T_\gamma}$ and $\boldsymbol{r}$ is a selector, i.e., a column vector whose entries are all null but one, equal to 1, which corresponds to the considered component of γ. Let us rewrite (18.1) as

$$
\dot{\boldsymbol{x}} = \boldsymbol{f}(\boldsymbol{x}(t)) + \boldsymbol{r}[y_R \cos(\omega_p t) + y_I \cos(\omega_p t)] \tag{18.15}
$$

and assume that, before perturbing the system, $y_R = y_I = 0$ and $\omega_p = {}^{2\pi}/_{T_p}$ holds a generic positive value. Let us apply the following modeling choices:

- a coefficient of a component of $\boldsymbol{\gamma}$, e.g., v_R, is *slightly* perturbed;
- y_I and T_γ are kept constant;
- v_I is free to vary.

Other choices of perturbed, constant, and free variables are possible, but these do not alter the results of the PIT.§ An adjusted y_R can be identified such that (18.15) admits a periodic solution that can be viewed as a small perturbation of γ. In practice, the periodic trajectory can be obtained by solving a BVP defined by (18.15) augmented by

$$\dot{\kappa}_R = \frac{2}{T_p}\cos(\omega_\gamma t)\boldsymbol{r}^{\mathrm{T}}\boldsymbol{x}(t) \tag{18.16}$$

with $y_I = 0$, $\omega_p = \omega_\gamma$ and with the boundary conditions $\boldsymbol{x}(T_\gamma + t_0) = \boldsymbol{x}(t_0)$ and $\kappa_R(t_0 + T_\gamma) - \kappa_R(t_0) = v_R + \Delta v_R$.

Equation (18.16) together with the last mentioned boundary condition yield the integral constraint originally reported as (1) in [25]. The formulation of this integral constraint in differential form allows us to solve (18.16) along with (18.15) and obtain

$$\kappa_R(t_0 + T_p) - \kappa_R(t_0) = \frac{2}{T_p}\int_{t_0}^{t_0+T_p} \cos(\omega_\gamma t)\boldsymbol{r}^{\mathrm{T}}\boldsymbol{x}(t)dt. \tag{18.17}$$

The aforementioned BVP, whose unknowns are $\boldsymbol{x}_0$ and y_R, can be solved through the TDSM. According to Section 18.2, the Jacobian matrix at the ith step of the Newton iterative scheme used to numerically solve the BVP (18.15) and (18.16) is

$$\begin{bmatrix} \boldsymbol{B} & \boldsymbol{\rho}_{y_R} \\ \boldsymbol{\chi}_{v_R} & \alpha_{y_R} \end{bmatrix} \tag{18.18}$$

where

$$\begin{aligned}
\boldsymbol{B} &= \boldsymbol{R}_{\boldsymbol{x}(t_0)} \\
\boldsymbol{\rho}_{y_R} &= \frac{\partial \boldsymbol{\varphi}(\boldsymbol{x}(t_0), t_0, t_0 + t, y_R, y_I, \omega_p)}{\partial y_R} \\
\boldsymbol{\chi}_{v_R} &= \boldsymbol{r}^{\mathrm{T}}\frac{2}{T_p}\int_{t_0}^{t_0+T_p} \boldsymbol{\Phi}^i(t_0 + T_p, t_0)\cos(\omega_p\tau)d\tau \\
\alpha_{y_R} &= \boldsymbol{r}^{\mathrm{T}}\frac{2}{T_p}\int_{t_0}^{t_0+T_p} \boldsymbol{\rho}_{y_R}\cos(\omega_p\tau)d\tau
\end{aligned} \tag{18.19}$$

evaluated for $t = t_0 + T_p$, $\boldsymbol{x}_0 = \boldsymbol{x}_0^i$, and $y_R = y_R^i$.

As far as the computation of $\boldsymbol{\rho}_{y_R}$ is concerned, evaluating the partial derivative of (18.15) with respect to y_R, leads to

$$\dot{\boldsymbol{\rho}}_{y_R} = \boldsymbol{f_x}\boldsymbol{\rho}_{y_R} + \boldsymbol{r}\cos(\omega_p t) \tag{18.20}$$

Thus, since $\boldsymbol{\rho}_{y_R}(t_0) = \frac{\partial \boldsymbol{x}_0(y_R)}{\partial y_R} = \boldsymbol{0}$, $\boldsymbol{\rho}_{y_R}$ can be obtained by solving (18.20) along with (18.15).

In this example, the v_R parameter is *perturbed*, y_R is thus *adjusted*, ω_p and y_I are *fixed* and v_I is left *free*. In principle, one can identify two parameter sets, i.e.,

§The PIT performance, basically in terms of converge rate of the algorithm, is sensitive to the choice of $\boldsymbol{s}$ and to the parameters to be altered, fixed, let free and adjusted. A detailed discussion on these aspects is beyond the scope of this work. The interested reader can refer to [39].

$\{v_R, v_I, \omega_p\}$ and $\{y_R, y_I\}$. The two parameters to be *perturbed* and *adjusted* cannot belong to the same set. Depending on how they are chosen, $\boldsymbol{\rho}$, $\boldsymbol{\chi}$ and α will differ from those reported in (18.19) (see [25] for details). The role played by the remaining three parameters is summarized in Table 18.1 of Appendix E.1.

The PIT is particularly useful to determine unstable $\boldsymbol{y}$ limit cycle of (18.15) with $y_R = y_I = 0$, starting from a limit cycle with non-null y_R and/or y_I. To explain this point, let $\dot{\boldsymbol{x}} = \boldsymbol{f}(\boldsymbol{x})$ exhibit a $\boldsymbol{y}$ unstable limit cycle. The PIT is applied as follows.

- `Step 1`. Identify an initial guess for ω_p. For limit cycles originated by a Hopf bifurcation of an equilibrium E_0 (assume, for simplicity, $E_0 = 0$), the eigenvalues of the $\boldsymbol{f_x}$ system Jacobian at E_0 identify the angular frequencies related to the imaginary part of these eigenvalues. Such frequencies are good initial guesses for ω_p.
- `Step 2`. Select the component of $\boldsymbol{x}$ to be used to insert the forcing term ruled by y_R, y_I and ω_p, and set $\boldsymbol{s}$ accordingly. Mechanisms to identify a suitable component of $\boldsymbol{x}$ were suggested in [25, 39].
- `Step 3`. Set $y_I = 0$, $v_R = \Delta v_R$ (with $|\Delta v_R| \ll 1$), and $y_R^{\text{old}} = 0$.
- `Step 4`. Solve the PIT-BVP to find a value of y_R and a set of initial conditions $\boldsymbol{x}_0$ that correspond to a periodic solution of $\dot{\boldsymbol{x}} = \boldsymbol{f}(\boldsymbol{x}(t)) + \boldsymbol{r} y_R \cos(\omega_p t)$ with $v_R = \frac{2}{T_p}\int_{t_0}^{t_0+T_p} \cos(\omega_p \tau)\boldsymbol{r}^{\mathrm{T}}\boldsymbol{x}(\tau)d\tau$.
- `Step 5`. If $y_R^{\text{old}} y_R < 0$ fix $\bar{v}_R = v_R$ and go to `Step 6`. Otherwise, set $y_R^{\text{old}} = y_R$, $v_R = v_R + \Delta v_R$ and go to `Step 4`.
- `Step 6`. Fix $y_R = 0$ and solve the PIT-BVP to find out a value of ω_p and a set of initial conditions $\boldsymbol{x}_0$ that correspond to a periodic solution of $\dot{\boldsymbol{x}} = \boldsymbol{f}(\boldsymbol{x}(t))$ with $\bar{v}_R = \frac{2}{T_p}\int_{t_0}^{t_0+T_p} \cos(\omega_p \tau)\boldsymbol{r}^{\mathrm{T}}\boldsymbol{x}(\tau)d\tau$. In this final step, a proper $\boldsymbol{\rho}_{\omega_p}$ and α_{ω_p} must be used in (18.18) instead of $\boldsymbol{\rho}_{y_R}$ and α_{y_R}, respectively.[‖]

After completing `Step 6`, $\boldsymbol{B}$ in (18.18) exhibits one null eigenvalue and vectors $\boldsymbol{\rho}_{\omega_p}$ and $\boldsymbol{\chi}_{v_R}$ represent a bordering of the Jacobian of the Newton iterative scheme used to apply the PIT.

We recall that since the PIT is applied to identify a limit cycle of power systems, the BVP inherent to the PIT must be further augmented by using the bordering based on the centre of inertia as detailed in Sections 18.3.3 and 18.3.4.

[‖] An efficient implementation of the PIT should compute simultaneously $\boldsymbol{\rho}_{y_R}$, $\boldsymbol{\rho}_{y_I}$, $\boldsymbol{\rho}_{\omega_p}$, $\boldsymbol{\chi}_{v_R}$, $\boldsymbol{\chi}_{v_I}$, α_{y_R}, α_{y_I} and α_{ω_p}. This can be easily done by solving the forward sensitivity problem [40], associated with (18.15) and (18.16) and

$$\dot{\kappa}_I = \frac{2}{T_p}\sin(\omega_p t)\boldsymbol{r}^{\mathrm{T}}\boldsymbol{x}(t) \ , \tag{18.21}$$

not only with respect to the initial conditions (i.e., the variational equation) but also with respect to the parameters y_R, y_I and ω_p. This is useful to design sweeping strategies other than that described in `Step 1`–`Step 6`.

Information on the stability of the limit cycle detected by using the PIT can be derived by computing a scalar G defined as:

$$G = \alpha_{yR}^{-1}[1 - \chi_{vR}(\rho_{yR}\chi_{vR} - \alpha_{yR}B)\rho_{yR}] \tag{18.22}$$

If $G < 0$ after completing `Step 6`, then the limit cycle is stable, otherwise it is unstable. The interested reader can refer to [25] for further details on this point.

18.3 Revisited power system model

This section presents a transient stability model of power systems in a formal and slightly unconventional notation. This notation allows applying the TDSM described in Section 18.2 and poses the basis for the PIT discussed in Section 18.2.4.

18.3.1 Outlines of standard power system models

The standard electromechanical equation of the rotor speed of synchronous machines can be written as follows:

$$\boldsymbol{M}\,\dot{\boldsymbol{\omega}} = \boldsymbol{p}_m(\boldsymbol{\omega}, \boldsymbol{z}, \boldsymbol{y}) - \boldsymbol{p}_e(\boldsymbol{\delta}, \boldsymbol{z}, \boldsymbol{v}, \boldsymbol{\theta}, \boldsymbol{y}) - \boldsymbol{D}(\boldsymbol{\omega} - \omega_0) \tag{18.23}$$

where the dependence on time is dropped to gain compactness. Being n_G the number of machines, symbols in (18.23) have the following meaning:

- $\boldsymbol{\omega}(t) \in \mathbb{R}^{n_G}$ are the rotor speeds of the machines;
- $\omega_0 \in \mathbb{R}$ is the reference synchronous frequency;
- $\boldsymbol{\delta}(t) \in \mathbb{R}^{n_G}$ are the rotor angles of the machines;
- $\boldsymbol{M} \in \mathbb{R}^{n_G \times n_G}$ is a diagonal matrix whose entries model the inertia constants of the machines;
- $\boldsymbol{z}(t) \in \mathbb{R}^{n_z}$ are the n_z state variables of the PSM ($\boldsymbol{\omega}$ and $\boldsymbol{\delta}$ excluded) that may influence the dynamics of the machines,
- $\boldsymbol{y}(t) \in \mathbb{R}^S$ are all the algebraic variables of the PSM but $\boldsymbol{v}$ and $\boldsymbol{\theta}$;
- $\boldsymbol{D} \in \mathbb{R}^{n_G \times n_G}$ is a diagonal matrix whose entries d_{jj} (for $j = 1, \ldots, n_G$) model the damping factor of the machines;
- $\boldsymbol{v}(t) \in \mathbb{R}^{n_B}$ and $\boldsymbol{\theta}(t) \in \mathbb{R}^{n_B}$ are bus voltages and phases, respectively, where n_B are the number of buses;
- $\boldsymbol{p}_m(\boldsymbol{\omega}, \boldsymbol{z}, \boldsymbol{y}) \in \mathbb{R}^{n_G}$ is the mechanical power regulated by controllers depending on $\boldsymbol{\omega}$, $\boldsymbol{z}$, and $\boldsymbol{y}$;
- $\boldsymbol{p}_e(\boldsymbol{\delta}, \boldsymbol{z}, \boldsymbol{v}, \boldsymbol{\theta}, \boldsymbol{y}) \in \mathbb{R}^{n_G}$ is the electrical power exchanged by machines.

The DAEs ruling the dynamics of the whole PSM are defined as:

$$\begin{aligned}
\dot{\boldsymbol{\delta}} &= \Omega\,(\boldsymbol{\omega} - \omega_0) \\
\boldsymbol{M}\dot{\boldsymbol{\omega}} &= \boldsymbol{p}_m(\boldsymbol{\omega}, \boldsymbol{z}, \boldsymbol{y}) - \boldsymbol{p}_e(\boldsymbol{\delta}, \boldsymbol{z}, \boldsymbol{v}, \boldsymbol{\theta}, \boldsymbol{y}) - \boldsymbol{D}(\boldsymbol{\omega} - \omega_0) \\
\dot{\boldsymbol{z}} &= \boldsymbol{\phi}\,(\boldsymbol{\delta}, \boldsymbol{\omega}, \boldsymbol{z}, \boldsymbol{v}, \boldsymbol{\theta}, \boldsymbol{y}) \\
\boldsymbol{0} &= \boldsymbol{g}\,(\boldsymbol{\delta}, \boldsymbol{\omega}, \boldsymbol{z}, \boldsymbol{v}, \boldsymbol{\theta}, \boldsymbol{y})
\end{aligned} \tag{18.24}$$

where $\boldsymbol{\phi}$ and $\boldsymbol{g}$ complete the model of the machines, Ω is the base synchronous frequency in rad/s. $\boldsymbol{\phi}$ accounts for regulators and other dynamics included in the system, while $\boldsymbol{g}$ models algebraic constraints such as lumped models of transmission lines, transformers and static loads. Note that the standard power flow equations are a subset of $\boldsymbol{g}$ and are used to compute a stationary solution. In particular, power flow equations involve $\boldsymbol{v}$ and $\boldsymbol{\theta}$ only.

The initial equilibrium point of (18.24) can be determined using a two-step approach, namely, first solving power flow equations and then initializing state variables given voltage profile and power injections at network buses. A detailed description of the procedure to initialize differential equations based on the power flow solution can be found in [29, 41]. Hereinafter the subscript 0 will denote the initial equilibrium point of (18.24).

To better understand the discussion provided later on in this section, it is important to note that $\boldsymbol{p}_e(\cdot)$ does not singularly depend on $\boldsymbol{\theta}_0$ and $\boldsymbol{\delta}_0$ but on their difference $\boldsymbol{\theta}_0 - \boldsymbol{\delta}_0$ only. This means that a shifted set of variables $\boldsymbol{\delta}' = \boldsymbol{\delta}_0 + \bar{\delta}$ ($\bar{\delta} \in \mathbb{R}$) still satisfy (18.24).

From the viewpoint of dynamical systems theory, the consideration above means that the stationary PSM solution is not an *isolated equilibrium* but it is embedded in a *continuum of equilibria* [42]. This is confirmed by an *always null eigenvalue* of the Jacobian matrix of the PSM linearized at any equilibrium point of (18.24) [43]. This should not surprise. The classical PSM aims at representing the envelope of the *actual* system dynamics through a steady-state solution. In other words, the *periodic steady-state solution* at the fundamental frequency with a *constant* envelope is actually represented as a *constant* steady-state solution playing the role of a *stationary solution*. All together, the admissible shifted equilibria represent a one-dimensional manifold Γ (parametrized by $\bar{\delta}$) in the phase space.¶ A projection of this manifold in the n_G-dimensional subspace of the state variables $\boldsymbol{\delta}$ is the locus of point $\boldsymbol{\delta}' = \boldsymbol{\delta}_0 + \bar{\delta}$.

The statement above on the *always null eigenvalue* is easily exemplified through a two-machine system example shown in Figure 18.3. Classical machine models, i.e., constant voltage source behind the transient reactance and constant mechanical power, and lossless transmission system are assumed. While oversimplified, such a

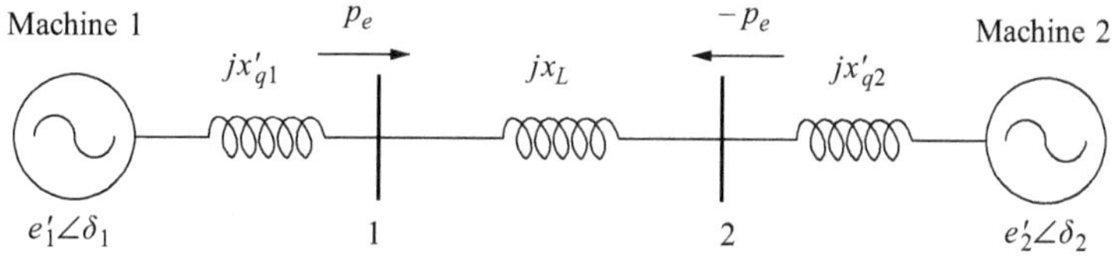

Figure 18.3 Two-machine system

¶The *phase space* is the space in which all possible states of a dynamical system are represented. For a DAE the dimension of the phase space is the given by the number of differential variables. In this case, such a dimension is $2n_G + n_z$.

system is commonly used for illustrating inter-area oscillations and transient stability analysis issues in power systems. The equations that describe the dynamic behaviour of the system are:

$$\begin{aligned} \dot{\delta}_1 &= \Omega(\omega_1 - \omega_0) \\ M_1\dot{\omega}_1 &= p_m - p_e(\delta_1, \delta_2) - D_1(\omega_1 - \omega_0) \\ \dot{\delta}_2 &= \Omega(\omega_2 - \omega_0) \\ M_2\dot{\omega}_2 &= -p_m + p_e(\delta_1, \delta_2) - D_2(\omega_2 - \omega_0) , \end{aligned} \tag{18.25}$$

with

$$p_e = \frac{e_1' e_2'}{x_{\mathrm{eq}}} \sin(\delta_1 - \delta_2) , \tag{18.26}$$

where $x_{\mathrm{eq}} = x'_{q1} + x_L + x'_{q2}$. The equilibrium point is thus $\omega_{10} = \omega_{20} = \omega_0$ and $\delta_{10} - \delta_{20} = \mathrm{asin}\left(\frac{p_m x_{\mathrm{eq}}}{e_1' e_2'}\right)$ from where it is apparent that shifting by the same amount $\bar{\delta}$ both δ_{10} and δ_{20} will satisfy the steady-state conditions of (18.25).

At the equilibrium, the state matrix of (18.25) is:

$$\boldsymbol{A} = \begin{bmatrix} 0 & \Omega & 0 & 0 \\ -k/M_1 & -D_1/M_1 & k/M_1 & 0 \\ 0 & 0 & 0 & \Omega \\ k/M_2 & 0 & -k/M_2 & -D_2/M_2 \end{bmatrix} , \tag{18.27}$$

where $k = \frac{e_1' e_2'}{x_{\mathrm{eq}}} \cos(\delta_{10} - \delta_{20})$. It is straightforward to show that matrix $\boldsymbol{A}$ defined in (18.27) has a null eigenvalue independently from system parameters. In fact, using the Laplace expansion to compute the determinant, one has:

$$\begin{aligned} \det(\boldsymbol{A}) &= -\Omega \det\left(\begin{bmatrix} -k/M_1 & k/M_1 & 0 \\ 0 & 0 & \Omega \\ k/M_2 & -k/M_2 & -D_2/M_2 \end{bmatrix}\right) \\ &= \Omega^2 \left(k^2/(M_1 M_2) - k^2/(M_1 M_2)\right) = 0 . \end{aligned} \tag{18.28}$$

The null eigenvalue is a consequence of the fact that the equilibrium point depends on $\delta_1 - \delta_2$ and, hence, an equation expressing the time derivative of machine rotor angles is redundant. The singularity of the state matrix can be removed by using a rotor angle of the two machines as phase reference. For example, if machine 1 is the reference, say $\delta_{10} = \bar{\delta}$, (18.25) becomes:

$$\begin{aligned} M_1\dot{\omega}_1 &= p_m - p_e(\bar{\delta}, \delta_2) - D_1(\omega_1 - \omega_0) \\ \dot{\delta}_2 &= \Omega(\omega_2 - \omega_1) \\ M_2\dot{\omega}_2 &= -p_m + p_e(\bar{\delta}, \delta_2) - D_2(\omega_2 - \omega_0) , \end{aligned} \tag{18.29}$$

with state matrix, at the equilibrium point:

$$\tilde{\boldsymbol{A}} = \begin{bmatrix} -D_1/M_1 & k/M_1 & 0 \\ -\Omega & 0 & \Omega \\ 0 & -k/M_2 & -D_2/M_2 \end{bmatrix} , \tag{18.30}$$

whose determinant, namely,

$$\det(\tilde{A}) = -\Omega k \frac{D_1 + D_2}{M_1 M_2} \neq 0\,, \tag{18.31}$$

is not null if $D_1 + D_2 \neq 0$. Note that the formulation for which a machine rotor angle is used as phase reference, i.e., (18.29), is actually never used in software tools for power systems analysis. The main reasons to prefer the formulation with a redundant equation, i.e., (18.25), are to maintain the modularity of generator models and to prevent losing the reference in case some section of the system is islanded.

18.3.2 From polar to rectangular coordinates

Since the objective is the location of periodic steady-state solutions of the PSM *periodic orbits must be represented as such*. This apparently trivial statement is actually not generally satisfied by the standard formulation of PSM in polar coordinates. A recast of (18.24) in rectangular coordinates is required since the TDSM is based on the formulation of a BVP with periodic constraints. In fact, it can happen that a limit cycle in rectangular coordinates is a diverging trajectory in polar coordinates having a phase angle arbitrary increasing in time. To clarify this important point, let us consider the following autonomous polar oscillator example

$$\begin{aligned} \dot{\rho} &= \rho_0 - \rho \\ \dot{\delta} &= \Omega\,(\omega - \omega_0) \end{aligned} \tag{18.32}$$

whose steady-state solution, i.e., $\rho(t) = \rho_0$ and $\delta(t) = \Omega\,(\omega - \omega_0)\,t$ is unbounded.** However, if (18.32) is recast using the following:

$$\begin{aligned} c &= \rho\cos(\delta) \\ s &= \rho\sin(\delta) \end{aligned} \tag{18.33}$$

i.e., using rectangular coordinates, the system exhibits a trivial periodic solution.

PSM can exhibit a similar behaviour as (18.32). In fact, there can exist steady state solutions of (18.24) such that $\boldsymbol{\omega} - \omega_0 = \Delta\boldsymbol{\omega} \neq \mathbf{0}$ is a *constant* vector and $\dot{\boldsymbol{\omega}} = 0$. This can happen, for example, as a consequence of the droop control of primary frequency regulation. In this case, as t increases, $\boldsymbol{\delta}(t)$ linearly tends to ∞ and so does $\boldsymbol{\theta}(t)$. Nevertheless, the rectangular coordinates of voltage phasors, namely, $\boldsymbol{v}_{\mathrm{q}}(t)$ and $\boldsymbol{v}_{\mathrm{d}}(t)$ components of $(\boldsymbol{v}, \boldsymbol{\theta})$, are periodic time varying functions describing a close trajectory. This means that the trajectory is divergent in the polar coordinate frame and periodic in rectangular coordinates.

To avoid the divergent behaviour of polar coordinates, the first equation of (18.24) can be recast using a rectangular reference frame. Each component δ_j of $\boldsymbol{\delta}$ is projected onto the unit circle in $\mathbb{R}^2$ using the transformation (18.33) with $\rho = 1$, i.e., $(\delta_j, 1) \rightarrow (c_j, s_j)$. Since there are two new unknowns for each entry of δ, the first equation in (18.24) is replaced by n_G pairs of equations. Introducing the vectors $\boldsymbol{c}$ and $\boldsymbol{s}$ storing the

**The initial condition $\delta(t_0) = 0$ is assumed.

state variables c_j and s_j (for $j = 1, ..., n_G$), the other equations of (18.24) are properly modified and the new set of PSM equations

$$\begin{aligned}
\dot{\boldsymbol{c}} &= -\Omega\,(\boldsymbol{\Delta}_\omega - \omega_0)\,\boldsymbol{s} \\
\dot{\boldsymbol{s}} &= \Omega\,(\boldsymbol{\Delta}_\omega - \omega_0)\,\boldsymbol{c} \\
\boldsymbol{M}\dot{\boldsymbol{\omega}} &= \boldsymbol{p}_m(\boldsymbol{\omega}, \boldsymbol{z}, \boldsymbol{y}) - \boldsymbol{p}_e(\boldsymbol{c}, \boldsymbol{s}, \boldsymbol{z}, \boldsymbol{v}, \boldsymbol{\theta}, \boldsymbol{y}) - \boldsymbol{D}(\boldsymbol{\omega} - \omega_0) \\
\dot{\boldsymbol{z}} &= \boldsymbol{\phi}\,(\boldsymbol{c}, \boldsymbol{s}, \boldsymbol{\omega}, \boldsymbol{z}, \boldsymbol{v}, \boldsymbol{\theta}, \boldsymbol{y}) \\
\boldsymbol{0} &= \boldsymbol{g}\,(\boldsymbol{c}, \boldsymbol{s}, \boldsymbol{\omega}, \boldsymbol{z}, \boldsymbol{v}, \boldsymbol{\theta}, \boldsymbol{y})
\end{aligned} \tag{18.34}$$

is obtained, where $\boldsymbol{\Delta}_\omega$ is a diagonal matrix whose entries $\boldsymbol{\Delta}_{\omega_{jj}} = \omega_j$, for $j = 1, .., n_G$.[††]

For example, based on (18.34), (18.25) of the two-machine system can be recast as follows:

$$\begin{aligned}
\dot{c}_1 &= -\Omega(\omega_1 - \omega_0)s_1 \\
\dot{s}_1 &= \Omega(\omega_1 - \omega_0)c_1 \\
M_1\dot{\omega}_1 &= p_m - p_e(c_1, c_2, s_1, s_2) - D_1(\omega_1 - \omega_0) \\
\dot{c}_2 &= -\Omega(\omega_2 - \omega_0)s_2 \\
\dot{s}_2 &= \Omega(\omega_2 - \omega_0)c_2 \\
M_2\dot{\omega}_2 &= -p_m + p_e(c_1, c_2, s_1, s_2) - D_2(\omega_2 - \omega_0)\,,
\end{aligned} \tag{18.35}$$

with

$$p_e = \frac{e'_1 e'_2}{x_{\text{eq}}}(s_1 c_2 - s_2 c_1)\,. \tag{18.36}$$

To conclude this subsection, let us recall the one-dimensional manifold Γ introduced in Section 18.3.1. When dealing with PSM equations in rectangular coordinates, the inherent periodicity of $\boldsymbol{\delta}$ angles makes Γ a closed one-dimensional curve in $\mathbb{R}^{3n_G+n_z}$, being $3n_G + n_z$ the new dimension of the phase space. Note that, in practice, $n_G \ll n_z$, hence the computational burden of (18.34) is similar to that of (18.24).

18.3.3 On the unit multipliers of the power system model periodic orbits

If the PSM exhibits a periodic (un)stable steady-state solution $\boldsymbol{\gamma}$, originated, for instance, from a Hopf bifurcation of the equilibrium point of (18.24), such a *limit cycle* has to be handled with care. As a matter of fact, as it happens for the equilibria, $\boldsymbol{\gamma}$ is not an isolated limit cycle but it is embedded in a *continuum of limit cycles* [44]. This occurs since the PSM aims at isolating one of the infinite possible solutions $\boldsymbol{\gamma}$, exactly as it does for the equilibria. Each one of these infinite periodic solutions represents the *periodic modulation* of the constant envelope of the periodic steady-state solution at the fundamental frequency represented as a constant equilibrium.

These modulated envelopes differ only by a constant shift of their $\boldsymbol{\delta}$ components and a continuum of periodic solutions is thus often considered as an isolated limit

[††] Further details concerning numerical aspects related to the described variable transformation are provided in Appendix E.2.

cycle. Among the drawbacks of this modeling approach, if $\boldsymbol{\gamma}$ is a self-sustained limit cycle, i.e., the system is not periodically forced, its monodromy matrix has two unit multipliers. One of them is *trivial*, as discussed in Section 18.2.1, but the other one is a direct consequence of the limit cycle *isolation procedure* which is inherent to the PSM. The non-trivial unitary Floquet multiplier is the equivalent of the always null eigenvalue of the Jacobian matrix of the PSM that can be found for any equilibrium of (18.24) solution [43]. If the system is non-autonomous, i.e., there is a periodic external signal forcing it, only the non-trivial null multiplier is observed in the monodromy matrix of $\boldsymbol{\gamma}$.

These considerations pose issues if one aims at locating $\boldsymbol{\gamma}$ through the TDSM since, as shown in Section 18.2.3, unit multipliers correspond to rank deficiency of the Jacobian matrix involved in the Newton scheme used to solve the periodic BVP.‡‡

18.3.4 Bordering based on the centre of inertia

As a first step, in (18.34), a constant $c_0 \in \mathbb{R}$ is added to c_j in all the n_G pairs of equations ruling the machine dynamics thus obtaining§§:

$$\begin{aligned} \dot{\boldsymbol{c}} &= -\Omega\left(\boldsymbol{\Delta}_\omega - \omega_0\right)\boldsymbol{s} \\ \dot{\boldsymbol{s}} &= \Omega\left(\boldsymbol{\Delta}_\omega - \omega_0\right)(\boldsymbol{c} + c_0) \end{aligned} \tag{18.37}$$

where the c_0 constant is used only in the detection of γ limit cycles through the TDSM and it is set to 0 when the equilibrium is solved.‖‖ When the TDSM is applied, the PSM is integrated along with its variational equations (see Section 18.2.1) thus achieving the sensitivity of its trajectories with respect to their initial conditions.

If the system is non-autonomous, the period T_γ is imposed by the time-varying periodic driving function and a phase condition is not needed. If the system is autonomous, a phase condition is introduced to remove the trivial unit multiplier of the monodromy matrix and to identify T_γ.

In both cases, to remove the non-trivial singularity of the Jacobian, due to the second unit multiplier of the monodromy matrix, a bordering equation is added taking the cue from the COI expression [29] in rectangular coordinates. In particular, the following equation is used¶¶:

$$q_c(\underbrace{\boldsymbol{c}, \boldsymbol{s}, \boldsymbol{\omega}, \boldsymbol{z}}_{x}, c_0) = \frac{\sum_j M_j \left(c_j(t) + c_0\right)}{\sum_j M_j} = 0 \tag{18.38}$$

where M_j is the inertia constant of the jth machine, and c_0 is updated during the iterations of the Newton scheme used in the TDSM to compute the steady state

‡‡Attention should be paid also if the periodic solution is searched in the frequency domain by resorting to harmonic balance techniques [45]. As a matter of fact, the singularities of the Jacobian of the Newton scheme adopted in the harmonic balance approach are exactly the same found in the TDSM [11].

§§An equivalent result is obtained by adding a constant $s_0 \in \mathbb{R}$ to the s_j variables.

‖‖In this way, the conventional structure of machine models is preserved.

¶¶If in (18.37) a constant s_0 is added to $\boldsymbol{s}$ instead of adding c_0 to $\boldsymbol{c}$, one should use:

$$q_s(\boldsymbol{c}, \boldsymbol{s}, \boldsymbol{\omega}, \boldsymbol{z}, s_0) = \frac{\sum_j M_j \left(s_j(t) + s_0\right)}{\sum_j M_j} = 0$$

solution. At converge, (18.38) isolates a limit cycles among the infinite equivalent ones as discussed in Section 18.3.3.

For example, for the 2-machine system, the bordering equation (18.38) becomes:

$$q_c(c_1, c_2, c_0) = \frac{M_1\,(c_1 + c_0) + M_2\,(c_2 + c_0)}{M_1 + M_2} = 0 \tag{18.39}$$

As far as conditions C1 and C2 are concerned, in general, it is not trivial to verify that a chosen bordering equations fulfils them, since $\boldsymbol{\Psi}$ depends on the dynamical equations governing the system and on the specific limit cycle $\boldsymbol{\gamma}$ the monodromy matrix is referred to. We have tested (18.38) for several power systems, including but not limited to the case studies discussed in Section 18.4. The proposed expression for $q_c(\boldsymbol{x}(t_0), c_0)$ revealed suitable in all cases.

18.4 Case studies

In this section, the TDSM is applied at first to two well-known *smooth* test systems: (i) the IEEE 14-bus system, which shows a stable limit cycle; (ii) the WSCC 9-bus system, which shows an unstable limit cycle coexisting with a stable equilibrium point. In these two examples, a preliminary parametric analysis to determine the occurrence of Hopf bifurcations is solved. With this aim, load power consumption as well as generation active power productions are parametrized by means of a scalar loading level μ. PQ load power consumption is defined as:

$$\begin{aligned} \boldsymbol{p}_L &= \boldsymbol{p}_L^0\,(1+\mu) \\ \boldsymbol{q}_L &= \boldsymbol{q}_L^0\,(1+\mu) \end{aligned} \tag{18.40}$$

where $\boldsymbol{p}_L^0$ and $\boldsymbol{q}_L^0$ are load base-case active and reactive powers, respectively. In a similar way, PV generator active powers are defined as:

$$\boldsymbol{p}_G = \boldsymbol{p}_G^0\,(1+\mu) \tag{18.41}$$

where $\boldsymbol{p}_G^0$ are the generator base-case active power productions. The models above are used to define the power flow solution, then machine and regulator variables are initialized according to the operating point.

Then the TDSM is applied to locate a stable limit cycle in two power system models described by a set of *hybrid* DAEs: (i) the 2-area system with inclusion of a bank of mechanically switched capacitors; (ii) a radial network in which two ULTCs and a switching capacitor bank interact causing the onset of periodic oscillations. All examples were solved in the simulator PAN [46].

18.4.1 IEEE 14-bus test system

The proposed TDSM was first applied to the IEEE 14-bus test system. Generators are described by fifth- and sixth-order models with voltage regulators (IEEE type I) and turbine governors. To force the occurrence of Hopf bifurcations and following limit cycles, no power system stabilizer is included in the system. All static and dynamic data can be found in [29].

A sweep of the loading level parameter μ shows that a Hopf bifurcation occurs for $\mu = \mu^* \approx 0.2802$. For $\mu = 0.4$, the eigenvalues lay "well inside" the instability region and a time domain analysis evidences the existence of a stable limit cycle. Since stable limit cycles can be determined using a conventional time domain simulation, the use of the PIT is not strictly necessary. However, the time domain simulation only provide a qualitative information. Then, matrix bordering is still necessary as discussed in Section 18.2.3. By observing the imaginary part of the pair of eigenvalues generating the Hopf, we chose a frequency of 1.442 Hz as initial guess for TDSM. At convergence of the TDSM, the working frequency was 1.430 Hz, that is very close to the initial value. Figure 18.4 shows the projection of the orbit on the (v_1, ω_1) plane, where v_1 is the voltage phasor amplitude at bus 1 and ω_1 is the working angular frequency of the generator connected to bus 1.

Finally, we consider the Floquet multipliers, e.g., the eigenvalues of the monodromy matrix, of the obtained stationary solution. By computing and sorting in descending order the moduli of Floquet multipliers, the two largest ones are equal to 1, as expected, since they are due to the autonomous nature of the power system and to the PSM structure as detailed in Section 18.3.3. The third largest Floquet multiplier is equal to 0.986, thus confirming the stability of the periodic orbit shown in Figure 18.4.

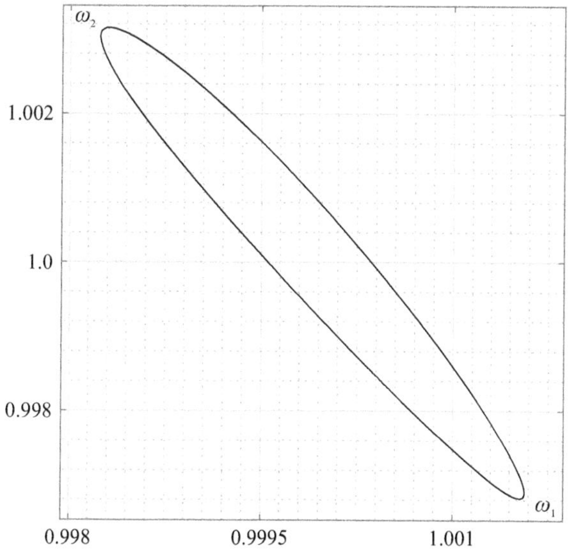

Figure 18.4 A projection of the stable limit cycles on the plane (ω_1, ω_2) obtained for the overloaded IEEE 14-bus test system ($\mu = 0.4$). ω_1 and ω_2 are the working angular frequency in pu of the generator connected to bus 1 and bus 2, respectively

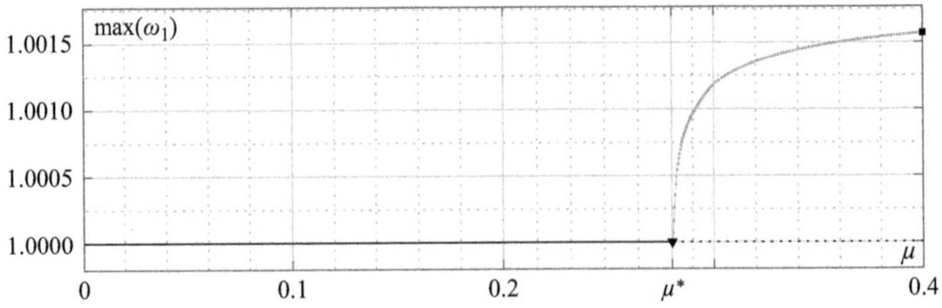

Figure 18.5 Bifurcation diagram for the IEEE 14-bus test system for $\mu \in [0,0.4]$. The triangle represents the super-critical Hopf bifurcation of the equilibrium point that is stable for $\mu < \mu^*$ (black solid line) and unstable for $\mu > \mu^*$ (black dotted line). Grey dots represent limit cycles originated from the Hopf bifurcation. The square corresponds to the limit cycle depicted in Figure 18.4

The TDSM and the computation of the eigenvalues of the principal matrix required 720 ms of CPU time; 8 iterations of the Hybrid-Newton method [47] were required to obtain the steady state solution.

To validate that the limit cycle obtained by resorting to the TDSM for $\mu = 0.4$ is actually originated by the Hopf bifurcation of the equilibrium point, we decreased μ from 0.4 to μ^* moving on a discrete grid of values of this parameter and the TDSM was applied at all of these values. Figure 18.5 shows the bifurcation diagram of ω_1, i.e., the angular speed of the generator connected to bus 1, for each λ value both for the equilibrium point and the limit cycle. The diagram confirms that the Hopf bifurcation is super-critical and the limit cycle obtained for $\lambda = 0.40$ is originated from such a bifurcation.

18.4.2 WSCC 9-bus test system

In this case study, we consider the WSCC 9-bus 3-machine system described in [41]. Generators are modeled using the *d–q* axis model and automatic voltage regulators (IEEE type I). The parametric analysis shows that for $\mu = \mu^* \approx 1.0571$ the equilibrium point becomes unstable and undergoes a sub-critical Hopf bifurcation. For $\mu \in [0, \mu^*)$, a stable equilibrium point of the PSM can be found. This means that for $\mu \in [0, \mu^*)$ an unstable limit cycle coexists with the equilibrium [26]. In this case, the time domain simulation does not help as trajectories cannot fall on the unstable limit cycle.

For the sake of example, let $\mu = 1.0250 < \mu^*$. Then, the steps of the PIT described in Section 18.2.4 are applied after having recast the PSM in rectangular coordinates as discussed in Section 18.3.2. The external signal forcing the system has been added to the equations describing the dynamics of the *c* variable (see (18.34)) of the machine connected to bus 2. The algorithm starts with $v_R = 10^{-2}$ and $\omega = 4\pi/5$. In `Step 4`, a first limit cycle of the forced system is located corresponding to $y_R = -0.0837$ (see the curve *a* in Figure 18.6).

The procedure continues iterating from `Step 5` to `Step 4` until $v_R = 0.0259$ when y_R becomes positive (*c* curve in Figure 18.6). At each iteration, v_R is increased

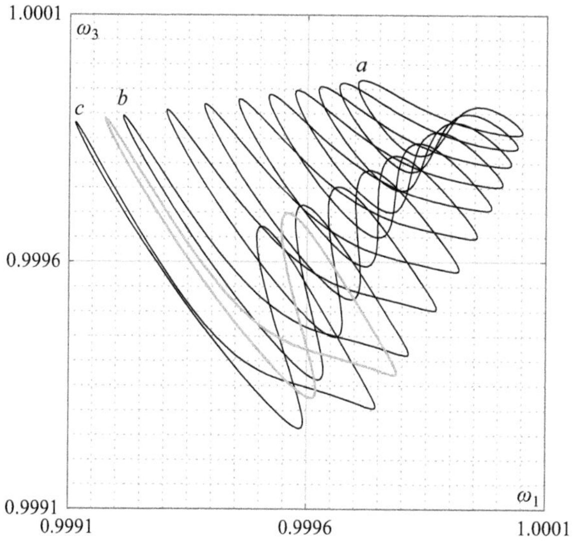

Figure 18.6 A projection of the limit cycles obtained with the PIT on the plane (ω_1, ω_3). ω_1 and ω_3 are the rotor speeds in pu of the machines of the WSCC 9-bus test system connected to bus 1 and 3, respectively. The first limit cycle obtained for $v_R = 10^{-2}$ is labeled with a. The label c corresponds to the last limit cycle obtained for $v_R = 0.0259$. y_R changes sign from negative to positive from the b limit cycle and the c one. The light grey limit cycle is detected at `Step 6` *and it corresponds to an unstable periodic steady-state solution of the autonomous systems, i.e., $y_R = y_I = 0$*

by 10%. Then `Step 6` is solved and a limit cycle of the autonomous system is obtained for $y_R = y_I = 0$, $v_R = 0.0214$, $v_I = 0.0381$, and $\omega = 2.5424$ (light grey curve in Figure 18.6). Evaluating (18.22), one can find $G = 31.4833 > 0$, i.e., the limit cycle is unstable. This is also confirmed by computing the Floquet multipliers of the periodic orbit. In fact, the magnitude of the largest multiplier is 2.2347.

As discussed in [19, 21], it is important to determine unstable limit cycles as they provide the boundary of the region of attraction of the stable equilibrium point. With respect to other techniques proposed in the literature, the PIT is able to define the exact trajectory of the limit cycle, which, as shown in Figure 18.6 is not necessarily a "simple" curve.

The PIT took 14.43 s CPU time; this time is greater than that of the IEEE 14-bus case since several TDSM simulations were performed during `Step 4` and `Step 6` of the PIT.

18.4.3 A switching 2-area PSM

The vast majority of studies on limit cycles in power systems focuses on continuous DAEs. However, discontinuities and switching devices, including discrete controllers,

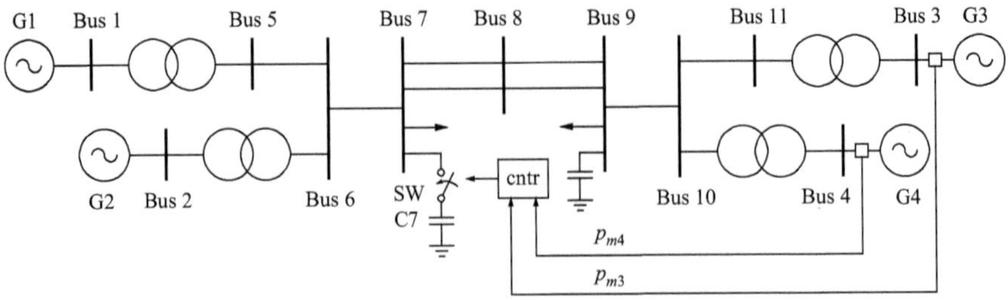

Figure 18.7 The schematic of the 2-area test circuit used as test

are common in power systems. In this section, we present an example of a periodic steady-state solution originated by the interaction of turbine governors and the control of a bank of mechanically switched capacitors. With this aim, we properly modify the 2-area test PSM (see Figure 18.7) originally proposed and described in in [48, p. 813]. The differences with respect to the original version are as follows.

- A switch (`sw`) that connects/disconnects the `C7` capacitor bank to/from `bus7`.
- The *digital control* block (labeled as `cntr`) that governs the `sw` switch. Having fixed the parameters $p_{\min}$ and $p_{\max} > p_{\min}$, the switch is closed when $p_{m3} + p_{m4} > p_{\max}$ and it is opened when $p_{m3} + p_{m4} < p_{\min}$. The $p_{\max}$ and $p_{\min}$ thresholds implement 2 different manifolds, $\mathcal{H}_1$ and $\mathcal{H}_2$, respectively, as described by (18.7). Each time one of these manifolds is traversed according to the logic implemented in `cntr` a saltation matrix is computed during the shooting analysis that determines the steady-state solution, as described in (18.8). These manifolds depend on the state variables of the turbine governors connected to the synchronous machines.
- A model of the COI as we did for the IEEE 14-bus and WSCC 9-bus test systems.

The control proposed in this section is a kind of wide area control system (WACS), which are more and more common today. It is important to stress that this is just an example. While it does not represent any real-world implementation, we think that such an example well illustrates possible undesired interactions that arise among different controllers in non-linear systems.

The `cntr` logic block is modelled in a very compact and efficient way through the *verilog-rtl* formal language, that is largely used in microelectronics design modelling [49]. In the literature, this kind of analog/digital simulations is known as *analog mixed signal* (AMS). The two $\mathcal{H}_1$ and $\mathcal{H}_2$ manifolds depend on the mechanical power set point of the `tg3` and `tg4` turbine governors (type I) that drive `G3` and `G4`. The model of these governors is:

$$
\begin{aligned}
p_{in} &= p_{\text{ref}} + \frac{\omega_{\text{ref}} - \omega}{R} \\
T_s\,\dot{x}_{g1} &= p_{in} - x_{g1} \\
T_c\,\dot{x}_{g2} &= \left(1 - \frac{T_3}{T_c}\right) x_{g1} - x_{g2}
\end{aligned}
$$

$$T_5\,\dot{x}_{g3} = \left(1 - \frac{T_4}{T_5}\right)\left(x_{g2} + \frac{T_3}{T_c}x_{g1}\right) - x_{g3}$$
$$p_m = x_{g3} + \frac{T_4}{T_5}\left(x_{g2} + \frac{T_3}{T_c}x_{g1}\right)$$

where p_m is the mechanical power set point (more details can be found in [29]). For simplicity, saturations are omitted in the expression above, but are considered in the model used for solving the numerical simulations. The manifolds are located by zeroing the scalar functions:

$$h_1(p_{m3}, p_{m4}) = p_{m3} + p_{m4} - p_{\max}$$
$$h_2(p_{m3}, p_{m4}) = p_{m3} + p_{m4} - p_{\min}$$

where p_{m3} and p_{m4} are the mechanical power set points of `tg3` and `tg4`. The zeros of $h_1(p_{m3}, p_{m4})$ and $h_2(p_{m3}, p_{m4})$ are considered when it passes from negative (positive) to positive (negative) values.

We perform a power-flow analysis that finds an equilibrium point of the system and initializes the synchronous machines, voltage regulators and turbine governors. Then, we increase the active power of the `Lo9` load from 1.767 pu to 2 pu. We approximately expect that total power by `G3` and `G4` increases above $p_{\max}$ causing `sw` to close. This improves power transfer from `G1` and `G2` to `Lo9` and leads to a damped power oscillation if `sw` is kept closed. The magnitude of this oscillation is large enough to cause the opening of `sw` after some time delay, thus triggering an oscillatory solution where `sw` periodically opens and closes. We look for a stable limit cycle and thus adopt only the shooting method.

This simple test circuit behaves like an autonomous system and thus there are 2 Floquet multipliers equal to 1 as described in Section 18.3.3. To perform the shooting analysis with a well-formed (bordered) Jacobian matrix, we inserted an extra element in the model of the 2-area system, i.e., a COI, as we did for the IEEE 14-bus and WSCC 9-bus systems described above. Note that, since the primary frequency regulation does not provide a perfect tracking of the frequency, the average values of the normalised rotor speeds of the synchronous machines decrease below 1 pu. As a consequence, a modulation of $\boldsymbol{c}$ and $\boldsymbol{s}$, i.e. of the $\boldsymbol{\delta}$ phases, occurs. To avoid such a modulation, $\boldsymbol{\Delta}_\omega$ are referred to the ω_c of the COI. Equation (18.42) describes how equations of the synchronous machines change if the ω_c angular frequency of the COI is used:

$$\dot{\boldsymbol{c}} = -\Omega\,(\boldsymbol{\Delta}_\omega - \omega_c(t))\,\boldsymbol{s} \tag{18.42}$$
$$\dot{\boldsymbol{s}} = \Omega\,(\boldsymbol{\Delta}_\omega - \omega_c(t))\,(\boldsymbol{c} + c_0)$$
$$\omega_c(t) = \frac{\sum_{j=1}^{4} M_j\omega_j(t_0)}{\sum_{j=1}^{4} M_j}\ .$$

The introduction of the $\omega_c(t)$ COI reference frequency leads to a solution where each entry of $\boldsymbol{c}$ and $\boldsymbol{s}$ has periodic variations around a fixed value.

Some simulation results by the proposed shooting method are reported in Figure 18.8. As it can be seen the `sw` switches open/close once per working period; at

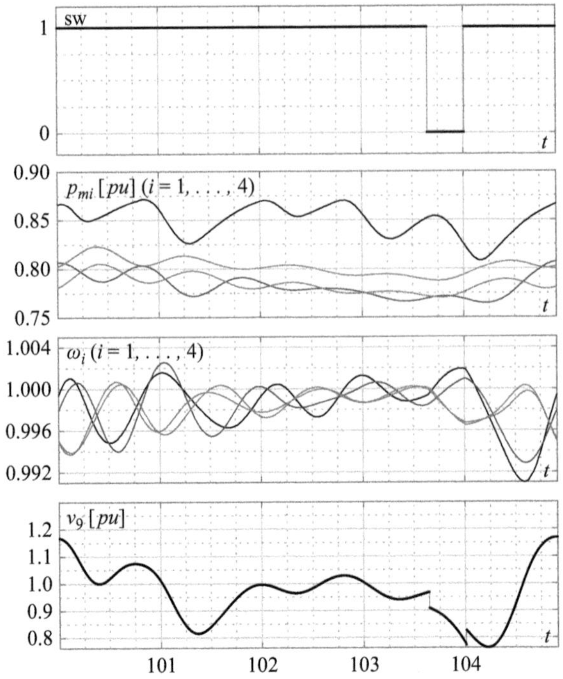

Figure 18.8 Simulation results obtained with the proposed shooting method performed on the 2-area test system shown in Figure 18.7. From upper to lower panel: the digital control signal that drives `sw`*: high level means switch closed; low level means switch open; instantaneous mechanical power of the four synchronous machines; instantaneous angular frequencies of the four synchronous machines; the magnitude of the* v_9 *voltage of* `bus9`*. In the upper and lowest panels, the grey solid lines serve only as a guide to the eye. In the second and third panels, curves from black to lightest grey refer to the first, second, third and fourth machines, respectively. Time t is expressed in seconds and the period of the oscillations is approximately 4.9118 s*

each closing/opening time instant there is an instantaneous jump of the voltage at `bus9` (and also at the other buses). Note that the voltage can jump because it is modelled as an algebraic variable. By observing Figure 18.8, we can easily see that average values of $\mathbf{\Delta}_\omega$ entries are lower than 1.

The first three Floquet multipliers with largest modulus are 1.001, 1.000, and 0.906, respectively. The first one is practically equal to 1 and corresponds to the eigenvector tangent to the trajectory; it is thus due to the autonomous nature of the circuit. The second multiplier is due to the PSM model. The third as well as all other remaining Floquet multipliers are well inside the unit circle in the complex plane, thus showing that the steady-state solution is stable.

18.4.4 Cascaded ULTC transformers

In the Introduction we cited [5], which shows a simple limit cycle originated by the poor design of the discrete voltage regulators of two under-load tap changers (ULTCs) in a radial distribution network. This kind of oscillations is relatively uncommon and can be straightforwardly eliminated through a proper set-up of the ULTC voltage regulation bands. More sophisticated is the study of the effect of regulator hard limits and power electronic switches on power systems modelled as hybrid DAEs that is proposed in [50–52].

The references above are likely the most relevant works on bifurcation analysis of hybrid power system models to the date. This topic is thus an open research field. However we implemented a test bench similar to that used in [51, Figure 7] to show the onset of periodic oscillations due to the interaction among three different actors, namely two ULTCs and a switching capacitor bank in a radial network. Note that the test system in [51], we use here, was in turn derived from that in [52]. The schematic of our test system is reported in Figure 18.9. We adopted a detailed model of the ULTCs; they are implemented by digital state machines described through the *verilog* formal language. The target of the ULTCs in Figure 18.9 is to keep voltages in the [0.99, 1.01] p.u. deadband by adjusting the transformer ratios in the [0.88, 1.20] range over 33 discrete positions. Thus the ratio varies by 0.01 from one position to the next.*** When the load voltage leaves a deadband at time t_0, the first tap change may take place at time $t_0 + \tau_1$ if at the end of this time interval, the voltage is still outside the current deadband. The subsequent changes may occur after the same τ_1 delay. The tap-changing action is dropped if voltage has reentered to or jumped from one side to the other of the deadband. This means that if the bus voltage oscillates too much and/or in a fast way with respect to the τ_1 delay, the tap does not move at all. These delays are differently set for the two ULTCs to avoid unrealistic synchronization.†††

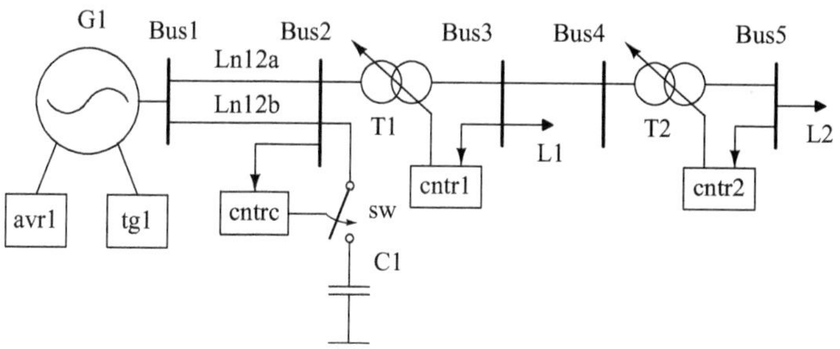

Figure 18.9 The schematic of the test system using two ULTCs and a switching capacitor bank

***The model of the ULTCs is similar to that described in [53]. The number of positions and ratio can vary in our simulations as specified.

†††The delays of `cntr1` and `cntr2` are set to 35 s and 40 s, respectively.

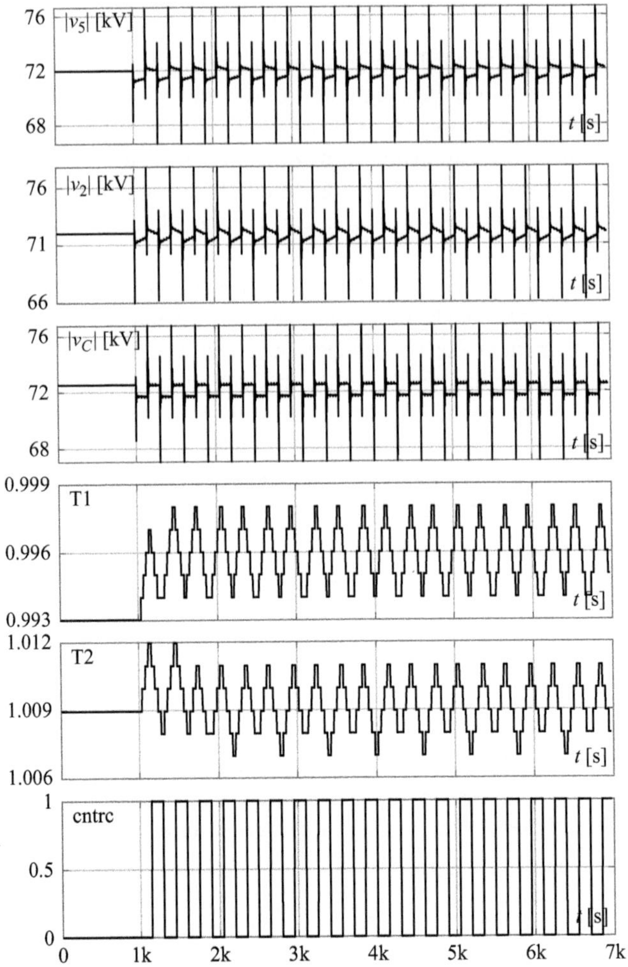

Figure 18.10 Simulation results obtained by long-term transient stability analysis. From upper to lower panel: the modulus of the `bus3` *voltage, it is used as input of the* `cntr1` *controller of the* `T1` *ULTC (kV) (upper panel). The modulus of* `bus5` *voltage, it is used as input of the* `cntr2` *controller (kV) (second panel from top). The modulus of* `bus2` *voltage, it is used as input of the* `cntrc` *controller that switches on/off the capacitor bank (kV) (third panel from top). The tap ratio of the* `T1` *transformer (third panel from the lowest one). The tap ratio of the* `T2` *transformer (second panel from the lowest one). The digital on/off signal of the capacitor bank (lowest panel). X-axis is time (s)*

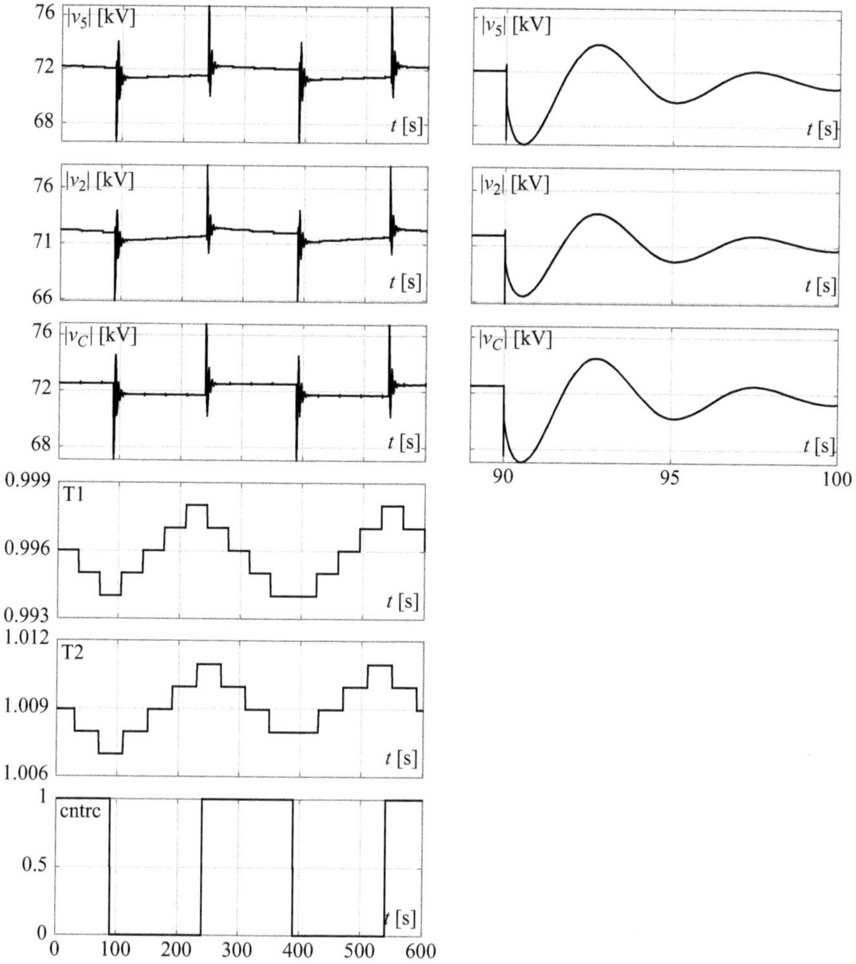

Figure 18.11 *Simulation results obtained by the TDSM method. The meaning of the waveforms reported in each panel is identical to that described in the caption of Figure 18.10*

The capacitor bank is inserted if the modulus of the voltage at `bus1` falls below a lower threshold and it is disconnected when the same voltage goes above an upper threshold. These two thresholds form a deadband. As for the ULTCs, when the voltage modulus exists this deadband, the on switching of capacitor bank may take place after the τ_2 delay and only if the voltage is still outside the dead band and has not crossed it.[‡‡‡] The switch off occurs when the voltage comes back inside the deadband and is driven by the same mechanism.

[‡‡‡]The delay is set to 150 s.

Consider the `cntr1` block in Figure 18.9. It senses the modulus of the `bus3` voltage and changes the tap position of `T1` as described before. The other two `cntr2` and `cntrc` controllers act in an identical way.

We used dynamic model of the lines and transformers, loads are constant impedance. In Figure 18.10, we report some results from a transient stability analysis. After 2,000 s from the beginning of the simulation, the `Ln12b` line is opened. This causes a sag in the voltage at `bus2`, `bus3`, `bus4`, and `bus5`. After the specified delays, the `cntr1` and `cntr2` controllers start changing tap positions to restore voltage magnitudes. The voltage moduli at `bus3` and `bus5` are shown in the first and second panel from top of Figure 18.10, respectively. The corresponding tap factors are shown in the third and the second panel from the last one, respectively.
However, as we can see, too much time passes without a full restore of moduli in their respective deadbands. The `cntrc` thus inserts the capacitor bank, since the voltage modulus at `bus2` is too low and outside the corresponding deadband. Insertion is visible since voltage moduli show a positive spike. Capacitor insertion causes the ULTCs to revert their action and to lower tap ratios. Also, this action takes too much time and being the voltage modulus at `bus2` still too high, after the τ_2 delay the capacitor bank is disconnected. The listed sequence of actions repeats indefinitely and periodically in time, that is, we have an oscillation.

This periodic steady-state oscillation can be computed through the described TDSM keeping in mind that we are dealing with a *hybrid dynamical* system characterized by *delayed events* as shown in Figure 18.2. Saltation matrices are computed as detailed in (18.11). In Figure 18.11, we report the periodic waveforms corresponding to those already shown in Figure 18.10. We see that the period of the oscillation is quite long being almost 600.47 s. This system instability can be classified as *long term*. The system shows several different orbits with different periods with respect to small parameter variations. In Figure 18.10, we have shown one among those. All those can be eliminated by suitably selecting the parameters (mainly delays and deadband magnitudes) of the controllers of the ULTC and capacitor bank.

18.5 Conclusions

This chapter shows an application of TDSM and PIT for the determination of limit cycles in power systems. The main advantage of the proposed technique is the ability to determine both stable and unstable periodic orbits in a unique framework. The proposed technique can also cope with hard limits and/or discontinuities, such as switched capacitor banks, on the right hand side of differential equations. Moreover, the technique shows a lower computational burden than other techniques proposed in the literature, e.g., [21].

The chapter also discusses an unconventional formulation of the power system model to cope with the requirements of the TDSM. The main feature of the proposed model concerns the representation of the speed reference of synchronous machines. This model is basically a generalized centre of inertia and involves a recast of the variables to avoid aperiodic drifting of machine angles. The proposed generalization of the centre of inertia indicates that a proper reformulation of synchronous machine equations allows applying techniques that are well-assessed in circuit analysis. Rethinking power system models based on rigorous formalism appears as a challenging field of research.

Bibliography

[1] Seydel R. *Practical Bifurcation and Stability Analysis*, 3rd ed. New York, NY: Springer; 2010.

[2] Abed EH, Varaiya PP. Nonlinear oscillations in power systems. *International Journal of Electrical Power and Energy Systems*. 1984;6:37–43.

[3] Alexander JC. Oscillatory solution of a model system of nonlinear swing equation. *International Journal of Electrical Power and Energy Systems*. 1986;8:130–136.

[4] Mitani Y, Tsuji K. Bifurcations associated with sub-synchronous resonance. *IEEE Transactions on Power Systems*. 1998;13(1):139–144.

[5] Hiskens IA. Stability of Limit cycles in hybrid systems. In: *Proceedings of the 34th Hawaii International Conference on System Sciences*; 2001. p. 1–6.

[6] Nomikos BM, Potamianakis EG, Vournas CD. Oscillatory stability and limit cycle in an autonomous system with wind generation. In: *Power Tech, 2005 IEEE Russia*; 2005. p. 1–6.

[7] Dobson I, Alvarado F, DeMarco CL. Sensitivity of Hopf bifurcation to power system parameters. *IEEE Decision and Control*. 1992;3:2928–2933.

[8] Cañizares CA, Hranilovic S. Transcritical and Hopf bifurcation in AC/DC Systems. In: *Proceedings of the Bulk Power System Voltage Phenomena III – Seminar*. Davos, Switzerland; 1994. p. 105–114.

[9] Joshi SK, Srivastava SC. Estimation of closest Hopf bifurcation in electric power system. In: *12th Power System Computational Conference*; 1996. p. 195–200.

[10] Lautenberg MJ, Pai MA, Padiyar KR. Hopf bifurcation control in power system with static var compensators. *International Journal of Electrical Power and Energy Systems*. 1997;19(5):339–347.

[11] Nastov O, Telichevesky R, Kundert K, White J. Fundamentals of fast simulation algorithms for RF circuits. *Proceedings of the IEEE*. 2007;95(3):600–621.

[12] Suarez A. Analysis and design of autonomous microwave circuits. *Wiley Series in Microwave and Optical Engineering*. New York, NY: John Wiley & Sons; 2009.

[13] Traversa FL, Bonani F. Improved harmonic balance implementation of Floquet analysis for nonlinear circuit simulation. *AEU – International Journal of Electronics and Communications*. 2012;66(5):357–363.

[14] Wang B, Ngoya E. Integer-N PLLs verification methodology: large signal steady state and noise analysis. *Circuits and Systems I: Regular Papers, IEEE Transactions*. 2012;59(11):2738–2748.

[15] Abed EH, Varaiya PP. Nonlinear oscillations in power systems. *International Journal of Electrical Power & Energy Systems*. 1984;6(1):37–43.

[16] Ajjarapu V, Lee B. Nonlinear oscillations and voltage collapse phenomenon in electrical power system. In: *Power Symposium, 1990. Proceedings of the Twenty-Second Annual North American*; 1990. p. 274–282.

[17] Laufenberg MJ, Pai MA, Padiyar KR. Hopf bifurcation control in power systems with static var compensators. *International Journal of Electrical Power & Energy Systems*. 1997;19(5):339–347.

[18] Howell F, Venkatasubramanian V. Transient stability assessment with unstable limit cycle approximation. *IEEE Transactions on Power Systems*. 1999;14(2):667–677.

[19] Li J, Venkatasubramanian V. Study of unstable limit cycles in power system models. In: *IEEE Power Engineering Society Summer Meeting*, vol. 2; 2000. p. 842–847.

[20] Chow JH, Wu FF, Momoh J. Applied mathematics for restructured electric power systems. In: Chow JH, Wu FF, Momoh J, editors. *Applied Mathematics for Restructured Electric Power Systems. Power Electronics and Power Systems*. New York, NY: Springer US; 2005. p. 1–9.

[21] Venkatasubramanian V, Li Y. Computation of unstable limit cycles in large-scale power system models. In: *IEEE International Symposium on Circuits and Systems (ISCAS 2006)*; 2006. p. 735–738.

[22] Aprille TJ, Trick TN. Steady-state analysis of nonlinear circuits with periodic inputs. *Proceedings of the IEEE*. 1972;60:108–114.

[23] Aprille T, Trick T. A computer algorithm to determine the steady-state response of nonlinear oscillators. *IEEE Transactions on Circuits and Systems*. 1972;19(4):354–360.

[24] Bizzarri F, Brambilla A, Gruosso G, Gajani GS. Time domain probe insertion to find steady state of strongly nonlinear high-Q oscillators. In: *2013 IEEE International Symposium on Circuits and Systems (ISCAS)*; 2013. p. 1865–1868.

[25] Bizzarri F, Brambilla A, Gruosso G, Gajani G. Probe based shooting method to find stable and unstable limit cycles of strongly nonlinear high-Q oscillators. *IEEE Transactions on Circuits and Systems – I: Regular Papers*. 2013;60(7):1870–1880.

[26] Kuznetsov YA. *Elements of Applied Bifurcation Theory*, 3rd ed. New York, NY: Springer-Verlag; 2004.

[27] Farkas M. *Periodic Motions*. New York, NY: Springer-Verlag; 1994.

[28] Ben-Israel A, Greville TNE. *Generalized Inverses: Theory and Applications. CMS Books in Mathematics*. New York, NY: Springer; 2003.

[29] Milano F. *Power System Modelling and Scripting*. New York, NY: Springer; 2010.

[30] Ascher UM, Petzold LR. *Computer Methods for Ordinary Differential Equations and Differential-Algebraic Equations*, vol. 61. Philadelphia, PA: Siam; 1998.

[31] Peters K, Parlitz U. Hybrid systems forming strange billiards. *International Journal of Bifurcation and Chaos in Applied Sciences and Engineering*. 2003;13(9):2575–2588.

[32] Di Bernardo M, Budd CJ, Champneys AR, Kowalczyk P. *Piecewise-smooth Dynamical Systems, Theory and Applications*. New York, NY: Springer-Verlag; 2008.

[33] Nordmark AB. Non-periodic motion caused by grazing incidence in an impact oscillator. *Journal of Sound and Vibration*. 1991;145(2):279 – 297.

[34] Bizzarri F, Brambilla A, Storti Gajani G. Extension of the variational equation to analog/digital circuits: Numerical and experimental validation. *International Journal of Circuit Theory and Applications*. 2013;41(7):743–752.

[35] Mees AI. *Dynamics of Feedback Systems. A Wiley-Interscience Publication*. New York, NY: J. Wiley; 1981.

[36] Parker TS, Chua LO. *Practical Numerical Algorithms for Chaotic Systems*. New York, NY: Springer-Verlag; 1989.

[37] Ngoya E, Suárez A, Sommet R, Quéré R. Steady state analysis of free or forced oscillators by harmonic balance and stability investigation of periodic and quasi-periodic regimes. *International Journal of Microwave and Millimeter-Wave Computer-Aided Engineering*. 1995;5(3):210–223.

[38] Gourary M, Ulyanov S, Zharov M, Rusakov S, Gullapalli KK, Mulvaney BJ. Simulation of high-Q oscillators. In: *1998 IEEE/ACM International Conference on Computer-Aided Design, 1998. ICCAD 98*. Digest of Technical Papers; 1998. p. 162–169.

[39] Brambilla A, Gruosso G, Gajani GS. Robust harmonic-probe method for the simulation of oscillators. *IEEE Transactions on Circuits and Systems – I: Regular Papers*. 2010;57(9):2531–2541.

[40] Petzold L, Li S, Cao Y, Serban R. Sensitivity analysis of differential-algebraic equations and partial differential equations. *Computers & Chemical Engineering*. 2006;30(10):1553–1559.

[41] Sauer PW, Pai MA. *Power System Dynamics and Stability*. Hoboken, NJ: Prentice Hall; 1998.

[42] Aulbach B. *Continuous and Discrete Dynamics Near Manifolds of Equilibria*. Lecture Notes in Mathematics. New York, NY: Springer-Verlag; 1984.

[43] Sauer PW, Pai MA. Power system steady-state stability and the load-flow Jacobian. *IEEE Transactions on Power Systems*. 1990;5(4):1374–1383.

[44] Aulbach B. Behavior of solutions near manifolds of periodic solutions. *Journal of Differential Equations*. 1981;39(3):345–377.

[45] Nakhla M, Vlach J. A piecewise harmonic balance technique for determination of periodic response of nonlinear systems. *IEEE Transactions on Circuits and Systems*. 1976;23(2):85–91.

[46] Bizzarri F, Brambilla A, Storti Gajani G, Banerjee S. Simulation of real world circuits: extending conventional analysis methods to circuits described by heterogeneous languages. *IEEE Circuits and Systems Magazine*. 2014;14(4):51–70.

[47] Powell MJD. A Hybrid Method for Nonlinear Equations (Chap 6, pp. 87–114) and A Fortran Subroutine for Solving systems of Nonlinear Algebraic Equations (Chap 7, pp. 115–161) in Numerical Methods for Nonlinear Algebraic Equations. P. Rabinowitz, editor. Gordon and Breach; 1970.

[48] Kundur P. *Power System Stability and Control. The EPRI power system engineering series*. London: McGraw-Hill; 1984.

[49] Lee JM. Verilog® quickstart: a practical guide to simulation and synthesis in verilog. In: *The Springer International Series in Engineering and Computer Science*. New York, NY: Springer; 2002. Available from: https://books.google.it/books?id=9XPFGjP8MlwC.

[50] Donde V, Hiskens I. Shooting methods for locating grazing phenomena in hybrid systems. *International Journal of Bifurcation and Chaos*. 2006;16(3):671–692.

[51] Donde V, Hiskens IA. Analysis of tap-induced oscillations observed in an electrical distribution system. *IEEE Transactions on Power Systems*. 2007;22(4):1881–1887.

[52] Sakellaridis N, Karystianos M, Vournas C. Local and global bifurcations in a small power system. *International Journal of Electrical Power & Energy Systems*. 2011 09;33:1336–1347.

[53] Van Cutsem T, Papangelis L. Description, Modeling and Simulation results of a Test System for Voltage Stability Analysis; University of Liége, Internal Report, 2013. Available: https://orbi.uliege.be/bitstream/2268/141234/1/Nordic_test_system_V6.pdf.

[54] Brambilla A, Storti-Gajani G. Frequency warping in time domain circuit simulation. *IEEE Transactions on Circuits and Systems I: Fundamental Theory and Applications*. 2003 July;50:904–913.

Chapter 19
Stability assessment in advanced DC microgrids

Daniele Bosich[1], *Massimiliano Chiandone*[1] *and Giorgio Sulligoi*[1]

The DC technology appears as promising in enabling new advanced microgrids. The reason is to be sought in the distributed implementation of PV renewable energy sources and battery storage systems, which actually operate by a DC distribution. When a widespread use of power converters guarantees the DC microgrids functionality, an attentive evaluation is to be carried out on the DC stability matter. The present chapter wants to investigate the methods to assess the stability in isolated DC distribution systems. As the latter can be lost when high are the converters control bandwidths, analytical developments are proposed to correlate requested control performance and poles positioning. If the methodology is initially conceived for a classical DC radial distribution, a final study will transfer it on complex DC Zonal Electrical Distribution Systems.

19.1 Introduction

In recent years, the impressive advancements in power electronics have launched the DC technology as decisive in the modernization of electrical power distribution. Indeed, nowadays the latter not only has to guarantee the proper loads supply, but also it has to ensure the requirements of efficiency and sustainability. Only the smart implementation of a totally controlled grid can be the answer when pursuing these upgraded features. As a matter of fact, the two requirements are now fundamental to get the transition in electric power generation towards the green era. The desired transformation in the way of managing the energy is mainly supported by the adoption of Distributed Energy Resources (DER) and storage. The available technologies for DER (e.g. photovoltaic plants, wind farms, hydropower plants, etc.) and storage (e.g. batteries, pumped-hydro plants, flywheels, etc.) are revolutionizing the concept of electrical distribution grid into the new one of controlled microgrid. Thanks to the wide utilization of power conversion, the latter is able to address the power flows in real time, while at the same time fostering the exploitation of aleatory sources

[1]Department of Engineering and Architecture, University of Trieste, Italy

thanks to the storage. Although the microgrid concept has been conceived for the low-voltage AC grids [1], the adoption of the DC technology appears now feasible and effective for the microgrids implementation [2–4]. Not only the DC distribution can promote the sustainability in the residential context [5,6], but other important benefits can make the DC microgrids convenient when feeding the onboard users of electric ships [7–12]. In land applications, the DC microgrids are well regarded being simplified the integration of DC subsystems, like PV generation and battery storage. Differently, in shipboard context the direct current can play an important role when chasing the space-weights reduction in power systems [13,14]. Indeed, when adopting the DC technology to reduce the space for hosting the electrical installations, new onboard areas are made available to increase the ship pay-load [15,16]. If on one side the DC distribution is powerful to increase the system flexibility (with all the related advantages), on the other such a feature is made possible thanks to a widespread use of controlled power converters. In a DC distribution, the latter are responsible to interface sources-loads to the DC bus (i.e. radial distribution [13]) or to interconnect a specific source-load area (i.e. zonal distribution [17]). Whatever the distribution is implemented, a large use of controlled power converters is undeniable for enabling a smarter management of sources and loads. Evidently, each converter must be equipped with a proper filtering stage to ensure the power quality requirements on DC distribution [13]. On the other hand, a not optimized match between controlled converter and filter can induce voltage oscillations and even the instability if the DC grid is isolated. By considering the importance of system stability [18], the present chapter wants to investigate on the destabilizing phenomena in advanced DC microgrids. First, Section 19.2 provides the basis for modeling islanded DC distributions, with a particular attention on simplifying assumptions (i.e. Constant Power Load (CPL) or controlled load's model) and system topologies (i.e. radial and zonal). After an introduction on the possible destabilizing resonance in DC grid, great importance is put on the control systems, which actually can be responsible for the instability in bad-designed filtered systems. Once defined the scenario, Section 19.3 is conversely oriented on the methodology to assess the DC stability. By starting from the mathematical models of DC controlled systems, a convenient procedure is proposed to finally find the system poles, thus verifying the stability requirement. Finally, Section 19.4 shows an example where the proposed methodology is able to predict the DC grid's behavior.

19.2 DC power systems modeling

The future power distribution will be based on a large employment of electronics converters to proficiently address the power flow, while at the same time ensuring controllability, flexibility and sustainability, both in land and in marine applications [4, 10]. Among the different distributions, the one based on DC appears to be a convenient possibility, especially when integrating subsystems naturally operating in DC current (i.e. PV, storage). When the DC technology is adopted to distribute the power towards the loads, unstable behaviors can occur when the LC filtering arrangements are not properly tuned on the DC–DC converters' control bandwidths. This phenomenon is made evident in an isolated DC microgrid feeding a high-performance controlled load. In such a case, a perturbation on the bus (i.e. load connection) can potentially trigger the system instability, thus the protections intervention. If the DC microgrid operates in islanded configuration, the blackout results consequent as well as the negative outcomes related to microgrid switch-off. During the last 25 years, a great effort has been spent in analyzing this destabilizing resonance, firstly assuming the condition of infinite bandwidth on the load converter control, thus the so-called CPL model [19–22]. Although this nonlinear representation does not constitute the worst case scenario [23], a large bibliography has been developed on the control systems to compensate for the CPL instability. Not only several contributions have considered the CPL infinite bandwidth's assumption in the near past [24–28], but also nowadays the same hypothesis is conventionally adopted to synthesize the stabilizing control in DC microgrids [29–33]. In the cases discussed in bibliography, the CPL modeling results convenient for two main reasons: on the one hand, it is representative of a well-recognized critical case (i.e. small/negative damping factor when a filtered DC–DC converter feeds a CPL), on the other hand, the no-dynamics modeling of a CPL is simple and straightforward. A different approach has been proposed by the authors to overcome the nonlinear CPL modeling, then investigating in detail the reason of DC system instability. Particularly, the works [34–37] have demonstrated how the unstable behavior is strictly related to the control performance on load converter. In other words, the infinite bandwidth of CPL is now made finite, where its magnitude greatly influences the stability behavior for a given LC filter on the DC load. A not integrated and consequential design of filtering stage (i.e. resonance frequency ω_f) and control (i.e. bandwidth ω_c) is the cause of DC instability, as observed in the optimization process in [38].

From this background, this section discusses about two islanded DC distributions on which study the system stability in the presence of perturbations. As seen before, the applied perturbation (e.g. load step, generating converter disconnection, etc.) can provoke unstable behavior on the DC filtered system in the presence of large controlled bandwidths on the loads side. Therefore, the models to represent the DC distributions must enable the possibility to manage/change the control bandwidths on controlled DC–DC converters. To do this, evidently the CPL assumption is overtaken, whereas the simplified modeling in [34–37] constitute the effective way to take into account the load converter dynamics and its effect on stability. In this regard, detailed considerations on the microgrid controls are in Section 19.2.3. For what concerns the possible DC distributions, the authors have concentrated their attention on the marine context. This particular scenario is noteworthy when discussing about the DC stability. Indeed, the complex DC shipboard grids are designed to supply large power (MW) to the loads, the LC filters are possibly downsized for space reasons while the system is islanded by definition [7,10,39]. All the three features make the shipboard DC microgrid a challenging test-bed on which assessing the system stability. In this chapter, the two electrical distributions on which evaluate the destabilizing effects are therefore collected from the IEEE Standard 1709 [13]. Such a standard is of paramount importance, as it provides a precise overview on controlled Medium Voltage DC (MVDC) power systems on ships. Although this standard is therefore focused on the marine applications, most of the guidelines are anyway transferable to a generic DC microgrid. Thus, the methodology to assess the stability is first conceived on complex MVDC shipboard grids. Later it can be applied on controlled Low Voltage DC–Medium Voltage DC (LVDC–MVDC) distribution systems to establish the stability performance in islanded microgrids on land.

19.2.1 Radial distribution

The system stability is firstly evaluated on the DC distribution proposed in Figure 19.1. In this radial topology [13], the generating part is depicted on the left, while the controlled loads are on the right. The system is supposed to be powered by a Diesel engine and a gas turbine, then the buffering function is guaranteed by a dedicated storage system (e.g. capacitor bank, batteries, fuel cells). Also, a shore connection platform on the right can feed the DC onboard system when the ship is mooring at the port. Clearly, several components are necessary to make available the generating/stored power. The cascade of AC generator (i.e. synchronous machine) and generic AC–DC rectifier is employed to make available the power from diesel/gas prime movers. The same power conversion's solution also for exploiting the landed power from the shore connection [40–42]. Each AC–DC interface can be uncontrolled or controlled. In the first case, a standard 6-pulses diodes bridge is able to impose an unregulated DC voltage, whose average value is proportional to the AC line–line Root Mean Square (RMS) voltage at the interface input (i.e. 1.35 is the ratio). Conversely, in the second case, Insulated Gate Bipolar Transistor (IGBT) switches are used to implement a totally controlled Active Front End converter to regulate voltage/current at the output of AC–DC power conversion stage. When the AC–DC power conversion is uncontrolled, an additional DC–DC buck-boost converter is conventionally added at the diode rectifier output to control the DC bus voltage. Differently, such a stage is not necessary if the AC–DC stage is able to regulate its voltage/current outputs. Other DC–DC interface is finally used to enable the charge/discharge functionality on battery modules. For what regards the power scheme on the right, several DC–DC converters work as interfaces towards the onboard loads. As the standard is focused on marine shipboard systems, also the loads are typical of the marine context. Thus, it is possible to enumerate the drives for moving the ship propellers, high-demanding loads such as the radar and the pulsed one and finally generic ship service load centers. Evidently, the widespread utilization of power converters on the one hand increases the flexibility in managing the onboard power toward the loads, on the other hand, enables high-performing dynamics performance on controlled loads. A similar complexity in electronics interfaces is also expectable in terrestrial DC microgrids working in islanded mode, where conversely the power conversion is mainly devoted in ensuring the optimization in power management, then pursuing the target of sustainability.

19.2.2 Zonal distribution

The distribution in Figure 19.2 is the zonal case where assessing the DC system stability. This innovative structure is envisaged in the IEEE Std 1709 [13], where the so-called Zonal Electrical Distribution System (ZEDS) is proposed as a flexibility

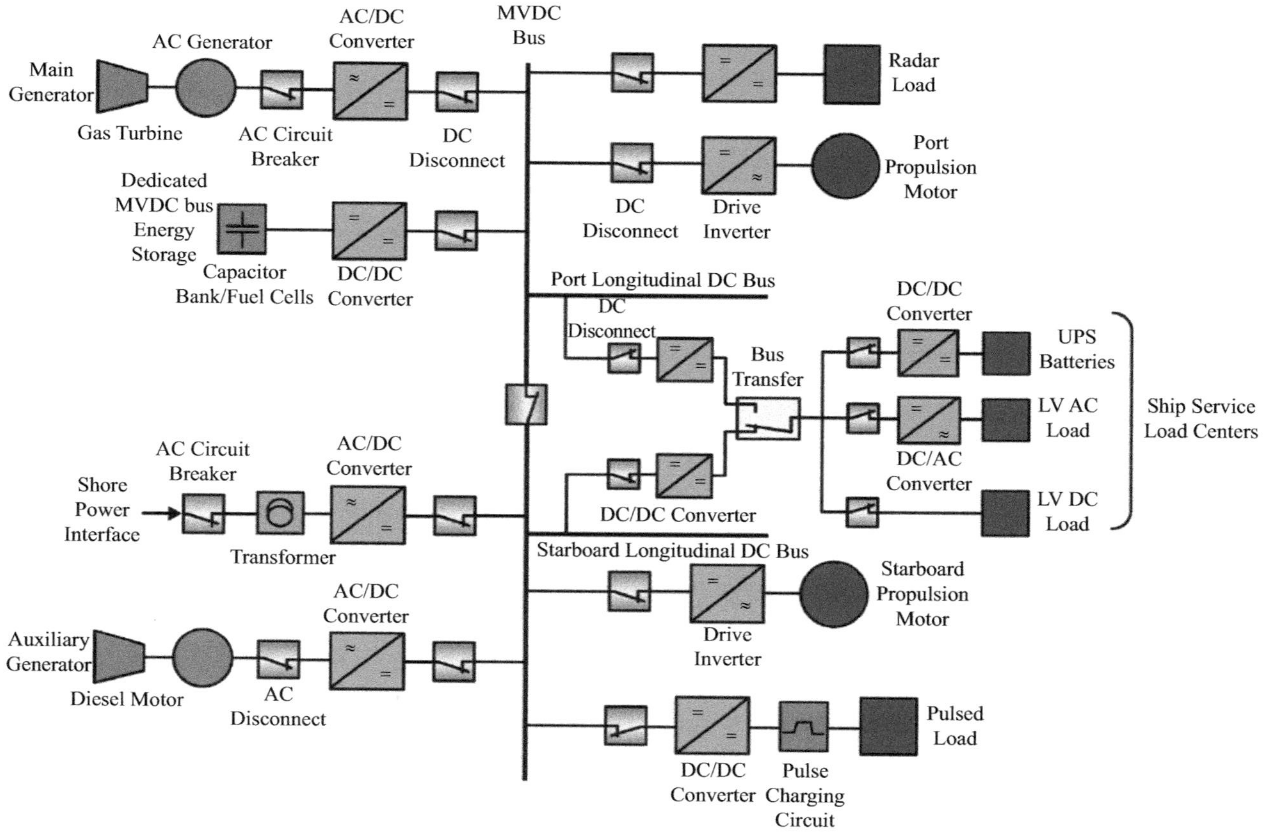

Figure 19.1 Radial DC marine power system [14]

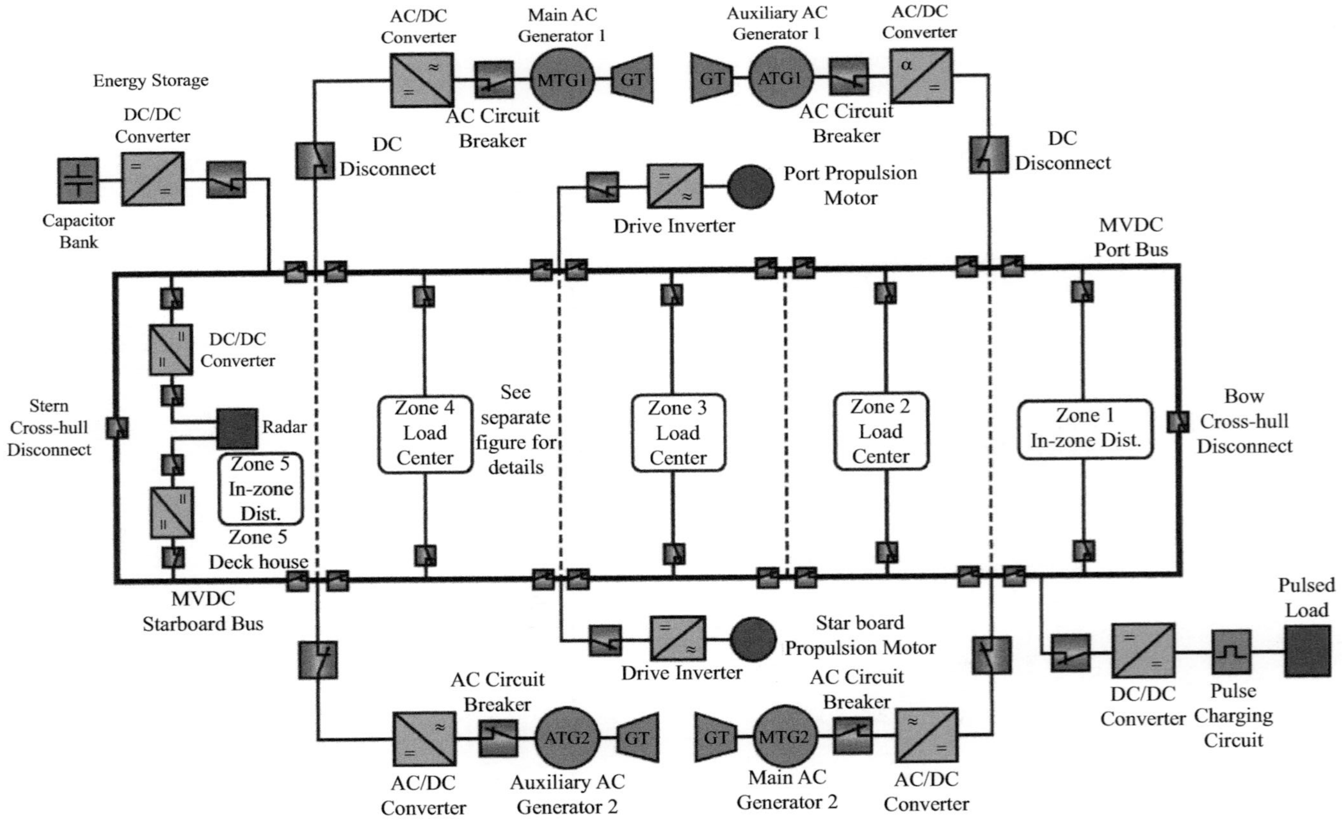

Figure 19.2 Zonal DC marine power system [14]

enabler for marine applications. The ZEDS topic is introduced in the IEEE Std 1826 [17], which identifies the zonal distribution as logical transition from the previous concepts and guidelines introduced in the IEEE Std 1662 [43] and the IEEE Std 1676 [44]. In this regard, the IEEE Std 1826 introduces the ZEDS as a power electronics open system interface with a power rated above 100 kW. Albeit the ZEDS topic has been proposed for navy shipboard power systems since 20 years [45], a recent contribution [46] has clarified the control systems and protections as they are essential in the management of such an islanded flexible grid. In such a zonal system, the ZEDS is configured to feed a group of loads as conceived in [17]. Particularly, ZEDS and its supplied loads are part of a larger set called Zone. This corresponds to the smallest logical-physical grouping of generating units, storage systems, and consumption (Figure 19.2). By means of a limited number of power/control interfaces, one or more external power systems (or other Zones) are interconnected to each Zone. The potentiality in reconfiguring each Zone appears evident and summarized in its intrinsic features: (a) it contains one or more independent power device, (b) it operates as an integral part of a larger system (normal operation), (c) it is capable of working independently for a short time (special operating conditions). As in [17,45–48], the ZEDS plays the role of linking block among the several grids, while showing some characteristics that are fundamentals in the marine case (i.e. redundancy, reconfigurability, fault resilience, and high efficiency). Such important advantages are guaranteed in the presence of a smart control system [49–51], which is designed and optimized to implement the master–slave strategy (i.e. centralized control), the independent/communicating controls and intelligent devices (i.e. distributed control), or the global control without communication (i.e. autonomous).

The ZEDS hierarchical control architecture is shown in Figure 19.3, where the function layers are depicted. To give a complete treatise about this architecture [17,46], in the following, the main functions are presented: (1) the external-to-bus conversion block to manage the power flow between in/out of a ZEDS; (2) the in-zone distribution bus to similarly ensure the power exchange among the ZEDS subsystems; (3) the in-zone energy storage system to guarantee the PQ/QoS requirements; (4) the in-zone generation element to support the power production inside the ZEDS; (5) the bus-to-internal conversion element to properly adapt the output supply; (6) the faults prevention in the the conversion system; (7) the distribution panel to interface the final devices. All the elements here described are connected through power electronics interfaces to avoid interruption in power supply during the transition from one interface to the next one. To ensure this functionality, three control functional layers are adopted: (a) multi-zone control (time constant above 100 ms) to coordinate the system mission/duties; (b) zonal control (time constant above 10 ms) to manage the in-zone control in order to impose the zone mission; (c) in-zone control (time constant smaller than 10 ms) to regulate the actions of power electronics components. Once the time constants are defined in Figure 19.3, the first-order assumption on the controls is useful to easily find the control bandwidths, then consequently provide guidelines for getting the dynamics decoupling in ZEDS controls. Then, by mathematically inverting the time constants, the consequent control bandwidths are found as 10–100–1000 rad/s. The decoupling is an important requirement to be attained when designing

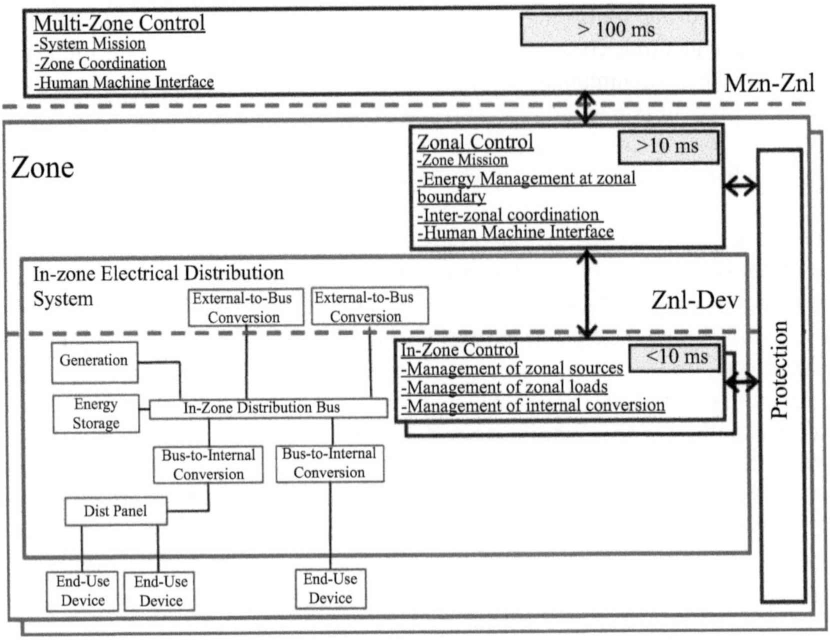

Figure 19.3 ZEDS hierarchical control architecture [46]

the control loops. To prevent destabilizing interactions among controls, a convenient repositioning [46] in accordance to the IEEE Std 1826 [17] configure three increasing bandwidths, where the one decade distance is imposed. The bandwidth on multi-zone control bandwidth is put equal to 5 rad/s. The zonal and in-zone bandwidths are, respectively, 50 and 500 rad/s.

19.2.3 DC voltage control

Both in radial (Figure 19.1) and in zonal (Figure 19.2) distributions, the DC–DC interface converters are the actuators to guarantee the system operation on stable equilibrium points. As in [52], three are the main control strategies (i.e. centralized, decentralized, and distributed) to properly manage the power electronics in DC microgrids. In general, the interfaces based on power electronics support the chosen strategy by performing the current or the voltage controls as output targets [53,54]. In the first case, a PI current loop is used to regulate the operation of a DC–DC interface converter (i.e. buck, boost, buck-boost topology). Such an action is aimed at imposing a certain current (i.e. reference) at the converter output. Differently, the PI voltage control on DC–DC converter (again step-up or step-down system) wants to calibrate the duty cycle to ensure the desired voltage at the converter output stage. As the voltage-controlled converters force the energization on DC grid, their importance results apparent. In a DC distribution with several interfaces, some DC–DC converters are usually current-controlled while the others are in charge of imposing the reference DC voltage on the bus. As multiple converters govern the same bus voltage, thus droop functionality is mandatory to ensure a bus voltage near the rated value (i.e. small droop coefficients) while sharing the power among voltage-controlled converters [10]. Hierarchical control strategies for DC systems are in [55–57].

19.3 Methodology to assess the DC stability

This section wants to propose a new methodology to investigate the stability in islanded DC microgrids, like the ones adopted in marine applications. As these power systems are isolated from the infinite-power point (e.g. the PCC, Point of Common Coupling), the system stability can be easily put in danger. The absence of this fundamental requirement means uncontrolled voltage oscillations in the presence of perturbations (i.e. load reconfigurations), definitely the protections interventions and the system blackout. The stability relevance is even more evident in the DC controlled grids where destabilizing resonances can be more common especially when control and LC filters arrangements are not conscientiously tuned in a stability-oriented system design [38]. In this regard, the authors have proposed a complete study on DC stability assessment in [58]. This paper follows a first research interest, where pursuing the definition of an aggregated method to understand how multiple-differently controlled converters impact on the system stability of a DC radial grid. By posing on important assumptions and initial hypotheses, this important work proposes a method to largely simplify the stability analysis while ensuring the analytical study even when the controlled loads are numerous. Different research trend the one followed in this section, where conversely the system complexity is maintained and the poles definition is achieved by digital tools. To gets the numerical evaluation on DC stability performance, a multi-model methodology is here proposed as in [53]. This advanced method is based on the determination of several subsequent models. Their development is actually necessary to ensure the logical transition from the physical DC power system to the system matrix, whose study leads to the poles positioning in the Gauss plane. As well known, the poles position (i.e. right/left part of real axis) is representative of the stability performance (i.e. unstable/stable behavior). The determination of these positions deserves therefore a notable importance when investigating how a controlled DC power system plays in terms of stability.

The methodology of Section 19.3.1 is depicted in Figure 19.4 where different models are consequently obtained to reach the final numerical determination of poles positioning, thus the stability performance as an outcome. The flowchart proposed in the following is necessary to explain the methodology. In such a flowchart, a particular attention is put on models by providing remarks on their definition and validation.

19.3.1 Stability assessment flowchart

In the past, important contributions like [59] have demonstrated how impedance and Eigenvalue Based Method (EBM) are equivalent in studying the DC stability, thus returning comparable results. By starting from this consideration, the proposed methodology is devoted in applying the eigenvalues' study [60–63], albeit opening to the possibility of others equivalent methodologies if adopting the impedance method.

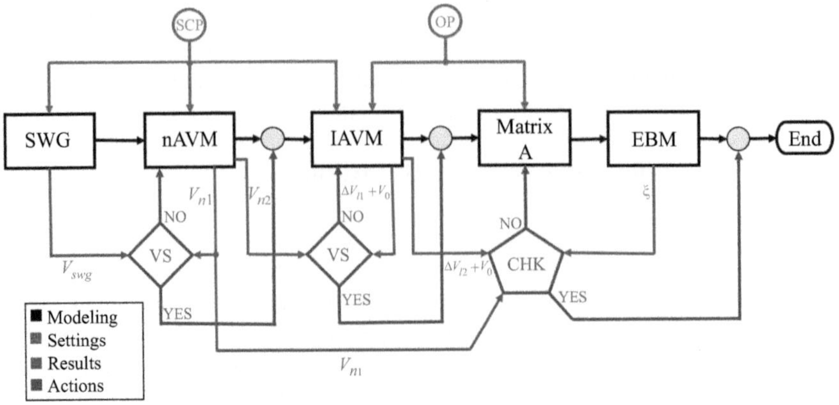

Figure 19.4 Stability assessment flowchart [53]

As known, the eigenvalue method is based on the definition of the analytical state-space matrix of the system. Consequently, the first assumption is related to the modeling data: the complete knowledge about the system is the starting point prior the mathematical modeling. In this work, all the system parameters are sufficiently known by hypothesis, therefore the state space matrix can be definable with a convenient margin of error. Basing on this, the flow chart in Figure 19.4 wants to provide the step-by-step path from the system definition towards the final poles determination. The procedure [53] to investigate the stability behavior relies on transients comparisons, where only a sufficient correspondence can enable the next stage of modeling. In the proposed flowchart, three colors are adopted to express the multi-model methodology. The subsequent models or methods are highlighted in black squares, where for example SWG stands for switching model or EBM represents the Eigenvalue Based Method to find the poles. The green color is used to express the system settings and data, then subdivided in system and control parameters (SCP*) and operating points (OP). The blue color shows the numerical simulations, which provides the data for performing the comparison (VS*) in red blocks. In red also, the final check (CHK*) prior the flowchart end. To cover the flowchart, it is possible to observe the following steps. From the SCP (i.e. filters parameters, power converters characterization, control constants, distribution topology), the switching model (SWG) is built as a circuital power scheme. In this regard, the proposed study utilizes the potentiality offered by PSCAD™ environment. Evidently other dynamics simulations tools can provide equivalent results. The two models named nonlinear average value model (nAVM) and linearized average value model (lAVM) are direct consequence of SWG modeling. Although focusing on the only average values (i.e. the power converters' switching is neglected), the nAVM is aimed at clearly representing the nonlinear interactions which are at the basis of SWG functioning. Then, the linearized lAVM is

*SCP, VS and CHK are acronyms of 'system and control parameters', 'comparison' and 'final check', respectively.

the direct consequence, when the equations of nAVM are linearized in the nearby of the equilibrium point (OP). Each modeling step is feedback verified by contrasting the blue output. When the red blocks (VS) confirm the conformity, it is possible to overcome the yellow circles checkpoints. Finally, the state-space matrix and the EBM eigenvalues are the final outputs to conclude about the system stability. Models, checkpoints, and outputs are explained in the following.

19.3.2 Models for stability study

As a matter of fact, different are the possibility for modeling a controlled DC grid. Each path in modeling is strictly dependent on the final target. The proposed path is the one for reaching the identification of system poles. Other goals are conversely obtained by following other alternatives in modeling. This section explains all the consequent steps to ensure the transition from the circuital model to the linearized system on which identify the matrix. The initial topology of a DC power system (in the next sections some examples will be discussed in detail) is the case which develops the circuital model. In authors experience, the latter can be obtained in PSCAD environment, being this simulation software able to accurately model the power converters by taking also into account switching phenomena. This SWG model is the closest to the effective power system behavior, therefore it constitutes the reference for the subsequent verifications/validations on the resulting nAVM model (Figure 19.4). As the dynamics under study is usually distant (i.e. at least one decade for the separation in bandwidths) from the AC control on generation side, ideal voltage generators are usually considered as sources of the interfaces to the DC bus. This assumption is quite common in the modeling of islanded DC microgrids, indeed other examples on this issue are in [25,27,58]. Differently, the SWG representation can model the action of storage systems, when their interface converters react with a dynamics comparable to the one achieved on DC distribution power electronics. Once modeled the power grid and the control on power converters, the SWG model can provide the first transient of interest. When the DC power grid under study presents a radial topology, the main bus voltage transient V_{swg} after a perturbation is the output to be considered from the SWG model. Conversely, in zonal topology, two approaches are possible. The first one declares one bus as dominant, so the related voltage transient constitutes the output from the SWG model. Instead, a second approach identifies several bus voltages as important in the zonal distribution. In this case, the comparisons to validate the models are to be replied on different V_{swg_k} inputs, then evidently complicating the analysis.

As in Figure 19.4, the SCPs are the initial inputs for the creation of the nAVM model. These parameters describe the only beginning point that builds the mathematical development. Second important issue in the nAVM representation is the one related to the AVM hypothesis. As suggested in [13], also this methodology wants to disregard the switching behavior of power converters. By putting this assumption, the converters' dynamics operations can be synthesized in mathematical equations thus opening the curtain toward the analytical modeling. Being the presence of nonlinear switching hardly to be modeled in mathematical equations, therefore this assumption plays a crucial role when pursuing a convenient mathematical representation of the investigated phenomena. From the knowledge about system distribution (i.e. power system's components, components' topology, power converters, control laws), a set of nonlinear differential equations can be written to represent the dynamics operation of the DC microgrid. This set of equations is conveniently defined by considering two aspects. A certain number of equations is directly obtainable by solving the power scheme thanks to the Kirchhoff's circuit laws. Conversely, a second group depends on the control laws, therefore the regulating signals implemented by the power electronics actuators. By combining these two groups, a final set of nonlinear equations describes the nAVM model. In authors experience, the Simulink® is a convenient tool for modeling the nAVM, especially when the integrators are made explicit to easily managing the differential equations. By forcing the same load step in the numerical model on Simulink as well as in the SWG on PSCAD, the nAVM dynamic response on the bus voltage V_{n1} is put in comparison (Figure 19.4) with the V_{swg} from the SWG model. When the nAVM transient corresponds to the average value of the switching one, then the nAVM model is validated so it can behaves as new reference for the next verification with linearized lAVM model. If this check is not verified, then the nAVM must be corrected up to provide a transient that corresponds to the mean value of the PSCAD evolution.

Finally, the lAVM model is the third outcome. As in Figure 19.4, such a linearized model is developed by considering the nonlinear equations of nAVM and the operating points (OP). To linearize the nAVM equations, the classical method foresees a sequence of partial derivatives in a set of stable operating points. The latter are constituted by the state variables (e.g. voltages, currents, duty cycles) in steady-state conditions. In the nAVM Simulink model, they represent the steady-state values that are visible at the integrators' outputs once the transient is concluded. In this regard, a consideration is consequent when taking into account the system perturbation (e.g. load step). To propose an accurate model, the equilibrium points are to be calculated in two scenarios, both before and after the perturbation. To perform the linearization, the nonlinear equations of nAVM are initially gathered as $\frac{d}{dt}x = f(x, u)$. In last equation, x are the state variables while u is utilized to represent the system inputs (e.g. voltage references, ideal voltages to supply the DC interfaces). When the nonlinear equations are linearized in the stable equilibrium points, the resulting model equations become effective in small-signal conditions (i.e. small perturbations around the steady-state condition) and they take the new linearized representation as in $\frac{d}{dt}\Delta x = \tilde{A} \cdot \Delta x + B \cdot \Delta u$. In last equation, $\tilde{A}$ and B are the matrices to precisely describe the linearized system. Being constant the inputs on the system, the B matrix is null. After the linearized lAVM model is defined, an important test is now requested to verify its correctness then proceeding in the multi-model stability assessment. In particular, an evaluation wants to compare two transients from nAVM and lAVM. As the validity of nAVM model has been already proven by the SWG correspondence, the nAVM can play as benchmark for the comparison with lAVM in terms of dynamics transients. To check the lAVM model against the nAVM one, the same small-perturbation (-10% on bus voltage) acts on both models implemented on Simulink tool. Particularly, by changing the initial condition on system's aggregated integrator (i.e. cumulative capacitance in radial distribution), the starting voltage value on the main bus voltage is accordingly modifiable in both models. As in Figure 19.4, the nAVM and the lAVM voltage transients (i.e. V_{n2} and $\Delta V_{l1} + V_0$) are now compared in the CHK checkpoint. When the transients are sufficiently similar, the two models behave as equivalent in the nearby of operating points. In this condition, the linearized state matrix is attainable from the lAVM for the stability study. Differently, the lAVM model needs a review and possibly some modifications.

19.3.3 State-space matrix

The final step of the multi-model methodology is related to the output of linearization process, thus the state-space $\tilde{A}$ matrix . The importance of this matrix is notable, as the study on $\tilde{A}$ can identify the poles position, so finally the stability performance of the DC system under investigation. As expressed before, the $\tilde{A}$ matrix is to be defined in two different conditions. Indeed, as the operating points are modified by the perturbation, consequently also the matrix's parameters change. In other words, one matrix is to be intended for the operation before the perturbation (i.e. load step), a second one for the analysis after the load increase. Once the matrices are ready to be studied, the EBM block numerically extracts the eigenvalues, thus finally defining the system poles. The analysis of the eigenvalues of these two matrices is effective to investigate the small-signal stability of the system before and after the perturbation. As well known from control theory, all the real part of eigenvalues must be negative for ensuring stable evolutions in the nearby of a precise operating point (i.e. the one in which the $\tilde{A}$ is calculated). In other words, a single eigenvalue with a positive real part is sufficient for an unstable behavior. The importance of these eigenvalues is therefore prominent, as well as the logical steps to find them. Additionally, the EBM at the final step is also able to identify the ξ damping factor, by observing the real part of the most critical poles, thus the ones more oriented on the right (unstable) plane.

Besides these important conclusions about stability, an additional confirmation is necessary to have the certainty about $\tilde{A}$ definition. A cross check between the information from EBM and the dynamics evolutions of both the nAVM and lAVM is therefore performed as visible in Figure 19.4. Also in this case, a step down on initial bus voltage condition (i.e. −10%) is adopted to close the discussion on nAVM, lAVM and $\tilde{A}$. Therefore, the stability assessment on $\tilde{A}$ eigenvalues is compared to the dynamic information coming from the nAVM and lAVM models, thus V_{n1} and $\Delta V_{l2} + V_0$ transients. Again, if the matrix eigenvalues have always negative real part, then both nAVM and lAVM models must experience a stable evolution toward the new operating point. When only one eigenvalue is characterized by a positive real part, thus the unstable evolutions must be made evident in nAVM and lAVM transients. In such a comparison, the ξ damping factor is also convenient for quickly glimpsing unstable conditions (i.e. negative damping factor). As said before, if this final verification is not positive, then certainly the construction of $\tilde{A}$ matrix or EBM algorithm are failed as the other models (i.e. nAVM and lAVM) are already verified thanks to the step-by-step methodology. The identification of system eigenvalues is at the end capable of giving the final check on system stability. On the other hand, the numerical assessment on the dynamics transients provides the proof about the modeling accuracy. Evidently, the choice of building partial-consecutive models is successful being fostered the error identification if necessary. The multi-model methodology flowchart is applicable to several DC controlled distributions. When modifying the system topology (e.g. moving from radial to zonal distribution), the only modification regards the mathematical model which ever must be sufficiently accurate to correctly approximate the real system behavior.

19.4 Application on DC microgrids

The present section is aimed at applying the multi-model methodology [53] on the DC distribution in Figure 19.5. The proposed power scheme is related to a controlled MVDC microgrid to be implemented on ships. Again, the marine context imposes a particular attention on the stability issues giving the intrinsic isolated operation. In DC systems, the large presence of controlled interface converters and relative filters actually constitute the perfect environment in which a small perturbation can trigger the destabilizing oscillations. Such an effect depends on the filtering arrangements and on the control performance to be ensured by the converters. This section considers both aspects when locating the poles by EBM.

19.4.1 Power system design

The advanced control of DC microgrids is achieved only in presence of a large employment of power electronics. For this reason, the radial distribution in Figure 19.5 shows

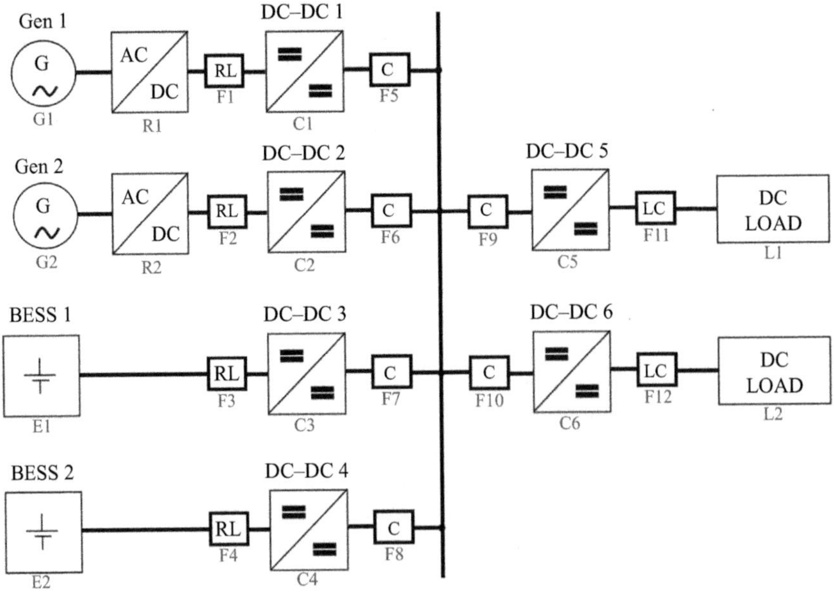

Figure 19.5 Radial multiconverter DC microgrid on ships [53]

several converters to interface the different elements to the common DC bus [53]. In the generating section on the left, two cascades of diode rectifiers (R1–R2) and DC–DC boost converters (C1–C2) are utilized to supply the grid from diesel generators (G1–G2). The storage function is given by E1–E2, the two battery energy storage systems (BESS). Their output voltage is adapted to the DC bus value thanks to two DC–DC buck-boost converters (C3–C4). On the load side, two DC–DC step-down buck converters (C5–C6) are in charge of feeding the two equivalent DC loads. These resistive terms are sufficiently accurate in aggregating the onboard loads. The data for the power converters design [53] are in Table 19.1, where the k subscript identifies the k-quantity or parameter for each component of the system. Then, the f_s is the switching frequency, V_{ACk} is the synchronous machine line–line output voltage, V_{DCk} represents the battery rated output and V_k the DC load rated voltage. To complete the information, P_{nk} is the converter rated power, V_n is the rated bus voltage, and I_{nk} the rated inductor current. Finally, R_{nk} models the rated load, while the converter control signal in nominal condition is named u_{nk}. To ensure the power quality requirements [13], each converter is equipped with its own input (F1–F4, F9–F10) and output (F5–F8, F11–F12) filter. An ohmic-inductive filter (R_{fk} and L_{fk}) is put at the input of diode rectifiers and batteries. For the C1–C2 converters, this filter resistor takes into account the converter losses. Differently, it models the battery internal resistance for C3–C4. In the DC system, each power converter has its own capacitor facing the bus. All the capacitors are thus parallel connected then summed in a single C_{BUS} capacitance. For the load converters (C5–C6), an LC filter is installed at the output, then C_{fk} and L_{fk} are the parameters once neglected the filter resistance. The onboard filters are in Table 19.2, while the design guidelines are in [64–67]. Attention is to be spent on the last two lines of the table. By assuming that a single converter is supplying its rated power to a CPL, the ω_{0fk} natural frequency and ξ_{fk} damping factor are two indices (Table 19.2) to weigh how the single filter impact on stability. Although the CPL is a pejorative assumption and this simplified evaluation considers each single converter as independent, a very negative damping factor (e.g. less than −0.5) anyway means the need of actions. An increment in capacitive filter or the control by linearizing functions can be crucial solutions.

Table 19.1 Design of DC–DC power converters [53]

	C1/C2	C3/C4	C5	C6
f_s [Hz]	3000	3000	3000	3000
V_{Ack} [V]	850	–	–	–
V_{Dck} [V]	–	1200	–	–
V_k [V]	–	–	1300	1300
P_{nk} [MW]	3	1	3	5
V_n [V]	1500	1500	1500	1500
I_{nk} [A]	2500	833	2308	3846
R_{nk} [Ω]	0.75	2.25	0.563	0.338
u_{nk}	0.2	0.2	0.867	0.867

Table 19.2 Design of filtering arrangements [53]

	C1/C2	C3/C4	C5	C6
Type	Boost	Buck-boost	Buck	
R_{fk} [mΩ]	35	85	–	–
L_{fk} [mH]	0.2	0.2	0.103	0.062
C_{fk} [mF]	–	–	0.360	0.560
C_{DCk} [mF]	6.0	6.0	6.0	6.0
ω_{0fk} [rad/s]	730	730	5193	5366
ξ_{fk}	–0.032	0.24	–0.37	–0.49

Table 19.3 Design of control bandwidths [53]

	I	II
Current controller C1–C2 [rad/s]	100	125
Bus voltage controller C3–C4 [rad/s]	10	220
Load voltage controller C5–C6 [rad/s]	1000	1000

19.4.2 Control design

In DC microgrids, the coordination of control tasks is essential and each converter must provide a specific work. In Figure 19.5, C3–C4 are indeed responsible for the bus voltage regulation in droop mode while ensuring an equal power sharing (i.e. $R_{dp} = 0.01\Omega$). This value is sufficient to have both dynamic decoupling and limited voltage drop in full power condition. Then, C1–C2 are current controlled and C5–C6 regulate the voltages on their own loads. To make the system equations more manageable, the control loop of each converter is based on the only action of an integral regulator. Although this assumption can appear simplistic, this control can properly command the converters action in voltages/currents regulation. The integral gain of each loop is individually tuned to ensure the requested control bandwidth. Then, the dynamics coherence is verified by identifying the time constants on reference step tests. Table 19.3 shows two combinations of control bandwidths. The first (I) is imposed to ensure the stability after the perturbation. In this case, the three control loops are dynamically decoupled. The second combination (II) is conversely conceived to experiment an unstable condition. In this case, not only the decoupling is lost but also the system poles are voluntarily put on the y-axis in the Gauss plane. This constitutes a sort of test for understanding the performance limit before dealing with instability. By performing the tests on these combinations, it is possible to verify the effective correspondence between poles location and simulated transients:

$$\begin{aligned}\frac{d}{dt}V_{DC}(t) = \frac{1}{C_{BUS}} \cdot \big[\big(1 - D_1(t)\big) \cdot I_1(t) + \big(1 - D_2(t)\big) \cdot I_2(t) + \\ + \big(1 - D_3(t)\big) \cdot I_3(t) + \big(1 - D_4(t)\big) \cdot I_4(t) + \\ - D_5(t) \cdot I_5(t) - D_6(t) \cdot I_6(t)\end{aligned} \tag{19.1}$$

$$\frac{d}{dt}I_1(t) = \frac{1}{L_1} \cdot \left[V_{RT1} - \left(1 - D_1(t)\right) \cdot V_{DC}(t) - R_1 \cdot I_1(t)\right] \tag{19.2}$$

$$\frac{d}{dt}D_1(t) = K_{i1} \cdot \left[I_1{}^* - \left(1 - D_1(t)\right) \cdot I_1(t)\right] \tag{19.3}$$

$$\frac{d}{dt}I_2(t) = \frac{1}{L_2} \cdot \left[V_{RT2} - \left(1 - D_2(t)\right) \cdot V_{DC}(t) - R_2 \cdot I_2(t)\right] \tag{19.4}$$

$$\frac{d}{dt}D_2(t) = K_{i2} \cdot \left[I_2{}^* - \left(1 - D_2(t)\right) \cdot I_2(t)\right] \tag{19.5}$$

$$\frac{d}{dt}I_3(t) = \frac{1}{L_3} \cdot \left[V_{BT3} - \left(1 - D_3(t)\right) \cdot V_{DC}(t) - R_3 \cdot I_3(t)\right] \tag{19.6}$$

$$\frac{d}{dt}D_3(t) = K_{i3} \cdot \left[V_{DC}{}^* - V_{DC}(t) - R_{dp} \cdot \left(1 - D_3(t)\right) \cdot I_3(t)\right] \tag{19.7}$$

$$\frac{d}{dt}I_4(t) = \frac{1}{L_4} \cdot \left[V_{BT4} - \left(1 - D_4(t)\right) \cdot V_{DC}(t) - R_4 \cdot I_4(t)\right] \tag{19.8}$$

$$\frac{d}{dt}D_4(t) = K_{i4} \cdot \left[V_{DC}{}^* - V_{DC}(t) - R_{dp} \cdot \left(1 - D_4(t)\right) \cdot I_4(t)\right] \tag{19.9}$$

$$\frac{d}{dt}V_5(t) = \frac{1}{C_5} \cdot \left[I_5(t) - \frac{V_5(t)}{R_{L1}}\right] \tag{19.10}$$

$$\frac{d}{dt}I_5(t) = \frac{1}{L_5} \cdot \left[D_5(t) \cdot V_{DC}(t) - V_5(t)\right] \tag{19.11}$$

$$\frac{d}{dt}D_5(t) = K_{i5} \cdot \left[V_5{}^* - V_5(t)\right] \tag{19.12}$$

$$\frac{d}{dt}V_6(t) = \frac{1}{C_6} \cdot \left[I_6(t) - \frac{V_6(t)}{R_{L2}}\right] \tag{19.13}$$

$$\frac{d}{dt}I_6(t) = \frac{1}{L_6} \cdot \left[D_6(t) \cdot V_{DC}(t) - V_6(t)\right] \tag{19.14}$$

$$\frac{d}{dt}D_6(t) = K_{i6} \cdot \left[V_6{}^* - V_6(t)\right] \tag{19.15}$$

19.4.3 Average value models

Once neglected the converters switching, (19.1)–(19.15) models the grid in Figure 19.5. This model preserves the nonlinear behavior, but it considers the only mean value of electrical quantities. The equations of each converter (C1–C6) are at the basis of the model. The common DC bus voltage is named V_{DC}, while the single current in the converter inductance is I_k. Each DC–DC converter is regulated by a duty cycle named D_k. In all the control loops acting on converters, the only integral coefficient K_{ik} is adopted, whereas the terms with asterisk are the references. To simplify the modeling, the V_{RTk} output of rectifiers is constant like the V_{BTk} output of batteries. The voltage on R_{Lk} load is named V_k. By combining (19.1)–(19.15), the nAVM numerical model is built in Simulink. From nAVM and equilibrium points $[V_{DC0}, I_{k0}, D_{k0}, V_{k0}]$, the linearization by partial derivatives defines the lAVM as in [54]. Also, this model is then implemented in Simulink.

19.4.4 Poles location

The small-signal stability of the controlled DC grid (Figure 19.5) is assessed by locating the system poles in the $\mathbb{C}$ plane [53,54]. Such an evaluation wants to verify how the control bandwidths impact on poles positioning. In this test, the DC grid parameters are in Tables 19.1 and 19.2. From the initial condition where the C5–C6 high-bandwidth (i.e. 1000 rad/s) converters provide 2 MW and 1.5 MW, respectively, an additional step request of 2 MW from C6 is the way to perturb the system. The performed analysis demonstrates how this perturbation cannot move the poles in the right plane when the system is controlled with tight and decoupled bandwidths (I). Conversely, the same perturbation triggers the instability if high performance (i.e. 220 vs. 10 rad/s) is requested on the bus voltage dynamics as in (II) combination. Thanks to the multi-model methodology and its consequent verifications on transients (Section 19.3.1), four lAVM models are defined to analyze the four cases. Thus, two lAVM models where the bandwidths are configured on combination (I), before and after the perturbation. Other two linearized models by setting on the (II) combination, in the first equilibrium point and in the second after the load step. From the lAVM models, the four matrices are analytically defined whereas the related eigenvalues are revealed by means of a numerical calculus. In the resulting representation in the Gauss plane, the complex conjugate poles are relevant for the stability matter, as they can experiment the shifting in the right half plane after the load increase. Conversely, the real poles are always on the stable plane, whichever the control bandwidths are. Figure 19.6 shows the poles with the (I) combination of bandwidths (100-10-1000 rad/s). Both before and after the step, all the poles have negative real parts, thus disclosing system stability. Instead, the poles with the (II) bandwidths combination (125-220-1000 rad/s) are in Figure 19.7. The rise in voltage-current control bandwidths force a pair of complex poles in passing the *y*-axis. Particularly, the focus is on the complex poles related to the bus capacitance. Before the load step, the stability is confirmed by their negative (−7) real part. Differently, the instability arises when their real part becomes positive (6) after the perturbation.

19.4.5 Numerical verification

The poles location is the method to envisage the possible reasons for microgrid instability (e.g. high-performance control, not-integrated filters design). Once identified the cause, the DC grid can be redesigned (e.g. reduction in control performance, increase in capacitive terms) to host the poles in the left half plane, thus ensuring the stability. As the location of poles is the result of a complex analytical-numerical process, the errors in modeling are quite common. A validation is thus unavoidable to verify the coherence between poles position and simulated transients. This section proposes the transients in Figures 19.8–19.13 to check the poles location. To provide an effective verification, each figure has the switching transients from SWG models in PSCAD and also the ones by running the nAVM models in Simulink. The results are in per-unit notation. The rated bus voltage (i.e. 1500 V) and the total generating

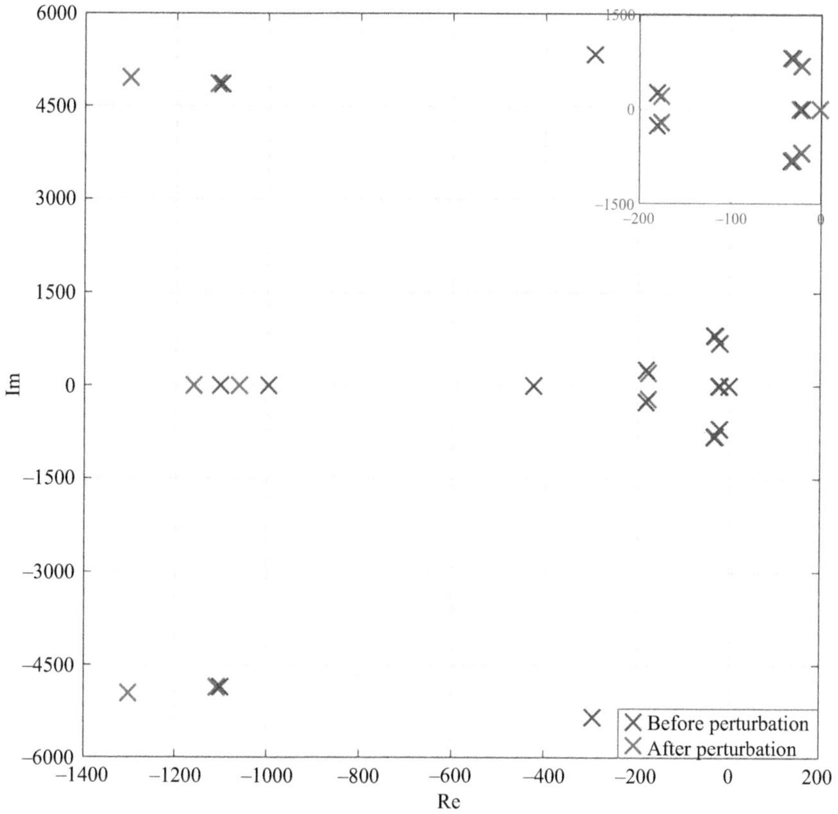

Figure 19.6 Poles location in the stable case, (I) bandwidths combination [53]

power (i.e. 8 MW) are the main bases, while the current is rescaled as a consequence (Table 19.1). For more clarity, only the per-unit of load voltage is referred to its own rated value (1300 V). To validate the EBM analysis on poles, Figures 19.8 and 19.9 give the DC bus voltage responses to the load increase at 0.7 s. The (I) combination is adopted in the transients of Figure 19.8, while the destabilizing combination (II) configures the systems simulated in Figure 19.9. As foreseen by poles location, when the control loops are set on (I) bandwidths, the stability is guaranteed even after the perturbation (Figure 19.8). Here, the voltage drop after the load increase is near –9%, then validating the small-signal assumption. Then, at steady-state the voltage settles at 0.995 p.u. due to the droop control. On the other hand, the unstable behavior predicted by the poles location is made evident also in the dynamics transients. Indeed, Figure 19.9 displays how the instability is triggered by the same perturbation that is responsible for the complex poles' repositioning in the right half plane (Figure 19.7). The good correspondence between switching (SWG) and average results (AVM) in Figures 19.8 and 19.9 is an additional verification on the validity of the multi-model methodology. To complete the study, also the power/current transients are provided

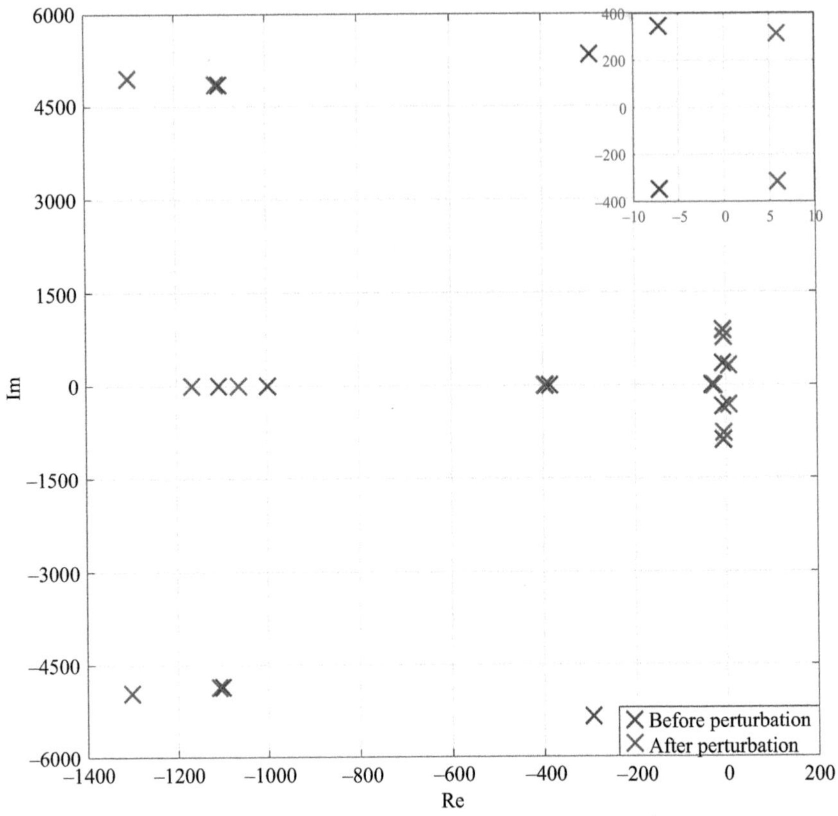

Figure 19.7 Poles location in the unstable case, (II) bandwidths combination [53]

for the stable case on (I) bandwidths configuration. Figure 19.10 shows the generating converters power, before and after the perturbation. At the initial steady-state, the C1–C2 current-controlled converters supply 3.5 MW (i.e. almost 0.45 p.u.) to the loads. This power is shared on the generating converters basing on their current references. Talking about the C3–C4 on BESS, at the beginning they do not furnish power being only responsible of controlling the bus voltage. Their overlapped transient shows no-power till the perturbation. At 0.7 s, the power from C6 (Figure 19.11) increases from 1.5 MW to 3.5 MW. As the bandwidth on current control is higher than the one on voltage control, then C1 and C2 are the first to react to this demand. Then the power shortage is supplied by C3–C4 as in Figure 19.10. As they control the bus voltage in droop mode, a slight reduction on this voltage is expectable (Figure 19.8) as well as a consequent small decrease in the powers from C1–C2 current-controlled converters. Therefore, C3–C4 must finally feed a little more than 1 MW each (i.e. 0.13 p.u.). Lastly, the C6 input current and output voltage are in Figures 19.12 and 19.13. In the last figure, the high-performance of C6 is evident where indeed the voltage is restored in about 2.5 ms after the load step.

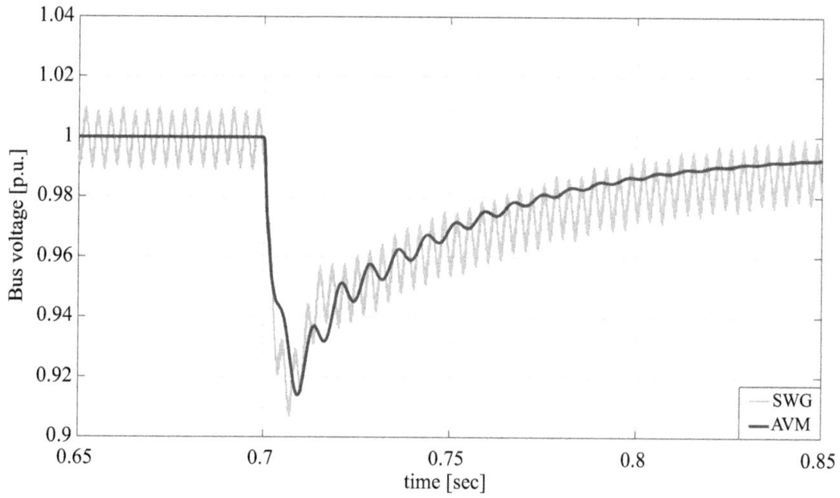

Figure 19.8 Bus voltage transient after perturbation, stable case [53]

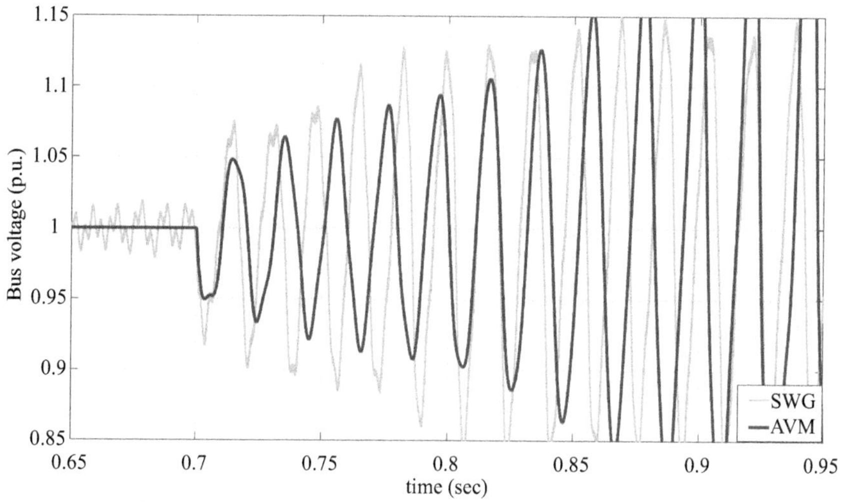

Figure 19.9 Bus voltage transient after perturbation, unstable case [53]

19.4.6 Considerations on stability assessment in zonal distribution

Since the early 2000, the DC technology has been the enabler for the zonal power systems on ships [45]. Nowadays, this distribution is the smartest way to manage the marine grids if seeking higher efficiency and redundancy [17,46]. As in Section 19.2.2, other advantages force towards the ZEDS implementation, for example

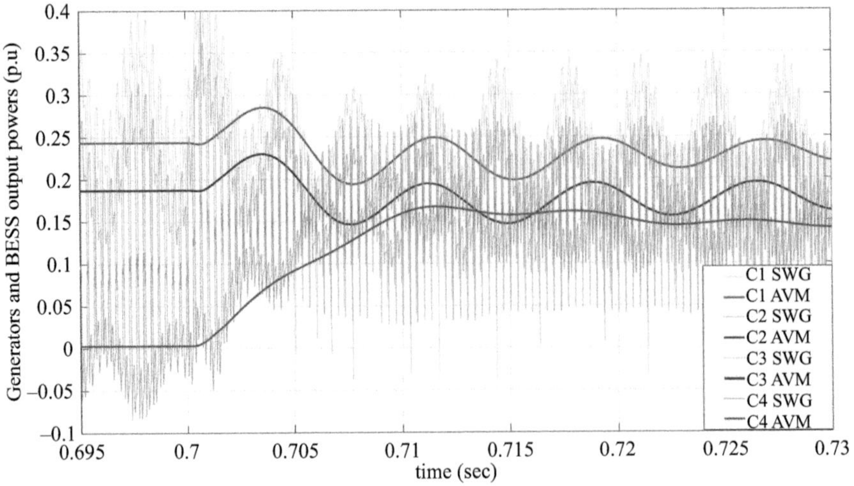

Figure 19.10 Output powers of generating converters [53]

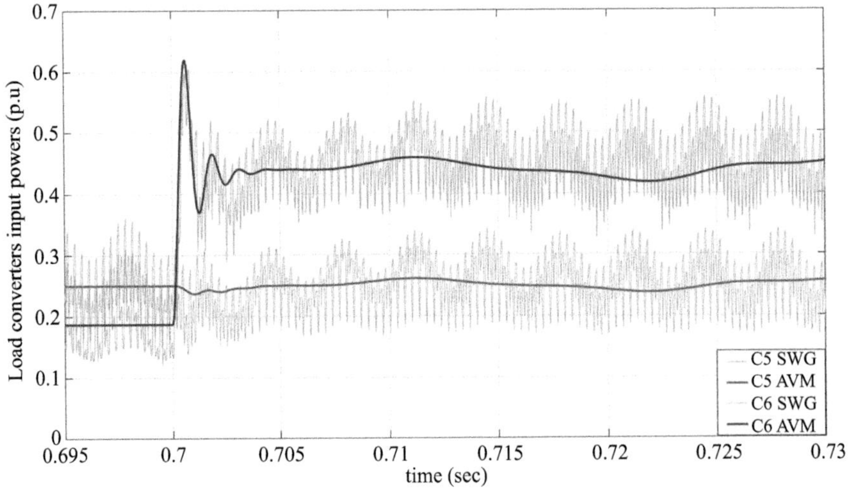

Figure 19.11 Input powers of load converters [53]

high reliability and advanced power management. Regarding the first, the controlled topology of ZEDS can avoid the disturbances propagation between distinct electrical zones. Considering the decoupling action of interface converters, the disturbances are not propagated between zones, neither when adjacent. The absence of propagation thus leads to reliability and feasibility. Second, the wide employment of power

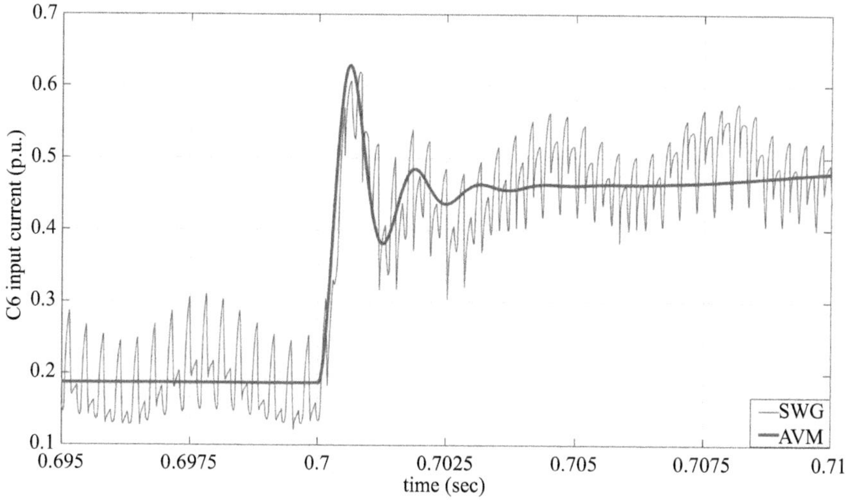

Figure 19.12 Input current of C6 converter [53]

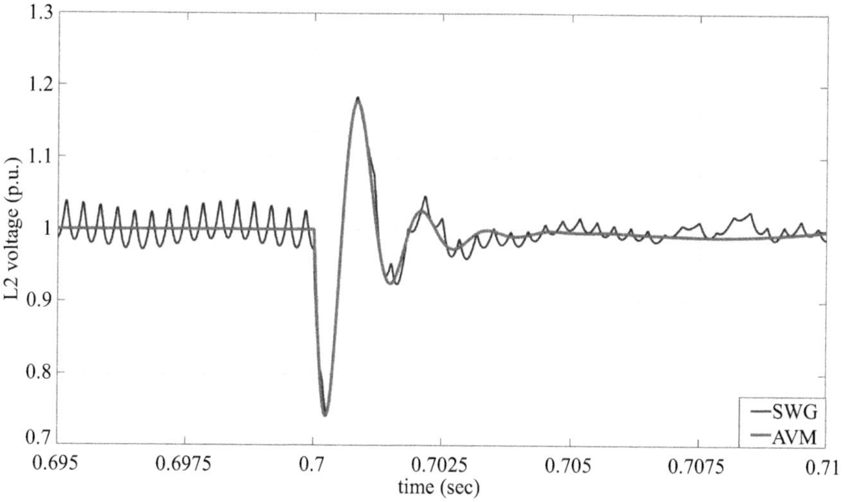

Figure 19.13 Voltage on L2 load [53]

converters improve the power flows between zones. Not only the convenient power sharing from the generating converters (the ones that interface the sources) is guaranteed, also the transient power from storage systems is optimized in order to ensure loads feeding and system stability, even in adverse conditions. A ZEDS is a power-enabler to maximize the operational capability, both in standard-operative and in extreme-faulty scenarios. As in [17,46], the ZEDS acts as a controlled distribution to feed the different sections of a system. Evidently, the main components of a ZEDS (i.e.

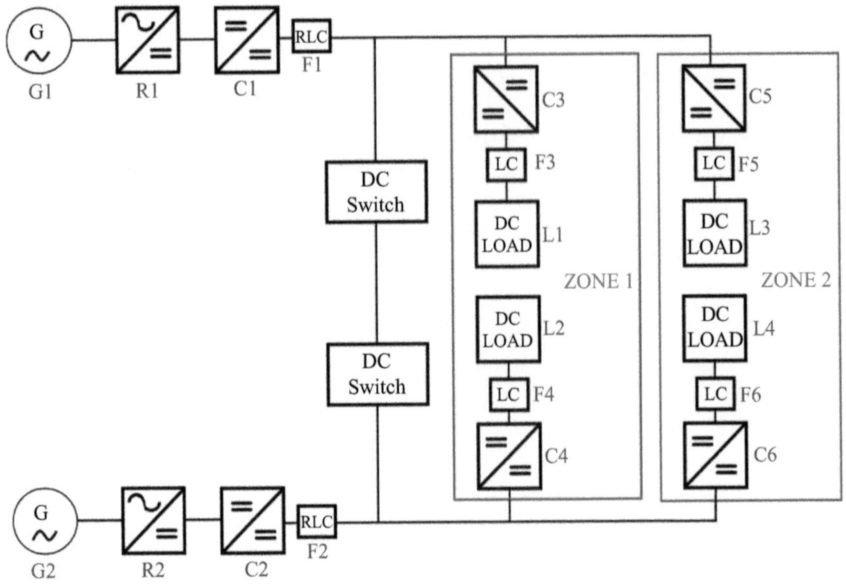

Figure 19.14 Zonal multiconverter DC microgrid on ships, disjointed configuration

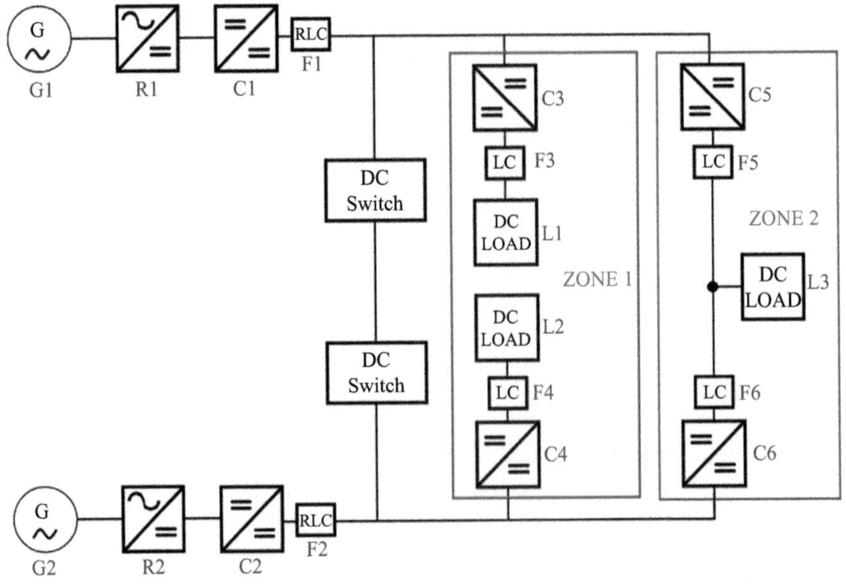

Figure 19.15 Zonal multiconverter DC microgrid on ships, parallel configuration

power converters, control systems, generation-storage systems and cables) ensures its power delivering to the loads. In general, a ZEDS does not contain its supplied loads. Conversely, the ZEDS and its loads constitute the so-called Zone, which is interfaced to the external by converters. Albeit the ZEDS does not include the loads, anyway it must ensure power quality and quality of service on the systems it supplies. As a ZEDS does not contain its loads [17,46], also the DC zonal systems on which evaluate the stability can have different configurations. By assuming the coherence with the guidelines in Section 19.2.2, two systems are identified to discuss about ZEDS implementation and stability study. The first case in Figure 19.14 is a disjointed configuration acting as zonal. The two loads in Zone 1 (L1–L2) are supplied from different converters (C3–C4), identical considerations for the loads in Zone 2. The four converters manage the power balance in the two zones, but the stability issue is after all correspondent to the one of a radial system. If the switches are opened, then L2–L4 are the paralleled loads of C2 and L1–L3 are the ones of C1. If the switches are closed the system remains radial, but all the load converters (C3–C4–C5–C6) are simultaneously fed by the two generating converters (C1–C2) in parallel. When approaching the stability analysis on the disjointed configuration, the same considerations of Sections 19.4.1–19.4.4 can be tailored on the zonal (radial) grids in Figure 19.14. Different remarks if the parallel configuration is adopted for the intrinsic DC zonal distribution (Figure 19.15). When the DC switch is open, L1–L2 are the radial loads for C1–C2, while L3 is paralleled fed by C5–C6. On the other hand, when the switch is closed, the unique bus supplies the radial loads L1–L2 and also the C5–C6. In turn, these converters supplies the L3 in parallel thus constituting a second distribution, whose voltage is controlled by C5–C6. The stability study on the distribution in Figure 19.15 thus needs a particular attention. Particularly, equations similar to (19.7) and (19.9) are to be written and integrated into the model in order to take into account both the two droop controlled distribution.

19.5 Conclusions

This chapter proposes a multi-model methodology to assess the system stability in advanced DC microgrids. As destabilizing behaviors are possible and dangerous in these LC filtered grids, a complete analysis based on poles location becomes pivotal, especially when the DC microgrids are islanded as in the marine context. Indeed, in these systems even small destabilizing perturbations on the bus can provoke the ship blackout caused by the intervention of over-under voltage protections. The problem is also significant in the DC microgrids on land, where the protections action means outages therefore loss of business. For these reasons, important efforts are invested on studying the DC stability, thus identifying both methods to investigate the nonlinear phenomenon and control techniques to compensate for the instability. This chapter is mainly oriented on the first target, thus it wants to provide a methodology to study the controlled grid and its criticality on the stability matter. To do this, the chapter describes a new way to assess the DC stability by means of an analytical–numerical tool. The latter is capable to foresee the stability performance without the need of time-consuming emulations, thus a great advantage for whom is in charge of designing the microgrid. After a detailed description on radial and zonal DC distributions, the chapter introduces a flowchart to achieve the stability assessment of DC grids. The proposed methodology is mainly based on three consequent models. First, the switching model (SWG) is the circuital representation of the controlled system. As it faithfully reproduces the dynamics behavior of the actual DC grid, it is mainly used to verify the correspondence of next models. Once neglected the switching behavior on SWG, the nonlinear average value model (nAVM) is thus defined. Although this model is only capable of providing the mean value of electrical quantities, it results sufficiently accurate when investigating the system stability in presence of perturbations. The lAVM is then found by linearizing the nAVM in the stable equilibrium point in which the system operates in steady-state condition. Then, the state-space matrix is defined by taking into account the mathematical equations that are behind the lAVM model. By a numerical analysis on this matrix, the poles on which the system operates are localized in the Guass plane. When all the poles have negative real parts, the system stability is guaranteed. Conversely, even a single pole with positive real part means system instability. Dynamics tests on PSCAD and Simulink can finally validate the multi-model methodology by confirming its capability in predicting the instability. In conclusion, a radial converters-based DC grid is the study-case on which the proposed method verifies the influence of control bandwidths on stability. Additional considerations about DC zonal distribution are at the end necessary to enable the stability assessment on totally controlled flexible-resilient grids.

Acknowledgment

The authors wish to thank Prof. Giovanni Giadrossi and PhD student Andrea Alessia Tavagnutti for their valuable contribution in preparing this chapter.

Bibliography

[1] N. Hatziargyriou, *Microgrids: Architectures and Control*, New York, NY: Wiley, 2014.

[2] T. Dragicevic, J. C. Vasquez, J. M. Guerrero, and D. Skrlec, Advanced LVDC electrical power architectures and microgrids: a step toward a new generation of power distribution networks, *IEEE Electrification Magazine*, vol. 2, no. 1, pp. 54–65, 2014.

[3] L. E. Zubieta, Are microgrids the future of energy? DC microgrids from concept to demonstration to deployment, *IEEE Electrification Magazine*, vol. 4, no. 2, pp. 37–44, 2016.

[4] M. Nasir, H. A. Khan, N. A. Zaffar, J. C. Vasquez, and J. M. Guerrero, Scalable solar dc micrigrids: on the path to revolutionizing the electrification architecture of developing communities, *IEEE Electrification Magazine*, vol. 6, no. 4, pp. 63–72, 2018.

[5] B. T. Patterson, DC, come home: DC microgrids and the birth of the internet, *IEEE Power and Energy Magazine*, vol. 10, no. 6, pp. 60–69, 2012.

[6] E. Rodriguez-Diaz, J. C. Vasquez, and J. M. Guerrero, Intelligent DC homes in future sustainable energy systems: when efficiency and intelligence work together, *IEEE Consumer Electronics Magazine*, vol. 5, no. 1, pp. 74–80, 2016.

[7] G. Sulligoi, A. Tessarolo, V. Benucci, A. Millerani Trapani, M. Baret and F. Luise, Shipboard power generation: design and development of a medium-voltage dc generation system, *IEEE Industry Applications Magazine*, vol. 19, no. 4, pp. 47–55, 2013.

[8] P. Cairoli and R. A. Dougal, New horizons in DC shipboard power systems: new fault protection strategies are essential to the adoption of DC power systems, *IEEE Electrification Magazine*, vol. 1 no. 2, pp. 38–45, 2013.

[9] M. R. Banaei and R. Alizadeh, Simulation-based modeling and power management of all-electric ships based on renewable energy generation using model predictive control strategy, *IEEE Intelligent Transportation Systems Magazine*, vol. 8, no. 2, pp. 90–103, 2016.

[10] Z. Jin, G. Sulligoi, R. Cuzner, L. Meng, J. C. Vasquez, and J. M. Guerrero, Next-generation shipboard DC power system: introduction smart grid and dc microgrid technologies into maritime electrical networks, *IEEE Electrification Magazine*, vol. 4, no. 2, pp. 45–57, 2016.

[11] D. Paul, A history of electric ship propulsion systems, *IEEE Industry Applications Magazine*, vol. 26, no. 6, pp. 9–19, 2020.

[12] G. Chang, Y. Wu, S. Shao, Z. Huang, and T. Long, DC bus systems for electrical ships: recent advances and analysis of a real case, *IEEE Electrification Magazine*, vol. 8, no. 3, pp. 28–39, 2020.

[13] IEEE Std 1709–2018 (Revision of IEEE Std 1709–2010), *IEEE Recommended Practice for 1 kV to 35 kV Medium-Voltage DC Power Systems on Ships*, pp. 1–54, 2018.

[14] D. Bosich, A. Vicenzutti, R. Pelaschiar, R. Menis, and G. Sulligoi, Toward the future: the MVDC large ship research program, *Proceedings of the 2015 AEIT International Annual Conference (AEIT)*, 2015, pp. 1–6.

[15] G. Sulligoi, A. Vicenzutti, and R. Menis, All-electric ship design: from electrical propulsion to integrated electrical and electronic power systems, *IEEE Transactions on Transportation Electrification*, vol. 2, no. 4, pp. 507–521, 2016.

[16] V. Bucci, U. la Monaca, D. Bosich, G. Sulligoi, and A. Pietra, Integrated ship design and CSI modeling: A new methodology for comparing onboard electrical distributions in the early-stage design, *Proceedings of the NAV2018 International Conference on Ship and Shipping Research*, Trieste, Italy, 20–22 June 2018; p. 124.

[17] IEEE Std 1826-2020 (Revision of IEEE Std 1826-2012), *IEEE Standard for Power Electronics Open System Interfaces in Zonal Electrical Distribution Systems Rated Above 100 kW*, pp. 1–44, 25 Nov. 2020.

[18] P. Kundur, *Power System Stability and Control*, McGraw-Hill, New York, NY, 1994.

[19] S. D. Sudhoff, K. A. Corzine, S. F. Glover, H. J. Hegner, and H. N. Robey, DC link stabilized field oriented control of electric propulsion systems, *IEEE Transactions on Energy Conversion*, vol. 13, no. 1, pp. 27–33, 1998.

[20] C. Rivetta, G. A. Williamson, and A. Emadi, Constant power loads and negative impedance instability in sea and undersea vehicles: statement of the problem and comprehensive large-signal solution, in *Proceedings of the IEEE Electric Ship Technologies Symposium*, 2005, pp. 313–320.

[21] C. H. Rivetta, A. Emadi, G. A. Williamson, R. Jayabalan, and B. Fahimi, Analysis and control of a buck DC–DC converter operating with constant power load in sea and undersea vehicles, *IEEE Transactions on Industry Applications*, vol. 42, no. 2, pp. 559–572, 2006.

[22] A. Kwasinski and C. N. Onwuchekwa, Dynamic behavior and stabilization of DC microgrids with instantaneous constant-power loads, *IEEE Transactions on Power Electronics*, vol. 26, no. 3, pp. 822–834, 2011.

[23] M. Cupelli, L. Zhu, and A. Monti, Why ideal constant power loads are not the worst case condition from a control standpoint, *IEEE Transactions on Smart Grid*, vol. 6, no. 6, pp. 2596–2606, 2015.

[24] A. M. Rahimi and A. Emadi, Active damping in DC/DC power electronic converters: a novel method to overcome the problems of constant power loads, *IEEE Transactions on Industrial Electronics*, vol. 56 no. 5, pp. 1428–1439, 2009.

[25] G. Sulligoi, D. Bosich, G. Giadrossi, L. Zhu, M. Cupelli, and A. Monti, Multiconverter medium voltage DC power systems on ships: constant-power loads instability solution using linearization via state feedback control, *IEEE Transactions on Smart Grid*, vol. 5, no. 5, pp. 2543–2552, 2014.

[26] X. Lu, K. Sun, J. M. Guerrero, J. C. Vasquez, L. Huang, and J. Wang, Stability enhancement based on virtual impedance for DC microgrids with constant

power loads, *IEEE Transactions on Smart Grid*, vol. 6, no. 6, pp. 2770–2783, 2015.

[27] D. Bosich, G. Sulligoi, E. Mocanu, and M. Gibescu, Medium voltage DC power systems on ships: an offline parameter estimation for tuning the controllers' linearizing function, *IEEE Transactions on Energy Conversion*, vol. 32, no. 2, pp. 748–758, 2017.

[28] X. Chang, Y. Li, X. Li, and X. Chen, An active damping method based on a supercapacitor energy storage system to overcome the destabilizing effect of instantaneous constant power loads in DC microgrids, *IEEE Transactions on Energy Conversion*, vol. 32, no. 1, pp. 36–47, 2017.

[29] M. A. Hassan and Y. He, Constant power load stabilization in DC microgrid systems using passivity-based control with nonlinear disturbance observer, *IEEE Access*, vol. 8, pp. 92393–92406, 2020.

[30] H. Wu, V. Pickert, M. Ma, B. Ji, and C. Zhang, Stability study and nonlinear analysis of DC–DC power converters with constant power loads at the fast timescale, *IEEE Journal of Emerging and Selected Topics in Power Electronics*, vol. 8, no. 4, pp. 3225–3236, 2020.

[31] M. S. Sadabadi and Q. Shafiee, Scalable robust voltage control of DC microgrids with uncertain constant power loads, *IEEE Transactions on Power Systems*, vol. 35, no. 1, pp. 508–515, 2020.

[32] M. Srinivasan and A. Kwasinski, Control analysis of parallel DC–DC converters in a DC microgrid with constant power loads, *International Journal of Electrical Power and Energy Systems*, vol. 122, p. 106207, 2020.

[33] Y. Gui, R. Han, J. M. Guerrero, J. C. Vasquez, B. Wei, and W. Kim, Large-signal stability improvement of DC–DC converters in DC microgrid, *IEEE Transactions on Energy Conversion*, vol. 36, no. 3, pp. 2534–2544, 2021.

[34] S. Pastore, D. Bosich, and G. Sulligoi, Influence of DC–DC load converter control bandwidth on small-signal voltage stability in MVDC power systems, in *Proceedings of the 2016 International Conference on Electrical Systems for Aircraft, Railway, Ship Propulsion and Road Vehicles & International Transportation Electrification Conference (ESARS-ITEC)*, 2016, pp. 1–6.

[35] S. Pastore, D. Bosich, and G. Sulligoi, Analysis of small-signal voltage stability for a reduced-order cascade-connected MVDC power system, in *Proceedings of the IECON 2017 – 43rd Annual Conference of the IEEE Industrial Electronics Society*, 2017, pp. 6771–6776.

[36] S. Pastore, D. Bosich, and G. Sulligoi, A frequency analysis of the small-signal voltage model of a MVDC power system with two cascade DC-DC converters, in *Proceedings of the 2018 IEEE International Conference on Electrical Systems for Aircraft, Railway, Ship Propulsion and Road Vehicles & International Transportation Electrification Conference (ESARS-ITEC)*, 2018, pp. 1–6.

[37] S. Pastore, D. Bosich, and G. Sulligoi, An analysis of the small-signal voltage stability in MVDC power systems with two cascade controlled DC–DC converters, in *Proceedings of the IECON 2018 – 44th Annual Conference of the IEEE Industrial Electronics Society*, 2018, pp. 3383–3388.

[38] D. Bosich and G. Sulligoi, Stability-oriented filter design optimization in cascade-connected MVDC shipboard power system, *Proceedings of the 2020 IEEE Power & Energy Society General Meeting (PESGM)*, 2020, pp. 1–5.

[39] U. Javaid, F. D. Freijedo, D. Dujic, and W. van der Merwe, MVDC supply technologies for marine electrical distribution systems, *CPSS Transactions on Power Electronics and Applications*, vol. 3, no. 1 pp. 65–76, 2018.

[40] IEC/IEEE International Standard – utility connections in port, Part 1: high voltage shore connection (HVSC) systems – general requirements, in *IEC/IEEE* 80005-1:2019 , pp. 1–78, 18 March 2019.

[41] IEC/IEEE International Standard – utility connections in port, Part 2: high and low voltage shore connection systems – data communication for monitoring and control, in *IEC/IEEE* 80005-2 Edition 1.0 2016-06, pp. 1–116, 27 June 2016.

[42] G. Sulligoi, D. Bosich, R. Pelaschiar, G. Lipardi, and F. Tosato, Shore-to-ship power, in *Proceedings of the IEEE*, vol. 103, no. 12, pp. 2381–2400, 2015.

[43] IEEE Std 1662-2016 (Revision of IEEE Std 1662-2008), *IEEE Recommended Practice for the Design and Application of Power Electronics in Electrical Power Systems*, pp. 1–68, 9 March 2017.

[44] IEEE Std 1676-2010, *IEEE Guide for Control Architecture for High Power Electronics (1 MW and Greater) Used in Electric Power Transmission and Distribution Systems*, pp. 1–47, 11 Feb. 2011.

[45] J. G. Ciezki and R. W. Ashton, Selection and stability issues associated with a navy shipboard DC zonal electric distribution system, *IEEE Transactions on Power Delivery*, vol. 15, no. 2, pp. 665–669, April 2000.

[46] G. Sulligoi, D. Bosich, A. Vicenzutti, and Y. Khersonsky, Design of zonal electrical distribution systems for ships and oil platforms: control systems and protections, *IEEE Transactions on Industry Applications*, vol. 56, no. 5, pp. 5656–5669, 2020.

[47] P. Kankanala, S. C. Srivastava, A. K. Srivastava, and N. N. Schulz, Optimal control of voltage and power in a multi-zonal MVDC shipboard power system, *IEEE Transactions on Power Systems*, vol. 27, no. 2, pp. 642–650, 2012.

[48] M. M. Biswas, T. Deese, J. Langston, *et al.*, Shipboard zonal load center modeling and characterization on real-time simulation platform, in *Proceedings of the 2021 IEEE Electric Ship Technologies Symposium (ESTS)*, 2021.

[49] T. V. Vu, D. Gonsoulin, D. Perkins, B. Papari, H. Vahedi, and C. S. Edrington, in *Distributed Control Implementation for Zonal MVDC Ship Power Systems, Proceedings of the 2017 IEEE Electric Ship Technologies Symposium (ESTS)*, 2017.

[50] D. Perkins, T. Vu, H. Vahedi, and C. S. Edrington, Distributed power management implementation for zonal MVDC ship power systems, in *Proceedings of the IECON 2018 – 44th Annual Conference of the IEEE Industrial Electronics Society*, 2018.

[51] S. Chen, J. Daozhuo, L. Wentao, and W. Yufen, An overview of the application of DC zonal distribution system in shipboard integrated power system, in *Proceedings of the 2012 Third International Conference on Digital Manufacturing & Automation*, 2012, pp. 206–209.

[52] M. Ahmed, L. Meegahapola, A. Vahidnia, and M. Datta, Stability and control aspects of microgrid architectures – a comprehensive review, *IEEE Access*, vol. 8, pp. 144730–144766, 2020.

[53] A. A. Tavagnutti, D. Bosich, and G. Sulligoi, A multi-model methodology for stability assessment of complex DC microgrids, in *Proceedings of the 2021 IEEE Fourth International Conference on DC Microgrids (ICDCM)*, 2021, pp. 1–7.

[54] A. A. Tavagnutti, D. Bosich, and G. Sulligoi, Active damping poles repositioning for dc shipboard microgrids control, in *Proceedings of the 2021 IEEE Electric Ship Technologies Symposium (ESTS)*, 2021, pp. 1–8.

[55] L. Che and M. Shahidehpour, DC microgrids: economic operation and enhancement of resilience by hierarchical control, *IEEE Transactions on Smart Grid*, vol. 5, no. 5, pp. 2517–2526, 2014.

[56] F. Gao, R. Kang, J. Cao, and T. Yang, Primary and secondary control in DC microgrids: a review, *Journal of Modern Power Systems and Clean Energy*, vol. 7, no. 2, pp. 227–242, 2019.

[57] Y. Han, X. Ning, P. Yang, and L. Xu, Review of power sharing, voltage restoration and stabilization techniques in hierarchical controlled DC microgrids, *IEEE Access*, vol. 7, pp. 149202–149223, 2019.

[58] D. Bosich, G. Giadrossi, S. Pastore, and G. Sulligoi, Weighted bandwidth method for stability assessment of complex DC power systems on ships, *Energies*, vol. 15, p. 258, 2022.

[59] M. Amin and M. Molinas, Small-signal stability assessment of power electronics based power systems: a discussion of impedance- and eigenvalue-based methods, *IEEE Transactions on Industry Applications*, vol. 53, no. 5, pp. 5014–5030, 2017.

[60] G. O. Kalcon, G. P. Adam, O. Anaya-Lara, S. Lo, and K. Uhlen, Small signal stability analysis of multi-terminal VSC-based DC transmission systems, *IEEE Transactions on Power Systems*, vol. 27, no. 4 pp. 1818–1830, 2012.

[61] J. Beerten, S. DArco, and J. A. Suul, Identification and small-signal analysis of interaction modes in VSC MTDC systems, *IEEE Transactions on Power Delivery*, vol. 31, no. 2, pp. 888–897, 2016.

[62] L. Guo, S. Zhang, X. Li, Y. W. Li, C. Wang, and Y. Feng, Stability analysis and damping enhancement based on frequency-dependent virtual impedance for DC microgrids, *IEEE Journal of Emerging and elected Topics in Power Electronics*, vol. 5, no. 1, pp. 338–350, 2017.

[63] A. Maulik and D. Das, Stability constrained economic operation of islanded droop-controlled DC microgrids, *IEEE Transactions on Sustainable Energy*, vol. 10, no. 2, pp. 569–578, 2019.

[64] D. Bosich, M. Gibescu, and G. Sulligoi, Large-signal stability analysis of two power converters solutions for DC shipboard microgrid, in *Proceedings of the*

2017 IEEE Second International Conference on DC Microgrids (ICDCM), 2017, pp. 125–132.

[65] D. Bosich, A. Vicenzutti, and G. Sulligoi, Robust voltage control in large multi-converter MVDC power systems on ships using thyristor interface converters, in *Proceedings of the 2017 IEEE Electric Ship Technologies Symposium (ESTS)*, Arlington, VA, 2017, pp. 267–273.

[66] M. H Rashid, *Power Electronics Handbook: Devices, Circuits, and Applications*, Burlington, MA: Academic, 2006.

[67] N. Mohan, T. M. Undeland, and W. P. Robbins, *Power Electronics, Converters, Applications and Design*, New York, NY: John Wiley and Sons, Inc., 2003.

Chapter 20

Scanning methods for stability analysis of inverter-based resources

Younes Seyedi[1], Ulas Karaagac[2], Jean Mahseredjian[3] and Houshang Karimi[1]

Recently, attempts have been made to understand and analyze the dynamic interactions between the power grids and the inverter-based resources (IBR) such as wind and solar photovoltaic (PV) parks. These adverse incidents that mainly stem from the controller interactions can lead to unwanted oscillations in sub- or super-synchronous frequency ranges, and thus jeopardize the reliable operation of power systems. To address such stability analysis issues, frequency-dependent impedance scanning techniques based on small-signal perturbations have been developed. This chapter deals with detailed explanation of different scanning methods that can be employed for predicting the stability issues and characterizing the sub- or super-synchronous oscillations. Implementation, computational burden, accuracy, and stability criteria for different scanning methods are also discussed. Stability assessment based on the scanning methods is investigated in three practical benchmark systems that involve full size converter (FSC) and doubly-fed induction generator (DFIG) wind parks. The positive-sequence, the dq and the $\alpha\beta$ scans are applied to stable and unstable cases in each benchmark and the results are verified by the electromagnetic transient (EMT) simulations.

20.1 Introduction

Modern power systems are experiencing unprecedented increase in the use of grid-connected IBRs that is mostly driven by large-scale integration of wind and solar power plants. Each of these integrated parts is a complex subsystem, with their own complicated characteristic and dynamic behavior, and they are designed and manufactured separately [1]. Co-existence and cooperation of these sub-systems in an interconnected power grid with other subsystems such as compensated transmission lines or HVDC systems give rise to some interactions among their controllers, which

[1]Department of Electrical Engineering, Polytechnique Montreal, Canada
[2]Department of Electrical Engineering, The Hong Kong Polytechnic University, China
[3]Faculty of Electrical Engineering, Polytechnique Montréal, Canada

can jeopardize the stability and safe operation of the system and even bring about severe damages to the grid assets.

Numerous interaction and undamped oscillation incidents involving grid-connected inverters of wind plants have been reported in recent years around the globe. The 2007 sub-synchronous oscillation (SSO) event of South Central Minnesota Wind Plant Substation in the United States is probably the first reported oscillation event involving DFIGs. In that incident, the interaction of a 100 MW wind farm involving DFIGs with a highly compensated 345 kV transmission line led to high magnitude sub-synchronous oscillations in the frequency ranges of 9–13 Hz and 37–43 Hz, which caused damages to the wind turbines [2]. The 2009 Event in ERCOT in Texas, USA is another example of the interactions between DFIGs and compensated transmission lines which damaged the involved wind turbines and also the series capacitor [3, 4]. Similar interaction event is reported to have happened in 2012 in Hebei Province in China involving DFIG-based wind farms and series compensation in the transmission system [5].

Oscillations due to interactions between the grid-connected inverters and other components in power systems are not limited to low frequency or sub-synchronous cases. The reported high or super-synchronous oscillations in range of 400–500 Hz due to interaction between an offshore wind farm and the BorWin1 HVDC System in Germany are an example of possible interactions in the super-synchronous region. The mentioned oscillations occurred due to the resonances between the converter filters and the grid [6]. Sub- and super-synchronous undamped oscillations are also reported to involve inverters in solar PV plants of different sizes [7].

Detection of aforementioned interactions between the grid-connected inverter-based resources and other components of power systems is not a straightforward task. The 2017 events in ERCOT in Texas, USA gives a valuable insight to the complexity of the interaction detection in power systems. Although procedures of planning and integration of wind farms were carefully reviewed and updated after the previously mentioned 2009 event in ERCOT [3, 8, 9], they failed to prevent three following similar incidents in the same power system due to the interaction between compensated transmission systems and DFIGs [10].

Within the context explained above, sophisticated analyzing methods are of great need and importance to investigate and detect any possible interactions in power systems with high penetration of the IBRs.

20.2 State-space methods

State-space modeling and subsequent modal analysis is a well-established and powerful tool to investigate the dynamic characteristics and stability of complex engineering systems including power systems [11, 12]. State-space modeling and modal analysis are widely used to examine the stability and interaction issues in modern power systems with high penetration of inverter-based resources [13–16].

The procedure to develop the state-space model of the investigated power systems and the advantages and drawbacks of utilizing the state-space method for stability analysis is presented in the upcoming subsections.

20.2.1 Development of the state-space models and stability analysis

The state-space model of a power system can be obtained through the following steps:

- Breaking the system into its subsystems and components.
- Driving the nonlinear differential equations that govern the dynamic of each component.
- Determining the state-variables for each component.
- Linearizing the governing equations at the operating point.
- Forming the matrices representing the state-space model of the component following the well-known standard procedures.
- Combining the state-space model of the components and subsystems according to their connection configuration.
- Developing the complete state-space model of the whole interconnected power system.
- Investigating the system stability through eigenvalues, participation factors, and sensitivity analysis.

20.2.2 Advantages and shortcomings of the state-space methods

State-space methods can provide deep insight into the dynamic behavior of the investigated system through modal analysis. Their advantages can be summarized as follows:

- Provides insight to the frequency of oscillations and their damping characteristics.
- Enable detection of most critical components through participation factors and facilitate the mitigation process.
- Facilitate investigation of the effects of different parameters on stability and dynamic interactions through providing analytical sensitivity analysis.
- Enable accurate assessment of stability margins.

Despite the above advantages, application of the state-space methods is restricted due to the following shortcomings:

- They are hard to obtain particularly for large and complex systems.
- They require linearization around an operating point which may decrease the accuracy of stability predictions.
- They cannot be used with black-box models, i.e., they are not applicable when detailed dynamic models of sub-systems are not available due to confidentiality issues, which is the case for many inverter-based components in power systems.

To overcome these limitations in the application of the state-space methods, more practical scanning methods are proposed to investigate the stability of the IBRs which are discussed in the next section.

20.3 Impedance-based methods

Impedance-based models are another class of methods which have been proposed for evaluation of the controller interactions between the IBRs and the grids [17, 18]. d–q frame impedance modeling [19] and the sequence-domain impedance modeling [20] are common approaches for deriving the IBR's impedance.

Once the IBR's impedance is determined, small-signal or large-signal stability (with several assumptions) issues due to the interactions between the grid and the IBR can be studied [21]. The small-signal analysis methods can only deal with the linearized dynamics of a nonlinear system around a specific operating point. The large-signal impedance of a component is related to its response for different perturbation magnitudes injected at its terminals and can deal with different operating conditions [22, 23].

An advantage of the impedance-based methods is that the models can be obtained analytically or through measurements. Moreover, the interpretation of the dynamical behavior based on the impedance models might be easier as the concept is physically related to the electrical circuits.

20.3.1 Analytical approaches

When the detailed structure of the IBRs and the grid as well as their control parameters are known, it might be possible to employ an accurate analysis method, and explicitly derive the impedance/admittance of the components [18, 24]. After obtaining the impedance/admittance, the closed-loop model of the entire network is constructed which characterizes the network's response seen from the point of connection. If the Nyquist stability criterion is employed, the number of encirclements of the eigenloci is counted and the stability of the equivalent network is assessed.

However, in many scenarios, the grid is a complex and large network, and the use of analytical approaches is not feasible. Moreover, the details of many commercial IBRs may not be available to the network designers at the early stages of design and planning, hence, the analytical approaches, similar to the state-space methods, may not be the preferred method in many practical scenarios.

To overcome the aforementioned challenges, measurement-based methods have been developed that can reliably estimate the impedance/admittance of the IBRs for stability assessment. Moreover, the measurement-based methods can be implemented in real-world power systems using the available sensors such as digital relays.

20.3.2 Measurement-based approaches

Recently, measurement-based impedance modeling and the associate stability criteria have gained attention in many research projects [25–27]. In contrast with the analytical methods, measurement-based methods do not require the detailed parameters of the IBRs' components and do not provide explicit expressions for the impedance. In fact, the measurement-based methods provide small-signal or large-signal impedance responses of the IBRs and the grid for a specified frequency range. Below are the main advantages of the measurement-based methods:

1. The EMT simulation tools in conjunction with the discrete Fourier transform (DFT) can efficiently provide accurate samples of the impedance/admittance that match the actual models with less effort.
2. When some electrical components are modeled as black boxes, in the sense that their detailed control structure or parameters are not known, then the measurement-based methods outperform the analytical methods.

Due to the above advantages, the measurement-based methods are often preferred in real-life and practical scenarios, and many utilities adopt them to perform planning and design studies related to the IBR integration.

20.4 Scanning methods

In light of fast and accurate EMT simulation tools, it is possible to extract the measurement-based impedance models for stability prediction. Hereinafter, the computational methods that calculate the impedance/admittance based on the EMT simulations are referred to as scanning methods.

The basic premise of the scanning methods is to apply small/large signal perturbations with some predetermined properties and capture the electrical responses of the components at their connection points or terminals.

The scanning methods can be classified in three groups depending on the domain of the applied perturbations [20, 19, 28, 29]:

1. Positive-sequence scan
2. dq scan
3. $\alpha\beta$ scan

It is common that the IBRs and the grid components have at least one non-linear device, e.g., PLLs. Under such circumstances, the best practice is to use EMT simulations and apply perturbations with given frequencies. The measured electrical signals are normally the corresponding voltages/currents at the terminals of the components.

20.4.1 Positive-sequence scan

The positive-sequence scan developed has been extensively used for stability prediction of wind parks in series-compensated transmission grids [30]. The basic premise is that the frequency-dependent positive-sequence impedances of the grid and the IBR are extracted from the measurements of voltage and current at the point of connection (PoC). The estimated impedances are further analyzed to assess the stability of the entire system.

The positive-sequence scan has two separate subroutines:

- Converter-side positive-sequence scan
- Grid-side positive-sequence scan

Figure 20.1 illustrates the procedure of the converter-side scan. The converter-side scan aims to extract the positive-sequence impedance of the IBR based on the voltage perturbation injection at the PoC. V_a, V_b, and V_c denote the instantaneous voltages at the operating point under the steady-state conditions and with the nominal frequency. $\tilde{V}_a$, $\tilde{V}_b$, and $\tilde{V}_c$ represent the instantaneous perturbation voltages with the desired frequency.

Figure 20.2 shows the generic procedure of the grid-side scan which aims to calculate the positive-sequence impedance of the grid that is connected to the IBR. In Figure 20.2, current perturbations are injected at the PoC for impedance calculation. I_a, I_b, and I_c denote the instantaneous currents at the operating point under the steady-state conditions and with the nominal frequency. $\tilde{I}_a$, $\tilde{I}_b$, and $\tilde{I}_c$ show the instantaneous perturbation currents with the desired frequency that are applied on top of steady-state currents I_a, I_b, and I_c, respectively.

The main steps of the combined scan are described below, where a subsystem can be either a transmission system, a grid, or an IBR:

1. Balanced sinusoidal current/voltage perturbations with small amplitude are applied to the subsystems. A recent study shows that single-tone sinusoidal

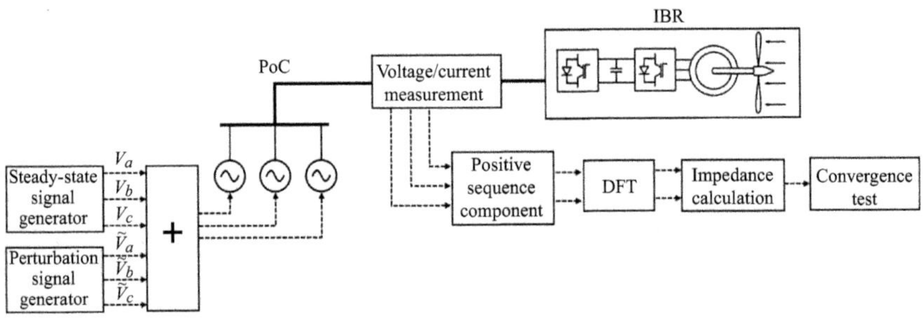

Figure 20.1 The procedure of the positive-sequence scan with voltage perturbations: converter-side

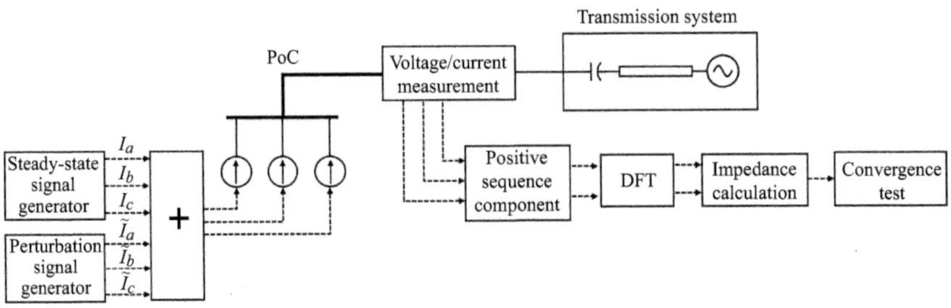

Figure 20.2 The procedure of the positive-sequence scan with current perturbations: grid-side

perturbations provide more accurate representation of the manufacturer's specific models [31].

2. Once the subsystems enter a steady state, the voltages and currents are measured at the PoC. The positive-sequence components of the measured signals are then extracted.
3. The discrete Fourier transform (DFT) of the positive-sequence components of the measured voltages and currents are obtained.
4. The positive-sequence impedance is calculated by dividing the DFT of the positive-sequence component of the voltage by the DFT of the positive-sequence component the current.

It should be emphasized here that the phasor-domain scanning can be used for the grid subsystem. If the grid subsystem contains converters (such as IBRs, HVDC stations), they can be represented with their frequency dependent impedance tables for the associated operating conditions. The frequency-dependent impedance tables of converters can be obtained through EMT-level frequency scanning for various operating conditions as shown in Figure 20.1. The phasor domain tools available in EMT-type simulators can quickly determine the frequency-dependent input impedance of the grid subsystem. Hence, the EMT-level positive-sequence scan is required only for the converter subsystem. The phasor domain scanning has negligible computational burden compared with the EMT-level scanning and allow determination of the frequency-dependent input impedance of very large-scale grids in a very short times.

20.4.1.1 The stability criterion:

It is known that the impedances of the subsystems in a sequence domain provide useful information for stability assessment and system design. For example, the Bode diagrams for the impedances can be employed to evaluate the stability margins, e.g., phase and gain margins, as well as the oscillation frequency. However, it is shown that assessing stability based on the phase difference at the frequency of unity magnitude can lead to false predictions in some converter systems [32].

The impedance ratio in the positive-sequence domain is given by:

$$Z_{ps}(f) = \frac{Z_{grid}(f)}{Z_{conv}(f)} \tag{20.1}$$

Where the input impedances of the subsystems are

$$Z_{grid/conv}(f) = \frac{DFT(V_{ps,m})}{DFT(I_{ps,m})} \tag{20.2}$$

the $DFT(.)$ function denotes the DFT of the operand. $V_{ps,m}$ and $I_{ps,m}$, respectively, represent the positive-sequence components of the measured voltages and currents at the PoC.

Z_{ps} can be regarded as the frequency-dependent open-loop gain of the system. Therefore, the Nyquist stability criterion can be adopted to reliably assess stability of the interconnected system. In this method, the system is stable if the Nyquist diagram does not encircle the critical point, $(-1, j0)$ given that the grid and the converter subsystems do not have any right half plane poles.

20.4.2 *dq-frame scan*

The *dq*-frame impedance modeling was introduced in 1997 to provide the impedance of three-phase systems in the synchronously rotating *dq*-frame [33]. The *dq*-frame scan aims to derive the input admittance and impedance matrices by processing the measured currents and voltages at a given operating point [28]. The *dq* impedance of three-phase power systems is represented as a 2-by-2 transfer matrix, hence, the generalized Nyquist criterion for multiple-input, multiple-output (MIMO) systems should be used for the stability assessment.

Similar to the combined scan, the *dq* scan involves two subroutines:

- Converter-side *dq* admittance scan
- Grid-side *dq* impedance scan

Figure 20.3 illustrates the procedure of the converter-side scan in *dq*-frame. The converter-side scan aims to extract the *dq* admittance of the IBR based on the voltage perturbation injection at the PoC. V_d and V_q, respectively, denote the *d*-axis and *q*-axis voltages at the operating point under the steady-state conditions. $\tilde{V}_d$ and $\tilde{V}_q$ show the *d*-axis and *q*-axis voltage perturbations with the desired frequency, respectively.

In each scan, the voltage of one axis is perturbed, i.e., $\tilde{V}_d \neq 0, \tilde{V}_q = 0$ or $\tilde{V}_q \neq 0, \tilde{V}_d = 0$. The *dq* to abc transformation is employed to convert the generated *dq* voltages into their three-phase counterparts for the EMT simulation. Next, the voltages and currents are measured at the PoC, and the measurements are transformed back to the *dq*-frame. It is noted that the same phase angle, ω_0, is used as the reference for the transformations.

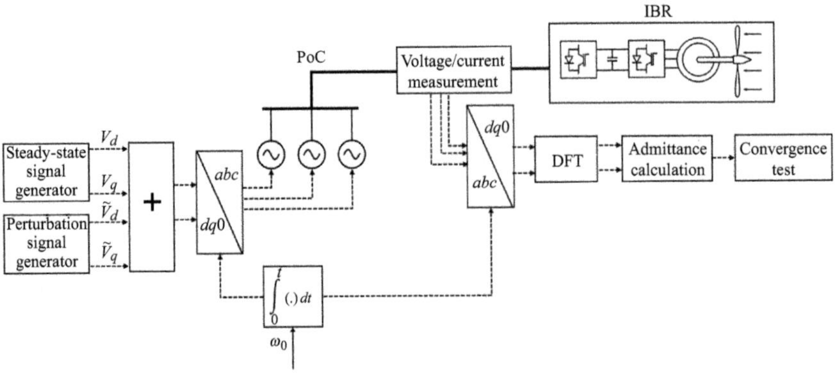

Figure 20.3 The procedure of the dq scan with voltage perturbations: converter-side

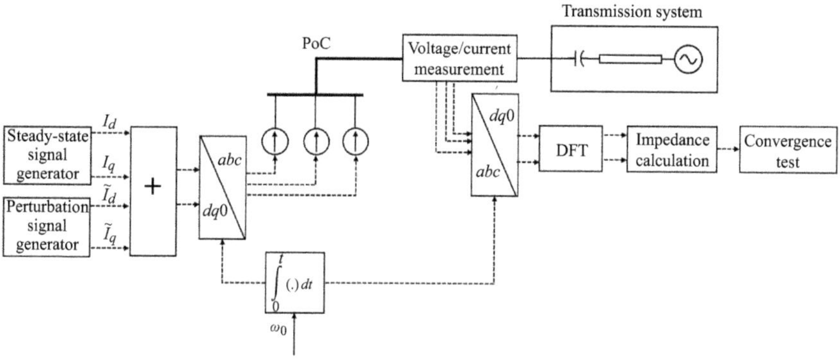

Figure 20.4 The procedure of dq scan with current perturbations: grid-side

The DFT is used to extract the admittance matrix of the converter as follows:

$$Y_{dq,conv}(f_i) = \begin{pmatrix} Y_{d,d}(f_i) & Y_{d,q}(f_i) \\ Y_{q,d}(f_i) & Y_{q,q}(f_i) \end{pmatrix} \tag{20.3}$$

where f_i denotes the perturbation frequency and the frequency-dependent admittances are calculated as:

$$Y_{d,d}(f_i) = \frac{I^d_{d,m}(f_i)}{V^d_{d,m}(f_i)} = \frac{DFT\left(I_{d,m}; \tilde{V}_q = 0\right)}{DFT\left(V_{d,m}; \tilde{V}_q = 0\right)} \tag{20.4}$$

$$Y_{d,q}(f_i) = \frac{I^q_{d,m}(f_i)}{V^q_{q,m}(f_i)} = \frac{DFT\left(I_{d,m}; \tilde{V}_d = 0\right)}{DFT\left(V_{q,m}; \tilde{V}_d = 0\right)} \tag{20.5}$$

$$Y_{q,d}(f_i) = \frac{I^d_{q,m}(f_i)}{V^d_{d,m}(f_i)} = \frac{DFT\left(I_{q,m}; \tilde{V}_q = 0\right)}{DFT\left(V_{d,m}; \tilde{V}_q = 0\right)} \tag{20.6}$$

$$Y_{q,q}(f_i) = \frac{I^q_{q,m}(f_i)}{V^q_{q,m}(f_i)} = \frac{DFT\left(I_{q,m}; \tilde{V}_d = 0\right)}{DFT\left(V_{q,m}; \tilde{V}_d = 0\right)} \tag{20.7}$$

In the above equations, $I_{d,m}$ and $I_{q,m}$ are the d-axis and q-axis components of the measured currents that flow at the PoC, respectively. Similarly, $V_{d,m}$ and $V_{q,m}$ represent the d-axis and q-axis voltages measured at the PoC, respectively. It is noted that, in general, $Y_{dq,conv}(f_i)$ is a complex matrix.

Figure 20.4 shows the procedure of the grid-side scan which aims to extract the dq impedance of the transmission grid. In this case, the current perturbations are injected at the PoC for the impedance calculation. I_d and I_q, respectively, denote the d-axis and q-axis voltages at the operating point under the steady-state conditions. $\tilde{I}_d$

and $\tilde{I}_q$ represent the d-axis and q-axis perturbation currents that are superimposed on the steady-state I_d and I_q, respectively.

In this part of the scan, the impedance matrix of the transmission grid is obtained:

$$Z_{dq,grid}(f_i) = \begin{pmatrix} Z_{d,d}(f_i) & Z_{d,q}(f_i) \\ Z_{q,d}(f_i) & Z_{q,q}(f_i) \end{pmatrix} \tag{20.8}$$

The parameter f_i shows the perturbation frequency, and the frequency-dependent impedances are calculated as:

$$Z_{d,d}(f_i) = \frac{V^d_{d,m}(f_i)}{I^d_{d,m}(f_i)} = \frac{DFT\left(V_{d,m}; \tilde{V}_q = 0\right)}{DFT\left(I_{d,m}; \tilde{V}_q = 0\right)} \tag{20.9}$$

$$Z_{d,q}(f_i) = \frac{V^q_{q,m}(f_i)}{I^q_{d,m}(f_i)} = \frac{DFT\left(V_{q,m}; \tilde{V}_d = 0\right)}{DFT\left(I_{d,m}; \tilde{V}_d = 0\right)} \tag{20.10}$$

$$Z_{q,d}(f_i) = \frac{V^d_{d,m}(f_i)}{I^d_{q,m}(f_i)} = \frac{DFT\left(V_{d,m}; \tilde{V}_q = 0\right)}{DFT\left(I_{q,m}; \tilde{V}_q = 0\right)} \tag{20.11}$$

$$Z_{q,q}(f_i) = \frac{V^q_{q,m}(f_i)}{I^q_{q,m}(f_i)} = \frac{DFT\left(V_{q,m}; \tilde{V}_d = 0\right)}{DFT\left(I_{q,m}; \tilde{V}_d = 0\right)} \tag{20.12}$$

Similar to the admittance matrix, $Z_{dq,grid}(f_i)$ is a complex matrix.

The study in [28] shows that it normally suffices to carry out the *dq* scan for the frequency range [1,100] Hz (for 50 Hz power systems) or [1,120] Hz (for 60 Hz systems) in order to detect sub- or super-synchronous oscillations due to interconnection of wind parks.

The *dq*-frame impedances obtained through EMT-level frequency scanning cannot be converted to stationary (such as sequence, abc, $\alpha\beta$) frame impedances (or vice versa). Hence, the phasor-domain scanning tool of an EMT simulator cannot be used for grid side impedance scan. As EMT-level grid side scanning is essential in this method, it is not well-suited for large-scale grid applications.

20.4.2.1 The stability criterion

The partitioning of the system into the grid-side and converter-side subsystems allows us to use the equivalent circuit that relates the *dq*-frame voltages and currents, as shown in Figure 20.5.

The stability of the equivalent circuit can be evaluated by means of the system return ratio in the *dq*-frame which is obtained as the product of the grid impedance and IBR admittance matrices:

$$L_{dq}(s) = Z_{dq,grid}(s)Y_{dq,conv}(s) \tag{20.13}$$

If the *dq*-scan is performed with a fine frequency step, then the scan results can be processed to procure the characteristic loci of the system. Moreover, suppose that the

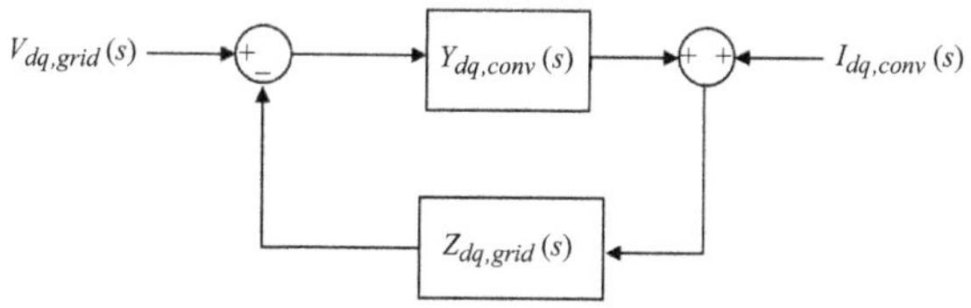

Figure 20.5 The block diagram of the equivalent circuit for stability assessment in dq-frame

transfer matrices $Z_{dq,grid}(s)$ and $Y_{dq,conv}(s)$ do not have any open-loop uncontrollable or unobservable modes in the right-half plane.

Under such circumstances, the system in Figure 20.5 is stable if and only if the net sum of counter-clockwise encirclements of the critical point by the set of characteristic loci of $L_{dq}(s)$ is equal to the total number of right-half plane poles of the matrices $Z_{dq,grid}(s)$ and $Y_{dq,conv}(s)$ [34].

The study in [17] explains the stability criteria for the individual subsystems. It is known that when the following two conditions are met, the *dq*-frame impedance and admittance matrices do not have any poles in the right-half plane:

- The transmission grid is stable before connection of the IBR.
- The IBR is stable when connected to an ideal voltage source.

In many practical scenarios, the above conditions are satisfied, and the generalized Nyquist criterion points out that the system is stable if and only if the net sum of counter-clockwise encirclements of the critical point by the set of characteristic loci of $L_{dq}(s)$ is zero.

20.4.3 $\alpha\beta$-frame scan

The $\alpha\beta$-frame impedance modeling aims to extract the impedances of three-phase power systems in the stationary $\alpha\beta$-frame. The $\alpha\beta$-frame scan finds the input admittance and impedance matrices by processing the measured currents and voltages at a given operating point. Conceptually, the $\alpha\beta$ impedance of three-phase power systems is represented as a 2-by-2 transfer matrix, thus, similar to the *dq*-frame scan, the generalized Nyquist criterion for MIMO systems can be used for the stability assessment.

Similar to the procedure for the *dq* scan, the $\alpha\beta$ scan involves two subroutines:

- Converter-side $\alpha\beta$ scan
- Grid-side $\alpha\beta$ scan

Figure 20.6 illustrates the procedure of the converter-side scan in $\alpha\beta$-frame. The converter-side scan extracts the $\alpha\beta$ admittance of the converter subsystem based on the voltage perturbation injection at the PoC. V_α and V_β, respectively, denote the α-axis and β-axis voltages at the operating point under the steady-state conditions. $\tilde{V}_\alpha$

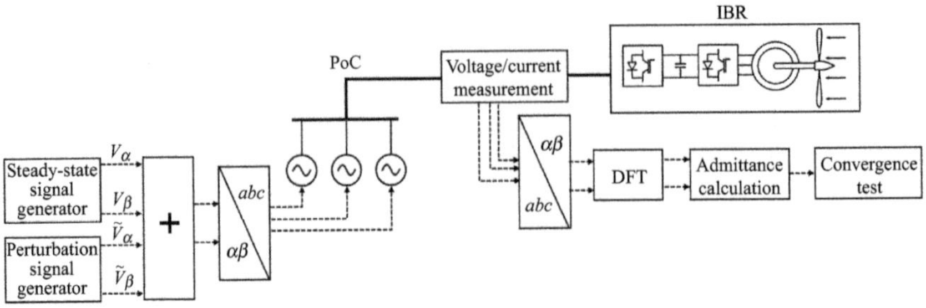

Figure 20.6 The procedure of αβ scan with voltage perturbations: converter-side scan

and $\tilde{V}_\beta$ show the α-axis and β-axis voltage perturbations with the desired frequency, respectively.

In each EMT simulation, the voltage of one axis is perturbed, i.e., $\tilde{V}_\alpha \neq 0,\ \tilde{V}_\beta = 0$ when the α-axis is perturbed and $\tilde{V}_\beta \neq 0, \tilde{V}_\alpha = 0$ when the β-axis is perturbed. The $\alpha\beta$ to abc transformation is used to convert the generated $\alpha\beta$ voltages into their three-phase counterparts for the EMT simulations. In the next step, the voltages and currents are measured at the PoC, and the measurements are transformed back to the $\alpha\beta$-frame.

The DFT is used to extract the admittance matrix of the converter as:

$$Y_{\alpha\beta,conv}(f_i) = \begin{pmatrix} Y_{\alpha,\alpha}(f_i) & Y_{\alpha,\beta}(f_i) \\ Y_{\beta,\alpha}(f_i) & Y_{\beta,\beta}(f_i) \end{pmatrix} \tag{20.14}$$

where f_i denotes the perturbation frequency and the frequency-dependent admittances are calculated as:

$$Y_{\alpha,\alpha}(f_i) = \frac{I^{\alpha}_{\alpha,m}(f_i)}{V^{\alpha}_{\alpha,m}(f_i)} = \frac{DFT\left(I_{\alpha,m};\tilde{V}_\beta = 0\right)}{DFT\left(V_{\alpha,m};\tilde{V}_\beta = 0\right)} \tag{20.15}$$

$$Y_{\alpha,\beta}(f_i) = \frac{I^{\beta}_{\alpha,m}(f_i)}{V^{\beta}_{\beta,m}(f_i)} = \frac{DFT\left(I_{\alpha,m};\tilde{V}_\alpha = 0\right)}{DFT\left(V_{\beta,m};\tilde{V}_\alpha = 0\right)} \tag{20.16}$$

$$Y_{\beta,\alpha}(f_i) = \frac{I^{\alpha}_{\beta,m}(f_i)}{V^{\alpha}_{\alpha,m}(f_i)} = \frac{DFT\left(I_{\beta,m};\tilde{V}_\beta = 0\right)}{DFT\left(V_{\alpha,m};\tilde{V}_\beta = 0\right)} \tag{20.17}$$

$$Y_{\beta,\beta}(f_i) = \frac{I^{\beta}_{\beta,m}(f_i)}{V^{\beta}_{\beta,m}(f_i)} = \frac{DFT\left(I_{\beta,m};\tilde{V}_\alpha = 0\right)}{DFT\left(V_{\beta,m};\tilde{V}_\alpha = 0\right)} \tag{20.18}$$

In the above equations, $I_{\alpha,m}$ and $I_{\beta,m}$ are the α-axis and β-axis components of the measured currents that flow at the PoC, respectively. Moreover, $V_{\alpha,m}$ and $V_{\beta,m}$ represent the α-axis and β-axis voltages measured at the PoC, respectively.

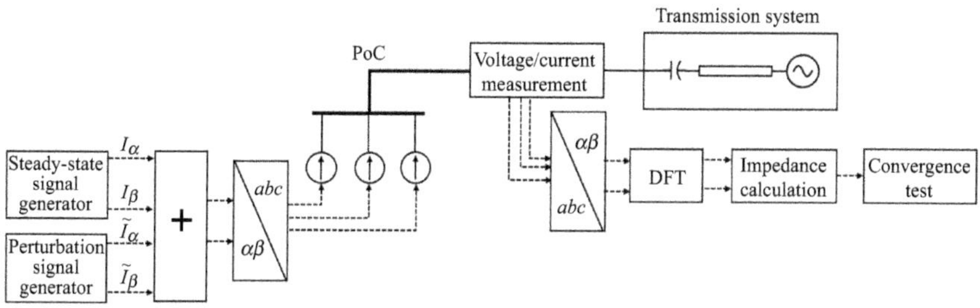

Figure 20.7 The procedure of αβ scan with current perturbations: grid-side scan

Figure 20.7 demonstrates the procedure of the grid-side $\alpha\beta$ scan which aims to extract the $\alpha\beta$ impedance of the grid subsystem. In this scan, the current perturbations are injected at the PoC for the impedance calculation. I_α and I_β, respectively, denote the α-axis and β-axis voltages at the operating point under the steady-state conditions. $\tilde{I}_\alpha$ and $\tilde{I}_\beta$ represent the α-axis and β-axis perturbation currents that are superimposed on the steady-state currents I_α and I_β, respectively.

In this part of the scan, the impedance matrix of the grid subsystem is obtained:

$$Z_{\alpha\beta,grid}(f_i) = \begin{pmatrix} Z_{\alpha,\alpha}(f_i) & Z_{\alpha,\beta}(f_i) \\ Z_{\beta,\alpha}(f_i) & Z_{\beta,\beta}(f_i) \end{pmatrix} \tag{20.19}$$

The parameter f_i shows the perturbation frequency, and the frequency-dependent impedances in the $\alpha\beta$-frame are found as:

$$Z_{\alpha,\alpha}(f_i) = \frac{V^{\alpha}_{\alpha,m}(f_i)}{I^{\alpha}_{\alpha,m}(f_i)} = \frac{DFT\left(V_{\alpha,m}; \tilde{V}_\beta = 0\right)}{DFT\left(I_{\alpha,m}; \tilde{V}_\beta = 0\right)} \tag{20.20}$$

$$Z_{\alpha,\beta}(f_i) = \frac{V^{\beta}_{\beta,m}(f_i)}{I^{\beta}_{\alpha,m}(f_i)} = \frac{DFT\left(V_{\beta,m}; \tilde{V}_\alpha = 0\right)}{DFT\left(I_{\alpha,m}; \tilde{V}_\alpha = 0\right)} \tag{20.21}$$

$$Z_{\beta,\alpha}(f_i) = \frac{V^{\alpha}_{\alpha,m}(f_i)}{I^{\alpha}_{\beta,m}(f_i)} = \frac{DFT\left(V_{\alpha,m}; \tilde{V}_\beta = 0\right)}{DFT\left(I_{\beta,m}; \tilde{V}_\beta = 0\right)} \tag{20.22}$$

$$Z_{\beta,\beta}(f_i) = \frac{V^{\beta}_{\beta,m}(f_i)}{I^{\beta}_{\beta,m}(f_i)} = \frac{DFT\left(V_{\beta,m}; \tilde{V}_\alpha = 0\right)}{DFT\left(I_{\beta,m}; \tilde{V}_\alpha = 0\right)} \tag{20.23}$$

Many cases involve the interactions of wind parks with transmission grids. Under such scenarios, detection of the oscillations requires the $\alpha\beta$-scan for the frequency range $[1,2f_0]$ Hz where f_0 is the fundamental system frequency.

As $\alpha\beta$ is a stationary reference frame, it is possible to convert $\alpha\beta$ impedances to sequence impedances (or vice versa). Hence, the phasor-domain scanning tool of EMT simulators can be used for the grid subsystem.

20.4.3.1 The stability criterion

The $\alpha\beta$ scan paves the way for multi-variable stability assessment of the system. The partitioning of the system into the grid-side and converter-side subsystems allows us to use the equivalent circuit that relates the $\alpha\beta$-frame voltages and currents, as shown in Figure 20.8.

Based on the equivalent circuit model, the stability can be evaluated by means of the system return ratio in the $\alpha\beta$-frame which is given as the product of the grid impedance and the converter admittance matrices:

$$L_{\alpha\beta}(s) = Z_{\alpha\beta,grid}(s)Y_{\alpha\beta,conv}(s) \tag{20.24}$$

If the $\alpha\beta$-scan is performed with a fine frequency step, then the scan results can be processed to obtain the eigenloci of the system. Moreover, it is assumed that the transfer matrices $Z_{\alpha\beta,grid}(s)$ and $Y_{\alpha\beta,conv}(s)$ do not have any open-loop uncontrollable or unobservable modes in the right-half plane. Under such circumstances, the system is stable if and only if the net sum of counter-clockwise encirclements of the critical point by the set of eigenloci of $L_{\alpha\beta}(s)$ is equal to the total number of right-half plane poles of the matrices $Z_{\alpha\beta,grid}(s)$ and $Y_{\alpha\beta,conv}(s)$. Normally, in power systems that incorporate wind parks, the subsystems do not have right-half plane poles, hence, the system is stable if the eigenloci of $L_{\alpha\beta}(s)$ do not encircle the critical point.

20.4.4 Convergence test

In EMT simulations, it is often helpful to verify the convergence of the impedance calculation. The convergence test can reduce the simulation time required for each perturbation frequency since the measurement noise and harmonic distortions in the voltage/current measurements result in time-varying changes in the impedances. The objective of the convergence test is to verify that the variations of the impedance magnitude and phase angle are negligible. Once the convergence test is passed the estimated impedance parameters are saved for post-processing.

An effective convergence test is based on the analysis of the short-term residual signal. Suppose that the variable $x(t)$ denotes either of the impedance magnitude or phase angle estimated at the time t. Moreover, let the parameters T and ε denote the averaging period and the convergence tolerance, respectively. The convergence test involves the following steps:

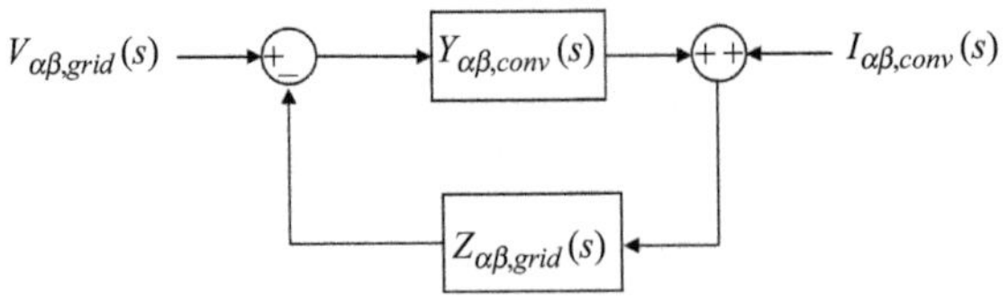

Figure 20.8 The block diagram of the equivalent circuit for stability assessment in $\alpha\beta$-frame

First, the moving average value of $x(t)$ is calculated as:

$$x_a(t) = \frac{1}{T}\int_{t-T}^{t} x(\tau)d\tau \tag{20.25}$$

Next, the mean absolute residual signal is obtained as:

$$x_r(t) = \frac{1}{T}\int_{t-T}^{t} |x(\tau) - x_a(\tau)|d\tau \tag{20.26}$$

The convergence of $x(t)$ is achieved if the following condition is met:

$$\left|\frac{x_r(t)}{x_a(t)}\right| < \varepsilon \tag{20.27}$$

The convergence test is passed when both impedance magnitude and phase angle meet the condition in (20.27).

20.4.5 Comparison of the scanning methods

It is noted that either of the previously discussed scanning techniques can be used in conjunction with the EMT simulations to estimate the frequency-dependent impedance/admittance of the subsystems for stability assessment. However, in general, different scanning techniques result in different accuracies, and their computational burden may not be identical. This subsection explains the key items for comparison of the scanning techniques.

- Stability criterion
  The scanning methods should be used with a proper stability criterion to provide reliable predictions. In general, the frequency-dependent impedance datasets are used to plot the Nyquist diagrams that can be analyzed for stability assessment. The positive-sequence scan deals with the single-variable impedance modeling and thus a single Nyquist diagram can determine the stability of a system. On the other hand, the dq and the $\alpha\beta$ scans deal with multi-variable stability impedance modeling and therefore the generalized Nyquist diagrams should be used for stability assessment.
  Inherently, the dq and $\alpha\beta$ scans provide two sets of eigenlocus and thus more than one Nyquist diagram should be analyzed. Under such circumstances, instability is confirmed if at least one eigenlocus encircle the critical point. It is concluded that the stability assessment based on the positive-sequence scan is easier and faster since it is associated with a single Nyquist diagram.
- Prediction reliability
  A scan is reliable if it does not result in false prediction results. The reliability of different scan methods mainly depends on the accuracy of their corresponding Nyquist diagrams specially where they approach the critical point. For the positive-sequence scan, the prediction reliability depends on the accuracy of the

impedance ratio for a range of frequencies. For the dq and $\alpha\beta$ scans, the reliability depends on the accuracy of the eigenvalues of $L_{dq}(s)$ and $L_{\alpha\beta}(s)$ for a range of frequencies. In the dq and $\alpha\beta$ scans, the accuracy of the dominant eigenlocus should be evaluated.

- Stability margin estimation
  The stability margins can be evaluated to obtain the degree to which the system oscillates in the presence of a voltage/current perturbation. Intuitively, the stability margins provided by different scanning methods can be compared with the EMT simulation results that indicate the extent and amplitude of the oscillations. Stability margins such as phase and gain margins can be estimated by analyzing the Nyquist diagrams of the impedance (return) ratio. In the dq and $\alpha\beta$ scans, the stability margins should be evaluated based on the dominant eigenlocus of $L_{dq}(s)$ and $L_{\alpha\beta}(s)$, respectively.
- Oscillation frequency
  The frequency of oscillations can be estimated based on the Nyquist diagrams. If the positive-sequence scan is used, the frequency at which the Nyquist diagram of $Z_{ps}(s)$ intersects with the unit circle gives an estimate of the oscillation frequency. If the dq or $\alpha\beta$ scans are used, the frequency at which the dominant eigenlocus of $L_{dq}(s)$ or $L_{\alpha\beta}(s)$ intersects with the unit circle is an estimate of the oscillation frequency.
- Computational burden
  The computational burden is another important aspect of the scanning methods. The stability criterion, i.e., the Nyquist diagrams, can be readily evaluated by post-processing of the scan results and such evaluations do not require significant computations. In contrast, EMT simulations that deal with the input impedance estimation for the grid and converter subsystems may impose significant computational burden.

The EMT simulators are capable of saving the voltage and current signals for parametric/statistical simulations. By virtue of this feature, the steady-state conditions of the subsystems are simulated only once, and the results are saved for subsequent simulations that involve the perturbation. Regardless of the domain/frame in which the perturbations are applied to the subsystems, the runtime of a scan is governed by the time interval required for the impedance estimation. Motivated by this fact, convergence time is defined as the time interval during which the impedance convergence is achieved after the perturbation inception. The impedance convergence is achieved when the magnitude and phase angle variations become negligible. Different scans may differ in their convergence times since they deal with different perturbation frequencies or domains. Finally, it should be noted that the convergence time is different from the computer simulation time for example, if the processor speed changes the convergence time does not change. In other words, the convergence time is independent from the simulation platform.

The DFT window size can also affect the convergence time in the scanning methods. In general, a larger DFT window results in a better spectral resolution which in turn can increase the accuracy of the impedance parameters. However, as the DFT window size increases, the convergence time can increase. This gives rise to a trade-off between the estimation accuracy and the computational burden in the scanning methods.

In addition, the number of EMT simulations in scanning of sub- and super-synchronous frequency ranges plays an important role in the computational burden. For the positive-sequence scan, a single simulation run is required for each perturbation frequency, and thus the number of simulations for scanning a subsystem equals the number of perturbation frequencies. However, for the dq and $\alpha\beta$ scans, two separate simulation runs are required to yield the impedance parameters for each perturbation frequency. Hence, the total number of simulations per subsystem is twice that of the positive-sequence scan.

The main advantage of positive-sequence and $\alpha\beta$ over the dq scanning method is that it has substantially lower computational burden where the phasor-domain scanning is applicable to the grid subsystem. As the computational burden of the phasor domain scanning is negligible compared with the EMT-level scanning, both positive-sequence and $\alpha\beta$ scanning methods can be used for stability assessment of large-scale grids.

20.5 Simulation results

In this section, the wind systems are implemented based on their average value model, and the magnetization branch of all transformers are excluded. The scanning techniques have been implemented in EMTP [35] based on the procedures that are explained in Section 4. Each time-domain EMT simulation deals with a single perturbation frequency. The scanning of each perturbation frequency is completed when the impedance magnitude and phase angle are converged. The time step in EMT simulations is 50 μs, and the convergence threshold is 1%. The maximum simulation time is equal to 100 s, hence, if the impedance parameters do not satisfy the condition in (20.27), the last estimate of the impedance at the time $t = 100\,\text{s}$ is used for the stability assessment.

The default perturbation amplitude is 1% of the nominal voltage/current at the PoC regardless of the scanning technique. The DFT window size in EMT simulations is fixed and equals 10 s. The stability criteria are evaluated when all the perturbation frequencies in the range $[1, 120]$ Hz are successfully scanned. The scan frequency is 1 Hz. The results of the frequency scans are saved in datasets for offline post-processing and stability assessment by MATLAB.

Stability prediction under different scenarios with DFIG and FSC wind turbines have been investigated. The results of the scans are presented and compared for both stable and unstable cases under each scenario.

20.5.1 Scenario I: FSC wind park connected to series-compensated transmission lines

In the first scenario, a wind park that consists of FSC wind turbines is connected to HV networks via two parallel transmission lines. The benchmark system is adopted from [30] with the single-line diagram depicted in Figure 20.9.

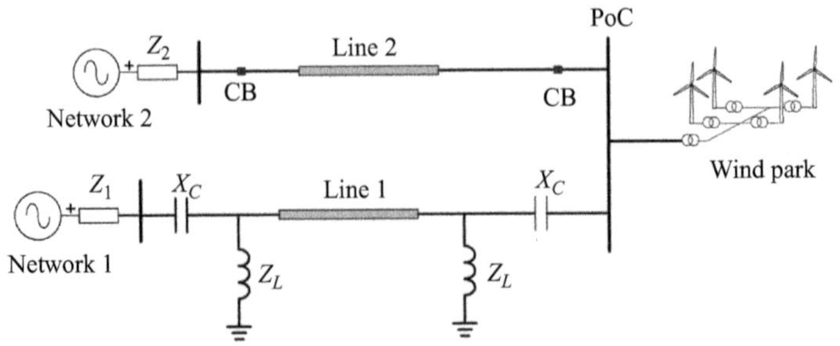

Figure 20.9 The benchmark system for interaction of wind parks with series compensated transmission lines

Table 20.1 The simulation parameters of the FSC wind park in scenario I

Parameter	Value
Number of wind turbines	400
Wind turbine rated power (MW)	1.5
Rated power of wind turbines (MVA)	1.66
Rated voltage of turbines (kV)	0.575
Rated voltage of the collector (kV)	34.5
Rated voltage of the grid	500
Reactive power of filter (kVAr)	75
Collector grid resistance (Ω)	0.014
Collector grid inductance (mH)	0.43
Collector grid capacitance (μ F)	62.2
Grid-side converter rise time (ms)	20
Machine-side converter rise time (ms)	20
Vdc regulator time constant (ms)	40
Prop. Gain of park controller – V control	2
Prop. Gain of park controller – Q control	0
Integral gain of park controller – Q control	0.15
Wind speed (m/s)	11.24

The transmission networks operate at the rated voltage of 500 kV. The shunt reactors (Z_L) each provide 230 MVAr reactive power (equivalent to 75% shunt compensation), while the series capacitor banks (X_C) at the ends of line 1 are employed for series compensation of 50%. The simulation parameters of the wind park are given in Table 20.1.

Stable case:

In the stable case, the wind park is connected to the HV Networks 1 and 2 via the transmission lines 1 and 2, respectively. The real power generated by the wind park is shown in Figure 20.10. The Nyquist diagrams obtained from different scans are

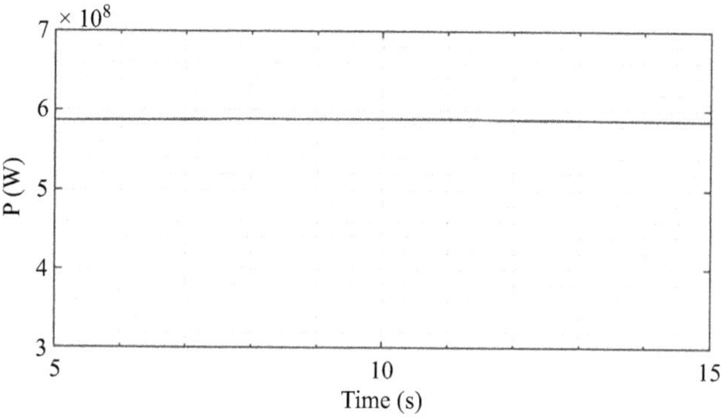

Figure 20.10 The real power injected by the wind park under the stable case of scenario I

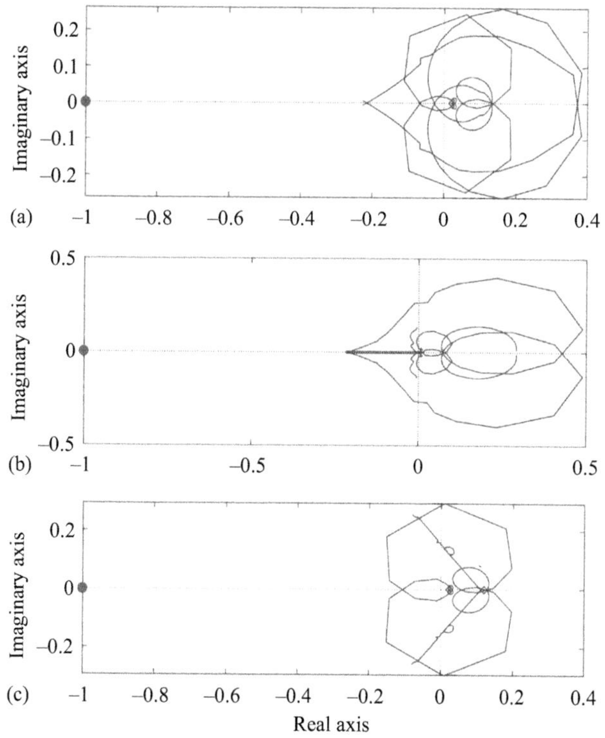

Figure 20.11 The Nyquist diagrams obtained for the stable case of scenario I: (a): of scenario I. (a) Using the positive-sequence scan. (b) Using the dq scan. (c) Using the αβ scan

demonstrated in Figure 20.11. It can be seen that the scans can successfully predict stability as the critical point is not encircled in Figure 20.11(a)–(c).

Unstable case:

In this case, line 2 trips at the time instant $t = 5$s to simulate a contingency that results in unstable voltage. After the line trip, the wind park interacts with the series capacitor and consequently oscillations emerge as shown in Figure 20.12. The spectrum of the voltage (measured at the PoC) reveals that the super-synchronous component is dominant (has the largest relative magnitude), and the oscillation frequency is 83.2 Hz (Figure 20.13). The Nyquist diagrams are shown in Figure 20.14, and they encircle the critical point. It is concluded that the scanning techniques can

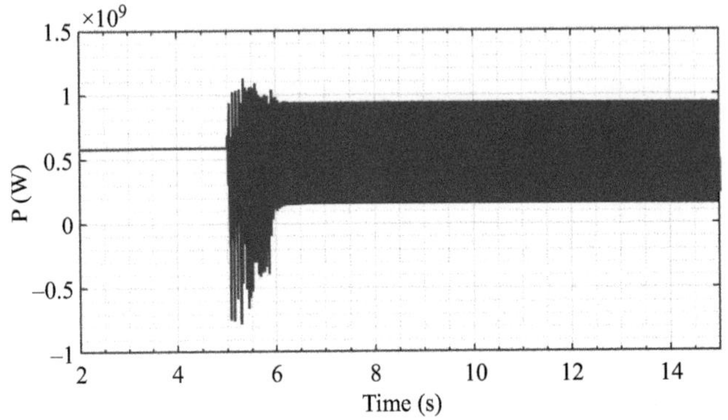

Figure 20.12 The real power injected by the wind park under the unstable case of scenario I

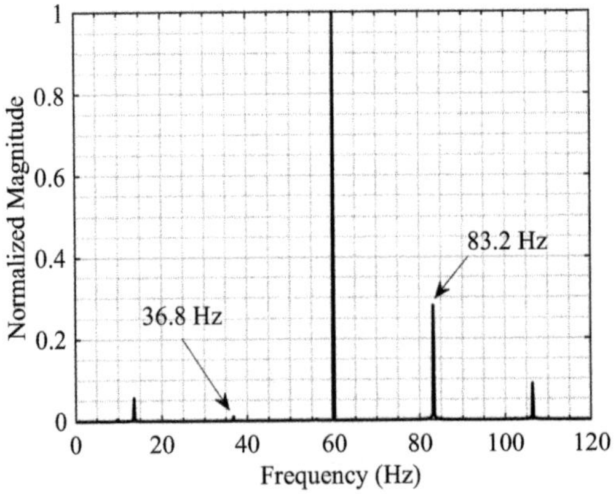

Figure 20.13 The normalized spectrum of the PoC voltage for the unstable case of scenario I

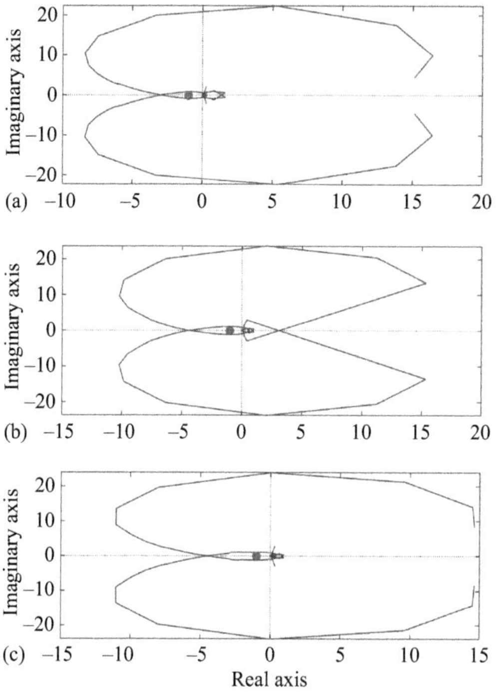

Figure 20.14 The Nyquist diagrams obtained for the unstable case of scenario I. (a) Using the positive-sequence scan. (b) Using the dq scan. (c) Using the $\alpha\beta$ scan

properly predict instability in this case. Moreover, the oscillation frequency is approximately 87 Hz based on the Nyquist plot in Figure 20.14(a) that is obtained by the positive-sequence scan. The *dq* and $\alpha\beta$ scans both estimate the oscillation frequency as 85 Hz which is closer to the actual oscillation frequency given by the DFT.

20.5.2 Scenario II: DFIG wind park connected to series compensated transmission lines

In the second test scenario, the wind park consists of DFIG wind turbines and is connected to the same transmission network as in scenario I. The parameters of the transmission network and the compensation level are identical to those in scenario I. The simulation parameters of the wind park are given in Table 20.2.

Stable case:

Similar to the first scenario, the stable case is achieved when the wind park is connected to both transmission lines. As can be seen in Figure 20.15, the real power injected by the wind park does not show any oscillatory behavior. The Nyquist diagrams produced by the frequency scans, shown in Figure 20.16, also confirm stability in this case as they do not encircle the critical point.

Table 20.2 The simulation parameters of the DFIG wind park in scenario II

Parameter	Value
Number of wind turbines	400
Wind turbine rated power (MW)	1.5
Rated power of wind turbines (MVA)	N1.66
Rated voltage of turbines (kV)	N0.575
Rated voltage of the collector (kV)	34.5
Rated voltage of the grid	500
Reactive power of filter (kVAr)	75
Collector grid resistance (Ω)	0.013
Collector grid inductance (mH)	0.04
Collector grid capacitance (μF)	70
Grid-side converter rise time (ms)	10
Rotor-side converter rise time (ms)	20
Prop. Gain of rotor controller – V control	2
Prop. Gain of rotor controller – P control	1
Integral gain of rotor controller – P control	10
Prop. Gain of park controller – V control	2
Prop. Gain of park controller – Q control	0
Integral gain of park controller – Q control	0.75
Wind speed (m/s)	9

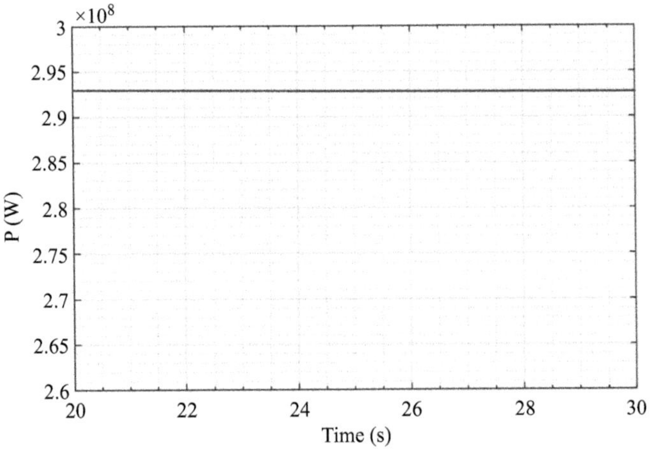

Figure 20.15 The real power generated by the DFIG wind park under the stable case of scenario II

Unstable case:

After the contingency, i.e., tripping of line 2 at $t = 5$ s, the system becomes unstable. The DFIG wind park interacts with the series capacitor that brings about oscillations as shown in Figure 20.17. In this case, the spectrum of the PoC voltage,

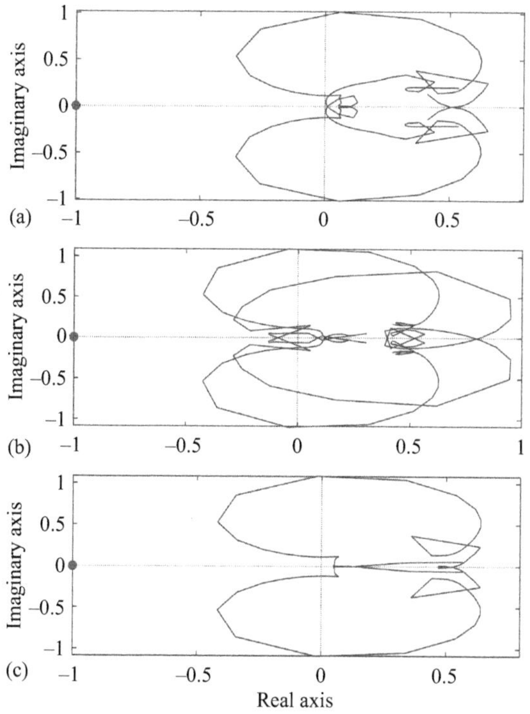

Figure 20.16 The Nyquist diagrams obtained for the stable case of scenario II: (a) using the positive-sequence scan. (b) Using the dq scan. (c) Using the $\alpha\beta$ scan

demonstrated in Figure 20.18, indicates that the sub-synchronous component is the dominant mode, and the oscillation frequency is 25.8 Hz. The corresponding Nyquist diagrams are shown in Figure 20.19, and evidently, they encircle the critical point. Therefore, the scanning techniques in conjunction with the Nyquist stability criterion can reliably predict instability in this case. Moreover, the diagrams in Figure 20.19 (a)–(c) provide the estimated oscillation frequency as 26 Hz. It turns out that the positive-sequence, the *dq* and the $\alpha\beta$ scans result in the same estimated oscillation frequency which is very close to the actual oscillation frequency given by the DFT.

Scenarios I and II investigate the super- and sub-synchronous oscillations where the grid subsystem (transmission network) does not have any converter. However, the frequency scanning techniques can be adopted for stability assessment in more complicated scenarios, where both the grid and the converter subsystems involve inverters.

20.5.3 Scenario III: interactions of wind parks with large-scale power systems

The third scenario is a realistic and large-scale power system (based on the Hydro-Quebec network) that utilizes five DFIG-based wind parks, HVDC system, and

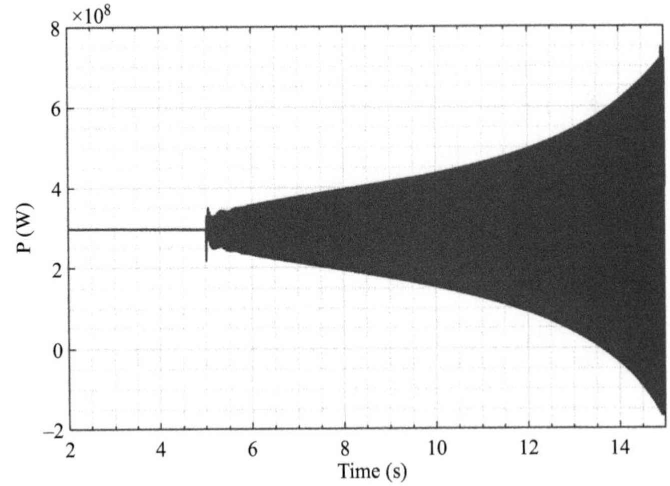

Figure 20.17 The real power generated by the wind park under the unstable case of scenario II

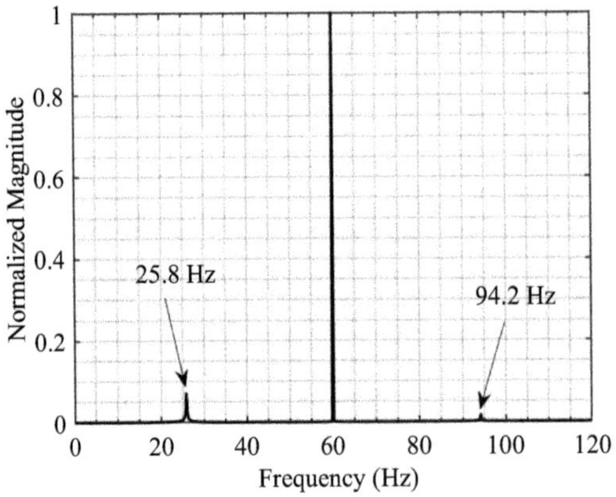

Figure 20.18 The normalized spectrum of the PoC voltage for the unstable case of scenario II

distributed voltage sources. The single-line diagram of the large-scale system and the locations of the wind parks (denoted by WP1,..., WP5) are depicted in Figure 20.20. The parameters of the wind parks are given in Table 20.3.

Stable case:

In this case, the wind parks operate under their nominal conditions and the system is stable. For instance, the real powers generated by the wind parks 1 and 5 are shown in Figure 20.21. To further investigate stability, the system is partitioned into a converter and a grid subsystem at the nodes $N_1, \ldots, N_5$ where the positive-sequence scans are carried out.

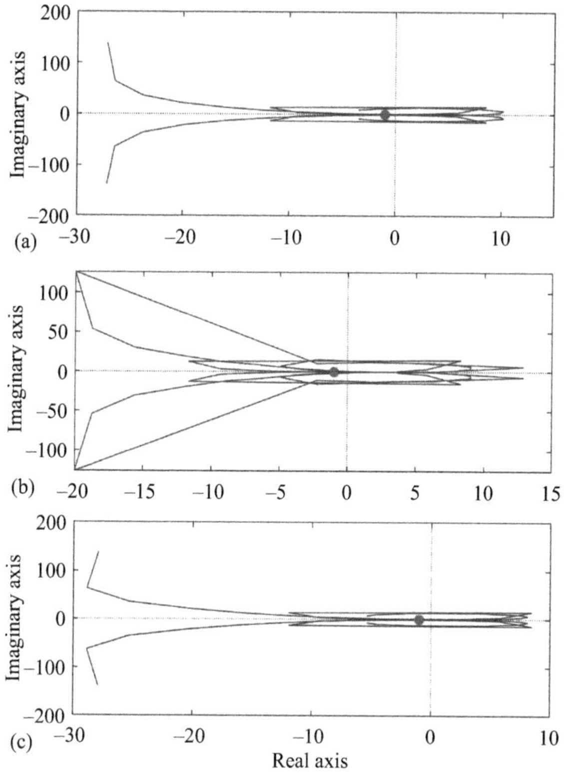

Figure 20.19 The Nyquist diagrams for the unstable case of scenario II: (a) using the positive-sequence scan. (b) Using the dq scan. (c) Using the αβ scan

The Nyquist diagrams obtained based on the positive-sequence scanning of WP 1 and 5 are shown in Figure 20.22. It is seen that the Nyquist diagrams do not encircle the critical point. The Nyquist diagrams for the rest of the wind parks are not shown in this section, however they confirm stability, and thus the system is considered stable.

Unstable case:

In this case, WP 5 is connected to the system via a series-compensated transmission line with a compensation level of 60%. The lines that connect the rest of the wind parks are not compensated. The Nyquist diagram obtained by scanning WP 5 is depicted in Figure 20.23. Since the Nyquist diagram encircles the critical point, the system is deemed to be unstable.

The time-domain simulations validate the scan result. For example, the instantaneous voltage measured at the PoC of WP 5 is demonstrated in Figure 20.24. Moreover, the corresponding voltage spectrum indicates that sub-synchronous oscillations emerge with the frequency 22 Hz, as shown in Figure 20.25. It is obvious that

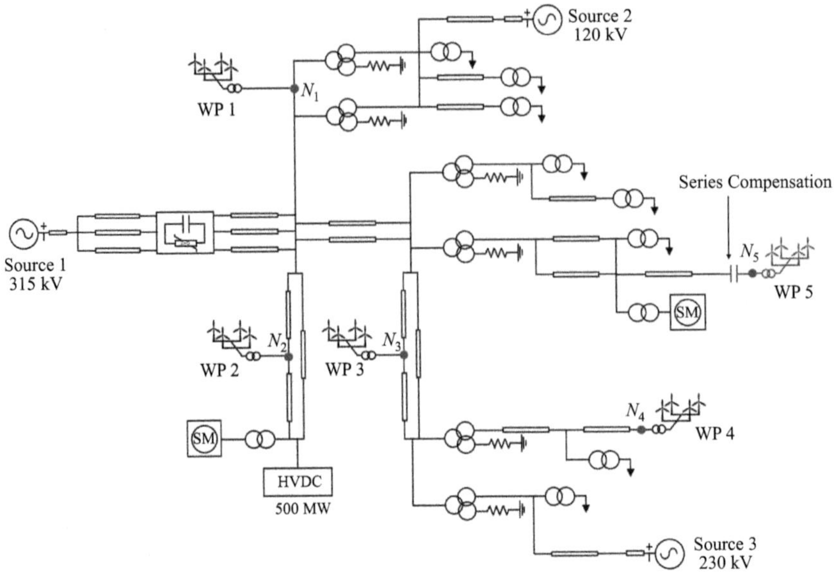

Figure 20.20 The single-line diagram of the large-scale system with DFIG-based wind parks

Table 20.3 The simulation parameters of the DFIG wind parks in the large-scale benchmark system

Parameter	**Value WP 1, 2, 3**	**Value WP 4, 5**
Number of wind turbines	266	200
Wind turbine rated power (MW)	1.5	1.5
Rated power of wind turbines (MVA)	1.66	1.66
Rated voltage of turbines (kV)	0.575	0.575
Rated voltage of the collector (kV)	34.5	34.5
Rated voltage of the grid	315	230
Reactive power of filter (kVAr)	75	75
Collector grid resistance (Ω)	0.02	0.0267
Collector grid inductance (mH)	0.6	0.8
Collector grid capacitance (μ F)	47	35
Grid-side converter rise time (ms)	10	10
Rotor-side converter rise time (ms)	20	20
Prop. Gain of rotor-side – V control	2	2
Prop. Gain of rotor-side – P control	1	1
Integral gain of rotor-side – P control	10	10
Vdc regulator time constant (ms)	100	100
Prop. Gain of park controller – V control	2	2
Prop. Gain of park controller – Q control	0	0
Integral gain of park controller – Q control	0.15	0.15
Wind speed (m/s)	7	7

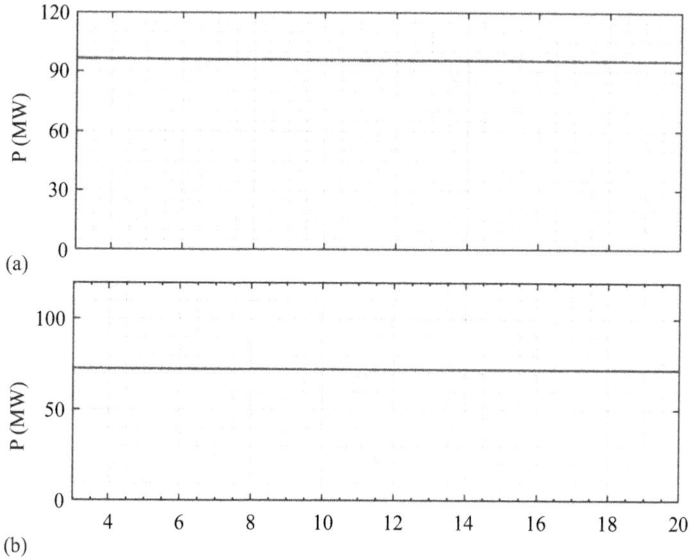

Figure 20.21 The real powers injected by the wind turbines under stable case of scenario III. (a) WP 1. (b) WP 5

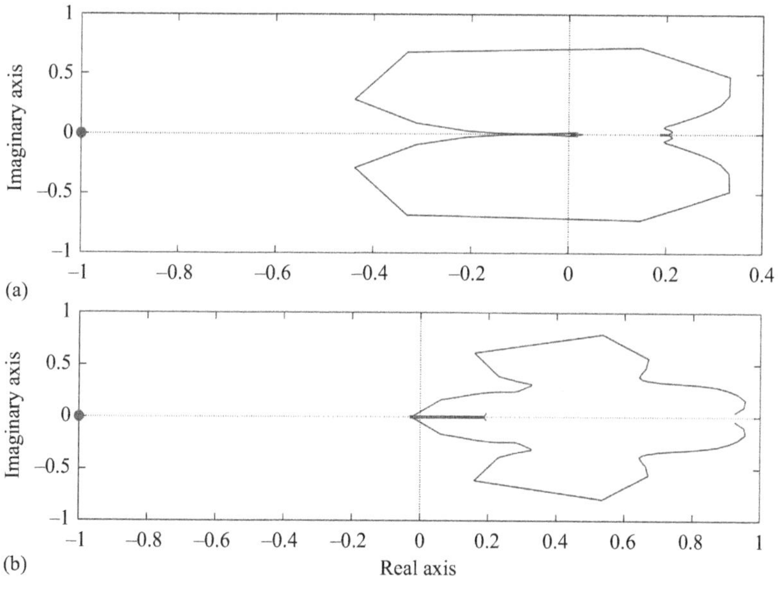

Figure 20.22 The Nyquist diagrams of the impedance ratio under the stable case of scenario III. (a) WP 1. (b) WP 5

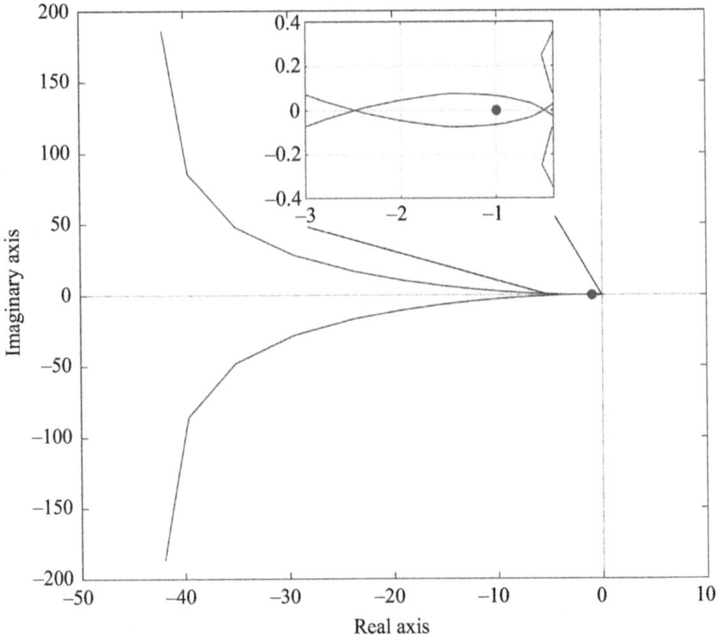

Figure 20.23 The Nyquist diagram of the impedance ratio for WP 5 under the unstable case of scenario III

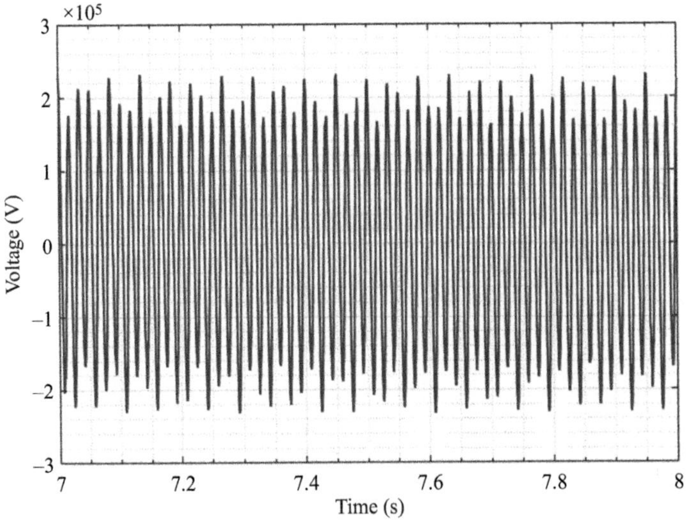

Figure 20.24 The instantaneous voltage at the PoC of WP 5 for the unstable case of scenario III

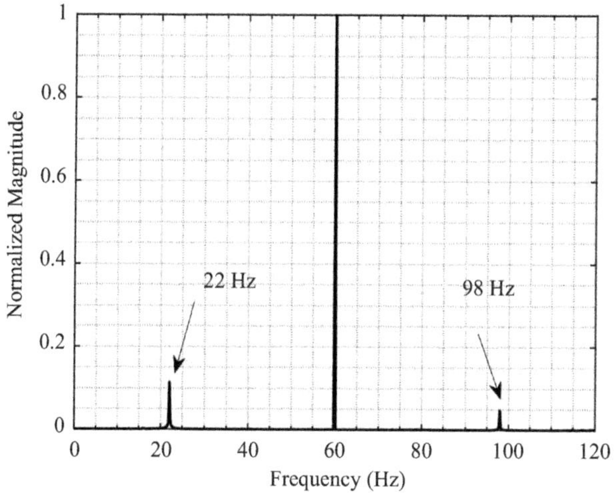

Figure 20.25 The normalized spectrum of the voltage measured at the PoC of WP 5 under the unstable case of scenario III

the complementary super-synchronous component (with the frequency of 98 Hz) has a smaller magnitude. In this case, the positive-sequence scan estimates the oscillation frequency as 22 Hz which is identical to the actual oscillation frequency given by the DFT.

The time-domain simulation results and the scan results match very well under both stable and unstable cases in the three investigation scenarios. The results corroborate the effectiveness of the frequency scanning methods and the usefulness of the Nyquist criterion for stability prediction in power systems that incorporate multiple inverter-based resources.

20.6 Conclusions

The stability assessment is indispensable for planning and reliable operation of power systems that incorporate IBRs such as wind and solar PV parks. Measurement-based stability assessment methods extract the frequency-dependent impedances of the grid and the converter subsystems based on the injection of small-signal perturbations at the point of connections. Hence, they are highly effective tools when the system under study involves non-linear or black-box components. In this chapter, the positive-sequence, the dq-frame, and the $\alpha\beta$-frame scans along with their associate stability criteria are discussed and compared. In general, either of the above scanning methods in conjunction with the Nyquist diagrams of the impedance ratio can be employed for the small-signal stability assessment. However, the scanning methods differ in their computational burden and accuracy. Three practical benchmark systems are implemented in EMTP where DFIG and FSC wind parks interact with their transmission networks. Both stable and unstable cases of each benchmark are assessed, and the results confirm that the positive-sequence scan can reliably predict stability issues and estimate the oscillation frequency as accurate as dq-frame and the $\alpha\beta$-frame scans, and also impose a lower computational burden especially compared to dq-frame scan. The dq-frame scanning method is not well suited for large-scale systems as it requires EMT-level grid-side scanning.

Bibliography

[1] A. S. Trevisan, M. Fecteau, Â. Mendonça, R. Gagnon, and J. Mahseredjian, "Analysis of low frequency interactions of DFIG wind turbine systems in series compensated grids," *Electric Power Systems Research,* vol. 191, pp. 1–8, 2021.

[2] A. Mulawarman and P. G. Mysore, "Detection of undamped sub-synchronous oscillations of wind generators with series compensated lines," in *Minnesota Power Systems Conference*, 2011.

[3] NERC, "Online Report: Lessons learned – Sub-synchronous interaction between series-compensated Transmission Lines and Generation," Atlanta, 2011.

[4] G. D. Irwin, A. K. Jindal, and A. L. Isaacs, "Sub-synchronous control interactions between type 3 wind turbines and series compensated AC transmission systems," in *IEEE Power and Energy Society General Meeting*, Detroit, 2011.

[5] L. Wang, X. R. Xie, Q. R. Jiang, *et al.*, "Investigation of SSR in practical DFIG-based wind farms connected to a series-compensated power system," *IEEE Transactions on Power Systems,* vol. 30, no. 5, pp. 2772–2779, 2015.

[6] C. Buchhagen, C. Rauscher, A. Menze, *et al.*, "BorWin1 – first experiences with harmonic interactions in converter dominated grids," in *International ETG Congress 2015; Die Energiewende – Blueprints for the New Energy Age*, Bonn, 2015.

[7] C. Li, "Unstable operation of photovoltaic inverter from field experiences," *IEEE Transactions on Power Delivery,* vol. 33, no. 2, pp. 1013–1015, 2018.

[8] Y. Cheng, M. Sahni, D. Muthumuni, and B. Badrzadeh, "Reactance scan crossover-based approach for investigating SSCI concerns for DFIG-based wind turbines," *IEEE Transactions on Power Delivery,* vol. 28, no. 2, pp. 742–751, 2013.

[9] Y. Cheng, S. H. (Fred) Huang, J. Rose, V. A. Pappu, and J. Conto, "Sub-synchronous resonance assessment for a large system with multiple series compensated transmission circuits," *IET Renewable Power Generation,* vol. 13, no. 1, p. 27–32, 2019.

[10] IEEE PES Task Force, "Wind energy systems sub-synchronous oscillations: events and modeling," IEEE Technical Report, 2020.

[11] P. L. Dandeno and P. Kundur, "Practical application of eigenvalue techniques in the analysis of power system dynamic stability problems," *Canadian Electrical Engineering Journal,* vol. 1, no. 1, pp. 35–46, 1976.

[12] P. J. Nolan, N. K. Sinha, and R. T. H. Alden, "Eigenvalue sensitivities of power systems including network and shaft dynamics," *IEEE Transactions on Power Apparatus and Systems,* vol. 95, no. 4, pp. 1318–1324, 1976.

[13] R. Majumder, "Some aspects of stability in microgrids," *IEEE Transactions on Power Systems,* vol. 28, no. 2, pp. 3243–3252, 2013.

[14] N. Pogaku, M. Prodanovic, and T. C. Green, "Modeling, analysis and testing of autonomous operation of an inverter-based microgrid," *IEEE Transactions on Power Electronics,* vol. 22, no. 2, pp. 613–625, 2007.

[15] A. Ostadi, A. Yazdani, and R. K. Varma, "Modeling and stability analysis of a DFIG-based wind-power generator interfaced with a series-compensated line," *IEEE Transactions on Power Delivery,* vol. 24, no. 3, pp. 1504–1514, 2009.

[16] H. Saad, J. Mahseredjian, S. Dennetiere, and S. Nguefeu, "Interactions studies of HVDC–MMC link embedded in an AC grid," *Electric Power Systems Research,* vol. 138, p. 202–209, 2016.

[17] J. Sun, "Impedance-based stability criterion for grid-connected inverters," *IEEE Transactions on Power Electronics* , vol. 26, no. 11, pp. 3075–3078, 2011.

[18] C. Zhang, M. Molinas, A. Rygg, and X. Cai, "Impedance-based analysis of interconnected power electronics systems: impedance network modeling and comparative studies of stability criteria," *IEEE Journal of Emerging and Selected Topics in Power Electronics,* vol. 8, no. 3, pp. 2520–2533, 2020.

[19] B. Wen, D. Boroyevich, R. Burgos, P. Mattavelli, and Z. Shen, "Small-signal stability analysis of three-phase AC systems in the presence of constant power loads based on measured d-q frame impedances," *IEEE Transactions on Power Electronics,* vol. 30, no. 10, pp. 5952–5963, 2015.

[20] M. Cespedes and J. Sun, "Impedance modeling and analysis of grid-connected voltage-source converters," *IEEE Transactions on Power Electronics,* vol. 29, no. 3, pp. 1254–1261, 2014.

[21] A. S. Trevisan, A. A. El-Deib, R. Gagnon, J. Mahseredjian, and M. Fecteau, "Field validated generic EMT-type model of a full converter wind turbine based on a gearless externally excited synchronous generator," *IEEE Transactions on Power Delivery,* vol. 33, no. 5, pp. 2284–2293, 2018.

[22] S. Shah and L. Parsa, "Impedance-based prediction of distortions generated by resonance in grid-connected converters," *IEEE Transactions on Energy Conversion,* vol. 34, no. 3, pp. 1264–1275, 2019.

[23] S. Shah, P. Koralewicz, V. Gevorgian, *et al.*, "Large-signal impedance-based modeling and mitigation of resonance of converter-grid systems," *IEEE Transactions on Sustainable Energy,* vol. 10, no. 3, pp. 1439–1449, 2019.

[24] X. Wang, F. Blaabjerg, and P. C. Loh, "An impedance-based stability analysis method for paralleled voltage source converters," in *International Power Electronics Conference (IPEC-Hiroshima 2014 – ECCE ASIA)*, Hiroshima, 2014.

[25] L. Fan and Z. Miao, "Admittance-based stability analysis: bode plots, Nyquist diagrams or eigenvalue analysis?," *IEEE Transactions on Power Systems,* vol. 35, no. 4, pp. 3312–3315, 2020.

[26] L. Fan and Z. Miao, "Time-domain measurement-based DQ-frame admittance model identification for inverter-based resources," *IEEE Transactions on Power Systems,* vol. 36, no. 3, pp. 2211–2221, 2021.

[27] Y. Xu, M. Zhang, L. Fan and Z. Miao, "Small-signal stability analysis of type-4 wind in series-compensated networks," *IEEE Transactions on Energy Conversion,* vol. 35, no. 1, pp. 529–538, 2020.

[28] A. S. Trevisan, Â. Mendonça, R. Gagnon, J. Mahseredjian and M. Fecteau, "Analytically validated SSCI assessment technique for wind parks in series compensated grids," *IEEE Transactions on Power Systems,* vol. 36, no. 1, pp. 39–48, 2021.

[29] X. Wang, L. Harnefors, and F. Blaabjerg, "Unified impedance model of grid-connected voltage-source converters," *IEEE Transactions on Power Electronics* , vol. 33, no. 2, pp. 1775–1787, 2018.

[30] U. Karaagac, J. Mahseredjian, S. Jensen, R. Gagnon, M. Fecteau, and I. Kocar, "Safe operation of DFIG-based wind parks in series-compensated systems," *IEEE Transactions on Power Delivery,* vol. 33, pp. 709–718, 2018.

[31] M. Lwin, R. Kazemi, and D. Howard, "Frequency scan considerations for SSCI analysis of wind power plants," in *IEEE Power & Energy Society General Meeting (PESGM)*, Atlanta, 2019.

[32] Y. Liao and X. Wang, "General rules of using bode plots for impedance-based stability analysis," in *IEEE 19th Workshop on Control and Modeling for Power Electronics*, Padua, 2018.

[33] S. Shah, P. Koralewicz, V. Gevorgian, and H. Liu, "Impedance methods for analyzing stability impacts of inverter-based resources: stability analysis tools for modern power systems," *IEEE Electrification Magazine*, vol. 9, no. 1, pp. 53–65, 2021.

[34] A. G. J. MacFarlane and I. Postlethwaite, "The generalized Nyquist stability criterion and multivariable root loci," *International Journal of Control*, vol. 25, no. 1, pp. 81–127, 1977.

[35] J. Mahseredjian, S. Dennetière, L. Dubé, B. Khodabakhchian, and L. Gérin-Lajoie, "On a new approach for the simulation of transients in power systems," *Electric Power Systems Research,* vol. 77, no. 11, pp. 1514–1520, 2007.

Part IV

Appendices

Appendix A

Stochastic process fitting procedure

The fitting procedure presented in Section 4.2.2 involves fitting the function in (4.23) to the Autocorrelation Function (ACF) of the data as well as identifying a Probability Density Function (PDF) that best captures the probability distribution of the data. The procedure is demonstrated below with an example measured wind speed data set. The data set includes 3 years of data sampled and averaged hourly. The measurement location is Mace Head, Galway, Ireland [12].

A.1 Find the ACF parameters

Figure A.1 shows the autocorrelation for the example data set. To capture this ACF using the method presented in Section 4.2.2, (4.23) has to be fitted to the autocorrelation. This can be done with any typical curve fitting algorithm. In this work, a non-linear least squares method, included in the Python package SciPy, is utilized. The number of decaying exponential and/or damped sinusoidal functions used to fit the ACF can most often be estimated visually or, if not, by trial and error. In this case, three exponentially decaying functions are considered. In order to determine the ideal number of exponentially decaying functions, the chi-squared test is used and the results are presented in Table A.1. The chi-square results indicate that it is sufficient to use two components as adding the third component does not significantly improve the fit.

In Figures A.1 and A.2, the fit of the 3-component models are compared to the ACF of the example data set. The 1-component decaying exponential function only

Table A.1 The chi-squared test results for fitting a weighted sum of 1, 2, and 3 exponentially decaying functions to the ACF of Data Set 1

Decaying exponential #	***p*-value**
1	563.05113902
2	0.1422387958
3	0.1422342338

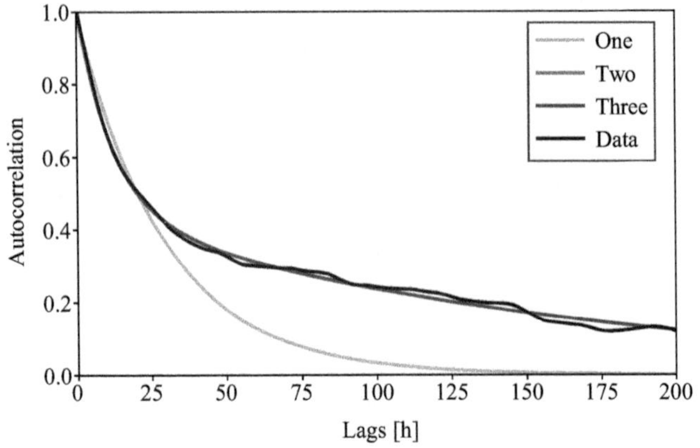

Figure A.1 *The ACF of the example data set and the fitted sum of 1–3 component exponentially decaying functions.*

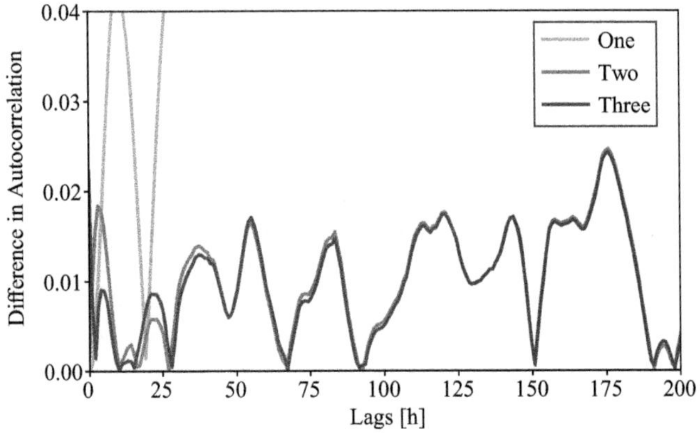

Figure A.2 *The difference in the ACF of the example data set and the fitted sum of 1–3 component exponentially decaying functions*

manages to capture the decay over the first 24 h. To capture the correlation after that more components are needed. In Figure A.1, the 2 and 3 component exponentially decaying processes are indistinguishable. To further examine the fit, the difference in the fitted processes from the actual ACF of the example data set is shown in Figure A.2. Component 2 performs slightly worse but the error for both components 2 and 3 is less than 3 %. The difference between the 2- and 3-component models is minimal in this case. Therefore, the simpler model is chosen, i.e., the 2-component model.

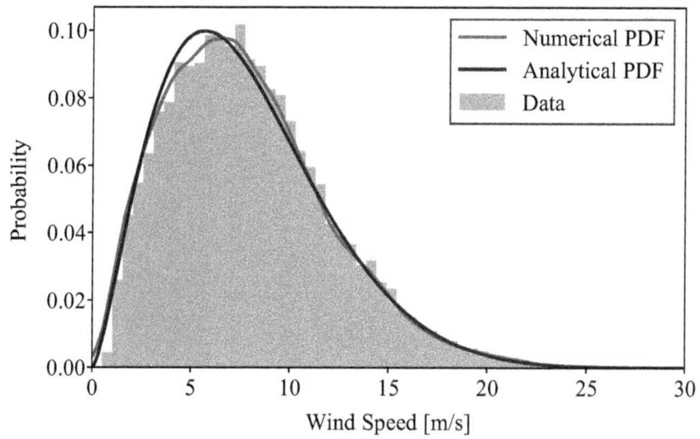

Figure A.3 *The histogram for the example data set and the fitted numerical and analytical PDF*

The fitted ACF function of the example data set is thus approximated as:

$$R_1(\tau) = w_1 \exp(-\alpha_1 \tau) + w_2 \exp(-\alpha_2 \tau), \tag{A.1}$$

where $w_1 = 0.55$, $w_2 = 0.45$, $\alpha_1 = 0.0811$, and $\alpha_2 = 0.00648$. Since no periodical behavior is present in this data set, $\omega_1 = \omega_2 = 0$.

The σ_1 and σ_2 parameters are defined from the α_1 and α_2 parameters, respectively, and the set standard deviation, σ of the two OU processes. The standard deviation can be set to any value as long as this value is used to define the Gaussian CDF $\Phi(\cdot)$ in (4.15). In this case, σ_η is set to be 1 and thus, $\sigma_1 = \sqrt{2\sigma^2\alpha_1} = 0.4027$, and $\sigma_2 = \sqrt{2\sigma^2\alpha_2} = 0.1138$.

A.2 Find the PDF parameters

Figure A.3 shows the probability distribution of the example data set. The PDF can be defined in two ways, analytically or numerically. The latter approach is to be preferred if the wind speed distribution is irregular or has two or more peaks.

A.2.1 Analytical PDF

Table A.2 shows the results of the Kolmogorov–Smirnov tests for six PDFs applied to the example data set. The Kolmogorov–Smirnov test is used to compare the analytical Cumulative Distribution Function (CDF) to the empirical cumulative distribution function (ECDF) of the data and allows deciding which analytical PDF best captures the probability distribution of the data set. The six PDFs considered in Table A.2 have all been used in the literature to model the probability distribution of wind speed data.

Table A.2 The Kolmogorov–Smirnov test results for 6 analytical PDFs fitted to Data Set 1

PDF	p-value
3-parameter Beta	$1.0688 \cdot 10^{-13}$
3-parameter Gamma	$1.3787 \cdot 10^{-77}$
2-parameter Inverse Gaussian	$3.5201 \cdot 10^{-166}$
2-parameter Lognormal	No fit
1-parameter Rayleigh	$4.7572 \cdot 10^{-14}$
2-parameter Weibull	$1.0849 \cdot 10^{-84}$

Based on Table A.2, the 3-parameter Beta PDF has the highest p-value and thus, provides the best fit. The 3-parameter Beta function is defined as:

$$p_{\mathrm{B}}(\eta) = \begin{cases} \dfrac{1}{\lambda_3 \, B[\lambda_1, \lambda_2]} \left(\dfrac{\eta}{\lambda_3}\right)^{\lambda_1 - 1} \left(\dfrac{\lambda_3 - \eta}{\lambda_3}\right)^{\lambda_2 - 1}, & \text{if } \eta > 0 \\ 0, & \text{if } \eta \leq 0 \end{cases} \tag{A.2}$$

where $B[\cdot, \cdot]$ is the Beta function, the shape parameters are $\lambda_1 = 2.6006$ and $\lambda_2 = 10.1357$ and the noncentrality parameter is $\lambda_3 = 38.0838$, for Data Set 1. The fitted analytical PDF is shown in Figure A.3.

A.2.2 Numerical PDF

Among the many possible numerical techniques to approximate the probability distribution of a set of measurements, the ECDF is considered here. It is a non-parametric estimator of the underlying Cumulative Distribution Function (CDF) of a stochastic process. The ECDF is found by first sorting the N data points from smallest to largest. Each data point is assigned a probability of $1/N$. The result is a step function that increases by $1/N$ for each data point. The ECDF is a discrete numerical approximation of the CDF of the data set. An estimation of the underlying continuous CDF function can be found from the ECDF by using interpolation.

The equivalent inverse CDF for the fitted numerical and/or analytical PDF is then used for imposing the probability distribution of the data set using the memoryless transformation, as discussed in Section 4.2.2. Figure A.3 shows the PDF defined through the ECDF for the example data set.

Appendix B
Modeling correlated stochastic processes

B.1 Two correlated Wiener processes

The correlation of two Wiener processes, say $W_1(t)$ and $W_2(t)$, can be achieved by defining a third Wiener process, say $V(t)$, which is independent from $W_1(t)$. The correlation between $W_1(t)$ and $W_2(t)$ is defined through a parameter $\rho_W(t)$. This can be constant or time varying. In the latter case, $\rho_W(t)$ is modeled as another stochastic process. Then $W_2(t)$ is defined as the following adapted Wiener process:

$$dW_2(t) = \rho_W(t)dW_1(t) + \sqrt{1-\rho_W(t)^2}dV(t). \tag{B.1}$$

B.2 Two correlated Poisson jump processes

The jump times of two jump diffusion SDEs as presented in (4.24) can also be correlated. To model the correlation between two Poisson distributed jumps, $J_1(t)$ and $J_2(t)$, three Poisson distributed jumps $n_1(t)$, $n_2(t)$ and $n_3(t)$, with jump rates λ_1, λ_2 and λ_3 are used. $J_1(t)$ and $J_2(t)$ are defined as:

$$J_1(t) = n_1(t) + n_3(t), \tag{B.2}$$

$$J_2(t) = n_2(t) + n_3(t). \tag{B.3}$$

Thus, the mean of the two Poisson jump processes is:

$$(\lambda_i + \lambda_3)t, \quad i = 1, 2. \tag{B.4}$$

Their covariance is λ_3 and their correlation is:

$$\rho_J = \frac{\lambda_3}{\sqrt{(\lambda_1+\lambda_3)(\lambda_2+\lambda_3)}}. \tag{B.5}$$

B.3 Correlated stochastic load model

The model in (4.53) is modified to include the correlation between the load consumption of d buses:

$$\begin{aligned}
\boldsymbol{p}_{\mathrm{L}}(t) &= (\boldsymbol{p}_{\mathrm{L}0} + \eta_p(t))\boldsymbol{v}_p(t)\,, \\
\boldsymbol{q}_{\mathrm{L}}(t) &= (\boldsymbol{q}_{\mathrm{L}0} + \eta_q(t))\boldsymbol{v}_q(t)\,, \\
\begin{bmatrix} d\eta_p(t) \\ d\eta_q(t) \end{bmatrix} &= \begin{bmatrix} \boldsymbol{a}_p(\eta_p(t)) \\ \boldsymbol{a}_q(\eta_q(t)) \end{bmatrix} dt + \begin{bmatrix} \boldsymbol{b}_p(\eta_p(t)) \\ \boldsymbol{b}_q(\eta_q(t)) \end{bmatrix} \mathbf{C} \begin{bmatrix} d\boldsymbol{W}_p(t) \\ d\boldsymbol{W}_q(t) \end{bmatrix},
\end{aligned} \tag{B.6}$$

where $\boldsymbol{p}_{\mathrm{L}} \in \mathbb{R}^d$ and $\boldsymbol{q}_{\mathrm{L}} \in \mathbb{R}^d$ represent the active and reactive power consumption at load buses, respectively; $\boldsymbol{p}_{\mathrm{L}0} \in \mathbb{R}^d$ and $\boldsymbol{q}_{\mathrm{L}0} \in \mathbb{R}^d$ represent the nominal active and nominal reactive power consumption at load buses, respectively; $\boldsymbol{v}_p \in \mathbb{R}^d$ and $\boldsymbol{v}_q \in \mathbb{R}^d$ represent vectors whose elements are calculated as:

$$\begin{aligned}
v_{p,i}(t) &= \left(v_i(t)/v_{0,i}\right)^{\gamma_{p,i}}\,, \quad i = 1, \ldots, d\,, \\
v_{q,i}(t) &= \left(v_i(t)/v_{0,i}\right)^{\gamma_{q,i}}\,, \quad i = 1, \ldots, d\,,
\end{aligned}$$

respectively; and $\boldsymbol{a}_p$, $\boldsymbol{a}_q$, $\boldsymbol{b}_p$ and $\boldsymbol{b}_q$ are all d-dimensional vectors with same meanings as in (4.30). Matrix $\mathbf{C} \in \mathbb{R}^{2d \times 2d}$ and is obtained as the Cholesky decomposition of a correlation matrix $\mathbf{R}$ with the following structure:

$$\mathbf{R} = \begin{bmatrix} \mathbf{R}_{p,p} & \mathbf{R}_{p,q} \\ \mathbf{R}_{q,p} & \mathbf{R}_{q,q} \end{bmatrix}, \tag{B.7}$$

where $\mathbf{R}_{q,p} = \mathbf{R}_{p,q}^T$ and:

$$\mathbf{R}_{p,p} = \begin{bmatrix} 1 & r_{p1,p2} & \cdots & r_{p1,pd} \\ r_{p2,p1} & 1 & \cdots & r_{p2,pd} \\ \vdots & \vdots & \ddots & \vdots \\ r_{pd,p1} & r_{pd,p2} & \cdots & 1 \end{bmatrix},$$

$$\mathbf{R}_{p,q} = \begin{bmatrix} r_{p1,q1} & r_{p1,q2} & \cdots & r_{p1,qd} \\ r_{p2,q1} & r_{p2,q2} & \cdots & r_{p2,qd} \\ \vdots & \vdots & \ddots & \vdots \\ r_{pd,q1} & r_{pd,q2} & \cdots & r_{pd,qd} \end{bmatrix},$$

$$\mathbf{R}_{q,q} = \begin{bmatrix} 1 & r_{q1,q2} & \cdots & r_{q1,qd} \\ r_{q2,q1} & 1 & \cdots & r_{q2,qd} \\ \vdots & \vdots & \ddots & \vdots \\ r_{qd,q1} & r_{qd,q2} & \cdots & 1 \end{bmatrix}.$$

B.4 Correlated stochastic power flow equations

To ensure a secure operation of the grid, it is required that generation and demand are balanced at all times. The power balance at ith bus is given by the well-known power flow equations, which in polar form is written as:

$$
\begin{aligned}
0 &= p_{G,i}(t) - p_{L,i}(t) \\
&\quad - \hat{v}_i(t) \sum_{j=1}^{n_B} [\hat{v}_j(t) B_{ij} \sin(\hat{\theta}_i(t) - \hat{\theta}_j(t)) \\
&\quad + \hat{v}_i(t) G_{ij} \cos(\hat{\theta}_i(t) - \hat{\theta}_j(t))] , \quad i = 1, \dots, n_B , \\
0 &= q_{G,i}(t) - q_{L,i}(t) \\
&\quad - \hat{v}_i(t) \sum_{j=1}^{n_B} [\hat{v}_j G_{ij} \sin(\hat{\theta}_i(t) - \hat{\theta}_j(t)) \\
&\quad - \hat{v}_j(t) B_{ij} \cos(\hat{\theta}_i(t) - \hat{\theta}_j(t))] , \quad i = 1, \dots, n_B ,
\end{aligned}
\tag{B.8}
$$

where $p_{G,i}$ and $q_{G,i}$ represent the sum of the active power generation, and the sum of reactive power generation at the ith bus, respectively. Similarly, $p_{L,i}$ and $q_{L,i}$ are the sum of the active power consumption, and the sum of the reactive power consumption at the ith bus, respectively. n_B is the total number of buses of the grid. G_{ij} and B_{ij}, respectively, are the real and imaginary part of the (i,j) element of the system admittance matrix.

In [18], noise is included in the bus voltage phasor to account for possible sources of volatility and fluctuations not modelled in the set of DAEs for transient stability analysis, e.g., the effects of harmonics, nonlinearities, load unbalances, and electromagnetic transients, etc. In the same vein, the stochastic processes in (B.8) are included through the variables $\hat{v}_i$ and $\hat{\theta}_i$, which are the bus voltage magnitude and the voltage phase angle, respectively, and are obtained as n_B-dimensional correlated SDEs as follows:

$$
\begin{aligned}
\hat{\boldsymbol{v}}(t) &= \boldsymbol{v}(t) - \eta_v(t) , \\
\hat{\theta}(t) &= \theta(t) - \eta_\theta(t) , \\
\begin{bmatrix} d\eta_v(t) \\ d\eta_\theta(t) \end{bmatrix} &= \begin{bmatrix} \boldsymbol{a}_v(\eta_v(t)) \\ \boldsymbol{a}_\theta(\eta_\theta(t)) \end{bmatrix} dt + \begin{bmatrix} \boldsymbol{b}_v(\eta_v(t)) \\ \boldsymbol{b}_\theta(\eta_\theta(t)) \end{bmatrix} \mathbf{C} \begin{bmatrix} d\boldsymbol{W}_v(t) \\ d\boldsymbol{W}_\theta(t) \end{bmatrix} ,
\end{aligned}
\tag{B.9}
$$

where $\boldsymbol{v} \in \mathbb{R}^{n_B}$ and $\theta \in \mathbb{R}^{n_B}$ are the noise-free components of the voltage magnitude and phase angles, respectively, at network buses; and $\boldsymbol{a}_v$, $\boldsymbol{a}_\theta$, $\boldsymbol{b}_v$ and $\boldsymbol{b}_\theta$ are all n_B-dimensional vectors with same meanings as in (4.30). $\mathbf{C} \in \mathbb{R}^{2n_B \times 2n_B}$ is calculated based on the correlation matrix, using (4.29), using a correlation matrix $\mathbf{R}$ that contains the correlation values between voltage magnitudes and voltage angles. The structure of **R** is similar to that of (B.7), namely:

$$\mathbf{R} = \begin{bmatrix} \mathbf{R}_{v,v} & \mathbf{R}_{v,\theta} \\ \mathbf{R}_{\theta,v} & \mathbf{R}_{\theta,\theta} \end{bmatrix}, \tag{B.10}$$

where $\mathbf{R}_{\theta,v} = \mathbf{R}_{v,\theta}^T$ and:

$$\mathbf{R}_{v,v} = \begin{bmatrix} 1 & r_{v_1,v_2} & \cdots & r_{v_1,v_d} \\ r_{v_2,v_1} & 1 & \cdots & r_{v_2,v_d} \\ \vdots & \vdots & \ddots & \vdots \\ r_{v_d,v_1} & r_{v_d,v_2} & \cdots & 1 \end{bmatrix},$$

$$\mathbf{R}_{v,\theta} = \begin{bmatrix} r_{v_1,\theta_1} & r_{v_1,\theta_2} & \cdots & r_{v_1,\theta_d} \\ r_{v_2,\theta_1} & r_{v_2,\theta_2} & \cdots & r_{v_2,\theta_d} \\ \vdots & \vdots & \ddots & \vdots \\ r_{v_d,\theta_1} & r_{v_d,\theta_2} & \cdots & r_{v_d,\theta_d} \end{bmatrix},$$

$$\mathbf{R}_{\theta,\theta} = \begin{bmatrix} 1 & r_{\theta_1,\theta_2} & \cdots & r_{\theta_1,\theta_d} \\ r_{\theta_2,\theta_1} & 1 & \cdots & r_{\theta_2,\theta_d} \\ \vdots & \vdots & \ddots & \vdots \\ r_{\theta_d,\theta_1} & r_{\theta_d,\theta_2} & \cdots & 1 \end{bmatrix}.$$

B.5 Correlated wind fluctuations

The spatial and temporal correlation between different wind turbines within a power plant, as well as among power plants can be modeled as a set of correlated wind speeds. Such a model, extending the wind speed model in (4.54), is written as:

$$\begin{aligned} \boldsymbol{v}_{\text{wind}}(t) &= \boldsymbol{v}_c + \eta_w(t)\,, \\ d\eta_w(t) &= \boldsymbol{a}_w(\eta_w(t))dt + \boldsymbol{b}_w(\eta_w(t))[\mathbf{C}\,d\boldsymbol{W}_w(t)]\,. \end{aligned} \tag{B.11}$$

where $\boldsymbol{v}_c \in \mathbb{R}^{n_W}$ is the vector of uncorrelated wind speeds; $\mathbf{C} \in \mathbb{R}^{n_W \times n_W}$ is calculated from a wind correlation matrix **R** using (4.29); and other variables and parameters have the same meaning as in (4.30).

Appendix C

Data of lines, loads and distributed energy resources

C.1 IEEE 34-bus distribution test feeder data

The per unit (pu) of length resistance, reactance and susceptance of the used line configuration are given in Tables C.1(a) and C.1(b), respectively, and are represented by three-phase matrices. They correspond to the configuration #300 of Reference 1. Table C.2 gives the line data in terms of connected buses and line length.

In order to avoid confusion, Table C.3 shows the correspondence of the bus numbering between Reference 1 and Figure 6.2.

Table C.1(a) Per unit of length resistance R and reactance X for the used configurations of the IEEE 34-bus distribution test feeder

Configuration	***R* (Ω/mile)**			***X* (Ω/mile)**		
IEEE_34 #1	1.3368	0.2101	0.2130	1.3343	0.5779	0.5015
	0.2101	1.3238	0.2066	0.5779	1.3569	0.4591
	0.2130	0.2066	1.3294	0.5015	0.4591	1.3471

Table C.1(b) Per unit of length susceptance B for the used configurations of the IEEE 34-bus distribution test feeder

Configuration		***B* (uS/mile)**	
IEEE_34 #1	5.3350	−1.5313	−0.9943
	−1.5313	5.0979	−0.6212
	−0.9943	−0.6212	4.8880

Table C.2 Line data for the IEEE 34-bus distribution test feeder

Begin bus	End bus	Length (ft.)
1	2	2,580
2	3	1,730
3	4	32,230
4	5	5,804
4	6	37,500
6	7	29,730
7	8	10
8	9	310
9	10	1,710
10	11	48,150
11	12	13,740
9	13	10,210
13	14	3,030
13	15	840
15	16	20,440
16	17	520
17	19	36,830
17	18	23,330
19	20	10
20	21	10
21	22	10,560
20	23	4,900
23	24	1,620
23	25	5,830
25	26	280
26	27	1,350
27	28	3,640
28	29	530
25	30	2,020
30	31	2,680
31	34	860
31	32	280
32	33	4,860

C.2 IEEE 13-bus distribution test feeder data

The lines in the IEEE 13-bus distribution test feeder are unbalanced and correspond to the #602 line configuration of Reference 1. The pu of length values of the resistance, reactance and susceptance are given in Tables C.4a and C.4b. Table C.5 gives the line data in terms of connected buses and lengths. In order to avoid confusion, Table C.6 shows the correspondence of the buses' numbering between Reference 1 and Figure 8.3.

Table C.3 Correspondence of bus numbering between Reference 1 and Figure 8.2

Bus number in Reference 1	Bus number in Figure 8.2
800	1
802	2
806	3
808	4
810	5
812	6
814	7
850	8
816	9
818	10
820	11
822	12
824	13
826	14
828	15
830	16
854	17
856	18
852	19
832	20
888	21
890	22
858	23
864	24
834	25
842	26
844	27
846	28
848	29
860	30
836	31
862	32
838	33
840	34

Table C.4a Per unit of length resistance R and reactance X for the IEEE 13-bus distribution test feeder

Configuration	*R* (Ω/mile)			*X* (Ω/mile)		
IEEE_13 1	0.7526	0.1580	0.1560	1.1814	0.4236	0.5017
	0.1580	0.7475	0.1535	0.4236	1.1983	0.3849
	0.1560	0.1535	0.7436	0.5017	0.3849	1.2112

Table C.4b Per unit of length susceptance B for the IEEE 13-bus distribution test feeder

Configuration		*B* (uS/mile)	
IEEE_13 1	5.6990	−1.0817	−1.6905
	−1.0817	5.1795	−0.6588
	−1.6905	−0.6588	5.4246

Table C.5 Line data for the IEEE 13-bus distribution test feeder

Begin bus	End bus	Length (ft.)
1	2	2,000
2	3	500
3	4	300
2	5	500
5	6	300
2	7	2,000
7	8	300
8	9	500
7	10	300
10	11	300
10	13	800
7	12	1,000

Table C.6 Correspondence of bus numbering between Reference 1 and Figure 8.3

Bus number in Reference 1	Bus number in Figure 8.3
650	1
632	2
633	3
634	4
645	5
646	6
671	7
692	8
675	9
684	10
611	11
680	12
652	13

C.3 IEEE 39-bus transmission test system data

Table C.7 gives the line data in terms of connected buses and resistance, reactance and susceptance. The transformers are treated with their equivalent impedance. Table C.8 provides the data related to the transformers, namely, their location and the assumed voltage levels.

Table C.7 Line data for the IEEE 39-bus transmission test system

Begin bus	End bus	*R* (pu)	*X* (pu)	*B* (pu)
1	2	0.0035	0.0411	0
1	39	0.001	0.025	0.75
2	3	0.013	0.0151	0.2572
2	25	0.007	0.0086	0.146
3	4	0.0013	0.0213	0.2214
3	18	0.0011	0.0133	0.2138
4	5	0.0008	0.0128	0.1342
4	14	0.0008	0.0129	0.1382
5	6	0.0002	0.0026	0.0434
5	8	0.0008	0.0112	0.1476
6	7	0.0006	0.0092	0.113
6	11	0.0007	0.0082	0.1389
7	8	0.0004	0.0046	0.078
8	9	0.0023	0.0363	0.3804
9	39	0.001	0.025	1.2
10	11	0.0004	0.0043	0.0729
10	13	0.0004	0.0043	0.0729
13	14	0.0009	0.0101	0.1723
14	15	0.0018	0.0217	0.366
15	16	0.0009	0.0094	0.171
16	17	0.0007	0.0089	0.1342
16	19	0.0016	0.0195	0.304
16	21	0.0008	0.0135	0.2548
16	24	0.0003	0.0059	0.068
17	18	0.0007	0.0082	0.1319
17	27	0.0013	0.0173	0.3216
21	22	0.0008	0.014	0.2565
22	23	0.0006	0.0096	0.1846
23	24	0.0022	0.035	0.361
25	26	0.0032	0.0323	0.513
26	27	0.0014	0.0147	0.2396
26	28	0.0043	0.0474	0.7802
26	29	0.0057	0.0625	1.029
28	29	0.0014	0.0151	0.249
12	11	0.0016	0.0435	0
12	13	0.0016	0.0435	0
6	31	0	0.025	0

(Continues)

Table C.7 (Continued)

Begin bus	End bus	*R* (pu)	*X* (pu)	*B* (pu)
10	32	0	0.02	0
19	33	0.0007	0.0142	0
20	34	0.0009	0.018	0
22	35	0	0.0143	0
23	36	0.0005	0.0272	0
25	37	0.0006	0.0232	0
2	30	0	0.0181	0
29	38	0.0008	0.0156	0
19	20	0.0007	0.0138	0

Table C.8 Transformer data for the IEEE 39-bus transmission test system

Transformer number	Connected buses	Nominal voltage ratio
1	31_6	380 kV/230 kV
2	11_12	230 kV/125 kV
3	13_12	230 kV/125 kV
4	19_20	230 kV/125 kV
5	19_33	230 kV/15 kV
6	2_30	230 kV/15 kV
7	25_37	230 kV/15 kV
8	29_38	230 kV/15 kV
9	10_32	230 kV/15 kV
10	22_35	230 kV/15 kV
11	23_36	230 kV/15 kV
12	20_34	125 kV/15 kV

Bibliography

[1] Group IDPW. "Radial distribution test feeders". *IEEE Transactions on Power Systems*. 1991;6:975–985.

Appendix D

Proofs and tools for DDAEs

D.1 Determination of A_0, A_1 and A_2

This appendix describes how (17.8)–(17.10) are determined based on (17.5)–(17.6). From (17.6), one obtains:

$$\Delta \boldsymbol{y} = -\boldsymbol{g}_y^{-1}\boldsymbol{g}_x \Delta \boldsymbol{x} - \boldsymbol{g}_y^{-1}\boldsymbol{g}_{x_d} \Delta \boldsymbol{x}_d \tag{D.1}$$

Substituting (D.1) into (17.5), one has:

$$\Delta \dot{\boldsymbol{x}} = (\boldsymbol{f}_x - \boldsymbol{f}_y \boldsymbol{g}_y^{-1}\boldsymbol{g}_x)\Delta \boldsymbol{x} + (\boldsymbol{f}_{x_d} - \boldsymbol{f}_y \boldsymbol{g}_y^{-1}\boldsymbol{g}_{x_d})\Delta \boldsymbol{x}_d + \boldsymbol{f}_{y_d}\Delta \boldsymbol{y}_d \tag{D.2}$$

In (D.2), one has still to substitute $\Delta \boldsymbol{y}_d$ for a linear expression of the actual and/or of the retarded state variable. To do so, consider the algebraic equations $\boldsymbol{g}$ computed at $(t-\tau)$. Since algebraic constraints have to be always satisfied, the following steady-state condition must hold:

$$\boldsymbol{0} = \boldsymbol{g}(\boldsymbol{x}(t-\tau), \boldsymbol{x}_d(t-\tau), \boldsymbol{y}(t-\tau)) \tag{D.3}$$

Then, observing that $\boldsymbol{x}_d = \boldsymbol{x}(t-\tau)$, $\boldsymbol{y}_d = \boldsymbol{y}(t-\tau)$, and $\boldsymbol{x}_d(t-\tau) = \boldsymbol{x}(t-2\tau)$, differentiating (D.3) leads to:

$$\boldsymbol{0} = \boldsymbol{g}_x \Delta \boldsymbol{x}_d + \boldsymbol{g}_{x_d}\Delta \boldsymbol{x}(t-2\tau) + \boldsymbol{g}_y \Delta \boldsymbol{y}_d \tag{D.4}$$

In steady-state, for any instant t_0, $\boldsymbol{x}(t_0) = \boldsymbol{x}(t_0-\tau) = \boldsymbol{x}(t_0-2\tau) = \boldsymbol{x}_0$ and $\boldsymbol{y}(t_0) = \boldsymbol{y}_d(t_0) = \boldsymbol{y}_0$. Hence, the Jacobian matrices in (D.4) are the same as in (17.6). Equation (D.4) can be rewritten as:

$$\Delta \boldsymbol{y}_d = -\boldsymbol{g}_y^{-1}\boldsymbol{g}_x \Delta \boldsymbol{x}_d - \boldsymbol{g}_y^{-1}\boldsymbol{g}_{x_d}\Delta \boldsymbol{x}(t-2\tau) \tag{D.5}$$

and, substituting (D.5) into (D.2), one obtains:

$$\begin{aligned}\Delta \dot{\boldsymbol{x}} =& (\boldsymbol{f}_x - \boldsymbol{f}_y \boldsymbol{g}_y^{-1}\boldsymbol{g}_x)\Delta \boldsymbol{x} \\ &+ (\boldsymbol{f}_{x_d} - \boldsymbol{f}_y \boldsymbol{g}_y^{-1}\boldsymbol{g}_{x_d} - \boldsymbol{f}_{y_d}\boldsymbol{g}_y^{-1}\boldsymbol{g}_x)\Delta \boldsymbol{x}_d \\ &+ (-\boldsymbol{f}_{y_d}\boldsymbol{g}_y^{-1}\boldsymbol{g}_{x_d})\Delta \boldsymbol{x}(t-2\tau)\end{aligned} \tag{D.6}$$

which leads to the definitions of $\boldsymbol{A}_0$, $\boldsymbol{A}_1$, and $\boldsymbol{A}_2$ given in (17.8), (17.9), and (17.10), respectively.

D.2 Chebyshev's differentiation matrix

Chebyshev's differentiation matrix $\boldsymbol{D}_N$ of dimensions $(N+1) \times (N+1)$ is defined as follows. First, one has to define $N+1$ Chebyshev's nodes, i.e., the interpolation points on the normalized interval $[-1, 1]$:

$$x_k = \cos\left(\frac{k\pi}{N}\right), \quad k = 0, \ldots, N. \tag{D.7}$$

Then, the element (i,j) differentiation matrix $\boldsymbol{D}_N$ indexed from 0 to N is defined as

$$\boldsymbol{D}_{(i,j)} = \begin{cases} \frac{c_i(-1)^{i+j}}{c_j(x_i - x_j)}, & i \neq j \\ -\frac{1}{2}\frac{x_i}{1-x_i^2}, & i = j \neq 1, N-1 \\ \frac{2N^2+1}{6}, & i = j = 0 \\ -\frac{2N^2+1}{6}, & i = j = N \end{cases} \tag{D.8}$$

where $c_0 = c_N = 2$ and $c_2 = \cdots = c_{N-1} = 1$. For example, $\boldsymbol{D}_1$ and $\boldsymbol{D}_2$ are:

$$\boldsymbol{D}_1 = \begin{bmatrix} 0.5 & -0.5 \\ 0.5 & -0.5 \end{bmatrix}, \quad \text{with } x_0 = 1,\ x_1 = -1\ .$$

and

$$\boldsymbol{D}_2 = \begin{bmatrix} 1.5 & -2 & 0.5 \\ 0.5 & 0 & -0.5 \\ -0.5 & 2 & -1.5 \end{bmatrix}, \text{ with } x_0 = 1,\ x_1 = 0,\ x_2 = -1\ .$$

D.3 Kronecker's product

If $\boldsymbol{A}$ is a $m \times n$ matrix and $\boldsymbol{B}$ is a $p \times q$ matrix, then Kronecker's product $\boldsymbol{A} \otimes \boldsymbol{B}$ is an $mp \times nq$ block matrix as follows:

$$\boldsymbol{A} \otimes \boldsymbol{B} = \begin{bmatrix} a_{11}\boldsymbol{B} & \cdots & a_{1n}\boldsymbol{B} \\ \vdots & \ddots & \vdots \\ a_{m1}\boldsymbol{B} & \cdots & a_{mn}\boldsymbol{B} \end{bmatrix} \tag{D.9}$$

For example, let $\boldsymbol{A} = \begin{bmatrix} 1 & 2 & 3 \\ 3 & 2 & 1 \end{bmatrix}$ and $\boldsymbol{B} = \begin{bmatrix} 2 & 1 \\ 2 & 3 \end{bmatrix}$. Then:

$$\boldsymbol{A} \otimes \boldsymbol{B} = \begin{bmatrix} \boldsymbol{B} & 2\boldsymbol{B} & 3\boldsymbol{B} \\ 3\boldsymbol{B} & 2\boldsymbol{B} & \boldsymbol{B} \end{bmatrix} = \begin{bmatrix} 2 & 1 & 4 & 2 & 6 & 3 \\ 2 & 3 & 4 & 6 & 6 & 9 \\ 6 & 3 & 4 & 2 & 2 & 1 \\ 6 & 9 & 4 & 6 & 2 & 3 \end{bmatrix}$$

Note that $\boldsymbol{A} \otimes \boldsymbol{B} \neq \boldsymbol{B} \otimes \boldsymbol{A}$.

Appendix E

Numerical aspects of the probe-insertion technique

E.1 Parameters of the probe-insertion technique

In Table E.1, a possible role played by the parameters v_R, v_I, y_R, y_I and $\omega_p\}$ is reported. The combinations in which either y_R or y_I are *perturbed* and v_R or v_I are *adjusted* have been omitted as it is worth verifying that they provide $\rho = 0$ since φ (see (18.19)) depends on neither v_R nor v_I.

E.2 Integration of (18.34)

The projection of each component δ_j of δ on the unit circle in $\mathbf{R}^2$, using the transformation (18.33) with $\rho = 1$, i.e., $(\delta_j, 1) \to (c_j, s_j)$, must be handled with care if one aims at numerically solving (18.34). As a matter of fact linear multi-step integration

Table E.1 The possible role played by the parameters $\{v_R, v_I, \omega_p\}$ and $\{y_R, y_I\}$ in the probe-insertion technique

Perturbed	Adjusted	Fixed	Free
v_R	y_R	y_I, ω_p	v_I
v_R	y_I	y_R, ω_p	v_I
v_I	y_R	y_I, ω_p	v_R
v_I	y_I	y_R, ω_p	v_R
ω_p	y_R	y_I, v_R	v_I
ω_p	y_I	y_R, v_R	v_I
ω_p	y_R	y_I, v_I	v_R
ω_p	y_I	y_R, v_I	v_R
y_R	ω_p	y_I, v_R	v_I
y_R	ω_p	y_I, v_I	v_R
y_I	ω_p	y_R, v_R	v_I
y_I	ω_p	y_R, v_I	v_R

methods, which are largely used to solve non-linear stiff DAEs warp time and introduce additional damping factors [54]. This means that, from a numerical point of view, focusing on the jth machine, the dynamics of c_j and s_j is ruled by the following equations:

$$\begin{cases} \dot{c}_j + \lambda_j^c c_j + s_j \Omega \left(\omega_j - \omega_0\right) = 0 \\ \dot{s}_j + \lambda_j^s s_j - c_j \Omega \left(\omega_j - \omega_0\right) = 0 \end{cases}$$

where λ_j^c and λ_j^s are real parameters depending on the chosen integration method. It is worth realizing that, at the equilibrium, i.e., for $\omega_j = \omega_0$, the state variables c_j and s_j do not remain constant but collapse toward $(0, 0)$ as far as the simulation time increases. To avoid this one can use the dynamical equation $\dot{\rho}_j + \rho_j - 1 = 0$. In this way, the numerical integration procedure provides the steady-state solution $\rho_j = 1$ and δ_j is guaranteed to belong to the unit circle, thus satisfying $c_j^2 + s_j^2 = 1$. In other words, ρ_j is *not assumed* to be 1 but it is *forced* to be so. If this is done the equation ruling the dynamics of δ_j becomes

$$\begin{cases} \dot{c}_j + c_j + s_j \Omega \left(\omega_j - \omega_0\right) - c_j \left(c_j^2 + s_j^2\right)^{-1/2} = 0 \\ \dot{s}_j + s_j - c_j \Omega \left(\omega_j - \omega_0\right) - s_j \left(c_j^2 + s_j^2\right)^{-1/2} = 0 \end{cases}$$

and the PSM equations (18.34) must be updated.

As far as (18.37) is concerned, if the constant c_0 is adopted, the equations ruling the jth machine read

$$\begin{cases} \dot{c}_j + (c_j + c_0) + s_j \Omega \left(\omega_j - \omega_0\right) + \\ \qquad\qquad -(c_j + c_0) \left[(c_j + c_0)^2 + s_j^2\right]^{-1/2} = 0 \\ \dot{s}_j + s_j - s_j \Omega \left(\omega_j - \omega_0\right) - s_j \left[(c_j + c_0)^2 + s_j^2\right]^{-1/2} = 0 \end{cases}$$

Index